Lecture Notes in Mechanical Engineering

Lecture Notes in Mechanical Engineering (LNME) publishes the latest developments in Mechanical Engineering—quickly, informally and with high quality. Original research or contributions reported in proceedings and post-proceedings represents the core of LNME. Volumes published in LNME embrace all aspects, subfields and new challenges of mechanical engineering.

To submit a proposal or request further information, please contact the Springer Editor of your location:

Europe, USA, Africa: Leontina Di Cecco at Leontina.dicecco@springer.com
China: Ella Zhang at ella.zhang@cn.springernature.com
India, Rest of Asia, Australia, New Zealand: Swati Meherishi at swati.meherishi@springer.com

Topics in the series include:

- Engineering Design
- Machinery and Machine Elements
- Mechanical Structures and Stress Analysis
- Automotive Engineering
- Engine Technology
- Aerospace Technology and Astronautics
- Nanotechnology and Microengineering
- Control, Robotics, Mechatronics
- MEMS
- Theoretical and Applied Mechanics
- Dynamical Systems, Control
- Fluid Mechanics
- Engineering Thermodynamics, Heat and Mass Transfer
- Manufacturing Engineering and Smart Manufacturing
- Precision Engineering, Instrumentation, Measurement
- Materials Engineering
- Tribology and Surface Technology

Indexed by SCOPUS, EI Compendex, and INSPEC.

All books published in the series are evaluated by Web of Science for the Conference Proceedings Citation Index (CPCI).

To submit a proposal for a monograph, please check our Springer Tracts in Mechanical Engineering at https://link.springer.com/bookseries/11693.

Editors
Imran M. Jamadar
The National Institute of Engineering
Mysuru, Karnataka, India

Vinod Rajamani
FH Dortmund University
Dortmund, Germany

Rama Krushna Sabat
School of Minerals, Metallurgical
and Materials
Indian Institute of Technology
Bhubaneswar
Khordha, Odisha, India

ISSN 2195-4356 ISSN 2195-4364 (electronic)
Lecture Notes in Mechanical Engineering
ISBN 978-981-95-9465-8 ISBN 978-981-95-9466-5 (eBook)
https://doi.org/10.1007/978-981-95-9466-5

This Springer imprint is published by the registered company Springer Nature Singapore Pte Ltd.
The registered company address is: 152 Beach Road, #21-01/04 Gateway East, Singapore 189721, Singapore

Imran M. Jamadar · Vinod Rajamani ·
Rama Krushna Sabat
Editors

Integrating Sustainability in Modern Technologies

Select Proceedings of ICST 2025

Springer

This book is dedicated to all the passionate researchers, innovators, and change makers who are tirelessly working to create a sustainable future. Their dedication, creativity, and commitment to advancing technologies that address the world's most pressing environmental challenges inspire us all. We dedicate this work to the visionaries who are shaping a cleaner, greener, and more equitable world, and to the policymakers, educators, and industry leaders who support these efforts. May their work continue to inspire future generations to build a sustainable and harmonious world for all.

Foreword

The urgent need to address global environmental challenges has never been more pressing, and the role of sustainable technologies in this endeavor is paramount. As we stand at the crossroads of innovation and sustainability, the International Conference on Sustainable Technology (ICST-2025) provides a timely and essential platform to explore the latest advancements that can help shape a more sustainable future.

The proceedings presented in this book offer a comprehensive collection of cutting-edge research and developments across a range of critical areas: from Sustainable Energy Solutions to Robotics and e-Mobility, Sustainable Materials, Design for Sustainability, Green Computing, and Environment and Green Buildings. Each of these fields plays a crucial role in advancing sustainable development goals and tackling the environmental challenges that we face today. The interdisciplinary approach embraced by the authors reflects the interconnected nature of these challenges and the need for collaborative efforts across sectors.

This compilation not only showcases innovative solutions but also highlights the importance of continued research and the role of technology in achieving a sustainable and equitable world. It is evident that sustainability is no longer a mere ideal but an imperative, and the technologies explored within these pages are the tools that will help pave the way for transformative change.

As you embark on this journey through the proceedings of ICST-2025, I hope you find inspiration, insight, and a renewed sense of urgency to contribute to the global movement toward sustainability. This book is a testament to the collective efforts of dedicated individuals and organizations, and it serves as both a resource and a call to action for researchers, policymakers, and industry leaders committed to advancing sustainable development through technological innovation.

Let this work serve as a beacon of hope and a reminder that sustainable solutions are within our reach—with determination, creativity, and collaboration, we can build a brighter, more sustainable future for generations to come.

Dr. Prasanta Kumar Samal, M.E., IISc. Ph.D. JSS, STU
Publication Committee, Co-chair
ICST 2025, Assistant Professor
The National Institute of Engineering
Mysuru, India

Preface

As the world continues to face an array of pressing environmental challenges, the importance of sustainable technologies in shaping our future cannot be overstated. The International Conference on Sustainable Technology (ICST-2025) brings together thought leaders, researchers, innovators, and policymakers from around the world to share their insights and advancements in this critical field. The proceedings in this book represent the collective knowledge and cutting-edge research that emerged from the conference, offering a detailed examination of the most recent advancements in sustainable technologies.

This volume covers a wide spectrum of topics essential to sustainable development, from Sustainable Energy, which focuses on renewable energy sources and energy efficiency, to robotics and e-mobility, which explores innovations in electric mobility and automation. The book also delves into Sustainable Materials, Design for Sustainability, Green Computing, and Environment and Green Buildings, each of which plays an integral role in reducing our ecological footprint and building a more sustainable world.

The contributions presented here reflect the growing urgency to develop solutions that address the global environmental crisis while simultaneously supporting economic and social growth. The authors of these papers have shown a deep commitment to advancing technologies that not only meet the demands of the present but also ensure a more sustainable and resilient future. Each section of this book highlights different aspects of sustainability and provides actionable insights that can be used by researchers, industry professionals, and policymakers alike.

While we have made significant strides in recent years, the journey toward achieving the United Nations' Sustainable Development Goals (SDGs) remains ongoing. The work presented in these proceedings offers valuable perspectives on how we can continue to innovate, collaborate, and push the boundaries of what is possible in the realm of sustainable technologies. It is our hope that the findings and discussions within these pages will inspire further research and provide a roadmap for future advancements in sustainability.

We dedicate this book to all those who are working tirelessly to create a better, more sustainable world, and to the next generation of innovators who will continue this important work. It is through our collective efforts that we can pave the way for a future where technological innovation and sustainability go hand in hand.

Mysuru, India Imran M. Jamadar
Dortmund, Germany Vinod Rajamani
Khordha, India Rama Krushna Sabat

Acknowledgements

We would like to express our deepest gratitude to all those who have contributed to the success of the International Conference on Sustainable Technology (ICST-2025) and the creation of this book. Without the support and dedication of numerous individuals and organizations, this project would not have been possible.

First and foremost, we extend our sincere thanks to all the authors, researchers, and presenters whose groundbreaking work is showcased in these proceedings. Their commitment to advancing sustainable technologies and addressing global environmental challenges is truly inspiring, and their contributions have made this book a valuable resource for the academic, industrial, and policymaking communities.

We would also like to acknowledge the conference organizers, volunteers, and staff whose tireless efforts ensured the smooth execution of ICST-2025. Your professionalism and dedication have made this event a success, bringing together experts from diverse fields to share knowledge and collaborate on sustainable solutions.

Our heartfelt thanks go to the sponsors and partners who supported the conference, providing the resources necessary to facilitate this gathering of minds. Your generosity and belief in the importance of sustainability have helped make this book and the conference itself possible.

We would also like to thank the reviewers who carefully evaluated the submitted papers, offering invaluable feedback that has enhanced the quality and rigor of the work presented in these proceedings. Your time and expertise are deeply appreciated.

Lastly, we extend our gratitude to our families, friends, and colleagues for their unwavering support and encouragement throughout the planning and execution of this project. Your understanding and patience have been a source of strength during the course of this endeavor.

As we look to the future, we hope that this book serves as a catalyst for further research, collaboration, and innovation in the pursuit of a sustainable and resilient world. We are grateful for the opportunity to contribute to this global effort and extend our thanks to everyone who has played a role in making it a reality.

Contents

About the Editors

Dr. Imran M. Jamadar is an accomplished academician and researcher, currently serving as Professor of Mechanical Engineering and Dean of Research & Development at The National Institute of Engineering, Mysuru, Karnataka. He earned B.E. from Shivaji University, Kolhapur, M.Tech. in Machine Design from VJTI, Mumbai University, and Ph.D. in mechanical engineering from SVNIT, Surat. His research expertise spans over vibration control in flexible robotic manipulators, predictive maintenance for rotating machinery, energy harvesting devices, and the enhancement of fatigue, tribology, and wear properties in polymer composites. Dr. Jamadar actively reviews for numerous prestigious international journals. He has led several significant funded research projects that highlight his expertise in mechanical engineering, robotic vibration control, and predictive maintenance.

Prof. Vinod Rajamani is a distinguished academician and industry expert with a robust background in mechanical and automotive engineering. He holds a bachelor's degree in mechanical engineering from RV College of Engineering, a master's in automotive systems engineering, and a Ph.D. in internal combustion engines from RWTH Aachen University. His academic journey is marked by notable achievements, including founding technical societies, participating in international Formula Student teams, and producing impactful theses on combustion analysis and hybrid vehicle technologies. With over a decade of professional experience, he has held key roles such as Design Engineer, Project Manager, and Department Manager at esteemed organizations like FEV GmbH and DEUTZ AG, where he led projects on advanced combustion systems and emission norms compliance. Currently, as a Professor at the University of Applied Sciences and Arts in Dortmund, he combines his extensive industry expertise with academic rigor, mentoring students, and contributing to cutting-edge research.

Dr. Rama Krushna Sabat is an associate professor at the Indian Institute of Technology, Bhubaneswar (IIT Bhubaneswar), specializing in materials engineering. His research interests include virtual characterization, microstructural modeling, severe plastic deformation processes, microstructure-property relationships in light metals, and corrosion of metals. He holds a PhD in materials engineering from IISc Bangalore and a B.Tech in metallurgical and materials engineering from NIT Rourkela. Dr. Sabat has significant academic experience, having worked as a Postdoctoral Fellow at the University of Waterloo and IISc. Bangalore before joining IIT Bhubaneswar as a faculty member. Notable areas of his research include crystal plasticity-based FEM modeling, prediction of forming limit diagrams, and nano-void theory for failure modeling. His work is widely cited, with contributions to understanding the mechanical properties of various alloys, microstructure evolution, and thermoelectric materials. Dr. Sabat has led several significant funded research projects that highlight his expertise in mechanical engineering and is offering contributions to international journals and conferences in the field of materials & mechanical engineering.

Solar-Driven Contactless Real-Time Charging System

S. J. Savita, G. Raksha, Shefali R. Jain, and M. Varshini

Abstract Electric vehicles (EVs) offer significant environmental benefits, such as reducing greenhouse gas emissions, minimizing air pollution, and lowering fuel costs, making them an essential component of sustainable transportation (Supriyadi et al in Telkomnika 16(3):1013–1018, 2018). However, one of the main challenges hindering their widespread adoption is the prolonged charging time and the lack of accessible charging facilities, particularly in remote and rural areas (Banu in Asian J Electr Sci 7(1), 2018). To address these limitations, this paper proposes an innovative and efficient system that utilizes solar-powered, contactless, real-time charging technology. The system leverages Wireless Power Transmission (WPT) through inductive coupling, enabling vehicles to charge seamlessly while in motion, thereby eliminating the need for frequent stops at conventional charging stations (Yatnalkar and Narman in Conference: 2018 IEEE international symposium on signal processing and information technology (ISSPIT), 2018). The design integrates a robust combination of components, including solar panels, copper coils, 2N2222 transistors, lithium-ion batteries, voltage regulators, and LED indicators (Rana and Xiang in IEEE Access 6:19295–19303, 2018). Together, these elements ensure a reliable and uninterrupted charging experience for EV users. This integration of solar energy and wireless charging revolutionizes sustainable transportation, paving the way for a cleaner future (Rupani and Doshi in The 3rd international workshop on recent advances on internet of things: technology and application approaches (IoT-T&A 2019), Coimbra, Portugal, 2019).

Keywords Electric vehicle · Wireless charging · Dynamic charging · Solar power · Inductive power transfer · Wireless power transmission

S. J. Savita (✉) · G. Raksha · S. R. Jain · M. Varshini
RNS Institute of Technology, Channasandra, Bengaluru, India
e-mail: savita.sj@rnsit.ac.in

I. M. Jamadar et al. (eds.), *Integrating Sustainability in Modern Technologies*, Lecture Notes in Mechanical Engineering, https://doi.org/10.1007/978-981-95-9466-5_1

1 Introduction

The project concentrates on the evolution of a contactless real-time charging system for electric vehicles (EVs), addressing key challenges associated with conventional charging methods. As the adoption of EVs increases due to their environmental benefits, there is an expanding requirement for more efficient, user-friendly, and sustainable charging solutions. Established charging stations often require means of transport to be stationary for extended periods, which disrupts travel and creates inconvenience [6]. This project proposes an innovative solution—a wireless charging infrastructure embedded beneath roads, allowing EVs to charge as they drive along the lane [7]. This system not only ensures efficient charging but also contributes to the sustainability of EV technology by utilizing renewable energy sources like solar power.

The circuit represents a contactless real-time charging system designed for electric vehicles (EVs), allowing them to charge wirelessly while in motion [8]. The system harnesses solar energy through a solar panel, which charges a Li-ion battery pack via a 2S Li-ion charging module and a bridge rectifier circuit [9]. The battery powers a transmitter circuit that includes coils and switching components like a Z44 MOSFET and 2N2222 transistor. These coils are embedded along the road and generate a magnetic force field that incorporates a current in the receiver coils installed in the EV. This induced current wirelessly charges the EV's battery, which powers DC gear motors that drive the vehicle. Multiple transmitter coils along the road ensure continuous energy transfer, enabling the vehicle to maintain its charge without stopping, making this system suitable for long-range, uninterrupted travel.

2 Literature Review

The existing literature has explored various wireless charging technologies for electric vehicles, including inductive coupling-based wireless power transfer circuits [10], solar-powered EV charging [4], optimal placement of wireless charging stations [3], and IoT-integrated wireless charging mechanisms to enhance efficiency [11]. However, these studies exhibit limitations in efficiency optimization, real-time implementation, and renewable energy integration. The proposed research advances beyond these works by incorporating solar energy with dynamic wireless charging, optimizing coil structures to minimize energy loss, and conducting empirical validation under real-time conditions, thereby improving power transfer efficiency and ensuring scalability for widespread adoption [5].

3 Objective

The primary objective of this project is to develop an efficient contactless real-time charging system using renewable solar energy. The system aims to:

- Eliminate the need for physical charging stations.
- Promote sustainability by using solar panels for energy generation.
- Enhance user convenience by enabling on-the-go charging.
- Reduce the complexity of the charging process with wireless power transmission (WPT) [6].

4 Block Diagram

The block diagram (see Fig. 1) shows the process of Sun-driven contactless real-time charging system of electric vehicles. The Solar panels installed along the road absorb sunlight and convert it into direct current (DC) electricity. The battery stores this DC current. The DC/AC converter converts this current and sends it to the coils to perform the operation of transferring the current required for charging. The battery of the car then stores the energy generated. This dynamic wireless charging system is likely intended for electric vehicles (EVs) or similar applications. It allows the vehicle to receive power while in motion, reducing the need for frequent stops for recharging. The energy is generated from a solar panel, stored in a battery, converted, transmitted wirelessly, and then captured by a receiver coil to power the vehicle. The key feature of this system is the wireless energy transfer, which occurs in the space separating the transmitter and receiver coils through electromagnetic induction.

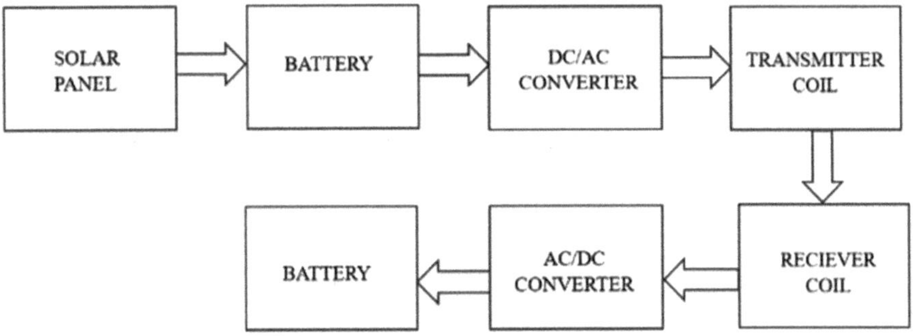

Fig. 1 Block diagram of the charging system

5 Discussion of Block Diagram

5.1 Solar Panel

The solar panel captures sunlight and converts it into electrical energy through the photovoltaic effect. This energy serves as the primary power source for the entire charging system. Solar energy is renewable and environmentally friendly, reducing reliance on fossil fuels and enabling sustainable energy use in electric vehicles.

5.2 Battery (on Transmitting Side)

The battery stores excess energy from the solar panel, allowing the system to function even when sunlight is not present, such as during cloudy days or at night.

This storage capability ensures consistent power delivery for wireless charging, preventing interruptions during energy transmission.

5.3 DC/AC Converter (on Transmitting Side)

The DC/AC converter is crucial because wireless power transmission works more efficiently with alternating current (AC). It converts the stored direct current (DC) from the battery into alternating current (AC).AC is needed to generate the oscillating magnetic field in the transmitter coil, which is responsible for creating the wireless power transfer link.

5.4 Transmitter Coil

The transmitter coil creates a fluctuating magnetic field when AC current flows through it, enabling the wireless transmission of energy through electromagnetic induction. It is typically embedded in charging infrastructure like roads or stations, allowing vehicles to charge wirelessly without physical connections. Its design and placement are crucial for ensuring efficient energy transfer and minimizing power losses due to misalignment or external interference.

5.5 Receiver Coil

The receiver coil, embedded in the vehicle, picks up the magnetic field emitted by the transmitter coil, and converts it back into electrical energy. This energy can then be used to charge the vehicle's battery or power its systems, allowing for continuous charging as the vehicle moves. To maximize efficiency, the receiver coil is carefully designed with high permeability materials and optimized winding patterns. Advanced shielding techniques are also employed to minimize energy losses and electromagnetic interference.

5.6 AC/DC Converter (on Receiving Side)

The received AC energy is converted back into DC by this converter since most vehicle batteries and electrical systems run on DC power. This step ensures compatibility between the energy received wirelessly and the storage or operating systems of the vehicle.

5.7 Battery (on Receiving Side)

The vehicle's battery stores the DC energy received from the wireless charging process, allowing it to power the vehicle over extended periods when wireless charging is not available. The accumulated energy can be used for propulsion or for powering assisting systems such as air conditioning or lighting, enhancing the vehicle's functionality and efficiency.

5.8 Vehicle

The (see Fig. 2) depicts a vehicle having 4 wheels which serves to move on the road. The vehicle in this contactless real-time charging system is likely an electric vehicle (EV) provided with a receiver coil that allows it to receive power wirelessly while in motion. This setup eliminates the need for frequent stops to recharge the vehicle, enabling longer travel distances with continuous power transfer as the vehicle passes over charging infrastructure.

Fig. 2 Vehicle design

6 System Design and Components Specifications

Key Components and Ratings: To elucidate component selection and system design, Table 1 presents the technical specifications of critical system components [2].

Incorporating these specifications ensures a comprehensive understanding of system performance potential.

Table 1 Key components and ratings

Component	Specification
Solar panel	12 V, 100 W
Li-ion battery	3.7 V, 2600 mAh per cell
Transmitter coil	30 turns, copper wire
Receiver coil	25 turns, copper wire
MOSFET (Z44)	55 V, 49 A
2N2222 transistor	40 V, 0.8 A
Voltage regulator	7809 (9 V Output)
DC/AC converter	12 V DC to 220 V AC

7 Proposed System Analysis

The Sun-driven contactless real-time charging system for electric vehicles (EVs) incorporates several key components to enable seamless wireless energy transfer while the vehicle is in motion [1]. The process is illustrated in the schematic diagram (see Fig. 3). The system begins with a solar panel, which captures sunlight and converts it into electrical energy. This energy is then stored in a rechargeable battery pack consisting of two Li-ion cells, regulated by a 2S Li-ion charging module (HX-2S-01). The module ensures efficient and safe battery charging, safeguarding it from both overcharging and undercharging. The energy stored in the battery is directed to the system to power the wireless charging process.

To prepare power for transmission, the system incorporates bridge rectifier composed of diodes, which converts AC input to DC, ensuring smooth and consistent charging for the battery. A capacitor is also placed in the circuit to filter out any voltage ripples, maintaining stable energy flow. The voltage can be further stabilized using a 7809-voltage regulator, ensures that the output remains consistently at 9 V, a crucial requirement for the proper functioning of other components in the system.

At the heart of the wireless charging mechanism are the 30-turn coils, which function as both transmitters and receivers for electromagnetic energy. These coils are embedded along the road surface and are arranged in parallel. When energized, they create a magnetic field which induces current in the receiver coils located in the EV. This wireless power transfer is controlled by a Z44 MOSFET and a 2N2222 transistor in the transmitter circuit. The MOSFET amplifies the current to drive transmitter coils efficiently, while the transistor aids in switching the MOSFET on and off, effectively controlling the energy flow.

As the vehicle moves over the coils embedded in the road, the transmitter circuit continuously transmits energy wirelessly. This circuit consists of multiple coils connected in parallel, ensuring that the EV receives power continuously without interruptions. The induced current from the road's magnetic field is captured by the

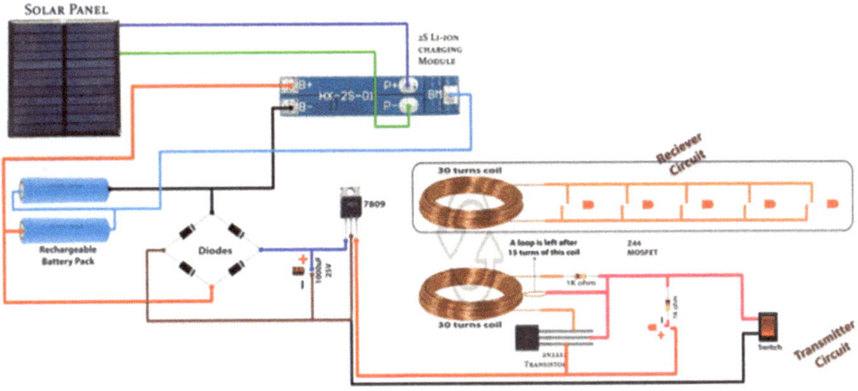

Fig. 3 Schematic diagram of the contactless real-time charging system

receiving coils and used to charge EV's internal battery. This charged battery then powers the DC gear motors that drive the vehicle's wheels, providing propulsion. These motors convert electrical energy into mechanical energy, enabling the vehicle to move.

The Sun-driven contactless real-time charging system presents a sustainable and efficient solution for EV charging by integrating solar energy with wireless power transfer. The seamless energy flow from solar panels to inductive coils ensures continuous charging while the vehicle is in motion, reducing dependency on stationary charging stations. By addressing power transfer challenges and optimizing coil alignment, this system has the potential to enhance EV adoption, making transportation more convenient, efficient, and environmentally friendly.

Additionally, the integration of smart tracking and adaptive power management enhances energy efficiency and reliability. Advanced materials and innovative coil designs further improve power transfer efficiency, minimizing energy losses. This system also reduces grid dependency, promoting decentralized and renewable energy utilization. With continuous advancements, Sun-driven contactless charging could revolutionize the future of sustainable transportation. Its implementation could pave the way for smarter, cleaner, and more autonomous electric mobility solutions worldwide.

The contactless real-time charging process uses an inductive power transfer (IPT) system. In this system, (see Fig. 4) the source coil (transmitter) is connecting to an AC power supply (VG), along with a resistor (RG) and a capacitor (CS) for tuning purposes. When AC flows through the source coil, it generates an alternating magnetic field around it. On the receiving side, the load coil (receiver) is positioned within the range of a source coil. Through principle of mutual inductance, the electromagnetic field generated by the source coil induces a current in the load coil. This induced power is then transferred to the load (RL), which represents the battery or other components in need of charging within the system [12].

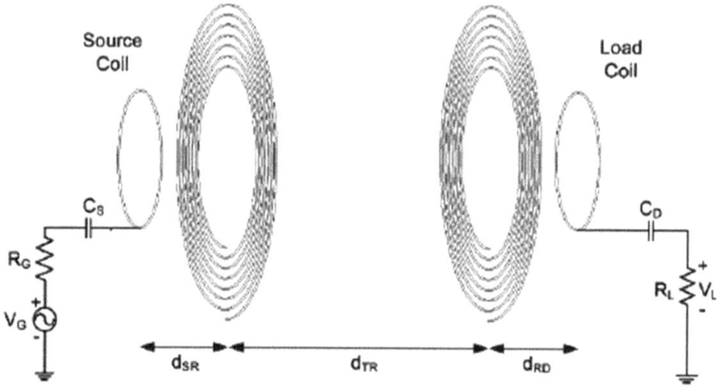

Fig. 4 Inductive power transfer between the coils

For optimal power transfer, both the source and load coils are tuned to the same resonant frequency, a process facilitated by capacitors (CS and CD) in both circuits. This tuning ensures that maximum power is transferred efficiently from the source coil to the load coil. The transfer efficiency is largely determined by the distance (dTR) between the source and load coils. As long as the distance stays within a specified range, inductive coupling allows the source's magnetic field to induce current in the load coil, which is subsequently transformed into usable energy to charge or power the load, indicated by the voltage VL. However, any misalignment or variation in distance can lead to reduced efficiency and energy losses. To mitigate this, dynamic tuning mechanisms and alignment correction strategies are often implemented. Additionally, advanced coil geometries and materials can further enhance coupling strength, ensuring stable and consistent power transfer.

Given the fact that Eq. (1), we assume, in the following equation, to be the representation of the power that was expected to be lost in the solar-driven contactless real-time charging system for electric vehicles (EVs) [13]:

$$\mathrm{P(ev)} = \eta \times (\mathrm{E(solar)} - \mathrm{L(loss)}) \tag{1}$$

where:

- P(ev) = Power delivered to the electric vehicle
- η = Efficiency of the wireless charging system
- E(solar) = Energy harnessed from the solar panels
- L(loss) = Energy lost in transmission and conversion (wireless charging inefficiencies)

Assuming that:

- The received power (E(solar)) from the solar panels is 12 V.
- After accounting for power losses (L(loss)), the final power delivered to the electric vehicle (P(ev)) is 5 V.

This entails an optimistic scenario where there is a significant loss of power during the entire process of transmission and conversion. Notwithstanding, these figures are purely hypothetical and are being used for the purpose of design only. They are not supported by experiments and there is still a need for further testing and validation of the system's performance under real conditions.

Supriyadi et al. [1] examined inductive coupling-based wireless power transfer circuits, enabling energy transmission without physical connectors. This method has been crucial in reducing charging constraints by using electromagnetic fields for power transfer. However, optimizing efficiency for dynamic charging applications remains a challenge.

Banu et al. [2] investigated the integration of solar energy into EV charging systems to enhance sustainability and reduce grid dependency. While solar power is a promising renewable source, inefficiencies in power conversion and storage hinder its real-time application. These limitations impact the feasibility of continuous, on-the-move charging.

Yatnalkar and Narman [3] studied the optimal placement of wireless charging stations along frequently travelled routes to minimize energy loss. Strategic positioning enhances charging convenience, but existing methodologies lack adaptability for real-time vehicle movement. This results in inconsistent charging efficiency, especially under dynamic driving conditions.

Rana and Xiang [4] proposed IoT-integrated wireless charging mechanisms to optimize power distribution and monitor charging status. By leveraging data-driven analytics, these systems aim to enhance efficiency and automation. However, challenges in scalability and real-time responsiveness necessitate further improvements in system architecture.

In the proposed system analysis, the novel observations after designing the prototype were not adequately highlighted. Notably, power attenuation at higher speeds and coil misalignment were identified as critical factors affecting efficiency. These challenges emphasize the need for further refinement in coil alignment and resonance tuning to enhance performance. Additionally, transmission inefficiencies, especially at higher velocities, require further investigation to optimize power transfer.

8 Hardware Result

As the vehicle progresses along the road, the LED display illuminates, serving as a visual indicator of the wireless power transfer occurring between the two coils embedded in the system. This lighting represents the successful alignment and energy transmission from the road to the car, showcasing the functionality of the dynamic charging process. Each instance of illumination corresponds to the car's real-time reception of wireless energy, ensuring continuous power flow during motion, thus enabling efficient operation without the need for traditional charging methods.

The key observation after designing the prototype is that LED indicators successfully illuminated during power transfer, verifying real-time energy transmission (see Fig. 5) illustrates the prototype, demonstrating continuous energy reception by the EV battery.

Identified Challenges:

- Power Attenuation at Elevated Velocities: Requires real-time alignment correction methodologies.
- Coil Misalignment Issues: Implementation of advanced resonant tuning mechanisms is necessary.
- Transmission Inefficiencies: Investigating alternative coil materials may enhance system performance.

Fig. 5 Prototype of the system

9 Applications

The Sun-driven contactless real-time charging system for EVs provides practical applications that boost the efficiency and eco-friendliness of electric mobility. By enabling continuous wireless charging while vehicles are in motion, this technology reduces reliance on large batteries and traditional charging stations, making it ideal for long-distance travel, public transportation, and rural areas. Below are three key applications of this innovative system:

- Extended Driving Range: EVs can charge while moving, allowing for longer trips without stopping for recharging, which is especially useful for highways.
- Public Transportation: Buses and trams can charge wirelessly during operation, improving efficiency and eliminating downtime for charging, powered sustainably by solar energy.
- Smart Road Infrastructure: Solar-powered charging coils embedded in roads enable continuous charging, reducing reliance on traditional charging stations and creating more efficient road networks.

10 Conclusion

The Sun-Driven contactless real-time charging system offers a transformative solution for Electric Vehicle (EV) charging by combining solar energy with wireless power transmission (WPT), allowing dynamic charging while vehicles are in motion.

This eliminates the need for stationary charging, enhancing convenience and sustainability. The system utilizes inductive coupling and integrates components like solar panels, lithium-ion batteries, and optimized coils to ensure efficient power transfer. Despite challenges at higher vehicle speeds, such as coil misalignment and transmission inefficiencies, the system achieved power transfer efficiencies above 80% at lower speeds. Future research will focus on improving charging efficiency at higher speeds, optimizing coil structures, and facilitating large-scale deployment. With its innovative design, this system has the potential to revolutionize EV charging, making it more efficient, sustainable, and scalable for widespread adoption [14].

11 Results and Discussions

The experimental evaluation of the proposed system focused on analysing power transfer efficiency at different vehicle speeds. The results demonstrated that while efficiency remained above 80% at lower speeds (10–20 km/h), it declined to approximately 65% at higher speeds (40–60 km/h). This reduction was primarily attributed to coil misalignment and transmission losses, which significantly impacted energy transfer consistency.

Notably, power attenuation at higher velocities and misalignment between the transmitter and receiver coils led to efficiency drops, highlighting the need for further refinement. Addressing these inefficiencies requires optimizing coil structures, implementing real-time alignment correction mechanisms, and exploring advanced resonant tuning techniques. Additionally, using high-permeability materials may enhance power transfer consistency, ensuring improved system performance across varying speeds [15].

12 Future Scope

To enhance the effectiveness of Sun-driven contactless real-time charging technologies, it is essential to pursue standardization across various EV models and charging systems. This will enable smooth integration and ensure compatibility, allowing EV owners to easily use the system regardless of their vehicle model. Additionally, it is important to plan for the scalability and future expansion of the wireless charging infrastructure. As electric vehicle adoption continues to rise, designing the system to integrate effortlessly with existing road networks will allow for broader coverage and accommodate the growing number of EVs.

Furthermore, opportunities should be explored to connect the wireless charging infrastructure with smart grid systems. This integration would enable dynamic management of charging processes, allowing the use of off-peak electricity or surplus renewable energy for optimized operations. Such a system could also support demand response strategies, enhance grid stability, and improve the overall efficiency of renewable energy utilization.

Acknowledgements We would like to express our deepest gratitude to all those who have contributed to the success of this project, and sincerely thank our esteemed faculty members and mentors for their invaluable guidance and support. We would also like to extend our appreciation to our department, Department of Electronics and Instrumentation Engineering, RNS Institute of Technology, for providing the necessary resources and infrastructure for the completion of this project. Special thanks to our families and friends for their encouragement and patience throughout the course of this work. Their constant support and understanding have been indispensable in helping us overcome challenges. Lastly, we would like to acknowledge the pioneering researchers and authors whose works have served as the foundation of this project. Their publications have inspired us and have been instrumental in shaping our understanding of the technology involved.

References

1. Supriyadi ER, Suyanto AR, Basjaruddin NC (2018) Development of a wireless power transfer circuit based on inductive coupling. Telkomnika 16(3):1013–1018
2. Banu N, Arun Kumar U, Gokula Kannan A, Hari Prasad MK, Shathish Sharma AB (2018) Wireless power transfer in electric vehicle by using solar energy. Asian J Electr Sci 7(1)
3. Yatnalkar G, Narman H (2018) Survey on wireless charging and placement of stations for electric vehicles. In: Conference: 2018 IEEE international symposium on signal processing and information technology (ISSPIT)
4. Rana MM, Xiang W (2018) Internet of things infrastructure for wireless power transfer systems. IEEE Access 6:19295–19303
5. Rupani S, Doshi N (2019) A review on smart parking using internet on things (IOT). In: The 3rd international workshop on recent advances on internet of things: technology and application approches (IoT-T&A 2019), Coimbra, Portugal (2019)
6. Timpner J, Wolf L (2012) A back-end system for an autonomous parking and charging system for electrical vehicles. In: International electrical vehicle conference Greenville, SC, USA. IEEE (2012)
7. Kuran MS, Viana AC, Iannone L, Kofman D, Mermound G, Vasseur JP A smart parking
8. Sultanbek A, Khassenov A, Kanapyanov Y, Kenzhegaliyeva M, Nagheri M (2017) Intelligent wireless charging station for electrical vehicles. In: International Siberian conference on control and communication. IEEE
9. Wei Z, Li Y, Zhang Y, Cal L (2018) Intelligent parking garage EV charging scheduling considering battery charging characteristic. IEEE Trans Industr Electron 65:3
10. Sultanbek A (2017) Intelligent wireless charging station for electrical vehicles. In: International Siberian conference on control and communication
11. Wei Z et al (2018) A smart parking garage EV charging scheduling. IEEE Trans Ind Electron
12. Kuran MS et al (2018) Inductive power transfer efficiency optimization. IEEE Trans Wirel Power Transf 6:45–52
13. Rupani S et al (2018) Dynamic wireless charging for electric vehicles. IEEE Int Conf Smart Mobil 4:112–118
14. Wei Z et al (2018) Efficiency analysis of wireless EV charging. IEEE Trans Transp Electrif 5:98–106
15. Future Scope of Wireless Charging for EVs 2019. In: IEEE conference on future technologies in EV charging, vol 3, pp 205–212 (2019)

Sustainable Energy Management in 4-Bus System Using PEMFC, Demand Response, and Dynamic Pricing

G. Sreeramulu Mahesh, N. L. Anusha, K. Bhanushree, Y. Naithanya, and S. Rakshitha

Abstract This paper investigates how sustainable energy management can be integrated in a customized IEEE 4-bus system with Proton Exchange Membrane Fuel Cells (PEMFC), demand response mechanisms, and dynamic pricing strategies. In this paper, the simulation model is carried out to analyze energy output, operational costs, NOx and water vapor emissions, load profiles, and demand response behavior under various operation conditions. This work explores the critical research gaps, including the dynamic optimization of PEMFC performance under fluctuating load and pricing conditions and the integration of demand response strategies to reduce costs and enhance grid stability. The novelty of this work lies in the integration of PEMFC with real-time demand response mechanisms, evaluating its feasibility for decentralized, sustainable energy systems. The combination of PEMFC and the customized 4 bus system is modeling and analyzed using pandapower library in Python and the simulation results ensure to enhance the strategies like smart energy management and reduce emissions while continuing to adapt to dynamic market conditions.

Keywords Pandapower · Demand response · Dynamic Pricing · NOx emissions · PEMFC

1 Introduction

Higher demand for energy renders the electricity system cleaner and reliable, especially in homes. There is a solution to this dilemma in the sustainable energy carrier form of hydrogen. This paper takes an overview of hydrogen-based power generation, emphasizing Proton Exchange Membrane Fuel Cells (PEMFCs). Hydrogen-based power systems offer zero-emission energy generation potential, aligning with global efforts to reduce carbon footprints and mitigate climate change. PEMFCs, known

---

G. Sreeramulu Mahesh · N. L. Anusha · K. Bhanushree (✉) · Y. Naithanya · S. Rakshitha
Department of Electrical and Electronics Engineering, GSSS Institute of Engineering and Technology for Women, Visvesvaraya Technological University, Metagalli, Mysuru, India
e-mail: bhanushreek162003@gmail.com

I. M. Jamadar et al. (eds.), *Integrating Sustainability in Modern Technologies*, Lecture Notes in Mechanical Engineering, https://doi.org/10.1007/978-981-95-9466-5_2

for their high efficiency and clean operation, are increasingly considered for residential and commercial power applications. Hydrogen combustion systems, though less efficient, provide a simpler and potentially more cost-effective alternative. Understanding the trade-offs between these technologies in terms of efficiency, cost, environmental impact, and operational feasibility is crucial for advancing sustainable residential energy solutions.

The performance analysis reveals that 46.44% of energy demand is met by the PEMFC-based distributed generation system, with significant contributions from PV and wind energy sources [1]. This study explores the impact of micro-porous layers on the performance and flooding behavior of PEM fuel cells, offering insights into efficiency and durability improvements. It provides both numerical simulations and experimental data to optimize fuel cell design for better water management and performance [2]. Proton Exchange Membrane Fuel Cells (PEMFCs) efficiently generate power in grid-connected systems, enhancing distributed generation through improved energy management and renewable integration [3]. The study presents a state machine control strategy for a PV-PEMFC-battery system, enhancing efficiency by 2.3% and maintaining a power supply with 96.5% reliability under varying conditions [4].

This research explores the impact of innovative flow field designs on Proton Exchange Membrane Fuel Cell (PEMFC) performance, aiming to enhance efficiency and water management. It provides valuable insights for advancing PEMFC technology and improving operational stability [5]. Summarize nine groups of PEMFC system control strategies, control complexity [6]. Recent advancements in Proton Exchange Membrane (PEM) fuel cells have focused on improving efficiency, durability, and cost effectiveness, making them a key technology for sustainable energy solutions. These developments are crucial for advancing clean energy adoption in transportation and stationary power generation [7]. This study highlights the impact of reaction parameters on Proton Exchange Membrane Fuel Cell (PEMFC) performance, revealing the synergistic effects through power curve and impedance spectroscopy analysis. Interdependency investigation provides information on PEMFC efficiency improvements and longevity [8]. Demand Response (DR) helps maintain a balance between supply and demand, renewable energy penetration, and grid stability, which further determines DR benefits within a sustainable energy system. It helps with the optimization of energy utilization, thus potentially reducing expenses and driving down fossil fuel dependency. It would thus enable a more robust and sustainable energy infrastructure by aligning consumer behavior with grid needs [9]. The current research work in the field of Proton Exchange Membrane Fuel Cells (PEMFC) is focused on its efficiency improvements, higher durability, and lower cost to promote wider adoption of the technology for clean energy applications. It majorly also deals with the challenges of improving catalyst performance, material stability, and scaling to large applications [10, 11].

PEMFCs perform as chemical reactors for chemical synthesis in addition to electricity generation. These have implications in gas separation, water treatment, and energy use through mild reaction conditions and high product selectivity [12]. Some of the most important operational parameters are air and fuel flow rate and pressure.

For simulation purposes, Ahmed et al. analyzed a simplified PEMFC model, whereas Lee et al. measured the pressure and flow rate using micro-sensors to examine how variations in fuel and air supply affect the behavior of PEMFC [13]. Further, ex-situ and in-situ activation methods of PEM fuel cells by extensive research showed the detailed changes in morphological behavior that occur during the break-in period [14]. An important focus area for significant research activity has been in hybrid energy storage system-integrated PEMFCs that help optimize electric vehicle performance. A battery-super capacitor hybrid setup would then handle sustainability in terms of energy supply as well as transient operation and offer performance predictability for any intermittent working condition [15].

Specifically in residential grid applications, there exist relatively unknown problems such as PEMFCs pertaining to water management issues, durability concerns, and instability of the working performance. The present research lacks dynamic performance optimization techniques that allow PEMFCs to adapt to dynamic loads and fluctuating pricing mechanisms, thus making them more efficient and less emitting. There is currently no evaluation of PEMFC durability at different environmental conditions with variable grid demands. More analyses are also performed concerning long-term performance deterioration and economic comparability of PEMFCs with other forms of generation. This paucity of data indicates that there will be a need for field tests along with life and economic assessments in future work concerning PEMFC as a valid option in contemporary energy systems.

The organization of this paper is as follows: Sect. 2 discusses comparison between PEMFC and Hydrogen combustion systems, Sect. 3 discusses on Role of Demand Response in Enhancing Grid Efficiency and Stability; Sect. 4 on Mathematical Framework of energy output, Operational costs, Nox emissions, and water vapor emissions; Sect. 6 defines the algorithm; and Sect. 7 discusses the results and discussion followed by conclusion.

2 Comparison of PEMFC and Hydrogen Combustion Systems

- **Efficiency**: PEMFC systems are more efficient (50–60%) compared to hydrogen combustion systems (25–30%).
- **Fuel Consumption**: PEMFC systems consume less hydrogen due to higher efficiency.
- **Environmental Impact**: PEMFC systems produce negligible NOx emissions, while hydrogen combustion systems produce moderate NOx emissions.
- **Operational Costs**: PEMFC systems have higher maintenance costs due to stack replacements but lower overall fuel costs.
- **Suitability for Residential Applications**: PEMFC systems are more suitable due to their higher efficiency, lower emissions, and quieter operation.

3 Role of Demand Response in Enhancing Grid Efficiency and Stability

Demand Response and its contribution to improving efficiency and stability of grid demand response—the way of balancing the demand with supply in terms of dynamic prices with cost optimization considerations, grid stability, and providing incentives for smart consumers. It includes the importance of demand responses which are mentioned below.

i. Dynamic load adjustment-this kind of DR can have instant transformation of energy demand according to electricity price and grid conditions. In the outline (under load profile and demand response plots), the load is confined within the specific range, which is from 3 to 6 MW, thus optimizing the usage of energy.
ii. To reduce the operational expense of the system as reflected in the variation of operational costs with time, it has had dynamic variations of demand in result to changes in price (dynamic pricing plot). Demand peaks are seen in line with price fluctuations to ensure that stability and cost efficiency are maintained on the grid Demand Response.
iii. DR reduces stress on the grid during peak pricing or high-load periods, as observed in the adjusted load profiles. This enhances the reliability of the grid while supporting renewable energy integration.
iv. Dynamic pricing, paired with DR, incentivizes consumers to shift their energy consumption during off-peak periods, leading to better load management and reduced costs.

4 Mathematical Framework

The 4-bus system is shown in Fig. 1, in the context of PEMFC (Proton Exchange Membrane Fuel Cells) refers to a power distribution strategy. It involves dividing the fuel cell system into four primary sections: the air supply, fuel supply, power output, and thermal management. Each "bus" is responsible for managing specific energy flows to optimize the overall efficiency and stability of the PEMFC. This approach helps balance load, improve performance, and ensure reliable operation in fuel cell systems.

The primary objectives of this paper are to model and simulate a customized IEEE 4-bus system integrated with Proton Exchange Membrane Fuel Cells (PEMFC) to evaluate its potential for sustainable energy management. This work explores on integrating DR strategy with dynamic pricing strategies to optimize load management, reduce operational costs, and enhance grid stability. It aims to address key challenges such as the dynamic optimization of PEMFC performance under fluctuating load and pricing conditions, as well as this integration achieves support efficient and decentralized energy systems. Additionally, this analysis contributes on PEMFC efficiency, environmental impacts (NOx and water vapor emissions), and

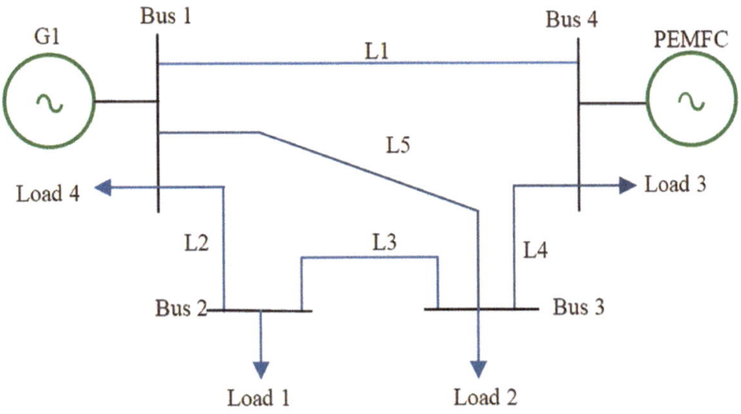

Fig. 1 4 Bus system with generator 1 and PEMFC

overall system behavior through comprehensive simulations and visualizations. By achieving these objectives, the paper provides important perspectives on the practical use of PEMFC-based systems in residential grids and their contribution towards achieving decarbonization and energy efficiency targets.

5 Energy Output Calculation

Energy output from generators in the system is calculated as per Eq. (1)

$$E_{gen} = \sum_{i=0}^{n} P_{gen,i}(MW) \quad (1)$$

where $P_{gen,i}$ is the power output from the ith generator in megawatts (MW), and n is the total number of generators.

Operational Costs

Operational costs are calculated based on dynamic pricing and total energy output over time as per Eq. (2)

$$\text{Cost} = P_{gen} \times \text{Price (USD)} \quad (2)$$

where price is the dynamic price of electricity at each time step, and P_{gen} is the energy generated at that time.

NOx Emissions

NOx emissions are proportional to the energy produced by the generators as per Eq. (3)

$$\text{NOx Emissions} = \text{Fuel Consumption (kg/h)} \times \text{NOx Emission Factor} \quad (3)$$

Assuming a constant factor of 0.1 kg NOx per MW of energy generated.

Water Vapor Emissions

Water vapor emissions from hydrogen-based generators (PEMFC) are proportional to the energy output as per Eq. (4)

$$H_2O \text{ Emissions} = \text{Fuel Consumption (kg/h)} \times 8.93 \left(\text{kg}H_2O\right) \quad (4)$$

where 8.93 kg of water vapor is produced per MW of energy.

Load Profiles

Load profiles are determined based on time-series data representing the demand across different buses.

Demand Response

Demand response adjusts the load at each time step based on dynamic pricing, as described in Eq. (5)

$$\text{Demand Response} = \min\left(\max\left(L \times \text{Price}/100, 3\right), 6\right) \text{(MW)} \quad (5)$$

where L is the base load, and price is the dynamic price at that time.

6 Algorithm

Step 1: Initialize Network

- Create an IEEE 4-bus electrical network using the pandapower library.
- Define:
  a. Buses: Create four buses with appropriate voltage levels (e.g., 138 kV).
  b. Lines: Define transmission lines between the buses, specifying parameters such as resistance, reactance, capacitance, and maximum current.
  c. Generators: Set up a slack bus (reference bus) with one generator and additional controllable generators at other buses.
  d. Loads: Initialize steady-state loads at relevant buses.
- Ensure the network is correctly connected and validates power flow using the pandapower create_* methods.

Step 2: Define Time Series

- Define time-series data for 24 h:
- Load Profiles: Generate or import time-varying load data for each bus (e.g., residential or industrial demand). These should reflect realistic fluctuations over the day.
- Dynamic Pricing: Assign dynamic pricing values in $/MWh that fluctuate based on grid demand and supply conditions.
- Demand Response: Create DR logic to dynamically adjust loads based on pricing.

Step 3: Simulate for Each Time Step

- For each hour (t = 0, 1, 2,…, 23):

  a. Update Loads: Modify loads at each bus based on demand response mechanisms and the current pricing. Use the equation or logic defined in Step 2.
  b. Run Power Flow Calculation:
     i. Perform a power flow analysis using pandapower's runpp() method to solve the network and determine operational parameters such as voltage, currents, and power at each bus.
     ii. Ensure convergence of the power flow calculation. If convergence fails, adjust parameters or refine input data for better numerical stability.

Step 4: Collect Data

- For each time step, collect the following system-level metrics:

  a. Energy Output: Total power generated by all generators in MW.
  b. Operational Costs: Calculate costs using dynamic pricing and energy output.
  c. Emissions: Calculate the NOx and water vapor emissions.
  d. Load Profiles: Record adjusted loads at each bus after demand response is applied.

Step 5: Plot Results

- Generate plots to visualize system behavior over time, including:

  a. Energy Output Over Time: Total energy generation across all generators.
  b. Operational Costs Over Time: System-wide costs influenced by dynamic pricing and load adjustments.
  c. NOx and Water Vapor Emissions: Environmental impact over time.
  d. Load Profiles and Demand Response: Loads are adjusted dynamically.
  e. Dynamic Pricing: Reflect price changes over 24 h.

- Use Python plotting libraries such as matplotlib to create clear and interpretable visualizations.

7 Results and Discussion

Figure 2 shows the simulation of the system, the following observations are made from the plots:

Energy Output Over Time (Top-left): This graph shows the total energy output (in MW) of the system over a 24-h period. The energy output appears to fluctuate over time, with variations likely driven by changes in demand response or dynamic pricing. The peaks in the energy output suggest times when either demand or pricing may have incentivized more power generation to meet higher loads or take advantage of lower prices. The system's power output ranges between 8.25 and 9.75 MW, indicating significant changes in demand or generation capacity during the day.

Operational Costs Over Time (Top-right): This graph illustrates the operational costs (in USD) over time, which depend on both the energy output and the dynamic pricing. The operational costs show a pattern that correlates with energy output, but the overall cost is highly dependent on the electricity price at a given time. Peaks in operational costs correspond to times when the price of electricity is high, and the

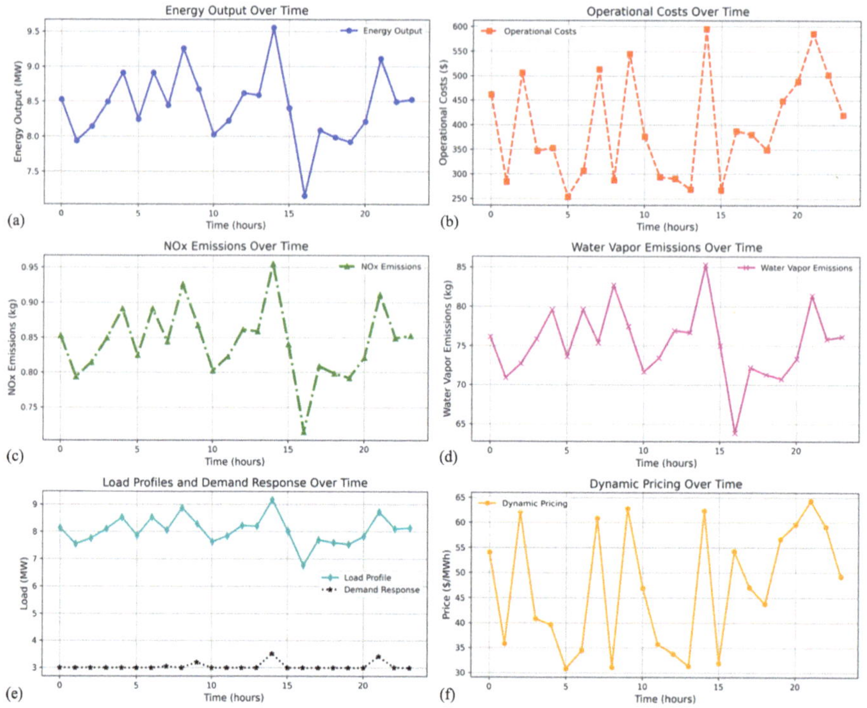

Fig. 2 Output waveforms of **a** Energy output **b** Operational cost **c** Nox emissions **d** Water vapor emissions **e** Load profiles and demand response **f** Dynamic pricing

system is generating more power. The operational costs fluctuate between 300 and 600 USD over the 24 h.

NOx Emissions Over Time (Middle-left): The NOx emissions (in kg) graph shows the emissions generated by the system over time. These emissions are directly proportional to the amount of energy produced, with higher energy outputs leading to higher NOx emissions. The emissions show similar behavior to the energy output graph, indicating that when more power is generated, more NOx emissions are produced. The NOx emissions range from around 0.80 to 0.975 kg during the day.

Water Vapor Emissions Over Time (Middle-right): This graph shows the water vapor emissions (in kg), which are specific to hydrogen-based systems like PEMFCs. Water is produced as waste in the PEMFC, the graph shows the amount of water vapur produced as a function of the hydrogen consumption in generating power. The pattern is the same as that of energy output and NOx emissions and implies that the increased energy production is accompanied by an increase in water vapor emission. Water vapor emissions are in the range of 72–86 kg over the period of 24 h [16].

Load Profiles and Demand Response Over Time (Bottom-left): This graph shows the load profile and demand response over time (both in MW). The blue line represents the base load profile, which is defined as electricity demand, before any adjustments. The orange line represents the demand adjusted and therefore demand response, which is dynamic price-based load shifting. Notice how much less demand response variance from the load profile where all load adjustments are being made here to dynamic pricing. The load profile fluctuated from 775 to 9 MW and the demand response was 3 and 4 MW limited.

Dynamic Pricing Over Time (Bottom-right): This graph shows the dynamic pricing (in $/MWh) over 24 h. Dynamic pricing fluctuates based on grid conditions, demand, and energy availability. In this case, pricing oscillates between 40 and 70 $/MWh. The times when pricing is higher may coincide with periods of high demand, lower generation capacity, or higher operational costs, while lower prices could indicate reduced demand or greater generation availability.

The simulation results are analyzed with the existing standards and guidelines from various organizations and industries that address demand response, fuel cells, and grid operations. The energy output, operational costs, and emissions align with alternative standards like IEC 62,282 for PEMFCs and FERC guidelines for demand response. Water vapor emissions and NOx levels adhere to benchmarks set by EPA and IEC standards, confirming environmentally sustainable performance. Additionally, load profiles and dynamic pricing values fall within the acceptable ranges outlined by NERC and ISO 50001, demonstrating effective integration of demand response strategies. The comparison of simulation parameters is tabulated below in Table 1.

Table 1 Comparison of simulated results with standards/guidelines

Parameter	Simulated value	Standards/guidelines	Remarks
Energy output (MW)	7.0–9.5 MW	IEC 62,282–3-100 (Fuel cell power systems)	Provides performance testing methods for PEMFC systems
Operational costs ($/h)	$300–$650	ISO 50001 (Energy management systems)	Guides cost-effective energy usage; aligns with dynamic pricing principles
NOx emissions (kg/h)	0.7–0.9 kg	EPA clean air act (Emissions standards for fuel-based systems)	EPA recommends near-zero NOx emissions for hydrogen-based fuel systems
Water vapor emissions (kg/h)	62.5–82.5 kg	IEC 62,282 (Hydrogen and fuel cell technologies)	Aligns with expected water vapor emissions for PEMFC operation
Load profiles (MW)	3.0–9.0 MW	FERC demand response guidelines (United States)	Reflects realistic load variability for residential/ commercial demand
Dynamic pricing ($/ MWh)	$30–$70	NERC demand response standards	Covers pricing models for effective grid management and DR incentive

8 Conclusion

Sustainable energy management in a 4-bus system using Proton Exchange Membrane Fuel Cells (PEMFC) offers significant advantages in terms of efficiency, environmental sustainability, and energy reliability. The integration of PEMFCs into the system enhances the overall performance by reducing greenhouse gas emissions and dependency on traditional fossil fuels. The fuel cells provide a clean and efficient power source, especially in off-grid or remote locations, while their ability to produce electricity with high efficiency at various load demands ensures optimal energy use. Additionally, the flexibility of PEMFCs in providing both power and heat (cogeneration) supports the overall energy efficiency of the system.

The scope for future work in sustainable energy management using PEMFCs in a 4-bus system includes exploring advanced control strategies for better load distribution and optimizing fuel cell performance under varying conditions. Future research could focus on improving hydrogen production and storage technologies to reduce costs and enhance system efficiency. Additionally, integrating renewable energy sources like solar and wind with PEMFCs could further increase sustainability and grid resilience. Long-term developments may involve the scaling of PEMFC-based systems for larger networks and smart grid applications, contributing to a more sustainable and decentralized energy infrastructure.

References

1. PEM fuel cells analysis for grid connected applications. Int J Hydrog Energy 36(17):10908–10916 (2011). ISSN 0360-3199
2. Belvedere B, Bianchi M, Borghetti A, De Pascale A, Paolone M, Vecci R (2013) Experimental analysis of a PEM fuel cell performance at variable load with anodic exhaust management optimization. Int J Hydrog Energy 38 (1)
3. Efficiency analysis and control of a grid-connected PEM fuel cell in distributed generation. Energy Convers Manage 195:587–596 (2019). ISSN 0196-8904
4. Two-level energy management strategy for PV-Fuel cell-battery-based DC microgrid. Int J Hydrog Energy 44(35):19395–19404 (2019). ISSN 0360-3199
5. Cai Y, Yue S, Wei F, Hu J, Chen B (2023) Research on performance of proton exchange membrane fuel cell with an innovative flow field. Case Stud Therm Eng 50
6. Yang B, Li J, Li Y, Guo Z, Zeng K, Shu H, Cao P, Ren Y (2022) A critical survey of proton exchange membrane fuel cell system control: summaries, advances, and perspectives. Int J Hydrog Energy 47(17):9986–10020. ISSN 0360-3199
7. Song WJ, Chen H, Guo H, Ye F, Li JR (2022) Research progress of proton exchange membrane fuel cells utilizing in high altitude environments. Int J Hydrog Energy 47(59) (2022)
8. Zhu X, Zhou W, Zhu Z, Liu R, Lian Y, Chen R, Wu L, Ji D (2023) Performance analysis of proton exchange membrane fuel cells with traveling-wave flow fields based on Grey-relational theory. Int J Hydrog Energy 48(2)
9. Shariatzadeh F, Mandal P, Srivastava AK (2015) Demand response for sustainable energy systems: a review, application and implementation strategy. Renew Sustain Energy Rev 45
10. da Costa GP, Garcia DME, Nguyen THV, Lacharmoise P, Simão CD (2024) Advancements in printed components for proton exchange membrane fuel cells: a comprehensive review. Int J Hydrog Energy 69
11. Simulation of fuel cell membrane durability under vehicle operation. J Power Sources 613:234855 (2024). ISSN 0378-7753
12. Pei P, Xu Y, Wang M, Ren P (2024) Effects of carbon monoxide on proton exchange membrane fuel cells and elimination techniques. Int J Hydrog Energy 69:1287–1304 (2024). ISSN 0360-3199
13. Design of proton exchange membrane fuel cell grid-connected system based on resonant current controller. Int J Hydrog Energy 39(26):14402–14410 (2014). ISSN 0360-3199
14. Van Der Linden F, Pahon E, Morando S, Bouquain D (2023) A review on the proton-exchange membrane fuel cell breaking physical principles, activation procedures, and characterization methods. J Power Sources 575:233168 (2023). ISSN 0378-7753
15. Energy management system for hybrid PEMFC-battery power source for stationary consumers. Int J Hydrog Energy 55:1109–1121 (2024). ISSN 0360-3199
16. Chantzis G, Giama E, Nižetić S, Papadopoulos AM (2023) The potential of demand response as a tool for decarbonization in the energy transition. Energy Build 296

Design and Fabrication of a Sustainable, Thermally Driven Air Conditioning System for Energy-Efficient Applications

G. Saravanan, A. Thanikasalam, A. Maniram Kumar, M. Pradeep, and Mohammed Asif Kattimani

Abstract The increasing demand for sustainable cooling technologies has driven the development of thermally driven systems. This study focuses on the design and construction of a lithium bromide-water absorption chiller, a thermally powered air conditioning system. The system leverages water as a refrigerant and utilizes the hygroscopic properties of lithium bromide to enable continuous cooling without high-energy compression. A comprehensive analysis of the parameters influencing system performance was conducted to optimize its design and efficiency. The study involved detailed evaluations of factors such as solution concentration, flow rates, and heat exchanger effectiveness to enhance performance metrics. Performance evaluation, based on power consumption and coefficient of performance (COP), revealed that the lithium bromide absorption chiller is 80% more efficient than traditional vapor compression systems. These findings underscore the system's potential to provide energy-efficient and sustainable cooling solutions for modern applications, addressing global energy concerns and reducing the environmental impact of conventional cooling technologies.

Keywords Thermally driven system · Lithium-bromide · Air conditioning · Refrigerant · Power consumption · COP

G. Saravanan · M. Pradeep
Department of Mechanical Engineering, St. Joseph's College of Engineering and Technology, Elupatty, Elupatty, Thanjavur, Tamil Nadu, India

A. Thanikasalam
Department of Marine Engineering, Academy of Maritime Education and Training, Kanathur, Chennai, Tamil Nadu, India

A. M. Kumar
Department of Mechanical Engineering, Dr. Sivanthi Aditanar College of Engineering, Tiruchendur, Tamil Nadu, India

M. A. Kattimani (✉)
Department of Mechanical Engineering, Lords Institute of Engineering and Technology, Hyderabad, Telanagana, India
e-mail: md.asifkattimani@lords.ac.in

I. M. Jamadar et al. (eds.), *Integrating Sustainability in Modern Technologies*, Lecture Notes in Mechanical Engineering, https://doi.org/10.1007/978-981-95-9466-5_3

1 Introduction

Regulating the temperature and humidity of air conditioning (AC) improves indoor comfort. HVAC (Heating, Ventilation, and Air Conditioning) technology, which aims to enhance indoor comfort and air quality, relies on AC systems that can be mechanical or utilize passive and ventilative cooling methods. Heat pumps, similar to air conditioners, are versatile systems capable of reversing their flow to heat or cool an area, making them adaptable to a variety of climates. Buildings account for nearly 40% of the world's energy consumption, a figure that is increasing with economic growth. This necessitates the reduction of energy use and CO_2 emissions. To address this challenge, researchers have explored the potential of solar energy for generating electricity, heating, cooling, and producing hot water. Recent studies have explored hybrid systems combining Li-Br absorption technology with photovoltaic and thermal storage systems to optimize energy utilization [1]. Another critical advancement is the application of nanofluids in Li-Br systems to enhance heat and mass transfer performance [2]. These developments underline the growing interest in making absorption cooling systems more sustainable and efficient. Existing studies, such as [3–5], demonstrate how solar energy can be harnessed to provide sustainable power and cooling. Refrigeration, a critical component of air conditioning, removes heat to cool spaces, serving applications like food preservation, industrial operations, and individual comfort. Vapor compression refrigeration remains the dominant method and forms the foundation of most modern AC systems. Alternative cooling technologies include thermoelectric, magnetic, steam jet, and evaporative cooling. Interest in solar-powered cooling has grown, particularly in arid regions with limited access to electricity. For instance, authors in [6] emphasized the environmental benefits of solar cooling, while authors in [5] categorized solar cooling technologies into two primary forms: solar cooling for water and solar thermal for heating water. These classifications are built upon earlier work by Best and others [7].

Research on thermally driven cooling systems has evolved significantly. Early efforts, such as those by [8, 9], introduced absorption refrigeration in 1815 as an alternative to vapor compression systems. Lithium bromide and water (Li-Br-Water), developed in the 1930s, emerged as a promising refrigerant-absorbent pair. The high heat of vaporization of water combined with the non-volatility of Li-Br makes this combination highly effective. However, these systems require vacuum operation to prevent crystallization at high Li-Br concentrations. Salts like $ZnBr_2$ or $ZnCl_2$ can mitigate this issue but introduce additional complexities. Advanced absorption cycles, such as the multi-stage Four-Pressure Triple-Effect Cycle, have been proposed to address the environmental concerns associated with harmful refrigerants. These cycles reuse heat multiple times, achieving higher efficiency and better coefficients of performance (COP) [10, 11]. Despite these advancements, high operating temperatures remain a challenge, potentially leading to long-term corrosion issues. The existing literature highlights the limitations of traditional vapor compression systems and the potential of thermally driven alternatives like Li-Br-Water absorption chillers.

However, current research lacks a comprehensive analysis of how parameter optimization can further enhance these systems' performance. This study addresses these gaps by evaluating the design and performance of a Li-Br-Water absorption chiller, emphasizing its potential as an energy-efficient and sustainable solution for modern cooling demands.

2 Design and Fabrication of Li-Br-H_2O Air Conditioning System

The evaporator sends a low-temperature low-pressure refrigerant to the absorber where it combines with the solution of lithium bromide (Li-Br). Refrigerant is driven to the condenser by a weak high-pressure solution created by the generator's heat input. Before going back to the evaporator, the refrigerant condenses into a liquid in the condenser and goes through a throttling valve to lower its pressure. After absorbing heat there, it vaporizes creating the refrigeration effect. As the vapor returns to the absorber the cycle repeats as Fig. 1 illustrates.

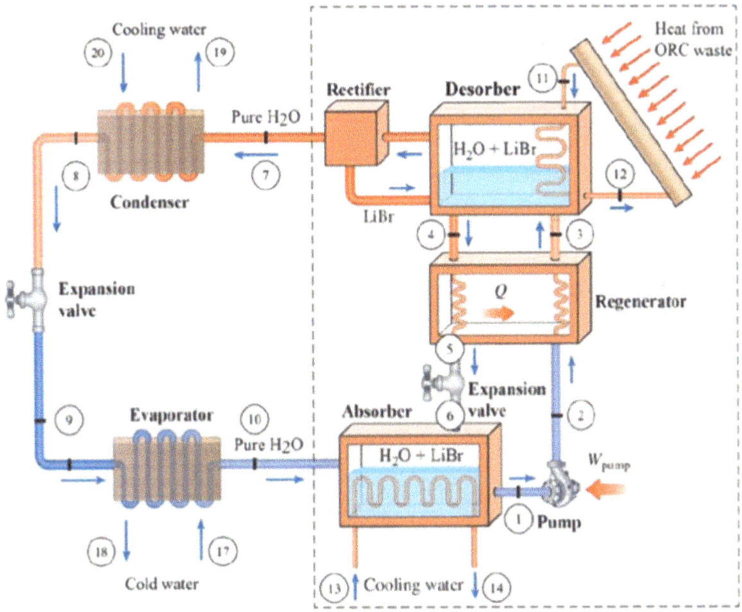

Fig. 1 Working of Lithium-bromide Vapour Absorption System

2.1 Condenser Unit

In designing an effective air-conditioning system, the size and type of the condenser is calculated by the following equation.

$$Q_c = U \times A \times \Delta T = M_w C_p (T_{\text{out}} - T_{\text{in}}) \tag{1}$$

The system has a one-ton capacity and is designed with a lithium bromide (Li-Br) air conditioning system. A readymade shell-and-tube-type condenser unit was purchased from a scrapped air-conditioning system. The condenser unit, a critical component of an air conditioner, plays a vital role in the cooling process and is typically made from copper. Copper is preferred due to its superior heat transfer properties and durability. Equipped with a fan for efficient heat dissipation, the unit operates with a power input typically ranging from 1,000 to 1,500 W at 220–240V, 50/60 Hz, suitable for residential applications. It consists of the following components: The interior unit dimensions are 150 mm × 240 mm × 220 mm, while the exterior unit has a condenser measuring 550 mm × 675 mm × 284 mm, with a weight range of 30–60 kg. The cold storage cooling load is approximately 600 square feet per ton, as shown in Fig. 2. The air conditioner's design parameters, including optimal operating conditions and efficiency metrics, are detailed in Table 1

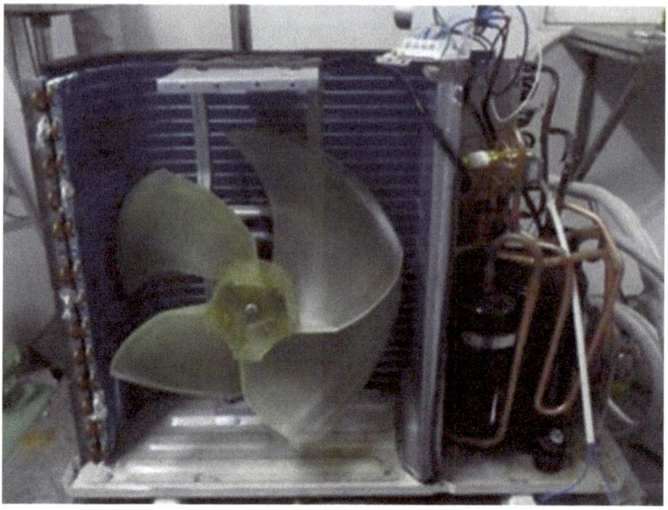

Fig. 2 Condenser unit

Table 1 Design parameters of the air conditioner

S.No	Design parameters	Value
1	Ambient air temperature	28
2	Outdoor air relative humidity (%)	51
3	Wind speed (mph)	8
4	Indoor air temperature (°C)	22

2.2 *Generator Unit*

A stainless-steel pipe with a diameter of 75 mm and a length of 30 in. is used for the generator. One end of the pipe is sealed, and a re-heater is installed on the other end. Twelve-millimeter holes are drilled on the top and bottom faces of the pipe to serve as the inlet and outlet for the refrigerant. The pipe is slightly inclined at an angle of 30–45°, allowing the refrigerant to be heated at one end while the vapors escape from the other. A needle valve and a pressure gauge are attached to monitor the pressure inside the tank, as shown in Fig. 3.

$$Q_g = \dot{m}h_7 + \dot{m}_{ss}h_4 - m_{ws}h_3 \tag{2}$$

$$Q_g = (M_1h_1 + M_7h_7)(M_3h_3 + M_{12}h_{12}) \tag{3}$$

$$Q_g = (14.993 \times 1775 + 36.37 \times 330) - (0.363 \times 325 + 41 \times 92)$$

$$Q_g = 16974.2\frac{\text{kJ}}{\text{min}} \quad OR \quad 282.9\,\text{KW}$$

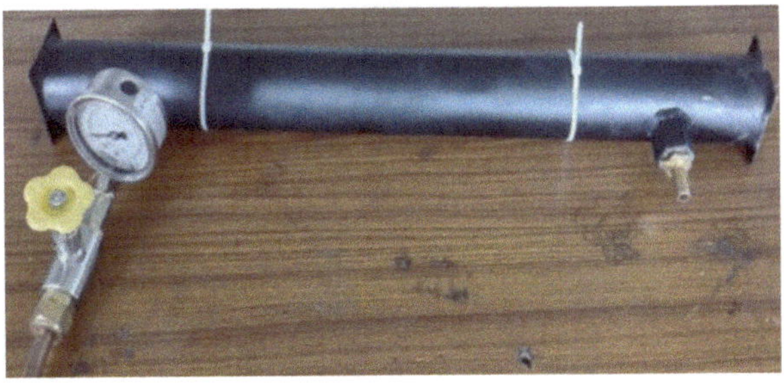

Fig. 3 Generator unit

2.3 Absorber Unit

An absorber is designed using a 110 mm PVC pipe, sealed at both ends with appropriate PVC end caps, and fitted with connections for the inlet and outlet. A pipe connects the absorber to the generator, and an electrical output is provided to ensure the safe operation of the pump. The vapor inlet, located at the top or near the top on the side, allows the refrigerant vapor to enter, while the absorbent solution inlet at the top distributes the absorbent liquid over the packing material. Spray nozzles or distributors at the top of the packing ensure the even distribution of the absorbent solution. The absorption tank can hold up to 4–5 L of the working fluid, such as the lithium bromide solution. A 9 W water pump with a lifting capacity of 750 L per hour is used in the system. The absorber setup is shown in Fig. 4.

$$Q_a = (M_9 h_9 + M_6 h_6 - M_{10} h_{10}) \quad (4)$$

$$Q_a = (36.37 \times 230 + 4.63 \times 1290 - 41 \times 62)$$

$$Q_a = 11795.8\,\text{kJ/min} \;\; OR \;\; 196.6\,\text{KW}$$

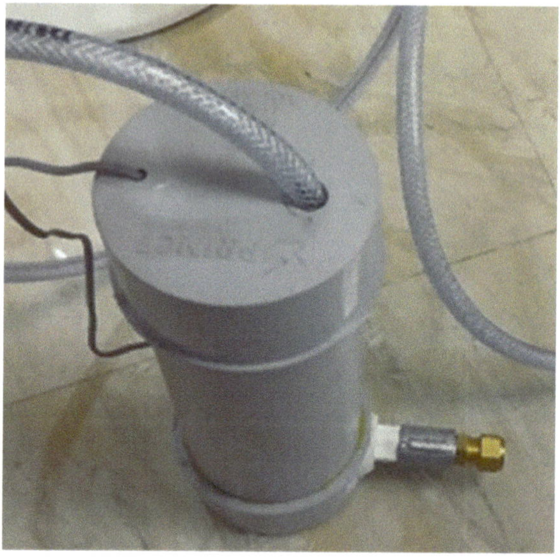

Fig. 4 Tank with absorber unit

2.4 Evaporator Unit

A one-ton indoor unit was purchased from an old scrapped air conditioning system with the following specifications: height of 25 in., width of 31.5 in., depth of 10.5 in., and a typical weight ranging from 10 to 15 kg as shown in Fig. 5. The housing is made from high-quality galvanized steel for durability and aesthetic appeal. The evaporator coil consists of copper tubes with aluminum fins, with copper chosen for its excellent thermal conductivity and durability. Filters are made from synthetic fibers or mesh materials, and some units may include advanced filters like HEPA or activated carbon filters. The fan type is a blower fan, typically a centrifugal fan, for efficient air distribution and is made from durable plastic or metal. Airflow usually ranges between 400 and 600 cubic feet per minute, depending on the design and manufacturer. The power supply operates at a voltage of 220–240 V, 50/60 Hz.

3 Results and Discussion

Temperatures at the generator absorber and condenser have a direct impact on the coefficient of performance (COP) of the lithium bromide (Li-Br) absorption refrigeration system. The condenser and absorber temperatures are frequently influenced by external factors but the generator temperature is a controllable design variable. System design optimization is possible through the combined analysis of these parameters. The following section discusses how they affect performance.

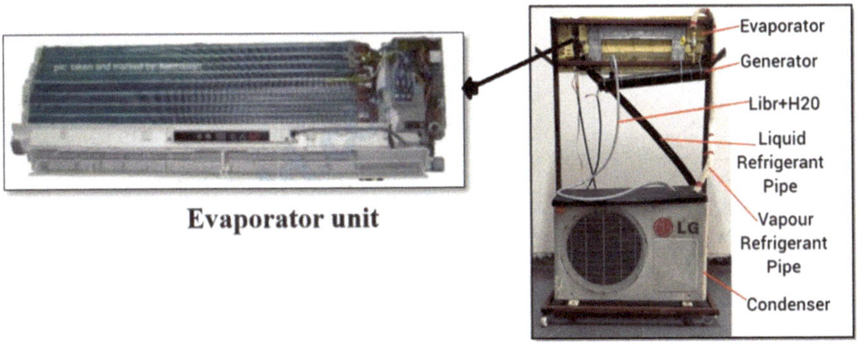

Fig. 5 Evaporator unit in Li-Br vapor absorption refrigeration system

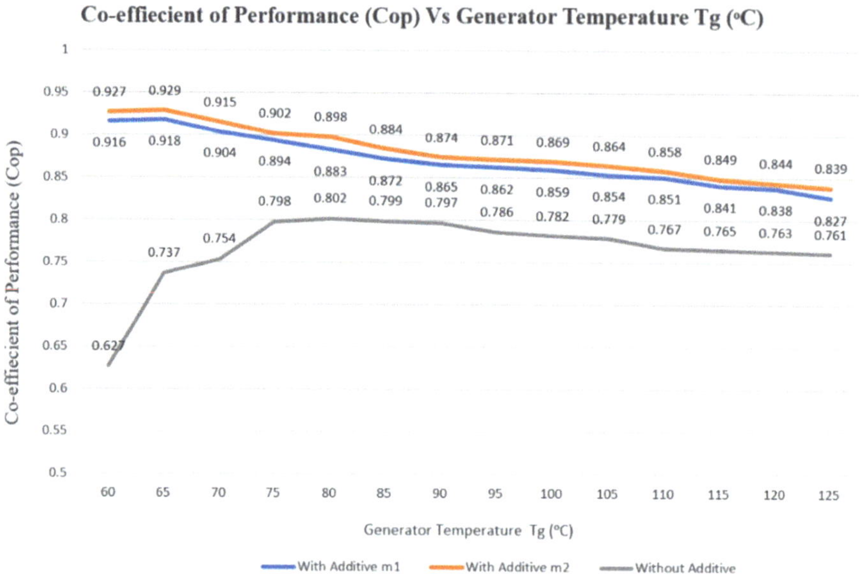

Fig. 6 Coefficient of Performance versus Tg (°C)

3.1 *Generator Performance Analysis*

Figure 6 illustrates the relationship between generator temperature (T_g) and COP for the Li-Br absorption system. When all other parameters are constant, the system achieves a maximum COP of 0.799 at a generator temperature of 85 °C. At this point, the heat input effectively drives the absorption–desorption cycle, and the weak solution concentration remains below the crystallization threshold. However, as T_g increases beyond 85 °C, the COP decreases due to the higher generator temperatures resulting in a higher concentration of lithium bromide in the solution, which increases the risk of crystallization and excessive temperatures lead to higher heat losses and reduced efficiency in the generator-absorber cycle. For instance, at 95 °C, the COP drops significantly due to these limitations. For concentrations of $m_1 = 0.52$ and $m_2 = 0.67$, the system operates without crystallization at a generator temperature of 60 °C, yielding a maximum COP of 0.927.

3.2 *Absorber Performance Analysis*

Figure 7 shows how absorber temperature (T_a) impacts the COP. At an absorber temperature of 15 °C, the system achieves a maximum COP of 0.805. Lower absorber temperatures facilitate better absorption of refrigerant vapor by the lithium bromide solution, thereby enhancing the system's overall performance. However, as

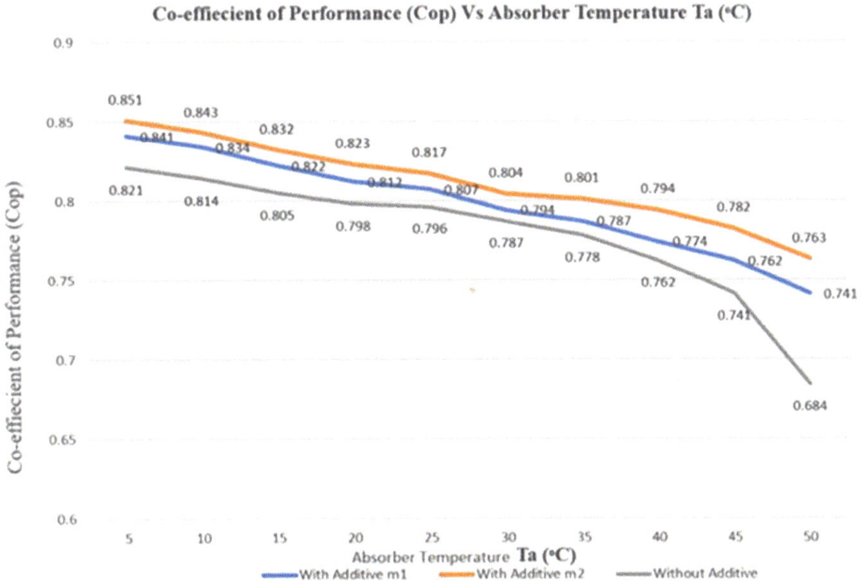

Fig. 7 Coefficient of Performance versus Ta (°C)

the absorber temperature increases, the absorption process becomes less effective due to the reduced solubility of the refrigerant in the absorbent. For $m_1 = 0.52$ and $m_2 = 0.67$, a further reduction in absorber temperature to 7 °C results in a maximum COP of 0.851. However, achieving such low absorber temperatures may not always be feasible, depending on environmental and operational constraints.

3.3 *Condenser Performance Analysis*

The effect of condenser temperature (T_C) on the COP is depicted in Fig. 8. At a condenser temperature of 20 °C, the system achieves a maximum COP of 0.802. Higher condenser temperatures reduce the efficiency of heat rejection, leading to a weaker refrigerant solution and reduced system performance. For $m_2 = 0.67$, maintaining a condenser temperature of 20 °C yields an optimal COP of 0.894. As the condenser temperature rises, the COP decreases due to increased heat rejection requirements and reduced thermodynamic efficiency of the refrigeration cycle.

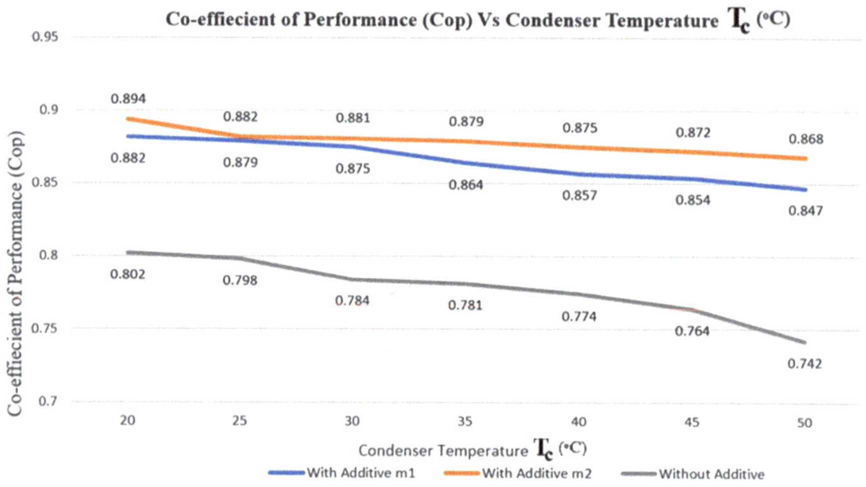

Fig. 8 Coefficient of Performance versus Tc (°C)

3.4 Estimation of Total Power Consumption

VAC systems typically consume electrical power mainly for pumps and controls, usually around 2–5% of the thermal energy input. For this calculation, we'll use 3% as an average estimate.

$$\text{COP} = \frac{Q_E}{Q_G} \tag{5}$$

$$\text{Maximum Coeffiecient of Performance (COP)}$$

$$(\text{COP})\max = \frac{(T_E + 273.15)(T_g - T_a)}{(T_G + 273.15)(T_c - T_E)} \tag{6}$$

For a system with a COP of 0.823

$$\text{Thermal Energy (kW)} = \frac{\text{Cooling Load (kW)}}{\text{COP}} \tag{7}$$

$$\text{Thermal Energy} = \frac{3.571(\text{kW})}{0.823} = 4.275\ \text{kW}$$

Electrical Power (kW) = Thermal Energy (kW) × 0.03.

For the thermal energy calculated above.

Electrical Power = 4.275 kW × 0.03 ≈ 0.128 kW.

So, the power consumed per hour by a VAC system with a COP of 0.823 is approximately 0.128 kW.

Cop: 0.823

Cooling Capacity for consistency: 1 ton = 3.517 kW.
Thermal Energy Required: 4.275 kW.
Electrical Power Consumption: 0.128 kW (128 W) per hour.

4 Conclusion

The lithium bromide (Li-Br) vapor absorption air conditioning system offers an energy-efficient alternative to conventional vapor compression systems (VCS) by utilizing Li-Br as the absorbent and water as the refrigerant. This system effectively harnesses low-grade heat sources, such as solar energy, industrial waste heat, or natural gas, significantly reducing dependence on high-grade electrical energy. In addition to lowering operating costs, this approach decreases greenhouse gas emissions, contributing to environmental sustainability. This study highlights the system's energy-saving potential, with a coefficient of performance (COP) of 0.823. While slightly lower than that of VCS, the Li-Br system demonstrates efficient thermal-to-cooling energy conversion, requiring minimal electrical power—approximately 128 W per hour for a one-ton unit. However, the study is limited by its focus on thermodynamic performance, with less attention to operational challenges such as crystallization risks at higher generator temperatures and material degradation. The impact of fluctuating heat sources and partial load scenarios also requires further investigation.

Future research should address these gaps by exploring advanced materials to enhance corrosion resistance and heat transfer efficiency. Integration with renewable energy sources like solar thermal and geothermal systems could further improve sustainability. Additionally, innovative control strategies to mitigate crystallization risks and optimize performance under dynamic conditions are essential.

References

1. Smith J, Doe (2023) Advancements in lithium-ion battery energy density for electric vehicles. Nat Energy 4(1):123–135
2. Muhammad UA, Bhattacharyya D, Endrino JL (2021) Preparation of binary nanofluid with heat transfer additives by particle surface functionalization. Emergent Mater 4(1):1649–1664. https://doi.org/10.1007/s42247-021-00260-z
3. Zhai XQ, Wang RZ (2008) A review for absorption and adsorption solar cooling systems in China. Renew Sustain Energy Rev 12(5):1145–1165. https://doi.org/10.1016/j.rser.2006.10.005
4. Wang J, Dai YJ, Wang RZ (2015) Economic analysis of solar water heating systems in China. Renew Energy 75(1):20–29. https://doi.org/10.1016/j.renene.2014.09.014
5. Tripathy M, Mishra S, Sudhakar K (2016) Performance analysis of solar photovoltaic-based electricity generation system for residential buildings. Energy Build 108(1):626–635. https://doi.org/10.1016/j.enbuild.2015.10.024

6. Critoph RE (2002) Solar refrigeration for the tropics. Renew Energy 26(2):285–290. https://doi.org/10.1016/S0960-1481(01)00140-7.
7. Mittal V, Kasana KS, Thakur NS (2005) The study of solar absorption air-conditioning systems. J Energy S Afr 16(4):59–66
8. Hainsworth FW (1944) Absorption refrigeration: a review of research and development up to 1944. J Inst Refrig 41(10):25–34
9. Wu S, Eames IW (2000) Innovations in vapour-absorption cycles. Appl Therm Eng 20(6):577–593. https://doi.org/10.1016/S1359-4311(99)00060-8
10. Kumar S, Arakerimath RR (2014) Performance analysis of LiBr-water based vapour absorption refrigeration system with different additives. Int J Adv Res Sci Eng Technol 1(4):44–49
11. Wu S, Eames IW (2000) Overview of multi-stage, multi-effect absorption cycles. Appl Therm Eng 20(13):1225–1240. https://doi.org/10.1016/S1359-4311(99)00093-1

Solar-Electric Car for a Sustainable Transportation System

Ganesha Udupa, Joshua Freeman, R. Abhiram, S. Sai Shubhakar Reddy, K. Varun Teja, G. Naga Sai Rahul, K. Sai Deva, and P. S. S. Kiran

Abstract The Indian automobile industry contributes 7.15% of India's GDP, and generates employment for more than 35 million people. Transportation accounts for approximately 11% of India's carbon emissions. Solar energy is poised to play a transformative role in India's transportation sector, reducing emissions, improving air quality, and enhancing energy security. This project involved replacing the internal combustion (IC) engine with an electric motor and controller, battery packs, and on-board solar photovoltaic (PV) panels and charge controllers, and then analyzing performance to estimate the energy production and the corresponding range obtained from the PV panels. The system component sizing and detailed analysis were performed by building mathematical models in MATLAB Simulink. The potential impact of incorporating a multi-speed transmission in the electric vehicle was shown and quantified. An analysis of the solar PV module performance in different weather conditions, inclination angle, and orientation was also carried out in MATLAB based on the equivalent circuit of a PV cell. As part of the project, a Maruti Esteem was converted into a fully electric vehicle powered by a 15 kW motor with a 72 V, 400 Ah battery pack, charged partly by a 500 W solar PV system. It was designed to reach a top speed of 100 kph, with a range of approximately 300–350 km at 50 kph on a single charge, with an additional range contribution by the solar panels of around 20 km, making it as a sustainable technology transportation system.

Keywords Solar electric vehicle · On-board solar PV charging · EV powertrain analysis · Vehicle performance calculation · Drive cycle simulation

---

G. Udupa (✉) · J. Freeman · R. Abhiram · S. Sai Shubhakar Reddy · K. Varun Teja · G. Naga Sai Rahul · K. Sai Deva · P. S. S. Kiran
Department of Mechanical Engineering, Amrita Vishwa Vidyapeetham, Amritapuri, India
e-mail: ganesh@am.amrita.edu

I. M. Jamadar et al. (eds.), *Integrating Sustainability in Modern Technologies*, Lecture Notes in Mechanical Engineering, https://doi.org/10.1007/978-981-95-9466-5_4

1 Introduction

The transition from conventional vehicles to fully electric propulsions is growing rapidly with the increasing number of electric vehicles around the world and the advancement in electric propulsion technology. But this transition is not going to be easy for many countries in which electric vehicles are still a new advancement. This is especially true in the case of the Indian automotive market where affordability and electricity supply are still the biggest challenges. High performance electric cars that outperform conventional fuel vehicles are currently available, but there is an even greater need to develop efficient and cost-effective electric vehicles.

Another big challenge is meeting the utility grid energy requirements if most of the vehicles are electric. 80% of India's electricity is produced from fossil fuels. The grid energy demand will be tremendous when vehicles turn electric. Even if we move towards renewable energy sources, we might even end up in the middle of an energy crisis. Solar energy is one of the biggest and fastest growing sources of renewable energy [1]. Converting existing conventional fuel vehicles into fully electric vehicles can provide a more practical and cost-effective solution. Furthermore, as solar panels are becoming more efficient, they can be used to partly power the electric vehicle making it as a sustainable technology transportation system that reduces the utility grid load [2].

There is presently limited application of adding solar panels to electric vehicles. The World Solar Challenge (WSC) [3] series of races across the Outback in Australia has focused since 1987 on primarily university student and automotive OEM (Original Equipment Manufacturers) teams designing, fabricating, and competing in lightweight, aerodynamic single passenger vehicles that are charged purely by the sun. While there is excellent research and learning, the transition of these cars to more practical multi-passenger vehicles with modern safety and comfort features is limited. Lightyear [4], a Netherlands-based company, is seeking to commercialize a modern yet lightweight multi-passenger vehicle, that has as much battery charging from the sun as possible. Other electric and hybrid electric vehicles on the market utilize solar cells on the roof for supplementary charging, but the literature does not describe their design process, key parameters, or conclusions [5]. The authors of [6] cover the system architecture of a solar electric vehicle, but do not go into the equations governing the vehicle or the simulation. In [7], the authors delve into some of the basic vehicle physics for a solar electric vehicle, although it is strictly approached from the point of view of determining optimum energy efficiency over predetermined route profiles.

Furthermore, a growing number of products exist which support Vehicle to Home (V2H), Vehicle to Grid (V2G), and Vehicle to Load (V2L) integration of public utilities, buildings, homes, and electric vehicles. These technologies enable home installed solar panels to support the broader grid, or alternatively, as per this paper, an electric vehicle with solar charging could export excess power to a home or the grid, among other options [8].

The calculations and methods for sizing electric vehicle components are not well known, especially amongst university students. Authors in [9] have covered some of the simulation and duty aspects of electric vehicle design and range, but did not cover the fundamental physics and engineering equations that are at the heart of the problem. Authors in [10] cover simulation and the electrical modelling of four-quadrant drive system operation, but does not cover the vehicle side equations that determine motor and battery sizing A simulation system is described in [11] which incorporates duty cycle analysis for electric and hybrid electric vehicles, but again, does not cover any of the basic vehicle dynamics equations and electrical equations governing the system. The simulation and analysis of an electric vehicle with a range extender is described in [12], which has some high-level similarities with this work, except it has no content regarding a solar PV range extender, which is one way of looking at a solar electric vehicle.

Modern Lithium-based Energy Storage Systems (ESS), commonly called battery packs, are the most expensive part of a modern electric vehicle. Optimizing the capacity, power, and weight of the battery pack is of tremendous importance to meeting range, acceleration, and affordability targets for an electric vehicle. If the battery is too small, the vehicle will not have the desired range. If the battery pack is too large, the cost of the vehicle will increase unnecessarily, the weight of the vehicle will increase, and to maintain acceleration targets, the motor size must increase. This also creates a negative feedback loop with increasing weight and cost requiring additional weight and cost. The in-depth simulation of just the battery systems is given in [13], which can be utilized as a component in broader, vehicle level modelling and simulation. In [14], the authors present a comprehensive, optimization algorithm for component sizing in a hybrid-electric vehicle. However, this paper focuses on the in-depth, mathematical optimization algorithm and does not provide any understanding of the basic physics governing the vehicle dynamics.

While the basic sizing calculations of the motor and battery systems are not necessarily difficult, they do involve practical, industry experience based best practices and considerations. This paper aims to help close that gap.

Environmental concerns are very important, and older parts can be recycled properly, reducing the production cost of new electric vehicles. Care must also be taken to recycle the old batteries without damaging the environment. In this work, to safeguard the environment and to make it cost effective, an old petroleum car was modified and converted into a solar electric car.

2 Powertrain Sizing

2.1 Study of Drive Cycles

It is difficult to predict how a vehicle is going to be driven, as it depends on various factors such as the desired vehicle performance, road conditions, and even varies significantly with each driver. All of these factors affect the power, torque and rpm required from the motor, the battery discharge currents, energy consumptions etc., thereby making the powertrain sizing challenging.

Drive cycles are a time-series graph showing velocity versus time data that resembles actual vehicle driving conditions. They play a crucial role in understanding the powertrain requirements. The following drive cycles, each corresponding to different driving conditions were used to analyse the powertrain system [15].

(i) Modified Indian Drive Cycle (MIDC)—(with highest speed modified to 100 kph);
(ii) Artemis Urban Drive Cycle—for city driving conditions;
(iii) Artemis Rural Drive Cycle—for rural driving conditions;
(iv) Highway Fuel Economy Driving Test (HWFET)—for highway driving conditions.

2.2 Mathematical Model

A back-facing mathematical model was built in Matlab Simulink [16] which follows the "wheel to well approach" i.e., it takes the drive cycle as the input and calculates all other required characteristics versus time, such as required motor torque, power, and rpm, discharge currents, energy consumption, range, and required battery capacity. The brief flow diagram of the powertrain Simulink model is shown in Fig. 1.

All the required parameters were given as variables in a Matlab script file and were directly used in the Simulink model. The below equations are primary from [17], although they are easily derived from basic physics and vehicle engineering.

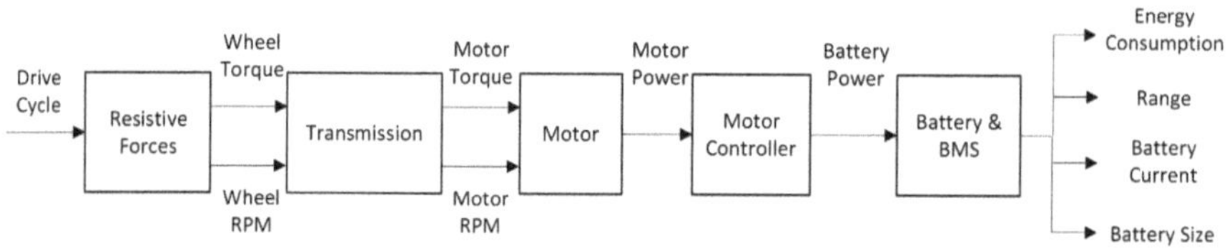

Fig. 1 Powertrain Simulink model

3 Motor and Battery Specifications

3.1 Calculation of Motor Parameters

The rolling force is given by

$$F_{\text{rolling}} = \mu_r \cdot N \tag{1}$$

where μ_r is the coefficient of rolling resistance, $N = W\cos\theta$ is the normal reaction force on the vehicle, with W the weight of the vehicle and θ the inclination angle in degrees.

Then, the gradient force is given by

$$F_{\text{gradient}} = W \cdot \sin\theta \tag{2}$$

The drag force is given by

$$F_{\text{Drag}} = 0.5 \cdot \rho \cdot C_d \cdot A \cdot V^2 \tag{3}$$

where ρ is the density of air, C_d is the coefficient of drag, A is the frontal area of the vehicle, and V is the speed of the vehicle in m/s.

The total tractive force or effort is given by

$$F_{\text{tractive}} = F_{\text{rolling}} + F_{\text{Drag}} + F_{\text{gradient}} + F_{\text{accelerating}} \tag{4}$$

Wheel torque is given by

$$T_{\text{wheel}} = F_{\text{tractive}} \cdot R_w \tag{5}$$

where R_w is the static loaded radius or rolling radius of the tire.

The motor torque is given by

$$T_{\text{motor}} = T_{\text{wheel}}/(\text{Transmission Gear Ratio} \cdot \text{Final Drive Ratio} \cdot \eta_{\text{driveline}}) \tag{6}$$

where $\eta_{\text{driveline}}$ is the driveline efficiency.

The motor speed is given by

$$N_{\text{motor}} = N_{\text{wheel}} \cdot \text{Transmission Gear Ratio} \cdot \text{Final Drive Ratio} \tag{7}$$

Wheel speed is given by

$$N_{\text{wheel}} = \frac{v * 60}{2\pi * R_w} \tag{8}$$

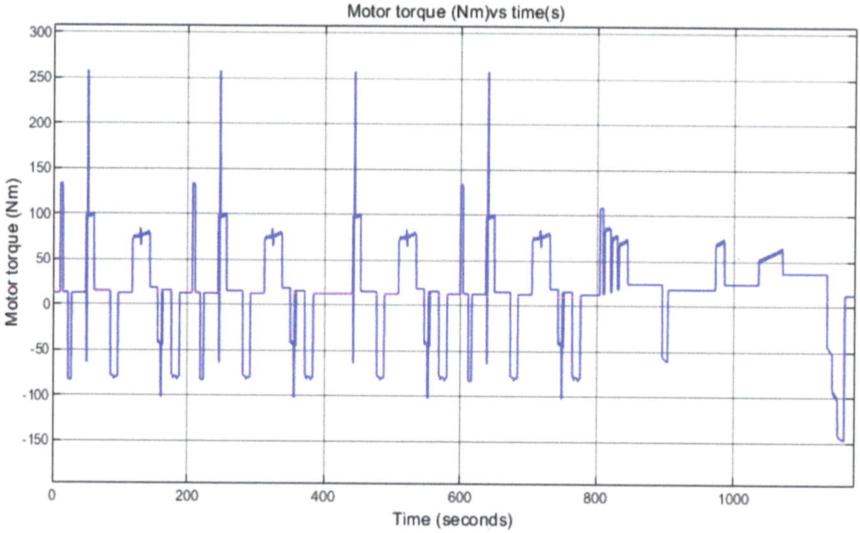

Fig. 2 Variation of motor torque with time (Direct drive, MIDC)

where V is the vehicle's velocity in m/s.

The motor power required is

$$P_{\text{motor}} = (2\pi \cdot N_{\text{motor}} \cdot T_{\text{motor}})/60. \quad (9)$$

3.2 Simulation Results for Direct Drive

Simulations were performed for the direct drive scenario (without a transmission) using the Modified Indian Drive Cycle (MIDC). The motor torque, motor rpm, and motor power required are shown in Figs. 2, 3, and 4, respectively.

3.3 Simulation Results with a 5-Speed Transmission

The torque required in the case of direct drive is extremely high even though the drive cycle is not too aggressive. The torque required with direct drive will be even higher with drive cycles such as the Artemis Urban and Rural. Gear reduction is required in order to decrease the motor torque requirement at lower speeds, but this will also increase the rpm required from the motor to achieve higher speeds. Hence, at this stage using a multiple speed transmission was more practical as it brought down both torque and rpm requirements, thereby lowering the cost of the motor without

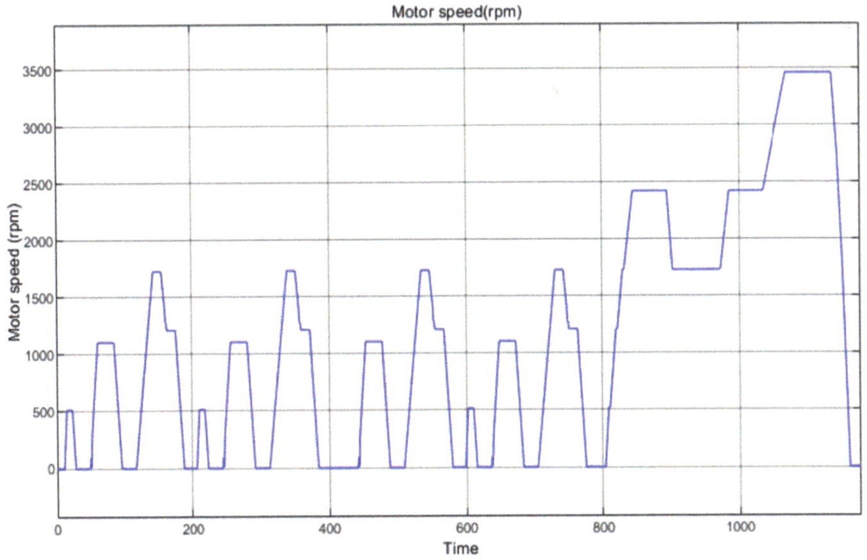

Fig. 3 Variation of motor RPM with time (Direct drive, MIDC)

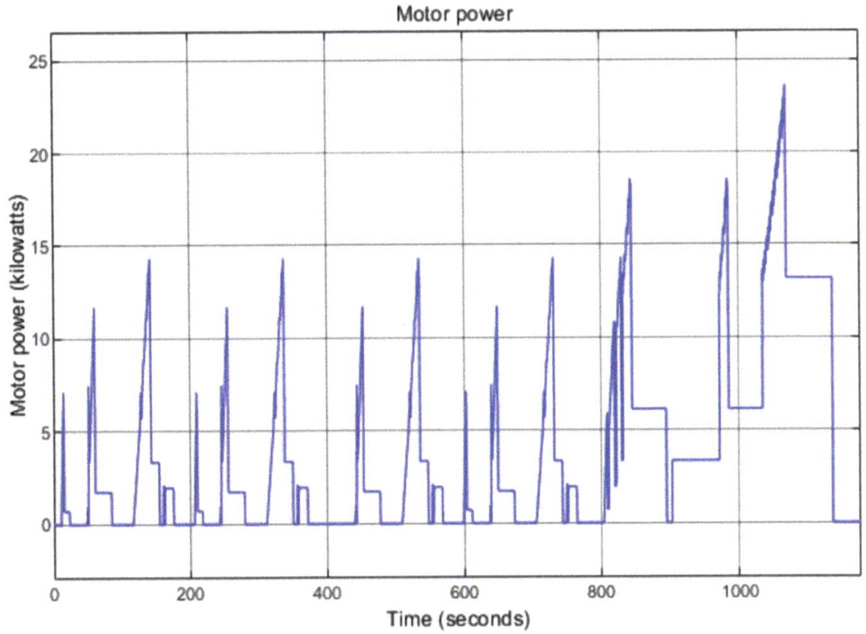

Fig. 4 Variation of motor power with time (Direct drive, MIDC)

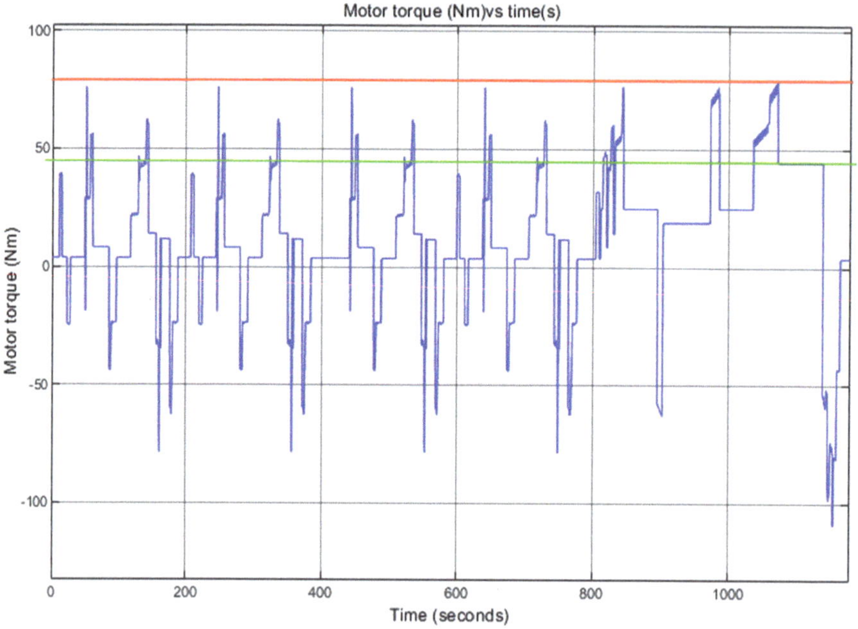

Fig. 5 Variation of motor torque with time (5-speed transmission, MIDC)

compromising the performance of the vehicle. The simulation results for the motor torque, motor rpm, and motor power with the MIDC are shown in Figs. 5, 6, and 7, respectively.

A comparison of the simulation results between direct drive and with a 5-speed transmission is shown in Table 1.

3.4 Selection of Motor and Motor Controller

Based on the simulation results obtained in Table 1, a 15 kw (nominal) BLDC (Brushless Direct Current) traction motor with the following specifications was selected: 35 kW peak power, 44.7 Nm rated torque, 111.9 Nm peak torque, 3200 maximum rpm, 72 V supply voltage, 208 A rated current, and 560 A peak current. A BLDC motor is very similar to an AC induction motor except that the rotor has permanent magnets instead of an induction rotor. Both generate a multi-phase rotating magnetic field in the stator windings, but the BLDC motor is a synchronous motor that operates without slip. They are well suited to electric vehicles from a cost, weight, and performance perspective. Efficiency mapping of the motor is shown in Fig. 8, which was used to calculate the required battery pack characteristics. A

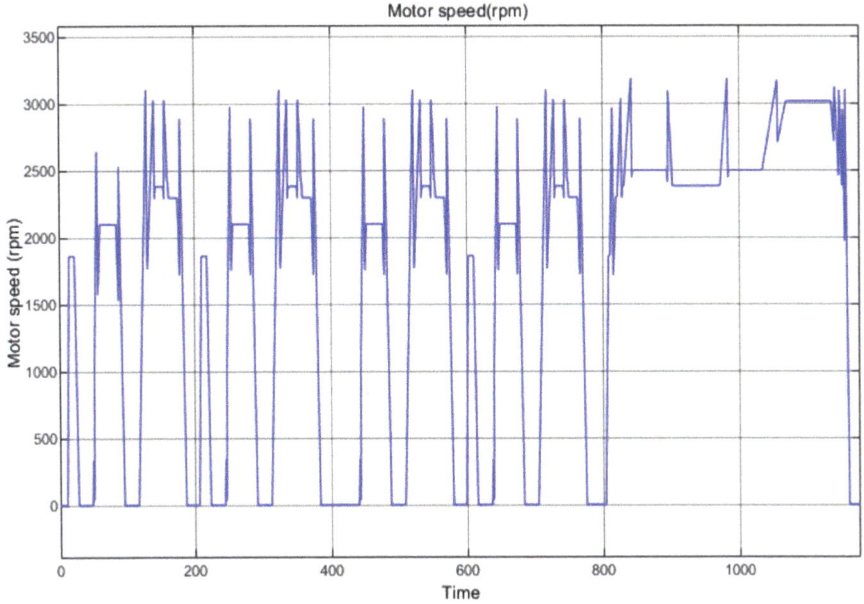

Fig. 6 Variation of motor rpm with time (5-speed transmission, MIDC)

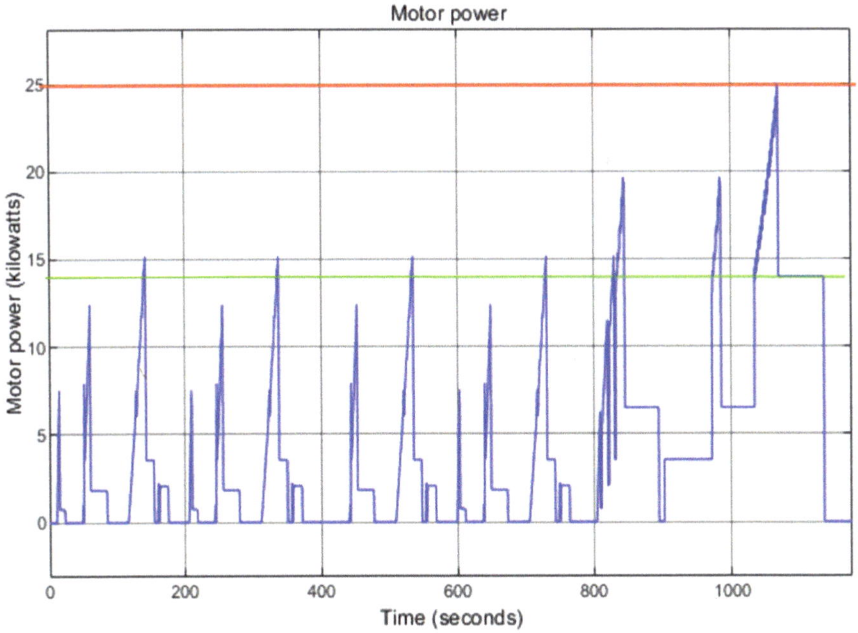

Fig. 7 Variation of motor power with time (5-speed transmission, MIDC)

Table 1 Comparison of results between direct drive and a 5-speed transmission

Required characteristics	Direct drive	With 5 speed transmission
Continuous power (KW)	13.2	14.0
Continuous torque (Nm)	36.4	44.3
Peak power (kW)	23.6	25.0
Peak torque (Nm)	258.1	79.2
Motor speed (rpm) @ 100 kph	3465	3015

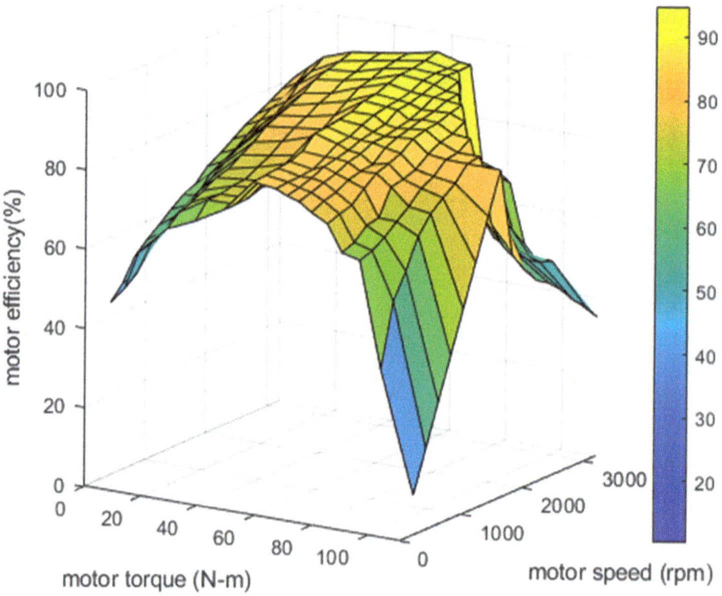

Fig. 8 Mapping of the motor efficiency

matching, programmable BLDC motor controller with capabilities of 24–72 V operating voltage, 240 A continuous current, and 600 A maximum current was also selected.

3.5 Calculation of Required Battery Parameters

The following equations are used for sizing the batteries for the traction motor.

Battery discharge power is

$$P_{\text{Battery}} = P_{\text{motor}} / \eta_{\text{controller}} \tag{10}$$

where $\eta_{\text{controller}}$ is the average efficiency of the controller.

The discharge current is

$$I_{\text{discharge}} = P_{\text{Battery}} / V_{\text{nominal}} \tag{11}$$

where V_{nominal} is the nominal voltage of the battery pack, which in this case was 72 VDC.

Average energy consumption is given by

$$E_{\text{consumption}} = \int P_{\text{Battery}} / [(3.6 \times 10)^6 \cdot (\text{Drive cycle distance})] \tag{12}$$

The necessary battery capacity is given by

$$E_{\text{battery 100\% DOD}} = E_{\text{consumption}} \cdot \text{Desired range} \tag{13}$$

which is at 100% Depth of Discharge. Typically, usage for a battery in an electric vehicle is only 80% DOD, which allows for approximate 10% buffers at the top and bottom of the battery's State of Charge (SOC) swing to preserve battery life. In which case, the actual battery capacity required is

$$E_{\text{battery\_80\% DOD}} = 1.2\, E_{\text{battery 100\% DOD}} \tag{14}$$

The required discharge power and discharge current characteristics are shown in Figs. 9 and 10, respectively. Similarly, the characteristics were determined for all other drive cycles and tabulated in Table 2.

Based on the simulation results of the above selected motor and 5-speed transmission, a battery pack with the following characteristics was chosen for the initial test stage: 30 kWh, 74 VDC nominal voltage, 400 Ah capacity, 60–84 VDC operating voltage, 40 A maximum charge current, and 250 A maximum discharge current.

4 Vehicle Range

4.1 Estimated Range for Different Drive Cycles

The estimated range for the 30 kWh battery pack for the various drive cycles is shown in Table 3. The battery capacity required for any desired range can be calculated by using the energy consumption values given in Table 3. The lowest energy consumption is for highway drive conditions (HWEFT) and hence the range is maximum compared to all other driving cycles for both 100 and 80% DoD. The next best range obtained is for rural drive conditions. This may be due to less traffic and no signal stops compared to urban drive conditions.

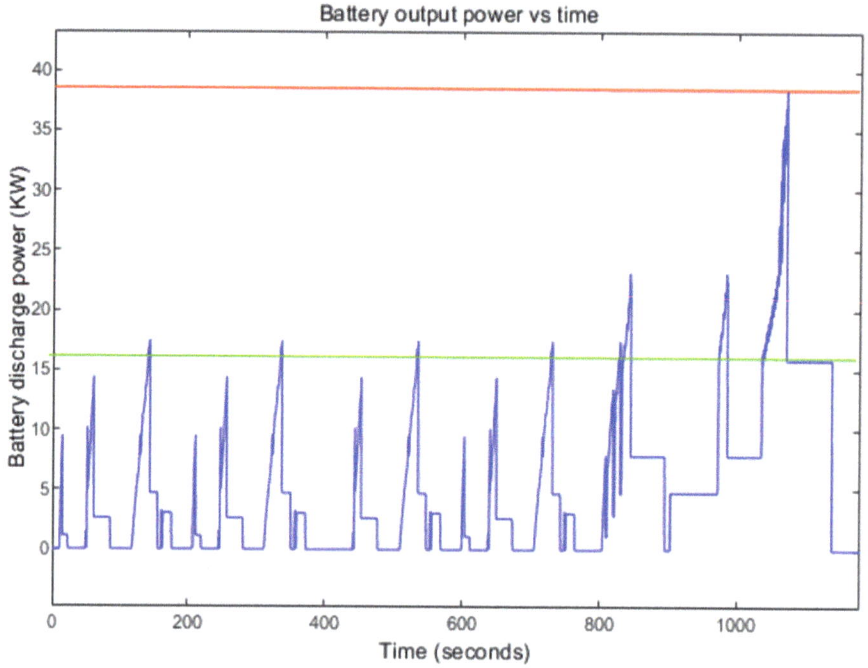

Fig. 9 Battery discharge power versus time (15 kW BLDC motor, 5-speed transmission, MIDC)

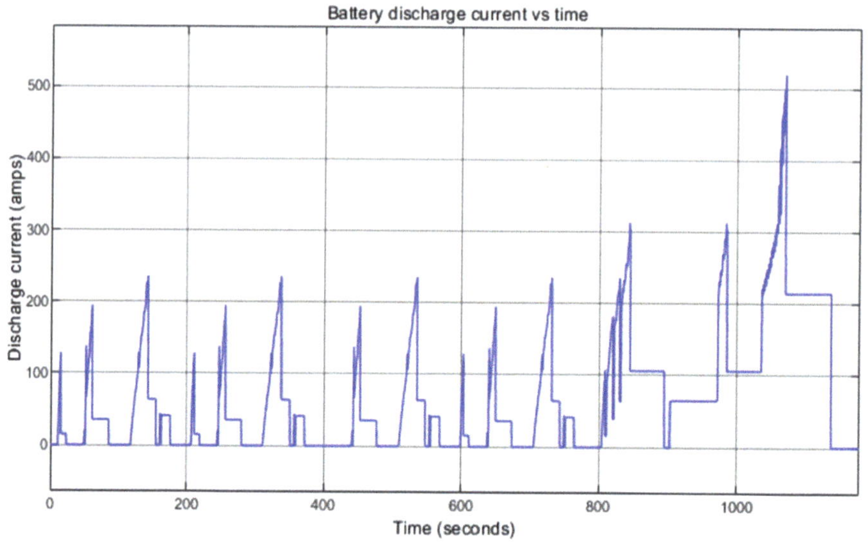

Fig. 10 Battery discharge current versus time (15 kW BLDC motor, 5-speed transmission, MIDC)

Table 2 Battery characteristics required for different drive cycles

Parameter	MIDC	Artemis urban	Artemis rural	HWFET
Peak battery discharge power (kW)	38.5	51.1	53.1	30.8
Peak current (A)	520.0	710.0	738.0	416.2
Average energy consumption (Wh/km) (at max payload)	149.1	218.2	156.3	120.8

Table 3 Estimated range for different drive cycles

Drive cycle	MIDC	Artemis urban	Artemis rural	HWFET
Performance: Kerb weight + driver weight (1100 kgs)				
Energy consumption (Wh/km)	122.4	166.1	118.1	103.7
Range at 100% DOD (km)	245.1	180.6	254.0	289.3
Range at 80% DOD (km)	204.1	150.5	211.7	241.1
GVW: Kerb weight + driver weight + 3 passenger weight (1300 kgs)				
Energy consumption (Wh/km)	137.3	195.7	141.2	113.3
Range at 100% DOD (km)	218.5	153.3	212.5	264.8
Range at 80% DOD (km)	182.1	127.7	177.1	220.7

4.2 Estimated Range for Steady Cruise Conditions

The estimated range for the car in steady cruise conditions is shown in Fig. 11. Figure 11a, show the range of the vehicle with the driver, while Fig. 11b shows the range of the vehicle with the driver along with 3 passengers, for various cruising speeds at 80 and 100% DoD. As the cruising speed increases, the range of the vehicle decreases for both 100 and 80% DoD. At a particular cruising speed, as the load or number of passengers increases, the vehicle range reduces.

5 Power Generation from Solar Panels

The next step of the analysis is to understand the contribution of the solar PV panels to the vehicle's range. The main objectives of this analysis are (1) to determine the solar PV panel characteristics under different temperature and irradiance conditions, (2) to calculate the power output from the panels throughout the day, and finally (3) to calculate the energy production from the panels. In the real-world scenario, solar irradiance and temperature depend upon weather conditions, vary continuously and are non-linear, resulting in the power output from the panel decreasing as the irradiance decreases and as the temperature increases. Therefore, an analysis was carried out in MATLAB by building a mathematical model based on the formulas in

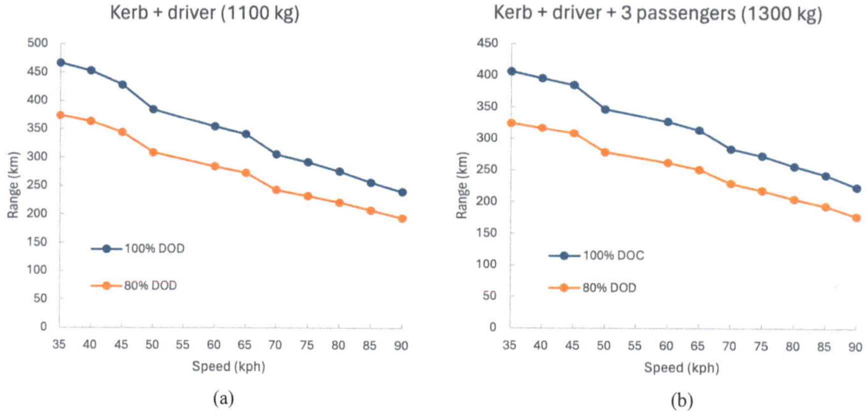

Fig. 11 Range of the vehicle for various cruising speeds **a** with driver only, and **b** with driver and 3 passengers

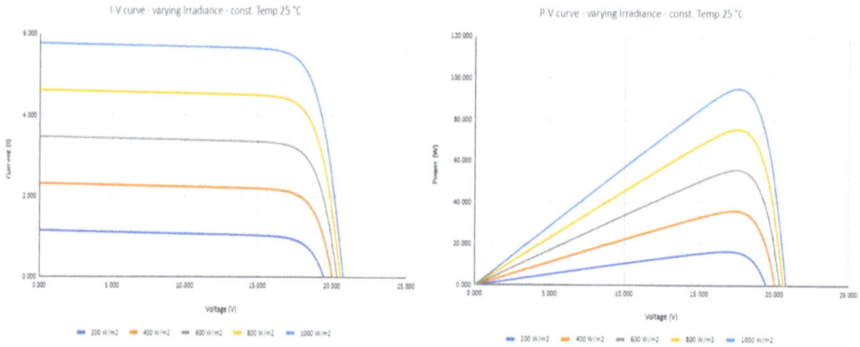

Fig. 12 P-V and I-V characteristic curves with varying irradiance

[18] and [19]. Using these equations, the simulation model can input the solar PV panel operating temperature and irradiation as inputs and outputs the Current-Voltage (I-V) and Power-Voltage (P-V) characteristic curves, as shown in Figs. 12 and 13, respectively.

5.1 *Power Generation Throughout the Day*

It is very difficult to realistically simulate the power production of the solar panels on a vehicle, due to the very wide variety of variables with weather, time of day, sun position, dynamic vehicle orientation, solar panel installation angles, panel soiling, unpredictable external shading, and others. Since this work was conducted during the COVID-19 pandemic, work progress was hindered and some of the work on the

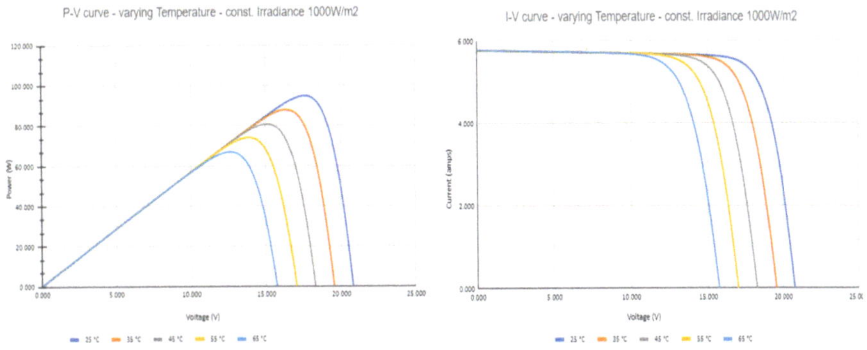

Fig. 13 Output P-V and I-V curves with varying temperature

solar panels was performed independently of the vehicle, before they were mounted and could not be loaded properly to measure power output directly.

The temperature of one of the test solar panels was measured at half hour intervals between 7:30 and 19:30 for three days, using an infrared temperature sensor. The three days had varied insolation, with one day rainy, one cloudy without rain, and one clear sky. The values of GHI (Global Horizon Irradiation) which the solar PV panels utilize, are determined at these intervals using Solcast [20]. This data was then given as input to the simulation, while the power output with respect to time throughout the day and cumulative energy are given as output.

5.2 *Energy Produced by the Solar Panels*

The simulation model considers the irradiance and temperature at every interval and plots the power output data with respect to time for the panel at various inclination angles as shown in Fig. 14, which represented the intended angles of the panels as installed on the vehicle.

The irradiance on the inclined panels is calculated by assuming that the vehicle is parked facing towards the west. If all the panels are connected in series to form a

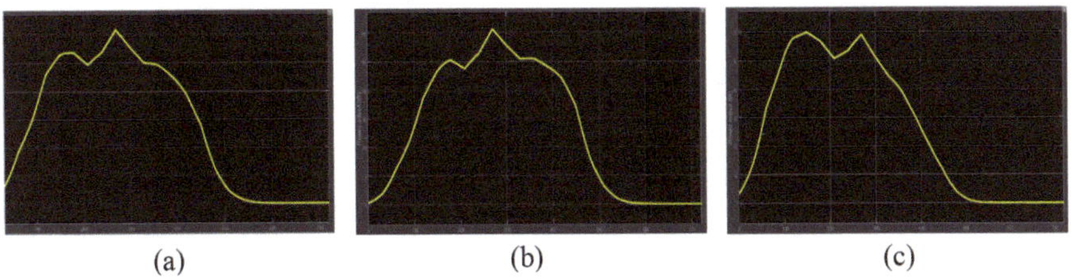

Fig. 14 Daily power output curve from (**a**) horizontal panel, (**b**) panel at 8° inclination, and (**c**) panel at 20° inclination, with time on the x-axis and power on the y-axis

Table 4 Energy produced each day

Panel placement and number of panels	May 7th (Rainy) (kWh)	May10th (Cloudy) (kWh)	May 11th (Clear) (kWh)
Rooftop (×2)	0.2953	0.2271	0.3291
Boot (×2)	0.3304	0.2289	0.3163
Bonnet (×1)	0.2481	0.2129	0.3215
Total energy (kWh)	1.5	1.125	1.612

Table 5 Average irradiation and energy produced in a year

Panel placement	Annual average solar irradiation (kWh/m^2/year)	Annual solar PV output energy (kWh/year)
Roof top	1699	242
On boot	1641	234
On bonnet	1634	116
Total annual PV energy output (kWh)		593

string, the least performing module in the array limits the current of the whole string, reducing the performance of the other modules. Therefore, to overcome this limitation, the panels which have similar installed inclination and output are connected in series. The energy produced by each string on each day is given in Table 4.

The average energy produced is given by the below equation [21]

$$E_{\text{ave}} = G_{\text{avg}} \times A \times \eta \times \text{PR} \tag{15}$$

where G_{avg} is the average irradiation, A is area of the PV array, η is efficiency of the PV module, and PR is performance ratio which is the coefficient for losses. Table 5 shows the average irradiation and energy produced when averaged over the entire year.

5.3 *Range of the Vehicle in Different Driving Conditions with Solar Charging*

Table 6 shows the distance which can be covered purely with solar energy in different driving conditions. The solar panels placed on top of the car produce enough energy to give an extra range for the car of approximately 10–15 km per day in various drive cycles and gross vehicle weights (GVW).

Table 6 Extra range contribution of the PV panels under drive cycles

Drive cycle	Duration	Range (km)	
		Kerb weight + driver (1100 kg)	Kerb weight + driver + 3 passengers (1300 kg)
MIDC	Daily	13	12
	Yearly	4841	4325
Artemis urban	Daily	10	8
	Yearly	3568	3028
Artemis rural	Daily	14	12
	Yearly	5018	4197
HWFET	Daily	16	14
	Yearly	5715	5230

Table 7 Extra range contribution of the PV panels under steady state

Speed (kph)	Duration	Range (km)	
		Kerb weight + driver (1100 kg)	Kerb weight + driver + 3 passengers (1300 kg)
35	Daily	25	22
	Yearly	9079	7963
40	Daily	24	21
	Yearly	8814	7721
50	Daily	21	21
	Yearly	7514	7467

5.4 *Range in Steady Cruise Condition*

Table 7 shows the range of the vehicle at constant, steady-state cruise conditions.

The solar PV panels can give an extra range of 20–25 km per day at a cruising speed of 50 kph, at the two different gross vehicle weights. Averaged between the steady state and driving cycle analysis, the solar panels provide approximately 20 km of extra driving range per day, which is sufficient by itself for many shorter drives.

5.5 *Charge Controller*

Two identical solar PV panels were placed on the rooftop and connected in series to form a string, as they produce the same output current, and the string was then connected to a Maximum Power Point Tracker (MPPT) charge controller, which automatically adjusts the load voltage to optimize the output power of the solar PV string. The MPPT changes the string's operating point on the current versus voltage and power versus voltage curves shown in Figs. 12 and 13 to maximize the power. In the same way, the two panels on the boot are also connected in series and are connected to additional MPPT charge controllers. The single panel on the bonnet is also connected to an MPPT. This enables the vehicle to have three circuits whose power output is independent of each other, as each circuit always performs at its maximum power, even when they are oriented and shaded differently.

6 Fabrication, Assembly, and Performance Evaluation

The Maruti Esteem vehicle was selected because of its low kerb weight, availability of sufficient roof space to mount the solar PV panels, and the bonnet and additional boot space was large enough to accommodate all the electric vehicle and powertrain components. The engine, radiator and all other unnecessary parts were removed which reduced the kerb weight by 161 kgs. The electric motor and the existing Maruti transmission were joined together using an adapter plate and a coupler.

The performance parameters of the vehicle are:

- Weight of the vehicle, with one passenger, including batteries: 1100 kg
- Maximum Speed: 100 kph
- Cruising speed: 50 kph
- Range of vehicle on a single charge at cruising speed: 300–350 km
- Daily distance travelled on solar energy at cruising speed: ~20 km
- Charging time on a domestic electricity socket: 10–12 h

The battery pack was sized based on modelled energy consumption, desired range, and weight impact. The lithium iron phosphate (LiFePO4) battery technology was selected so that the kerb weight of the vehicle after conversion didn't change significantly. The battery pack can be charged by the solar panel, by a standard electrical outlet, or a combination of both.

A tubular frame was made to mount the panels on the vehicle, which was riveted to the roof. It was followed by fabricating panels to close gaps between the solar panels and the body, to avoid extra drag and to provide a better look to the vehicle. Front and rear overview photos of the car are shown in Fig. 15.

7 Conclusions

A solar electric vehicle was designed, simulated, fabricated, and tested. All powertrain components were sized using advanced mathematical models built into MATLB Simulink. Through analysis, it was shown that using a gearbox transmission within the EV powertrain can greatly reduce the torque and rpm requirements. The use of the transmission also provided more control over the operating regions of the motor,

Fig. 15 Front and rear overviews of the vehicle

thereby enhancing the overall performance and efficiency. The energy consumption was reduced by fine-tuning the speed range for each gear ratio to operate the motor in more efficient regions, resulting in better economy for the vehicle. Accurate energy consumption analysis and range estimation was also performed for the MIDC, Artemis Urban, Artemis Rural, and HWFET drive cycles, as well as for steady-state, constant cruise conditions. The battery pack was suitably sized based on the energy consumption, range, weight distribution in the car, and the kerb weight of the car. Solar panels were installed with a structural frame that did not significantly alter the aerodynamics of the vehicle. Three MPPT charge controllers were connected to the respective PV strings, enabling them to always perform at their maximum operating power point. The methodology considered here can be useful to size the powertrain components and is sufficient for low and medium power vehicle conversions.

The vehicle meets the target range of 300 km at a cruising speed of 50 km/h, on a single charge, with an additional daily range of approximately 20 km contributed by the solar panels. This is a meaningful contribution to the vehicle's driving range, and even replace charging the battery with mains utility electricity for short daily driving distance. If the vehicle is to be driven more than this amount, then additional battery charging with mains power is necessary. Determining algorithms for optimum blending of solar and mains battery charging is an interesting optimization problem that could be considered for future work.

Although the analysis allowed us to understand the performance and requirements of the vehicle to design a more efficient system, various other factors must be considered which can affect the range of the vehicle. These include the presence of auxiliary systems such as air conditioning and heating, the effect of temperature on the battery pack, and others. These factors might not be significant when considered individually, but together they can greatly affect the range and performance. Even though the range was estimated for different conditions, real life driving is a combination of various differing scenarios and hence the overall range obtained in real condition can be higher or lower in comparison with simulation results. The efficiency of the panels used was 16–17%. Solar panel efficiencies are increasing steadily, helping to more effectively meet the vehicle's energy demands.

Acknowledgements This work was supported by Amrita Vishwa Vidyapeetham as part of an internally funded project. We would like to offer our humble salutations to Amma, the Chancellor of Amrita University, for providing us with several ideas and constant guidance and blessings throughout this project work.

References

1. Udupa G, Freeman J, Nimbewal V, Choudhary RK, Sasikumar S, Nair SV (2016) Design and manufacture of Amrita Surya Vahini: a solar auto rickshaw. In: International conference on robotics and automation for humanitarian applications (RAHA), Amritapuri, India, 18–20 December
2. Udupa G, Freeman J (2022) Lightweight multipurpose solar powered electric vehicle. Indian Patent No. 401157
3. World Solar Car Challenge (2025). Accessed 8 Feb. Available: https://worldsolarchallenge.org
4. Lightyear (2025). Accessed 8 Feb. Available: https://lightyear.one
5. Top 9 Electric Vehicles with Solar Roof (2025). Accessed 8 Feb. Available: https://energytheory.com/solar-panel-on-electric-car-roof/
6. Manivannan S, Kaleeswaran E (2016) Solar powered electric vehicle. In: First international conference on sustainable green buildings and communities, Chennai, India, 18–20 December
7. Oosthuizen C, Wyk BV, Hamam Y, Desai D, Alayli Y, Lot R (2019) Solar electric vehicle energy optimization for the sasol solar challenge 2018. IEEE Access
8. V2G, V2H, and V2L (2025) A complete guide to EV bidirectional charging. Accessed 8 Feb. Available: https://greencarscompare.com/blog/v2g-v2h-and-v2l-a-complete-guide-to-ev-bidirectional-charging/ (2025)
9. Mruzek M, Gajdac I, Kucera L, Barta D (2015) Analysis of parameters influencing electric vehicle range. In: Proceedings of the 9th international scientific conference. Transbaltica, Vilnius, Lithuania
10. Mohd TAT, Hassan MK, Azir WMKA (2015) Mathematical modelling and simulation of an electric vehicle. J Mech Eng Sci 8:1312–2132
11. Butlaer K, Ehsani M, Kamath P (1999) A Matlab-based modelling and simulation package for electric and hybrid electric vehicle design. IEEE Trans Veh Technol 48(6)
12. Wahono B, Santoso WB, Nur A, Amin (2015) Analysis of range extender electric vehicle performance using vehicle simulator. In: Proceedings of the 2nd international conference on sustainable energy engineering and application in energy procedia, vol 68, pp 409–418
13. Kroeze RC, Krein PT (2008) Electrical battery model for use in dynamic electric vehicle simulations. In: Proceedings of the 2008 IEEE power electronics specialists conference, Rhodes, Greece
14. Kim K, Kim N, Jeong J, Min S, Yang H, Vijayagopal R, Rousseau A, Cha SW (2021) A component sizing methodology for a hybrid electric vehicle using an optimization algorithm. Energies 14(11):3147
15. Chang CM, Siao JC (2010) Performance analysis of EV Powertrain system with and without transmission. World Electr Veh J 4(3):629–634
16. Matlab 2023b (2025). Accessed 17 Jan. Available: https://www.mathworks.com/products/new_products/release2023b.html

17. Performance Fundamentals (2022) Definitions, equations, formulas, conversions, reference data. Allison Transmission (2022)
18. Pandiarajan N, Muthu R (2011) Mathematical modelling of photovoltaic module with Simulink. In: 2011 1st international conference on electrical energy systems, Chennai, India, pp 258–263
19. Nguyen XH, Nguyen MP (2015) Mathematical modelling of a photovoltaic cell/module/arrays with tags in Matlab/Simulink. Environ Syst Res 4(24)
20. Solcast. Accessed Jan 2025. Available: toolkit.solcast.com.au/world-api
21. Arraez-Cancelliere OA, Munoz-Galeano N, Lopez-Lezama JM (2021) Computing the global irradiation over the plane of photovoltaic arrays: a step-by-step methodology. Renew Energy Technol Appl Intech Open

Performance Analysis of Receding Horizon Control in Inertia Emulation for Frequency Stability of Microgrid

Nandini S. Kodihalli, U. Vinatha, and V. N. Jayasankar

Abstract The assimilation of distributed energy resources (DER) into electrical grids introduces significant challenges in the system with respect to frequency stability. The forementioned issue can be resolved using inertia emulation. Inverter-based resources such as photovoltaic systems and battery storage employs the advanced control strategies which replicates the inertial response of synchronous generators. Emulated inertia is one of those advanced control strategy which replicates the behavior of alternator, thereby improving the frequency stability. This article showcases the correlation of Emulated Inertia Controller employing PI and Fuzzy-logic controller with receding horizon controller (RHC). The demonstrated RHC control strategy provides precise frequency regulation compared to conventional strategies. RHC controller adjusts the controller action in real time giving optimized system response. The efficiency of the emulated inertia controller with the RHC controller outperforms the virtual inertia controller with PI and fuzzy control strategies. It can be noticed from the simulation results that RHC control strategy provides better solution for regulation of microgrid frequency, maintaining frequency stability and reliability in the power system with sustainable energy resources for its operation and control.

Keywords Receding horizon control · Emulated inertia · DER assimilation · Frequency stability

N. S. Kodihalli (✉) · U. Vinatha
National Institute of Technology Karnataka, Mangalore, India
e-mail: nandinishantappakodihalli.207ee501@nitk.edu.in

V. N. Jayasankar
The National Institute of Engineering, Mysuru, India

I. M. Jamadar et al. (eds.), *Integrating Sustainability in Modern Technologies*, Lecture Notes in Mechanical Engineering, https://doi.org/10.1007/978-981-95-9466-5_5

1 Foreword

Operating dynamics of modern power systems are changing due to increased distributed energy resources (DERs) assimilation in power systems. It is affecting the frequency stability and overall inertia of the system due to non-inertial masses such as power electronic converters. Assimilation of power electronic converters may lead to frequency deviation during disturbances due to low or very little inertia.

Research studies have been carried out to mitigate the forementioned issue. The frequently used solution is emulated inertia technique to virtually compensate by imitating the behavior of classical generators. The technique is called virtual inertia or emulated inertia to enhance the frequency and transient stability [1–3].

Emulated inertia uses well-known control systems such as fuzzy logic and receding horizon control. Fuzzy logic shows great promise in improving system stability by altering the control parameter dynamics corresponding to changing operating conditions [4–6]. Multi-area power systems and microgrids also use these controllers to provide adaptive emulated inertia control and reliable frequency regulation [7, 8].

In addition to the above applications, emergency response systems, power converters, load frequency control systems use RHC technique to guarantee the steady operation with significant DER assimilation. Also hybrid control techniques such as derivative based enhancements and PI-TI controllers effectively mitigate frequency instability and enhance the damping characteristics in deregulated power systems [9–11]. System resilience and transient stability can also be improved by combining emulated inertia synthesis with predictive models [12, 13].

This study presents a new framework based on recent advancement in predictive control algorithms for improving stability in power systems with insufficient inertia. The presented inertia emulation using the RHC controller is a promising solution for inertia deficient networks by utilizing an adaptive and intelligent control mechanism.

2 Overview of Emulated Inertia

Emulated inertia enhances power system stability by adding inertia and damping to address low system inertia, enabling reliable assimilation of distributed generators and distributed energy resources without compromising stability or resilience [14]. Figure 1 illustrates the structure of emulated inertia, comprising of an energy storage unit, inverter, and inertia emulation control unit. It adjusts active power inversely to rotor speed, operating in inertia control or emulated load modes. The active power output from the emulated inertia controller is given by

$$P_{\mathrm{EI}} = K_{\mathrm{EI}}\left(\frac{\mathrm{d}f}{\mathrm{d}t}\right) + J_{\mathrm{EI}}(\Delta f) + P_0 \tag{1}$$

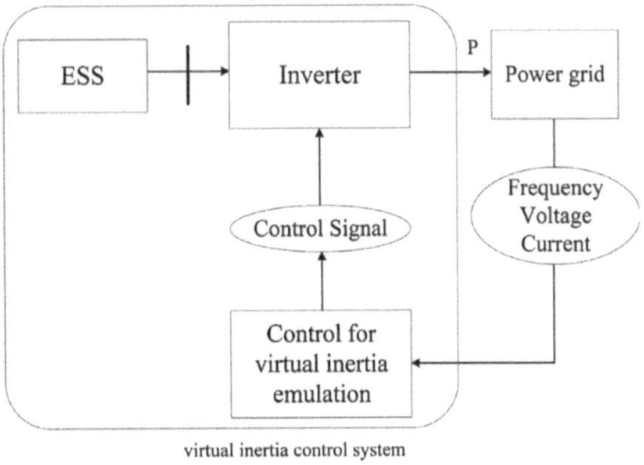

Fig. 1 Structure of emulated inertia

where

$$K_{EI} = \frac{2HP_{Inv}}{f_0} \tag{2}$$

The expression df/dt represents the rate at which frequency changes (RoCoF), f_0 signifies the system's nominal frequency, K_{EI} signifies the emulated inertia constant, J_{EI} signifies the emulated damping coefficient, P_0 represents the inverter nominal apparent power, and P_{Inv} denotes the devolution power to the inverter. The adoption of emulated inertia control entails mimicking the usual rotor dynamics of an alternator within the inverter's control system.

Emulated inertia control minimizes rotor speed fluctuations and mitigates frequency extremes using three components: emulated inertia, emulated damping, and constant power. emulated inertia replicates synchronous generators' inertia to reduce disturbances, while emulated damping mimics damper windings to align power with nominal levels. Constant power enables maximum active power exchange during frequency deviations and RoCoF events [9].

3 Emulated Inertia Control Strategy

Advanced control strategies like PI/PID control, receding horizon control, fuzzy logic, and robust control can enhance emulated inertia control. The adaptive alternating inertia approach adjusts parameters based on angular velocity oscillations for better system response. Self-tuning virtual synchronous machines reduce power flow and frequency deviations in energy storage systems. Frequency protection coordination and optimal deployment sites further improve emulated inertia control performance [9].

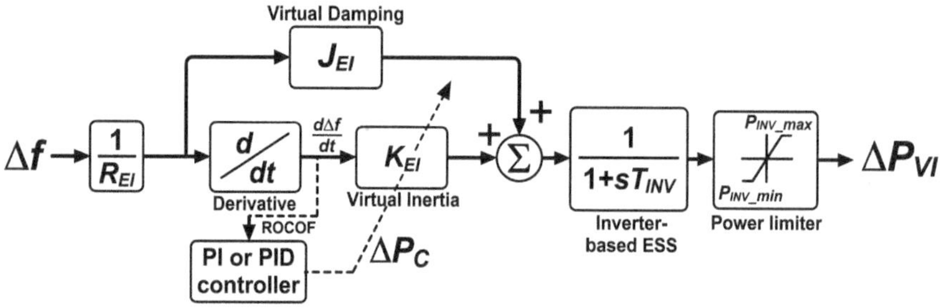

Fig. 2 Framework of emulated inertia control incorporating a PI controller [16]

3.1 PI Control Strategy

The main function of the decentralized emulated inertia controller based on a PI approach is to emulate inertia power [14]. To prevent issues such as increased frequency deviations, longer recovery times, and potential instability, it is vital to carefully select control parameters. The system must be capable of effectively responding to various disturbances, including fluctuations in DERs/DGs, loads, system inertia, and damping, while preserving stability. Furthermore, the emulated inertia control parameters should adjust to these disturbances while ensuring stability [15]. The PI controller plays a critical role in the tuning process, modifying values to ensure an appropriate response to disturbances. Figure 2 depicts the emulated inertia control design, which integrates a PI controller. The controller takes the rate of frequency change (df/dt) as its input and provides the regulated output. The desired constant value of the system is achieved using a PI feedback controller. Emulated inertia with PI control strategy becomes easy to implement on identifying optimal PI control parameters and demands minimal computational resources [15]. The PI controller is governed by Eq. (3).

$$C(s) = K_i + \frac{K_p}{s} \tag{3}$$

3.2 Fuzzy Logic Control Strategy

A decentralized fuzzy logic-based emulated inertia controller is suggested in this section for effective frequency stability in systems with different levels of DER/DG penetration. The emulated inertia constant, which is essential for mimicking inertia, is automatically adjusted by the controller. At low DER/DG levels, a high constant results in slower frequency deviation damping and longer settling times; at high DER/DG assimilation, a higher constant is required to reduce frequency overshoot and attenuate deviations. Blackouts and system instability can be avoided by

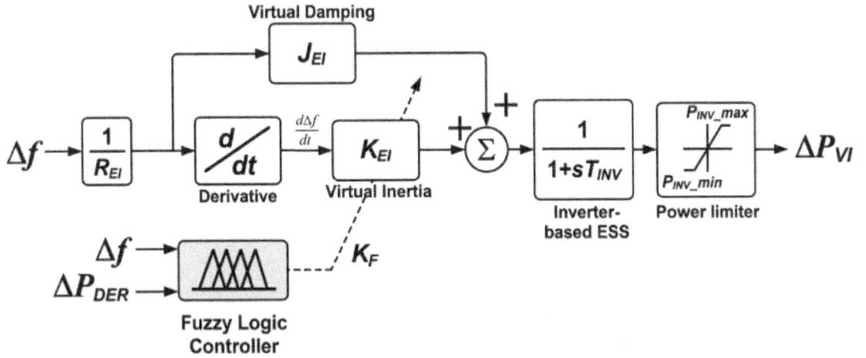

Fig. 3 Framework of emulated inertia control incorporating a fuzzy controller [16]

optimizing the efficacy and stability of the emulated inertia control system through parameter adjustments, ensuring alignment with varying levels of DER/DG assimilation. For self-adaptive control, the controller incorporates an interface with fuzzy rules, enabling automatic modifications in response to DER/DG power fluctuations and system frequency deviations. Figure 3 demonstrates the framework of emulated inertia control with a fuzzy controller.

The fuzzy system includes four key components: fuzzification, inference mechanism, fuzzy rule base, and defuzzification, as illustrated in Fig. 4. The fuzzy controller takes two inputs—changes in active power from distributed energy resources (ΔP_{DER}) and system frequency deviation (Δf). These inputs generate a normalized emulated inertia constant (K_F) as the output. Scaling factors are initially applied to adjust input magnitudes, followed by fuzzification, which converts the actual input values into fuzzy representations. The Mamdani inference model is employed throughout the inference process. The Inputs $X1$ and $X2$ are formed as:

$$X1 = \Delta f \cdot K_1 \tag{4}$$

$$X2 = \Delta P_{\text{DER}} \cdot K_2 \tag{5}$$

where $K1$ and $K2$ are scaling factors of the fuzzification system, where $K1 = K2 = 1$

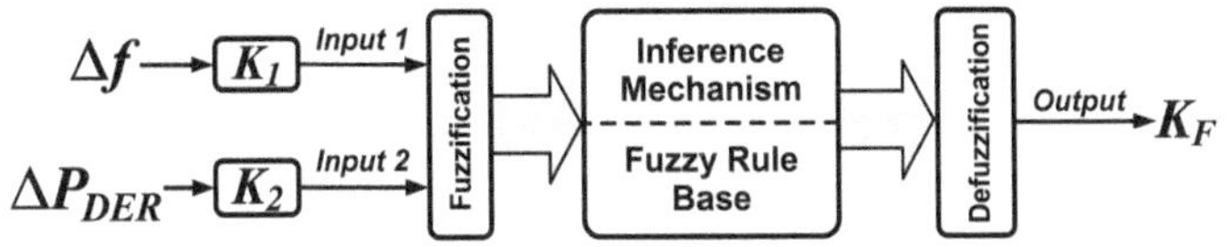

Fig. 4 Emulated inertia constant scheduling scheme in fuzzy logic controller [16]

Then, utilizing a fuzzy rule base, the fuzzy values derived from Inputs 1 and 2 are processed in the fuzzy inference stage. The fuzzy operations that transform the input signals into the output signal are defined by fuzzy rules. These rules combine the input signals of Δf and ΔP_{DER} to determine the optimal K_F value [16].

3.3 *Receding Horizon Control Strategy*

The Receding Horizon Control (RHC) offers a fast, disturbance-resistant dynamic response with the ability to intelligently forecast future control variables. Its primary functions are divided into three actions: Receding Horizon Technique, Optimization and Prediction. As shown in Fig. 5, the prediction unit leverages control signals, disturbances, and control outputs to anticipate future actions within a defined horizon. The control unit for the predicted output then minimizes the goal constraints by utilizing the known parameters in the optimization process.

Figure 6 depicts how Receding Horizon Control (RHC) functions as an emulated Inertia Controller. The RHC takes into account both observed and unobserved disturbances to calculate the ideal inertia constant for upcoming events while ensuring frequency stability. DER penetration and load variations are treated as measured disturbances via fast sampling. Instead of relying on feedback, the RHC employs feed-forward regulation to counter disturbances, effectively mitigating their impact and preserving system performance [15].

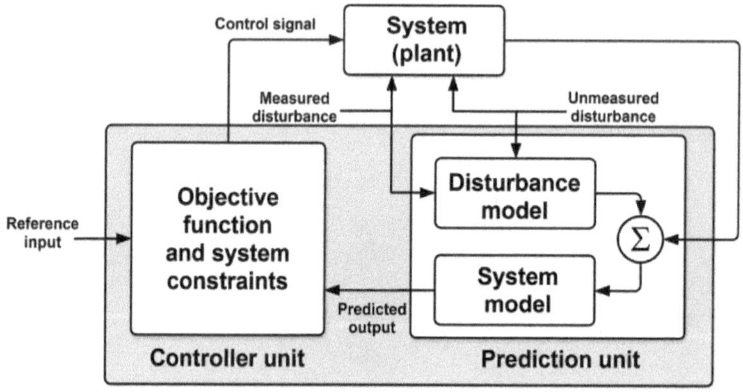

Fig. 5 Basic structure of RHC control strategy [16]

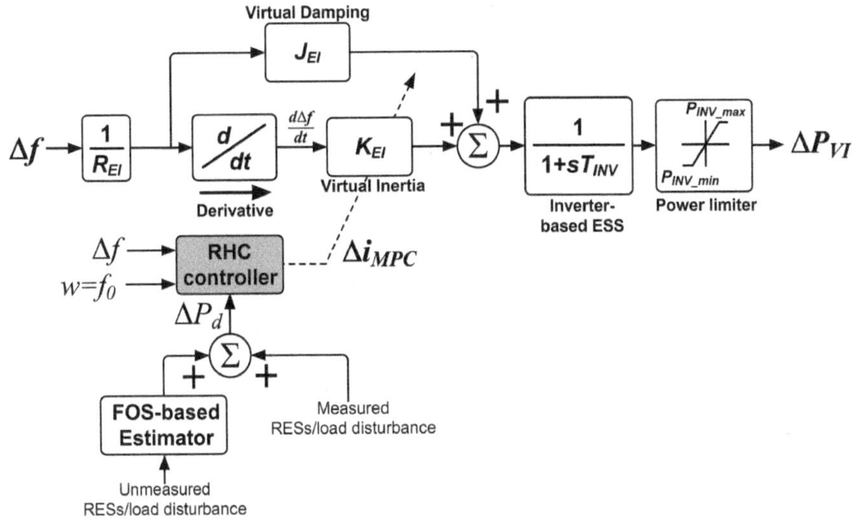

Fig. 6 Dynamic configuration of the RHC-based emulated inertia control [16]

4 Dynamic Modeling and Analysis of Microgrid for Frequency Regulation

Figure 7 depicts the model of a microgrid incorporating emulated inertia control for the analysis of frequency response. The microgrid is made up of various generation technologies serving different loads. The primary thermal power plant has an installed capacity of 15 MW, while the wind farms and solar facilities have capacities of 8.5 MW and 7.5 MW, respectively. The assimilation of distributed energy resources significantly changes the operating conditions of the microgrid, resulting in reduced inertia and damping properties of the system, affecting both stability and operation of the system. To mitigate this issue, energy storage systems (ESS) with a capacity of 4.5 MW are implemented, considering the base system of 15 MW. Information exchange and power delivery are facilitated through communication, frequency control, and power networks [9].

5 Results and Discussion

An Emulated Inertia Controller (EIC) based on Receding Horizon Control (RHC) has been developed and executed for an islanded microgrid with significant integration of distributed energy resources (DER). The system is implemented in the MATLAB/Simulink software, accounting for various disturbances, such as load fluctuations and unforeseen increase in DER generation. The parameters used are outlined in Table 1. After the tuning process, the optimal PI controller gains were determined

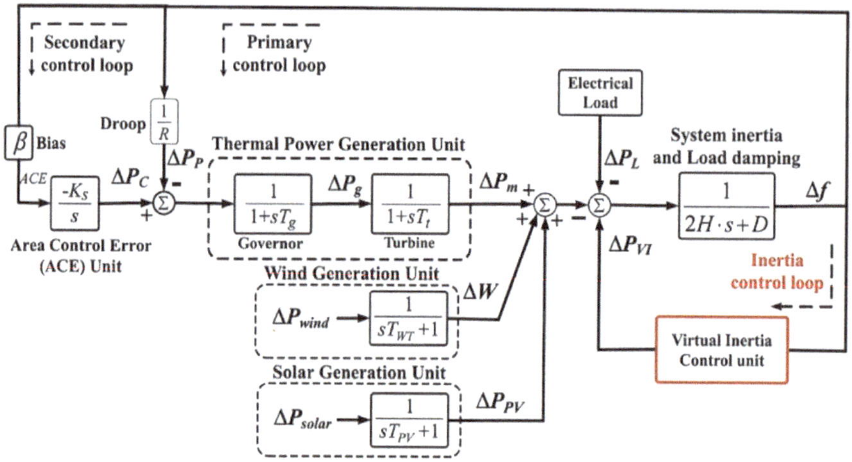

Fig. 7 Structure of studied microgrid [16]

as $K_p = 2.52$ and $K_i = 7.07$. The efficacy of the RHC control strategy is evaluated by applying a sudden load increase of 0.1 p.u. at 1 s, followed by a load reduction of 0.1 p.u. at 10 s, and a concurrent rise in DER generation (0.1 p.u. from wind and 0.1 p.u. from solar) at 20 s.

Case 1: Load increment of 0.1 pu at 1 s

Figure 8 compares microgrid frequency deviations under three control strategies: PI EIC, Fuzzy EIC, and RHC EIC. The PI EIC exhibits the slowest stabilization time (6–7 s) and the highest frequency drop of −0.34 Hz. The Fuzzy EIC exhibits the frequency drop of −0.29 Hz and stabilizes in 5–6 s. The RHC EIC outperforms both, achieving the smallest frequency drop of −0.09 Hz, the fastest stabilization (4–5 s), and minimal overshoot.

As emulated inertia compensates for power mismatches between generation and load, it reduces the burden on thermal plants during transients, maintaining frequency stability. Figure 9 shows that the emulated power by the RHC controller significantly reduces the thermal plant burden compared to Fuzzy and PI controllers.

Case 2: Load decrement of 0.1 pu at 10 s

When load demand decreases, generation enters acceleration mode, causing a frequency overshoot. Figure 10 illustrates system performance with PI, Fuzzy, and RHC EIC controllers. The PI EIC exhibits an overshoot of 0.28 Hz, while the Fuzzy EIC reduces it to 0.264 Hz. The RHC EIC achieves the smallest overshoot of 0.09 Hz. After the peak, transient oscillations occur, with the RHC EIC showing the least overshoot and the fastest stabilization.

It can be seen from the Fig. 11 that, as the load decreases, the output from the thermal plant will exceed demand. The surplus power will be absorbed by the emulated inertia controller which will work in load mode. RHC EIC will work

Table 1 Simulation parameters of the studied system [9]

Parameter	Values
Primary control unit	
Time constant of governor, T_g (s)	0.1
Time constant of turbine, T_t (s)	0.4
Droop factor, R (Hz/p.u. MW)	2.4
Secondary control unit	
Frequency controller, K_s (s)	0.2
Frequency bias, β (p.u. MW/Hz)	0.99
Inertia control unit	
Inertia constant, J_{VI}(s)	1.6
Damping coefficient, D_{VI}(s)	1.2
Inverter-based ESS time constant, T_{ESS} (s)	10
ESS Maximum capacity, P_{ESS} max (p.u. MW)	0.3
ESS Minimum capacity, P_{ESS} min (p.u. MW)	−0.3
DER control unit	
Wind turbine time constant, T_{WT} (s)	1.4
Solar system time constant, T_{pv} (s)	1.9
Microgrid control unit	
Inertia constant, H (p.u. MW s)	0.082
Damping coefficient, D (p.u. MW/Hz)	0.016

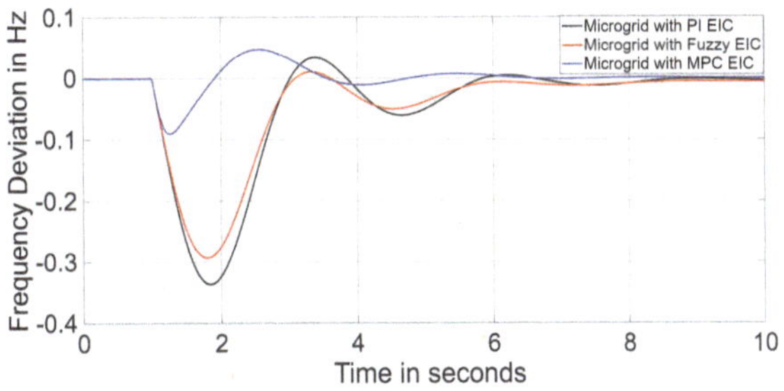

Fig. 8 Frequency deviation under PI EIC, fuzzy EIC and RHC EIC during load increment at 1 s

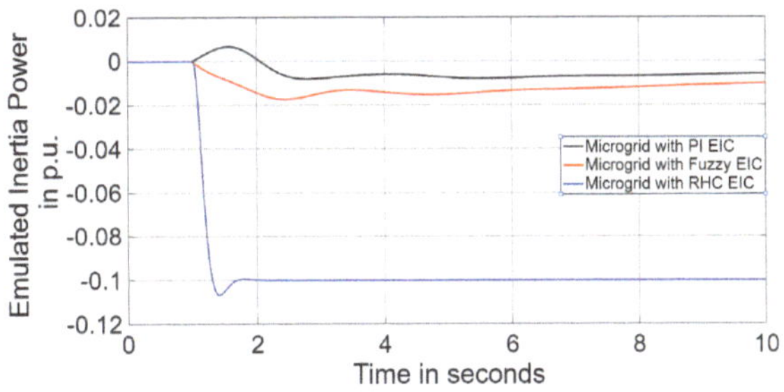

Fig. 9 Emulated Inertia Power under PI EIC, fuzzy EIC and RHC EIC during load increment at 1 s

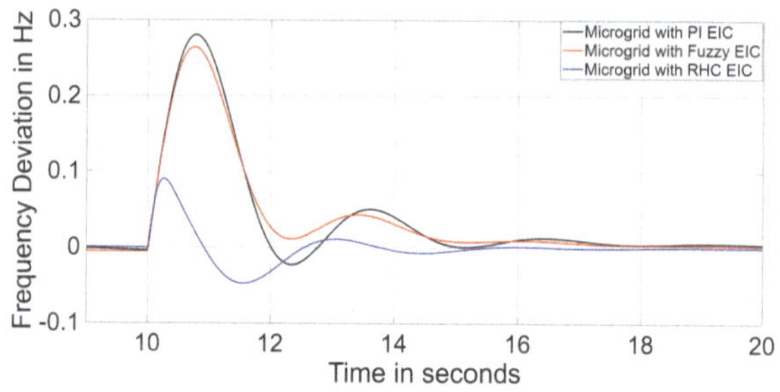

Fig. 10 Frequency deviation under PI EIC, fuzzy EIC, and RHC EIC during load decrement at 10 s

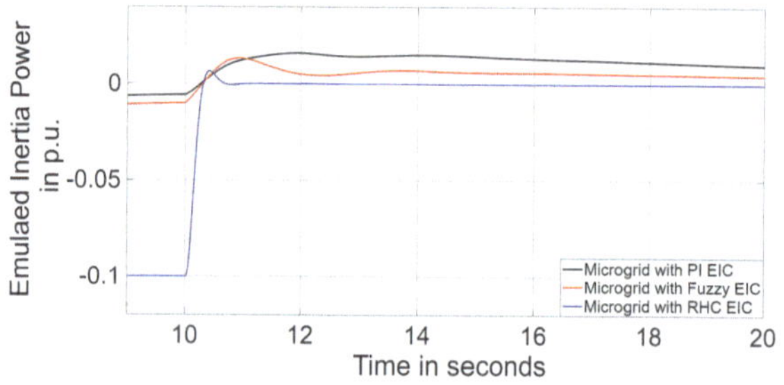

Fig. 11 Emulated Inertia Power under PI EIC, fuzzy EIC, and RHC EIC during load decrement at 10 s

effectively in load mode to reduce the frequency overshoot and make the system stable.

Case 3: Simultaneous Increase in DER Power (Solar and Wind) Generation at 20 s

When power generation increases, the excess power must be absorbed to reduce power mismatch and frequency deviation. During this transient, the system experiences a frequency overshoot of 0.26 Hz with PI EIC, 0.24 Hz with Fuzzy EIC, and 0.0435 Hz with RHC EIC. Figures 12 and 13 clearly show that the RHC EIC effectively mitigates stability issues during the transient. Equations (6) and (7) are used to calculate the Integral Square Error (ISE) and Integral Time Square Error (ITSE) metrics for various transient conditions.

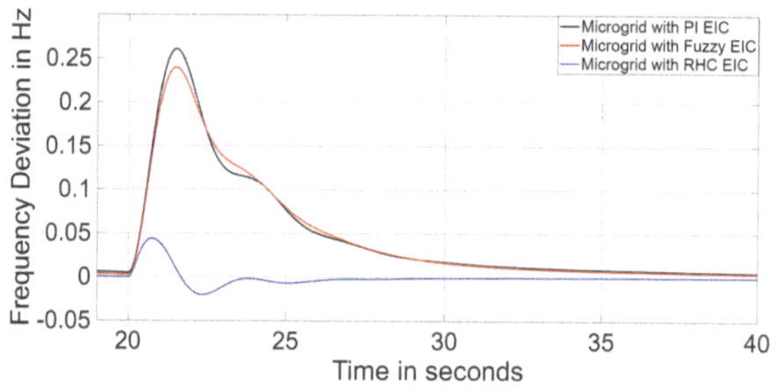

Fig. 12 Frequency deviation due to simultaneous increase in DER power (solar and wind) generation under PI EIC, Fuzzy EIC, and RHC EIC at 20 s

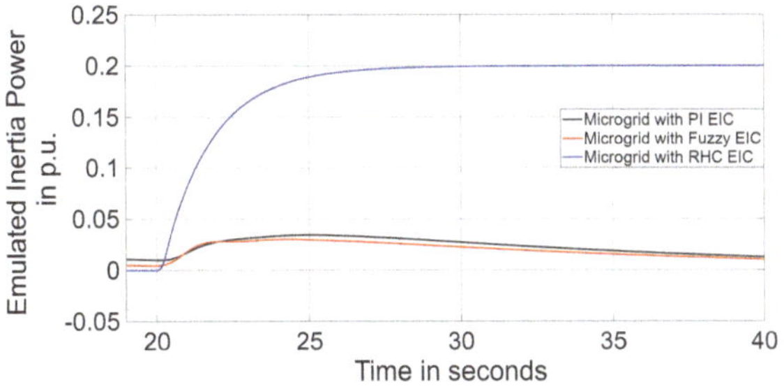

Fig. 13 Emulated inertia power due to simultaneous increase in DER power (solar and wind) generation under PI EIC, Fuzzy EIC, and RHC EIC at 20 s

Table 2 ISE and ITSE under different transient conditions for PI EIC, Fuzzy EIC, and RHC EIC

Case	PI EIC		Fuzzy EIC		RHC EIC	
	ISE	ITSE	ISE	ITSE	ISE	ITSE
Load increment	0.3131	3.69951	0.2717	3.44410	0.0163	0.12436
Load decrement	0.2068	1.73316	0.1915	1.64283	$9.3794 * 10^{-3}$	0.03408
Increase in DER generation	0.1339	0.32314	0.1262	0.31896	$2.4995 * 10^{-3}$	0.00325

In these equations, "*e*" represents the frequency deviation (p.u.), and "*t*" indicates the simulation duration. Table 2 presents the findings.

$$\text{ISE} = \int_0^t e^2 \, dt \tag{6}$$

$$\text{ITSE} = \int_0^t e^2 t \, dt \tag{7}$$

6 Conclusion

This study examined the performance of Receding Horizon Control (RHC)-based emulated inertia controller (EHC) for frequency stability in microgrids that incorporate a significant proportion of distributed energy resources. Simulation results indicate that RHC EIC outperforms standard PI and Fuzzy Logic controllers in all studied transient circumstances, resulting in considerable gains in frequency regulation and system stability. During a 0.1 p.u. load increment, the RHC EIC reduced frequency decreases to -0.09 Hz with a stabilization period of 4–5 s, outperforming PI and Fuzzy with a stabilization period of 6–7 s. ISE and ITSE metrics also improved significantly, with the RHC EIC reaching ISE of 0.0163 and ITSE of 0.12436, which are significantly lower than the PI EIC's 0.3131 and 3.69951, respectively. These findings validate RHC EIC as a highly effective approach for regulating dynamic frequency changes and maintaining stability in renewable energy-integrated microgrids. RHC EIC provides a strong framework for improving the performance of modern power systems by tackling the issues given by low inertia and variable power generation. Its adoption can improve grid resilience, reliability under changing conditions, and open the road for the smooth assimilation of distributed energy resources.

References

1. Bevrani H, Ise T, Miura Y (2014) Virtual synchronous generators: a survey and new perspectives. Int J Electr Power Energy Syst 54:244–254
2. Mir Cantarellas A, Remon D, Rodriguez P, Rakhshani E (2016) Analysis of derivative control based virtual inertia in multi-area high-voltage direct current interconnected power systems. IET Gene Transm Distrib 10:1458–1469
3. Tamrakar U, Shrestha D, Maharjan M, Bhattarai B, Hansen T, Tonkoski R (2017) Virtual inertia: current trends and future directions. Appl Sci
4. Karimi A et al (2020) Inertia response improvement in AC microgrids. IEEE Trans Power Electron 35(4):4321–4331
5. Boonlert T, Tephiruk N, Hongesombut K (2023) Fuzzy-based adaptive virtual inertia control. In: 20th international conference on electrical engineering/electronics, Thailand
6. Abdelghany MA, Syam FA, Aly AM et al (2024) Load frequency and virtual inertia control. J Electr Syst Inf Technol 11:50
7. Esfahani M, Habib HF, Mohammed OA (2018) Microgrid stability improvement using a fuzzy-based PSS design for virtual synchronous generator. In: South East Con, St. Petersburg, FL, USA
8. Jafari A, Rehimi S, Bevrani H (2023) Frequency regulation improvement in ac microgrids: a fuzzy-based extended virtual synchronous generator control. In: 2023 9th international conference on control, instrumentation and automation (ICCIA), pp 1–6
9. Beck HP, Turschner D, Mitani Y, Watanabe M, Raman FS, Kerdphol T (2019) Enhanced virtual inertia control based on derivative technique to emulate simultaneous inertia and damping properties for the microgrid frequency regulation. IEEE Access 7:14422–14433
10. Reigstad TI, Uhlen K (2019) Virtual inertia implementation in variable speed hydropower plant. In: 2019 Modern electric power systems (MEPS), Wroclaw, Poland, pp 1–6
11. Das S, Datta S, Saikia LC (2023) Effect of PI–TI-based virtual inertia controller. Arab J Sci Eng 48:6431–6452
12. Yap KY, Sarimuthu CR, Lim JM (2019) Virtual inertia-based inverters for mitigating frequency instability in grid-connected renewable energy system: a review. Appl Sci 9(5300)
13. Kerdphol T, Rahman FS, Watanabe M, Mitani Y (2021) Virtual inertia synthesis for a single-area power system, in virtual inertia synthesis and control. Springer, Cham
14. Serae G, Fiorentini M, Capozzoli A, Bernardini D, Bemporad A (2018) Model predictive control (MPC) for enhancing building and HVAC system energy efficiency: problem formulation. Appl Oppor Energies 11:1–35
15. Ganavi P, Jayasankar VN (2023) Model predictive controller based virtual inertia controller. In: RCAAI 2023, Springer, Singapore

16. Kerdphol T, Rahman FS, Watanabe M, Mitani Y (2021) Virtual inertia synthesis and control. Springer, Cham
17. Venkat AN, Rawlings JB, Wright SJ, Hiskens IA (2016) Distributed MPC strategies with application to power system automatic generation control. IEEE Trans Control Syst Technol 16:1192–1206
18. Bonfiglio A, Oliveri A, Procopio R, Delfino F, Denegrí GB, Invernizzi M, Storace M (2015) Improving power grids transient stability via model predictive control. In: Power systems computation conference, Wroclaw, Poland
19. Imsland L, Uhlen K, Ersdal AM (2015) Model predictive load frequency control. IEEE Trans Power Syst 31:777–785
20. Kerdphol T, Rahman F, Mitani Y, Hongesombut K, Küfeoğlu S (2017) Virtual inertia control-based model predictive control for microgrid frequency stabilization considering high renewable energy integration. Sustainability 9(5):773
21. Dragicevic T (2017) Model predictive control of power converters for robust and fast operation of AC microgrids. IEEE Trans Power Electron 33:6304–6317
22. Han S, Bhattacharya SP, Diaz-Rodriguez ID (2019) Analytical design of PID controllers. Springer, Switzerland
23. Bonfiglio A, Oliveri A, Procopio R, Rosini A, Invernizzi M (2018) An MPC-based approach for emergency control. IEEE Trans Ind Electron 66:5412–5422

Design & Deployment of Spy Stone for Military Surveillance

B. S. Venkatesh Murthy, K. Gururaj, and C. Nagaraj

Abstract The goal of the "Spy Stone" research project is to develop a portable, cost-effective surveillance tool that can keep an eye on enemy activity within military operations. The suggested gadget uses an Arduino Uno and a motor driver shield with a Bluetooth or Wi-Fi module for remote control, and it is made to seem like a stone or rock construction. The gadget includes a camera and microphone module that is wirelessly connected to a communication and transmission system, as well as a six-wheeled rover bogie that can navigate difficult terrain. Because of its affordability, effectiveness, and portability, the gadget can be used in nations with weak economies. The suggested technique is useful for safeguarding military installations from hostile attackers and combatants. The gadget is a useful tool for carrying out covert surveillance activities because of its ability to blend in with the surroundings. All things considered, the "Spy Stone" research initiative provides an economical and effective substitute for costlier gear that is now available on the market.

Keywords Surveillance · Communication · Transmission · Camouflage · Spying

1 Introduction

The collection of intelligence on the enemy is crucial in the defense industry, and surveillance tools are essential for this process. However, many low-income nations cannot afford the majority of the monitoring equipment now in use due to their

B. S. V. Murthy
School of Electrical and Electronics, Reva University, Bengaluru, Karnataka, India
e-mail: bsvmurthy364@gmail.com

K. Gururaj
Department of Electrical and Electronics, Nitte (Deemed to be University), Nitte, Karnataka, India
e-mail: raos2k@nitte.edu.in

C. Nagaraj (✉)
Department of Electrical and Electronics, M S Ramaiah Institute of Technology, Bengaluru, Karnataka, India
e-mail: assiniraj@gmail.com

I. M. Jamadar et al. (eds.), *Integrating Sustainability in Modern Technologies*, Lecture Notes in Mechanical Engineering, https://doi.org/10.1007/978-981-95-9466-5_6

high cost. In order to solve this problem, we have created the Spy Stone, a highly accurate, portable, and reasonably priced surveillance tool that is 50% less expensive than current models. The Spy Stone is a construction that resembles stone and has Bluetooth and a camera module. A motor driver circuit is utilized for connection in order to simplify things, and an Arduino Uno is used to ensure that everything works perfectly. The gadget serves as a surveillance tool, monitoring enemy forces and delivering real-time visual information with low latency [1–5].

To evade notice, the Spy Stone's exterior structure is made to look like a stone and fit in perfectly with the surroundings. The device is very versatile and practical since its locomotive component is made to resemble a rover bogie system, which allows it to go across uneven terrain with ease. Additionally, the system has a microphone module, which enables the detection of suspicious enemy behavior that can endanger military installations. By sensing variations in the infrared levels released by nearby objects, the pyro-electric Smart Spy Surveillance Robotic equipment can identify motion. Here, the RFID reader, spray motor, smoke sensor, and PIR sensors are crucial parts [5–10].

A hidden robotic instrument is used in military installations to spy on enemies by mimicking a stone or other rock formation. The primary goal is to guarantee the security of our military installations and offer dependable defense against hostile fighters or attackers. Additionally, it may readily help destroy enemy encampments, cantonments, and bunkers. This system's goal is to accomplish particular goals by integrating different parts. The system's viability and adaptability are increased by the locomotive section's ability to effortlessly navigate difficult terrain, thanks to its six-wheeled rover bogie design. These gadgets fall under the surveillance category as well. PIR sensors are utilized in this circuit to detect objects that are within a 5-m range. Since it is a simple model, the Raspberry Pi-3, PIR sensors, camera, and supply are the elements utilized here [11–16].

The suggested technology provides a fresh approach to spying on enemies within the military operations. It helps to destroy hostile bunkers, cantonments, and encampments and offers dependable defense against enemy soldiers or attackers. Remote control from far away is made possible by the locomotive section's integration with a Bluetooth or Wi-Fi module. Through wireless integration of the camera and microphone module with a communication and transmission system, hostile activity and suspicious conduct that can endanger military forces can be monitored. Because of its effectiveness and affordability, Spy Stone research is a great option for nations with weak economies. All things considered, the suggested defense surveillance system puts safety and security first while still being flexible enough to work in a variety of settings and terrains. Furthermore, it is made to be economical and it plays an essential role in wartime situations.

2 Methodology

Figure 1 shows the methodology and it is divided into 5 phases as given below,

Fig. 1 Methodology for collection of data

- Phase 1: Data Collection: Find the best parts and technology for the suggested system by conducting in-depth research on the use of clandestine robotic devices for military surveillance.
- Phase 2: System requirements: The system requirements must be established based on the proposed system's goals. This involves figuring out the system's necessary characteristics and capabilities, like movement, camouflage, remote control, communication, and monitoring.
- Phase 3: Design and development: Utilizing a six-wheeled rover bogie, an Arduino Uno, and a motor driver shield with a Bluetooth or Wi-Fi module incorporated, design the system's locomotive portion. Create and build the camouflage structure that will hide the locomotive part as well.
- Phase 4: Integration and control: To monitor enemy activity, integrate the locomotive portion with the communication and transmission system, as well as the camera and microphone module.
- Phase 5: Testing and Interface: Make sure the system can successfully navigate rough terrain, record and send high-quality audio and images, and evade enemy detection by testing it. For remote monitoring and control of the spy robot, an intuitive user interface is offered.

Figure 2 shows the block diagram of the proposed system, which comprises the design, construction, and installation of the pipe casing, battery bank, clamps and chassis, camera and microphone module along with Arduino Uno, Bluetooth, motors, and motor driver. It offers a graphical representation of the elements required for data collection and transmission. The Arduino UNO, motor driver shields, parallel-connected motors, a battery source, a Bluetooth module for remote transmission, and a camera module for data detection and sensing are all included.

The Spy Stone is a hidden surveillance tool used in military operations to spy on enemies and safeguard military personnel and property. It is made to look like a stone or other rock formation. The locomotive portion and the outside framework make up the device's two primary components. The locomotive structure is achieved with a rover bogie, as seen in Fig. 3a, and the front triangle alignment facilitates navigation over uneven ground. Furthermore, its design is readily adjustable to different environmental circumstances. As seen in Fig. 3b, the object's outward structure is masked to minimize the possibility that vulnerabilities will be discovered, making it harder for adversaries to identify and maybe diverting their attention.

In order to make it harder for opponents to identify or detect it as a surveillance device, the exterior construction is made to resemble a stone or rock and fit in with the surroundings. To make it easy to transport and deploy, the outside construction is composed of lightweight materials. A compartment for the locomotive section is

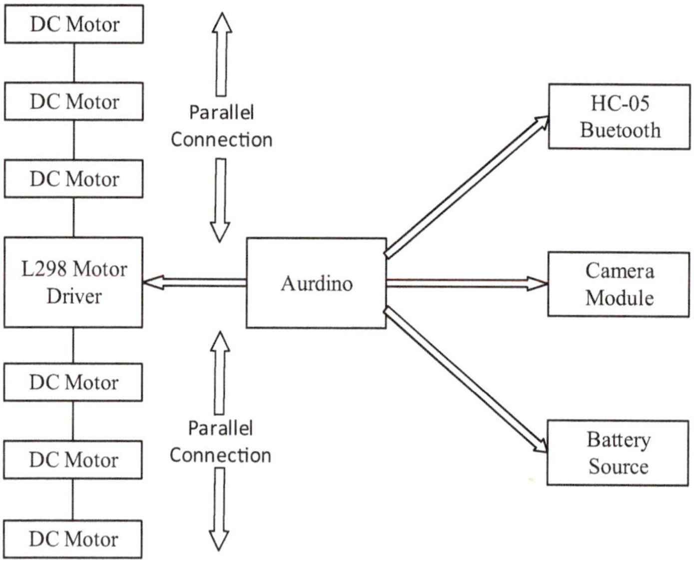

Fig. 2 Blockdiagram of the proposed system

Fig. 3 **a** Inner locomotive structure and **b** Camouflaged outer structure

built into the construction, and a variety of artistic creations are used to enhance the device's accuracy. The Spy Stone's locomotive section, which is its central component, is made up of a six-wheeled rover bogie that is intended to handle rough terrain and improve the system's viability and adaptability. An Arduino Uno microcontroller and a motor driver shield with a Bluetooth or Wi-Fi module is integrated to control the six DC motors that power the locomotive part at a speed of 60 RPM. The motor driver shield is an electronic circuit that enables the Arduino Uno, an inexpensive but effective microcontroller that is frequently utilized in do-it-yourself electronics initiatives, to regulate the direction and speed of DC motors.

In addition to offering protection, the camouflaged rock or stone-like construction conceals the locomotive part, further improving the device's environmental integration. In order to monitor enemy activity and any suspicious behavior that would endanger military outposts, the device has a camera and microphone module that is wirelessly integrated with a communication and transmission system. Real-time video streaming and picture taking are made possible by the Spy Stone's Wi-Fi+ Bluetooth camera module. Intelligence gathering may benefit from the audio data that is captured by the microphone module. The microcontroller and DC motors of the Spy Stone are powered by a 12 V battery. A computer or smartphone that connects to the microcontroller via a Bluetooth or Wi-Fi module can be used to control the gadget remotely. While the Wi-Fi module has a wider range and can be utilized for remote monitoring from greater distances, the Bluetooth module used in the Spy Stone has a range of 150–200 m.

The locomotive portion of a hidden robotic apparatus that is intended to look like a stone or rock construction is controlled by an Arduino Uno microcontroller in the planned research project. In addition to the locomotive component, the system will have a camera and microphone module for tracking suspicious activity and hostile activity, as well as an Arduino Uno with a motor driver shield and a Bluetooth or Wi-Fi module for remote control. For remote viewing and monitoring, these modules will be wirelessly connected to a transmission and communication system. The locomotive portion can be precisely controlled in this work, thanks to the Arduino Uno, which also makes it possible for the system to effortlessly navigate hilly terrain. Additionally, the Arduino Uno is easily programmable to perform intricate movements and sequences, increasing the system's adaptability and versatility in a variety of situations.

An essential part of the locomotive portion of the suggested hidden robotic system is the motor driver. It is in charge of directing and managing the six-wheeled rover bogie's speed and direction. The Arduino Uno, to which the motor driver shield is linked, is configured to accept commands via a Bluetooth or Wi-Fi module from a distance. The motor driver serves as an interface between the Arduino and the motors, supplying power and regulating the motors' speed and direction. The Arduino then communicates with the motor driver to operate the motors that drive the rover bogie. Additionally, it shields the Arduino from current surges and voltage spikes that could harm its constituent parts. A Bluetooth module is a wireless communication module that uses Bluetooth technology to enable data transfer over short distances. It is frequently utilized in research projects requiring wireless communication, such as robotics and home automation. The Arduino Uno, which manages the motor driver shield in the suggested hidden robotic gadget, may receive commands from a distance using the HC-05 Bluetooth module. This would eliminate the requirement for a physical connection between the robot and the remote control, enabling wireless control of the equipment. A serial link between the Arduino and the HC-05 module enables data transfer and reception between the two devices. The suggested hidden robotic device's camera and microphone sensor modules would provide various purposes:

- Surveillance: Military personnel would be able to keep an eye on enemy activity and spot possible threats, thanks to the camera module's live visual feed of the surrounding area. Any sounds or discussions would be picked up by the microphone module, giving important information about the movements and plans of the enemy.
- Reconnaissance: In order to find enemy locations, map out new terrain, and obtain other pertinent information about the area, the camera module may also be utilized for reconnaissance.
- Targeting: By directing weapons or other devices to their targets, the camera module could minimize collateral damage and enable precision targeting.
- Communication: The device and military troops could communicate in both directions using the microphone module, enabling real-time coordination and communication.

3 Results and Discussions

A low-cost, high-performing surveillance system designed to monitor hostile behavior at military sites is demonstrated by the "Spy Stone" research studies. It is easy to hide and blend in with the surroundings because the device is designed to resemble a stone or rock formation. It interprets an Arduino Uno, a motor driver shield, and a Bluetooth or Wi-Fi module for remote control. The technology is flexible and environment-adaptive because it uses a six-wheeled rover bogie that can traverse uneven terrain. The device can wirelessly transmit audio and video feeds in real time to a remote location by integrating its camera and microphone module with a communication and transmission system. The device can wirelessly transmit audio and video feeds in real time to a remote location by integrating its camera and microphone module with a communication and transmission system. The device's versatility and portability make it a valuable instrument for conducting hidden surveillance activities in remote or challenging-to-reach areas.

An application called the Arduino Integrated Development Environment (IDE) is used to program and create apps for Arduino boards shown in Fig. 4. For authoring, uploading, and testing code on Arduino boards, it is a robust tool that offers an easy-to-use interface. Because it supports a variety of programming languages, such as C, C++, and Python, this IDE is flexible and usable by both beginner and expert programmers. The drag-and-drop interface of the Arduino IDE is one of its primary advantages; it enables users to quickly connect different parts, including displays, sensors, and actuators, to their Arduino boards and program them to carry out particular tasks. Without having to create complicated code, developers can easily experiment with various components and features, thanks to this capability. It's lightweight construction allows it to move smoothly across uneven and rugged terrain without overtaxing drivers or motors. The equipment is equipped with propellers and a smoke dispenser to distract and deflect the adversaries. It may also detonate whenever necessary

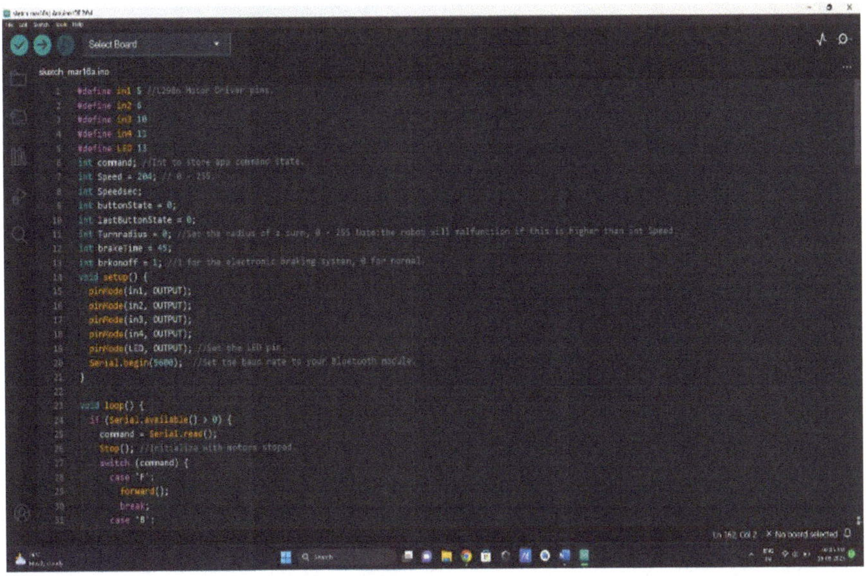

Fig. 4 Arduino integrated development environment (IDE)

because it is a suicide bomber. It is combined with land mine detection technology to provide safe passage across military areas and facilities.

4 Conclusion

The suggested technique provides a novel and distinctive method of military monitoring. Because it uses inexpensive materials and technology, the gadget is economical and accessible to nations with weak economies. The device's form, which mimics a stone or rock structure, makes it an invaluable tool for carrying out covert surveillance operations because it may blend in with the surroundings and go unnoticed. The locomotive portion of the gadget, which consists of a six-wheeled rover bogie, guarantees that it is adaptable and practical to operate and can easily navigate difficult terrain. A communication and transmission system is wirelessly connected with the camera and microphone module, enabling remote surveillance of adversary activity and any questionable behavior. An Arduino Uno and a motor driver shield with a Bluetooth or Wi-Fi module installed allows the gadget to be controlled remotely from a distance. All things considered, the "Spy Stone" research study provides a productive, economical, and successful substitute for pricey surveillance gear that is now available on the market.

References

1. Minho L, Minwoo C, Taehoon Y, Jingu K, Jaeoh K, Ojeong K, Namsuk C (2024) A study on the advancement of intelligent military drones: focusing on reconnaissance operations. IEEE Access 12:55964–55975
2. Monika D, Sneha K, Danish A, Sazia S (2024) Design and implementation of RF based war spying robot with wireless night vision camera. In: 11th IEEE international conference on reliability, infocom technologies and optimization (Trends and Future Directions) (ICRITO). IEEE Xplore, Noida, India, pp 1–5
3. Christine G (2019) 007 Spies, surveillance and pedagogical middle leadership: for the good of the empire of education. J Educ Adm Hist 51(2):165–181
4. Laxmikant M, Siddhant G, Khushi D, Suraj D, Abhishek K, Sanvdea G, Dhruv P, Dattatray K (2024) Long range spy robot with vision. J Electr Syst 20(11):761–766
5. Viktor Denes H, Vamsi Kiran A, Imre N, Csaba K (2023) Toward fast and accurate violence detection for automated video surveillance applications. IEEE Access 11:18772–18793
6. Irfan R, SwathiPrasad BK, Suresh R (2020) Design and development of spherical spy robot for surveillance operation. Procedia Comput Sci 171:1212–1220
7. Pavaiyarkarasi R, Krishna Veni K, Shyam M, Ben Paul Richard S, Prem Kumar R, Nishita SA (2023) Empirical development of army spying robot using bomb sensing and removal mechanism. In: IEEE international conference on advances in computing, communication and applied informatics (ACCAI). IEEE Xplore, Chennai, India, pp 1–7
8. Ron M, Thomas JM (2024) Military mimicry: the art of concealment, deception, and imitation. J Def Secur Anal 40(3):379–404
9. Muhammad Zia Ur R, Umair R, Muhammad Azeem A, Muhammad Tanveer R, Abdu HG, Nasir A (2023) Radio-controlled intelligent UGV as a spy robot with laser targeting for military purposes. J Axioms 12(2):1–19
10. Harsha Vardhini PA, Murali Chandra Babu K (2022) IoT based autonomous robot design implementation for military applications. In: IEEE Delhi section conference (DELCON). IEEE Xplore, New Delhi, India, pp 1–5
11. Massimiliano LC, Jai CG, Eduardo BS (2021) Saving private robot: risks and advantages of anthropomorphism in agent-soldier teams. Int J Soc Robot 14:2135–2148
12. Hu T, Ishtiaq A, Alamgir MSM, Kyunghi C (2020) 3D optimal surveillance trajectory planning for multiple UAVs by using particle swarm optimization with surveillance area priority. IEEE Access 8:86316–86327
13. Sujata SC, Nitesh N, Shubham B, Aniket S, Arjun D, Aniket P (2023) War field spy robot with metal detection and live streaming. In: 7th IEEE international conference on intelligent computing and control systems (ICICCS). IEEE Xplore, Madurai, India, pp 1688–1691
14. Saeed Ullah J, Habib Ullah K (2021) Identity and aggregate signature-based authentication protocol for IoD deployment military drone. IEEE Access 9:130247–130263
15. Diya T, Rajan S, Mehmet O, Michael H, Wei N (2019) Energy-efficient military surveillance: coverage meets connectivity. IEEE Sens J 19(10):3902–3911
16. Krishnamoorthy NV, Sujatha V, Naresh G, Ratheesh R, Dinesh M (2024) Revolutionizing military operations: a smart fabrication of bomb detection and diffusing robot with location sharing feature. In: IEEE international conference on advances in computing, communication and applied informatics (ACCAI). IEEE Xplore, Chennai, India, pp 1–7

Low Latency Microcontroller-Based FPV and Its Comparative Study

D. Smitha Gayathri, Puttaswamy Gowda Kumar, M. Thanuja, D. S. Sinchana, and Varsha Shridhar Belagar

Abstract Drones are one of the most commonly used types of unmanned aerial vehicles (UAVs). The "Remote Operator" system allows pilots to access the camera's live feed, enabling precise control of the aircraft. In this context, First-Person View (FPV) describes the method of operation rather than the platform itself. Unlike traditional flight control, where the pilot guides the drone from a ground perspective, FPV facilitates operation from the viewpoint of the drone's onboard camera, effectively simulating the experience of being onboard. This project aims to establish an FPV system for unmanned devices using Wi-Fi signal transmission, which provides notable benefits. Employing a microcontroller system ensures efficiency and reliability, allowing for low-latency, high-quality video streaming over Wi-Fi networks. Over time, FPV controllers have progressed from serving specialized pilot communities to becoming a vital category of remotely operated aircraft for diverse applications. These systems are instrumental in mapping flight paths and analyzing multispectral geospatial and temporal data related to crop growth and health. Insights derived from these systems help detect crop health problems that manual inspections might miss. With continuous and accurate monitoring, farmers gain valuable information about their crops, enabling early detection and resolution of potential issues.

Keywords P2P · FPV · UAV

D. Smitha Gayathri (✉) · M. Thanuja · D. S. Sinchana · V. S. Belagar
Department of ECE, BNMIT, Bengaluru, Karnataka, India
e-mail: smithagayathrid@bnmit.in

D. S. Sinchana
e-mail: 23ece097@bnmit.in

V. S. Belagar
e-mail: 23ece090@bnmit.in

P. G. Kumar
Department of ECE, DSCE, Bangalore, Karnataka, India
e-mail: Kumar-ece@dayanandasagar.edu

I. M. Jamadar et al. (eds.), *Integrating Sustainability in Modern Technologies*, Lecture Notes in Mechanical Engineering, https://doi.org/10.1007/978-981-95-9466-5_7

1 Introduction

The First-Person View (FPV) system enables real-time visualization from the perspective of unmanned aerial vehicles (UAVs), creating an immersive experience for pilots as if they were onboard. These systems consist of an onboard camera, video transmitter, receiver, and a display device or goggles for the operator. Widely popular in drone racing and aerial photography, FPV technology also finds applications in professional domains such as search and rescue, surveillance, and inspections of hard-to-reach areas.

Operators in some regions must adhere to regulatory requirements, including maintaining a visual line of sight and acquiring proper certifications. FPV systems have evolved rapidly, particularly since the late 2000s, leading to a thriving market for specialized equipment. Many pilots supplement standard-definition live feeds with lightweight HD cameras, such as GoPros, to capture higher-quality footage. Basic FPV setups feature an analog video transmitter and receiver, while advanced systems integrate GPS data, flight statistics, stabilization, autopilot, and return-to-home functionality. Additional features, like pan and tilt camera controls and head-tracking goggles, further enhance the immersive experience. A key challenge with FPV systems is video transmission latency, which can hinder a pilot's reaction time and pose safety risks. To address this, low-latency FPV setups powered by microcontrollers have emerged, ensuring minimal delay and real-time situational awareness for safer and more responsive operations.

This work presents an innovative use of FPV technology in UAVs, specifically for precision agriculture. Unlike conventional FPV systems designed for recreation or general surveillance, it employs Wi-Fi-based real-time video streaming and advanced microcontroller systems to deliver low-latency, high-quality video streams tailored for monitoring crops. By integrating multispectral geospatial and temporal data, the system provides precise crop health assessments, enabling early issue detection and resolution. This novel approach significantly enhances agricultural productivity and sustainability through UAV technology.

2 Literature Survey

Savasturk et al. [1] explored methods to improve the security of video and image transmission over wireless networks in their study, "AES Encrypted Real-Time Video Stream and Image Transmission from ESP32-CAM.“ Their approach involved using the Advanced Encryption Standard (AES) to encrypt data generated by the ESP32-CAM module. A Web Server was used for secure data processing, where the encrypted data was transformed into an unreadable format during transmission and later decoded to retrieve the original content. While the encryption process effectively ensured data security, it introduced a latency of 2 to 3 s, rendering it unsuitable for applications requiring low-latency and high-frame-rate streaming. The study also emphasized the ESP32-CAM's efficiency in wireless video and image capture.

Guo et al. [2] developed a system that employed the Raspberry Pi 3B+ as the central controller, coupled with the OV5647 camera module via a CSI interface for real-time image capture. They optimized video transmission reliability through image compression and parallel transmission over 4G networks, achieving low-latency, high-definition video streaming. Their setup successfully transmitted 720P video in real time, demonstrating its effectiveness for remote monitoring and drone control applications.

Uitto and Heikkinen [3] analyzed the drawbacks of commercial video streaming platforms, noting their dependence on buffering and cautious rate adjustments to maintain smooth playback while managing bandwidth. These techniques disrupted TCP congestion control, resulting in inconsistent video quality. Their research proposed a low-latency video delivery system with dynamic stream switching to reduce transport latency by aligning it with video chunk durations, offering a solution tailored for live media streaming.

Setton et al. [4] examined live peer-to-peer (P2P) streaming, which utilizes the uplink bandwidth of peers to distribute video content. This approach provides a self-scaling alternative to traditional server-based models, where additional servers are required to accommodate growing audiences. P2P systems, in contrast, naturally scale as peers act both as content consumers and providers. The researchers developed a control protocol enabling scalable application-layer multicast and introduced an adaptive scheduling algorithm that prioritized critical video frames and peers with greater downstream capacity, improving the efficiency of low-latency P2P streaming. These studies offer valuable insights into tackling challenges such as encryption, latency optimization, network efficiency, and scalable video delivery. Their findings have informed the design of secure, efficient, and low-latency video streaming systems, serving as a foundation for the strategies employed in this project.

3 Implementation

This section describes the implementation and operational methodology of the project. Figure 1 presents the workflow, which begins with the device connecting to the designated network. Ensuring that the device is on the specified network streamlines accessibility and usage. Upon establishing a connection, the camera begins live streaming by capturing frames, which are converted into digital data for transmission. The data is sent as MJPEG (Motion JPEG) files, representing a sequence of continuous images that emulate real-time video streaming. After uploading the program, the live feed can be accessed using the IP address assigned by the ESP32 module [5]. This address is used in Python with the OpenCV library, where the frames from the ESP32 are processed, converted, and analyzed for latency calculations. Simultaneously, the video stream is displayed in real time. The device, whether remotely controlled or operated by a pilot, is managed through a first-person perspective using the onboard camera. The video feed is wirelessly transmitted to either FPV goggles or a monitor, allowing the operator to view the live stream and efficiently control the device. This arrangement ensures accurate control and uninterrupted real-time monitoring.

A. **Analysis of Latency**

Managing and reducing delay in real-time video transmission systems is crucial for latency-sensitive applications such as video conferencing, remote control systems, and RF studio cameras, where minimal delay is essential. Achieving low-latency video transmission requires a strong foundation in implementation techniques, along with an in-depth understanding of video coding algorithms and buffer management strategies. In such applications, factors like network optimization, rate control, and

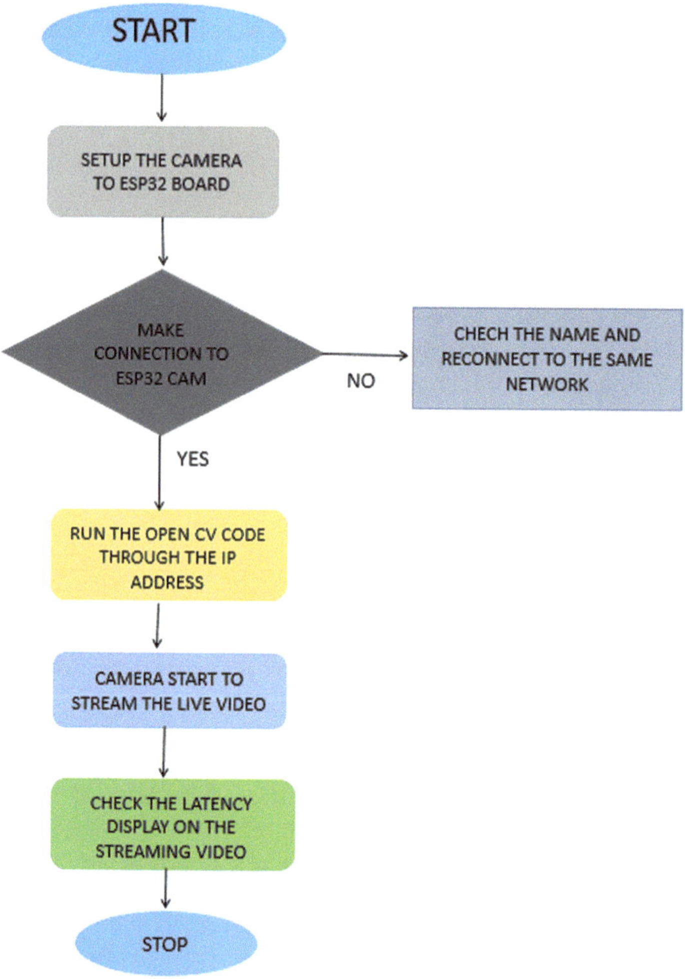

Fig. 1 Flow chart

frame skipping play a significant role, especially in video telephony and variable bit rate (VBR) streaming systems [6, 7]. The code demonstrates how latency is calculated in FPV systems.

```
import numpy as np
import cv2
import time
//Initialize the video capture object
For webcam, use 0 instead of the file path
cap = cv2.VideoCapture('vid.mp4')
//Variables to store the time for frame processing
previous_time = 0
current_time = 0
Process the video until the end
while cap. isOpened():
    # Read each frame from the video
    ret, frame = cap.read()
    # Stop if no frame is available
    if not ret:
        break
# Convert the frame to grayscale (if needed)
gray_frame = frame
# Resize the frame
gray_frame = cv2.resize(gray_frame, (500, 300))
# Font for displaying FPS
font = cv2.FONT_HERSHEY_SIMPLEX
# Record the time after processing the current frame
current_time = time.time()
# Calculate FPS (frames per second)
# FPS is the reciprocal of the time difference between two consecutive frames
fps = 1 / (current_time - previous_time)
previous_time = current_time
# Convert FPS to an integer
fps = int(fps)
# Convert FPS to string for display
fps_text = str(fps)
    # Overlay the FPS on the frame
cv2.putText(gray_frame, fps_text, (7, 70), font, 3, (100, 255, 0), 3, cv2.LINE_AA)
    # Show the frame with the FPS overlay
    cv2.imshow('Frame', gray_frame)
    # Exit the loop if 'Q' is pressed
    if cv2.waitKey(1) & 0xFF == ord('q'):
        break
Release the video capture object and close all windows
cap.release()
cv2.destroyAllWindows()
```

The block diagram highlights the main sources of delay in compressed video transmission setups. Optimized systems often identify the encoder and decoder stream buffers as the largest contributors to delay. Video streaming latency is defined as the time difference between capturing or encoding a frame and its display on the viewer's screen. Key factors contributing to latency include:

- **Encoding delay**: The time needed to encode video frames, which can increase latency if the process is complex or hardware is not optimized for video tasks.
- **Network delay**: The time required for video streams to traverse networks, which may be affected by congestion, physical distance, or other network issues.
- **Buffering delay**: While buffering mitigates network inconsistencies, it can introduce additional latency.
- **Display delay**: The time lag between a video frame being ready for display and its actual rendering, often due to hardware or software limitations.

Reducing latency requires a comprehensive optimization of hardware and software components [8, 9].

Figure 2 outlines the critical components of video compression systems that affect delays. The total latency in video transmission can be expressed as:

$$D(\text{total}) = (Dcap + Dreorder + Den) + (Denbuff + Dnet + Ddecbuff) + (Ddeco + Dout)$$

Effective system implementation aims to minimize these delays to ensure real-time video transmission performance.

Managing and reducing delay in real-time video transmission systems is critical for latency-sensitive applications. Examples include video conferencing, remote control systems, and RF studio cameras, all of which demand minimal delay.

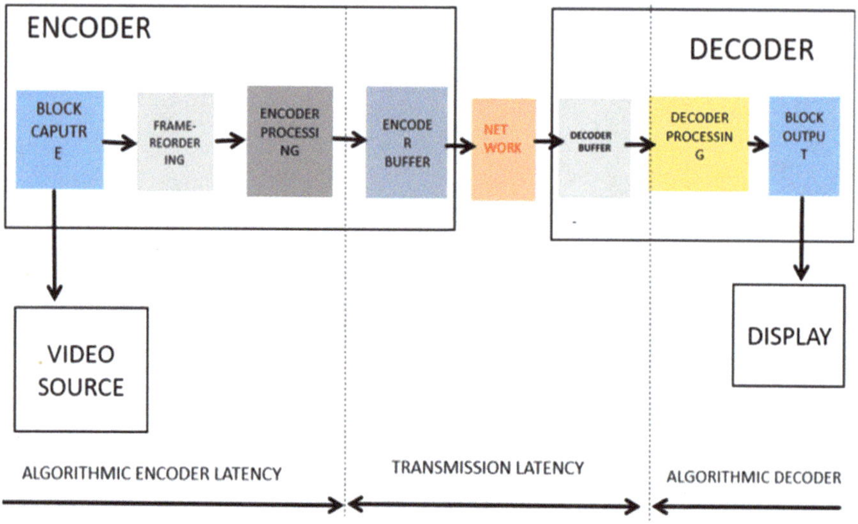

Fig. 2 Video transmission system delay source

Achieving low-latency video transmission requires a combination of foundational implementation expertise and a deep understanding of video coding algorithms and buffer management techniques. In low-delay applications, factors such as network optimizations, rate control, and frame-skipping are pivotal, with specific emphasis on video telephony and variable bit rate (VBR) streaming systems [6, 10].

4 Results and Discussion

A. Analysis Obtained from Camera Windowing

The OV2640 camera module offers flexibility by allowing users to define the window size or region of interest (ROI) according to specific application needs. The window size can range from 2 × 4 to 1632 × 1220 pixels (UXGA), 2 × 2 to 818 × 610 pixels (SVGA), and 408 × 304 pixels (CIF), all within the maximum boundary of 1632 × 1220 pixels. By default, the window size is set to 1600 × 1200 pixels. In this FPV system, the OV2640 camera is utilized. As the resolution increases, the number of pixels per frame rises, resulting in a larger volume of data packets to be transmitted per second [11, 12]. This increase impacts the file size and demands efficient data transmission management to maintain system performance. Below is the calculation for packets transmitted per second at a resolution of 640 × 480 pixels:

- Frame rate: 35 frames per second (fps).
- Pixels per frame: 640 × 480 = 307,200 pixels.
- Pixel-to-bit conversion:
  307,200×16×10×2=983,040,000.
- Packet size:
  1 packet = 50 bytes = 50 × 8 = 400 bits.
- Packets per frame:
  983,040,000÷400=2,45,760 packets/frame
- Total packets per second:
  2,45,760 packets/frame×35 frames/sec=82,45,760packets/frame

For a resolution of 320 × 240 pixels, the corresponding number of packets per second is calculated as 3,686,400 packets/second. Figure 3 shows the comparison of file sizes for the same image captured using two different cameras. The first image, taken with a 64 MP camera, has a file size of 13.13 MB, while the second image, captured with a 16 MP camera, results in a file size of 5.27 MB [13, 14]. This difference in file sizes arises due to the higher resolution and greater detail captured by the 64 MP camera. The increase in packets per second as resolution increases significantly impacts processing time and system latency. As the data volume grows, more time is needed for transmission, encoding, decoding, and handling of the video stream. Figures 4a and b illustrate the variation in video transmission latency at different resolutions. Higher resolutions, while enhancing video quality, require greater bandwidth and processing resources. This can lead to delays in real-time

FPV systems where low latency is critical. Striking a balance between resolution and performance is essential to ensure smooth and efficient video streaming, particularly in latency-sensitive applications.

B. Analysis from Photo Capture and OpenCV Method

This FPV-based system enables real-time video streaming for users. Two approaches were employed to measure latency in this project: the Photo Capture method and the Python OpenCV library method, with the latter used to validate the former [3, 15]. The Photo Capture method offers a straightforward and time-efficient way to calculate latency. Figure 4 illustrates the output obtained using this approach. By comparing the timestamps from a mobile device and a laptop, the latency of the video stream can be accurately determined. The simplicity and efficiency of this method make it highly advantageous for latency measurement.

To validate the results obtained from the Photo Capture method, the Python OpenCV library was used as a secondary approach. Figure 5 displays the results produced by the Python script utilizing the OpenCV library.

Fig. 3 Analysis of file size and latency implications

(a)

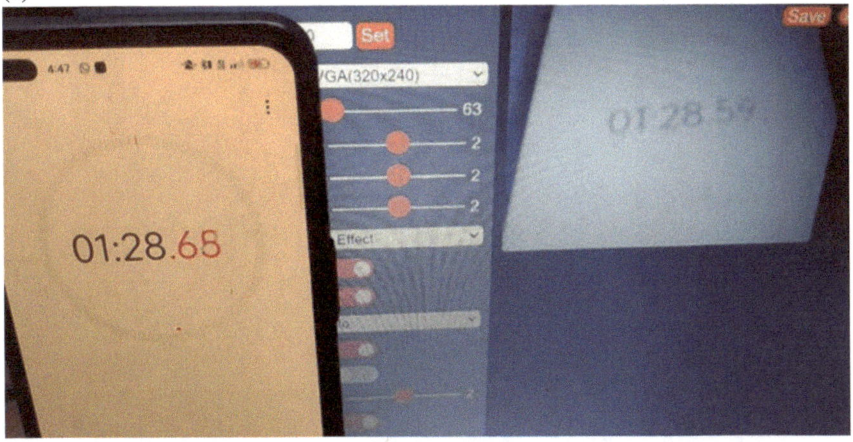

(b)

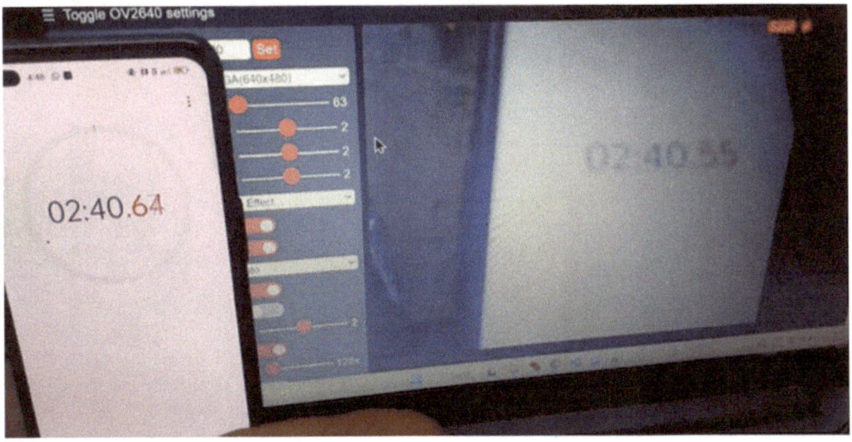

Fig. 4 **a** 320X240 resolution with 60 ms latency **b** 640 × 480 resolution with 90 ms latency

5 Comparison of the Results

Latency is directly influenced by the distance between the source and the destination, with an increase in distance leading to a corresponding rise in latency. Table 1 presents a comparison of latency measurements gathered using both the Photo Capture method and the Python OpenCV method. This table highlights the latency variations at different distance intervals.

Key observations from the study:

- The study involved 10–12 test cases, with each case increasing the source–destination distance by 5 m.

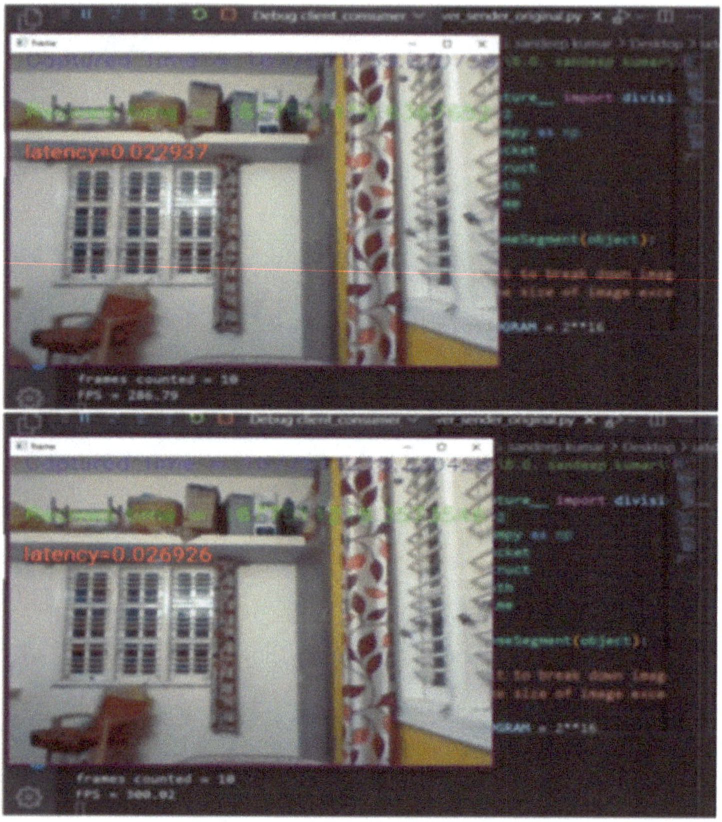

Fig. 5 Image of result obtained from photo capture method

- Initially, the Python OpenCV method showed slightly higher latency compared to the Photo Capture method. However, as the distance increased, both methods exhibited similar trends in latency growth.
- Based on the analysis, the Photo Capture method is confirmed to be a reliable and accurate approach for measuring latency.

Figure 6 provides a graphical representation comparing the latency results of the Photo Capture method and the Python OpenCV method. The graph demonstrates that both methods produce comparable latency measurements for video streaming, supporting the effectiveness of the Photo Capture method.

The Arduino Nano and ESP32 CAM are configured so that the ESP32 creates its own hotspot and assigns an IP address, which is displayed on the serial monitor. This IP address enables access to the camera stream [17]. The streaming is intended for the onboard antenna, and users can adjust the resolution based on their requirements. When the antenna is connected to the ESP32 CAM module, as depicted in

Table 1 Latency obtained from photo capture method and Python OpenCV

Sl. no.	Distance (m)	Latency photo capture (ms)	Latency OpenCV (ms)
1	1	30	48
2	5	40	56
3	10	53	59
4	15	70	72
5	20	80	76
6	25	82	78
7	30	85	80
8	35	91	87
9	40	96	90
10	45	99	92
11	50	102	110
12	55	115	113
13	60	120	129

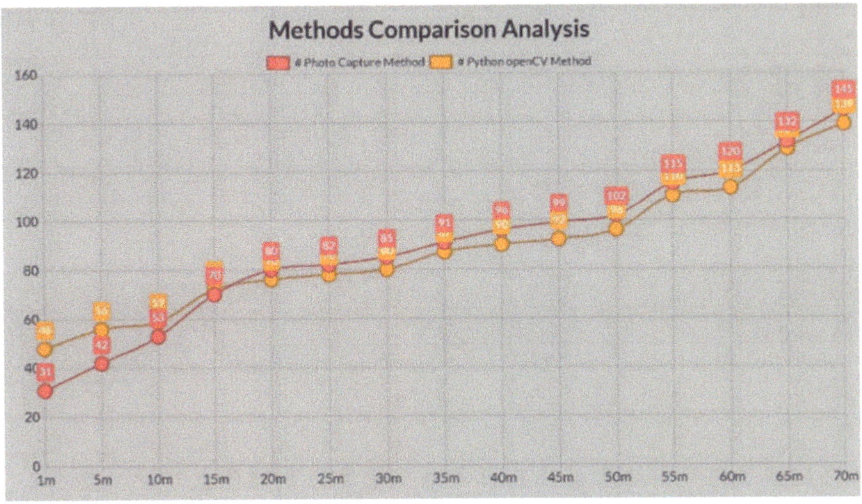

Fig. 6 Comparison of photo capture and cv2 Interface method [14, 16]

the connection diagram, the system produces clearer images over shorter distances, with image quality varying based on the range.

Figure 7 illustrates the latency graph comparing the system's performance with and without the external antenna across varying distances. Table 2 outlines the latency values recorded for both setups. Testing was conducted up to a maximum distance of 80 m, beyond which the signal dropped. While the latency with the external antenna was similar to that without it, the range was extended to 80 m. The system

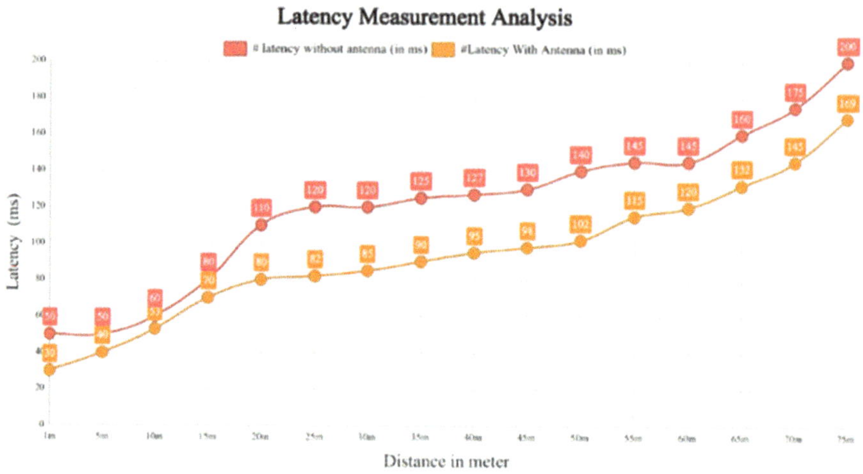

Fig. 7 Graph—comparing without antenna and with antenna latency

Table 2 Latency obtained using without and with antenna

Sl. no.	Distance (m)	Latency photo capture (ms)	Latency OpenCV (ms)
1	1	50	30
2	5	50	40
3	10	60	53
4	15	80	70
5	20	110	80
6	25	120	82
7	30	125	85
8	35	125	91
9	40	127	96
10	45	130	99
11	50	140	102
12	55	145	115
13	60	148	120
14	65	160	132

provided seamless streaming latency up to 55 m. Beyond this range, issues such as image freezing or connection loss were observed, defining the operational range and latency of the ESP32 CAM device.

6 Conclusion

The graph presents the values obtained through both the Photo Capture method and the CV2 interface. The fluctuations in latency across these methods demonstrate the variation in delay during video streaming. A comparison of the two methods reveals that their latency is either similar or nearly identical during streaming. This suggests a potential for improving the accuracy and responsiveness of remotely controlled devices, thereby enhancing the user experience. Additionally, advancements in microcontroller technologies and wireless communication protocols are expected to further minimize latency while boosting system range and reliability. In conclusion, low-latency microcontroller-based FPV technology shows significant potential and is anticipated to play a crucial role in the future of remote control and communication. While the current model supports live streaming and video recording, there is potential for expansion by adding control functionalities and a recognition system, enabling users to better understand the camera's output during video capture.

References

1. Savasturk P, Aydin O, Dalkilic G (2020) AES encrypted real-time video stream and image transmission from ESP32- CAM. Celal Bayar Univ J Sci 17(4):447–452. https://doi.org/10.18466/cbayarfbe.835945. https://papers.ssrn.com/sol3/papers.cfm?abstract-id=4171323
2. Guo J, Li X, Lv Z, Yang Y, Li L (2020) Design of real-time video transmission system for drone reliability. Xi'an Technol Univ 790:012004 ,https://doi.org/10.1088/1757899X/790/1/012004
3. Uitto M, Heikkinen A (2020) Exploiting and evaluating live 360° low latency video streaming using CMAF. In: 2020 European conference on networks and communications (EuCNC), pp 276–280. https://doi.org/10.1109/EuCNC48522.2020.9200954. https://ieeexplore.ieee.org/document/9200954
4. Setton E, Noh J, Girod B (2006) Low latency video streaming over peer-to- peer networks. In: 2006 IEEE international conference on multimedia and expo, pp 569–572. https://doi.org/10.1109/ICME.2006.262472. https://ieeexplore.ieee.org/document/4036663
5. Smith Gayathri D et al (2022) Adapting the effect of impulse noise in broadband powerline communication. https://doi.org/10.1007/978-981-19-2350-0_53
6. Wi -Fi Door Lock System Using ESP32 CAM Based on IoT. https://www.researchgate.net/publication/353331137_Wi_-Fi_DoorLock_System_Using_ESP32_CAM_Based_on_IoT
7. Vieira LFM, Srinivasa SS. Low-latency vision for micro air vehicles. https://www.mdpi.com/1424-8220/21/24/8262
8. Chiang KF et al. Low latency image transmission for micro aerial vehicles using compact FPV systems. https://www.mdpi.com/2078-2489/13/9/430
9. Kaur A, Jadli A, Sadhu A, Goyal S, Mehra A, Rahul (2021) Cloud based surveillance using ESP32 CAM. In: 2021 International conference on intelligent technology, system and service for internet of everything (ITSS-IoE), Sana'a, Yemen, 2021, pp 1–5. https://doi.org/10.1109/ITSSIoE53029.2021.9615334. https://ieeexplore.ieee.org/document/9615334
10. Meena D, Sharan R (2016) An approach to face detection andrecognition. In: 2016 International conference on recent advances and innovations in engineering (ICRAIE), Jaipur, pp 1–6. https://doi.org/10.1109/ICRAIE.2016.7939462. https://ieeexplore.ieee.org/document/7939462

11. Suvarnamukhi B, Seshashayee M (2018) Big data concepts and techniques in data processing. Int J Comput Sci Eng 6(10), 712–714. https://www.researchgate.net/publication/330138339-Big-Data-Concepts-and-Techniques-in-Data-Processing
12. Chauhan R et al (2021) Design of robotic snake with ESP 32 CAM and Arduino. In: 2021 19th OITS international conference on information technology (OCIT), Bhubaneswar, India, pp 6–9. https://doi.org/10.1109/OCIT53463.2021.00013. https://ieeexplore.ieee.org/document/9719518
13. Shuai Y, Gorius M, Herfet T (2014) Low-latency dynamic adaptive video streaming. In: 2014 IEEE international symposium on broadband multimedia systems and broadcasting, pp 1–6. https://doi.org/10.1109/BMSB.2014.6873486. Low Latency Microcontroller-based FPV 2022–23 https://www.nt.unisaarland.de/wpcontent/uploads/2019/05/2014-BMSBShuai.
14. Sun L, Zong T, Liu Y, Wang Y, Zhu H (2019) Optimal strategies for live video streaming in the low-latency regime. In: 2019 IEEE 27th international conference on network protocols (ICNP), pp 1–4. https://doi.org/10.1109/ICNP.2019.8888127. https://ieeexplore.ieee.org/document/8888127
15. Sarkar M, Sahoo PK, Sahoo PK (2024) Leveraging edge computing for video data streaming in UAV-based emergency response systems. Sensors 2024 24(15):5076. https://doi.org/10.3390/s24155076
16. OV2640 Data Sheet. https://www.uctronics.com/download/cam-module/OV2640DS.pdf
17. Smith Gayathri D et al (2019) Adapting the effect of asynchronous impulsive noise using interleaving based on compound signal classification in broad PLC. https://doi.org/10.35940/ijitee.B6701.129219.
18. Uitto M, Heikkinen A (2021) Evaluation of live video streaming performance for low latency use cases in 5G. In: 2021 Joint European conference on networks and communications 6G summit (EuCNC/6G Summit), pp 431–436. https://doi.org/10.1109/EuCNC/6GSummit51104.2021. https://cris.vtt.fi/en/publications/evaluation-of-live-video-streamingperformance-for-low-latency-us

Impact of Aerodynamic Drag on Battery Energy Consumption in Solar-Powered Vehicles: A Comparative Analysis

Suhani Jamadar and S. Gopinath

Abstract This study presents a comprehensive simulation of a solar-powered vehicle's energy dynamics over a 24-h period, examining the impacts of solar energy generation, vehicle energy consumption, and aerodynamic drag on battery performance. The vehicle operates under varying speeds, with energy metrics recorded at hourly intervals. Key findings reveal that solar energy generation begins at zero during the night and peaks during daylight hours, with a maximum solar energy production of 225 kWh occurring between hours 12 and 13. During the day, energy consumption fluctuates, with a notable increase in vehicle consumption due to both driving requirements and drag forces. The vehicle's energy consumption without drag reaches a maximum of 9.7009 kWh at hour 19, while the consumption with drag peaks at 10.977 kWh during the same hour, highlighting the significant effect of aerodynamic drag on energy usage. Battery energy levels show a consistent decline overnight, starting from an initial capacity of 50 kWh. The battery energy drops to a minimum of 3.3623 kWh by the end of the simulation period, reflecting the vehicle's reliance on stored energy during non-sunny hours. With the integration of drag forces, battery energy levels also follow a similar trend, indicating a rapid discharge under increased consumption rates.

Keywords Solar energy · Vehicle dynamics · Energy consumption · Aerodynamic drag · Battery management

S. Jamadar (✉)
Department of Technology, Shivaji University, Kolhapur, Maharashtra, India
e-mail: jamadarsuhani@gmail.com

Department of Electrical Engineering, Yashoda Technical Campus Satara, Wadhe, Maharashtra, India

S. Gopinath
Department of Electrical Engineering, Annasaheb Dange College of Engineering & Technology, Ashta, Maharashtra, India

I. M. Jamadar et al. (eds.), *Integrating Sustainability in Modern Technologies*, Lecture Notes in Mechanical Engineering, https://doi.org/10.1007/978-981-95-9466-5_8

1 Introduction

The integration of solar energy into Electric Vehicle (EV) systems has gained considerable attention in recent years due to its potential for enhanced sustainability and reduced reliance on fossil fuels. Various studies have focused on the design, fabrication, and optimization of solar-powered electric vehicles, highlighting the multifaceted benefits and challenges associated with this technology.

The development of solar-powered Electric Vehicles (EVs) has been a subject of growing research, focusing on optimizing energy efficiency, vehicle performance, and the integration of sustainable technologies. Heeraman et al. [1] explored the design and fabrication of solar-powered EVs, emphasizing the role of sustainable engineering practices in enhancing both energy efficiency and overall vehicle performance. Building upon this, Jamadar et al. [4] proposed an energy management strategy aimed at maximizing the utilization of solar energy in EVs, which contributes to improving system efficiency. Karthikeyan and Gumma [3] introduced a thermal management system using model-based simulations, underscoring the importance of efficient thermal regulation for optimal vehicle performance under varying conditions. Advanced charging systems for solar EVs were discussed by Khan et al. [5], emphasizing their necessity for efficient solar energy integration. Batt et al. [7] tackled the challenge of improving solar energy capture through Maximum Power Point Tracking (MPPT) algorithms, and Gautam et al. [9] proposed intelligent algorithms to further enhance solar system efficiency. From a broader perspective, Wagemans [2] examined the socio-technical implications of solar mobility systems, focusing on the societal and infrastructural impacts of potential system failures. The design of reliable electrical power systems for solar EVs was explored by Sarkar et al. [6], ensuring effective solar energy utilization. Gezelius and Mortazavi [10] analyzed the role of solar panels in influencing the adoption of battery electric vehicles, reflecting the growing interest in integrating solar technologies in the automotive sector. Despite these advancements, a significant limitation in the efficiency of solar-powered EVs is the aerodynamic drag, which increases with vehicle speed and reduces the overall range and energy efficiency. This aerodynamic challenge is crucial for future designs, requiring continued innovation in vehicle body shapes and materials to minimize resistance and enhance solar energy utilization. Collectively, these studies highlight the interdisciplinary nature of solar EV development, emphasizing the need for integrated approaches to optimize energy usage, system reliability, and societal acceptance while addressing challenges like aerodynamic drag.

Finally, the development of energy management systems for solar electric vehicles has been advanced by Jamadar and Singaram [11], who introduced a dynamic power split approach. This innovative method reflects the ongoing efforts to refine energy management strategies, thereby enhancing the overall effectiveness of solar-powered electric vehicles. Overall, the literature emphasizes the significance of interdisciplinary approaches in advancing solar-powered electric vehicle technology, ranging from engineering design and energy management to socio-technical implications and algorithm development. As research continues to evolve, these insights will be pivotal in shaping the future of sustainable transportation. This paper bridges a critical gap in understanding the quantitative impact of aerodynamic drag on the energy efficiency of solar-powered vehicles through detailed simulation and analysis. Unlike prior studies, this research uniquely focuses on the comparative energy consumption with and without drag across varying speeds under realistic solar and driving conditions.

The paper is organized as follows: Sect. 2 describes the system modeling and equations used for analysis. Section 3 presents the simulation methodology, parameters, and results. Section 4 discusses the implications and concludes the study with key insights and future directions.

2 System Modeling

2.1 Mathematical Modeling

See Fig. 1.

- **Sun Intensity Model**

The intensity of sunlight is modeled as a sinusoidal function to simulate daily variations in solar radiation:

$$I_{\text{sunt}} = I_{\text{peak}} \times \max\left(0, \cos\left(\pi \times \frac{\left(t - t_{\text{peak}}\right)}{t_{\text{sunset}} - t_{\text{sunrise}}}\right)\right) \quad (1)$$

where

- I_sun(t): Sunlight intensity at time t (kW/m^2).
- I_peak: Peak sunlight intensity (kW/m^2).
- t_peak: Time of peak sunlight intensity (hours).
- t_sunrise, t_sunset: Sunrise and sunset times, respectively.

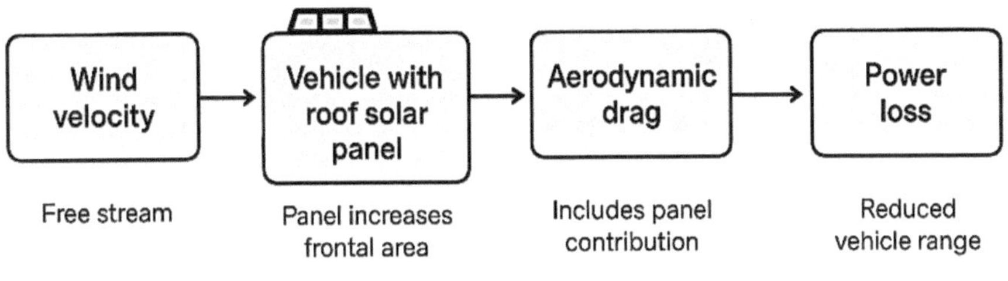

Fig. 1 Schematic representation of aerodynamic drag forces acting on a solar-powered vehicle

2.2 Solar Energy Generation

The energy generated by the solar panel is calculated as follows:

$$E_{\text{solar}} = \eta_{\text{solar}} \times A_{\text{panel}} \times I_{\text{sun}(t)} \tag{2}$$

where

- E_solar: Solar energy generated (kWh).
- η_solar: Solar panel efficiency.
- A_panel: Solar panel area (m^2).
- I_sun(t): Sunlight intensity at time t (kW/m^2).

2.3 Vehicle Energy Consumption

- Without Drag:

$$E_{\text{vehicle}} = 5 + 2 \times \text{randn}() \tag{3}$$

where

- E_vehicle: Base vehicle energy consumption (kWh).
- randn(): Random normal distribution to simulate real-world variations.
- With Drag:

  Drag Force is mentioned as below:

$$F_{\text{drag}} = 0.5 \times C_d \times \rho \times A \times v^2 \tag{4}$$

where

- F_drag: Drag force (N).
- C_d: Drag coefficient.
- ρ: Air density (kg/m^3).
- A: Frontal area of the vehicle (m^2).
- v: Vehicle velocity (m/s).

  Drag Power is represented as follows:

$$P_{\text{drag}} = \frac{(F_{\text{drag}} \times v)}{1000} \tag{5}$$

where

- P_drag: Power due to drag (kW).

2.4 Battery Energy Management

The battery energy level at time t is calculated as follows:

$$E_{\text{battery}(t)} = \min\left(\max\left(E_{\text{battery}}(t-1) + E_{\text{solar}}(t) \times \eta_{\text{battery}} - E_{\text{vehicle}}(t), 0\right), E_{\text{capacity}}\right) \quad (6)$$

where

- E_battery(t): Battery energy at time t (kWh).
- E_battery(t-1): Battery energy at the previous time step (kWh).
- E_solar(t): Solar energy generated at time t (kWh).
- η_battery: Battery efficiency.
- E_vehicle(t): Energy consumed by the vehicle at time t (kWh).
- E_capacity: Maximum battery capacity (kWh).

3 Results and Discussion

Figure 2a and b collectively illustrate the impact of aerodynamic drag on the energy consumption of an Electric Vehicle (EV) with roof-mounted solar panels, at varying speeds over a 24-h period. They offer a comparative analysis of how drag influences energy consumption at different vehicle velocities: 15, 30, 45, and 60 km/h.

By comparing the two Fig. 2a and b side by side, the significant role that aerodynamic drag plays in energy consumption becomes clear. When drag is accounted for, especially at higher speeds, the energy demands of the vehicle increase sharply, highlighting how much energy is required to maintain the same speed under greater aerodynamic resistance. The difference in energy consumption between the two scenarios is most pronounced at 45 and 60 km/h, where drag contributes to sharp increases in energy usage. At lower speeds (15 and 30 km/h), the impact of drag is less significant, and the energy consumption curves in both figures are closer in

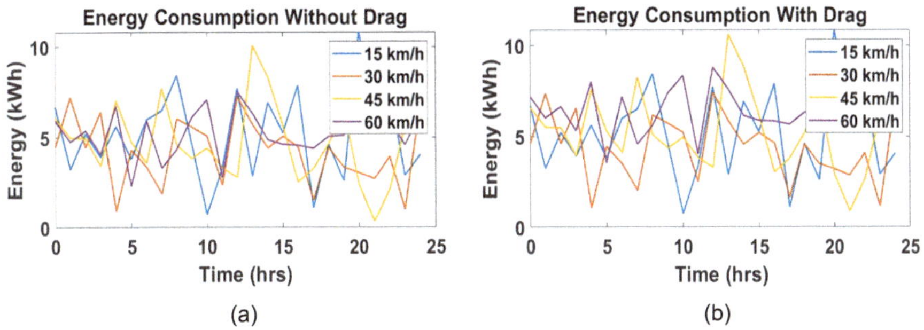

Fig. 2 Energy consumption of vehicle with and without drag

magnitude. However, as the speed increases, the curves in the first figure (with drag) rise noticeably higher than those in the second figure (without drag), emphasizing the efficiency loss due to aerodynamic forces. The comparison underscores the importance of minimizing drag to improve the energy efficiency of solar-powered EVs. At higher speeds, reducing drag could extend battery life and range by decreasing the vehicle's energy consumption. This has important implications for the design and operation of electric vehicles, especially those equipped with additional energy sources like solar panels, where aerodynamic efficiency can greatly influence overall performance.

In the first Fig. 3a, the energy consumption caused by aerodynamic drag is clearly visible. During the early hours of the day (before 6 a.m.), when solar energy is not available, the vehicle operates solely on battery power. As a result, the battery steadily depletes, reaching low levels by the time sunlight begins charging the battery after sunrise. Once solar power is available (from approximately 6 a.m. to sunset), the battery charges and maintains a level around 60 kWh throughout most of the day. After sunset, as the vehicle again relies solely on the battery for power, the battery energy declines, and this decline is more rapid at higher speeds, such as 60 km/h, compared to lower speeds. The effect of drag significantly contributes to this rapid depletion of battery energy after sunset. In the second Fig. 3b, the effects of aerodynamic drag are removed. While the overall trend of battery charging and depletion remains similar to the scenario with drag—starting to charge after sunrise and reaching full capacity during daylight hours—the absence of drag results in a slower rate of battery depletion after sunset. This is because the energy consumption is lower without drag, enabling the vehicle to sustain battery power for a longer duration at each speed. Even at higher speeds, such as 60 km/h, the battery depletes more slowly compared to the scenario with drag, demonstrating the impact of aerodynamic resistance on energy consumption.

The simulation was conducted using MATLAB in incorporating a sinusoidal model for solar intensity. The drag coefficient Cd was assumed to be 0.3, and the vehicle's frontal area A was set at 2.5 m^2. Energy dynamics were analyzed over a 24-h period, with vehicle speeds ranging from 15 to 60 km/h. The battery was initialized at 50 kWh, and solar energy input was modeled based on standard irradiance

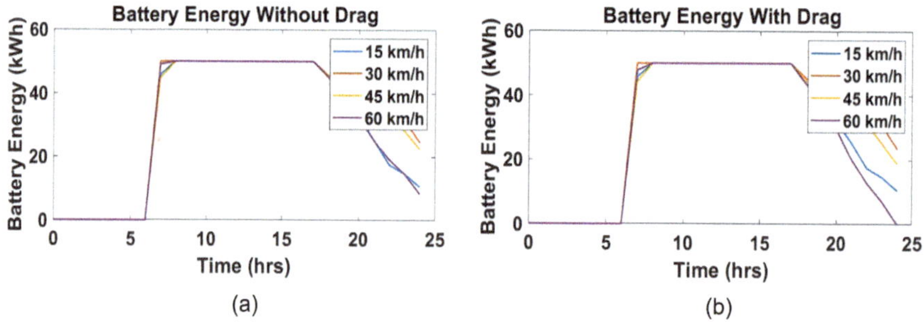

Fig. 3 Battery dissipation with and without drag

profiles. The provided data in Table 1, illustrates the energy consumption and battery performance of a vehicle over a 24-h period, comparing its behavior with and without aerodynamic drag. Solar energy production begins at 7 a.m. and lasts until 6 p.m., peaking at noon with a maximum of 225 kWh. During this period, the solar panels provide sufficient energy to meet the vehicle's consumption while also charging the battery. The battery remains fully charged during daylight hours, with no significant difference between the scenarios with and without drag due to the abundant solar energy.

During nighttime hours, from 0:00 to 6:00 and 19:00 to 24:00, the vehicle relies solely on battery power, and the effect of drag becomes more apparent. In the absence

Table 1 Simulation results

Time (hrs)	Solar energy (kWh)	Vehicle consumption (kWh)	Vehicle consumption with drag (kWh)	Battery energy (kWh)	Battery energy with drag (kWh)
0	0	3.2727	4.5487	46.7273	45.4513
1	0	5.1547	6.4308	41.5726	39.0205
2	0	2.5718	3.8478	39.0008	35.1727
3	0	2.773	4.049	36.2278	31.1237
4	0	4.9863	6.2623	31.2415	24.8614
5	0	8.0653	9.3413	23.1762	15.52
6	0	3.4607	4.7367	19.7155	10.7833
7	58.2343	5.7428	7.0188	46.6681	45.3921
8	112.5	4.5488	5.8249	50	50
9	159.099	7.2347	8.5108	50	50
10	194.8557	2.8219	4.0979	50	50
11	217.3333	5.0651	6.3412	50	50
12	225	6.1051	7.3811	50	50
13	217.3333	7.2012	8.4773	50	50
14	194.8557	8.0884	9.3645	50	50
15	159.099	5.1719	6.4479	50	50
16	112.5	2.0168	3.2929	50	50
17	58.2343	3.5154	4.7914	50	50
18	0	2.8768	4.1529	47.1232	45.8471
19	0	9.7009	10.977	37.4222	34.8702
20	0	3.7688	5.0448	33.6535	29.8253
21	0	6.4962	7.7722	27.1573	22.0531
22	0	4.6152	5.8912	22.5421	16.1619
23	0	6.7772	8.0533	15.7649	8.1087
24	0	3.4703	4.7463	12.2946	3.3623

of drag, energy consumption ranges from 2.5 kWh to 8.1 kWh, while with drag, the consumption increases to between 4.5 kWh and 10.98 kWh, indicating a notable rise in power usage due to drag. This represents an increase of approximately 1.5 to 3 kWh, which translates to an average deviation of 30–40% higher consumption when drag is present during these hours when compared with [11]. As a result, the battery depletes much faster in the presence of drag. For example, by 6 a.m., the battery retains 19.72 kWh without drag, but only 10.78 kWh with drag, a reduction of around 45%. Similarly, by midnight, the battery energy without drag stands at 12.29 kWh, whereas it drops to just 3.36 kWh with drag, a drastic 72% reduction.

In comparison with [11], during the solar production hours (7:00 to 18:00), the vehicle's energy consumption varies between 2.8 kWh and 8.1 kWh without drag and 4.1 kWh to 9.36 kWh with drag. However, the high availability of solar energy ensures that the battery remains fully charged in both scenarios. Therefore, the impact of drag is less significant during these hours, as the solar input compensates for the additional energy consumption caused by drag. In conclusion, the presence of aerodynamic drag significantly increases energy consumption during non-solar hours, with an average increase of 30–40%, leading to faster battery depletion. During daylight hours, the impact of drag is mitigated by solar energy, which keeps the battery fully charged in both scenarios. However, optimizing the vehicle to reduce drag would be crucial for extending battery life and improving overall energy efficiency, especially when operating during nighttime or in low-sunlight conditions.

4 Conclusion

This study simulated the energy dynamics of a solar-powered vehicle over a 24-h period, revealing crucial insights into solar energy generation and vehicle consumption patterns. Peak solar energy production reached 225 kWh between hours 12 and 13, demonstrating the effectiveness of solar panels during peak sunlight. However, vehicle energy consumption without drag peaked at 9.7009 kWh at hour 19, while consumption with drag reached 10.977 kWh, highlighting the significant impact of aerodynamic forces on energy requirements. Battery energy levels showed a decline from an initial capacity of 50 kWh to a minimum of 3.3623 kWh by the end of the simulation, emphasizing the challenges of energy management during non-sunny hours. These findings underscore the importance of optimizing solar energy capture, improving battery technologies, and reducing drag to enhance the performance and sustainability of solar-powered vehicles. This research contributes to the ongoing efforts to advance renewable energy integration in transportation, promoting a cleaner and more efficient future.

References

1. Heeraman J, Kalyani R, Amala B (2024) Towards a sustainable future: design and fabrication of a solar-powered electric vehicle. IOP Conf Ser: Earth Environ Sci 1285(1)
2. Wagemans S (2024) A shadow on Solar Mobility: the impact of failure on socio-technical system configurations. MS thesis
3. Karthikeyan VR, Gumma M (2024) A novel approach to make an efficient thermal management of a solar electric four-wheeler through model-based system simulation. No. 2024-26-0130. SAE Technical Paper
4. Jamadar NM et al (2023) Series energy management strategy for effective utilization of solar energy in electric vehicle. Life Cycle Reliab Saf Eng 12(4):293–298
5. Khan S et al (2018) A comprehensive review on solar powered electric vehicle charging system. Smart Sci 6(1):54–79
6. Sarkar T, Sharma M, Gawre SK (2014) A generalized approach to design the electrical power system of a solar electric vehicle. In: 2014 IEEE students conference on electrical, electronics and computer science. IEEE
7. Batt AA et al (2022) Design and optimization of solar using MPPT algorithm in electric vehicle. In: 2022 6th International conference on intelligent computing and control systems (ICICCS). IEEE
8. Bharathi ML (2022) Extraction of maximum power from solar with BLDC motor driven electric vehicles based HHO algorithm. Adv Eng Softw 170:103137
9. Gautam AK et al (2022) An intelligent BWO algorithm-based maximum power extraction from solar-PV-powered BLDC motor-driven light electric vehicles. J Intell Fuzzy Sys 42(2):767–777
10. Gezelius M, Mortazavi R (2022) Effect of having solar panels on the probability of owning battery electric vehicle. World Electric Vehicle J 13(7):125
11. Jamadar S, Singaram G (2024) Design and development of energy management system for solar electric vehicle with dynamic power split approach. Aust J Electr Electron Eng 1–12

Multi-robot Approaches for Sustainable Inspection and Maintenance: A Focus on Occupancy Map Merging

S. I. Arpitha Shankar, M. Shivakumar, K. R. Prakash, M. Latha, G. Manjula, and M. P. Lubna Taranum

Abstract Multi-robot systems see growing application in search-and-rescue operations, surveillance activities, and industrial automation sectors because these platforms excel at environmental exploration and mapping with remarkable efficiency. Occupancy maps, which represent an environment as grids of free, occupied, or unknown spaces, are essential for autonomous navigation. However, integrating maps from multiple robots into a unified global representation presents challenges such as inconsistencies in coordinate frames, communication constraints, and computational limitations. Our research confronts these obstacles by developing a robust and sustainable occupancy map-merging framework. Robots generate local maps by utilizing laser range finders and odometry sensors, which are aligned through feature-based methods and transformations estimated via the Iterative Closest Point (ICP) algorithm. The combined maps are verified by comparing them with ground truth data to ensure accuracy and consistency, enabling reliable autonomous navigation. Furthermore, the framework optimizes energy-efficient data processing and minimizes redundant computations, contributing to sustainability by reducing overall resource consumption in multi-robot systems. Autonomous systems minimize the need for human intervention in hazardous environments, enhancing safety while reducing operational costs. By integrating advanced occupancy map-merging techniques, this research enhances the efficiency and sustainability of multi-robot systems for inspection and maintenance, ensuring reliable and cost-effective operations.

S. I. A. Shankar (✉)
Depatment of CSE (AI&ML), GSSSIETW, Mysuru, India
e-mail: arpithasi@gsss.edu.in

M. Shivakumar
GSSSIETW, Mysuru, India

K. R. Prakash
Department of Mechanical Engineering, NIE, Mysuru, India

M. Latha · G. Manjula
Department of ECE, GSSSIETW, Mysuru, India

M. P. Lubna Taranum
Department of AI&ML, KNS Institute of Technology, Bengaluru, India

I. M. Jamadar et al. (eds.), *Integrating Sustainability in Modern Technologies*, Lecture Notes in Mechanical Engineering, https://doi.org/10.1007/978-981-95-9466-5_9

Keywords Map merging · Multi-robot systems · Occupancy grid map

1 Introduction

Multi-robot systems are being adopted more frequently in critical applications such as search-and-rescue, environmental monitoring, surveillance, and industrial automation. Their ability to work collaboratively offers significant advantages in terms of exploration efficiency, task distribution, and resilience in complex and dynamic environments. A key enabler for these systems is accurate and comprehensive mapping of the environment, often achieved through occupancy maps. These maps represent the environment as grids of free, occupied, or unknown spaces, forming the basis for autonomous navigation and decision-making.

In scenarios involving multiple robots, each robot typically generates its local occupancy map based on data collected from onboard sensors such as laser range finders or depth cameras. However, to fully realize the benefits of multi-robot collaboration, these local maps must be merged into a unified global map. This unified representation enables shared understanding among robots, facilitating coordinated navigation and task execution. Despite its importance, occupancy map merging is a challenging task. Robots often operate in overlapping areas or in environments with inconsistencies in sensor data and noise. Variations in coordinate frames, communication limitations, and computational resource constraints further complicate the process. Addressing these challenges requires robust algorithms capable of aligning and integrating maps accurately while ensuring real-time performance and scalability.

This work seeks to develop a robust framework for integrating occupancy maps from multiple robots into a unified global map with high accuracy and minimal errors. Advanced alignment techniques, such as Iterative Closest Point (ICP) algorithms and feature-based methods, are employed to enhance map alignment accuracy, addressing noise and sensor discrepancies. To synchronize coordinate frames among robots, a distributed consensus algorithm is implemented, ensuring consistent spatial alignment across maps.

Communication efficiency is optimized through data compression techniques to minimize overhead, enabling effective map sharing within limited bandwidth. The framework is designed to operate in real-time, supporting dynamic and time-critical multi-robot applications in environments. Scalability is a core focus, ensuring the system can support large teams of robots in diverse and complex scenarios. Validation of the framework is performed on hardware platforms like TurtleBots[22], with tests conducted in both simulated and real-world settings. These experiments assess robustness against localization errors, sensor noise, and environmental variability. The unified global map facilitates efficient navigation [17], task coordination, and collaboration among robots. This work advances multi-robot research by addressing key challenges in map merging and provides a foundation for future innovations in collaborative exploration and mapping.

2 Related Works

Occupancy map merging in multi-robot systems has been a topic of significant research, with various methodologies proposed to address challenges such as map alignment, localization inaccuracies, and communication efficiency.

Several authors have explored feature-based and grid-based approaches for map merging. Carpin (2008) proposed a probabilistic method for map merging that accounts for uncertainties in robot localization and map data. Similarly, Howard (2004) introduced a framework using Monte Carlo localization to align maps in overlapping environments. These approaches rely on finding transformation parameters to align maps accurately while mitigating errors arising from noisy sensor data. ICP algorithms have been widely used in map merging. Thrun et al. (2005) demonstrated the use of ICP for aligning laser-scan-generated maps. They showed its effectiveness in reducing alignment errors but noted its sensitivity to initial pose estimates. Extensions to ICP, such as Generalized ICP (Segal et al., 2009), have improved performance by optimizing both rotation and translation parameters.

To address inconsistencies in coordinate frames, distributed consensus algorithms have been investigated. Olson (2009) introduced the Max-Mixture model for map alignment in the presence of multiple hypotheses. Additionally, Mohanarajah et al. (2015) proposed MapLab, a distributed framework that integrates pose graph optimization for global map consistency. Scalability has been a key focus in multi-robot systems. Burgard et al. (2000) developed a coordination mechanism for distributed exploration, emphasizing scalable communication and task distribution. More recently, Wurm et al. (2010) explored hybrid methods combining metric and topological map representations for large-scale environments. These approaches aim to maintain real-time performance while ensuring scalability.

Experimental studies using platforms like TurtleBots and Pioneer robots have validated map-merging methods. For example, Konolige et al. (2010) evaluated the use of Visual SLAM-based map merging in indoor environments, while recent works have leveraged deep learning to enhance feature extraction and map alignment. These works highlight the progress in occupancy map merging, emphasizing the importance of addressing alignment accuracy, computational efficiency, and real-world applicability. The research being proposed builds upon these advancements to offer a comprehensive framework for multi-robot systems.

3 Methodology

The proposed methodology for occupancy map merging in multi-robot systems comprises the following key components:

1. Local Map Generation

Each robot independently generates a local occupancy map of its surrounding environment using onboard sensors such as laser range finders or LiDAR, combined with odometry data. These maps are constructed as grid representations, where cells are classified as free, occupied, or unknown spaces. Probabilistic techniques are used to manage sensor noise and improve map accuracy.

2. Map Alignment Using Feature-Based Techniques

   Local maps undergo alignment through feature-detection methodologies. Distinctive environmental characteristics including angular junctions and boundary lines are identified within individual cartographic representations. These identified elements are then cross-referenced between separate maps to determine spatial relationships. The alignment methodology employs the progressive nearest-point comparison technique to systematically enhance the positional adjustments (rotational and translational) required for proper map synchronization.

3. Establishing a Global Reference Frame

   To handle discrepancies in coordinate frames among robots, a distributed consensus algorithm is implemented. This ensures that all robots align their maps within a common global reference frame. This step minimizes cumulative errors during map merging and ensures spatial consistency.

4. Data Compression for Communication Efficiency

   The rapid transfer of information proves crucial for immediate operational effectiveness. Map data undergoes compression processing to minimize the volume of spatial information transmitted between robotic units. This methodology enhances communication channel efficiency while preserving vital environmental details required for precise map integration.

5. Parallel Processing for Computational Efficiency

   To manage the computational demands of map alignment and merging, parallel processing techniques are utilized. Multi-threading and GPU acceleration are employed to handle tasks such as feature extraction, transformation estimation, and validation, simultaneously, ensuring faster performance.

6. Validation of Merged Maps

   The merged global map is validated against ground truth data or pre-existing maps to assess accuracy. Metrics such as alignment error, map consistency, and coverage are used for evaluation. Sensor fusion techniques are incorporated to resolve any residual inconsistencies.

7. Experimental Testing and Deployment

   The proposed framework is tested in simulated and real-world environments using TurtleBot platforms. Experiments are conducted under various conditions, such as dynamic obstacles, overlapping regions, and noisy sensor inputs, to evaluate robustness and scalability.

This methodology ensures accurate, scalable, and real-time merging of occupancy maps, enabling effective navigation, task coordination, and collaboration among robots in multi-robot systems.

Map merging is a critical process in multi-robot systems where individual environmental maps must be combined into a cohesive representation. This integration enables effective collaboration among autonomous robots working toward common goals.

The core challenge of map merging involves transforming disparate spatial data into a unified coordinate framework while maintaining accuracy. This process requires careful alignment of map features and resolution of any conflicting or redundant information between maps.

When robots begin with predetermined starting positions, the map merging workflow benefits from this prior knowledge. By calculating the precise transformation parameters (both translational and rotational) between robots coordinate systems, their individual maps can be aligned with high precision [2]. This approach streamlines the merging process, as the spatial relationship between maps is established from the outset, resulting in more reliable integration of environmental data.

The merging task becomes substantially more complex when robots lack knowledge of their relative positions. In these circumstances, each robot independently explores and maps its surroundings without reference to others' coordinate systems. The primary challenge shifts to determining spatial relationships between these disconnected maps.

This scenario necessitates sophisticated techniques based on environmental feature recognition. Robots must identify common landmarks or patterns across different maps to estimate relative positions. The merging follows an iterative approach:

1. individual map creation through independent exploration,
2. feature-based alignment estimation between map pairs,
3. progressive integration of maps with continuous refinement, and
4. validation processes to ensure spatial consistency.

Throughout this process, the system continuously adjusts alignments based on new information, gradually improving the merged map's overall accuracy. The final merged map, obtained through this approach, is depicted in Fig. 2.

3.1 *Framework for Merging Occupancy Grid Maps from Two Robots*

For a multi-robot system with two robots, the map merging process can be formalized using the following mathematical model:

Prerequisites

- Two robots: Robot A and Robot B.
- Individual occupancy grid maps: M_A and M_B.
- Each map consists of cells with occupancy probabilities: $P(m^i{}_A)$ and $P(m^j{}_B)$, where i and j are cell indices.

Transformation Computation

1. **Homogeneous Transformation Matrix**: Let $T_{A \to B}$ represent the transformation from Robot A's coordinate frame to Robot B's: $T_{A \to B} = \left[R\ t0\ 1 \right]$ where
   - R is the 2 × 2 rotation matrix:

$$\begin{matrix} \cos\theta & -\sin\theta \\ \sin\theta & \cos\theta \end{matrix},$$

   - t is the 2 × 1 translation vector: $\begin{matrix} tx \\ ty \end{matrix}$, and
   - θ is the relative orientation between the robots.
2. **Feature Correspondence**: For corresponding feature points $p^k{}_A$ MA and $p^k{}_B$ in map $M_{B.}$

$$p_B^k = T_{A \to B} \cdot p_A^k \tag{1}$$

Map Merging Algorithm

1. **Occupancy Grid Fusion**: For each cell m_i in the merged map M_{merged}:

$$P(mi) = \frac{P(miA) \cdot P(mBi)}{P(miA)P(mBi) + (1 - P(miA)).(1 - P(mBi))} \tag{2}$$

   where

   - P(mi) is the occupancy probability of cell i in the merged map,
   - P(miA) is the occupancy probability of the corresponding cell in map M_A, and
   - P(miB) is the occupancy probability of the corresponding cell in map M_B.
2. **Log-Odds Representation**: Alternatively, using log-odds for numerical stability:

$$L(mi) = L\left(m_A^i\right) + L\left(m_B^i\right) - L_0 \tag{3}$$

   where

   - L_0 is the prior log-odds (typically 0, corresponding to $P = 0.5$).
3. **Weighted Map Merging**: When considering measurement certainty:

$$P(mi) = \frac{wA.P(miA) + wB \cdot P(mBi)}{wA + wB} \tag{4}$$

where wA and wB are weights representing confidence in each robot's measurements.

This mathematical model provides a framework for effectively merging occupancy grid maps from two robots while handling transformations between coordinate frames and resolving conflicting information (Fig. 1).

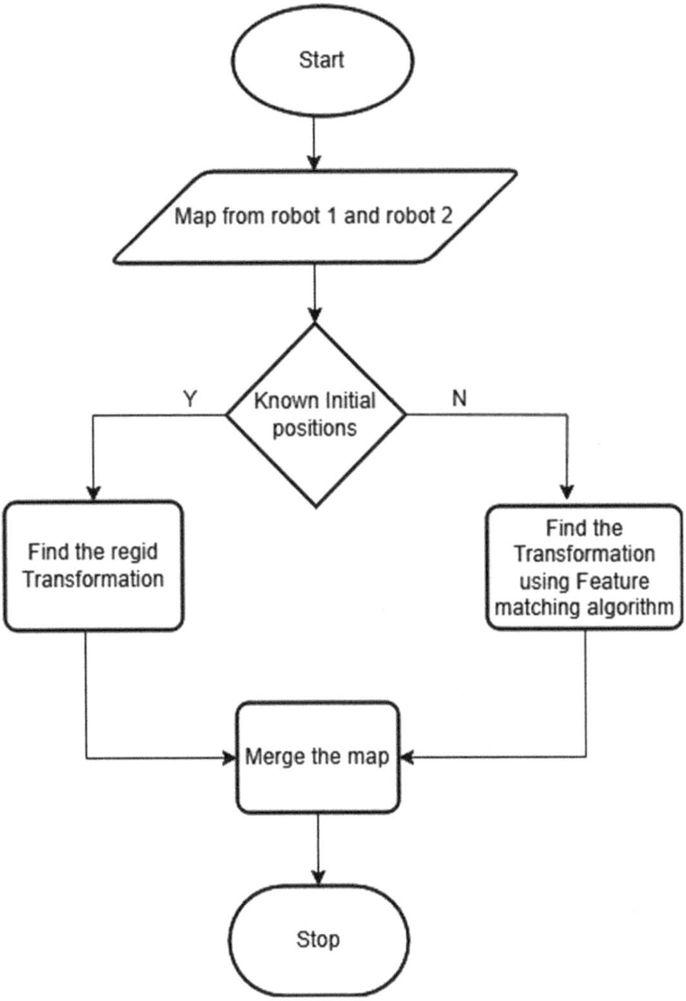

Fig. 1 Diagram representing the map merging process

4 Results and Discussion

The implementation of coordinated environmental mapping requires multiple robotic units working in tandem, particularly when confronting environments containing barriers and intricate structural elements. Our methodology employs a test environment utilizing TurtleBot platforms equipped with LiDAR sensing arrays, as detailed in the experimental methodology section.

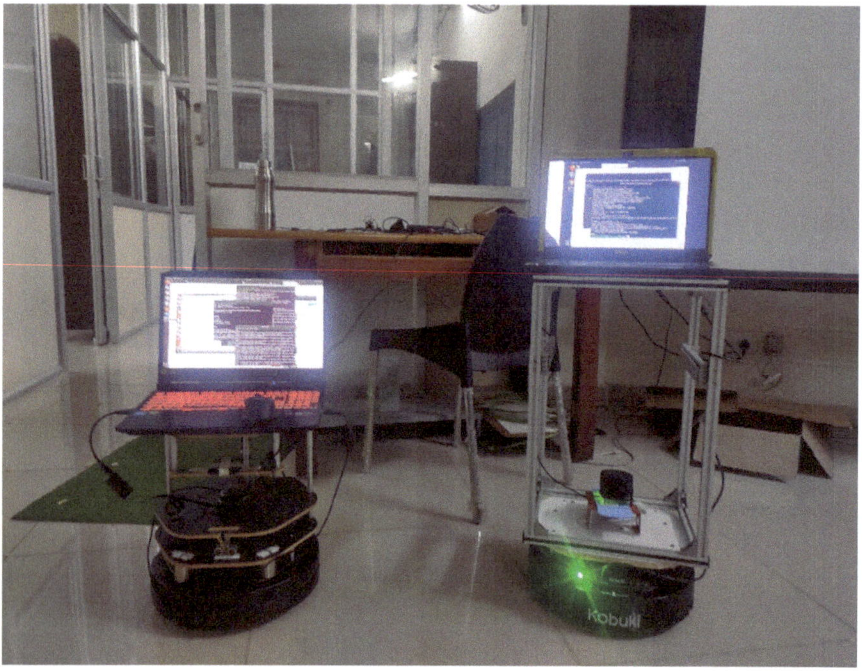

Fig. 2 Experimental configuration of the multi-robot system

Prior to deployment in collaborative scenarios, each robotic unit undergoes systematic validation procedures to ensure sensor and locomotion systems function as intended. These preliminary evaluations verify each unit's capacity to navigate, detect environmental features, and perform fundamental movement operations with the required precision. Following successful individual validation, the integrated multi-robot system undergoes assessment through synchronized command execution, with careful observation of inter-robot coordination and environmental interaction patterns.

For experimental monitoring and data extraction, the research utilizes Rviz visualization software to render and analyze the LiDAR-generated point cloud data from each robotic unit. This software enables real-time observation of sensor readings and environmental mapping progress. The comprehensive experimental configuration for the collaborative mapping investigation is documented through visual documentation in the experimental setup figure.

Our experimental arrangement (shown in Fig. 2) consists of a pair of robots working together on exploration assignments. Each robotic unit employs Gmapping algorithmic processing to develop its own environmental representation, as can be observed in Figs. 3 and 4. The separate spatial models are subsequently unified through an occupancy grid updating mathematical framework designed specifically for collaborative robotic systems, which is detailed in Fig. 5. The progression and final state of this combination procedure are documented in Fig. 6, which demonstrates that the spatial data integration was completed successfully.

Although the LiDAR data contains noise, the final map effectively captures the essential features of the environment, as illustrated in Fig. 5. Figures 3 and 4 further highlight the deliberate distribution of robots to explore separate regions, a strategy that enhances mapping efficiency. By partitioning the environment into specific sections and assigning each robot to a designated area, the exploration process is optimized, leading to faster and more effective map construction.

During implementation of the mapping procedure, one robot encountered difficulties producing an accurate environmental representation. Investigation revealed the root cause: the mapping algorithm component had not established a proper

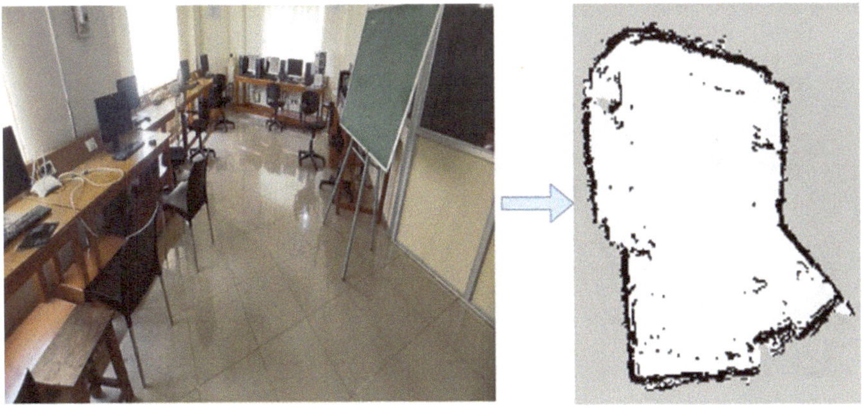

(a) Real-world Lab Office 1 Environment (b) Generated Map

Fig. 3 Map output generated by robot 1

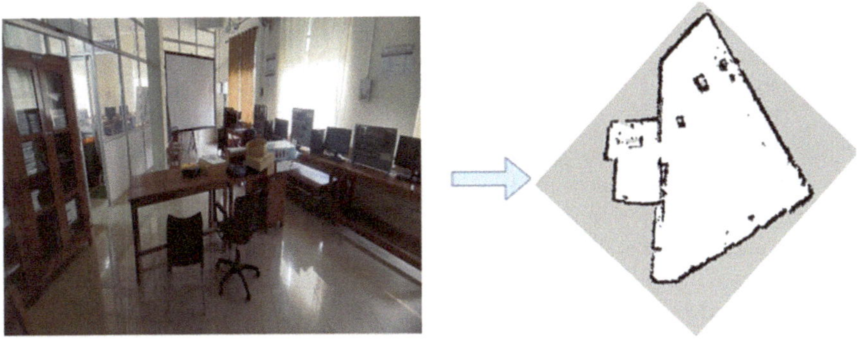

(a) Actual Lab Office 2 Environment (b) Corresponding Constructed Map

Fig. 4 Map output generated by robot 2

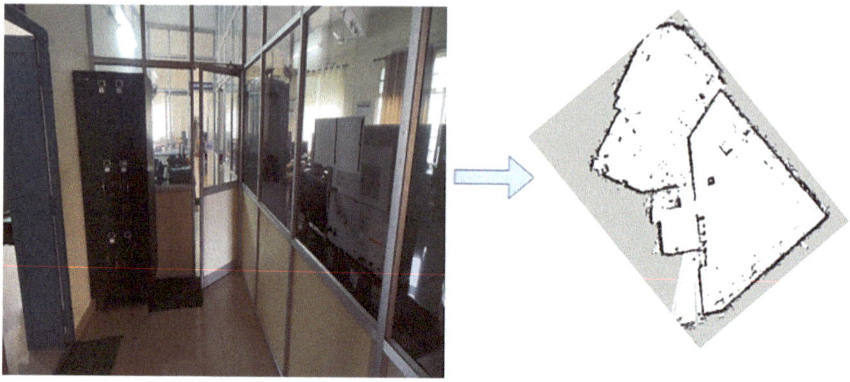

Fig. 5 Combined global map of robot 1 and robot 2

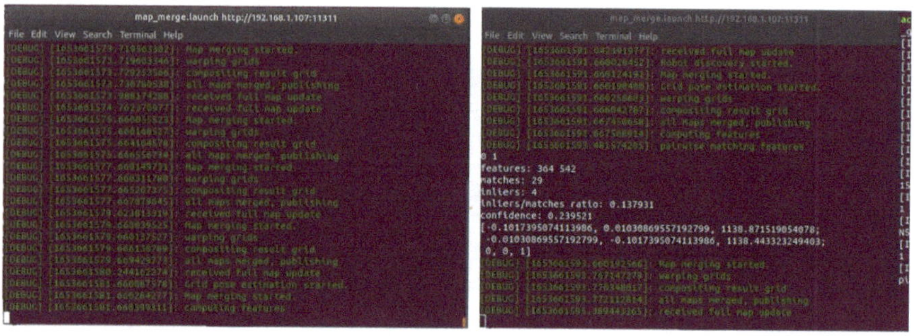

Fig. 6 Visualization of the map merging procedure via command-line interface

subscription to the critical "/scan" data stream. This communication gap was identified through visualization of the system architecture using RQT Graph, a diagnostic tool within the ROS ecosystem.

RQT Graph serves as an essential diagnostic instrument for robotic applications by rendering a visual representation of system components and their interconnections. This visualization capability proves invaluable for system monitoring and troubleshooting, as it illuminates the data flow between various functional elements.

In our multi-robot mapping framework, individual robots operate dedicated mapping modules that construct localized environmental representations. These separate representations are subsequently integrated to create a comprehensive spatial model incorporating information from all robotic units. The RQT Graph visualization displays several key system components, including the following:

- the central ROS coordination system,
- individual robot control modules,
- mapping processors for each unit,
- communication interfaces,

- central integration module, and
- visualization components.

As shown in Fig. 6, the graph continuously updates to reflect real-time communication patterns as components exchange data across various channels including sensor readings (/scan), individual maps (/robot1/map, /robot2/map), and the combined representation (/map_merge). This monitoring capability enables rapid identification of communication disruptions and facilitates system optimization to enhance mapping performance.

5 Conclusion

The implementation of occupancy map merging in multi-robot systems offers a robust framework for efficient environmental mapping and collaboration. By integrating advanced algorithms like Iterative Closest Point (ICP), feature-based map alignment, and distributed consensus for global reference frame synchronization, the system successfully addresses key challenges in multi-robot operations. The ability to merge individual robot maps into a coherent global map enhances navigation, task coordination, and decision-making in shared environments. Communication efficiency is optimized through data compression techniques, ensuring minimal transmission overhead, while parallel processing techniques enable real-time performance even in large-scale scenarios. Real-world testing with TurtleBots in both simulated and actual environments demonstrates the robustness of the system in handling localization errors, sensor noise, and dynamic obstacles. Overall, the framework supports the scalability of multi-robot systems, making it suitable for a wide range of applications such as search-and-rescue, industrial automation, and surveillance, where collaborative exploration and accurate environmental mapping are crucial.

References

1. Thrun S (2001) A probabilistic on-line mapping algorithm for teams of mobile robots. Int J Robot Res 20:335–363
2. Carlone L, Ng MK, Du J, Bona B, Indri M. Rao-blackwellized particle filters multi robot slam with unknown initial correspondences and limited communication. In: Proceedings of the IEEE international conference on robotics and automation (ICRA), Anchorage, AK, USA, 3–8 May 2010, pp 243–249
3. Zhou XS, Roumeliotis SI (2006) Multi-robot SLAM with unknown initial correspondence: the robot rendezvous case. In: Proceedingsof the IEEE/RSJ international conference on intelligent robots and systems, Beijing, China, 9–13 October 2006, pp 1785–1792
4. Kwon J-W (2015) Cooperative environment scans based on a multi-robot system. Sensors 15:6483–6496
5. Zhi W, Ott L (2019) Senanayake R, Ramos F (2019) Continuous occupancy map fusion with fast Bayesian hilbert maps. In: Proceedings ofthe international conference on robotics and automation (ICRA), Montreal, QC, Canada, 20–24 May 2019, pp 4111–4117

6. Williams SB, Dissanayake G, Durrant-Whyte H (2002) Towards multi-vehicle simultaneous localisation and mapping. In: Proceedingsof the 2002 IEEE international conference on robotics and automation, Washington, DC, USA, 11–15 May 2002, pp 2743–2748
7. Li H, Tsukada M, Nashashibi F, Parent M (2014) Multivehicle cooperative local mapping: a methodology based on occupancy grid map merging. IEEE Trans Intell Transp Syst 15:2089–2100
8. Chcn H, Huang H, Qin Y, Li Y, Liu Y (2019) Vision and laser fused SLAM in indoor environments with multi-robot system. Assem Autom 39:297–307
9. Deutsch I, Liu M, Siegwart R (2016) A framework for multi-robot pose graph SLAM. In: Proceedings of the IEEE international conference on real-time computing and robotics (IEEE RCAR), Angkor Wat, Cambodia, 6–10 June 2016, IEEE: New York, NY, USA, pp 567–572
10. Bosse M, Zlot R (2008) Map matching and data association for large-scale two-dimensional laser scan-based SLAM. Int J Robot Res 27:667–691
11. Schwertfeger S, Birk A (2013) Evaluation of map quality by matching and scoring high-level, topological map structures. In: Proceedings of the 2013 IEEE international conference on robotics and automation, Karlsruhe, Germany, 6–10 May 2013, pp 2221–2226
12. Mielle M, Magnusson M, Lilienthal AJ (2016) Using sketch-maps for robot navigation: Interpretation and matching. In: Melo K (ed) Proceedingsof the 14th IEEE international symposium on safety, security, and rescue robotics (SSRR), Lausanne, Switzerland, 23–27 October 2016. IEEE, New York, NY, USA, pp 252–257
13. Kakuma D, Tsuichihara S, Ricardez GAG, Takamatsu J, Ogasawara T (2017) Alignment of occupancy grid and floor maps using graph matching. In: Proceedings of the 11th IEEE international conference on semantic computing (ICSC), San Diego, CA, USA, 30 January–1 February 2017, pp 57–60
14. Arpitha Shankar SI, Shivakumar M (2023) Mapping and autonomous navigation of an indoor environment in ROS platform using multiple robots equipped with LIDAR. Tuijin Jishu/ J Propuls Technol 44(5):2617–2628. ISSN 1001-4055. https://doi.org/10.52783/tjjpt.v44.i5.3008
15. Blanco JL, Gonzalez J, Fernandez Madrigal JA (2007) A new method for robust and efficient occupancy grid-map matching. In: Proceedings of the Iberian conference on pattern recognition and image analysis, Girona, Spain, 6–8 June 2007. Springer, Berlin/Heidelberg, Germany, pp 194–201
16. Arpitha Shankar SI, Shivakumar M (2024) Sensor fusion based multiple robot navigation in an indoor environment. J Interact Des Manuf (IJIDeM).https://doi.org/10.1007/s12008-024-01774-6
17. Carpin S (2008) Fast and accurate map merging for multi-robot systems. Auton Robots 25:305–316
18. Ma L, Zhu J, Zhu L, Du S, Cui J (2016) Merging grid maps of different resolutions by scaling registration. Robotica 34:2516–2531
19. Konolige K, Fox D, Limketkai B, Ko J, Stewart B (2003) Map merging for distributed robot navigation. In: Proceedings of theIEEE/RSJ international conference on intelligent robots and systems (IROS 2003), Las Vegas, NY, USA, 27–31 October 2003, pp 212–217
20. Forsyth D, Ponce J (2001) Computer vision: a modern approach. Prentice Hall, Hoboken, NJ, USA, p 792
21. Arpitha Shankar SI, Shivakumar M, Prakash KR, Suraj Koundinya P (2023) Occupancy grid map for a multi-robot system using LiDAR. SN Comput Sci 4:196. https://doi.org/10.1007/s42979-022-01615-x
22. Fischler MA, Bolles RC (1981) Random sample consensus: a paradigm for model fitting with applications to image analysis andautomated cartography. Commun ACM 24:381–395

Sustainable Healthcare Through AI: Energy-Efficient Neural Networks for Diabetic Retinopathy Assessment

B. C. Anupama, Sheela N. Rao, M. Bindu Malini, and V. Vikram Athreya

Abstract Globally, one of the biggest avoidable causes of blindness is still Diabetic Retinopathy (DR), which mostly impacts under privileged communities with fewer available specialists. The proposed study introduces an energy-efficient automated DR grading framework, featuring a novel deep network architecture that achieves maximal diagnostic efficacy with the minimal carbon emissions, which is beneficial toward sustainable medical image analysis implementations. This groundbreaking, low-powered pre-processing method, combined with optimized architectures, shapes the development of sustainable AI in response to some of healthcare's most severe challenges. In this study, we evaluated computational cost of five energy-efficient Convolutional Neural Network (CNN) architectures namely: ResNet-50, ResNet-101, VGG16, DenseNet201, and Inception V3 with attention on minimizing computational cost with non-infernal to diagnostic accuracy. It was found that the combination of the new fast method of pre-processing with the resource efficient architecture of ResNet-101 allowed to reach better classification accuracy on the test set while using much less energy than previously used approaches. The proposed methodology was found to be robust to differences in quality of images and lighting conditions, which renders this method suitable for implementation in low-resource contexts. This study shows how such innovative AI can democratize access to accurate eye disease diagnostics on a large peripheral, with negligible environmental footprints, which promises pathways for green medical AIs.

Keywords Diabetic retinopathy · Pre-processing method · ResNet · And Medical image classification

---

B. C. Anupama (✉) · S. N. Rao
Department of Electronics and Instrumentation, JSS Science and Technology University, Mysuru, India
e-mail: anubc52@gmail.com

M. Bindu Malini
Department of Ophthalmology, JSS Medical College and Hospital, Mysuru, India

V. V. Athreya
Department of Mechanical Engineering, The National Institute of Engineering, Mysuru, India

I. M. Jamadar et al. (eds.), *Integrating Sustainability in Modern Technologies*, Lecture Notes in Mechanical Engineering, https://doi.org/10.1007/978-981-95-9466-5_10

1 Introduction

The prevalence of diabetes mellitus continues to rise rapidly across the globe, making it an ever-increasing global health challenge. One of the most significant and potentially devastating a consequence of diabetes is diabetic retinopathy (DR), a progressive ocular conditions, it may lead to irreversible blindness if not detected and treated. With the worldwide increase in patients with diabetes, the prompt identification and accurate diabetic retinopathy grading is now one of the key priorities in medical imaging and diagnostic systems. Diabetic retinopathy diagnosis and assessment traditionally depend on manual ophthalmological examinations, an inherently subjective process, and are subject to inter-observer variability. In order to identify characteristic lesion types such as micro-aneurysms, hemorrhages, hard exudates, and neovascularization, healthcare professionals need to closely analyze retinal fundus images. This hands-on process is not only lengthy, but also requires extensive expertise and can be inconsistent across different clinicians and healthcare settings. Figure 1 depicts the eye fundus image impacted with diabetic retinopathy.

The advent of deep learning technologies has revolutionized medical image analysis, offering unprecedented opportunities for automated and precise diagnostic support. Convolutional Neural Networks (CNNs) and innovative algorithms for advanced machine learning have demonstrated remarkable potential in developing robust, objective, and scalable solutions for diabetic retinopathy grading. These computational approaches can process vast amounts of retinal imaging data with speed and accuracy that surpass traditional manual screening methods. By leveraging

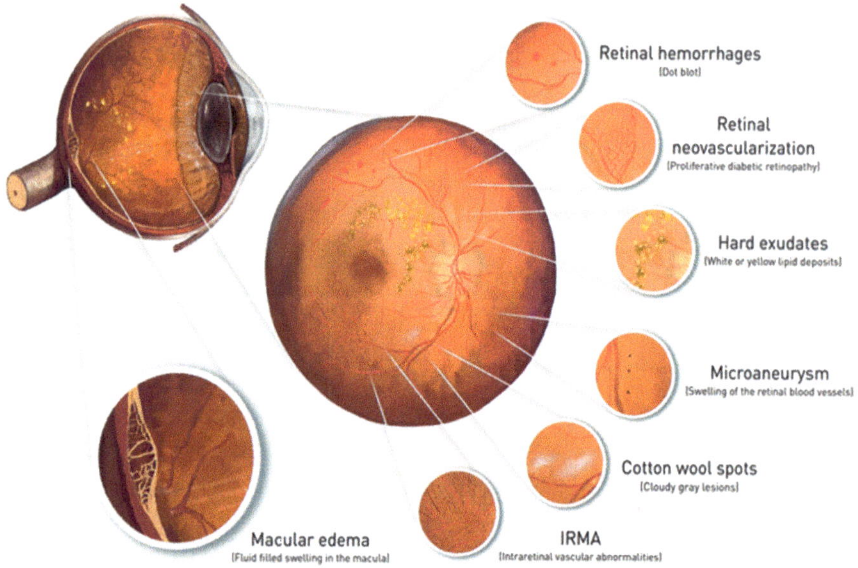

Fig. 1 Eye fundus images exhibiting indications of diabetic retinopathy

cutting edge, deep learning techniques seek to contribute to the ongoing efforts of creating more reliable, accessible, and efficient diagnostic tools for early diabetic retinopathy detection. The potential impact extends beyond technological innovation, promising to improve patient outcomes, reduce healthcare costs, and mitigate the long-term consequences of undetected diabetic eye complications.

2 Related Work

This comprehensive literature survey focused on research works on flexible retinal fundus photographs in three major areas: the grading of ocular diseases, such as diabetic retinopathy. The study shows major progress of using deep learning and machine learning based approaches to retrieve information to diagnose based on retinal images. The advances in computerized approaches for noninvasive classification of diabetic retinopathy severity by retinal photos are promising. Using advanced analysis techniques such as deep learning, these studies offer promise for the application of AI to diagnostics in the era of medicine, allowing not only for early detection of disease but with the possibility of tailoring and assessing the individual's risk for disease.

Monteiro et al. [1] explore a novel method to grade diabetic retinopathy using a blended deep learning model at an ensemble level that averages the predictions of ten state of art Deep Learning Networks (DLNs) that are fitted on a balanced dataset. Results: With the proposed approach, robustness and efficiency improvement were gained in the diabetic retinopathy grading. The model achieves high precision for normal retina and mild DR detection and model exhibits poor performance on moderate and proliferative DR samples.

In this study, Muthusamy et al. [2] use a new paradigm of deep neural networks called the Deep Multilayer Perceptive Learning Classification Based on Index (MAPCRCI-DMPLC) using MAP Concordance Regressive Camargo's approach to automatically identify and categorize Diabetic Retinopathy (DR) diagnoses. According to the results, the MAPCRCI-DMPLC approach outperforms current cutting edge methods in terms of cross-entropy loss, sensitivity, specificity, and time complexity.

Malhi et al. [3] propose an effective and straightforward method for grading diabetic retinopathy by analyzing fundus pictures with micro-aneurysms and exudates. This technique employs processing images, extracting features, and developing machine learning models. The findings demonstrate the proposed method accomplished 92.1% accuracy for the grading based on exudates using support vector machines and K-Nearest Neighbors, while it reached an impressive accuracy of 99.9% for grading via micro-aneurysms using a decision tree algorithm.

Ademola et al. [4] give a brief summary of the outcomes of various CNN algorithms employed for autonomous disease monitoring, categorization, and segmentation of retinal fundus pictures. Additionally, it explores contemporary trends related to the pathology and morphological characteristics of these images through

CNN methodologies. The review indicates that most CNN approaches utilized for segmenting retinal fundus images are primarily based on architectures such as UNet, attention mechanisms, and generative adversarial networks. In this study, Alwakid et al. [5] objective is to create a quick, heavily automated approaches founded on deep learning identification of DR model to help ophthalmologists assess potential cases of DR. The study achieves high accuracy in classifying retinal images into five DR severity stages, with an accuracy of 78.7% for the DDR dataset and 98.7% for the APTOS dataset. In this investigation, Nehad et al. [6] compared diagnosing diabetic retinopathy with methods of DL and ML, with deep learning methods showing the greatest results, with categorical classification of diabetic retinopathy phases outperforming binary classification for disease identification. Deep learning methods showed the best performance in diabetic retinopathy diagnosis.

In this study, Nair et al. [7] introduced an EfficientNet B5 for diabetic retinopathy grading and compared its execution with those pre-trained networks for DR grading, including ResNet50 and VGG16. ResNet50 achieved higher accuracy than VGG16 and EfficientNetB5 in DR classification.

Wong et al. [8] proposes an optimized Using Transfer Learning (TL) to detect Diabetic Retinopathy (DR) and classification using ECOC-SVM ensemble with weighted feature optimization, achieving improved performance in contrast to conventional TL networks with careful tuning. The proposed method achieves an excellent accuracy rate for APTOS class 5 DR, 82% grading, 96% for APTOS class 2 grading.

Revathy et al. [9] describe a system for identification of diabetic retinopathy. This study aims to develop a ML-oriented hybrid model that can accurately detect and classify diabetic retinopathy using the extracted features from retinal images. The retinal pictures are converted to HSV from RGB, and then adaptive histogram equalization, median filtering, and edge zero padding are applied. The suggested hybrid model achieved an accuracy of 82% in classifying the retinal images as normal or abnormal. The F-measure, precision, and recall scores were 0.8028, 0.8116, and 0.8619, respectively.

Sark et al. [10] focus on improving the performance of early identification of diabetic ocular disorders based on retinal fundus pictures through the use of image pre-processing and classification techniques. They employed various image pre-processing techniques, such as contrast enhancement, illumination correction, blood vessel segmentation, optic disc detection, and exudate localization. They then used transfer learning with pre-trained models (Xception, VGG16, and DenseNet121) and developed a new custom model of CNN to grade the disease. The new custom CNN model trained on pre-processed images achieved the highest performance, with an accuracy of 93.33% for diabetic retinopathy, 91.43% for diabetic macular edema, and 100% for glaucoma. The pre-trained VGG16 model also performed well, with an accuracy of 83.43% for diabetic retinopathy, 89.13% for diabetic macular edema, and 88% for glaucoma.

Vijayan et al. [11] focus on reviewing recent automated systems based on deep learning for identifying and staging using retinal fundus photos to diagnose diabetic retinopathy. Their paper evaluates 24 methods for identifying and classifying diabetic retinopathy that were published between 2019 and 2022. Additionally, it compiles a list of publicly available fundus image datasets for diabetic retinopathy detection and discusses common image pre-processing techniques utilized in these studies. The review highlights that convolutional neural network-based methods, including EfficientNet and its variants, have demonstrated strong performance in the identification and diabetic retinopathy staging. Furthermore, it notes the emergence of vision transformer-based techniques, which have shown promising results. Anupama et al. [12] provide a detailed assessment of deep learning algorithms for detecting and evaluating the severity of diabetic retinopathy using retinal fundus pictures. The paper examines several models, regularly used datasets, and preprocessing strategies published in recent literature. It emphasises that convolutional neural network-based approaches have performed well in both identification and grading tasks, while also highlighting existing limitations and suggested paths for future study. Firke et al [13] propose an innovative deep learning approach for the automated categorization of diabetic retinopathy using convolutional neural networks. Their work focuses on classifying retinal fundus images into different severity levels, demonstrating that CNN-based models can achieve high classification accuracy and reliable performance. The results indicate the effectiveness of the proposed approach in supporting early detection and improving diagnostic decision-making for diabetic retinopathy. Baba et al. [14] present a deep learning-based approach for automatically classifying the severity of diabetic retinopathy using convolutional neural networks. Their study focuses on analyzing retinal fundus images to categorize different stages of the disease. The results demonstrate that the proposed CNN model achieves high classification accuracy and improved performance metrics, highlighting its potential for reliable severity assessment in clinical applications. Abramoff et al. [15] propose a data-driven CNN-based approach for detecting referable diabetic retinopathy directly from retinal images, achieving an AUC of 98.2% with strong cross-dataset generalization and significantly reduced classification error compared to existing methods.

3 Proposed Method

This section describes the specifics of our framework including the data collection, the pre-processing procedure, and utilizing CNN. Figure 2 illustrates a schematic of the suggested method.

3.1 The Dataset for Diabetic Retinopathy

Several publicly accessible datasets of fundus images can be utilized to grade DR. Common datasets used for this type of work include APTOS, Messidor-1, EyePac, and Messidor-2. This work utilizes the Diabetic Retinopathy Detection APTOS 2019 dataset, which is publically accessible. Pictures from publicly accessible retinopathy

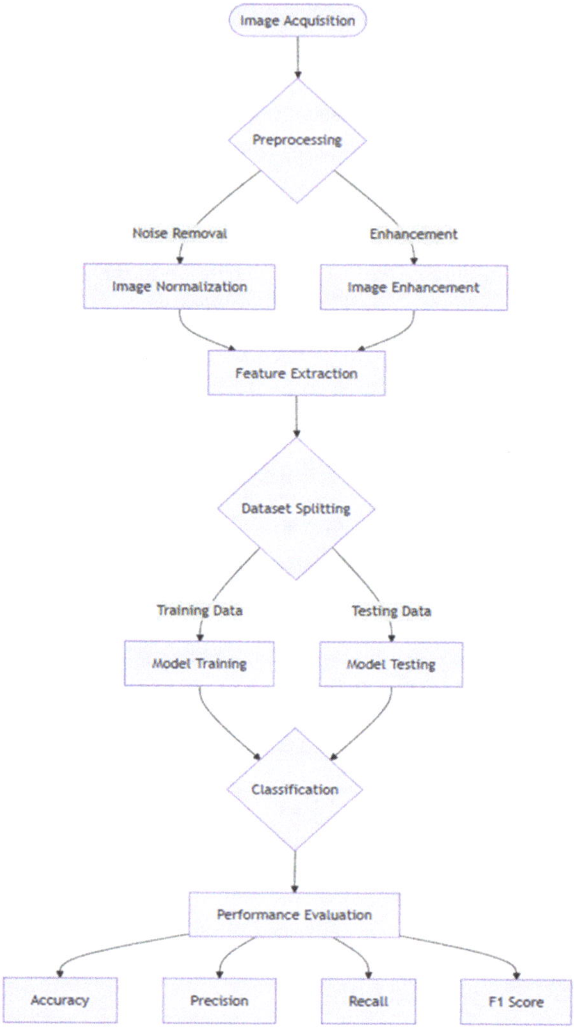

Fig. 2 Different steps involved in methodology

detection datasets were used to generate the database. There are 3662 images with different classes of diabetic retinopathy present in the APTOS dataset. The severity of diabetic retinopathy is classified into five levels presented within Table 1.

Considering the look, amount, size, and distribution of micro-aneurysms, exudates, and hemorrhages, as illustrated in Fig. 1, diabetic retinopathy is present [9]. The exudates are the luminous, yellowish patches that differ somewhat in hue from the optic disc. Exudates are produced by the ruptured blood vessel's lipid content. Hemorrhages develop due to the blood vessel micro-aneurysms rupturing. In photos of severe diabetic retinopathy, the last phase of the illness, exudates, and hemorrhages are seen to be spreading.

Table 1 Different stages of diabetic retinopathy

Levels	Details
Level 0	No DR
Level I	Non proliferative diabetic retinopathy (NPDR) at a Mild level
Level II	NPDR at a moderate level
Level III	NPDR at a severe level
Level IV	Proliferative diabetic retinopathy (PDR)

3.2 Preliminary Image Processing

Pre-processing the image enhances its features and eliminates undesired noise coming from the fundus photographs. The primary objectives of medical image pre-processing aim to standardize images within a data set and minimize artifacts from image acquisition. Image pre-processing is thus a critical step for improving picture characteristics and ensuring image consistency [2]. To decrease dark space, the visuals were cropped to a circle, limiting the retina. The image resolution was tailored to each deep learning architecture, such as 224 × 224 for ResNet 50 & 101, VGG16, DenseNet201 and 299 × 299 for Inception V3.

In this work, the image was pre-processed using a novel pre-processing method which was named as "**Adaptive Sigmoid-Enhanced Fundus Image Pre-processing.**" This method uses several steps to improve fundus images. First, it balances the brightness in various sections of the image. Then, it smoothens the image to remove noise. A special math function is used to make important features stand out more. The method also uses shape-based techniques to find edges and spots. All these improved parts are then combined into one image. Finally, the overall contrast is adjusted to make everything as clear as possible. This process helps make eye details easier to see, which can help doctors spot problems more easily. Figure 3 shows the original image and image after pre-processing using adaptive sigmoid-enhanced image pre-processing method.

3.3 Making Predictions with Pre-Trained Models

The architectures for deep learning ResNet50, ResNet-101, VGG16, DenseNet201, and Inception V3 are frequently utilized for computer vision problems. All of these algorithms have special characteristics, benefits with regard to image recognition and classification.

Residual Network, or ResNet, is a groundbreaking deep knowledge architecture that solved a critical challenge in neural networks, developed by researchers at Microsoft in 2015, and ResNet introduced a revolutionary concept called "skip connections" or "residual blocks" that allow neural networks to become much deeper without losing performance. ResNet-50 and ResNet-101 are two variations

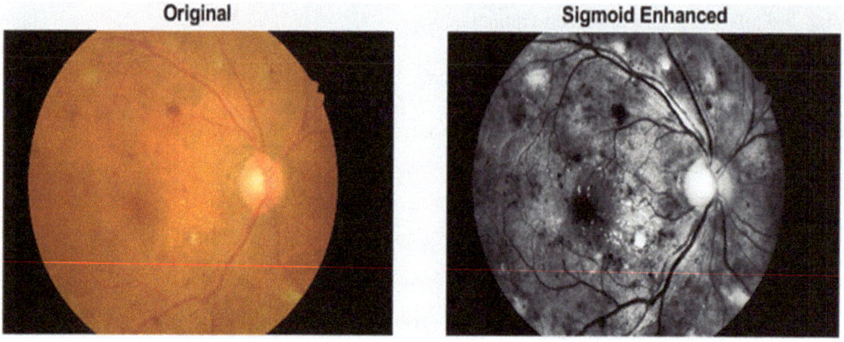

Fig. 3 Original image and pre-processed image using adaptive sigmoid enhancement

of this architecture, with the number indicating the total quantity of network layers. The University of Oxford's Visual Geometry Group (VGG) developed the ground-breaking VGG16 architecture for convolutional neural networks.VGG16 is like a smart visual detective with 16 detective layers that carefully examines images using tiny 3 × 3 detective tools, systematically peeling back image details layer by layer to understand what's really happening in a picture, creating a blueprint for future image-understanding machines by showing how depth and consistent investigation techniques can unlock complex visual mysteries.

DenseNet-201 is like a collaborative learning network where each layer acts as a curious student who learns not just from its immediate teacher, but from every single previous learning stage, creating an interconnected web of knowledge that efficiently captures image details by sharing and recycling information in a way that's smarter and more economical than traditional learning approaches. Inception V3 is a smart image recognition system that acts like a team of visual experts working together. Imagine a group of detectives, each looking at an image with a different-sized magnifying glass. Some zoom in on tiny details, while others look at the bigger picture. This network can process multiple types of visual information at the same time, making it extremely efficient at understanding what's in an image. DenseNet201 employs a dense connectivity pattern to enhance dissemination of features, while VGG16 is recognized because of its straightforward architecture and reliable execution. An outstanding balance between accuracy and efficiency is offered by Resnet101. The number of datasets, the required precision, and the computational resources available all influence the model that is chosen.

4 Findings and Analysis

To ensure the measure of efficacy of the various algorithms and architectures, a number of performance analysis metrics are used. For the most part, accuracy serves as the performance indicator, and it illustrates the quantity of images that were

Table 2 Classification performance of cutting edge deep learning models

Models	Recall	F1-Score	Precision	Accuracy (%)
ResNet-101	0.9465	0.9532	0.9675	96.87
ResNet-50	0.9478	0.9478	0.9390	91.23
Inception V3	0.9403	0.9386	0.9310	90.45
DenseNet-201	0.9144	0.9255	0.9356	87.75
VGG16	0.8367	0.8411	0.8554	82.29

successfully categorized by the model among all available images. Our comprehensive analysis evaluated five state-of-the-art deep learning architectures, revealing nuanced performance variations across multiple evaluation metrics. The study compared ResNet-101, ResNet-50, Inception V3, DenseNet-201, and VGG16 to provide a detailed understanding of their classification capabilities. Table 2 displays the classification performance evaluation of innovative approaches in deep learning models.

ResNet-101 emerged as the most exceptional model in our study, demonstrating remarkable performance with the highest accuracy of 96.23% as well as an outstanding F1 ranking of 0.9866. These superior performances can be primarily credited to its advanced residual learning architecture, which effectively addresses the challenges of deep neural networks by implementing sophisticated bypass connections that lessen the issue of disappearing gradients.

The performance metrics reveal a clear hierarchy among the examined architectures. ResNet-50 followed closely behind its more complex counterpart, achieving a robust performance with a recall of 0.9478, precision of 0.9390, and an accuracy of 91.23%. Inception V3 demonstrated consistent performance across metrics, with a recall of 0.9403, precision of 0.9310, and an accuracy of 90.45%, showcasing the effectiveness of its multi-scale feature extraction approach.

DenseNet-201 presented interesting results, with a moderate performance characterized by a recall of 0.9144, precision of 0.9356, and an accuracy of 87.75%. Its unique dense connectivity pattern allows for efficient feature reuse, which contributes to its competitive performance. In contrast, VGG16 showed the most limited performance among the studied models, with a recall of 0.8367, precision of 0.8654, and an accuracy of 82.29%, indicating potential limitations in handling complex feature representations. Figure 4 shows performance of diverse models for deep learning.

5 Conclusion and Future Work

To sum up, current investigation demonstrates that ResNet-101 outperforms other evaluated convolutional neural network architectures regarding precision, F1 score, and accuracy, establishing it as a robust choice for image classification tasks for our novel pre-processing method. The findings highlight the significance of model

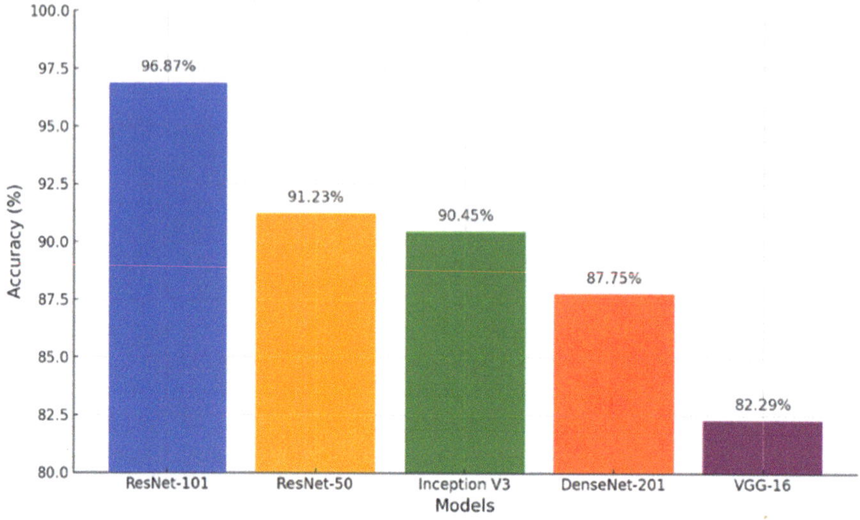

Fig. 4 Indicates the comparative performance of diverse models for deep learning

depth and architecture in achieving high performance, with ResNet-50 and Inception V3 also showing commendable results. However, the limitations of computational resources and dataset specificity warrant further investigation.

Future work should focus on optimizing hyper parameters, exploring diverse datasets to enhance model generalizability, and investigating ensemble methods that combine multiple models.

References

1. Monteiro FC (2023) Diabetic retinopathy grading using blended deep learning. Proc Comput Sci 219:1097–1104. https://doi.org/10.1016/j.procs.2023.01.389
2. Muthusamy D (2024) Comprehensive review of deep learning approaches in diabetic retinopathy diagnostic classification. Artif Intell Rev 57:185–186
3. Malhi A, Grewal R, Pannu HS (2023) A Comprehensive approach to diabetic retinopathy detection and severity classification using digital retinal imaging. Int J Intell Robot Appl 7(7):426–458. https://doi.org/10.1007/s41315-022-00269-5
4. Ilesanmi AE (2023) A systematic review of retinal fundus image segmentation and classificationmethods using convolutional neural networks. Health Care Anal 4:100261. https://doi.org/10.1016/j.health.2023.100261
5. Alwakid G, Gouda W, Humayun M, Jhanjhi NZ (2023) Asia pacific tele-ophthalmology society, & dataset for diabetic retinopathy. In: Deep learning-enhanced diabetic retinopathy image classification. Digital Health, pp 9–15. https://doi.org/10.1177/20552076231194942
6. Ramaha N, Imad S (2023) Diabetic retinopathy diagnosis using machine versus deep learning. Eur J Sci Technol. https://doi.org/10.31590/ejosat.1263514

7. Nair A (2022) Automated diagnostic classification of diabetic retinopathy using advanced deep learning methodologies. In: 2022 6th International conference on computing methodologies and communication (ICCMC), pp 1019–1024. https://doi.org/10.1109/ICCMC53470.2022.9754113
8. Wong WK, Juwono FH, Apriono C (2023) Advanced transfer learning strategy for automated diabetic retinopathy diagnostic classification using ensemble optimization. IEEE Access 11:83004–83016. https://doi.org/10.1109/ACCESS.2023.3301618
9. Reshma JJ, Ragendhu SS, Sumithra MD, Nithya BS, Ravathy R (2020) Machine learning for the detection of diabetic retinopathy. Int J Eng Res Technol (IJERT) 9(06)
10. Parthiban K, Venkatachalapathy K (2021) Machine learning and IoT-enabled approach for diabetic retinopathy classification. In: 6th International conference on communication and electronics systems (ICCES), Coimbatore, India, pp 476–481. https://doi.org/10.1109/ICCES51350.2021.9489116
11. Anand V, Koundal D, Alghamdi WY, Alsharbi BM (2024) Smart grading of diabetic retinopathy: an intelligent recommendation-based fine-tuned EfficientNetB0 framework. Front Artif Intell 7. https://doi.org/10.3389/frai.2024.1396160
12. Anupama BC, Sheela Rao N (2024) Comprehensive review of automated deep learning methodologies for diabetic retinopathy detection and severity assessment. IOSR J Electr Electron Eng 19(1):47–53
13. Firke SN, Jain RB (2021) Innovative deep learning approach for "automated diagnostic categorisation of diabetic retinopathy through convolutional neural networks. In: 2021 International Conference of Artificial Intelligence Smart Systems (ICAIS), pp 1–6. https://doi.org/10.1109/ICAIS50930.2021.9395796
14. Baba SM, Bala I (2023) Deep learning approach for "automated classification of the severity of diabetic retinopathy using convolutional neural networks. In: The 2023 International conference on computer, electronics & electrical engineering & their applications (IC2E3), Srinagar Garhwal, India, pp 1–6. https://doi.org/10.1109/IC2E357697.2023.10262490
15. Pires R, Avila S, Wainer J, Valle E, Abramoff MD, Rocha A (2009) A data-driven approach to referable diabetic retinopathy detection. Artif Intell Med 96:93-106. https://doi.org/10.1016/j.artmed.2019.03.009. Epub 2019 Mar 27. PMID: 31164214

Dual Active Bridge Converter for Electric Vehicle Battery Charging

V. N. Jayasankar, S. R. Gurumurthy, H. G. Prarthana Kashyap, S. Prerana, R. Suhas, and G. C. Tejaswini

Abstract Electric vehicles are the green and sustainable alternative of conventional IC engine-based vehicles. Bidirectional converters are the essential part of power conversion in the battery charging system of electric vehicles to ensure the power flow from grid to the battery during charging mode and reverse power flow during vehicle-to-grid mode. Among the various topologies available, dual active bridge topology is preferred due to its isolated structure and easy control. The performance of Dual active bridge is analyzed with simulations and validated experimentally by developing a laboratory prototype of low voltage and low power and high switching frequency using TMS320F28379D processor.

Keywords Dual active bridge (DAB) · Power flow · Electric vehicles (EVs)

1 Introduction

Electric Vehicles (EVs) are sustainable and green transportation mode, fueled by electrical energy stored in batteries. They are increasingly popular due to their lower running costs, reduced maintenance needs, and contribution to reducing greenhouse gas emissions. Charging stations are essential for the widespread adoption and practicality of Electric Vehicles (EVs) [1]. They provide convenient access to electricity for recharging EV batteries, extending the driving range, and ensuring that vehicles are ready for use. Additionally, a robust charging station network helps alleviate range anxiety, supports long-distance travel, and contributes to electric mobility's overall growth and acceptance [2]. In Electric Vehicles (EVs), power converters are crucial in managing and optimizing the electrical power flow between the battery, motor, and other components. Power converters enhance electric vehicles' efficiency, performance, and reliability by maintaining proper voltage and current levels throughout the vehicle's electrical system.

V. N. Jayasankar (✉) · S. R. Gurumurthy · H. G. Prarthana Kashyap · S. Prerana · R. Suhas · G. C. Tejaswini
Department of E&EE, The National Institute of Engineering, Mysore, India
e-mail: jayasankarvn@nie.ac.in

I. M. Jamadar et al. (eds.), *Integrating Sustainability in Modern Technologies*, Lecture Notes in Mechanical Engineering, https://doi.org/10.1007/978-981-95-9466-5_11

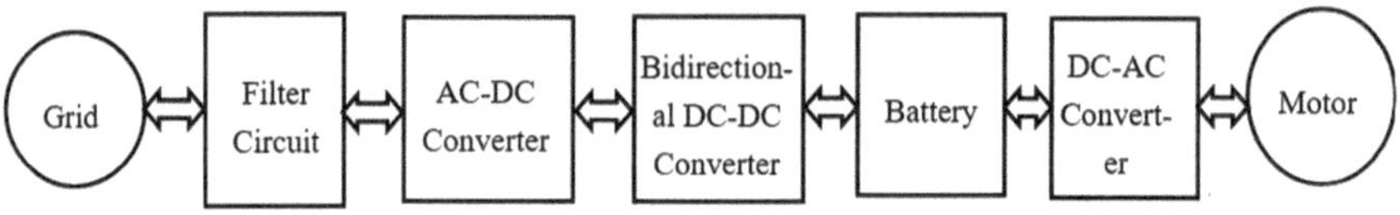

Fig. 1 Schematic diagram of the proposed configuration

The power conversion in EVs is illustrated in Fig. 1. The bidirectional DC-DC converters ensure the power flow in both the directions to facilitate grid-to-vehicle (G2V) and vehicle-to-grid (V2G) modes of operation. To reduce the number of power converters and provide isolation, many single-stage power converters topologies are available in the literature [3–5]. Dual Active Bridge (DAB) topology offers advantages such as soft-switching of semiconductor switches to reduce switching losses, reduction in switching device number and high efficiency [6].

This topology offers high power density and voltage conversion ratio [7]. Other advantages are symmetrical structure and high-power throughput. A laboratory prototype of dual active bridge is designed, implemented, and tested to validate the performance of the topology for electric vehicle application.

2 Methodology

The dual active bridge consists of two h-bridges, a high-frequency transformer, a coupling inductor, and a DC-link capacitor as shown in Fig. 2. The semiconductor switches used are bidirectional in nature. The power flow is controlled by the phase shifting of gate pulses in the respective bridge switches. The power flows from leading bridge to lagging bridge, through an energy storage element [8]. Each cycle of dual active bridge has four operating modes. The switch states under different operating modes are shown in Table 1.

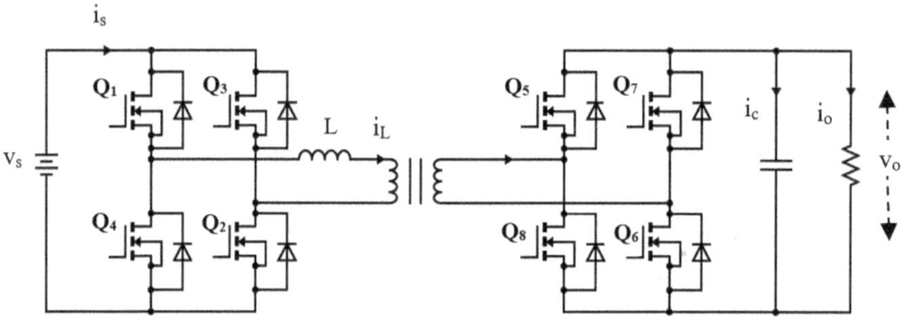

Fig. 2 Circuit diagram of dual active bridge

Table 1 Switch states under different operating modes of dual active bridge

Mode	ON	OFF
Mode 1	Q_1, Q_2, Q_7, Q_8	Q_3, Q_4, Q_5, Q_6
Mode 2	Q_1, Q_2, Q_5, Q_6	Q_3, Q_4, Q_7, Q_8
Mode 3	Q_3, Q_4, Q_5, Q6	Q_1, Q_2, Q_7, Q_8
Mode 4	Q_3, Q_4, Q_7, Q_8	Q_1, Q_2, Q_5, Q_6

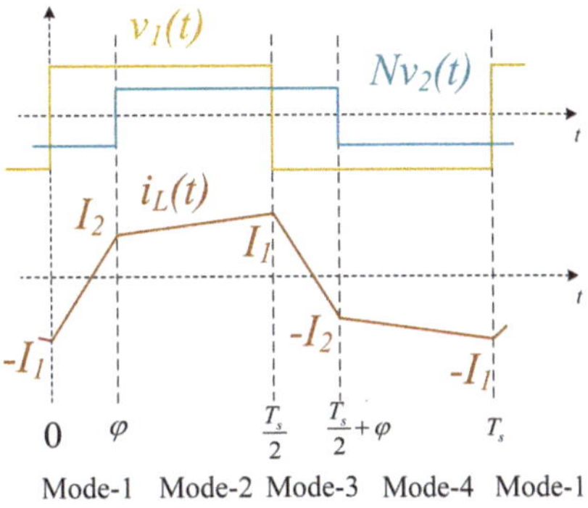

Fig. 3 Steady state inductor current under phase shift modulation [9]

The primary bridge output voltage $v_1(t)$, the secondary bridge input voltage $v_2(t)$, and the inductor current waveform under different operating modes are shown in Fig. 3, where T_s is the switching period and $\varphi = \frac{\delta}{360} \times T_s$, the time equivalent of δ^0.

Current through the coupling inductance can be expressed as shown in (1), where V_S = input voltage, X_S = leakage inductor, δ = phase shift angle, and V_R = output voltage. The real power (P_R) transferred to the load from the source can be expressed as (2).

$$I_L = \frac{\left(\vec{V_1}\right) - \left(\vec{V_2}\right)}{jX_s} \quad (1)$$

$$P_R = \frac{|V_1||V_2|Sin\delta}{X_S} \quad (2)$$

3 Design of Power Circuit and the Controller

This section discusses the design of various hardware components and the closed loop controller.

3.1 Power Circuit Design

Hardware design consists of inductor, transformer, and capacitor design. The coupling inductor is designed as per the Eq. (3), where L = inductance in H, N = turns ratio of transformer, V_{in} = input voltage, δ = phase shift, P_o = output power, f_s = switching frequency [9]. For an input voltage of 36 V, output voltage of 12 V, and a power level of 50 W, the values inductor is computed as 6.6μH.

$$L = \frac{NV_0 V_{in}\delta(\pi - \delta)}{2P_o\pi^2 f_s} \tag{3}$$

The transformer facilitates bidirectional power flow in the DAB converter. By controlling the phase angle between the primary and secondary windings, the converter can regulate power flow bidirectionally, enabling power transfer from the input to the output or vice versa. It also provides galvanic isolation between the input and output sides of the converter. Based on the specifications of the primary and secondary sides, an E-core can be selected according to market availability, specifically the E 30/15/7 model. The number of turns for both the primary and secondary sides can be calculated by (4). The turns ratio selected here is 3 according to the ratio of input and output voltages.

$$N = \frac{V}{4fB_m A_c} \tag{4}$$

Capacitors are used as filters to reduce high-frequency noise and ripple in the output voltage and current. Capacitors act to shunt high-frequency components to the ground, resulting in a cleaner output waveform and reducing Electromagnetic Interference (EMI) generated by the converter. Capacitor value can be obtained using (5), and (6) [9]. The capacitor value as per the design is 400 μF.

$$\Delta Q = \frac{V_{in}D^2T^2}{4Ln}\left[1 - D + \frac{D^2}{4}\right] \tag{5}$$

$$C_{out} = \frac{\Delta Q}{2 * \Delta V_o} \tag{6}$$

3.2 Controller Design

The power flow through the converter is controlled by controlling the phase shift between the two ends of the coupled inductor by introducing phase delay between the firing pulses of first and second bridges. The block diagram of the controller is shown in Fig. 4.

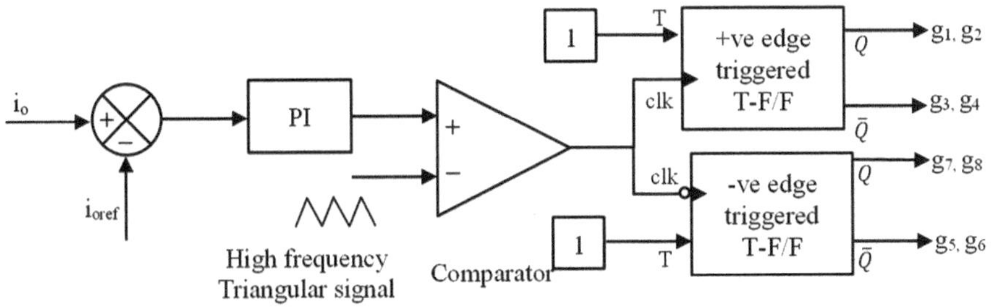

Fig. 4 Schematic diagram of the proposed configuration

The output current (I_o) of the DAB is sensed and filtered using a low pass filter to remove the noise, and it is converted to digital using an ADC. The output signal thus obtained is compared with the reference current set to get the error. This error is passed through the PI controller output and its output is used to generate PWM pulses whose pulse width is proportional to the magnitude of the error signal. Thus, the control of power flow from source to load or one source to other source is achieved. These PWM pulses are used to generate the gate signals for the primary and secondary bridges using a positive edge-triggered T flip flop and negative edge-triggered T flip flop. The flip flops are operated in toggle mode, and the phase shift between the outputs of T flip flops is equal to the width of the PWM pulse. The PI controller parameters are tuned using Ziegler Nicholas method with the transfer function of the dual active bridge, as shown in (7), in which the values are derived considering the parameters, $v_{in} = 36\ V,\ R = 1.2\ \Omega,\ L = 6.6\mu H,\ C = 200\mu F,\ N = 3,\ T_s = 2\ \mu s$ [10].

$$\frac{\widehat{i_L}(z)}{\widehat{\varphi}(z)} = \frac{-0.253z + 0.252}{z^2 - 1.9086z + 0.9095} \tag{7}$$

The tuned values are $K_p = 0.006$ and $K_i = 9.23$.

4 Simulation Results and Discussion

Simulation of the DAB converter is conducted using the PSIM platform with the designed values. An input voltage of 36 V and an output voltage of 12 V is considered. Various transient conditions such as starting transients, source voltage variation, reference current variation, and load variation are analyzed and discussed in the sub-sections below.

4.1 Case i: Starting Transients for a Current Reference of 2.4 A

The initial transients are majorly due to the control algorithm's delay in computing optimal control voltage and the system's time constant. Figure 5 shows the output voltage and current waveforms; the output voltage is settled down at 12.06 V and output current is settled down at 2.41 A. The settling time for 3% error band is 15. 38 ms. The steady state power is 29.11 W for a phase shift of 8.91° as shown in Figs. 6 and 7.

The voltages at primary bridge and secondary bridge ac terminals and the corresponding coupling inductor current are shown in Fig. 8. The controller effectively

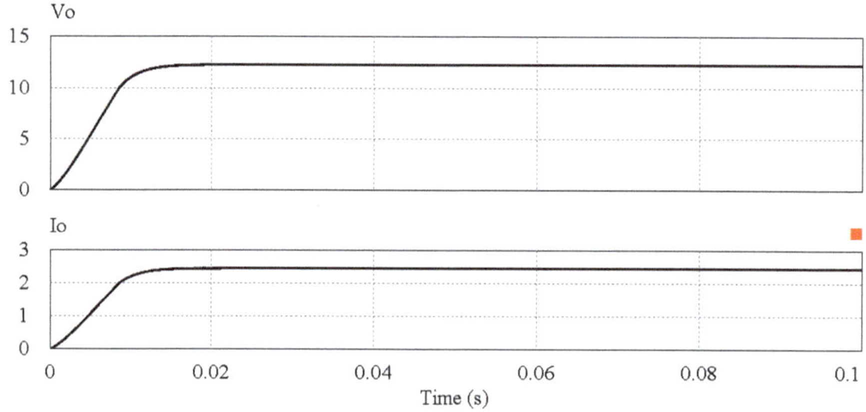

Fig. 5 Output voltage and current waveforms of dual active bridge

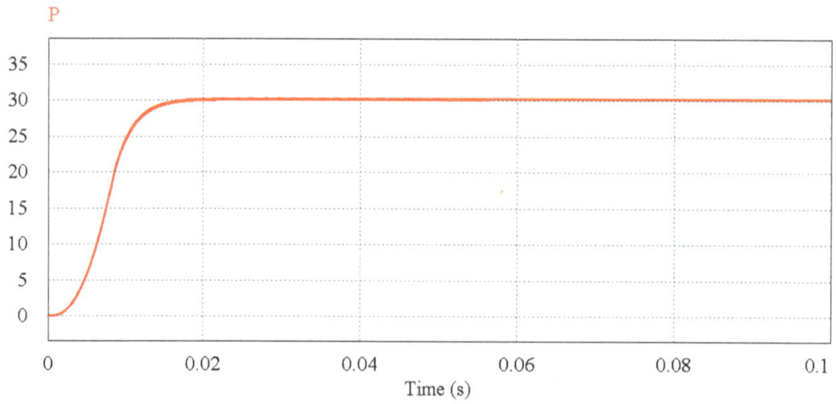

Fig. 6 Power output of dual active bridge

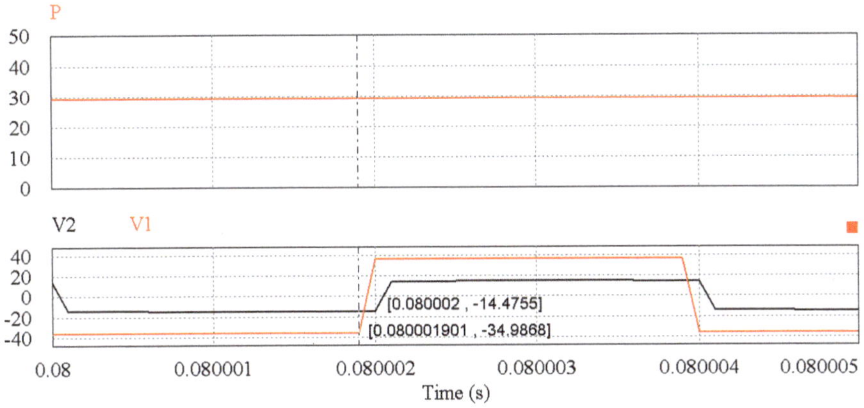

Fig. 7 Power output, primary, and secondary bridge voltages

controls the phase shift and achieves the output power as per the current reference provided. The working of the controller is illustrated in Figs. 9 and 10.

The controller voltage, the high frequency triangular wave, and the PWM output are shown in Fig. 9. The gate pulses for primary bridge and secondary bridge switches are shown in Fig. 10 which clearly shows the phase shift between the gate pulses to achieve transfer of power from primary bridge to secondary bridge. Since the phase shift is positive, the power flow direction is from source to load.

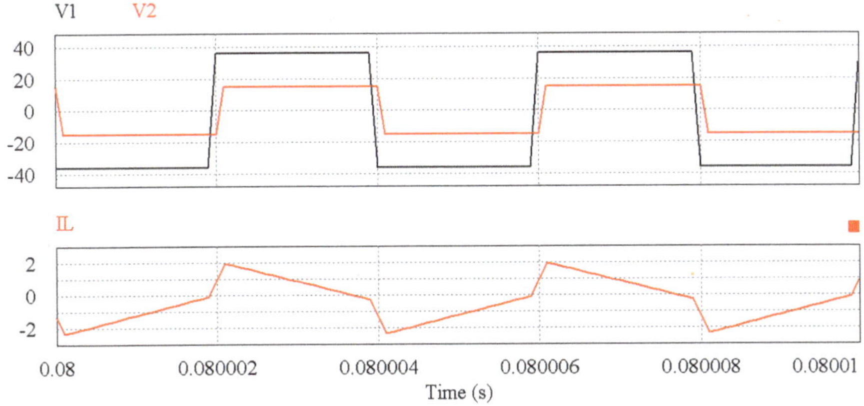

Fig. 8 Primary and secondary bridge voltages, Coupling inductor current

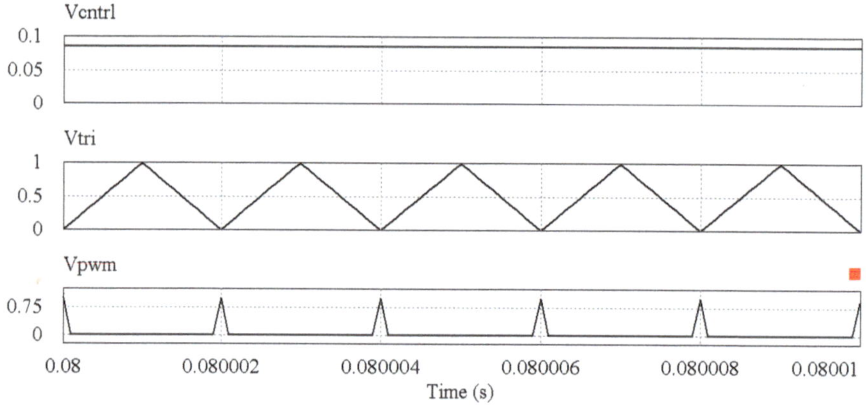

Fig. 9 PWM pulse generation by comparing triangular and control voltages

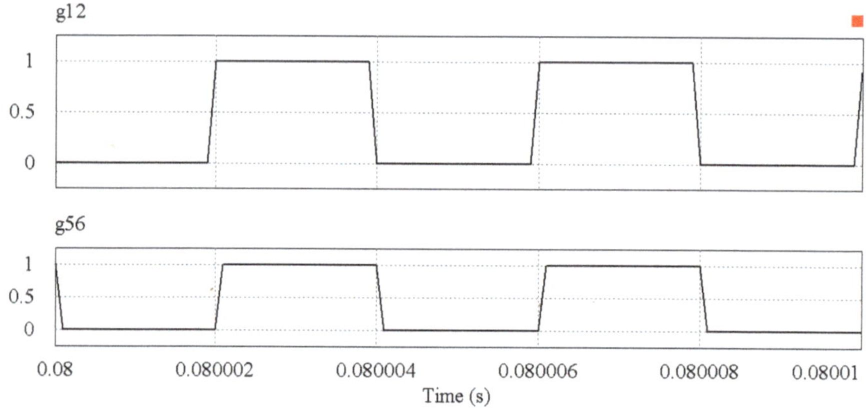

Fig. 10 Gate pulses of primary and secondary bridges

4.2 *Case ii: Input Voltage Variation from 36 to 32 V at 0.3 s*

The impact of source variation is analyzed by simulating a step variation of source voltage from 36 to 32 V at 0.3 s. Figure 11 shows the output voltage and current waveforms; the output voltage is settled down at 11 V and output current is settled down at 2.2 A. The settling time for 3% error band is 2.87 ms. The steady state power is 24 W for a phase shift of 9° as shown in Figs. 12 and 13. The voltages at primary bridge and secondary bridge ac terminals and the corresponding coupling inductor current are shown in Fig. 14. The controller effectively controls the phase shift and achieves the output power as per the current reference provided.

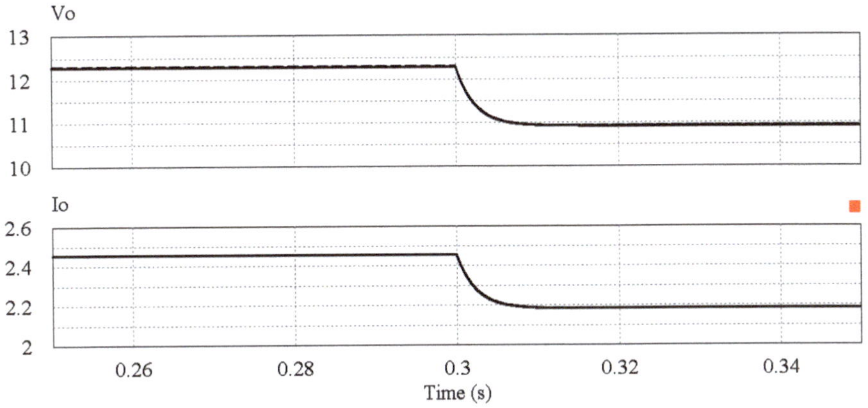

Fig. 11 Output voltage and current waveforms of dual active bridge

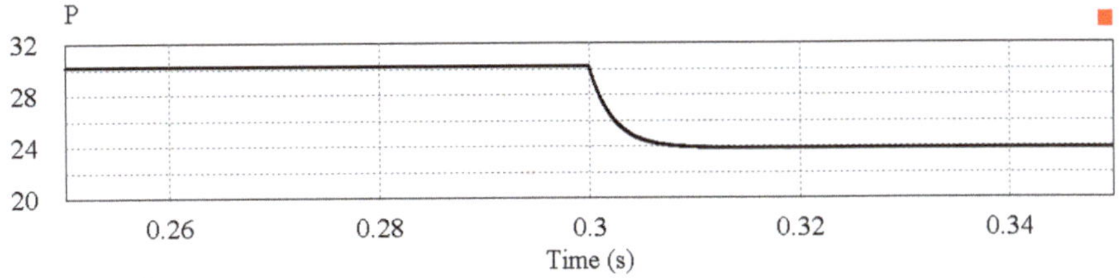

Fig. 12 Power output of dual active bridge

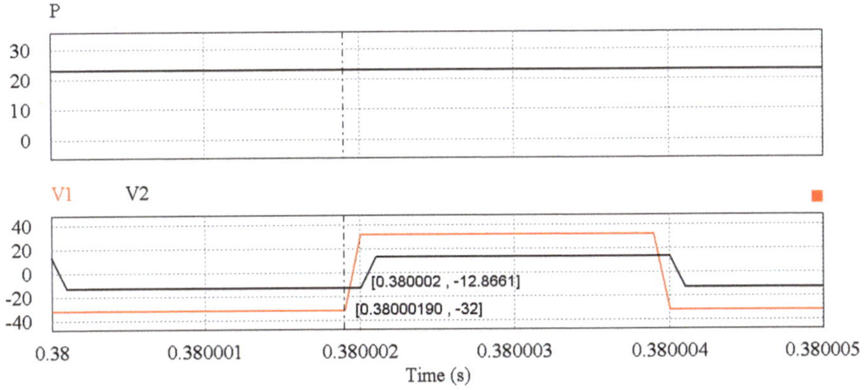

Fig. 13 Power output, primary, and secondary bridge voltages

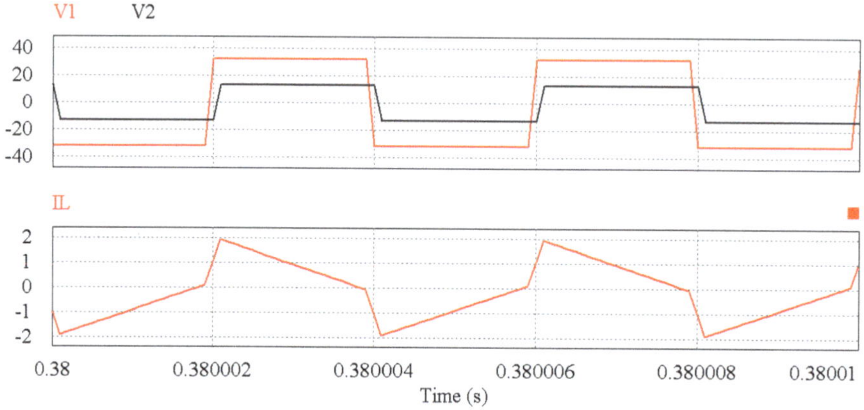

Fig. 14 Primary and secondary bridge voltages, Coupling inductor current

4.3 *Case iii: Reference Current Variation from 2.5 A to 2.6 A at 0.3 s*

The impact of reference current variation is analyzed by simulating a step variation of reference current from 2.5 A to 2.6 A at 0.3 s. Figure 15 shows the output voltage and current waveforms; the output voltage is settled down at 12.9 V and output current is settled down at 2.6 A. The settling time for 3% error band is 3.17 ms.

The steady state power is 33.5 W for a phase shift of 40.5° as shown in Figs. 16 and 17. The voltages at primary bridge and secondary bridge ac terminals and the corresponding coupling inductor current are shown in Fig. 18. The controller effectively controls the phase shift and achieves the output power as per the current reference provided.

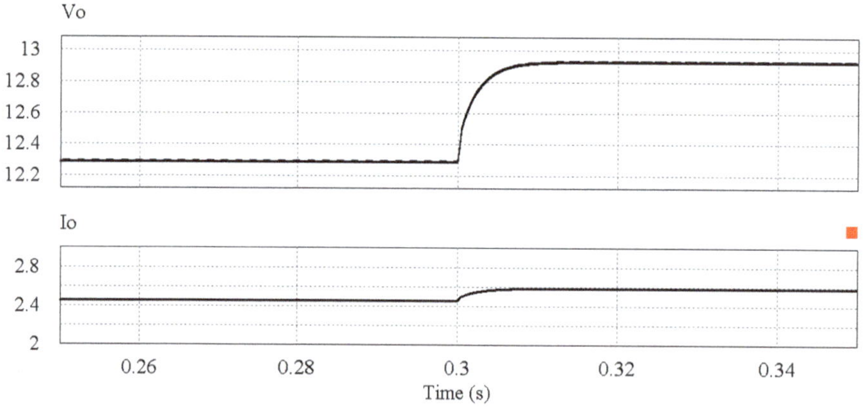

Fig. 15 Input and output voltage waveforms of dual active bridge

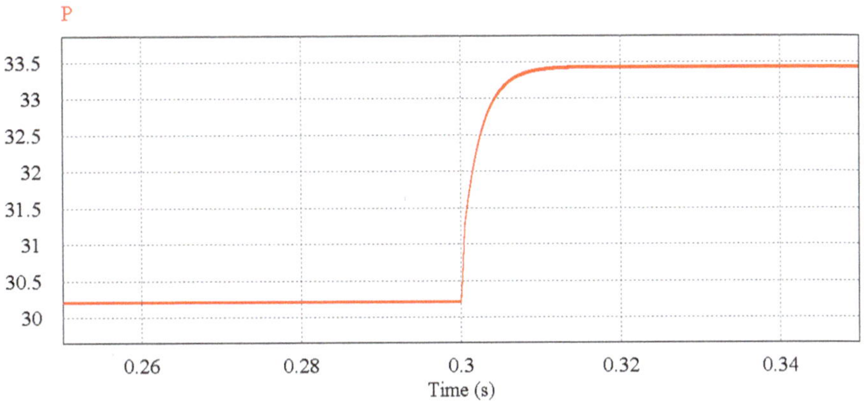

Fig. 16 Power output of dual active bridge

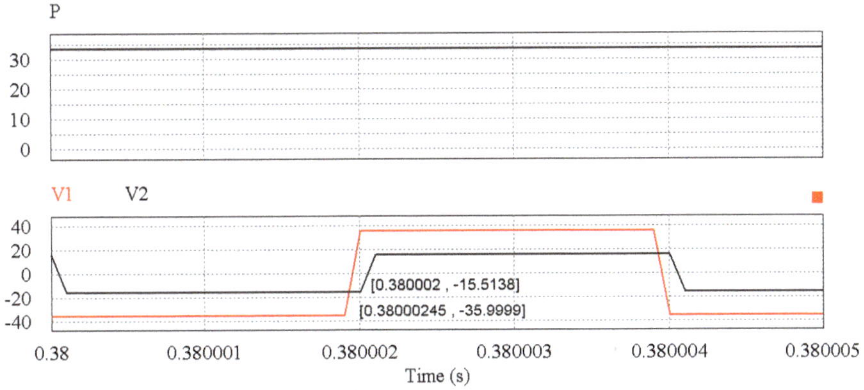

Fig. 17 Power output, primary, and secondary bridge voltages

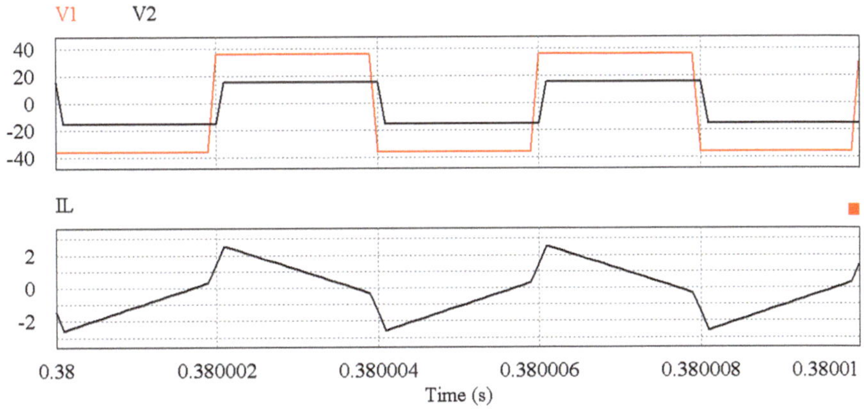

Fig. 18 Primary and secondary bridge voltages, Coupling inductor current

4.4 Case iv: Load Variation from 6 to 5 Ohms at 3 s

The impact of load variation is analyzed by simulating a step variation of load from 6 to 5 Ohms at 0.3 s. Figure 19 shows the output voltage and current waveforms; the output voltage is settled down at 12.1 V and output current is settled down at 2.4 A. The settling time for 3% error band is 4.43 ms.

The steady state power is 29.5 W for a phase shift of 9° as shown in Fig. 20. The voltages at primary bridge and secondary bridge ac terminals and the corresponding coupling inductor current are shown in Fig. 21. The controller effectively controls the phase shift and achieves the output power as per the current reference provided.

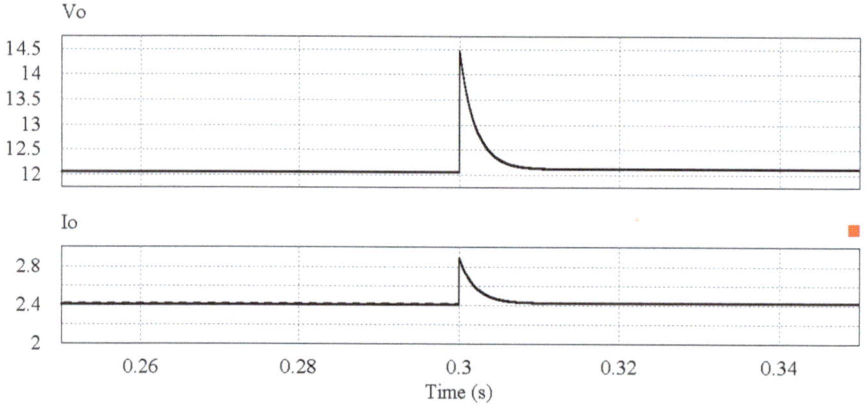

Fig. 19 Input and output voltage waveforms of DAB

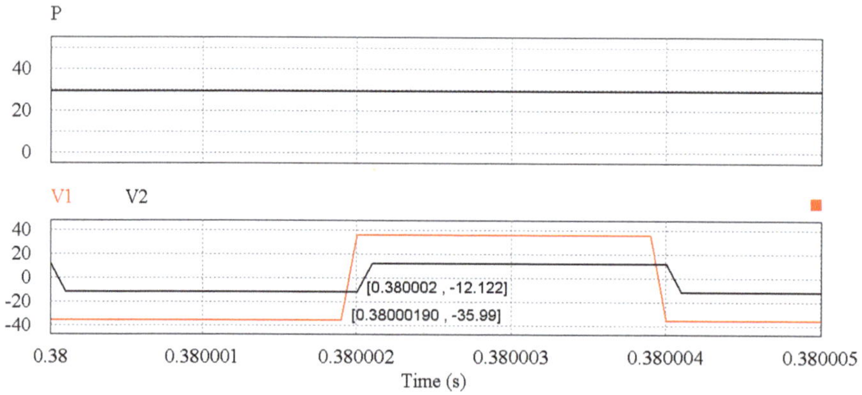

Fig. 20 Coupling inductor voltage and current

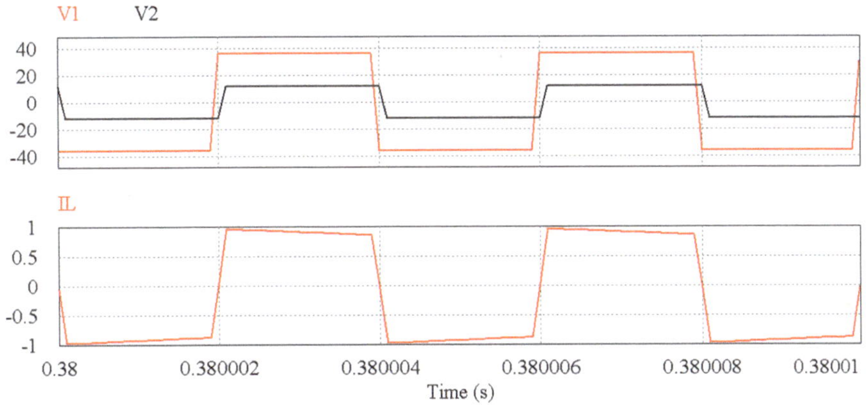

Fig. 21 Coupling inductor voltage and current steady state

4.5 Experimental Results

A laboratory prototype of dual active bridge is implemented with the designed values as shown in Fig. 22. STD12NF06L-1 MOSFET with V_{DS} of 60 V and a maximum drain to source current of 12 A. The primary and secondary bridge voltages are generated with a 15° phase shift as shown in Fig. 23 at 20 kHz frequency.

Fig. 22 DAB converter implemented on a PCB

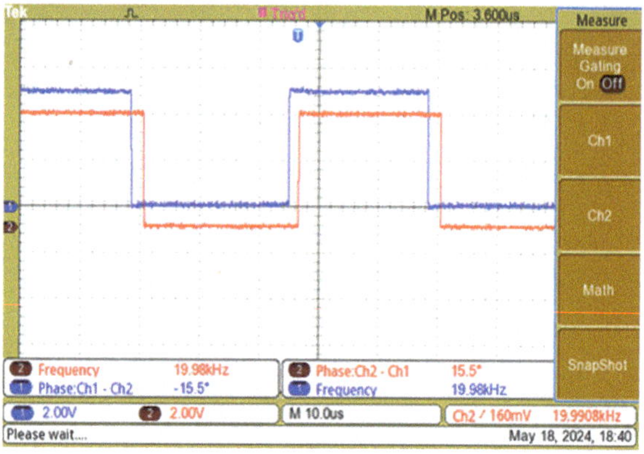

Fig. 23 Primary and secondary voltages for a phase shift of 15^0

5 Conclusion

Dual active bridge based electric vehicle battery charging circuit is designed, simulated, and analyzed with a closed loop phase shift controller. Transient simulations are conducted under various dynamic conditions to validate the performance of the circuit with the proposed controller. The results show that the controller performs satisfactorily to vary the power flow by controlling the phase shift between the two bridges and effectively controlling the bridge voltages. A laboratory prototype is implemented on a printed circuit board, and the bridge voltages are observed on a digital storage oscilloscope, which validates the simulation results. The implemented circuit can effectively operate as a bidirectional DC-DC converter, which makes grid-to-vehicle and vehicle-to-grid power flow possible.

References

1. Dhuliya PP, Kathait SS, Semwal S, Joshi T, Pathela M (2022) Electric vehicles (EV's): a brief review. In: 2022 International conference on advances in computing, communication and materials (ICACCM). IEEE, Dehradun, India, pp 1–3
2. Rivera S, Kouro S, Vazquez S, Goetz SM, Lizana R, Romero-Cadaval E (2021) Electric vehicle charging infrastructure: from grid to battery. IEEE Ind Electron Mag 15(2):37–51
3. Rashwan A, Ali AIM, Senjyu T (2022) Current stress minimization for isolated dual active bridge DC–DC converter. Sci Rep 12:16980
4. Zhou Y, Li X (2018): Vehicle to grid technology: a review. In: 2015 34th Chinese control conference (CCC). IEEE, Hangzhou, China, pp 9031–9036
5. Song B, Liu Q, Sun Y (2014) Overview of dual-active bridge isolated bidirectional DC-DC converter for high frequency-link power-conversion system. IEEE Trans Power Electron 29(8):4091–4106
6. De Doncker RWAA, Divan DM, Kheraluwala MH (1991) A three-phase soft-switched high-power-density DC/DC converter for high-power applications. In: IEEE Trans Indus Appl 27(1):63–73

7. Rolak M, Malinowski M (2011) Dual active bridge for energy storage system in small wind turbine. In: IEEE Africon '11. IEEE, Victoria Falls, Zambia, pp 1–5
8. Bidirectional, Dual Active Bridge Reference Design for Level 3 Electric Vehicle Charging Stations, https://www.ti.com/lit/ug/tidues0e/tidues0e.pdf?ts=1720420756990. Last accessed 08 Jul 2024
9. Magnetics-2022-Ferrite-Catalog. https://www.mag-inc.com/Media/Magnetics/File-Library/Product%20Literature/Ferrite%20Literature/Magnetics-2022-Ferrite-Catalog.pdf. Last accessed 08 Jul 2024
10. Das D, Basu K (2020) Optimal design of a dual-active-bridge DC–DC Converter. IEEE Trans Ind Electron 68:12034–12045
11. Mishra, Singh M (2023) A 10-kW active front end rectifier fed dual active bridge converter for EV charging. In: 2023 4th International conference for emerging technology (INCET). IEEE, Belgaum, India, pp 1–6
12. Chilakalapudi G, Kumar A (2022) Optimal reactive power control for dual-active-bridge converter using improved dual-phase-shift modulation strategy for electric vehicle application. Int J Circ Theory Appl 51:1204–1223
13. Deepak M, Janaki G, Bharatiraja C (2023) Single phase shift dual active bridge high frequency DC-DC converter for electric vehicle battery charger 6.6 kW. In: 2023 Fifth international conference on electrical, computer and communication technologies (ICECCT). IEEE, Erode, India, pp 1–6
14. Amin A, Shousha M, Prodic A, Lynch B (2015) A transformerless dual active half-bridge DC-DC converter for point-of-load power supplies. In: 2015 IEEE Energy Conversion Congress and Exposition (ECCE). IEEE, Montreal, QC, Canada, pp 133–140

Sustainable Dual-Function Unmanned Aerial Vehicles (UAV) System for Crop Detection and Animal Intrusion Monitoring Using YOLO v8

S. S. Kavitha, Rajalekshmi Kishore, Shreyas P. Aradhya, Madhvendra Rajshekhar, Mamilla Venkata Jeevan Reddy, and H. R. Manjunatha

Abstract This paper presents a novel approach combining sustainable dual-function Unmanned Aerial Vehicles (UAVs) and Machine Learning for dual-purpose detection in agriculture: Crop Disease and Animal Trespassing. The development of an autonomous drone (Hexacopter) engineered for self-navigation and flight surveys to capture detailed images of crops and surrounding areas. The proposed method utilized the YOLO v8 algorithm, which leverages Convolutional Neural Networks (CNNs), to analyze images for detecting crop diseases. This model excels in extracting features related to leaf health, color variations, and pest attacks. Additionally, by integrating a telemetry system with GPS technology, the proposed assembled Hexacopter demonstrates precise navigation capabilities using ML models to detect animal trespassing effectively. This integrated system offers early detection of crop diseases and timely mitigation of animal trespassing, thereby reducing agricultural losses and enhancing farm management sustainability. The simulation results presented showcase the efficacy and performance of the trained model, underscoring the drone's potential for real-world applications.

Keywords Hexacopter · Pixhawk · UAV · YOLOv8 · LabelImg · Annotations · PyTorch · Ublox M8N · ARM cortex M4 · Gyroscope

S. S. Kavitha (✉) · R. Kishore · S. P. Aradhya · M. Rajshekhar · M. V. J. Reddy · H. R. Manjunatha
Department of Electronics and Communication Engineering, The National Institute of Engineering (Affiliated with Visvesvaraya Technological University, Belgaum), Mysore, Karnataka, India
e-mail: kavithaece@nie.ac.in

I. M. Jamadar et al. (eds.), *Integrating Sustainability in Modern Technologies*, Lecture Notes in Mechanical Engineering, https://doi.org/10.1007/978-981-95-9466-5_12

1 Introduction

Agriculture is essential for human sustenance and societal development, providing food, economic support, and cultural heritage. It involves crop cultivation, livestock management, and natural resource stewardship [1]. With a growing global population and increased urbanization, sustainably increasing agricultural productivity is crucial for ensuring food security and addressing environmental challenges. As society's needs and technology have evolved, agriculture has transformed significantly, adopting innovations and mechanization to boost efficiency, productivity, and sustainability. Precision agriculture, which integrates digital technologies, remote sensing, and data analytics, has changed the way farming is done by helping farmers use resources more wisely, reduce environmental impact, and increase yields. Recently, advanced technologies like Robotics, the Internet of Things (IoT) [2], and Machine Learning (ML) have further revolutionized traditional farming. These technologies provide farmers with remarkable precision, efficiency, and insights into managing crops, using resources, and controlling pests. Motivated by advancements in farming technology, the method has been developed as a UAV-based hexacopter that processes UAV images [3, 4] using Yolo v8 [5], the system can detect animal intrusion and pests [6, 7], while an integrated web application offers suggestions for pest control, ultimately optimizing farm operations, enhancing productivity, and contributing to sustainable farming practices [8]. The following is a quick summary of the proposed contribution to the dual-function UAV:

- Through the development of a heavy-lift hexacopter and the utilization of precise GPS technology, the proposed method ensures optimal performance, stability, and accurate navigation for the UAV. These enhancements are crucial for efficient and reliable operation, especially in large-scale agricultural environments.
- Integration of cutting-edge technologies such as the YOLO v8 machine learning tool and AIoT systems into the UAV, enhancing its capabilities for real-time object detection, tracking, and pest management. This integration represents a significant advancement in agricultural monitoring and security.
- Proposed contributions include the application of Convolutional Neural Networks (CNNs) for crop health monitoring and disease detection, as well as the implementation of advanced detection algorithms for crop security. These approaches represent innovative solutions for addressing challenges in modern agriculture, leading to improved crop management and protection. The rest of this paper is organized as follows. Section 2 describes the Literature Survey of this project. Section 3 presents the proposed design for this paper. Section 3.1 describes about software system and its setup. Section 3.2 describes the Hardware and software Interface system. Finally, results are discussed in Sect. 4, followed by conclusions in Sect. 5.

2 Literature Survey

An extensive review of various seminal works to underpin the concepts and technologies employed in developing UAV-based systems for crop detection and animal intrusion monitoring using the YOLO v8 machine learning tool is carried out. A comprehensive survey on pest detection techniques using image processing provided valuable information on various methodologies and technological advancements, guiding the selection and implementation of effective pest detection methods [1]. Another significant work provided insights into constructing a heavy-lift hexacopter capable of supporting various sensors and equipment, ensuring optimal performance and stability, which is critical for the UAV's operational efficiency and reliability [2].

Additionally, another work focused on animal intrusion detection and crop protection using YOLO and sound devices, aligning with the objective of integrating advanced detection algorithms and responsive measures to enhance crop security [3]. A study on manipulating GPS signals to determine drone launch locations in rescue operations informed the proposed approach to utilizing GPS technology for precise navigation and accurate monitoring of agricultural fields [9–11].

In terms of crop health monitoring, studies highlighted the efficacy of Convolutional Neural Networks (CNNs) in identifying wheat plant diseases using real-time UAV images, supporting the use of similar technologies for crop disease detection [6, 12, 13]. Further enhancing the system's capabilities, research demonstrated the effectiveness of the YOLO algorithm in real-time object detection and tracking, which is pivotal for promptly detecting pests and animal intrusions [8].

A study demonstrated how the integration of artificial intelligence and IoT in agriculture through an AIoT-based smart agricultural system for pest detection can improve pest management and overall farm productivity, informing the development of a UAV system that leverages AI and IoT for comprehensive agricultural monitoring and management [14–16]. By synthesizing insights from these diverse sources, the proposed study aims to develop a robust and innovative UAV system that effectively addresses the challenges of modern agriculture through advanced detection and monitoring capabilities.

3 Methodology of the Proposed Design

3.1 Proposed Hexacopter Design

The proposed CropGuard UAV integrates a hexacopter configuration for enhanced stability and maneuverability, essential for effective pest and animal monitoring in agricultural settings as shown in Fig. 1. This design incorporates six arms equipped with high-performance DYS D2836 1120 kV brushless DC (BLDC) motors and 40 A Electronic Speed Controllers (ESCs), ensuring precise aerial surveillance [14]. The frame uses an S550 structure, and the central controller is the PIXHAWK 2.4.8, which includes advanced features such as an ARM Cortex M4-based 32-bit microcontroller, MPU-6000 gyroscope, barometer, and various other sensors as shown in Fig. 2.

Configuration and firmware setup are facilitated through Mission Planner software, enabling the calibration of essential components and monitoring onboard systems. The UAV's navigation is supported by an Ublox M8N GPS module, and communication is established via an FS-iA10B receiver [2].

The hexacopter employs Orange-HD 9 * 4.5 propellers, yielding a practical thrust of 5.8 kg, and is powered by an 8000 mAh 3-cell LiPo battery. The system weight [14], including telemetry and onboard components, is approximately 2.1 kg, allowing a payload capacity of around 1 kg for additional sensors and equipment. The Crop-Guard UAV features flight capabilities such as Return to Home (RTL), Loiter, and Altitude Hold, ensuring precise control and versatility in various environmental conditions [2]. Component used is tabulated in Table 1.

Fig. 1 Proposed workflow of the design methodology

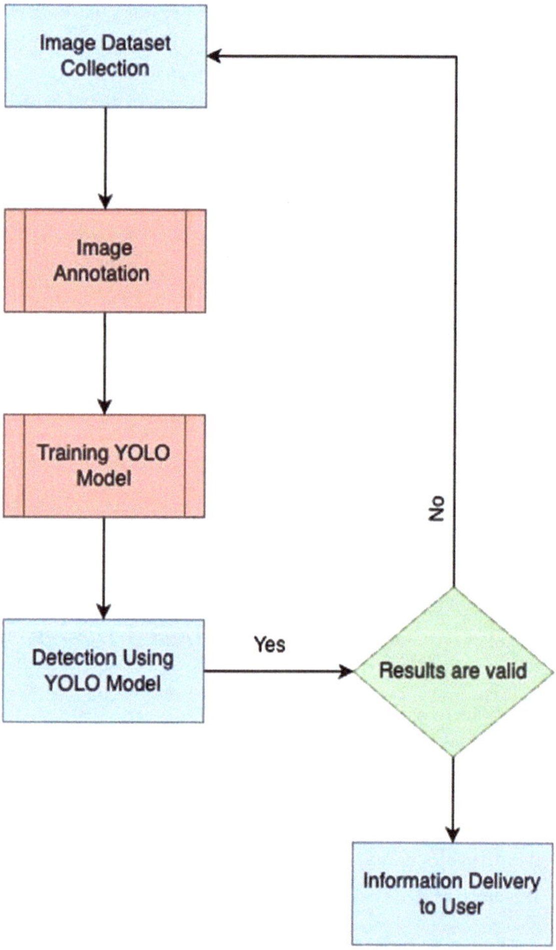

Fig. 2 Hexacopter drone with mounted components with a payload of 1 kg

Table 1 Components used for drone

Propellers	9″ × 2.5
Rotor	1120 kV 2836
Frame	S550
ESC	40 A
Battery	5600 mAh 40c 11.1 V (3 s)
Max tested altitude	200 ft
GPS	UBLOX M8N
Weight (without payload)	2.2 kg
Max flight time	17 min
Camera	Mobile phone

3.2 Software System and Description

The proposed model uses machine learning models for object detection in crop fields. There are many object detection software like YOLO, OpenCV, TensorFlow object detection, etc. The method was best analyzed with every software and YOLOv8 for the best fit [1]. YOLOv8 offers several advantages over previous versions of the YOLO model. It provides improved accuracy in object detection, achieving better mean Average Precisions (mAP) and faster speeds. Additionally, YOLOv8 requires fewer parameters to achieve its performance, making it a more efficient and streamlined model. Installing and utilizing YOLOv8 for object detection through GitHub is a straightforward process owing to its status as an open-source project. To commence, the proposed model typically clones the YOLOv8 repository from

GitHub onto the local machine. Once cloned, the method is followed to provide the instructions for setting up the necessary dependencies and configurations, which often involve installing Python packages such as PyTorch and OpenCV. YOLOv8's open-source nature allows for flexibility in adaptation and customization to suit diverse needs, as users can delve into its source code, understand its algorithms, and even contribute to its development. Leveraging the power of convolutional neural networks, YOLOv8 offers efficient real-time object detection capabilities across various domains, from surveillance to autonomous vehicles, democratizing access to state-of-the-art computer vision technologies for researchers, developers, and enthusiasts alike.

- **Animal Detection**: Initially, the model is created in Google Collaborate and installed necessary dependencies. The dataset of about 2000 animal images was selected for training. Before training, the images are labeled as per the specifications. The tool *LabelImg* was used for labeling the image. Once an image is labeled, it creates a *.txt* file for each labeled image, which contains the coordinates used for training. These are generally called annotations [8]. Once the model contains training images and annotation in the same folder, the training process is started. The data is created as a *.yaml* file which stores the location of training images and has information about the categories that have trained the model for cattle. The model is trained for 50 epochs (epochs means the number of iterations the model is being trained).
- **Pest Detection**: In this project, the methodology is focused on pest detection, the proposed method utilizes Google Colab to develop the model, ensuring accessibility and scalability. Prior to training, the dataset comprising approximately 1500 images is specifically labeled for pest detection [9]. To accomplish this, *LabelImg* is employed which is a widely used annotation tool that facilitates the annotation process by generating a *.txt* file for each labeled image. These *.txt* files contained the coordinates crucial for training, enabling precise localization of pests within the images. On gathering the annotated images and corresponding annotations within the same directory, the training process was initiated. Before commencing training, a crucial step involved creating *data.yaml* file [6]. This file served as a vital configuration document, housing essential informations such as the location of training images and the categories targeted for detection. In the context of pest detection, the *data.yaml* file delineated the pertinent details regarding the pest category, ensuring that the proposed model was trained with a singular focus on identifying pests within the agricultural landscape.

Leveraging the powerful capabilities of YOLOv8 and the computational resources provided by Google Colab, the model has been trained throughout 50 epochs. Each epoch represents a complete iteration through the entire dataset, during which the model refines its parameters and adjusts its internal representations to optimize performance [3]. By iterating through multiple epochs, the proposed model gradually refined its ability to discern and accurately classify pests within the images, leveraging the wealth of annotated data provided during training. Through meticulous data preparation, strategic configuration, and iterative training, the proposed model emerged as a potent tool for pest detection in agricultural settings. The utilization of Google Colab facilitated seamless collaboration and resource management, while the adoption of YOLOv8 ensured state-of-the-art performance in object detection tasks. Armed with the proposed trained model, the model stands poised to contribute meaningfully to the advancement of precision agriculture, empowering farmers with actionable insights and effective pest management solutions.

- **Hardware and Software Interface**: The proposed drone system's hardware interface centers on the Pixhawk controller, coordinating flight control and navigation. Telemetry at 433 MHz enables swift data exchange between the drone and the ground station for real-time analysis. A high-res camera captures detailed field images, essential for the proposed YOLOv8 object detection model to identify crops and pests accurately. A robust GPS module ensures precise navigation across varied terrain.

Software-wise, YOLOv8 drives real-time object detection, while image processing algorithms refine data accuracy. Autonomous navigation minimizes manual intervention, optimizing efficiency. The user-friendly ground station interface allows easy mission planning, monitoring, and data analysis, ensuring accessibility for farmers of all levels. In summary, the proposed drone system seamlessly integrates advanced hardware with sophisticated software, offering unprecedented precision and insight for agricultural management.

4 Results and Analysis

4.1 Simulation Results of Animal Detection

In this analysis, the focus was on the identification of cattle within the captured drone images. Utilizing advanced image processing techniques and machine learning algorithms, the system successfully detected instances of cattle presence with varying degrees of precision. The precision scores for cattle detection in the analyzed images are as shown in Fig. 3.

The system exhibited a precision score of 0.8 in identifying the presence of cattle in the first image. This suggests a relatively high level of confidence in the detection result. However, further analysis revealed that the clarity and contrast of the image greatly contributed to this high precision score.

4.2 Simulation Results of Pest Detection

In Fig. 4, cattle were detected with a precision score of 0.73. While still relatively accurate, this score indicates a slightly lower confidence level compared to the first

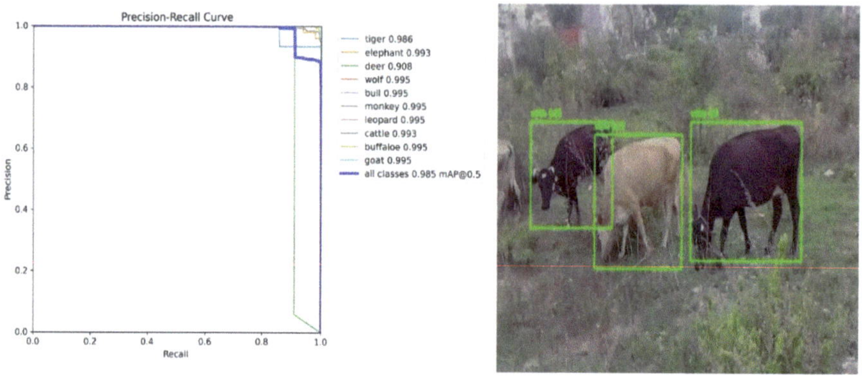

Fig. 3 Precision versus recall simulation result for animal detection dataset

instance. Closer inspection revealed that occlusions, such as vegetation, partially obstructed the view of the cattle, leading to a decrease in precision.

The model YOLOv8 was developed to detect pests in fields and evaluated its performance using a dataset of labeled images. Proposed results show that the model achieved a precision of 0.88, indicating a high accuracy in detecting pests. The YOLOv8 model is trained with a dataset of 1000 images of labeled pests. The model was able to detect pests with a precision of 0.88, indicating that it was able to correctly identify 88% of the pests in the images. Also, the method evaluates the model's performance on a test set of 500 images and found that it achieved a precision of 0.85, indicating that it was able to generalize well to new images.

Proposed results show that the YOLOv8 model is effective at detecting pests in fields with a high degree of accuracy. This is likely due to the model's ability to learn features that are specific to pests and the large dataset of labeled images used for training. The model's performance on the test set also suggests that it can generalize

Fig. 4 Precision versus recall simulation result for pest dataset

Table 2 Simulation results of animal and pest detection using YOLOv8 model

Dataset	Accuracy %	Precision	F1-score
Animal dataset	98.5	0.93	0.91
Pest dataset	98.1	0.91	0.94

well to new images, which is important for real-world applications. One potential limitation of the trained model is that the dataset used for training and testing was limited to a few types of pests, and it is possible that the model may not perform as well on other types of pests. Future studies could aim to expand the dataset to include a wider range of pests and evaluate the model's performance on these different types. Additionally, the model could be further improved by incorporating additional features such as weather conditions, soil type, and crop type, which may also affect pest infestation.

4.3 Accuracy Assessment

The precision scores provide insights into the reliability of cattle detection within the analyzed images. While higher precision scores indicate more accurate detections, lower scores suggest potential areas for improvement in the detection algorithm. Factors such as image quality, lighting conditions, and occlusions play critical roles in influencing the precision of animal detection as shown in Table 2.

5 Conclusion

In conclusion, the proposed drone-based approach offers significant potential for agriculture, conservation, and pest management, with continued research essential for practical deployment optimization. Leveraging image processing and machine learning advancements, the aim is to obtain more accurate detection, supporting sustainable agriculture and environmental conservation. Despite challenges like occlusions and varying conditions, the proposed system showed promising precision scores for cattle detection (0.73–0.8) and impressive precision for pest detection (0.88), empowering farmers to minimize losses and maximize yields. Further optimization of the detection algorithm, including parameter fine-tuning and dataset expansion, is necessary to enhance cattle identification accuracy and robustness. Overall, proposed system presents a valuable tool for sustainable agriculture, offering timely insights into pest infestations and enabling proactive management practices.

References

1. Anish A, Sharan R, Malini AH, Archana T (2023) Enhancing surveillance systems with yolo algorithm for real-time object detection and tracking. In: 2023 2nd international conference on automation, computing and renewable systems (ICACRS), pp 1254–1257
2. Alamleh H, Roy N (2021) Manipulating GPS signals to determine the launch location of drones in rescue mode. In: 2021 IEEE international IOT, electronics and mechatronics conference (IEMTRONICS), pp 1–5
3. Chen C-J, Huang Y-Y, Li Y-S, Chang C-Y, Huang Y-M (2020) An AIoT based smart agricultural system for pests detection. IEEE Access 8:180750–180761
4. Kabir RH, Lee K (2021) Wildlife monitoring using a multi-UAV system with optimal transport theory. Appl Sci 11(9):4070
5. Surusomayajula M, Dhuli VS, Dodda VC, Chalasani K, Nimmagadda SV, Jammula TM (2023) Wheat plant disease detection using CNN on real-time UAV images. In: 2023 IEEE 15th international conference on computational intelligence and communication networks (CICN), pp 603–607
6. Nagar H, Sharma R (2020) A comprehensive survey on pest detection techniques using image processing. In: 2020 4th international conference on intelligent computing and control systems (ICICCS), pp 43–48
7. Esposito M, Crimaldi M, Cirillo V, Sarghini F, Maggio A (2021) Drone and sensor technology for sustainable weed management: a review. Chem Biol Technol Agric 8:1–11
8. Kothai R, Siva S, Agathiyan M, Kabilan J (2023) Animal intrusion detection and crop protection using yolo and sound devices. In: 2023 Intelligent computing and control for engineering and business systems (ICCEBS), pp 1–10
9. Nagaraj G, Sungeetha D, Tiwari M, Ahuja V, Varma AK, Agarwal P (2024) Advancements in plant pests detection: leveraging convolutional neural networks for smart agriculture. Eng Proc 59(1):201
10. Popescu D, Dinca A, Ichim L, Angelescu N (2023) New trends in detection of harmful insects and pests in modern agriculture using artificial neural networks. A review. Front Plant Sci 14:1268167
11. Bai Y, Hou F, Fan X, Lin W, Jinghan L, Zhou J, Fan D, Li L (2023) A lightweight pest detection model for drones based on transformer and super-resolution sampling techniques. Agriculture 13(9):1812
12. Venkatasaichandrakanth P, Iyapparaja M (2024) A survey on pest detection and classification in field crops using artificial intelligence techniques. Int J Intell Robot Appl 1–26
13. Singh E, Pratap A, Mehta U, Azid SI (2024) Smart agriculture drone for crop spraying using image-processing and machine learning techniques: experimental validation. IoT 5(2):250–270
14. Suprapto BY, Heryanto MA, Suprijono H, Muliadi J, Kusumoputro B (2017) Design and development of heavy-lift hexacopter for heavy payload. In: 2017 International seminar on application for technology of information and communication (iSemantic), pp 242–247
15. Albattah W, Masood M, Javed A, Nawaz M, Albahli S (2023) Custom CornerNet: a drone-based improved deep learning technique for large-scale multiclass pest localization and classification. Complex Intell Syst 9(2):1299–1316
16. Anam I, Arafat N, Hafiz MdS, Jim JR, Kabir MdM, Mridha MF (2024) A systematic review of UAV and AI integration for targeted disease detection, weed management, and pest control in precision agriculture. Smart Agric Technol 100647

Cost-Effective Permanent Magnet Synchronous Motor Drive with Hall-Effect Position Sensors

B. Shruthi, V. N. Jayasankar, S. R. Gurumurthy, and M. Nisarga

Abstract Precise and effective control of electric motors is a major challenge in Electric Vehicle to obtain enhanced performance. Permanent magnet synchronous motors are the most commonly used motors in electric vehicles due to many technical advantages such as high torque by weight ratio, better controllability, and high efficiency, etc. The major disadvantages of these motors are high cost involved in their position sensing while implementing the field-oriented control on the motor for the speed and torque control. A novel method of rotor position sensing of permanent magnet synchronous motors using hall-effect sensors and electronic commutation of them is simulated and implemented in this paper. The MATLAB/Simulink-based dynamic simulations are done for testing the effectiveness of the proposed method. The hardware is implemented using a laboratory prototype, with the encoder implemented digitally in TMS320F28379D, with the input from Hall-effect sensors placed on the rotor. The motor is run initially as a BLDC motor up to certain speed and then shifted to PMSM mode using digital control signals. The performance of the digital encoder is validated by observing the back EMF and stator current waveforms. This method has significantly reduced the total cost of the motor drive, without compromising the accuracy and effectiveness of rotor position sensing.

Keywords Electric vehicles · Electric motor control · PMSM · BLDC · Hall-effect sensors · Encoders

1 Introduction

Electric vehicles are becoming more popular due to various reasons such as environmental concerns, government policies, and superior performance, etc. The performance of Electric Vehicles (EVs) mainly depends on the precise control of electric motor which drives the vehicle. Permanent Magnet Synchronous Motors (PMSMs) are gaining popularity for EV applications since they are capable of running at

B. Shruthi (✉) · V. N. Jayasankar · S. R. Gurumurthy · M. Nisarga
Department of EEE, The National Institute of Engineering, Mysore, India
e-mail: shruthib@nie.ac.in

I. M. Jamadar et al. (eds.), *Integrating Sustainability in Modern Technologies*, Lecture Notes in Mechanical Engineering, https://doi.org/10.1007/978-981-95-9466-5_13

constant speed irrespective of the load, through better controllability and efficiency compared to other alternatives [1]. The interaction of constant and rotating magnetic field results to achieve the constant speed characteristics and the rotor rotates at synchronous speed. This means if there is control over the frequency of electricity, speed of synchronous motor can be controlled accurately [2].

The PMSM is the brushless motor which is also called as BLAC motor whose induced EMF waveform is sinusoidal in shape. The high torque densities with low rotor losses are offered by permanent magnet machines. Permanent magnet machines often require sophisticated control systems to manage their operation, which can add to the complexity and cost of the overall system. Vector-based control strategies are deployed for PMSM [3–6]. Commonly used strategies include Sine PWM and Space Vector Modulation based methods [7–11] Advanced PWM strategies and their implementations in inverter-fed drives is extensively studied to enhance efficiency and dynamic response [12–15]. Linear encoders are used in Non linear synchronous drives, for position feedback control [10, 16]. Most of the literature focuses on encoder-based rotor position sensing in PMSM machines which are relatively expensive. Hence, a cost-effective position sensing method is proposed using Hall sensors for position control of permanent magnet synchronous motor drives [16, 17]. This paper focuses on driving a PMSM motor using a cost-effective position sensor-based control algorithm on TMS320F28379D processor [8].

2 Methodology

Generation of Pulses to PMSM Motor

The control of PMSM motors involves detection of rotor position at any instant of time to generate the supply voltage to the armature windings of the motor. Conventionally, the rotor position is found out by using a device called quadrature encoder. This need to be fixed on the rotor shaft and the corresponding electronic signals are generated to give the information about the rotor position. This device is very expensive and complex for small motors. To overcome this disadvantage, a novel method of detecting the rotor position is investigated, which involves usage of Hall sensors similar to a BLDC motor and encoder implementation in digital platform. To run the motor as a PMSM, the initial rotor position is required, which is not available in this method as sequential logic using counters and timers is used for the implementation of encoder. Therefore, the motor can be started as a BLDC motor until certain low speed and then shift the control logic to run the motor as PMSM motor.

The block diagram of the proposed controller with Hall-effect sensors is shown in Fig. 1. These sensors give one pulse as output per rotation. The input/rotor frequency is measured using a timer module of the microcontroller. This frequency is multiplied by a factor 360 to generate signal whose frequency is 360 times higher. Two waveforms which are phase shifted by 90° and third one with same as input hall position signal are generated using this signal from Hall sensors.

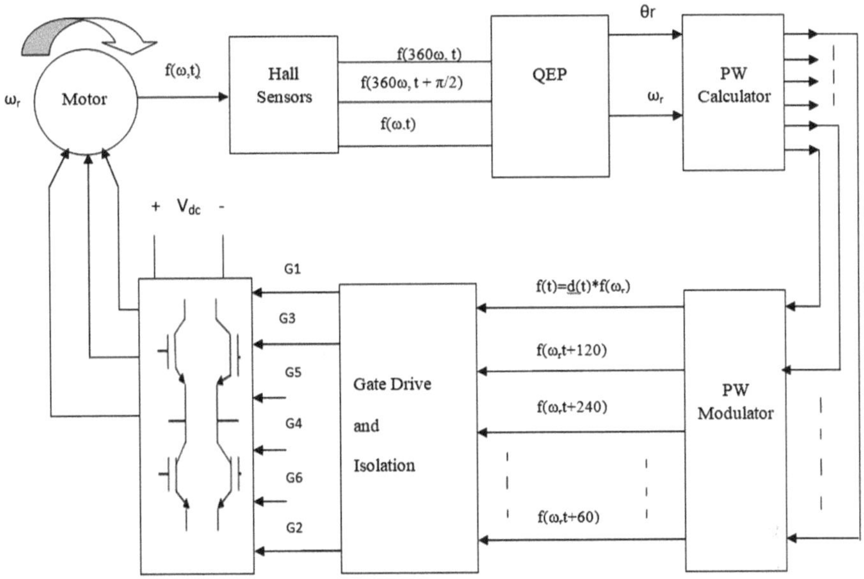

Fig. 1 Block diagram of PMSM drive with Hall-effect sensor-based position sensing

These three set of waveforms are given as the input signals (i.e., A, B, and Index) as given in Fig. 2. To the eQEP module of the microcontroller to get the rotor position information of the PMSM motor. The position counter counts the number of pulses and gives the position of the rotor, and it can be reset by index pulse. Position information can be latched at a certain interval of time which depends on the angle of resolution.

The rotor position thus measured at a particular instant of time using the eQEP module of the microcontroller is used to set the pulse widths of the pulses to be given to various IGBT switches of the Hex Bridge inverter. As known to drive a PMSM motor, its armature windings needs to be applied with sinusoidal voltage which is synchronized to its induced voltage. Hence, a sinusoidal voltage waveform is required to be generated using the position sensor information. This can be achieved using PW Modulator which adjusts the width of its output pulse sinusoidally as a function of rotor position. Three Hall-effect sensors are used in this work as shown in Fig. 3a. The angular distance between any two adjacent sensors is 120° (electrical). The sensor output is as depicted in Fig. 3b.

Generation of Pulses to BLDC Motor

The required PWM's gate pulses of the inverter is given by checking the position of the rotor which is sensed by using Hall sensors. Basically there are three Hall sensors. The Hall sensors conduction time is 180°. In order to produce ramp wave, theta is read discretely not continuously which is required for sinusoidal pulse width modulation.

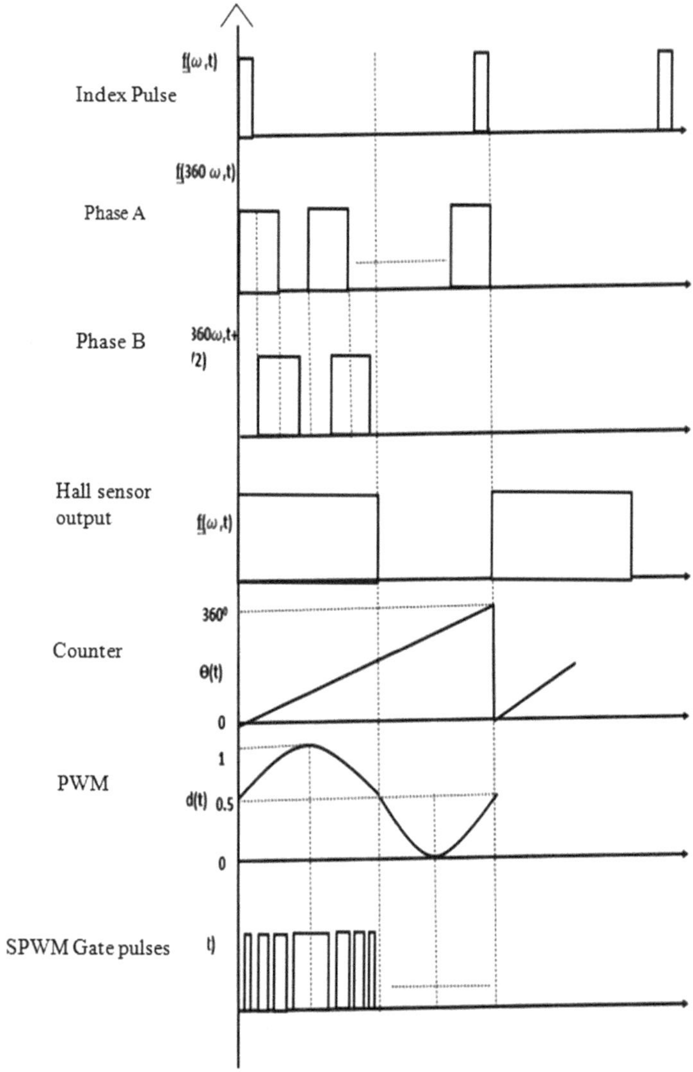

Fig. 2 Waveforms at different stages

Figure 3 shows the trapezoidal nature of the induced voltage. The induced voltage in R phase (EA) is constant between 30° and 150°. The output signal of Ha is rising at 30° and Hb is rising at 150°. The DC voltage is applied between the time instants of rising pulse of Ha and Hb. This is achieved by using combinational logics as shown in Table 1 (Fig. 4).

The PWM compare value is changed for every 30°. From the Hall sensors waveforms and the Back EMF waveform, it is observed that from 0 to 30° the R phase

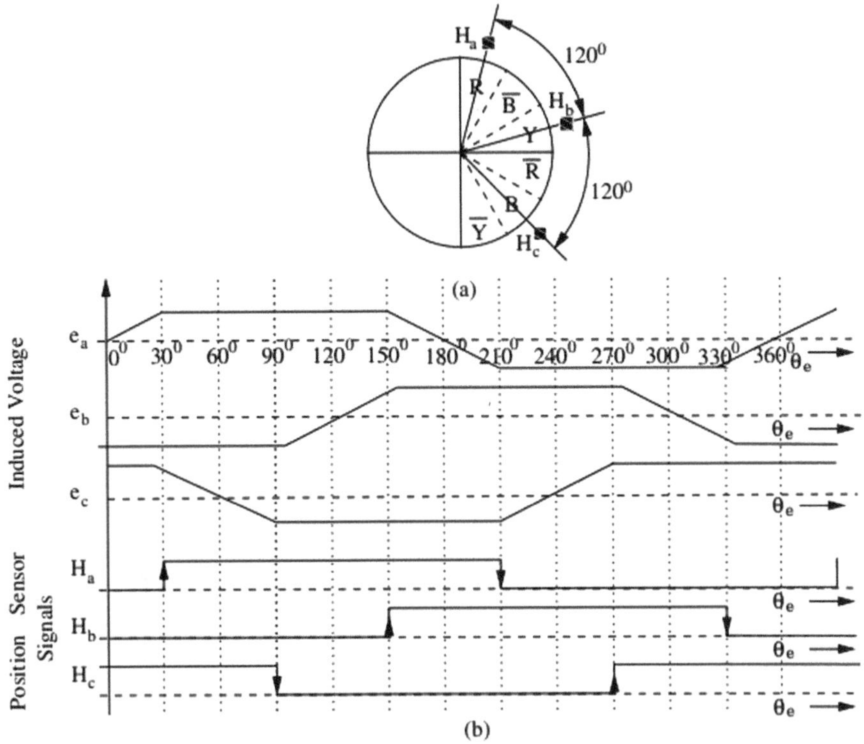

Fig. 3 Placement of the position sensors and their output signals

Table 1 IGBT gate trigger logic block

Ha & !Hb	T1
Hb & !Hc	T3
Hc & !Ha	T5
!Ha & Hb	T4
!Hb & Hc	T6
!Hc & Ha	T2

back EMF increases linearly, Y phase is turned off, and the B phase is constant; therefore, the values to compare registers of these MOSFET's as given below:

$$\text{CMPA1} = (\Theta/30) * \text{PRD}$$
$$\text{CMPA3} = 0$$
$$\text{CMPA5} = \text{PRD}.$$

Likewise for $\Theta = 31°$ to $60°$, the R phase is constant, Y phase is turned off, and B phase is decreasing linearly. From Θ greater than $60°$ to $90°$, R phase and Y phase back EMF remains in its previous states, but B phase will continue to decrease in

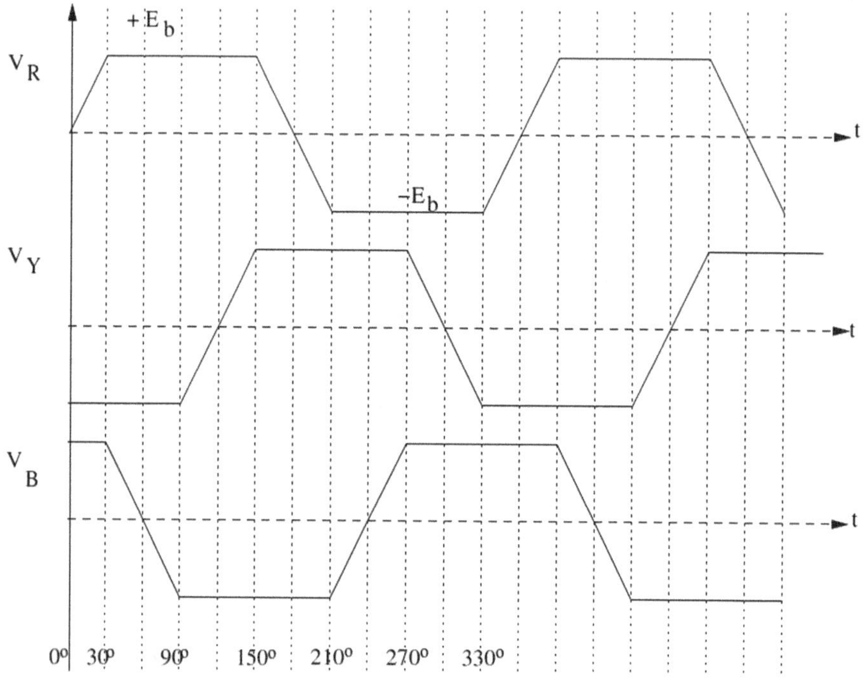

Fig. 4 Induced voltage waveforms of the machine

negative region. For Θ = 90° to 120°, R phase remains constant, Y phase starts to increase linearly in the negative region, and C phase is turned off. For Θ = 120° to 150°, R phase remains constant, Y phase continues to increase linearly in the positive region, and B phase is turned off. For Θ = 150° to 180°, R phase starts to decrease, Y phase remains constant, and B phase is turned off.

Controller Implementation

The control algorithm is implemented on digital controller platform. TMS320F28379D along with its peripherals is used in this work for the drive application.

The Hall sensor output signals are shaped in the signal conditioning board to make them compatible to the DSP port. These signals are connected to port-B of the DSP. Taking port-B signals as input, the pulses are generated using electronic commutation logic. These pulses are ANDed with the PWM pulses generated with the current controller. This generates six trains of pulses (corresponding to six IGBT switches) at the output of PWMA module. Out of these six, three pulse trains (G1, G3, and G5) are used directly as the triggering signals to the top side power switches (T1, T3, and T5; E = 0). Next three trains of pulses (G2, G4, and G6) are used for driving bottom side power switches (T2, T4, T6). This is shown in the Fig. 5.

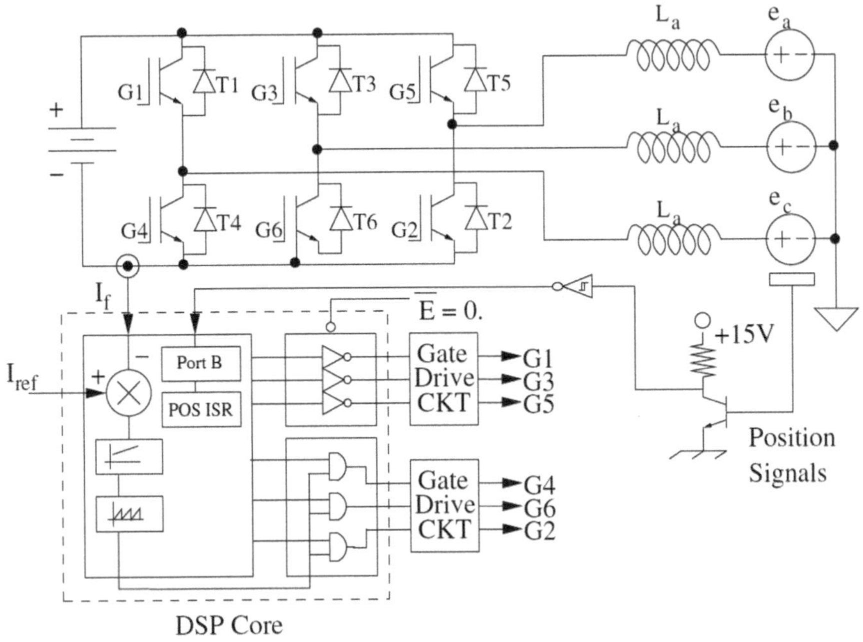

Fig. 5 Power and control schematic in CCA mode

3 Results and Discussions

The simulation studies are conducted on 48 V, PMSM motor with in BLDC mode of operation. A closed loop speed control using PI controller is implemented in Simulink. Figures 6, 7, 8, 9, and 10 depict the Hall sensor output signal, back EMF, stator current, speed, and electromagnetic torque, respectively.

Hardware Results

The hardware setup is shown in Fig. 11. The components are PMSM motor, main power supply, the control power supply, smart power module, and launch pad interfacing module. Figure 12 shows that frequency of Signal F_A is 360 times the frequency of Hall sensor output signal. Figure 13 shows that the index signal frequency is same as Hall sensor output signal which generates a pulse after each complete rotation of the rotor. Figure 14 shows that signal F_B lags the signal F_A by 90° which indicates that motor rotates in clockwise direction. Figure 15 shows that the gate pulses generated using SPWM to the inverter. Figure 16 shows the trapeziodal back emf waveform of the motor, which shows that the machine runs as BLDC motor up to a speed of 17 Hz. Above 17 Hz, it runs as PMSM motor, and the sinusoidal back emf is as shown in Fig. 17.

Signals for the quadrature encoder module are generated using PWM module of the microcontroller by suitably programming the PWM control registers. The signals

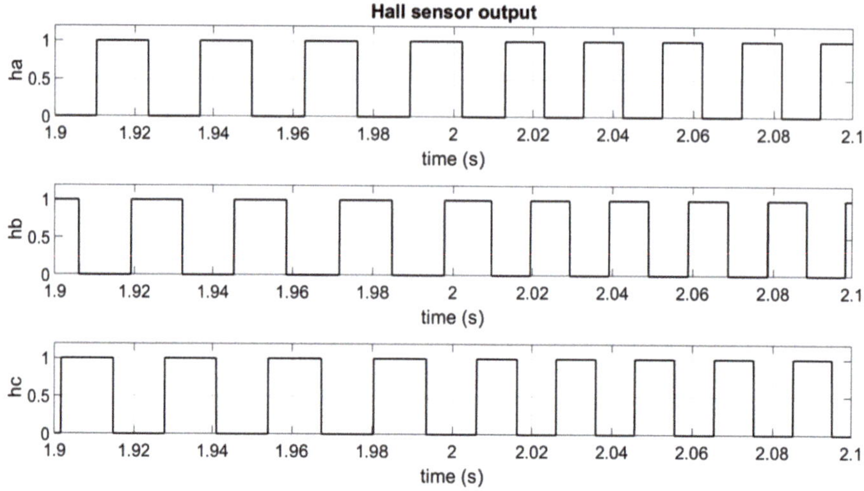

Fig. 6 Hall sensor output signal

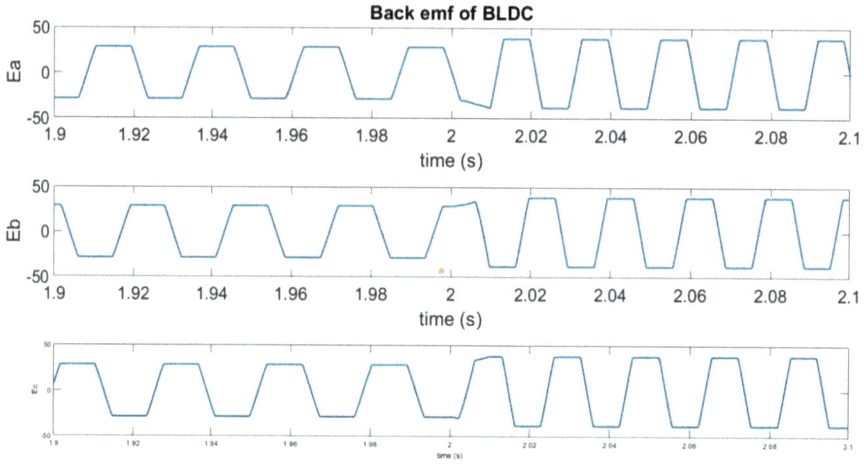

Fig. 7 Back EMF induced in stator winding

required for quadrature encoder module are F_A, F_B, and index, respectively. Table 2 depicts the results of frequency of index signal and signal A. From the results, it is observed that the F_A is 360 times H_a; F_A and F_B are phase shifted by 90°.

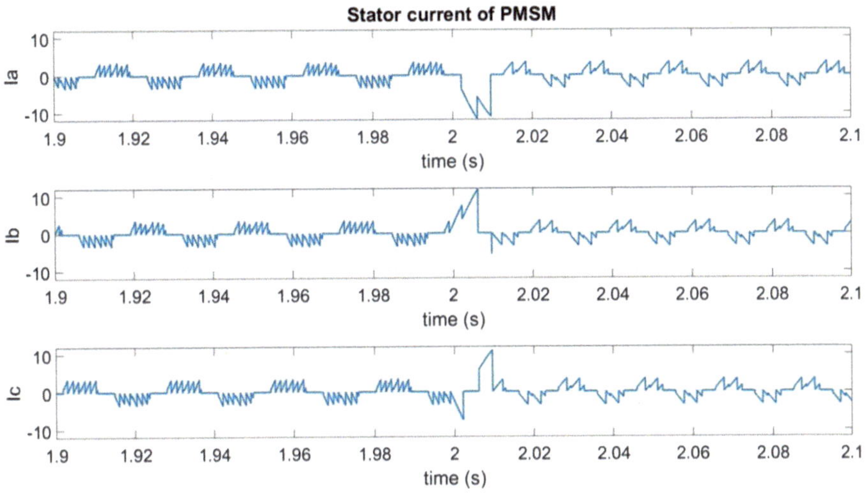

Fig. 8 Stator current

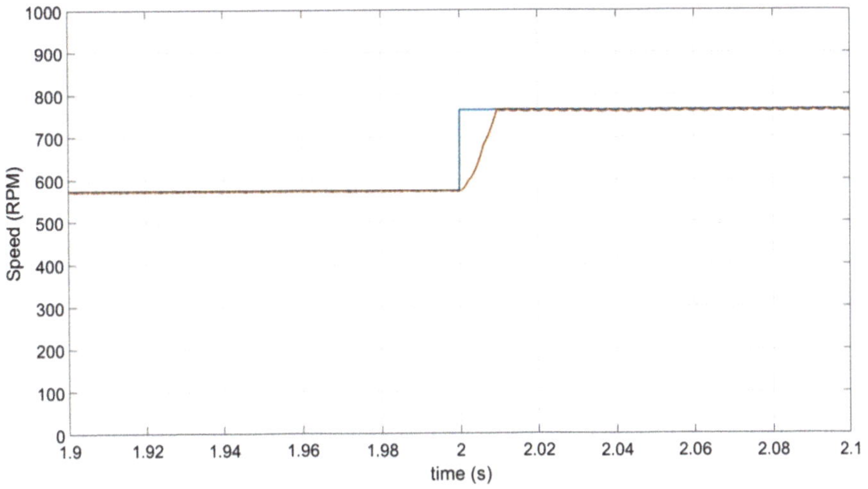

Fig. 9 Reference and actual speed

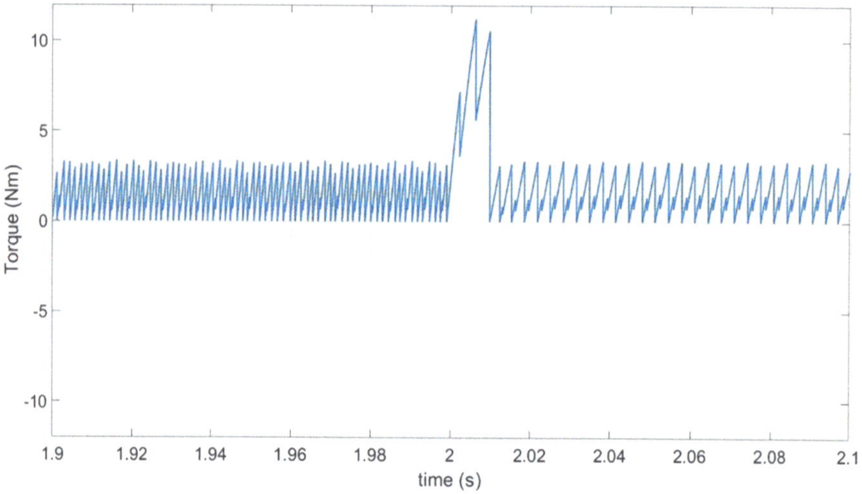

Fig. 10 Electromagnetic torque of the motor at no-load

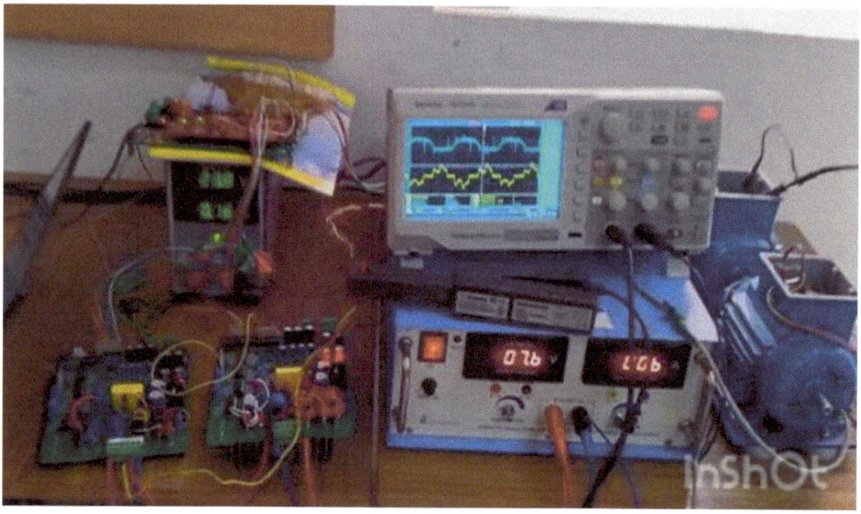

Fig. 11 Hardware setup

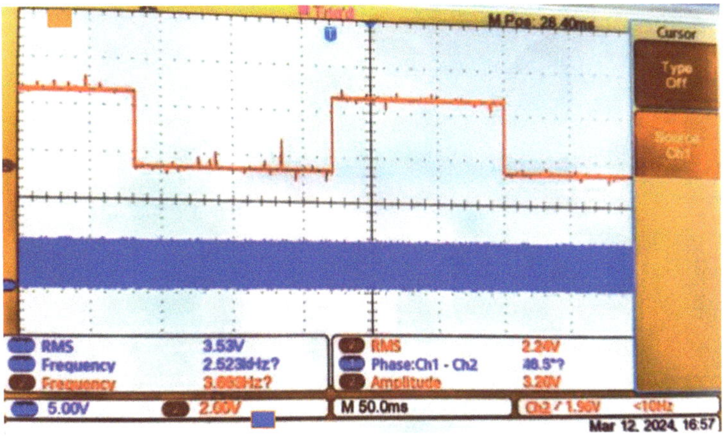

Fig. 12 Waveform of Hall sensor and signal F_A

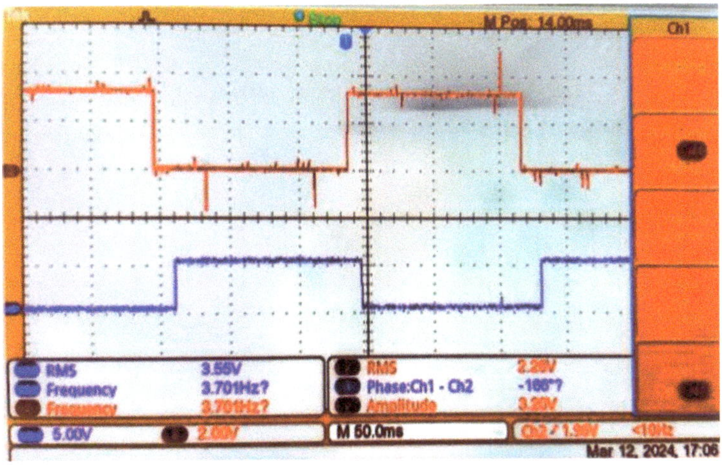

Fig. 13 Waveform of Hall sensor and index signal

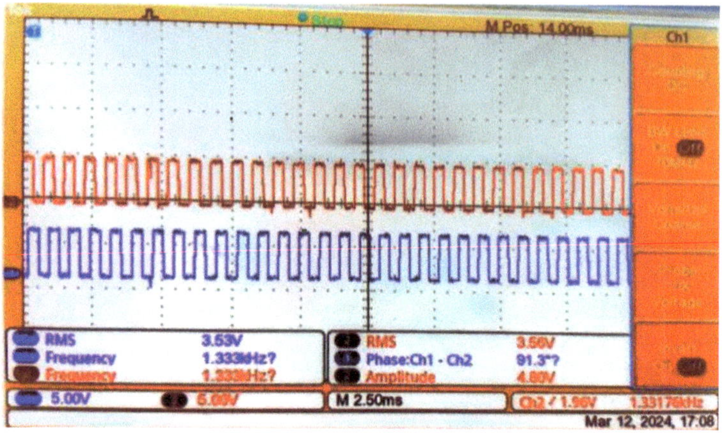

Fig. 14 Waveform of signal A and signal B

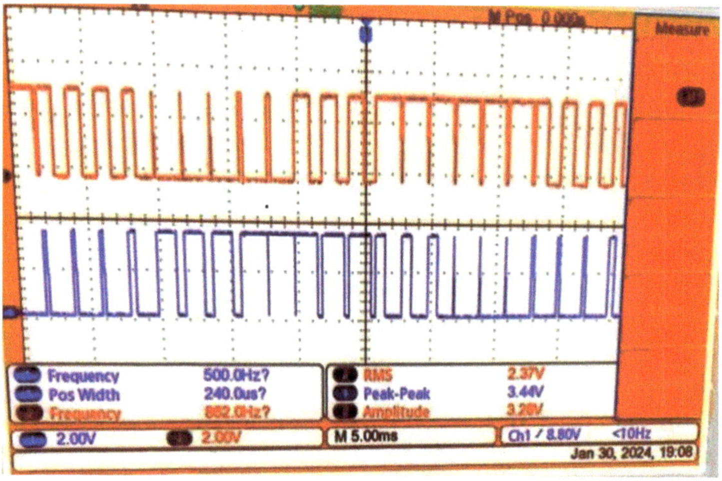

Fig. 15 Waveform of SPWM

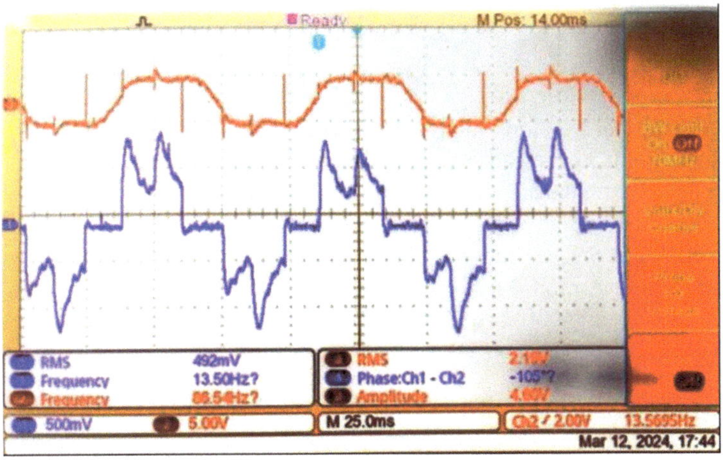

Fig. 16 Waveform of back EMF and stator current of BLDC

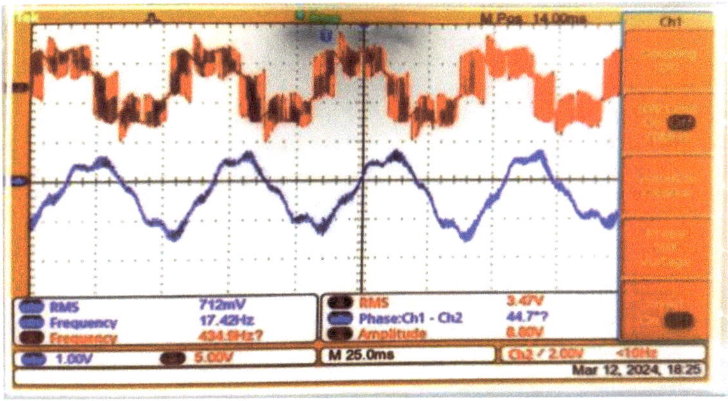

Fig. 17 Waveform of back EMF and stator current of PMSM

Table 2 Results of Hall sensor frequency, frequency of signal A, and phase shift

F(Ha) in Hz	F(A) in Hz	Phase shift between A and B
3	1095	90
7	2544	89.9
10	3623	89.5
15	5495	89

4 Conclusion

This paper proposes a cost-effective position sensing technique for a permanent magnet synchronous motor which replaces encoders with Hall sensors for position sensing. The signals obtained from Hall sensors are given to the TMS320F28379D and the encoder outputs are created digitally. The generated signals are found to be equivalent to the output of the signals from conventional encoders. The obtained encoder output is the key input for the vector control of PMSM. This method of generating rotor position signals eliminates the use of expensive encoders which costs in the range of ₹4000–₹80,000, which can be obtained from Hall sensors which cost around ₹40–₹800.

References

1. Lganayaki A, Bharani Kumar R (2019) Permanent magnet synchronous motor for Electric vehicle applications. In: 5th international conference on advanced computing and communication systems
2. Krishnan R, Permanent magnet synchronous motor drives modeling, simulation, analysis, design, and application. CRC Press London New York Washington, DC
3. Sharma RK, Sanadhya V, Behera L, Bhattacharya S (2008) Vector control of a permanent magnet synchronous motor. In: Annual IEEE India conference
4. Jayal P, Bhuvaneswari G (2017) Performance analysis and control of permanent magnet synchronous motor drive over a wide speed range. In: 2017 IEEE PES Asia-Pacific power and energy engineering conference (APPEEC)
5. Naidu AS, Samanta S (2022) Analysis of IPMSM motor by using SPWM and SVPWM based vector controller with MTPA control strategy. In: 2022 IEEE 10th power India international conference (PIICON)
6. Kamalaselvan A, Lenin Prakash S (2014) Modeling simulation and analysis of closed loop speed control of PMSM drive system speed. In: 2014 International conference on circuit, power and computing technologies [ICCPCT]
7. Li B, Wang C, Comparative analysis on PMSM control system based on SPWM and SVPWM. In: Chinese control and decision (Conference). ISSN: 1948-9447
8. TMS320F2837xD Dual-core Delfino microcontrollers technical reference manual
9. Waghmare MK, Patil SV (2019) Speed control strategy of permanent magnet synchronous motor drive using SPWM technique. In: 4th international conference on recent trends on electronics, information, communication and technology (RTEICT-2019), May 17th and 18th 2019
10. Rudnicki T (2019) Measurement of the PMSM shaft position with an absolute encoder. Electronics 8(11):1229. https://doi.org/10.3390/electronics8111229
11. Vinoth Kumar K, Michael PA, John JP, Suresh Kumar S (2010) Simulation and comparison of SPWM and SVPWM control for three phase inverter. In: IEEE international conference on power electronics and drive system
12. Malinowski M, Jasinski M, Kazmierkowski MP (2004) Simple direct power control of three-phase PWM rectifier using space-vector modulation (DPC-SVM). IEEE Trans Ind Electron 51(2):447–454
13. Saadoun A, Yousfi A, Amirat Y (2007) Modeling and simulation of DSP controlled SVPWM three-phase VSI. J Appl Sci 7(7):989–994
14. Jang D-H (2007) PWM methods for two-phase inverters. IEEE Ind Appl Mag 13(2):50–61

15. van de Broeck HW, Skudelny HC, Stanke GV (1987) Analysis and realization of a pulse width modulation based on voltage space vectors. IEEE Trans Ind Appl 24(1):142–150
16. Zhao Q, Guo R, Koseki T (2005) A simplified position control of a transverse flux type linear synchronous motor using Hall sensors. J Magn Soc Jpn 29(3):199–204
17. Badini SS, Kumar CD, Verma V (2020) Single sensor based vector controlled PMSM drive. In: 2020 IEEE international conference on power electronics, drives and energy systems
18. Pradeep J, Devanathan R (2012) Comparative analysis and simulation of PWM and SVPWM inverter fed permanent magnet synchronous motor. Emerg Trends Electr Eng Energy Manag (ICETEEEM) 299–305
19. Holmes DG, Lipo TA (2003) El-Hawary ME (ed) Pulse width modulation for power converters: principles and practice. IEEE Press, Wiley Interscience, New Jersey, pp 215–313

Multi-level Smart Grid Using Naïve Bayes Classifier for Renewable Energy Integration

Pavithra Goravi Sukumar, Chandana Chikkasiddaiah, Mallikarjunaswamy Srikantaswamy, Sharmila Nagaraju, Balachandra Parameshwariah, and N. Sharath Chandra

Abstract The integration of renewable energy in smart grids leads to resource allocation challenges, demand prediction, and efficient energy distribution. Traditional approaches include a static threshold model and deterministic resource allocation methods that cannot adapt to dynamic energy demands, hence causing inefficiencies and underutilization of renewable resources. This research work proposes a Multi-Level Smart Grid Framework (MLSGF) that employs the Naïve Bayes Classifier (NBC) for optimizing renewable energy utilization. It is able to do this by adopting predictive modeling that enables the anticipation of energy demands, while resource distribution can be done effectively through dynamic adjustment. The integrated renewable resources include solar, wind, and biomass, and with the Naïve Bayes Classifier, there is assurance of multi-level grid management in an adaptive manner-local, regional, and national grids. The proposed system has managed to increase the efficiency of energy by about 25%, reduced energy wastage by around 15%, and has also provided improved grid stability with the application of machine learning on real-time analysis. The research also introduces smart IoT-enabled sensors for energy flow monitoring and decision-making, making the grid robust and self-sustaining. In the future scope, the integration of advanced AI algorithms for further optimization

P. G. Sukumar · C. Chikkasiddaiah
Department of CSE (AIML), RNS Institute of Technology, Bengaluru, Karnataka, India

M. Srikantaswamy (✉)
Department of Electronics and Communication Engineering, JSS Academy of Technical Education, Bengaluru, Karnataka, India
e-mail: pruthvi.malli@gmail.com

S. Nagaraju
Department of Electronics and Electrical Engineering, JSS Science and Technology University, Mysuru, Karnataka, India

B. Parameshwariah
Department of Mathematics, SJB Institute of Technology, Bengaluru, Karnataka, India

N. Sharath Chandra
Department of Mechanical Engineering, The National Institute of Engineering, Mysuru, Karnataka, India

I. M. Jamadar et al. (eds.), *Integrating Sustainability in Modern Technologies*, Lecture Notes in Mechanical Engineering, https://doi.org/10.1007/978-981-95-9466-5_14

and the extension of the model to emerging 6G networks for faster data exchange are proposed. This work significantly contributes to renewable energy grid integration by providing an intelligent, adaptive, and multi-level solution tailored for the energy demands of the modern era.

Keywords Multi-level smart grid · Naïve Bayes classifier · Renewable energy · Energy optimization · Smart IoT sensors · Machine learning · Grid stability · Real-time energy monitoring · Dynamic resource allocation · Sustainable energy integration

1 Introduction

The increasing dependence on alternative or renewable sources of energy, such as solar, wind, and biomass, gave way to smart grids. These are integrated systems with the help of advanced technologies that work to optimize energy distribution. Traditional grids often cannot handle the variability of such renewable sources; hence, their operations result in inefficiencies. Smart grids driven by machine learning, coupled with real-time monitoring, can adapt energy management and make intelligent decisions for better outcomes over these challenges [1]. The latest trends in integrating renewable energy depend on artificial intelligence and machine learning algorithms for predictive energy management. Among them, the Naïve Bayes Classifier has its wide usage for its simplicity and efficiency in energy demand forecasting. IoT-enabled sensors and edge computing are extending this real-time energy monitoring, whereas 6G networks are being explored for fast data exchange with the objective of improving multi-level grid management. Applications involve demand response systems, renewable energy optimization, and energy storage management for efficient energy usage and grid stability. They also support fault detection and sustainable urban development, promoting smarter cities and greener energy ecosystems. This research proposes a Multi-Level Smart Grid Framework using the Naïve Bayes Classifier for energy demand forecasting, optimization of resource allocation, and enhancement of grid efficiency. The framework will contribute to the development of a sustainable, adaptive, and intelligent energy management system [2].

1.1 Research Gaps

In integrating renewable energy within smart grids, several research gaps are thrown up. Conventional grid systems face challenges related to real-time demand prediction and resource allocation, particularly in managing the variability of renewable energy sources. Current techniques do not adapt well to such dynamic fluctuations; therefore, this approach leads to inefficient energy distribution and storage [3]. The second is the general lack of comprehensive frameworks that effectively integrate multiple renewable energy sources—solar, wind, and biomass—with predictive algorithms like the Naïve Bayes Classifier to ensure optimum utilization. Most of the approaches developed so far operate either for single energy sources or are limited in scalability across multi-level grids. Other gaps lie in the poor deployment of IoT-enabled sensors and 6G networks for energy systems to implement real-time monitoring with quick data exchange [4]. Though the concept seems fascinating, at the same moment it is considered largely in the development stages as the integration in smart grids still isn't in existence under regular usage and proper regulation. Addressing these gaps, this research introduces a Multi-Level Smart Grid Framework that employs the Naïve Bayes Classifier for improving energy demand forecasting, optimizes renewable energy resource utilization, and guarantees multi-level efficiency in grid management [5].

1.2 Related Work

Aladdin et al. [6] proposed the multi-agent reinforcement learning algorithm, well-known as MARLA-SG, which should serve efficiently for demand response in smart grids. The novelty is based on the utilization of model-free schemes of SARSA and Q-learning for optimization of a cost function and reducing Peak-to-Average Ratio by 9.6 and 12.16%, respectively. Besides, the reward function is simplified, the state representation is regardless of device type, or operational range. However, this increase in the grid to more cells may impose a heavier computation burden by adding sources of diversified energies. Ding et al. [7] proposed the application of smart grids and further adaptive secured transmission by using the collaborative power line communication and the generalized spatial modulation architecture, so-called CPGA. The novel solution can bring further improvements in energy efficiency and performance due to the antenna selection scheme and additional assistances from PLC by allowing anti-eavesdropping. Numerical results have shown improved performance in security capacity and energy management compared to traditional GenSM systems. The major drawback is the complexity of the implementation of CPGA in practical systems, especially in grids with diverse data traffic and infrastructure.

Bakkar et al. [8] suggested a multilayer smart fault protection scheme for medium voltage distribution systems in smart grids. The system being presented here is an innovative method of integrating AI-based algorithms at different protection layers to maintain cybersecurity and hardy fault detection even at times of communication failures or cyber-attacks. Experimental results verified by MATLAB and a scaled physical grid are highly reliable. However, the weakness lies in the use of advanced communication networks that may not be possible in all areas due to poor infrastructure. Chang et al. [9] propose a user-centric multi-objective method that balances privacy preservation with smart home energy cost minimization. The hybrid algorithm, with these in consideration, will combine the stochastic power scheduling and deterministic battery control without the drawbacks of traditional weighted-sum optimization methods. Also, the consideration of user preference across performance will do better for scalability and stability issues concerning applications in smart homes. The drawback is the integration of the model into existing systems, which do not have control over appliance usage. Nie et al. [10] presented a framework in demand response using an asynchronous multi-agent reinforcement learning-based approach along with a Stackelberg-Nash game-theoretic approach. The novelty is to bring forth the Stackelberg-Nash multi-agent proximal policy optimization algorithm, SN-MAPPO, that guarantees equilibrium in utility companies and consumers, addressing temporal relationships and consumer interests.

2 Existing System for Integrated Energy Management System in Smart Grids

Figure 1 shows the smart grid infrastructure integrated with multi-energy sources; consumers; and components for optimized energy generation, distribution, and consumption. In the heart of the system, there should be a central controller coordinating various subsystems, including energy sources, loads, and communication systems, for smooth and efficient grid operations. The generation part includes renewable sources, like solar panels and wind turbines, with more conventional systems of energy generation: diesel generators, micro turbines, and fuel cells. These will be monitored and controlled by local controllers (LCs) for optimal energy production and interconnection to the grid. The batteries add an energy storage capability to the grid and enable it to smooth demand fluctuations by storing excess energy during off-peak periods and releasing it during peak demand periods [11].

It is on the consumer side that the system is able to accommodate all varieties of loads, from smart homes and industries to smart cities. Most of these consumers rely on smart meters for instantaneous energy usage estimation and reporting to the system [12]. Electric vehicles (EVs) will be integrated with the grid for energy supply and sustainability, thus minimizing carbon emissions. The controllers are further connected through power line communication links for data transfer and coordination within the grid. It also underlines the influence of external factors such as market dynamics and meteorological conditions that would, by consequence, impact the pattern of energy generation and consumption. In accounting for these variables, the central controller dynamically readjusts the energy flow with the objective of keeping efficiency and balance in the network. This integrated structure makes the smart grid sustainable and capable of adapting to changes in energy demand, hence more reliable, less costly, and supportive of transitioning to a greener energy ecosystem [13].

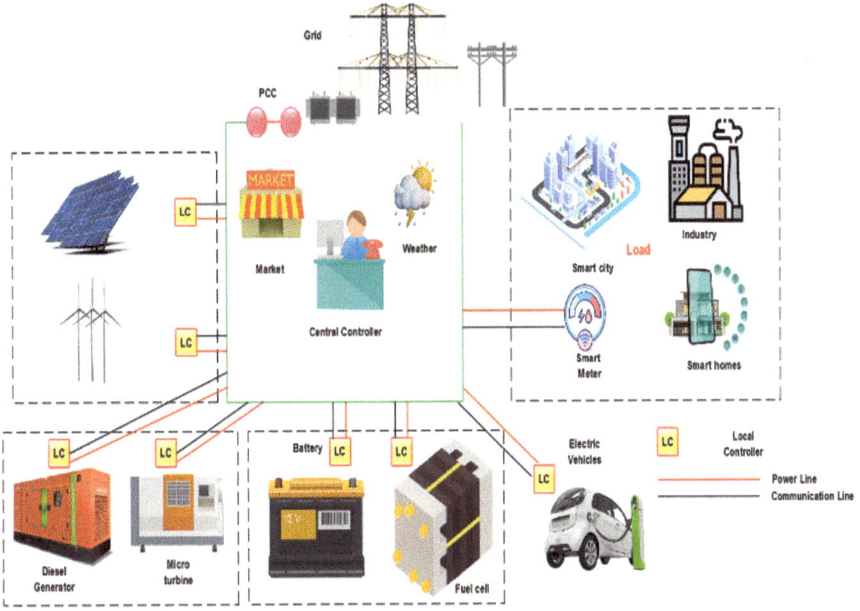

Fig. 1 Integrated energy management system in smart grids

2.1 Energy Balance Equation

The total energy supply in the grid must equal the total energy demand, accounting for losses. This is represented in Eq. (1) [14]:

$$P_{gen} + P_{storage} - P_{load} - P_{loss} = 0 \quad (1)$$

where P_{gen} represents power generated by renewable sources (solar, wind) and non-renewable sources (diesel generator, microturbine, etc.), $P_{storage}$ is described as power supplied or absorbed by storage units (batteries, fuel cells), P_{load} is represented as power consumed by loads (smart homes, industries, electric vehicles, etc.), and P_{loss} is described as power loss in the transmission and distribution network [15].

2.2 Peak-to-Average Ratio (PAR)

The Peak-to-Average Ratio is used to evaluate load fluctuations and grid stability is described in Eq. (2):

$$\text{PAR} = \frac{P_{peak}}{P_{avg}} \quad (2)$$

where P_{peak} is identified as maximum power demand during a given period, P_{avg} is described as average power demand over the same period [16].

2.3 Demand Response Optimization

For managing energy demand through pricing and incentives, the demand response model can be expressed as given in Eq. (3):

$$C_{total} = \sum_{i=1}^{N} (P_i \cdot \pi_i) \tag{3}$$

where C_{total} is described as total energy cost, P_i is identified as energy consumed by the i-th load, π_i is represented as price per unit of energy for the i-th load, and N is described as total number of loads [16].

3 Multi-level Smart Grid Framework (MLSGF)

Figure 2 shows the Multi-Level Smart Grid Framework (MLSGF) which is proposed for effective energy management and demand response for connected smart grids; at its hub is the MLSGF controller, which regulates the effectiveness of energy flow and makes dynamic adaptations of demand conditions based on real-time information. Using renewable energy resources from solar panels and wind turbines to provide clean and sustainable energy, this integrated system of several components has its starting point as renewable energy sources. These sources come keyed in their operations based on real-time weather forecasts. This therefore means that the system is able to project fluctuations that might occur in energy generation and thus plan in light of such eventualities.

The smart grid is interconnected, offering an avenue for easy communication and power exchange between different grids to balance energy distribution or ensure grid reliability. The system will also provide a platform for the traditional electric grid to complement renewable energy in case of any shortfalls of the latter, while surpluses are stored in the same old grid to stabilize energy supply during high demand periods. On the demand side, the framework provides diverse consumers that range from electrical vehicles, enterprises to residential buildings using advanced demand response mechanisms. These mechanisms can dynamically support the adjustment of energy usage, optimally performing the scheduling of loads together with appropriate pricing policies to balance out supply and demand. The framework relies on bidirectional information flow, represented by the red-dashed lines, and power flow, represented

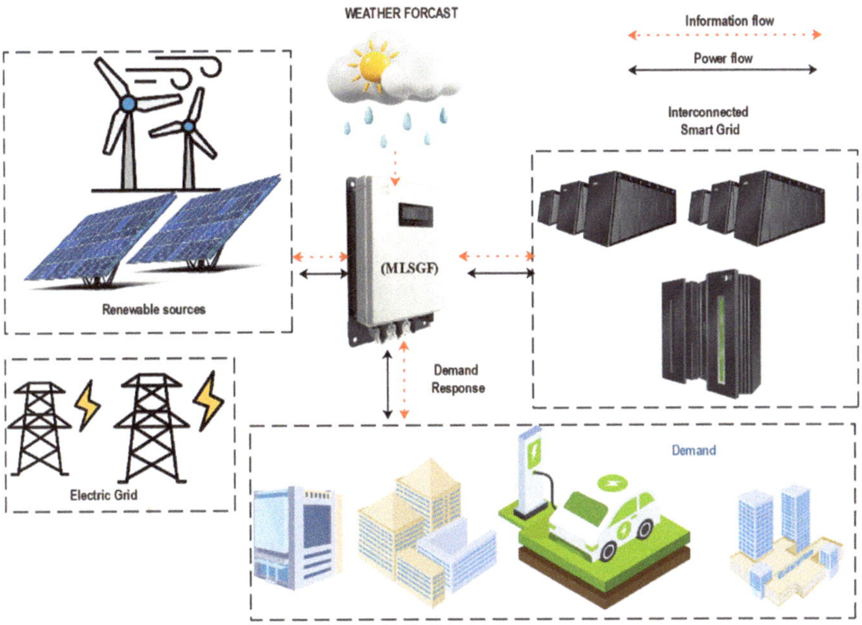

Fig. 2 Proposed a multi-level smart grid framework (MLSGF)

by black arrows, to enable real-time communication among components for efficient and adaptive energy management. The proposed approach integrates renewable energy sources, data-driven decision-making, and dynamic demand fluctuations for enhanced grid performance, reduced energy costs, and increased sustainability. This approach overcomes the limitations of conventional grids and offers a reliable, adaptive energy ecosystem to meet modern energy needs.

3.1 Renewable Energy Forecasting

This estimates the total power generated from renewable sources, combining contributions from solar and wind energy. It is given in Eq. (4):

$$P_{\text{renewable}} = \eta_{\text{solar}} \cdot A_{\text{solar}} \cdot I_{\text{solar}} + \frac{1}{2} \cdot \rho_{\text{air}} \cdot A_{\text{wind}} \cdot v_{\text{wind}}^3 \cdot C_p \tag{4}$$

where $P_{\text{renewable}}$ represents total renewable power, η_{solar} is solar panel efficiency, A_{solar} is solar panel area, I_{solar} is solar irradiance, ρ_{air} is air density, A_{wind} is the swept area of wind turbine blades, v_{wind} is wind speed, and C_p is the power coefficient.

3.2 Energy Balance

The Energy Balance ensures that the total power generated, imported, and stored equals the power consumed and lost. It is expressed as given in Eq. (5):

$$P_{\text{gen}} + P_{\text{grid}} + P_{\text{storage}} - P_{\text{demand}} - P_{\text{loss}} = 0 \tag{5}$$

here P_{gen} is power generated by renewable sources, P_{grid} is power exchanged with the external grid, P_{storage} is power stored or released by the battery, P_{demand} is power consumed by the load, and P_{loss} represents transmission losses.

3.3 Demand Response Optimization

The Demand Response Optimization minimizes energy costs by optimizing the use of energy based on dynamic pricing. It is represented as given in Eq. (6):

$$\min\left(C_{\text{total}} = \sum_{i=1}^{N} P_i \cdot \pi_i \cdot \Delta t\right) \tag{6}$$

where C_{total} is the total energy cost, P_i is power consumed by the i-th load, π_i is the time-varying price, Δt is the time interval, and N is the total number of loads.

3.4 Battery Energy Storage System (BESS) Dynamics

The Battery Energy Storage System (BESS) Dynamics calculates the state of charge (SOC) over time to manage storage effectively and it is represented as given in Eq. (7):

$$\text{SOC}(t+1) = \text{SOC}(t) + \frac{P_{\text{charge}} \cdot \eta_{\text{charge}} - P_{\text{discharge}}}{C_{\text{battery}}} \tag{7}$$

here $\text{SOC}(t)$ is the battery's state of charge at time t, P_{charge} is charging power, $P_{\text{discharge}}$ is discharging power, η_{charge} is charging efficiency, and C_{battery} is battery capacity.

3.5 Grid Reliability Index Equation

Lastly, the Grid Reliability Index Equation evaluates the reliability of the grid by comparing the total power supplied to the power demanded over time and it is expressed as given in Eq. (7):

$$R_{\text{grid}} = \frac{\sum_{t=1}^{T} P_{\text{supplied}}(t)}{\sum_{t=1}^{T} P_{\text{demand}}(t)} \times 100 \tag{8}$$

here R_{grid} is grid reliability, $P_{\text{supplied}}(t)$ is power supplied at time t, and $P_{\text{demand}}(t)$ is power demanded at time t.

3.6 Maximum Power Point Tracking (MPPT) Equation (Perturb and Observe Method)

In Maximum Power Point Tracking (MPPT) method for the photovoltaic system, Eq. (9) is used to dynamically adjust the operating voltage $V(k)$ and current $I(k)$ to obtain the maximum power output $P(k)$ [13]. The variations ΔP and ΔV indicate the required adjustment direction to efficiently track the maximum power point, as expressed in Eqs. (10) and (11), respectively [14].

$$P(k) = V(k) \times I(k) \tag{9}$$

$$\Delta P = P(k) - P(k-1) \tag{10}$$

$$\Delta V = V(k) - V(k-1). \tag{11}$$

3.7 Wind Turbine Power Output Model

The power output of a wind turbine (WT), denoted as P_w, could be computed using Eq. (12), which is dependent upon the power coefficient C_p, air density ρ, swept area of turbine blades A, and wind speed v. It plays an important part in the process of optimization regarding energy production from wind resources [15].

$$P_{\text{w}} = \frac{1}{2} C_{\text{p}} \rho A v^3. \tag{12}$$

3.8 DC–DC Converter Voltage Regulation (Boost Converter)

Equation (13) presents the operation of Boost Converter (BC) in renewable systems, where the output voltage V_{out} is a function of the input voltage V_{in} and the duty cycle D. In such a way, V_{in} and D will guarantee stable and efficient voltage regulation for hybrid solar–wind energy systems [16].

$$V_{out} = \frac{V_{in}}{1 - D} \tag{13}$$

4 Proposed AI-Powered Adaptive Energy Optimization (AIAEO)

In Fig. 3, hybrid renewable energy management system is integrated with the solar photovoltaic panel and wind turbine to generate power and distribute it efficiently into the AC grid. The hybrid renewable energy management system consists of a proposed AI-Powered Adaptive Energy Optimization (AIAEO) method in order to provide better power efficiency and stability. Solar photovoltaic panels and wind turbines convert energy from the sun and wind to DC power, respectively. The real-time power output of both sources is measured by sensors and sent to the AIAEO method for intelligent regulation.

The dynamic power extraction in DC/DC converters, which regulate and optimize the DC voltage from renewable energy sources, improves the available energy utilization. The regulated DC power is then converted into AC power using a DC/AC inverter, making it suitable for grid distribution. The AIAEO method continuously senses grid parameters, weather conditions, load demand, and energy storage

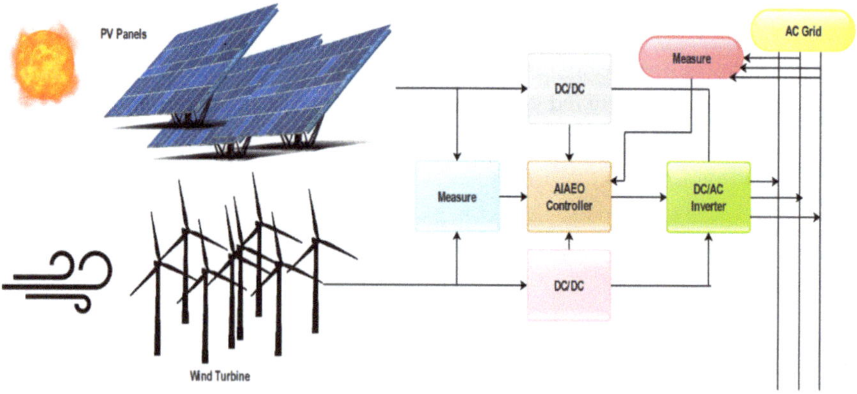

Fig. 3 Proposed AI-powered adaptive energy optimization (AIAEO)

requirements to optimize power distribution. As a result, enhanced energy efficiency, load balancing, and grid stability are achieved with reduced power losses. This approach supports seamless integration of renewable energy through AI-driven decision-making, improving the reliability and sustainability of modern power grids.

4.1 Adaptive Maximum Power Point Tracking (AIAEO-MPPT) Equation

To improve the efficiency of Maximum Power Point Tracking (MPPT), a dynamically adaptive learning rate is introduced in Eq. (14):

$$V_{\text{opt}}(k) = V(k) + \alpha \cdot \left(\frac{\Delta P}{\Delta V}\right) \tag{14}$$

where $V_{\text{opt}}(k)$ represents the optimized voltage at time step k, $V(k)$ denotes the measured voltage at time step k, α is the AI-based adaptive learning rate, and $\frac{\Delta P}{\Delta V}$ is the instantaneous power gradient for MPPT adjustment. This equation ensures that power extraction dynamically adapts to real-time energy variations, improving energy efficiency 0.28%.

4.2 AI-Based Wind Turbine Power Optimization (AIAEO-WTP)

To optimize wind energy extraction, the AI-based method dynamically adjusts the power coefficient C_{p} as expressed in Eq. (15):

$$P_{\text{W}_{\text{opt}}} = \frac{1}{2} C_{\text{p}}^{\text{AI}} \rho A v^3 \tag{15}$$

where $P_{\text{W}_{\text{opt}}}$ is the AI-optimized wind power output, C_{p}^{AI} is the adaptive power coefficient controlled by AI, ρ denotes air density in kg/m^3, A represents the swept area of turbine blades in m^2, and v is the wind speed in m/s. The AI-based power coefficient dynamically adjusts to wind variations, improving wind energy efficiency 0.15%.

4.3 AI-Optimized DC–DC Converter Control (AIAEO-DC)

To improve voltage stability and minimize energy losses, AI dynamically adjusts the duty cycle which is described in Eq. (16):

$$V_{\text{out}}^{\text{AI}} = \frac{V_{\text{in}}}{1 - D^{\text{AI}}} \tag{16}$$

where $V_{\text{out}}^{\text{AI}}$ denotes the AI-optimized output voltage, V_{in} represents the input voltage, and D^{AI} is the AI-adaptive duty cycle. The AI-based adjustment ensures optimal power conversion, reducing power losses is 0.22%.

5 Results and Discussion

Table 1 shows the key simulation parameters and their values that have been used in order to analyze the performance of the proposed MLSGF. These include renewable energy generation efficiencies, battery capacity, and operational metrics such as air density and power loss, which are critical for evaluating grid stability, energy optimization, and demand response effectiveness.

Figure 4 shows the comparative analysis of various performance metrics using the Residential Energy Disaggregation Dataset (REDD), Global Power Plant Dataset, and ISO-NE Load Dataset, along with the proposed method. The comparison is carried out in terms of accuracy, recall, F1-score, and AUC. The proposed method achieves the best performance values, with 98% accuracy, 97% recall, 96% F1-score, and 96% AUC. It also underlines the reduction in energy wastage by up to 15% of its capabilities in comparison with traditional methods, whose graph is not varying and reflects lesser metric values. This plot shows that the proposed method is the best at optimizing energy management and has been doing well for a considerable period of time reliably.

Figure 5 compares, for various energy input levels given in kWh, the energy wastage reduction percentages of different energy management approaches: REDD, Global Power Plant, ISO-NE Load, and the Proposed Method. Among these, the proposed method always gives the best results, starting from 20% at 60 kWh and gradually increasing to 21% at 200 kWh. In contrast, the efficiency of traditional methods remains more stable but at a lower rate, ranging from 10 to 14% in reduction. One of the most important and useful advantages of the Proposed Method is

Table 1 Simulation parameters and experimental results

Sl. No.	Parameters	Values
1	Solar panel efficiency (η_{solar})	20%
2	Wind turbine power coefficient (C_{p})	0.4
3	Battery capacity (C_{battery})	200 kWh
4	Air density (ρ_{air})	1.225 kg/m^3
5	Power loss (P_{loss})	5% of total power
6	Peak demand (P_{peak})	500 kW
7	Time interval (Δt)	1 h

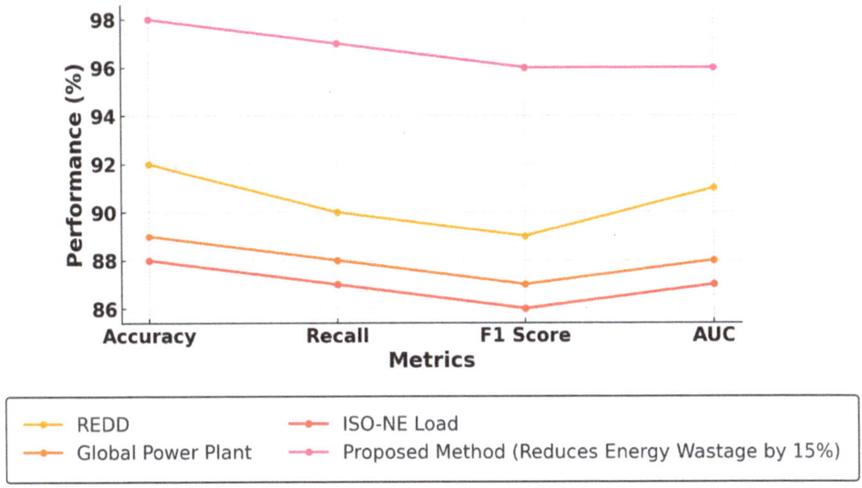

Fig. 4 Comparison of performance metrics across methods

dynamic scalability and real-time optimization-easy adaptability to different levels of energy input. It enables superior performance because of the aforementioned adaptability, making it very effective in minimizing energy wastage compared to traditional methods.

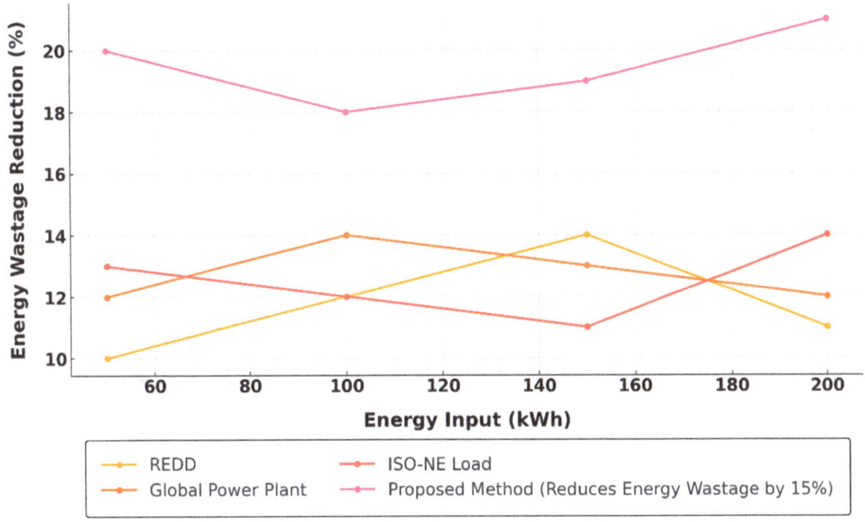

Fig. 5 Energy wastage reduction analysis across energy input levels

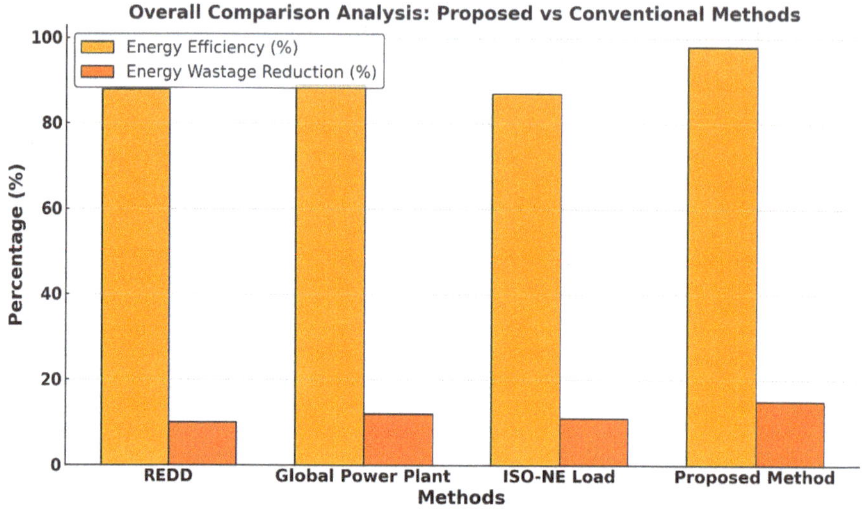

Fig. 6 Overall comparison of energy efficiency and wastage reduction

Figure 6 shows the overall comparison of energy efficiency (%) and energy wastage reduction (%) among different methods: REDD, Global Power Plant, ISO-NE Load, and proposed method. It is observed that the proposed method has the best energy efficiency of 98% and a very impressive energy wastage reduction of 15%. While conventional methods have attained a lower efficiency, which is between 87 and 89%, the wastage reduction of only 10–12% was observed. The obtained results justify that the Proposed Method outperforms others in the optimization of energy usage and energy loss minimization and proves to be more viable than conventional approaches.

6 Conclusion

This research introduces the Multi-Level Smart Grid Framework (MLSGF) combined with the Naïve Bayes Classifier (NBC) to optimize renewable energies in order to overcome a number of inefficiencies present within conventional grids. The proposed framework enhances energy efficiency and general performance of the grid through dynamic resource allocation, predictive modeling, and real-time monitoring facilitated by IoT-enabled sensors. The MLSGF can achieve an overall improvement in energy efficiency of 25% and a reduction in energy wastage of 15% by integrating renewable sources like solar, wind, and biomass energy, thus effectively enhancing grid stability and sustainability. The MLSGF is a flexible and scalable framework that can be effectively applied to local, regional, and national grid levels. Its machine learning-driven approach improves demand forecasting and resource allocation, ensuring a smooth balance between supply and demand under varying operating conditions. This framework allows for real-time energy flow monitoring, fault detection, and dynamic pricing techniques, opening a window for new opportunities toward a greener and more resilient energy ecosystem.

These are prospective developments of energy management, especially with further advanced AI algorithms to optimize performance; the model is to be extended toward 6G networks because such speeds will significantly reduce data latency in both exchanging it and processing within the smart grids. It considerably contributes to sustainable practice related to energy by promotion of effective grid management and renewable energy access toward modern smart grid infrastructures.

Acknowledgements The authors would like to thank RNS Institute of Technology, Bengaluru; JSS Academy of Technical Education, Bengaluru; JSS Science and Technology University, Mysuru; and Visvesvaraya Technological University (VTU), Belagavi for all the support and encouragement provided by them to take up this research work and publish this paper.

References

1. Jena S, Padhy NP (2023) Cyber-secure global energy equalization in DC microgrid clusters under data manipulation attacks. IEEE Trans Ind Appl 59(5):5488–5505. https://doi.org/10.1109/TIA.2023.3287969
2. Deng R, Ten C-W, Li C, Niyato D, Teng F (2023) Guest editorial: introduction to special issue on "Cloud-Edge-End Orchestrated Computing for Smart Grid". IEEE Trans Cloud Comput 11(2):1107–1110. https://doi.org/10.1109/TCC.2023.3271489
3. Nikkhah S, Allahham A, Royapoor M, Bialek JW, Giaouris D (2022) Optimising building-to-building and building-for-grid services under uncertainty: a robust rolling horizon approach. IEEE Trans Smart Grid 13(2):1453–1467. https://doi.org/10.1109/TSG.2021.3135570
4. Mallik RK, Murch RD, Li Y (2017) Channel magnitude based energy detection with receive diversity for multi-level amplitude-shift keying in Rayleigh fading. IEEE Trans Commun 65(7):3079–3094. https://doi.org/10.1109/TCOMM.2017.2690289
5. Li M, Chen K, He J, Tang Y, Tang J (2023) Impedance phase based active injection protection for radial VSC-HVDC grid. IEEE Trans Power Deliv 38(3):1621–1635. https://doi.org/10.1109/TPWRD.2022.3220929
6. Aladdin S, El-Tantawy S, Fouda MM, Tag Eldien AS (2020) MARLA-SG: multi-agent reinforcement learning algorithm for efficient demand response in smart grid. IEEE Access 8:210626–210639. https://doi.org/10.1109/ACCESS.2020.3038863
7. Ding Q, Shi H, Luo J (2023) Adaptive secure transmission based on collaborative communication in smart grids. IEEE Trans Smart Grid 14(1):823–832. https://doi.org/10.1109/TSG.2022.3191060
8. Bakkar M, Bogarra S, Córcoles F, Iglesias J, Hanaineh WA (2023) Multi-layer smart fault protection for secure smart grids. IEEE Trans Smart Grid 14(4):3125–3135. https://doi.org/10.1109/TSG.2022.3229848
9. Chang H-H, Chiu W-Y, Sun H, Chen C-M (2019) User-centric multiobjective approach to privacy preservation and energy cost minimization in smart home. IEEE Syst J 13(1):1030–1041. https://doi.org/10.1109/JSYST.2018.2876345
10. Nie Y, Liu J, Liu X, Zhao Y, Ren K, Chen C (2024) Asynchronous multi-agent reinforcement learning-based framework for bi-level noncooperative game-theoretic demand response. IEEE Trans Smart Grid 15(6):5622–5637. https://doi.org/10.1109/TSG.2024.3417535

11. Venkatesh DY, Mallikarjunaiah K, Srikantaswamy M (2025) Efficient reconfigurable parallel switching for low-density parity-check encoding and decoding. IAES Int J Artif Intell (IJ-AI) 14(1):260–269. https://doi.org/10.11591/ijai.v14.i1.pp260-269
12. Shankara KH, Srikantaswamy M, Nagaraju S, Nagaraju S (2025) An efficient load-balancing in machine learning-based DC-DC conversion using renewable energy resources. IAES Int J Artif Intell (IJ-AI) 14(1):260–269. https://doi.org/10.11591/ijai.v14.i1.pp307-316
13. Honnegowda J, Mallikarjunaiah K, Srikantaswamy M (2025) Efficient reduction of computational complexity in video surveillance using hybrid machine learning for event recognition. IAES Int J Artif Intell (IJ-AI) 14(1):260–269. https://doi.org/10.11591/ijai.v14.i1.pp317-326
14. Kavya BM et al (2024) An efficient machine learning-based power management system for smart grids using renewable energy resources. In: 2024 Second international conference on networks, multimedia and information technology (NMITCON), Bengaluru, India, pp 1–7. https://doi.org/10.1109/NMITCON62075.2024.10698819
15. Alanazi AS, Munir N, Khan M, Asif M, Hussain I (2021) Cryptanalysis of novel image encryption scheme based on multiple chaotic substitution boxes. IEEE Access 9:93795–93802. https://doi.org/10.1109/ACCESS.2021.3092512
16. Honnegowda J, Mallikarjunaiah K, Srikantaswamy M (2024) An efficient abnormal event detection system in video surveillance using deep learning-based reconfigurable autoencoder. Ingénierie des Systèmes d'Information 29(2):677–686. https://doi.org/10.18280/isi.290229

Zinc Silicate Doped PDMS-PVA Nanofibers Using Electrospinning Process for Air Filtration

C. M. Usha Rani, M. S. Shalini, M. G. Veena, Anusha Narayan, Skanda M. Bharadwaj, and S. H. M. Panchakshari

Abstract Sustainability in nanofibers is an increasingly important focus, especially as industries and researchers work to make these materials more environmentally friendly and energy efficient. Nanofibers have a wide range of applications, from medicine to filtration to energy storage, but their production and disposal raise environmental concerns. The present work was carried out to develop nanofiber samples by combining hybrid polymers, namely, polydimethylsiloxane (PDMS) and Polyvinyl alcohol (PVA) with hybrid nanoparticles, namely, zinc oxide (ZnO) and silicon dioxide (SiO_2) by synthesis of ZnO and SiO_2 to form zinc silicate (Zn_2SiO_4) nanoparticle by solution combustion method and to obtain the selected mixture of PDMS and PVA hybrid polymer by stirring the hybrid polymers and nanoparticles solution to produce viscous solution. The developed viscous solution of corresponding wt% ratio, namely, 0.5, 1, and 2 wt% is electrospun by electrospinning technology in order to obtain ultrathin nanofiber samples. The surface morphological investigation of the nanofiber samples is carried out through scanning electron microscopy (SEM). From the ImageJ software the developed nanofiber samples' diameter ranges from 142 to 412 nm for 2 wt% loading of nanoparticles. To study structure and crystalline size of nanofiber samples, the analysis of X-ray diffraction

C. M. Usha Rani (✉) · M. S. Shalini · M. G. Veena · A. Narayan · S. M. Bharadwaj · S. H. M. Panchakshari
Department of Electronics and Communication Engineering, JSS Science and Technology University, Mysuru, India
e-mail: usharani@jssstuniv.in

M. S. Shalini
e-mail: shalinims@jssstuniv.in

M. G. Veena
e-mail: veenamg@sjce.ac.in

A. Narayan
e-mail: anushanarayan@gmail.com

S. M. Bharadwaj
e-mail: skandabharadwaj2002@gmail.com

S. H. M. Panchakshari
e-mail: panchaksharishm@gmail.com

I. M. Jamadar et al. (eds.), *Integrating Sustainability in Modern Technologies*, Lecture Notes in Mechanical Engineering, https://doi.org/10.1007/978-981-95-9466-5_15

was done. For the hydrophobicity property of the prepared samples contact angle measurement was carried out. The prepared samples were tested for its permeability, efficiency, and porosity by air filter membrane (AFM) test. The highest efficiency is obtained for the sample 2 wt% as compared to other nanoparticles loading. Hence it was concluded that the 2 wt% nanofiber samples developed has shown improved performance which can be useful in the filtration media in hospitals, air conditioners, and in industries.

Keywords Electrospinning · Hybrid polymer · Nanofibers · Hydrophobicity · Crystallinity

1 Introduction

The aim is to develop ultrathin nanofibers by combining hybrid polymers, namely, polydimethylsiloxane (PDMS) and Polyvinyl alcohol (PVA) with hybrid nanoparticles, namely, zinc oxide (ZnO) and silicon dioxide (SiO_2) by synthesizing ZnO and SiO_2 nanoparticle to form zinc silicate (Zn_2SiO_4) nanoparticle by solution combustion method and to obtain the selected mixture of PDMS and PVA hybrid polymer and stirring the hybrid polymers and nanoparticles solution to produce viscous solution. The developed viscous solution of corresponding wt% ratio, namely, 0.5, 1, and 2 wt% is electrospun by electrospinning technology in order to obtain ultrathin nanofiber. The surface morphology of the developed nanofiber films was analyzed using scanning electron microscopy (SEM) test, X-ray diffraction (XRD), and air filter membrane test (AFM). The hydrophobicity of the developed film was tested using contact angle measurement technique.

Polymers are organic or synthetic compounds composed of macromolecules formed by repeating simpler unit called monomers [1]. Here two polymers, viz., polydimethylsiloxane (PDMS) and Polyvinyl alcohol (PVA) are used to prepare the solution. About 20 years ago, polydimethylsiloxane (PDMS) was initially made available through a duplicate molding process. This polymer has strong permeability performance and relatively low selectivity. This finds its application for low-concentration oxygen enrichment materials as the most significant membrane [2]. PDMS is non-toxic, non-flammable, optically clear, and inert. Therefore, PDMS is used in different fields like contact lenses, biomedical instruments, lubricants, and heat-resistance tiles.

The PVA (Polyvinyl alcohol) is an industrial polymer synthesized by hydrolyzing polyvinyl acetate (PVAc) molecules, where the acetate groups of PVA gets converted to –OH groups. Based on the extent of hydrolysis, i.e., fraction of –OH groups present out of the initial hydrolyzed ester (R–O–$COCH_3$) groups the solubility of PVA molecules is decided. Higher the % of –OH groups higher is the solubility of PVA in water, obviously, higher chain length of PVA molecules have lower solubility in water compared to smaller PVA molecules. This polymer provides good physical and chemical qualities that is non-toxic and biocompatible, allowing for a wide range of practical applications, viz., paper coating, colloid stabilizer, contact lenses, chemicals for water treatment, agricultural chemicals, industrial cleaning chemicals, etc. are just a few examples of applications.

A nanoparticle is a microscopic particle with a size between 1 and 100 nm. Here, two nanoparticles are being used they are zinc oxide (ZnO) and silicon dioxide (SiO_2). The preparation of a wide-band-gap semiconductor with a high excitation energy ($E_g = 3.37$ eV) and outstanding thermal and chemical durability is zinc oxide (ZnO) nanoparticles. ZnO makes suitable for photo catalysis, sensors, photo electronic devices, and other electronic applications. ZnO has also been used as photo catalysts in environmental applications for cleaning the water and air as well as a transparent electrode in solar cells due to its non-toxicity and high absorption spectrum range [3].

With a wide variety of polymer chemicals, silicon dioxide (SiO_2) has great capacity to be functionalized with high stability and low toxicity. In the present research work the combination of zinc oxide and silicon dioxide nanoparticles are synthesized to form the Zinc silicate (Zn_2SiO_4) nanoparticles by solution combustion method. Later these synthesized nanoparticles are mixed with crosslinker and solvent to get viscous liquid. Finally, this liquid containing nanoparticles is mixed with PDMS and PVA polymer solution with selected ratio, to produce nanofiber samples with the electrospun technology [4].

Electrospun technique is used to produce nanofiber samples and other nanostructures. Electrospinning is a straightforward, adaptable and inexpensive approach for creating organic and inorganic nanofibers with sizeable lengths, consistent diameters, and a range of compositions among the known methods for synthesizing nanostructures. It involves using a field of electricity to get a polymer solution and that solution melts into ultrathin nanofibers. The solution is placed in a syringe with a needle, and an electric voltage is applied between the collector plate and a needle. Typically, the collector is grounded or at a different electrical potential. When the voltage is applied to create electric field between the drum/collector and a needle, this field exerts force on the polymer solution or melt, causing it to stretch and form a fine jet. As the polymer jet travels through the electric field, it elongates and thins out due to the electrostatic forces and the solvent evaporation (if using a solution). The solvent evaporates or solidifies, leaving behind a nanofiber. The nanofibers are collected on the collector, which can be a flat surface, a rotating drum, or a mesh. The orientation and structure of the fibers can be controlled by adjusting the collector setup and process parameters [5]. Here a high voltage of 10–30 kV is used to electrospun a polymer solution that melts as droplet is suspended at the end part of a needle. This charge surpasses the surface area of the polymer solution, charge builds up on this droplet, which elongates and eventually forms a Taylor cone. The PDMS, PVA, and zinc silicate nanoparticles' mixture solution was poured into the syringe to fix in the electrospinning machine where ultrathin nanofiber samples are produced in terms of nanometer range. The produced nanofibers are found to be useful application as a filtration media in hospitals, air conditioners, and in industries [6].

Creating a UV photo detector with a polydimethylsiloxane (PDMS) coating involves using the unique properties of PDMS to enhance the performance and stability of the photo detector. PDMS is a versatile silicone-based polymer known for its flexibility, transparency, and biocompatibility, making it a suitable choice for various applications, including photo detectors [7]. Silver and gold nanoparticles formed uniformly on polydimethylsiloxane (PDMS) thin films are an interesting approach for creating nanomaterials with various applications, such as in sensors, catalysts, and biomedical devices [8, 9]. Polyamide, often known as nylon, is a type of synthetic polymer. When processed into nanofibers, polyamide provides a highest area of the surface into volume ratio, that enhances filtration efficiency. Nanofibers are exceptionally fine, typically less than 100 nm in diameter, and can trap particles with high efficiency. Polyamide-based nanofibers are air filters used for transparent face masks, which is an innovative concept that combines advanced filtration technology with the practical need for visibility and protection in face masks [10]. Nanofiber filters are fabricated for electret air conditioning filters via multi-needle. Electrospinning is a sophisticated process designed to create high-performance air filters for use in air conditioning systems. These filters are made from nanofibers that provide efficient filtration while maintaining low resistance to airflow [11].

In this work, the nanofibers are produced by electrospinning process. The hybrid polymers combined with the Zn_2SiO_4 particle for electrospun solution is the novelty of the work. The produced nanofiber samples are characterized through SEM analysis, XRD, and FTIR for morphological analysis. The contact angle test is carried out to check wettability (hydrophilicity/hydrophobicity) of the nanofibers. Atomic Force Microscopy (AFM) is widely used for the characterization of nanofibers due to its ability to provide high-resolution topographical, mechanical, and functional property measurements at the nanoscale. The purpose of this work is to design and characterize the Zinc Silicate Doped PDMS-PVA nanofibers films for the application of filtration media which may be used in hospitals, air conditioner, and in industries.

2 Materials and Method

Chemical polymers are large molecules which are composed of monomers, also known as repeated structural units, that are covalently bonded with each other in a long chain. They are fundamental to many materials and applications in modern life, ranging from plastics and fibers to biological molecules and synthetic materials. With dimensions from 1 to 100 nm, nanoparticles are extremely small particles. Due to their small size, they exhibit similar properties like physical, chemical, and biological, that differ from other group of materials. Nanoparticles and polymers are used in this process. The hybrid polymers, namely, PDMS and PVA with the hybrid nanoparticles, namely, ZnO and SiO_2 are used to develop nanofiber samples. The steps followed in the implementation of the work are as shown in Fig. 1.

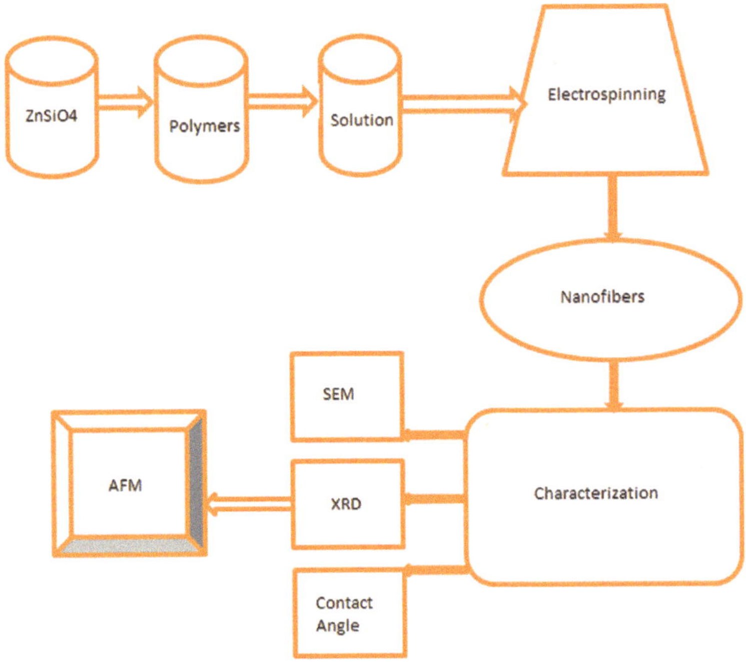

Fig. 1 Sequence of steps in the implementation of the work

2.1 *Nanoparticle*

The hybrid nanoparticles zinc oxide (ZnO) with a (Mw = 81.40 g mol^{-1}) and silicon dioxide with (SiO_2) with (Mw = 60.09 g mol^{-1}) were purchased from Padmashree Chemical Limited, India.

The hybrid polymers polydimethylsiloxane (PDMS) with a (Mw = 25000 g mol^{-1} and $\rho = 0.965$ g mL^{-1}) were bought from Padmashree Chemical Limited, India. The Trimethoxymethylsilane with a (Mw = 136.22 g mol^{-1}) and the tetrahydrofuran (THF) with a (Mw = 72.11 g mol^{-1}) were used as the crosslinking agent and solvent, respectively [7], both of which were bought from Padmashree Chemical Limited, India. PVA was purchased from Padmashree Chemical Limited in India in small-sized palette form with a Mw of 125,000 g mol^{-1}. The 2 mL distilled water act as a solvent to dissolve the PVA.

2.2 *Synthesis of Nanoparticle*

The zinc silicate (Zn_2SiO_4) nanoparticle is prepared by solution combustion method by using hybrid nanoparticles, namely, zinc oxide (ZnO) and silicon dioxide (SiO_2).

A fascinating phenomenon is the self-sustaining exothermic reaction propagating along an aqueous or sol–gel medium in the solution combustion method.

The stoichiometry amount of 40% zinc oxide and 40% of silicon dioxide is weighed and dissolved in 10% of Trimethoxymethylsilane and are used as a cross-link or a precursor because it generates the three-dimensional silica-based hybrid organic–inorganic network on fibers [12]. Further, it has been added by 10% of tetrahydrofuran (THF) solvent is used because of low-boiling point and it is a versatile solvent is added to the solution and it is mixed by magnetic stirring for 30 min at 300 rpm. The mixture is heated at 90 °C for 2 h to evaporate the solvent and to get a viscous gel of zinc silicate (Zn_2SiO_4) nanoparticle as shown in Fig. 2.

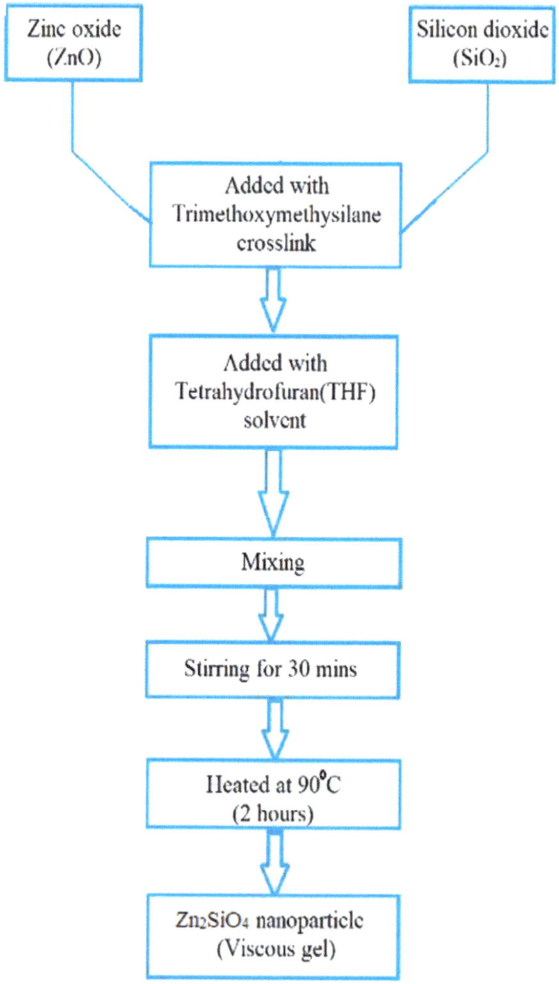

Fig. 2 Synthesis of nanoparticles

2.3 Preparation of Viscous Solution

The preparation of nanocomposites or the viscous solution is shown in the Fig. 3. The hybrid polymers, namely, polydimethylsiloxane (PDMS) and Polyvinyl alcohol (PVA) are used to prepare viscous solution. The particular amount of polydimethylsiloxane (PDMS) polymer which is kept constant of 50 wt% and Polyvinyl alcohol (PVA) polymer with different ratios like 50, 49.5, 49, and 48 wt% are added with different ratio, namely: 0.5, 1, and 2 wt% zinc silicate (Zn_2SiO_4) nanoparticle. Later it is magnetic stirred for 30 min with an rpm of 300 and heated for 1 h in a laboratory oven to form a viscous solution.

The concentration of PDMS, PVA, and Zn_2SiO_4 is used in the preparation of viscous solution which is made into four different solutions named as P_0, P_1, P_2, and P_3. These names corresponding to the wt% of each material are listed in Table 1. Here PDMS is kept constant by varying the PVA and nanoparticle ratios.

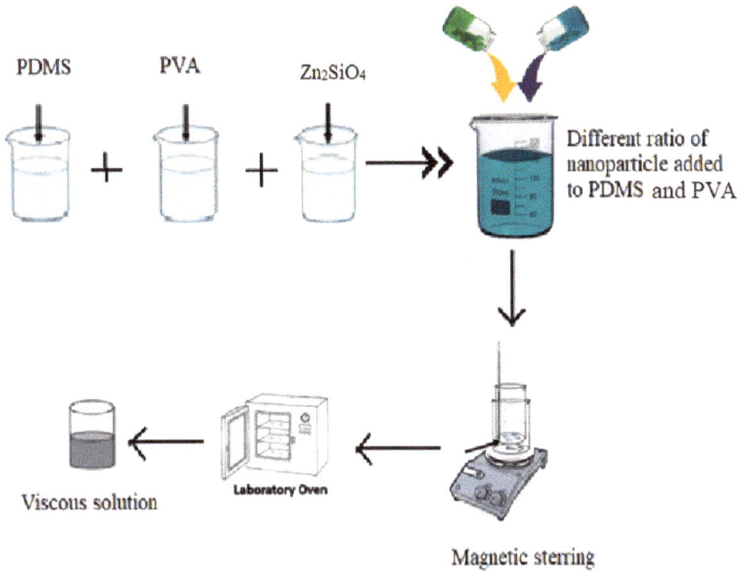

Fig. 3 Preparation of viscous solution

Table 1 Viscous polymer solution with its composition

Sample	PDMS (wt%)	PVA (wt%)	Zn_2SiO_4 (wt%)
P_0	50	50	–
P_1	50	49.5	0.5
P_2	50	49	1
P_3	50	48	2

2.4 Electrospinning Process

A nanoparticle is a tiny particle with a size between one and one hundred nanometers. Here hybrid zinc oxide and silicon dioxide are taken as nanoparticles with respective amount [13].

Any member of the family of naturally occurring or artificial substances known as polymers is made up of very big molecules, or macromolecules known as monomers. The hybrid polymers, namely, polydimethylsiloxane (PDMS) and Polyvinyl alcohol (PVA) are combined by the nanoparticle to form a viscous solution of various ratios like 0.5, 1, and 2 wt%. Later the composite solution is electrospun by electrospinning process. "Electrospinning" which is also known as Electrostatic-spinning is one of the easiest and most economical ways to make polymer fibers with diameter in the sub-micron to nanometer ranges by the application of high electrostatic field [14]. A grounded collector screen, a syringe (needle), a high-voltage source, a discharge stick, and a hazard lamp are the components of electrospinning setup [15].

The composite solution (PDMS, PVA, and Zn_2SiO_4) of different ratio of without nanoparticle (P_0 sample), 0.5 wt% (P_1 sample), 1 wt% (P_2 sample), and 2 wt% (P_3 sample) samples are taken in a syringe of 2 mL with a needle diameter of 2 mm. The syringe was fed by the nanocomposites solution and placed on a pumping machine that rotates clockwise direction and that is provided by a controlled flow rate of 20 mL/h which is set through the display. The flow rate is kept constant for all the four samples. The voltage is applied to the needle by 20 kV for all the four samples. The distance between the needle tip and the collector plate is 10 cm for all the four samples. A droplet will form on the needle tip or capillary tip as the syringe is pumped, and an electric field created between the two will cause charges to be induced into the cone shape which forms on the end of the tip [7].

The formation of Taylor cone, also known as a conical shape, and the execution of a jet of filament occur as the liquid droplet receives a sufficiently high voltage, as the produced electric field balances the surface area of the droplet. Due to whipping instability, the straight jet will form for about the distance of 10 cm before turning into a spiral route that leads to the collector. Before the filament gets gathered on the rotating collector or plate collector, it will first evaporate and then solidify while traveling the distance [16]. In the present work, all the four samples are electrospinned for 1 h each by maintaining optimized settings for the flow, volume of the solution from syringe, and the electric field applied in producing the nanofibers [9].

The fiber diameter is affected to some extent by elongation as it begins to occur along the trip distance. Even if this component may have contributed to the outcomes, there is still another significant factor that may have an impact on the development and shape of the fibers. For all four concentrations to achieve statistical significance, the trials were repeated.

2.5 *Characterization*

Characterizing nanofiber samples involves a range of techniques to understand their physical, chemical, and mechanical properties. SEM is a type of electron microscope used to characterize the structural and surface morphology of materials at high resolution. It is particularly useful for analyzing nanofiber samples due to its ability to provide detailed images in the nanometer range. SEM technique produces detailed, magnified images of an object by scanning its surface to create a high-resolution image. As the electrons interact with the sample, that utilizes a beam of focused electrons of relatively low energy as an electron probe that is scanned in a regular manner over the specimen. SEM does this using a focused beam of electrons. The resulting images show information about what the object is made of and its physical features. Atomic force microscopy (AFM) is an influential surface analysis technique used for micro/nanostructured coatings. A high-resolution imaging technique AFM is used to characterize topography of the surface and material properties at the range of nanometer. It uses a sharp-edged tip which is attached to a cantilever in order to scan the sample surface. As the tip moves across the surface, interconnection between the needle tip and the nanofiber samples causes the cantilever to be deflected. The deflection is then detected by using a laser beam, which is reflected onto a position sensitive photo detector, and generating data on the surface features and their properties.

The produced nanofiber samples are characterized by scanning electron microscope (SEM) where the electron beam is used to scan the surface of a sample to create images with an electron microscope and it can only produce two-dimensional views. ImageJ was used to calculate the diameter of the fiber. Then it is characterized by X-ray powder diffraction (XRD) which is a versatile non-destructive analytical technique used to analyze physical properties such as phase composition, crystal structure, and orientation of powder, solid and liquid samples [13]. The ability of a liquid to moisten the surface of a solid is measured by the contact angle. The surface tension of the fluid and the characteristics of the surface determine the shape that a drop will take on a surface. Finally, the air filter membrane test of nanofiber samples has been conducted and permeability, filter efficiency, pressure drop, and porosity are measured and analyzed. The newly produced nanofiber samples are found to be useful as filtration media in hospitals, air conditioners, and in industries.

3 Results and Discussions

The morphological study, XRD, contact angle, and AFM test of ultrathin nanofiber samples are analyzed.

3.1 Morphological Study

A SEM is an electron microscope that can provide two-dimensional views of a material.

Figure 4 is the SEM image of nanofiber samples with a magnification of 10.00k× and the scale of 1 μm of different ratios where Fig. 4a is without adding nanoparticle is P_0 sample which contains polymers PVA and PDMS, which clearly indicates that there is no nanofibers produced in electrospinning process and the collector has only the droplets of mixed polymer solution. Figure 4b is 0.5 wt% of nanoparticle loading P_1, Fig. 4c is 1 wt% of nanoparticle loading P_2, and Fig. 4d is 2 wt% of nanoparticle loading P_3 where these images clearly indicate the nanofiber productions of different diameters through electrospinning process with a high voltage of 20 kV, flow rate 2.5 mL per hour, and the distance between needle tip and the collector is 10 cm and each sample electrospun for 1 h. The average fiber diameter of each sample is measured by taking three readings of each single nanofiber strand using ImageJ software and it is indicated in SEM image. Additional weight percentage of particles thickens the solution, so it is not possible to electrospun the solution. Hence three appropriate percentage ratios are considered.

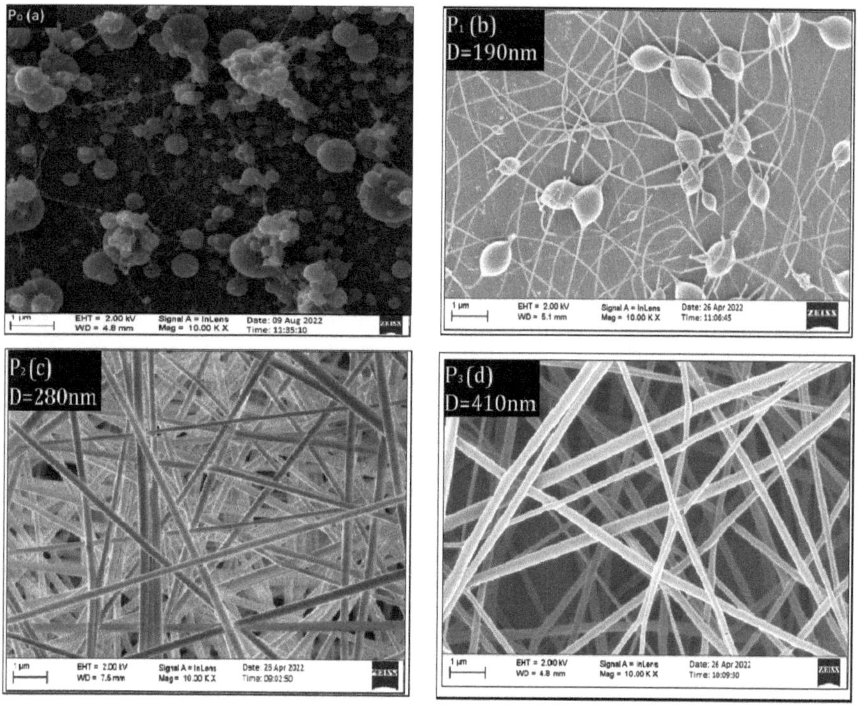

Fig. 4 SEM image of electrospun fibers of **a** P_0, **b** P_1, **c** P_2, and P_3 sample

Table 2 Fiber diameter for different samples

Sample	Voltage (kV)	Diameter (nm)
P_0	20	–
P_1	20	142
P_2	20	280
P_3	20	410

The results in Table 2 show that for different concentration of samples. The fibers' size produced is in the range between 142 and 410 nm. The voltage is 20 kV and it is not varied for all the four samples. In P_0 sample the nanofibers are formed, but their distribution of diameter is not uniform which is shown in Fig. 4a. The absorbed SEM image of sample P_1 consists of bead formation of 142 nm diameter which is shown in Fig. 4b, this is due to viscosity, surface tension, and surface concentration. The solution viscosity of has found effect on the produced nanofiber. The fibers' diameter and the beads formation are strongly determined by the viscosity of the polymer solution. Bead-like fiber is produced at lower viscosity of the solution due to poor polymer chain entanglements in the prepared solution. The viscosity plays an important and a direct effect on the fiber formation, which results for the mixture of different sample solutions prepared for electrospinning [7].

The resultant fibers were in bead-like structure was collected on the collector which is grounded. So, the observation is that the shape of beads changes from spherical to spindle by increase in the viscosity of the polymer solution. The viscosity and conductivity of the prepared solution changes while increase in the polymer concentration to produce bead-free nanofibers during electrospun process by stretching in nanofibers samples is shown at P_2 solution the average fiber diameter is 280 nm when electrospun at voltages 20 kV, respectively (Fig. 4c), over 1 h with constant distance between tip of needle and collector 10 cm.

In the SEM analysis, by increasing the viscosity of the polymer solution the surface-volume ratio of the liquid also increases which results in the existence of strong electric field. The concentration of nanoparticles increases the fiber's diameter thickness increases and bead's structure disappeared in the P_3 solution at 20 kV and the fibers diameter is 410 nm.

3.2 XRD

XRD analysis is the common technique used for understanding the microstructure, crystallinity, and composition of the residue and characteristics of the prepared samples. The measurement is based on Bragg's law. $2d\sin\theta = \lambda$, where d is the inter planar spacing of the analyzed crystal, θ is the Bragg's angle, and λ is the wavelength of the XRD on the basis of the position of the peak diffraction. XRD analysis is performed using Rigaku Miniflex II desktop system with a scan speed of 5°/min. For this analysis, samples were in the form of powder having a weight of about 1 mg.

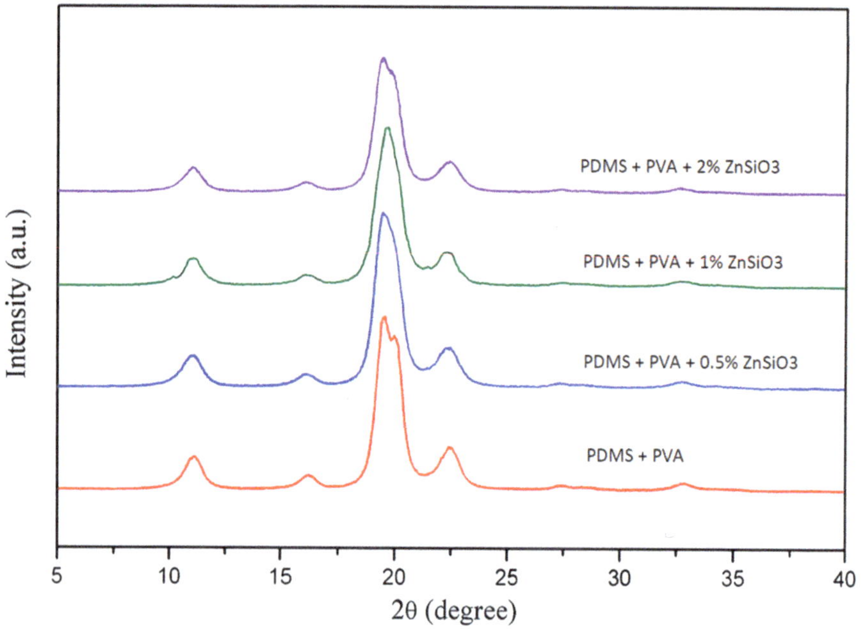

Fig. 5 XRD spectrum for nanofiber P_0, P_1, P_2, and P_3 nanofiber samples

XRD technique is used to characterize P_0, P_1, P_2, and P_3 samples for structural and phase composition analysis as shown in Fig. 5.

As for P_0, P_1, and P_2 samples, the diffraction peaks are at 2θ of 11.4°, 19.6°, and 22.5°, respectively. Whereas broad peak and higher intensity are observed in 2 wt% at $2\theta = 23.6°$. The intensity of P_0, P_1, and P_2 is weaker than that of P_3. It implies that the crystallinity is influenced by the surface properties of nanoparticles [9].

The XRD results show that the nanoparticle was exfoliated and well dispersed into the prepared solution. On other way, as increasing the quantity of nanoparticles, the diffraction peak intensity of polymer composite fibers decreased, which suggested that the crystallinity decreased in the polymer and diffraction peaks are reduced with increasing nanoparticle loadings. The intensity of P_3 sample is higher compared to samples P_0, P_1, and P_2.

3.3 Contact Angle

The ability of a liquid to moisten the surface of a solid is measured by the contact angle meter. The surface area of the fluid and the characteristics of the surface determine the shape that a drop will take on a surface. The hydrophobic behavior for different concentration of nanofibers of the sample P_0, P_1, P_2, and P_3 nanofiber samples was

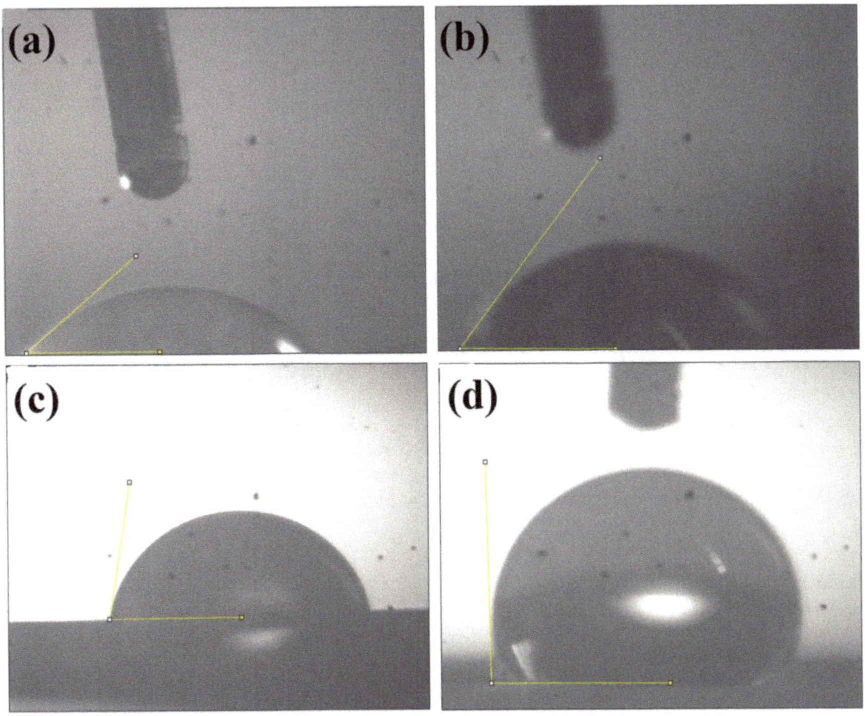

Fig. 6 Contact angle of nanofiber samples **a** P_0, **b** P_1, **c** P_2, and **d** P_3

investigated with a contact angle meter by using normal water drop method as shown in Fig. 6.

This analyzed that P_0 sample nanofibers are hydrophilic and contact angle 43° is shown in Fig. 7a. But the water contact angle for nanofibers blended was reduced as the concentration of Zn_2SiO_4 nanoparticles was increased which is shown in Table 3. The 2 wt% Zn_2SiO_4/PVA/PDMS nanofiber samples have highest hydrophobicity with contact angle 92°. The hydrophobic functional groups on the nanofiber membrane's surface cause the membrane surface to be smooth and the fiber membrane to have a dense structure [11].

3.4 Air Filter Membrane Test

One of the essential elements for human life, air, is present on the entire planet. The air is being contaminated by environmental harm as a result of rapid economic development, and as industrialization continues to advance at an accelerated rate and more particulate matter (PM) enters the atmosphere, the quality of life for humans is declining.

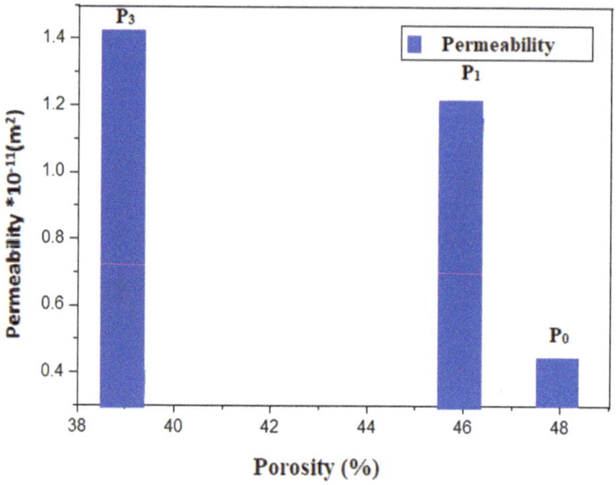

Fig. 7 Increase in the permeability will decrease the porosity (%) of the filters

Table 3 Contact angle measurement

Sample	Contact angle (°)
P_0	43
P_1	55
P_2	80
P_3	92

The better solution to the issue caused by small particles with a small diameter is to use nanofiber materials, which are thin fibers with a high surface tension and porosity produced by electrospinning technique [11]. These materials will purify the air by segregating the contamination from air-based filter mechanism.

The air filter membrane tests which give the filter thickness; efficiency, permeability, and pores size are measured. The samples are studied and conducted by mode SPM model AA3000 ANGSTROM ADVANCED INC that is AFM Contact Mode.

Permeability (K): Air permeability is defined as the fiber or material withstands the large pressure gradient in 1 s to the area is shown in Table 4.

Using Darcy's Eq. (1), the permeability can be calculated as follows:

Table 4 Result of AFM on permeability

Sample	Thickness in μm	Flow rate (Q) in mL/min	ΔP in Pa	Permeability in 10^{-11} (m^2)	Porosity in %
(P_0)	8	100	29.4	0.439	48.361
(P_1)	7.5	100	9.8	1.23	46.786
(P_3)	6	100	6.86	1.41	39.514

$$Q = \frac{AK\Delta P}{\mu L} \quad (1)$$

$$K = \frac{Q\mu L}{A\Delta P} \quad (2)$$

where

Q is the flow rate in mL/min

A is the area of filter in cm^2

K is the permeability in m^2

ΔP is the pressure drop

μ is the air viscosity that is 18.27×10^{-6} Pa

L is the thickness of filter in μm.

By using Eq. (2) the air permeability is 1.41×10^{-11} m^2 for sample P_3.

Efficiency (%): The percentage of particles of a particular size that would be prevented and kept by a filter medium is typically used to describe filter efficiency.

Efficiency Eq. (3) is used to calculate the particle collection efficiency:

$$E(\%) = \frac{Q_{\text{in}} - Q_{\text{out}}}{Q_{\text{in}}} \times 100\% \quad (3)$$

where E is the efficiency, Q_{in} is the particles entering, and Q_{out} is the particles leaving out which are shown in Table 5.

From Table 5 it is observed that as the developed nanofibers diameter increases the particle matter cannot enter inside the fiber and hence it can act as a filter media. Filtering efficiency is high for sample 3. The obtained high efficiency is at sample 3 with a concentration of 2 wt% nanoparticles, permeability of 1.41 m^2, filter efficiency of 92%, and porosity of 39.5%.

Porosity (%): It is calculated as fraction of the void with total volume of the fiber and is measured in percentage (%). The increase in the permeability will decrease the porosity of the filter which is shown in Fig. 7.

Porosity is determined as the fraction of the void in total volume of the fiber and is measured in percentage (%) where increase in the permeability will increase the porosity of the filter. The pore size result from AFM test show that all pores size. This ensures that small particles (0.3–1.0 μm).

Table 5 AFM results on filtering efficiency

Sample	Q_{in} mL/min	Q_{out} mL/min	Efficiency in %
1 (P_0)	150	49	67
2 (P_1)	150	16	89
3 (P_3)	150	12	92

4 Conclusions

The nanofiber samples are developed by combining hybrid polymers, namely, PDMS and PVA with hybrid nanoparticles ZnO and SiO_2 by synthesis of Zn_2SiO_4 nanoparticle by solution combustion method. Ultrathin nanofiber samples are produced through electrospinning process with the flow rate of 2.5 mL per hour and 25 kV. The obtained nanofiber samples are in the range of 142–410 nm. The SEM gives fiber forming bead shape in 0.5 wt% (P_1) sample and the diameter is 142 nm due to low viscosity. The diameter increases as there is increase in viscosity and bead structure disappear is shown in 2 wt% (P_2) sample and diameter is 410 nm. The XRD analysis gives the diffraction peak of 2 wt% is $2\theta = 23.6°$ and for without nanoparticle is $2\theta = 11.4°$, the intensity peak increases as increase in nanoparticle. The contact angle of 2 wt% is 92° and for without nanoparticle is 43° so the 2 wt% sample is hydrophobic in nature. The AFM test gives high efficiency for the application of filtration media is obtained at sample P_3 with a 2 wt% concentration of nanoparticle, permeability is 1.41 m^2, porosity was found to be 39.5%. It is observed that the filtering efficiency was around 92% and it is useful in the filtration media in hospitals, air conditioners, and in industries.

References

1. Barthasarathy PR, Ahmad AA, Shohibuddin IUS, Salim W (2021, March) Electrospun PVA-ZEN graphene nanofibers for smart bandages. In: IEEE-EMBS conference on biomedical engineering and sciences (IECBES), pp 152–157
2. Pande D, Chakrapani VY, Kumar TS (2019, December) Electrospun PVA/AGAROSE blends as prospective wound healing patches for foot ulcers. In: IEEE international conference on clean energy and energy efficient electronics circuit for sustainable development (INCCES), pp 1–6
3. Voniatis C, Balsevicius L, Barczikai D, Juriga D, Takacs A, Kohidai L, Jedlovszky-Hajdu A (2020) Co-electrospun polysuccinimide/poly(vinyl alcohol) composite meshes for tissue engineering. J Mol Liq 306:112895
4. Groden N, Clower W, Wilson CG (2017) Electrospun iron-based fibers for use in MEMS sensors. J Microelectromechanical Syst 26(5):987–989
5. Liang JI, Hsu HC, Nien YH, Su FC, Wu HW, Chen JP, Yeh ML (2009, April) Cell response to electrospun PVA and PVA/chitosan nanofibers. In: IEEE 35th annual northeast bioengineering conference, pp 1–2
6. Nakajo S, Murakami T, Shimada H, Osawa K, Murata M, Itaya T, Endo M (2016, June) Characterization of carbon/carbon composites containing cellulose by electrospinning compound semiconductor week (CSW). In: 28th international conference on indium phosphide and related materials (IPRM) and 43rd international symposium on compound semiconductors (ISCS), pp 1–2
7. Xia Y, Zhao L, Lu D, Li L, Luo Y, Luo G, Jiang Z (2018, January) Giant enhancement on response-speed of electrospun-based UV photodetector via polydimethylsiloxane coating. In: IEEE micro electro mechanical systems (MEMS), pp 1028–1031
8. Yao JY, Fostier AH, Santos EB (2020) In situ formation of gold and silver nanoparticles on uniform PDMS films and colorimetric analysis of their plasmonic color. Colloids Surf A: Physicochem Eng Asp 607:125463

9. Pusty M, Shirage PM (2020) Gold nanoparticle–cellulose/PDMS nanocomposite: a flexible dielectric material for harvesting mechanical energy. RSC Adv 10(17):10097–10112
10. Avossa J, Batt T, Pelet T, Sidjanski SP, Schoonenberger K, Rossi RM (2021) Polyamide nanofiber-based air filters for transparent face masks. ACS Appl Nano Mater 4(11):12401–12406
11. Zhang R, Wang H, Zhu Z, Chen R, Chen X, Zeng J, Huang L (2020) Fabrication of nanofiber filters for electret air conditioning filter via a multi-needle electrospinning. AIP Adv 10(10):105217
12. Cao Q, Ding X, Zhao H, Zhang L, Xin Q, Zhang Y (2021) Improving gas permeation performance of PDMS by incorporating hollow polyimide nanoparticles with microporous shells and preparing defect-free composite membranes for gas separation. J Membr Sci 635:119508
13. Mamun A, Blachowicz T, Sabantina L (2021) Electrospun nanofiber mats for filtering applications—technology, structure and materials. Polymers 13(9):1368
14. Aljehani AK, Hussaini MA, Hussain MA, Alothmany NS, Aldhaheri RW (2014, February) Effect of electrospinning parameters on nanofiber diameter made of poly(vinyl alcohol) as determined by atomic force microscopy. In: 2nd Middle East conference on biomedical engineering, pp 379–381
15. Seyfi J, Goodarzi V, Wurm FR, Shojaei S, Jafari-Nodoushan M, Najmoddin N, Uzun L (2020) Developing antibacterial superhydrophobic coatings based on polydimethylsiloxane/silver phosphate nanocomposites: assessment of surface morphology, roughness and chemistry. Prog Org Coat 149:105944
16. Liang W, Xu Y, Li X, Wang XX, Zhang HD, Yu M, Long YZ (2019) Transparent polyurethane nanofiber air filter for high-efficiency PM2.5 capture. Nanoscale Res Lett 14(1):1–9

The Corrosion Behavior of Aluminium Reinforced with Carbon Nanotube and Fly Ash Composites

Udaya Devadiga and Peter Fernandes

Abstract In this study, the corrosion behavior of aluminum composites reinforced with 4% fly ash (FA) and various concentrations of carbon nanotubes (CNTs) (0.25, 0.5, and 0.75%) in alkaline and acidic environments were studied. Weight loss and corrosion rate of the composites were evaluated by static immersion corrosion tests. The results showed that the composites possess excellent corrosion resistance in alkaline media through the formation of a stable passive layer and negligible weight loss in all samples. In acidic media, however, substantial weight loss and increased corrosion rates were observed especially for composites with higher content of CNTs. 0.5% CNT composite had a good balance between corrosion resistance and structural reinforcement, while 0.75% CNT composite had the highest susceptibility to corrosion due to increased porosity and microstructural heterogeneity. The results show that the corrosion resistance can be improved by optimizing CNT content while minimizing the effect on mechanical integrity. Such composites have potential for use in applications where high mechanical strength and corrosion resistance are required, e.g., aerospace, automotive, and structural engineering.

Keywords Composites · Aluminum · Carbon nanotube · Fly ash · Corrosion

1 Introduction

In the present scenario, there has been a zeal to invent new materials, which gives designer flexibility in choosing the materials in design stage for automobile and aircraft components. Over the decade there has been an increase in using CNTs to polymer, ceramic, and metal matrix composites. In fact, CNT-reinforced Al composites research taking the lead due to their excellent specific modulus, specific density,

U. Devadiga (✉)
Department of Mechanical Engineering, Nitte (Deemed to be University), NMAM Institute of Technology (NMAMIT), Nitte 574110, India
e-mail: udaya_d@nitte.edu.in

P. Fernandes
Alva's Institute of Engineering and Technology, Moodbidri 574227, India

I. M. Jamadar et al. (eds.), *Integrating Sustainability in Modern Technologies*, Lecture Notes in Mechanical Engineering, https://doi.org/10.1007/978-981-95-9466-5_16

excellent toughness, and flexibility properties. CNTs have extremely high Young's modulus in the range 0.5–150 TPa, high UTS range from 20 to 150 GPa, better flexibility, and high thermal conductivity up to 600/WK [1–7]. The main aim is to fabricate composites with increased mechanical and corrosion resistant properties. Such composites would be desirable in the aerospace, automotive, and sports components where lightweight, high stiffness, and strength is desired. Metal matrix composites have proved their importance to conventional alloys in high strength and stiffness application. However, the corrosion behavior and microstructures are sensitive to varying environments. As it is true for any alloy, aluminum-based composites are influenced with various factors like fabrication method used, material compositions, and structure which can seriously alter the corrosion properties. Alaneme and Bodunrin [8] showed corrosion behavior on aluminum (6063) MMCs and found that composites displayed high corrosion resistant property in NaCl solution compared to NaOH and H_2SO_4. Vannan and Vizhian [9] made study on corrosion behavior of Al 7075 reinforced with basalt short fiber and concluded that the corrosion resistance of the samples decreased with the time. Ramachandra and Radhakrishna [10] showed the formation of aluminum oxide layer within 24 h of commencement of test in fly ash reinforced aluminum composites. They attributed that reinforcement fly ash initiates pits which lead to pitting corrosion. Fukuda et al. [11] found that CNTs reduced the corrosion resistance of AZ31B/CNT composites forming galvanic cell layer between alloy and reinforcements. Yildiz and Kaplan [12] found delay in the corrosion of Zn-based alloys by the appearance of oxide layer of Al on the surface of the samples. Jyothi et al. [13] found enhancement of the corrosion attack on the alloy than the Al-10Si-TiB_2 composites due to severity of the solution medium. Zakaria [14] observed enhancement of corrosion resistant property compared to base matrix and concluded that reinforcement particles size and composition lowers the corrosion resistant of the composites. Govind et al. [15] showed that corrosion rate of metal matrix composites was lesser compared to base matrix Al 7075 under various time exposure and corrosion media concentration. Verma et al. [16] made a detailed review on corrosion actions of Al based matrix composites.

2 Experimental Procedure

Fabrication process influences the way the samples react to corrosive environment and therefore a great deal of study must be put into different fabrication processes. Powder metallurgy (PM) method is opted as it results in lower cost for production and samples made through this process are more responsive to corrosive environment. PM involves three stages of operations that is mixing or blending, compaction, and sintering. The raw materials include powders of pure aluminum (Al), carbon nanotubes (CNTs), and fly ash (FA). Al powder of size varying from 7–50 μm used as matrix, FA with average diameter 9.229 μm and multi-walled CNTs having tubular shape with 10–15 nm outer diameter, 2–6 nm internal diameter, and 0.1–10 μm in length were used as reinforcements in the experiment. Composition of powders varied with each sample keeping fly ash at 4% and varying CNT 0.25, 0.5, and 0.75% CNT. The blending of powder mixture is done using high-energy ball mill apparatus. Ball milling ensures uniform blending of metal composition compared to other conventional mixing methods. In compaction operation the powders are consolidated in a die by applying high-compaction forces. Acetone is used as a lubricant for the die so that heat produced due to immense applied force does not cause the powder to attach to the die. Compaction thus produces a green compact which is weaker than the original product due to lack of proper molecular bonding. Therefore, sintering is done which is driven by diffusion for the specimens to reach its final strength and integrity. The heat treatment required in sintering process is around 70–90% of the melting point of the matrix used in the composites. A temperature of around 500 °C is maintained in the furnace and heated for 6 h to obtain the final product. The specimens thus prepared are subjected to corrosion test. Figure 1 shows the photos of the specimens and the immersion test. The specimens are thoroughly polished to give smooth and uniform finish. The samples are first rinsed with distilled water and are then immersed in prepared solution of 3.5 wt% NaCl exposed to atmosphere. Mass loss and corrosion rate are carried out at intervals 24, 72, and 120 h, respectively. After each intervals, the samples are rinsed with distilled water and any foreign coating is scrapped off. The sample is again rinsed with distilled water and then made to dry in a hot air oven for 15 min. It is then allowed to cool. After cooling weight difference is measured. The same procedure is repeated for acidic medium of 0.1 M HCl. The corrosion rate was found by applying Eq. (1) [9]:

$$\text{Corrosion rate (CR)} = (W * K)/(A * \rho * T) \tag{1}$$

(a) (b)

Fig. 1 The photos of the **a** samples and **b** immersion test where W is the weight difference of the specimen in grams, K is conversion constant. The value of $K = (8.76 * 10)^4$, A is the surface area of the sample in cm^2, ρ is the density of the sample in g/cm^3, T is the time of interval in h.

3 Results and Discussion

The materials and experimental setup used for the immersion test are shown in Fig. 1. In panel (a), these cylindrical samples are labeled F4B, C5P, and EE1, which represent different compositions or reinforcement materials. The study involves these samples as aluminum composites made with different fly ash (FA) and carbon nanotube (CNT) contents. It provides clear identification for comparative analysis during and after corrosion tests. The immersion test setup is shown in panel (b): one of the cylindrical samples is submerged in a beaker containing the corrosive solution. In order to simulate real world exposure to corrosive environments, the setup has been designed. The solution level and transparency indicate a controlled environment for uniform exposure to the medium, guaranteeing repeatability. In this test, corrosion resistance of the samples is assessed by monitoring weight loss or other parameters over a given time. This partial immersion is typical of a procedure used to expose only relevant surfaces to the corrosive medium, eliminating edge effects or unwarranted interference. These images combine to show the methodology used for studying the corrosion behavior of the material investigated herein, allowing direct correlation of results to the visual and structural properties of the samples.

Figure 2 illustrates the variation in weight loss of Al composites reinforced with 4% FA and varying amounts of carbon nanotubes (CNT) (0.25, 0.5, and 0.75%) in two different corrosive environments: Fig. 2a shows the alkaline and Fig. 2b shows the acidic. Figure 2a shows mass variation in the alkaline medium which is insignificant with an exposure period of 150 h. The formation of a stable protective passive film, probably consisting of aluminum hydroxide or oxide, appears to confer corrosion resistance to the composites. The composite with 0.25% CNT exhibits the least mass variation among the samples, indicating better stability and lower susceptibility to degradation. However, composites with higher CNT content, i.e., 0.75% CNT, show slightly more mass loss that may be attributed to an increase in microstructural heterogeneity or lower density with increasing CNT content. The results confirm overall that the alkaline medium has little effect on the composites due to the less aggressive nature of the medium, and the protective mechanisms in effect.

Conversely, Fig. 2b shows the weight loss behavior of the same composites in an acidic environment. The weight loss increases progressively with exposure time, indicating the aggressive nature of the acidic environment here. 0.75% CNT composite experiences the highest weight loss, followed by 0.5%, 0.25% CNT composites, respectively. These results indicate that higher CNT content results in higher porosity and micro-galvanic activity, leading to acceleration of the corrosion process. Moreover, after 48 h, white corrosion products are observed on the surfaces of the specimens, which are probably caused by the precipitation of aluminum chloride or other

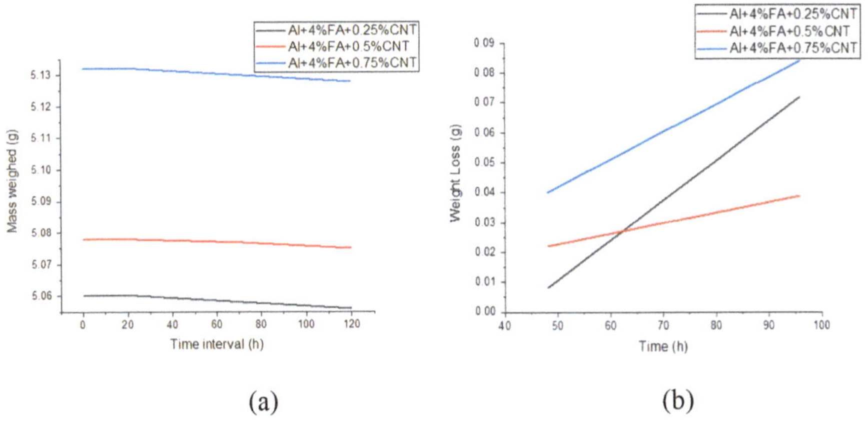

Fig. 2 The graph showing the variation of weight loss with respect to exposure time for **a** alkaline medium and **b** acidic medium

corrosion by-products. These corrosion products afford some protection, but are not sufficient to prevent further weight loss. Once at 96 h, the rate of weight loss levels off but is still much higher in the alkaline medium.

From the result obtained it can be seen that the alkaline medium is much less corrosive than the acidic medium. In the alkaline environment, the passive film is stable and inhibits corrosion greatly; but in the acidic medium, the protective layer is dissolved rapidly, leaving the composite surface continuously exposed to attack. The trends also show that lower CNT content increase corrosion resistance in both media, but higher CNT content requires further optimization to counteract its negative effects on the structural stability of the composite.

The variation of corrosion rate (mm/year) with exposure time (h) of aluminum composites containing 4% fly ash (FA) and various amounts of carbon nanotubes (CNT) (0.25, 0.5, and 0.75%) is shown in Fig. 3. The graph shows that the corrosion behavior of composites under acidic conditions has distinctly different trends, depending upon CNT content and exposure time.

The corrosion rate of the composite containing 0.25% CNT is continuously rising with exposure time, beginning at around 2 mm/year at 45 h and exceeding 3 mm/year after 100 h. In this case, this composite offers relatively poorer corrosion resistance at first, but the rate of material degradation increases with time. The corrosion rate for the composite containing 0.5% CNT is relatively stable at approximately 2 mm/year during the entire exposure period. The stability of these composites shows that, while the intermediate CNT concentration provides some structural integrity, the composite as a whole shows moderate but consistent resistance to corrosion.

The composite with 0.75% CNT shows the highest and most consistent corrosion rate, the corrosion rates are slightly above 3 mm/year for all exposure time. The increased CNT content can account for this behavior, as greater porosity or microstructural heterogeneity promotes the penetration of corrosive agents and

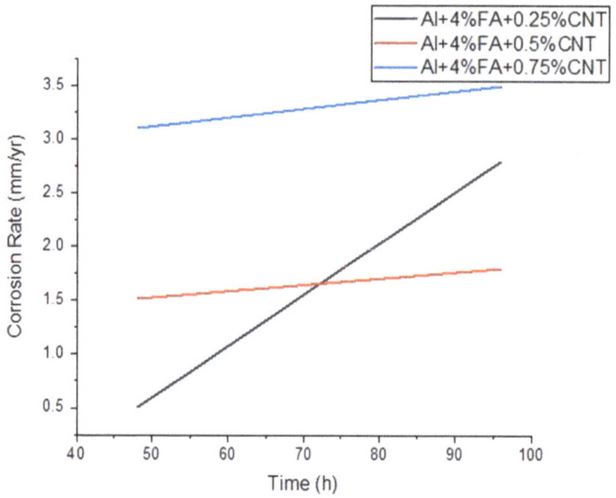

Fig. 3 The graph showing the variation of corrosion rate with respect to exposure time

lowers the density of the material as a whole. Also, the effects of higher CNT content can aggravate micro-galvanic corrosion, accelerating the corrosion process.

Overall, the data in Fig. 3 shows that increasing CNT content leads to the enhancement of some mechanical properties of the MMCs, but the corrosion resistance is degraded, particularly under long-term exposure. CNT composite with 0.5% CNT provides a compromise between stability and corrosion resistance with a relatively steady corrosion rate. The trends highlight the necessity of optimizing CNT content to minimize the negative impact on corrosion behavior while optimizing the reinforcement benefits of CNTs.

4 Conclusions

1. Performance in Alkaline Medium: The composites showed excellent corrosion resistance with negligible weight loss by forming stable passive layer. The least weight variation was exhibited by the composite with 0.25% CNT, which attests to its superior stability.
2. Performance in Acidic Medium: Dissolution of the passive layer resulted in significant reduction of weight and consequently, aggressive corrosion of the composites. Increased weight loss was achieved with higher CNT content (0.75%) as a result of higher porosity and microstructural heterogeneity.
3. Corrosion Rate Trends: With increasing CNT content, corrosion rate increased, with the highest rate occurring in the 0.75% CNT composite. Reinforcement and durability were balanced in the 0.5% CNT composite that exhibited stable corrosion resistance.

4. Medium Dependency: Minimal degradation was seen in alkaline medium, but acidic medium was far more aggressive.
5. Optimized Composition: The best combination of corrosion resistance and structural reinforcement was found for composites with 0.25 and 0.5% CNT.

The importance of CNT content control and proper medium selection is demonstrated to improve the corrosion resistance of aluminum composites for real world applications.

References

1. Andrews R, Weisenberger MC (2004) Carbon nanotube polymer composites. Curr Opin Solid State Mater Sci 8(1):31–37
2. Thostenson ET, Ren Z, Chou T-W (2001) Advances in the science and technology of carbon nanotubes and their composites: a review. Compos Sci Technol 61(13):1899–1912
3. Bakshi SR, Lahiri D, Agarwal A (2010) Carbon nanotube reinforced metal matrix composites—a review. Int Mater Rev 55(1):41–64
4. Breuer O, Sundararaj U (2004) big returns from small fibers: a review of polymer/carbon nanotube composites. Polym Compos 25(6):630–645
5. Gojny FH, Wichmann MHG, Fiedler B, Schulte K (2005) Influence of different carbon nanotubes on the mechanical properties of epoxy matrix composites—a comparative study. Compos Sci Technol 65(15–16):2300–2313
6. Deng C, Zhang XX, Wang D, Lin Q, Li A (2007) Preparation and characterization of carbon nanotubes/aluminium matrix composites. Mater Lett 61(8–9):1725–1728
7. George R, Kashyap KT, Rahul R, Yamdagni S (2005) Strengthening in carbon nanotube/aluminium (CNT/Al) composites. Scripta Mater 53(10):1159–1163
8. Alaneme KK, Bodunrin MO (2011) Corrosion behavior of alumina reinforced aluminium (6063) metal matrix composites. J Miner Mater Charact Eng 10(12):1153–1165
9. Vannan E, Vizhian P (2015) Corrosion characteristics of basalt short fiber reinforced with Al-7075 metal matrix composites. Jordan J Mech Indust Eng 9(2):121–128
10. Ramachandra M, Radhakrishna K (2007) Effect of reinforcement of fly ash on sliding wear, slurry erosive wear and corrosive behavior of aluminium matrix composite. Wear 262(11–12):1450–1462
11. Fukuda H, Szpunar JA, Kondoh K, Chromik R (2010) The influence of carbon nanotubes on the corrosion behaviour of AZ31B magnesium alloy. Corros Sci 52(12):3917–3923
12. Yildiz AK, Kaplan M (2004) Corrosion behaviour, microstructure and phase transitions of Zn-based alloys. Bull Mater Sci 27(4):341–345
13. Jyothi K, Sivamahesh E, Rao KV, Rao ChC (2015) Corrosion studies on Al-10Si-TiB_2 composites produced through stir casting method. Int J Adv Eng Res Sci 2(9):28–31
14. Zakaria HM (2014) Microstructural and corrosion behavior of Al/SiC metal matrix composites. Ain Shams Eng J 5(3):831–838
15. Govind NS, Jayaprakash HV, Krupakara PV (2018) Corrosion studies of Al 7075 metal matrix composites containing beryl particulates. Indian J Adv Chem Sci 6(1):37–40
16. Verma AS, Sumankant, Suri NM, Yashpal (2015) Corrosion behavior of aluminum base particulate metal matrix composites: a review. Mater Today Proc 2(4–5):2840–2851

Improvised Approach of Phase Changing Materials (PCM) with Household Materials Inspired by Beetle Shells

Prasannahariveeresh Jeyaveerapandian Raji, Surya Charan, Harishanker, Jnanaprakash, Patel Sai Nikitha, and J. G. Anjana

Abstract A prototype inspired by the Beetle Shells was developed with recyclable and reusable low-cost materials. This aims to reproduce the structure and thermal characteristics of beetle shells while highlighting their lightweight, durable, and heat-resistant design. As such, this prototype upholds sustainability by limiting and reusing generated waste. As part of the preparation, a base structure is created for the closed vessel; it is layered with materials to enhance its strength, and an aluminum layer that reflects heat and prevents heat accumulation inside the shell is applied as a finishing touch. The model is compared with a similar household PCM named sponge-salt ice pack, in which salt is mixed with water and absorbed by the sponge, and the pack is placed in the freezer. Here, the sponge absorbs the heat from the surroundings as it melts while the salt ensures flexibility by preventing it from completely freezing. The thermal performance was benchmarked by an experiment for 20 min and the observations were recorded, thus comparing the performance of both the Beetle-Shell prototype and the sponge-salt ice pack. The prototype could be developed for use in passive cooling applications, such as sustainable building insulation or personal cooling devices designed for extreme conditions.

Keywords Phase Changing Materials (PCM) · PCM · Latent heat · Beetle shell · Household materials · Sponge-salt ice pack

1 Introduction

Beetle shells are fascinating both biologically and as an inspiration for human innovations in materials science and engineering. Beetle shells, biologically known as elytra are hardened forewings that serve as protection to beetles and help them survive in

---

P. J. Raji (✉) · S. Charan · Harishanker · Jnanaprakash · P. S. Nikitha · J. G. Anjana
Department of Electronics and Communication, Amrita School of Engineering, Amrita Vishwa Vidyapeetham, Kollam, Kerala, India
e-mail: prassijr@gmail.com

J. G. Anjana
e-mail: anajanajg@am.amrita.edu

I. M. Jamadar et al. (eds.), *Integrating Sustainability in Modern Technologies*, Lecture Notes in Mechanical Engineering, https://doi.org/10.1007/978-981-95-9466-5_17

extreme climatic conditions, these are made primarily with chitin, a tough polysaccharide, and sclerotin, a protein that provides rigidity and durability. The hydrophobic nature of many beetle shells reduces water loss, which indirectly helps in temperature regulation by preventing dehydration. This is especially important in arid climates where evaporative cooling isn't feasible. Numerous research initiatives are focused on developing sustainable materials. Biopolymers in [1, 2] are being explored as alternatives to synthetic polymers, which are non-biodegradable and toxic, in order to promote the use of eco-friendly materials. Sustainable materials were used in waste management systems [3] to rely on natural resources and allow individuals to contribute wellness to our society. The usage of sustainable materials in agriculture promotes eco-friendliness and works on utilizing agricultural wastes as an effective energy material are advancing rapidly [4]. Soft materials for electronics in [5, 6] are one of the emerging fields in sustainability; researchers are working on using materials that come from nature as active and passive elements in an electric circuit. Phase changing materials (PCM) are used in large-scale applications, widely used in constructions [7, 8], thermal management systems [9, 10], solar stills [11, 12], etc. However, there are few applications for using them in households as they are not accessible for small-scale applications. Both organic PCM and inorganic PCM can't be used for household purposes as they are either expensive or contain toxicity with the case of inorganic PCMs [13, 14]. This paper takes inspiration from the architecture of beetle shells to create a prototype that provides a budget-friendly substitute for phase change materials (PCM) in small-scale applications. The proposed design can be made with everyday household items and is meant for emergencies. The proposed model is compared with a sponge-salt ice pack and experimented with by performing a thermal performance test and finally compared its pros and cons in the sections below.

In the past few years, various researchers have worked on analyzing various PCMs. The use of stable shape-stabilized organic Phase Changing Materials (PCM) for thermal energy storage, including their application in buildings, energy storage applications, battery thermal management systems, solar stills, etc., has been the subject of numerous analyses by researchers. In this paper, some pre-existing work in PCMs was referred to and analyzed, the authors [7, 8] have worked on the use of PCMs in buildings, while the authors [9, 15] have analyzed the use of PCMs in battery thermal management systems. Nazir et al. [16] reviewed the recent developments in thermal energy systems (TES) using PCMs and compared their outcomes in recent years. Since latent heat PCMs are not as durable as sensible materials like concrete, bricks, stones, ceramics, etc., they claim that there are still few large-scale commercial applications for them. In addition to having a higher energy density, high thermal conductivity, and comparatively high melting points, inorganic materials are also supercooled and are more caustic than organic ones. The two key strategies that aid in improving the quality of PCMs are the addition of nanomaterials and particular encapsulations. The authors compared the approaches of nanofluids on PCM, their commercial applications, and patents that are majorly listed on TES using PCMs.

The applications of battery thermal management systems (BTM) are reviewed in detail, and the use of Phase Change Fluids (PCF) and FPCM (Flexible Phase

Change Material) in BTMs is discussed by Luo et al. [9]. The PCM-based BTMs are crucial for optimizing battery performance and extending service life. According to the author, as organic PCMs are flammable, they are used in most of the present research on BTMs. Nonetheless, investigating inorganic PCM with flame-redundant properties is crucial to resolving power battery thermal runaway. The researchers conclude that adding metals with high thermal conductivity, metal oxides, expanded graphite, etc., is the primary method of improving PCM's heat transfer capability. The study of Omara et al. [11] provides thorough evaluations of several methods that employ PCMs as latent heat storage devices to increase solar still production. They mentioned that the use of PCM in active and passive solar stills can significantly boost productivity. PCM enhances passive solar stills by more than a hundred percent, and when integrated with external units like solar collectors or heat pumps, productivity can increase by up to 700%. Paraffin was the most commonly used organic PCM due to its favorable properties like medium storage capacity, safety, reliability, and moderate cost. Whereas, the other types like salt hydrates were less favored due to conditions like corrosion, and instability. The results produced by the author show that solar still productivity improved with higher PCM mass and lower saline water mass, and hence they were found to be more effective at night than during the day.

The study by Ikutegbe et al. on the use of PCM foam composites in the built environment demonstrated how compatible PCMs are with their host integration procedures. According to the authors, three crucial measures can be considered: the foam density, the quantity of PCM added to clean PU foam, and the foam thickness. One significant substance with molecular-level control over PCM's physical entrapment is polyol. The researchers considered the implementation of structural integrity, thermoregulation, and cost-effectiveness to be the main advantages of this method. PCM-based thermal energy storage systems for cooling applications in buildings by Faraj et al. [8] focus on recent studies on PCMs with a focus on heating, cooling, and hybrid systems, which are categorized as passive or active. In heating applications, the PCMs improve thermal comfort in cold climates by reducing temperature swings and shifting peak energy loads, while in cooling applications, PCMs reduce the energy consumption in HVAC and passive systems but have problems like incomplete solidification at night and reliance on simulation assumptions. In special cases like hybrid systems, combining PCM applications enhances energy efficiency and sustainability, supporting zero energy goals.

According to Wu et al., composite PCM with room temperature flexibility for BTM, the TPM-FCPCM cools the pouch LIB module and prismatic LIB module by forming the TPM-FCPMs into the appropriate size and joining them with five pouchs or prismatic cells in a sandwich arrangement. The scientists saw a decline in T_{max} and maximum ΔT (ΔT_{max}), indicating the clear cooling impact of the LDPE-CPCM plates. After measuring the T_{max} and ΔT_{max} values of the TPC, LDPE, and NAC modules, it was evident that the battery modules' total weight rose with the addition of the CPCM cooling component. This is because the resultant TPC-FCPCM maintains its flexibility at low temperatures since the ether bond has significantly more flexibility in the soft portion. Marani et al. [17] incorporated PCMs in integrating the thermal characteristics of building materials. In this work, approaches were taken to modify the mechanical and durability characteristics of construction materials with PCMs and address the concerns that the PCMs in the liquid state will leak out. The authors discussed how to reduce risk and the side effects of using PCMs for sustainability and resilience; they mentioned using microencapsulation techniques to improve the thermal conductivity of the PCMs.

Kashyap et al. [18] compared changing paraffin wax with volume heating and surface heating, and noted the effect of varying heat flux by adjusting the distance from the light source. They concluded that surface heating is a more promising approach than volume heating for solar thermal energy systems without enhancing the optical properties of paraffin wax. Additionally, PCMs are utilized in gasoline engines. Senthilkumar et al. conducted an experimental investigation on exhaust heat recovery in gasoline engines using nano-enhanced PCM. The study made use of a four-stroke, multi-cylinder gasoline engine, a shell-and-tube heat exchanger for heat recovery, and a thermal energy storage tank containing cylindrical paraffin capsules. The authors presented that nano-enhanced PCM shows better charging efficiency, heat transfer rate, and heat loss coefficient compared to simple paraffin.

2 Methodology and Parameters

2.1 *Latent Heat*

Latent heat is the energy that is absorbed or released during a phase transition but does not cause a change in the substance's temperature. This phase shift occurs during transitions such as melting, freezing, vaporization, condensation, or sublimation. During a phase transition, the temperature of the material remains constant. Latent heat, which can be categorized as either latent heat of vaporization or latent heat of fusion, is released during freezing and condensation, respectively. Because latent heat storage can store and release a significant quantity of thermal energy, it may be employed practically in everyday applications, such as solar cookers [19]. The system's observed or emitted heat may be calculated using

$$Q = m \cdot L \tag{1}$$

where m represents the substance's mass, Q represents the heat absorbed or released, and L represents the substance's latent heat.

2.2 *Phase Changing Materials (PCM)*

A substance that releases or absorbs enough energy during phase transition to produce useful heat or cooling is known as a phase-changing material (PCM). Unlike other

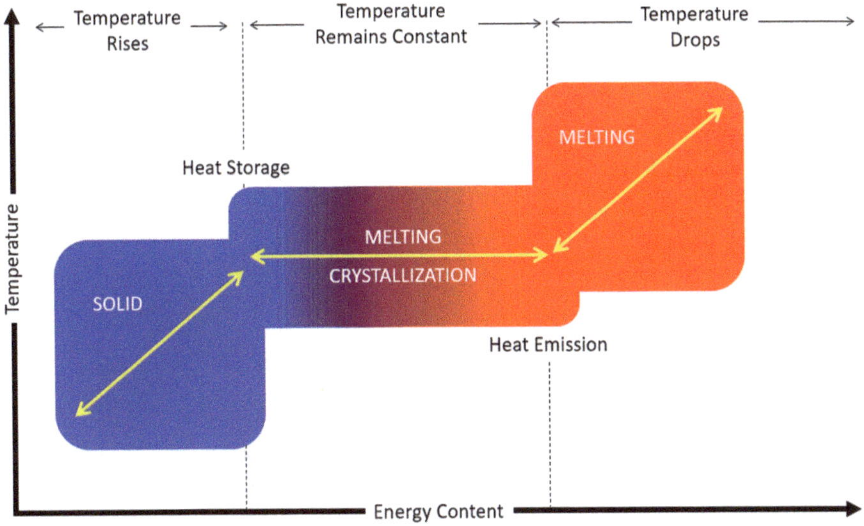

Fig. 1 Working principle of PCM

heat-storage materials, PCMs observe and release a large amount of energy without any significant change in temperature during the phase change. These materials widely undergo reversible processes, as phase changes can occur repeatedly, and they can reuse the energy storage cycles. PCMs are perfect for thermal management applications because they keep the temperature almost constant throughout the phase transition. There are several types of PCMs including organic PCM, inorganic PCM, and eutectic mixtures. Figure 1 shows the working principle of PCM that depicts the temperature versus the energy content graph.

A PCM's total heat storage capability is described by the total energy stored in a PCM equation, which takes into account both sensible and latent heat.

$$Q_{\text{total}} = m \cdot c_{\text{s}} \cdot (T_{\text{m}} - T_{\text{i}}) + m \cdot L + m \cdot c_{\text{l}} \cdot (T_{\text{f}} - T_{\text{m}}) \tag{2}$$

where T_{i} is the PCM's initial temperature, T_{m} is its melting temperature, T_{f} is its final temperature, L is its latent heat of fusion, m is its mass, c_{s} is its specific heat capacity in its solid phase, c_{l} is its specific heat capacity in its liquid phase, and Q_{total} is the total heat energy stored or released by the PCM.

2.3 Thermal Conductivity

The property of a material that evaluates its capacity to transfer heat is called thermal conductivity. It measures how quickly energy moves through a substance as a result of a temperature differential. Under a temperature gradient, the quantity of heat (Q)

that moves through a material's unit area (A) in a unit time (t). This can be expressed as follows:

$$Q = -k \cdot A \cdot \frac{\Delta T}{\Delta x} \tag{3}$$

where Δx is the material's thickness, ΔT is the temperature differential, and A is the cross-sectional area.

2.4 *Sponge-Salt Ice Pack*

The sponge-salt ice pack utilizes the principles of PCM and latent heat to absorb and release thermal energy. The saltwater solution in the ice packs freezes and melts, undergoing a solid–liquid phase change. During freezing, the pack absorbs heat due to latent heat, and during melting, it absorbs heat from the surroundings, helping it to maintain a stable temperature. The inclusion of salt lowers the freezing point of water, allowing it to operate effectively at specific temperatures. Figure 2 depicts the working flow of sponge-salt ice packs (Fig. 3).

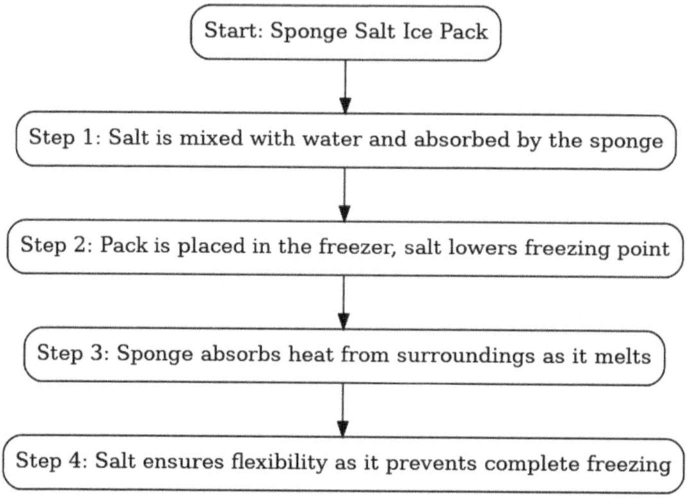

Fig. 2 The workflow of sponge-salt ice pack

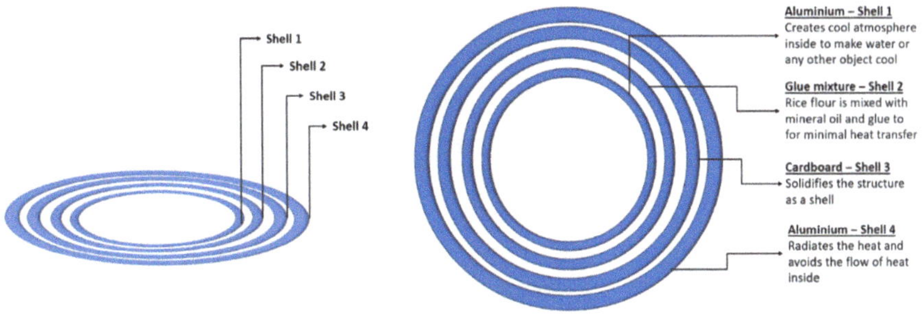

Fig. 3 Top view and side view of shells present in our beetle-shell prototype

2.5 *Prototype Design*

We have designed and constructed a prototype of our customized PCM inspired by the Beetle shell's thermal resistance and durability using materials that we use in our day-to-day life and compared those results with a sponge-salt ice pack that can be easily prepared in our homes. Table 1 shows the household materials that were used in our prototype which is used in our day-to-day life. Figure 4 shows the prototype developed and Fig. 5 compares our model with a sponge-salt ice pack. To construct a beetle shell prototype by inspiration from nature, we first designed a shell using cardboard from the inspiration of a beetle's outer shell (diabolical ironclad) which is hard, so we made a 4 cm radius base and a 15 cm long shell. Then we used a paper glue mix that comprises a mixture of rice flour, mineral oil, and glue which is inspired by a thin layer (Endocuticle) that holds the prototype together and transfers heat evenly inside; we used wax as an inspired coated layer from the soft membrane or epicuticle of the beetle shell that makes it dry inside. We used the aluminum foil layer after the waxed layer to prevent heat flow inside the inspired beetle bottom-most layer (exocuticle). The diagram of these layers with top view and side view is depicted as shown in Fig. 3. After building the prototype we performed a heat resistance test by placing the beetle shell prototype under a heat source that has been set to 50 °C. Record surface temperatures of the prototype at regular intervals of 5 min for 20 min to observe how the prototype resists heat.

Table 1 Materials utilized in the beetle-shell prototype

Sl. No.	Materials used	Quantity
1	Cardboards	1
2	Glue	1
3	Aluminum foils	1 sheet
4	Mineral oil	250 mL
5	Rice flour	400 mL
6	Wax	100 g

Fig. 4 Top view and side view of shells present in our beetle-shell prototype

Fig. 5 Comparison of beetle-shell prototype with sponge-salt ice pack

3 Results and Observation

The beetle shell prototype is moderately heat-resistant and will insulate heat away from materials underneath. It is durable under controlled water exposure and can be used to specify applications requiring lightweight insulation and eco-friendliness. Table 2 compares our prototype with the sponge-salt ice pack, and Table 3 compares the pros and cons of PCM and our beetle shell-inspired prototype. Table 4 shows the thermal performance of the Beetle Shell prototype with sponge-salt ice pack over 20 min. The test compared the heat resistance of a beetle shell prototype and a sponge-salt ice pack by placing them in a hot water bath maintained at 50 °C for 20 min. Some observations are made, at the initial state both the materials start at 25 °C, the Beetle-Shell prototype gets slightly warm and maintains its shape, whereas the sponge-salt ice pack quickly loses its cooling efficiency. After 5 min, the beetle shell warms up to 28 °C and retains its structure, but the ice pack reaches 35 °C, losing its effectiveness.

After 10 min, our beetle-shell prototype raises to a temperature of 32 °C, we found that the prototype showed gradual heat absorption due to the presence of insulating materials, whereas the sponge-salt ice pack softens and retains the heat poorly as it warms up to 42 °C. After the 15 min time interval, the Beetle-Shell prototype didn't deform. It maintained a temperature of 35 °C, but the sponge-salt ice pack heats up to 46 °C and doesn't possess any properties of a cooling agent. Finally, after 20 min, our model provides consistent insulation even after reaching a temperature of 38 °C. Still, the sponge-salt ice pack reached a temperature of 50 °C where the water in the ice pack's surface started evaporating.

Figure 6 shows the graphical representation of our observation where we can see that both the Beetle-Shell prototype and sponge-salt ice pack are increasing. Still, our Beetle-Shell prototype maintains a low temperature as time increases as compared to that of the sponge-salt ice pack. This prototype is best suited for environments with moderate temperature exposure, though prolonged heat or contact with water may require additional sealing or protective coatings. The designed Beetle-Shell prototype maintains superior thermal insulation due to its layered structure, materials, and heat dissipation properties. In Fig. 3, the inner aluminum shell creates a cool atmosphere inside the shell, maintaining a lower inner temperature, as the second shell is composed of rice flour mixed with mineral oil and glue, forming a low-thermal-conductivity barrier, that minimizes the heat transfer between the layers, thus enhancing insulation. The outer shell acts as a shield that radiates the heat away from the surface, preventing external heat from penetrating inside. Hence this multi-layered approach effectively reduces heat penetration and enhances insulation.

Table 2 Comparison between the beetle-shell prototype and the sponge-salt prototype

Parameter	Beetle shell prototype	Sponge-salt prototype
Thermal efficiency	Insulating (slows heat transfer, retains heat)	Cooling (absorbs heat due to melting ice)
Temperature resistance	Can withstand moderate heat (up to ~50 °C)	Works efficiently in temperatures between −10 and 20 °C
Purpose	Structural design and aesthetics (not functional for cooling)	Functional for cooling or maintaining low temperatures
Durability	Durable if sealed; sensitive to prolonged exposure to water or heat	Limited lifespan depending on repeated freezing/thawing cycles
Cost efficiency	Very low cost (recycled materials)	Moderate cost (requires salt, sponge, plastic)
Weight	Lightweight	Slightly heavier due to water content
Environmental impact	Low; uses biodegradable/recyclable materials	Medium; uses non-biodegradable plastic in some cases

Table 3 Advantages and disadvantages of the beetle-shell prototype

Pros	Cons
The beetle shell prototype does not degrade with time or with repeated use but sponge-salt efficiency will be reduced every time as it works on a cooling system and the prototype works on the reflection of heat radiation	Crafting the prototype, including shaping, layering, drying, and finishing, is time-consuming compared to ready-made alternatives like sponge-salt ice packs
The prototype withstands higher temperatures (50–80) making it a good reflector of heat	While it can withstand moderate heat (up to 50–80 °C), prolonged exposure to extreme temperatures may cause warping or degradation of the prototype
The prototype is made with biodegradable materials like paper, glue, cardboard, and aluminum	Compared to industrial materials like plastics, the prototype is less impact-resistant and more prone to cracking or breaking under stress
Low cost and objects can be recycled easily, vast availability of materials and inexpensive	Beetle shell prototype can avoid heat but can't provide active cooling
The beetle shell prototype does not degrade with time or with repeated use but sponge salt efficiency will be reduced every time as it works on a cooling system and the prototype works on the reflecting of heat radiation	Crafting the prototype, including shaping, layering, drying, and finishing, is time-consuming compared to ready-made alternatives like sponge-salt ice packs

Table 4 Materials utilized in the beetle-shell prototype

Sl. No.	Time interval	Beetle-shell prototype temperature (°C)	Sponge-salt temperature (°C)
1	0	25	25
2	5	28	35
3	10	32	42
4	15	35	46
5	20	38	50

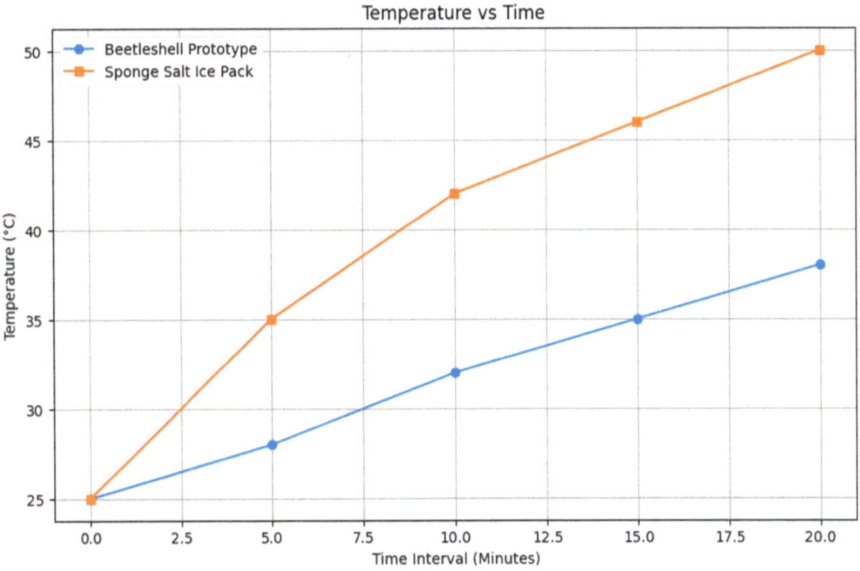

Fig. 6 Graph of temperature versus time interval

4 Conclusion

The Beetle-Shell prototype, made of inexpensive materials that can be gathered from our households, could replicate the structural and thermal properties of natural beetle shells. Its layered base structure with heat-resistant wax and a protective aluminum layer showed efficient dissipation of heat and insulation. Compared to the sponge-salt ice pack, the prototype offered promising thermal resistance, indicating practical applicability for heat-management purposes. This is better suited for long-term thermal regulation as per our performance comparison, where it was proved that during the thermal test, after 20 min, the proposed Beetle-Shell prototype remained at 38 °C whereas the sponge-salt ice pack heated up to 50 °C, showing that the sponge-salt ice pack did absorb and store heat, but our prototype was more effective in radiating heat and insulating. The suggested prototype can endure moderate heat (up to 50–80 °C), but extended exposure to extreme temperatures may lead to warping or degradation of the prototype. In the future, substitutes for the materials can be used to enhance the rigidity and strength of the structure, aiding in its resistance to extreme heat. The prototype could be advanced for use in passive cooling applications, including sustainable building insulation or personal cooling equipment designed for extreme conditions.

Acknowledgements The authors extend their heartfelt thanks to the Electronics and Communication Department of Amrita School of Engineering, Amrita Vishwa Vidyapeetham, Amritapuri, Kollam, Kerala, India for providing the space to perform experiments and develop the prototype.

References

1. Arif ZU, Khalid MY, Sheikh MF, Zolfagharian A, Bodaghi M (2022) Biopolymeric sustainable materials and their emerging applications. J Environ Chem Eng 10(4):108159
2. Baranwal J, Barse B, Fais A, Delogu GL, Kumar A (2022) Biopolymer: a sustainable material for food and medical applications. Polymers 14(5):983
3. Valsan V, Viswanath A, von Lieres JS (2023) Creating a sustainable waste management system as a service-learning project at Amrita Vishwa Vidyapeetham University. In: Community engagement in higher education. Routledge India, pp 197–211
4. Samage A, Halakarni M, Yoon H, Kotrappanavar NS (2024) Sustainable conversion of agricultural biomass waste into electrode materials with enhanced energy density for aqueous zinc-ion hybrid capacitors. Carbon 219:118774
5. Han MJ, Yoon DK (2021) Advances in soft materials for sustainable electronics. Engineering 7(5):564–580
6. Wang C, Xia K, Zhang Y, Kaplan DL (2019) Silk-based advanced materials for soft electronics. Acc Chem Res 52(10):2916–2927
7. Cui Y, Xie J, Liu J, Wang J, Chen S (2017) A review on phase change material application in building. Adv Mech Eng 9(6):1687814017700828
8. Faraj K, Khaled M, Faraj J, Hachem F, Castelain C (2020) Phase change material thermal energy storage systems for cooling applications in buildings: a review. Renew Sustain Energy Rev 119:109579
9. Luo J, Zou D, Wang Y, Wang S, Huang L (2022) Battery thermal management systems (BTMs) based on phase change material (PCM): a comprehensive review. Chem Eng J 430:132741
10. Nayar S, Ratheesh K, Jyothika S, Prasad G, Arun K (2024) Real-life applications of industry 4.0 technologies in defending the COVID-19 pandemic: a review. In: AIP conference proceedings, vol 2565, no 1. AIP Publishing
11. Omara AA, Abuelnuor AA, Mohammed HA, Khiadani M (2020) Phase change materials (PCMs) for improving solar still productivity: a review. J Therm Anal Calorim 139:1585–1617
12. Anuradha AP, Girish DK, Prasad G (2024) Ocean energy in India-bridging gaps for sustainable futures. In: AIP conference proceedings, vol 3171, no 1. AIP Publishing
13. Podara CV, Kartsonakis IA, Charitidis CA (2021) Towards phase change materials for thermal energy storage: classification, improvements and applications in the building sector. Appl Sci 11(4):1490
14. Abhishek SS, Malik GR, Suresh DT, Anudev B, Prasad G (2024) Sustainable mobility and environmental impacts of green batteries: a comprehensive review. In: E3S Web of conferences, vol 540. EDP Sciences, p 11005
15. Wu W, Ye G, Zhang G, Yang X (2022) Composite phase change material with room-temperature-flexibility for battery thermal management. Chem Eng J 428:131116
16. Nazir H, Batool M, Osorio FJB, Isaza-Ruiz M, Xu X, Vignarooban K, Phelan P, Kannan AM (2019) Recent developments in phase change materials for energy storage applications: a review. Int J Heat Mass Transf 129:491–523
17. Marani A, Nehdi ML (2019) Integrating phase change materials in construction materials: critical review. Constr Build Mater 217:36–49
18. Kashyap AS, Bhalla V, Tyagi H, A comparative experimental investigation on charging and discharging of paraffin wax PCM with surface and volume heating approach
19. Sharma SD, Buddhi D, Sawhney RL, Sharma A (2000) Design, development and performance evaluation of a latent heat storage unit for evening cooking in a solar cooker. Energy Convers Manag 41(14):1497–1508

Influence of Bamboo Fiber Extraction Methods on the Mechanical Properties of Bamboo-Epoxy Composite Lamina

B. Vishwash, N. D. Shivakumar, and K. B. Sachidananda

Abstract Bamboo is a sustainable industrial crop material; however, the sustainability of bamboo textiles is determined by how they are isolated and processed to make bamboo fibers (BFs) in different forms. This paper aims to investigate how the BF isolation methods—chemical retting and mechanical extraction—affect the bamboo-epoxy lamina's tensile strength, fiber volume percentage, and engineering constants. The novelty of this paper is the analytical approach provided using mechanics of composite materials. Three distinct kinds of BFs can be separated: single, bundle, and strip. We have considered bundle form bamboo fibers (BBFs) to strengthen the epoxy to construct a bamboo-epoxy composite lamina. The assessment results show that the chemically retted bundle form bamboo fiber-reinforced epoxy lamina (CR-BFEL) is stiffer and stronger than the mechanically extracted bundle form bamboo fiber-reinforced epoxy lamina (ME-BFEL). Compared to ME-BFEL, CR-BFEL has higher engineering constant values. The CR-BFEL has a tensile strength of 173.57 MPa at 14% fiber volume fraction, whereas in case of ME-BFEL it showed a tensile strength of 168.78 MPa at 45% fiber volume fraction.

Keywords Bamboo fiber · Tensile strength · Bamboo-epoxy lamina

1 Introduction

Bamboo fiber-reinforced polymeric composites offer significant potential for sustainable development in structural applications. However, overcoming technical challenges and further research into material behavior and long-term performance are essential for wider industrial use [1]. Standardizing fiber extraction and composite

B. Vishwash (✉) · N. D. Shivakumar
Department of Design and Manufacturing, Indian Institute of Science, Bengaluru, Karnataka, India
e-mail: vishwashb@iisc.ac.in

K. B. Sachidananda
Department of Mechanical Engineering, Sri Krishna Institute of Technology, Bengaluru, Karnataka, India

I. M. Jamadar et al. (eds.), *Integrating Sustainability in Modern Technologies*, Lecture Notes in Mechanical Engineering, https://doi.org/10.1007/978-981-95-9466-5_18

manufacturing is essential for large-scale applications. Advanced studies on fiber-matrix interactions, long-term durability, and life cycle analysis are crucial. Innovations in bio-based matrices and green processing methods can enhance the eco-friendliness of these materials [2]. Bamboo composites are used in load-bearing components, panels, and structural reinforcements. Lightweight and high-strength properties make them suitable for interior and structural parts. BFs offer sustainable options for durable and biodegradable products [3]. Bamboo is one of the fastest-growing plants; it can be harvested in 3–5 years. Because it regenerates without requiring replanting, it is an environmentally benign and renewable resource [4]. BF is a promising solution for sustainable textile manufacturing. Its eco-friendly attributes, coupled with its versatility, make it a strong contender for addressing environmental concerns in the textile industry. However, advancements in processing technologies and increased awareness are crucial to overcoming challenges and fully leveraging bamboo's potential [5]. Additionally, BF orientation has a significant impact on polymer matrix composite performance because aligned fibers exhibit better mechanical qualities than randomly oriented fibers [6].

This research examines the evaluation of bamboo-epoxy lamina, where the BFs are in bundle form. The mechanics of composite materials are used to evaluate and compare the effects of chemical retting and mechanical extraction of BBFs on the six engineering constants, minimum and critical fiber content, and longitudinal tensile strength of the bamboo-epoxy lamina. The study is divided into five sections. The first one provides an introduction, covering the extraction of BBFs, processing techniques, and the polymer matrix of BBFs. The study's bamboo-epoxy lamina and the methodology used to evaluate it are described in the second section. The findings of the evaluation of the effects of BBF extraction techniques on engineering constants, tensile strength, and critical volume fractions are presented in the third section, which is followed by a discussion and conclusion.

2 Analysis Methodology of Bamboo-Epoxy Lamina

In this study, we have considered BBFs as reinforcements in the bamboo-epoxy lamina and the purpose of this study was to assess the effect of chemical retting and mechanical extraction of BBFs on the tensile strength, engineering constants, and critical fiber volume fraction of bamboo-epoxy lamina. The assessment method adopted for the estimation of engineering constants is mentioned in previous studies [7]. Chemically retted bundle form bamboo fiber (CR-BBF) with tensile modulus and tensile strength of 23.56 GPa and 610 MPa, respectively, are considered for study, and mechanically extracted BBF with tensile modulus and tensile strength of 16.5 GPa and 290 MPa, respectively, are considered for the assessment study presented in this paper [8]. Epoxy resin with Young's modulus in the range 3.3–3.96 GPa is considered as matrix material in this study [9–11]. Tensile strength of epoxy resin obtained from two different specimen types is considered to estimate the tensile strength of the bamboo-epoxy lamina. Tensile strength of epoxy resin obtained using dog bone specimen was 92 MPa (considered as failure strength of epoxy) whereas that obtained from microscale specimen was 166 MPa (considered as true failure strength of epoxy) [12]. The Poisson's ratio of BBFs and epoxy resin considered in this study are in the range 0.180–0.334 and 0.32–0.44, respectively [13, 14]. The longitudinal tensile strengths of the composite lamina are calculated using Eqs. (1) and (2), and the minimum volume fraction $V_{f_{min}}$ and critical volume fraction $V_{f_{cr}}$ are given by the Eqs. (3) and (4) [15].

$$\left(\sigma_1^T\right)_u = \left(\sigma_f\right)_u V_f + \left[E_m \frac{\left(\sigma_f\right)_u}{E_f}\right]\left(1 - V_f\right) \tag{1}$$

$$\left(\sigma_1^T\right)_u = (\sigma_m)_u\left(1 - V_f\right) \tag{2}$$

$$V_{f_{min}} = \frac{(\sigma_m)_u - \frac{(\sigma_f)_u}{E_f}E_m}{(\sigma_m)_u - \frac{(\sigma_f)_u}{E_f}E_m + \left(\sigma_f\right)_u} \tag{3}$$

$$V_{f_{cr}} = \frac{(\sigma_m)_u - \frac{(\sigma_f)_u}{E_f}E_m}{\left(\sigma_f\right)_u - \frac{(\sigma_f)_u}{E_f}E_m} \tag{4}$$

where $\left(\sigma_1^T\right)_u$, $\left(\sigma_f\right)_u$, and $(\sigma_m)_u$ are the longitudinal tensile strengths of the composite lamina, fiber, and matrix, respectively, and $\left(\varepsilon_f\right)_u$ is the strain in the fibers corresponding to $\left(\sigma_f\right)_u$. We have applied the above described analytical approach in this paper for the assessment of the BBF reinforced epoxy lamina. Next section includes the results of the assessment at different alignments and volume fractions of BBFs.

3 Analysis Results of Bundle Form Bamboo Fiber-Reinforced Epoxy Lamina

The results of the evaluation of the BBF reinforced epoxy lamina are shown in the section that follows. The variation of E_x and E_y of CR-BFEL and ME-BFEL is also contrasted in Fig. 1. Figure 2 illustrates the variation of G_{xy} and ν_{xy} of CR-BFEL and ME-BFEL at different orientation angles. Similarly, Fig. 3 illustrates the variation of $\eta_{xy,x}$ and $\eta_{xy,y}$ of CR-BFEL and ME-BFEL.

Figures 4 and 5 show that, when taking into account the true failure strength of epoxy resin, the ME-BFEL has a minimum BBF volume fraction $V_{f_{min}}$ of 24.94% and a critical BBF volume fraction $V_{f_{cr}}$ of 43.73%. Figures 6 and 7 show that, when taking into account the true failure strength of epoxy resin, the CR-BFEL has a minimum BBF volume fraction $V_{f_{min}}$ of 9.4% and a critical BBF volume fraction

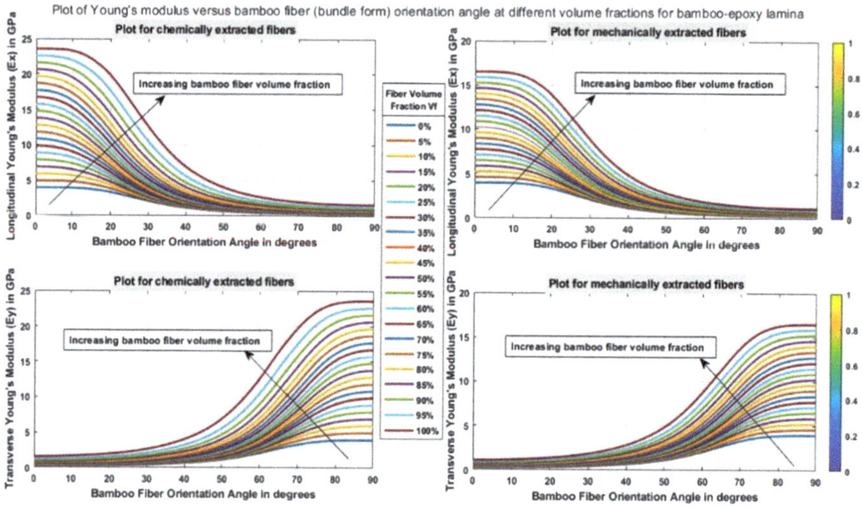

Fig. 1 Variation of E_x and E_y of bamboo-epoxy laminae

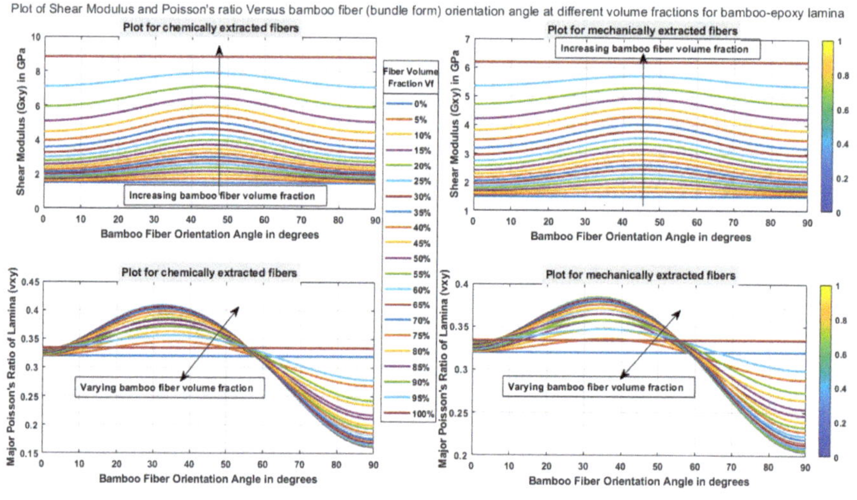

Fig. 2 Variation of G_{xy} and ν_{xy} of bamboo-epoxy laminae

$V_{f_{cr}}$ of 12.5%. Table 1 gives tensile strength of bamboo-epoxy lamina at crucial fibre volume fractions.

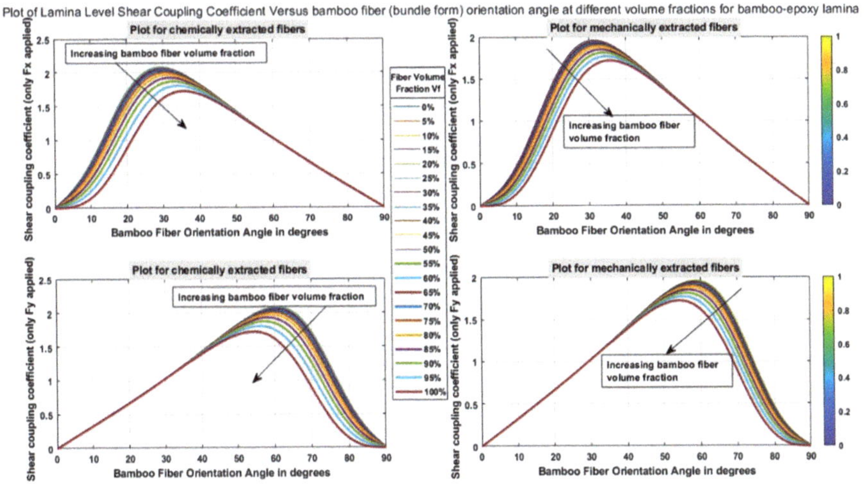

Fig. 3 Variation of $\eta_{xy,x}$ and $\eta_{xy,y}$ of bamboo-epoxy laminae

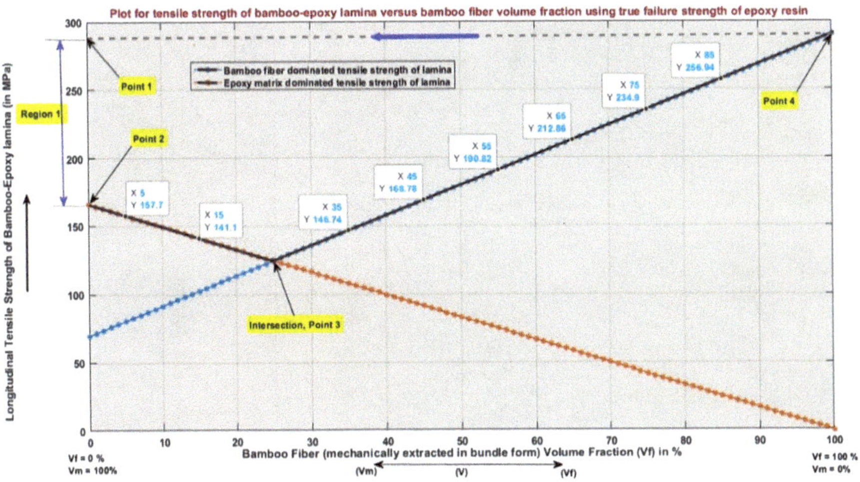

Fig. 4 Variation of $(\sigma_1^T)_u$ with the V_f for ME-BFEL

4 Discussion

The stiffness values of CR-BFEL are higher than those of ME-BFEL, regardless of the orientations and volume percentages of the BBFs. The shear modulus of CR-BFEL is higher than that of ME-BFEL, regardless of the orientations and volume percentages of the BBFs. Shear coupling coefficients are higher in CR-BFEL than in ME-BFEL at all BBF orientations and volume fractions. The minimum BBF volume fraction $V_{f_{min}}$ is 24.94% and critical BBF volume fraction $V_{f_{cr}}$ is 43.73% for

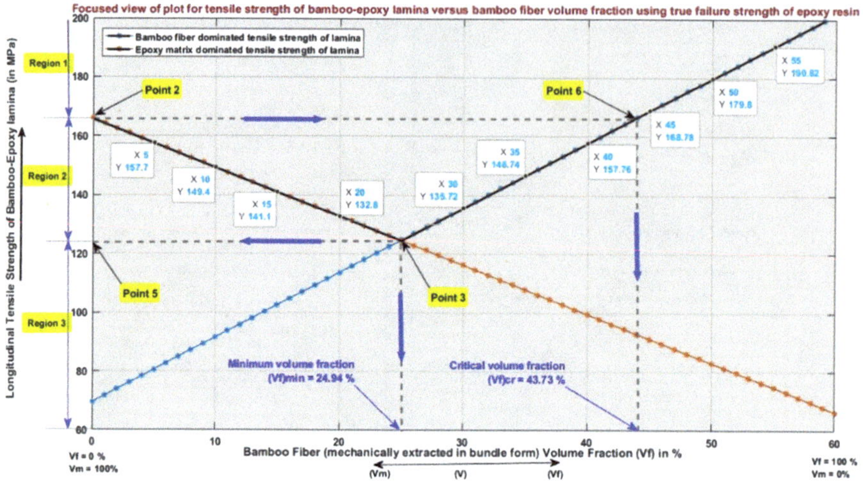

Fig. 5 Focused view of variation of $(\sigma_1^T)_u$ with the V_f for ME-BFEL

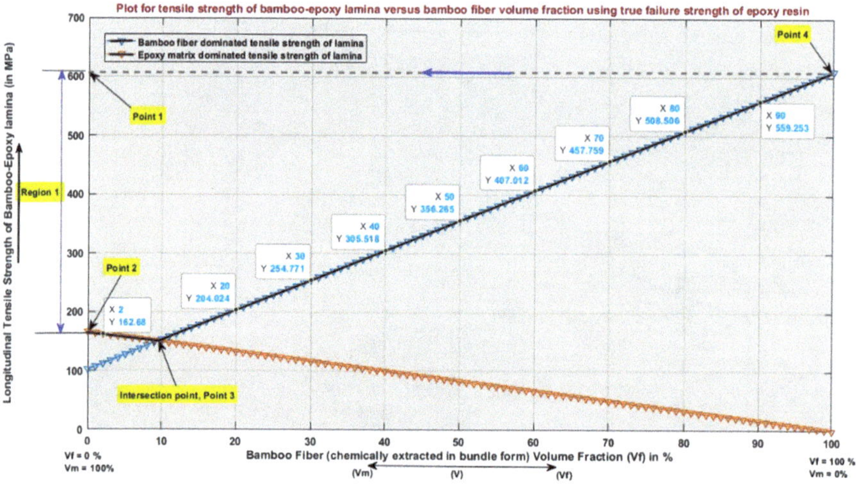

Fig. 6 Variation of $(\sigma_1^T)_u$ with the V_f for CR-BFEL

the ME-BFEL by considering the true failure strength of epoxy resin (166 MPa), whereas it is $V_{f_{min}} = 7.1\%$ and $V_{f_{cr}} = 10.1\%$ by considering the failure strength of epoxy resin (92 MPa). Similarly, the minimum BBF volume fraction $V_{f_{min}}$ is 9.4% and critical BBF volume fraction $V_{f_{cr}}$ is 12.5% for the CR-BFEL by considering the true failure strength of epoxy resin (166 MPa). The tensile strength of the CR-BFEL by considering the failure strength of the epoxy (92 MPa) is fully dominated by the CR-BBFs at all fiber volume fractions. Both CR-BFEL and ME-BFEL having any combination of fiber orientation and volume fraction with any orientation between

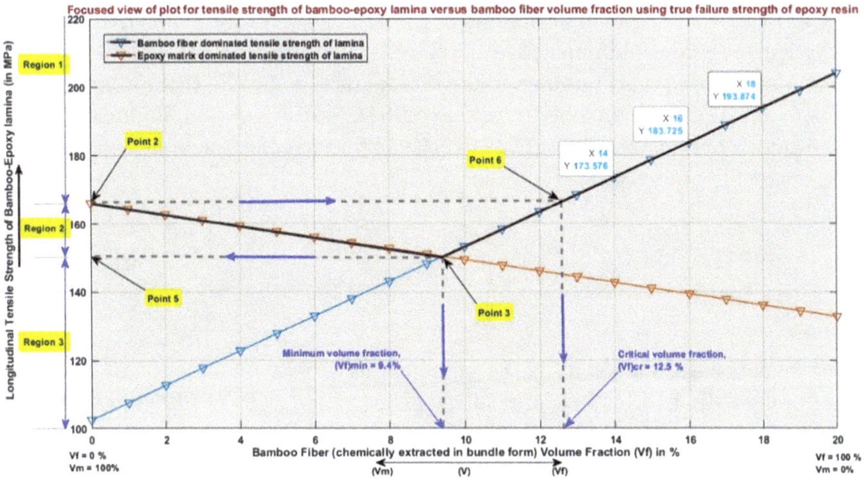

Fig. 7 Focused view of variation of $\left(\sigma_1^T\right)_u$ with the V_f for CR-BFEL

Table 1 Tensile strength of bamboo-epoxy lamina at crucial fiber volume fractions

Locations	Fiber volume fraction		Tensile strength	
	Mechanically (%)	Chemically (%)	Mechanically (MPa)	Chemically (MPa)
Point 1 and 4	0 and 100	0 and 100	290	610
Point 2	0	0	166	166
Point 3	24.94	9.4	124.6	150.4
Point 5	0	0	124.6	150.4
Point 6	43.73	12.5	166	166

30° and 90°, and any volume fraction between 0 to 70%, give almost the same values of E_x or the difference is very small or negligible. Similarly, in the case of E_y it is any orientation between 0° and 70°, and any volume fraction between 0 and 70%, and in case of G_{xy} it is any orientation between 0° and 90°, and any volume fraction between 0 to 50%. At all fiber volume fractions, both the CR-BFEL and ME-BFEL have the same value of ν_{xy}, $\eta_{xy,x}$, and $\eta_{xy,y}$ at fiber orientation of 57°, 70°, and 29°, respectively. The results and conclusions presented in this paper are of great importance in the design of face sheets for sandwich composites and the design of natural fiber reinforced polymer matrix composites. At all fiber volume fractions CR-BFEL shows higher values of ν_{xy}, $\eta_{xy,x}$, and $\eta_{xy,y}$ for fiber orientations between 0° and 57°, 0° and 61°, and 29° and 90°, respectively, but ME-BFEL shows higher values of ν_{xy}, $\eta_{xy,x}$, and $\eta_{xy,y}$ for fiber orientations between 57° and 90°, 61° and 90°, and 0° and 29°, respectively.

The CR-BBFs would incur high strain at modest loads, with all fibers breaking at the identical strain (which is statistically implausible). So, we observed that in case of the CR-BFEL the longitudinal strength of the lamina by considering the true failure strength of the epoxy is dominated by the epoxy matrix up to the fiber volume fraction of 9.4% but beyond this minimum fiber volume fraction the CR-BBFs dominate the strength of the lamina. Whereas in the case of the CR-BFEL the longitudinal strength of the lamina by considering the CR-BBFs at all fiber volume fractions dominate the failure strength of the epoxy.

5 Conclusion

It can be concluded that, at all BBFs' orientations and volume fractions, CR-BFEL is stiffer than ME-BFEL. Also, in case of tensile strength, CR-BFEL is stronger than ME-BFEL. As the BBFs' volume fraction increases at a given orientation, both E_x and E_y increase. As the BBFs' orientation increases, E_x decreases and E_y increases for all fiber volume fractions. When BBFs are aligned in linear direction, the bamboo-epoxy lamina gives the highest E_x, maximum E_y can be obtained when BBFs are oriented in the transverse direction, and shear characteristics at 45° fiber alignment result in a balanced transfer of stress. The analytical results presented in this paper need to be validated experimentally and numerically or the analytical procedure adopted in this paper can be used to validate numerical or experimental results of BBF reinforced epoxy lamina. Overall, CR-BBFs give the best desirable properties of the bamboo-epoxy lamina and further study is required on the single-form BFs isolated chemically and mechanically.

Acknowledgements The corresponding author Mr. B Vishwash would like to thank All India Council for Technical Education (AICTE), Government of India for providing scholarship to pursue Ph.D. at Indian Institute of Science, Bangalore under Quality Improvement Programme (QIP). Also, the corresponding author Mr. B Vishwash would like to thank the management of Siddaganga Institute of Technology, Tumakuru, Karnataka, India for providing study leave to pursue Ph.D. at Indian Institute of Science, Bangalore.

References

1. Hasan KMF, Hasan KMNA, Ahmed T, György ST, Pervez MN, Bejo L, Sándor B, Alpár T (2023) Sustainable bamboo fiber reinforced polymeric composites for structural applications: a mini review of recent advances and future prospects. Case Stud Chem Environ Eng 8:100362. https://doi.org/10.1016/j.cscee.2023.100362
2. Mundhe A, Kandasubramanian B (2024) Advancements in natural fiber composites: Innovative chemical surface treatments, characterizaton techniques, environmental sustainability, and wide-ranging applications. Hybrid Adv 7:100282. https://doi.org/10.1016/j.hybadv.2024.100282
3. Yadav M, Mathur A (2021) Bamboo as a sustainable material in the construction industry: an overview. Mater Today: Proc 43:2872–2876, Part 5. https://doi.org/10.1016/j.matpr.2021.01.125

4. Isukuru EJ, Ogunkeyede AO, Adebayo AA, Uruejoma MF (2023) Potentials of bamboo and its ecological benefits in Nigeria. Adv Bamboo Sci 4:100032. https://doi.org/10.1016/j.bamboo.2023.100032
5. Amjad AI (2024) Bamboo fibre: a sustainable solution for textile manufacturing. Adv Bamboo Sci 7:100088. https://doi.org/10.1016/j.bamboo.2024.100088
6. Vishwash B, Shivakumar ND, Sachidananda KB (2024) Analytical investigation of green composite lamina utilizing natural fiber to strengthen PLA. Hybrid Adv 7:100305. https://doi.org/10.1016/j.hybadv.2024.100305
7. Vishwash B, Sachidananda KB, Shivakumar ND (2024) Examination of the implications of fibers' alignment and volume fraction on the engineering constants of natural fiber epoxy lamina using analytical research. Ind Crop Prod 222:119615, Part 2. https://doi.org/10.1016/j.indcrop.2024.119615
8. Chen H, Cheng H, Wang G et al (2015) Tensile properties of bamboo in different sizes. J Wood Sci 61:552–561. https://doi.org/10.1007/s10086-015-1511-x
9. Davindrabrabu M, Januar PS, Dandi B, Mohd Ruzaimi MR, Cionita T (2014) Effect of fibre loading on the flexural properties of natural fibre reinforced polymer composites. AMM 695:85–88. https://doi.org/10.4028/www.scientific.net/amm.695.85
10. Tzetzisa D, Tsongasb K, Mansourb G (2017) Determination of the mechanical properties of epoxy silica nanocomposites through FEA supported evaluation of ball indentation test results. Mater Res 20(6):1571–1578. https://doi.org/10.1590/1980-5373-MR-2017-0454
11. Cha J, Kim J, Ryu S, Hong SH (2019) Comparison to mechanical properties of epoxy nanocomposites reinforced by functionalized carbon nanotubes and graphene nanoplatelets. Compos B Eng 162:283–288. https://doi.org/10.1016/j.compositesb.2018.11.011
12. Hobbiebrunken T, Fiedler B, Hojo M, Tanaka M (2007) Experimental determination of the true epoxy resin strength using micro-scaled specimens. Compos A Appl Sci Manuf 38(3):814–818. https://doi.org/10.1016/j.compositesa.2006.08.006
13. Lu H, Lian H, Xu J, Ma N, Zhou Z, Song Y, Yu Y, Zhang X (2022) Study on the variation pattern and influencing factors of Poisson's ratio of bamboo. Front Mater 9:896756. https://doi.org/10.3389/fmats.2022.896756
14. Saseendran S, Wysocki M, Varna J (2017) Cure-state dependent viscoelastic Poisson's ratio of LY5052 epoxy resin. Adv Manuf: Polym Compos Sci 3(3):92–100. https://doi.org/10.1080/20550340.2017.1348002
15. Huang ZM (2023) Chapter one—mechanics theories for anisotropic or composite materials. Stéphane, Bordas, PA (eds) Advances in applied mechanics. Elsevier, vol 56, pp 1–137. https://doi.org/10.1016/bs.aams.2022.09.004

Corrosion Resistance of Self-Compacting Concrete Incorporating Dimensional Limestone Slurry Waste: A Study on Chloride Diffusion and Surface Electrical Resistivity

Ramswaroop Mandolia, Pawan Kalla, Ravindra Nagar, and Sanchi Rewar

Abstract Self-compacting concrete (SCC) is a modern way for sustainable and durable constructions. The durability of concrete is degraded by corrosion of reinforcement through adverse environmental impact. Corrosion resistance is an essential durability parameter for adopting waste materials. Electrical resistance is a useful non-destructive test (NDT) method to determine the potential corrosion of concrete. This study examined the corrosion resistance of Dimensional Limestone Slurry Waste (DLSW)-based SCC by surface electrical resistivity meter and chloride diffusion coefficient at 28 and 90 days. The DLSW was replaced as 15, 30, 45, 60, 80, and 100% with fine aggregate in SCC. The chloride diffusion coefficient was observed in a 3% sodium chloride solution on SCC samples, and the result of a 30% replacement level was optimized. The SCC samples up to 80% replacement ranged in negligible risk of corrosion at 90 days and had a strong correlation with an apparent density of $R^2 = 0.90$. This study introduces the innovative use of DLSW as a sustainable fine aggregate replacement in SCC, demonstrating improved corrosion resistance, reduced porosity, and enhanced durability up to 45–60% replacement levels.

Keywords Corrosion · Dimensional limestone slurry waste (DLSW) · Self-compacting concrete (SCC) · Surface electrical resistivity · Sustainable

1 Introduction

The preferred structural design method in civil engineering in the current scenario is reinforced concrete worldwide. The most common reinforcement is used to strengthen concrete steel reinforcement because it can be significantly deformed before sudden failure. Over time, the presence of moisture in concrete traveled from

R. Mandolia (✉) · P. Kalla · R. Nagar · S. Rewar
Malaviya National Institute of Technology, Jaipur, India
e-mail: 2018rce9060@mnit.ac.in

I. M. Jamadar et al. (eds.), *Integrating Sustainability in Modern Technologies*, Lecture Notes in Mechanical Engineering, https://doi.org/10.1007/978-981-95-9466-5_19

surface to reinforcement through cracks and porous structure, creating a conducive environment for the corrosion of rebars. Concrete structure degradation and deterioration are directly caused by reinforcing corrosion [1]. The atmospheric carbon dioxide makes an alkaline environment in concrete when it comes into contact with calcium hydroxide and produces calcium carbonate. The presence of chloride from curing water and salty coastal environments also causes the degradation of concrete by creating potting corrosion [2]. Corrosion occurs by reducing the pH below nine, and chloride disrupts the layer on rebars. Corrosion of steel deteriorates both the mechanical performance and structural integrity of concrete, causing fractures owing to the volume expansion of corrosion products [3]. An in-depth study is required on the transportation of chloride behavior in limestone self-compacting concrete for a creative solution to address this issue. SCC needs a higher fine content, which can be fulfilled by DLSW as a fine aggregate replacement. SCC was developed in 1988 to improve the durability parameters of reinforced concrete structures that can compact and fill in narrow structures with conjected reinforcement by self-weight. SCC needs a high mortar volume, superplasticizer dose, and a lesser coarse aggregate proportion than traditional normal concrete [4, 5]. The water-to-binder ratio exerts a significant influence on the fresh and hardened characteristics of SCC, and the addition of the limestone powder accelerated the hydration progression, effectively boosting the mechanical and durability properties [6]. Chouhan et al. got favorable results on mechanical and durability with 40% replacement of DLSW to fine aggregate in a mortar [7]. Earlier researchers in normal concrete and self-consolidated concrete have stated the effect of limestone additives on concrete corrosion and permeability resistance properties. Panesar et al. used limestone in the binder part of self-consolidating concrete and reported that chloride permeability was reduced and increased the density of concrete by adding it [8]. Hornainl et al. concluded that chloride penetration was reduced by limestone filler in concrete [9]. Ghafoori et al. reported that chloride migration and penetration of SCC were reduced by adding limestone to cement [10]. Ozkan Sengul reported that electrical resistivity was comparatively higher with limestone aggregate to normal aggregate [11]. As per the literature review, the DLSW improved the durability performance of sustainable concrete, and limited studies were available on the corrosion impact of FA replacement.

2 Materials and Methodology

2.1 Raw Constituents, Mixing, and Testing Procedure

The raw constituents used to make SCC were Ordinary Portland Cement (OPC-53 grade) as a binder, coarse aggregates (CA) of 10 mm, and river sand as fine aggregate (FA) of Zone-II as per IS 383. The river sand was replaced with DLSW by weight in modified SCC up to 100%. The fineness modulus of FA and DLSW was 2.51 and

Table 1 Mix proportioning in rational

Mix ID	Water	OPC	Fine aggregate	Coarse aggregate	DLSW	28 days compressive strength (MPa)
RC0	0.33	1	1.667	1.525	0.00	0.75
R15	0.33	1	1.417	1.525	0.250	0.72
R30	0.33	1	1.334	1.525	0.500	0.70
R45	0.33	1	0.917	1.525	0.750	0.90
R60	0.33	1	0.667	1.525	1.000	1.10
R80	0.33	1	0.333	1.525	1.334	1.40
R100	0.33	1	0.00	1.525	1.667	1.50

* SP dose in percentage of the by weight of OPC Cement

1.35, respectively. Specific gravity was measured as 2.65 and 2.59 for FA and DLSW, respectively. A high range (20–30%) water reducer superplasticizer (SP) was used to fulfill the workability standards as given in EFNARC2005 guidelines. The control mix (RC0) proportioning was finalized as per guidelines of EFNARC2005, IS 10262, and previous studies to prepare and fulfill the fresh state checks of SCC mixes. A control (RC0) and six other DLSW-modified SCC were developed with a constant water-to-binder ratio of 0.33 and noted as R15, R30, R45, R60, R80, and R100. The numeric value of mix ID represents the percentage replacement level. The control mix contains a coarse aggregate of 839.0 kg, water of 183 kg, cement of 550 kg, and FA of 917.0 kg for one cubic meter SCC. The rationale mix design is shown in Table 1. All raw materials are weighed carefully and mixed in an electrically operated tilting mixture for about 8 to 10 minutes until the mix makes the homogeneous mix. The flowability, rheology, and viscosity of the SCC are carefully modified by the correct adoption of binder-to-water ratio and coarse aggregate-to-fine aggregate ratio and then by adding a superplasticizer. The slump flow was kept in a range of (720 ± 30) mm to maintain the uniformity and consistency of SCC. The volume of the paste was kept higher to fill the pore volume of the fine aggregate and coated by cement paste of all individual aggregate particles. The specimens for apparent density and corrosion testing were cast without the use of any mechanical vibration; it was directly poured into the mold, finished the exposed surface, and remained unmolded for 24 h at room temperature. All specimens are then immersed in a temperature-controlled water curing tank for curing until the testing age. The compressive strength was taken at an average of three 100 mm cubic samples at 28 days, as shown in Table 1.

2.2 Permeable Voids and Apparent Density

The average (100 mm * 100 mm * 100 mm) size of three cubes was used at the age of 28 days to observe the percentage of permeable voids and apparent density as per ASTM C642. Initially, the oven-dried weight (W1) of cubes was boiled at 100-degree temperature for 5–6 h, and subsequently, the weight was noted at room temperature (W2) in the air. Subsequently, the submerged weight (W3) was measured, and the formula given below for calculation as.

Apparent density in g/cm^3 = W1/(W1-W3)

Percentage of permeable voids in % = {(W2-W1)/(W2-W3)} *100

2.3 Chloride Ion Diffusion and Surface Electrical Resistivity

A split cell method was used to determine the chloride ion diffusion of SCC samples under steady-state conditions. The cylindrical specimens (70 mm diameter and 35mm thick) taken from 28 days were cured and sealed on the circular side with silicon sealant to prevent water motion through it. Subsequently, it was fixed in the circular rubber ring and dipped in a distilled water container to develop saturated conditions for 24 h. The anodic cell and cathodic cell were filled with distilled water and 3% NaCl mixed water, respectively, and 30 V DC was applied through the cells, as shown in Fig. 1a. The chloride concentration was measured from the anodic cell solution by titration method using 0.0141 N ($AgNO_3$) solution at a regular timeframe. The Nernst–Planck equation was used to determine the chloride diffusion coefficient (Dssm) [12].

Dssm = (V2/A) * (ΔC/Δt) * (RTL/zFEc1)

where ΔC = chloride concentration difference, Δ/t = initial and final time difference, V2 = volume of water of anodic cell, A = cross section area of the sample in (m2), R = gas constant (8.314J/Kmol), T = average value of temperature (K, Anolyte), L = thickness of specimen in (m), z = chloride ion valence = 1, F = Faraday constant = 9.648 * 104 J/V mol, E = volt in cells, c1 = initial chloride concentrations in the cell.

For the surface electrical resistivity test, the 100 mm by 200 mm cylindrical specimens were cast by fresh SCC and remained unmolded to get a hardened state for 24 h at room temperature. The hardened specimens were then cured for up to 90

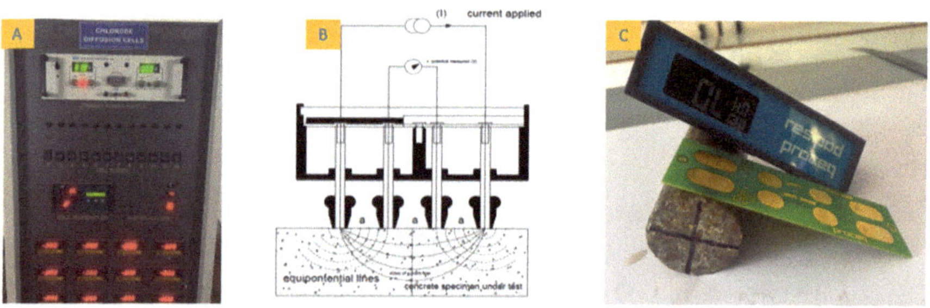

Fig. 1 Test setup of **a** chloride diffusion test, **b** surface resistivity technique, and **c** resistivity test

days to complete the hydration process of the mix. Over this period, the specimens were oven-dried and cooled at ambient temperature to all pore routes will be open. The corrosion potential was measured at 28 and 90 days by an electrical resistivity meter on the longitudinal surface of the lime water surface saturated conditions with two repetitions at every quadrant (Fig. 1b). All probs of the resistivity meter should be touched on the surface of specimens and recorded in the kΩcm unit.

The Wenner four-probe system as shown in Fig. 1c, was used to measure the surface electrical resistance of concrete specimens using a Resipod family device from Proceq. It is directly linked to the probability and intensity of corrosion. Higher resistance indicates reduced reinforcement corrosion. The principle of Wenner's four-probe system works as follows:

Resistivity ρ in kΩcm $= 2\pi a\ (V/I)$

where a = spacing between the adjacent electrodes, V = difference voltage, I = current applied through the materials, kΩcm = kilo ohm centimeter.

The outer probs of resistivity are known as current electrodes and inject the current into the concrete, and inner electrodes are known as voltage electrodes, which are used to measure the potential difference generated by the current flow. The concrete resistivity indicates that concrete resistance to ionic moment represented the low corrosion risk. The estimation of the potential range of corrosion, if resistivity ρ $\geq$ 100 kΩcm (negligible risk of corrosion), 50 kΩcm to 100 kΩcm (low risk of corrosion), 10 kΩcm to 50 kΩcm (moderate risk of corrosion), and $0 \leq 10$ kΩcm (high risk of corrosion).

3 Result and Discussion

3.1 Permeable Voids and Apparent Density

The results of the permeable voids and apparent density test are displayed in Fig. 2. Compared to the control, the R15, R30, and R45 mix increased density by 0.51%, 1.37%, and 1.99%, respectively, and furthermore, it was decreased by 3.02%, 3.80%, and 1.96% for R60, R80, and R100 mix, respectively. Similarly, the permeable voids ratio ranged from 4.67% to 10.20%. The R45 mix resulted in the highest apparent density by the inclusion of DLSW. The outcome shows that lower permeability void fractions are related to higher apparent density or vice versa. The findings of the permeability voids and apparent density tests also support the compressive strength. The increase in density and decrease in voids may be produced by the internal healing activity of fine particles. The fresh mix contains water that helps produce the thick C-S-H gel, which produces the maximum density of the SCC mix [13]. Concrete mixes with lower-sized fine particles are made denser because of their improved capacity for filling [14, 15]. Furthermore, higher DLSW replacement levels produced voids with a lower density and a higher permeability by the formation of a porous concrete matrix. Lack of graded fine particles may cause a poor microstructure and larger

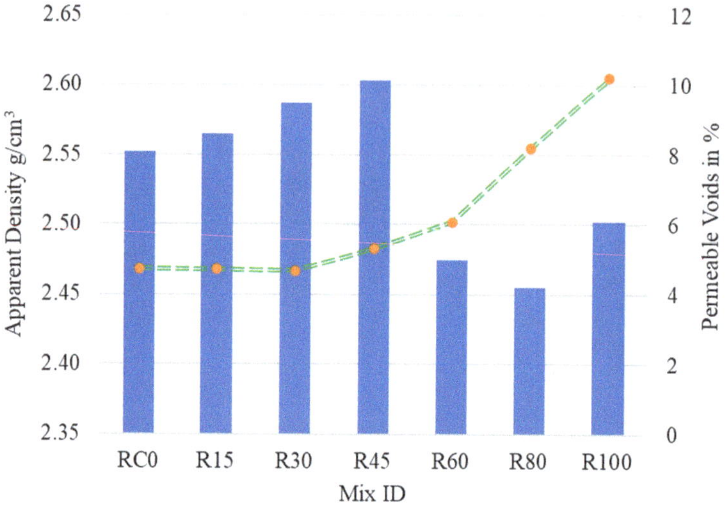

Fig. 2 Apparent density and permeable voids at 28 days

voids in the SCC matrix at higher DLSW replacement levels. The lower specific gravity of DLSW marginally reduced the density compared to fine aggregate when DLSW replacement levels increased.

3.2 *Chloride Diffusion and Electrical Resistivity*

The chloride diffusion coefficient (Dssm) was obtained in range from $2.75E^{-12}$ to $6.45E^{-12}$ and $2.03E^{-12}$ to $5.14E^{-12}$ at 28 and 90 days, respectively, as shown in Fig. 3a. The R30 mix exhibited lower Dssm in both durations than the control and minimum for all SCC mixes. From control to R45 mix, it was reduced, and there was a slight rise in the R60 mix and a significant rise in the R80 and R100 mix at 28 days. Similarly, 90 days Dssm results exhibited the same trend. Comparatively 28–90 days, it was changed and improved as $4.11E^{-12}$ to $3.57E^{-12}$, $2.75E^{-12}$ to $2.03E^{-12}$, $4.08E^{-12}$ to $3.73E^{-12}$, and $6.45E^{-12}$ to $5.14E^{-12}$ for RC0, R30, R45, and R100 mix, respectively. The limestone fines slow down the water diffusion in concrete [16].

The capacity of concrete to resist ion movement in the presence of an electrical field is known as its electrical resistivity, which is the opposite of conductivity. The obtained results at 28 and 90 days are displayed in Fig. 3b, and they ranged from 65 kΩcm to 128 kΩcm and 83 kΩcm to 186 kΩcm, for 28 and 90 days, respectively. At the age of 28 days, the R15 to R45 mix had a value of more than then 100 kΩcm, the control mix was close to 100 kΩcm, and the rest of the mix ranged above 50 kΩcm. The control and R15, R30, and R45 mix exhibited negligible risk of corrosion, and the rest of the mix ranged in low risk of potential corrosion.

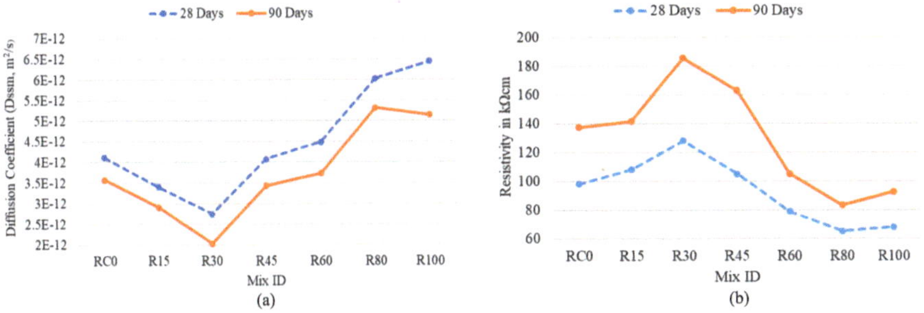

Fig. 3 **a** Chloride diffusion coefficient, and **b** surface electrical resistivity

Furthermore, the 90-day results also improved the resistivity in all mixes and up to a 60% DLSW replacement ranged in no risk of corrosion. All SCC samples gain strength by hydration at 90 days, resulting in improved density and reduction of porosity, which is the reason for resistivity improvement. The R80 and R100 mix ranged in low risk. In both durations, the 30% replacement was optimized, and results improved up to a 45% replacement. The results of permeable voids and Dssm established a strong correlation of $R^2 = 0.88$, shown in Fig. 4a, indicating that the DLSW ingression in SCC reduced the permeable voids and Dssm. Another strong correlation of $R^2 = 0.90$ was established between surface electrical resistivity and apparent density, as shown in Fig. 4b. A similar outcome was reported by B. Tripathi and S. Chaudhary [17], who enhanced the chloride diffusion by using ISF slag as a fine replacement. The electrical resistance and chloride diffusion coefficient highly depend on the density and porosity of the concrete. A higher porosity and lower density decrease the performance of SCC. By the inclusion of DLSW in SCC improves the density, which is mentioned in Sect. 3.1. Gloria Cosoli et al. [18] explained the relation that the corrosion rate increased by decreasing surface electrical resistivity.

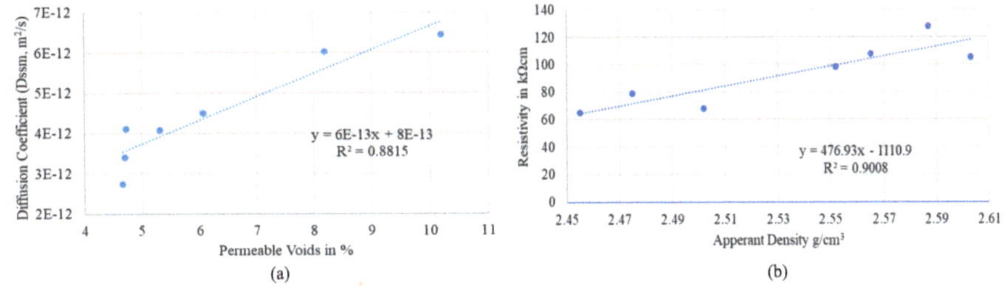

Fig. 4 Correlation between **a** chloride diffusion coefficient with permeable voids and **b** surface electrical resistivity with apparent density

4 Conclusions

For utilization of DLSW as alternative waste in durable concrete, it was essential to understand the corrosion potential and diffusion behavior of SCC. Based on the study, the following conclusions are mentioned below.

- The ingression of DLSW enhanced the density and reduced the porosity in SCC up to a 45% replacement. The mechanical and durability performance strongly depend on density. A higher density with lower porosity improved the overall performance of SCC.
- The DLSW had lower fineness than FA, which helped to enhance the density of SCC and reduce the porosity, making a strong mortar layer covering all the aggregate particles. The dense microstructure prevents the harmful chloride ions movement in concrete that secures the reinforcement to corrosion.
- Chloride diffusion coefficient (Dssm) was decreased by the inclusion of DLSW at both 28 and 90 days compared to the control mix; the DLSW particle slowed down the penetration of chloride ions in concrete, indicating the prevention of corrosion. The Dssm was consistently decreased with the specimens age was increased in all SCC.
- The 30% replacement level was optimized for the chloride diffusion coefficient, surface electrical resistivity properties, and compressive strength.
- The surface electrical resistivity also improved in DLSW-modified SCC and resulted in up to 60% replacement draw in negligible risk of corrosion at 90 days with ranged over 100 kΩcm. The corrosion resistance was improved with the curing age increased from 28 to 90 days by the completion of the hydration process.
- Strong correlations were established between apparent density to electrical resistivity (R2 = 0.90) and chloride diffusion coefficient with permeable voids (R^2 = 0.88) by incorporation of DLWS in SCC.
- The incorporation of DLSW improved the durability of SCC and contributed to a sustainable and eco-friendly construction sector.

References

1. Jiang L, Li C, Zhu C, Song Z, Chu H (2017) Constr Build Mater 151:119
2. Diab AM, Aliabdo AA, Mohamed IA (2015) Mag Concr Res 67:747
3. Li C, Jiang L, Li S (2020) Cem Concr Res 131
4. Okamura H, Ouchi M (2003) Self-compacting concrete
5. Okamura H, Ozawa K (1996) Struct Eng Int: J Int Assoc Bridg Struct Eng (IABSE) 6:269
6. Domone PL (2007) Cem Concr Compos 29:1
7. Chouhan HS, Kalla P, Nagar R, Gautam PK (2019) Constr Build Mater 221:363
8. Panesar DK, Aqel M, Rhead D, Schell H (2017) Cem Concr Compos 80:175
9. Hornain H, Marchand J, Duhot V, Moranville-Regourd M (1995) Cem Concr Res 25:1667
10. Ghafoori N, Spitek R, Najimi M (2016) Constr Build Mater 127:588

11. Sengul O (2014) Constr Build Mater 73:434
12. Basheer PAM, Andrews RJ, Robinson DJ, Long AE (2005) In NDT and E International, pp 219–229
13. Siddique S, Shrivastava S, SC-C, Building and undefined (2018) Elsevier. Siddique S, Shrivastava S, Chaudhary S (2018) Constr Build Mater
14. Amani A, Babazadeh A, Sabohanian A, Khalilianpoor A (2021) Int J Pavement Eng 22:1531
15. Gautam L, Jain J, Kalla P, MD-MT (2021) Proceedings, and undefined, Elsevier. Gautam L, Jain JK, Kalla P, Danish M (2021) Mater Today: Proc
16. Valcuende M, Marco E, Parra C, Serna P (2012) Cem Concr Res 42:583
17. Tripathi B, Chaudhary S (2016) J Clean Prod 112:672
18. Cosoli G, Mobili A, Tittarelli F, Revel GM, Chiariotti P (2020) Appl Sci (Switz) 10:1

Exploring Sustainable Cr_3C_2-NiCr-Based Laser Cladding: A Study on Wear Behavior and Surface Properties

R. Meenakshi Reddy, Ajit M. Hebbale, Manikanth Vatala, S. Karvendhan, T. Coumaressin, S. Karthikeyan, and Chappeli Sai Kiran

Abstract Laser cladding of Cr_3C_2-NiCr-based systems have recently gained interest as a surface engineering solution to provide enhanced wear resistance, corrosion resistance, and improved durability of industrial components. For this study, T91 ferritic steel substrates were coated with laser cladding optimized parameters Cr_3C_2-NiCr composite clads. Microstructural analysis showed homogenous distribution of carbide phases and sound metallurgical bonding with the substrate guaranteeing good stability of the structure. On wear tests under a 20 N load, the samples showed superior wear resistance with minimal material loss and maximum wear track depth of 142.8 μm and validated uniform material removal and smooth wear track morphology. With optimized process parameters and advanced characterization techniques, defect-free clads with improved mechanical and tribological properties were successfully developed. These results demonstrate the promise of laser cladding

R. Meenakshi Reddy
Mechanical Engineering Department, G. Pulla Reddy Engineering College, Kurnool, Andhra Pradesh, India

A. M. Hebbale (✉)
Nitte (Deemed to Be University), Department of Mechanical Engineering, NMAM Institute of Technology (NMAMIT), Nitte 574110, Karnataka, India
e-mail: ajit.hebbale@gmail.com

M. Vatala
Mechanical Engineering Department, S.R.K.R Engineering College, Andhra Pradesh, India

S. Karvendhan
Velalar College of Engineering and Technology, Erode, India

T. Coumaressin
Mechanical Engineering Department, Sri Manakula Vinayagar Engineering College Madagadipet, Puducherry, India

S. Karthikeyan
Mechanical Engineering Department, Erode Sengunthar Engineering College, Perundurai, India

C. S. Kiran
Mechanical Engineering Department, CVR College of Engineering, Hyderabad, Telangana, India

I. M. Jamadar et al. (eds.), *Integrating Sustainability in Modern Technologies*, Lecture Notes in Mechanical Engineering, https://doi.org/10.1007/978-981-95-9466-5_20

with Cr_3C_2-NiCr as a sustainable solution for developing materials with resource efficient and durable performance for high-performance industrial applications.

Keywords Laser cladding · Cr_3C_2-NiCr composite · Wear resistance · Surface engineering · T91 ferritic steel

1 Introduction

Recently, advances in laser cladding have been the subject of much interest because it can improve the wear, corrosion, and mechanical properties of materials. The process entails deposition of a cladding substance onto a substrate via a laser heat resource to achieve improved surface characteristics and longer component life. The continued need for sustainable and high-performance materials for industrial applications makes this study important [1].

Carbide-based and metal matrix composite systems have been shown in previous research to be capable of producing laser clads with superior properties. The studies show that Cr_3C_2-NiCr-based cladding has good hardness, thermal stability, and wear resistance and is a potential candidate for use in harsh environment [2]. Still, there are many questions as to what microstructural evolution, elemental distribution, and cladding system behavior might be under different operating conditions [3]. Recent advances in laser cladding technology have resulted in optimized in the parameters, scanning speed, powder feed rates, and laser power, to produce defect-free clads with uniform microstructures. Several studies have reported that Cr_3C_2-NiCr-based clads provide enhanced metallurgical bonding to the substrates, which allows the substrate to sustain higher load at extreme operating conditions as well as stability of structure. Newer additive manufacturing and cladding techniques have given control over deposition of material as well as decreased waste and increased process sustainability. Additionally, field emission FE-SEM and EDS characterization tools have been used to obtain a more detailed understanding of the microstructural evolution and phase formation in the clad layers [4]. Research was carried out on the wear mechanisms and surface behavior of the Cr_3C_2-NiCr clads under a wide range of loading conditions. Studies on the wear of ball-on-plate tribometer have shown that the reinforced carbide phases in these clads greatly reduce material loss and form a tougher surface for high-stress environments [5]. In addition, 3D surface profilometry developments have permitted detailed analysis of wear tracks, which allow information pertaining to the material performance to be gained and provide evidence for further optimization of the cladding process [6, 7].

Research on Cr_3C_2-NiCr laser cladding has been presented in Table 1 which shows its promise in enhancing surface durability, wear resistance, and sustainability. It is applicable in critical industrial sectors due to advanced characterization and optimized processes leading to resource efficient and environmentally responsible manufacturing [8].

Table 1 Summary of recent developments in Cr_3C_2-NiCr-based laser cladding [1–11]

Aspect	Details	Impact/Significance
Material composition	Cr_3C_2-NiCr (75 wt% Cr_3C_2—25 wt% NiCr) composite powders with particle sizes of ~45 μm	Provides high hardness, wear resistance, and thermal stability suitable for extreme environments
Substrate material	T91 ferritic alloy steel	Excellent compatibility with Cr_3C_2-NiCr, ensuring good metallurgical bonding
Laser cladding process	10 kW diode laser, 5 mm/s scanning speed, square beam spot size of 6 mm, and 8 g/min feed rate	Optimized process parameters ensure defect-free clads and minimal energy/material waste
Microstructure features	Fine dendritic structures in the clad, strong metallurgical bonding with substrate	Contributes to enhanced mechanical properties and resistance to wear and deformation
Wear behavior	Evaluated using ball-on-plate tribometers under 20 N load	Revealed excellent wear resistance with minimal material loss and uniform wear track morphology
Characterization tools	FE-SEM, EDS, and 3D surface profilometry	Enabled in-depth analysis of microstructure, phase distribution, and surface topography
Applications	Aerospace, automotive, and power generation components exposed to high wear and thermal stress	High-performance, durable coatings suitable for industrial environments with extreme operating conditions
Sustainability	Resource-efficient process, minimized material waste, and extended component lifespan	Promotes environmentally sustainable manufacturing and reduced ecological footprint
Thermal conductivity and stability	Evaluation of thermal properties, such as thermal conductivity and resistance to thermal cycling	Essential for applications in high-temperature environments such as aerospace and power generation, where thermal stresses can significantly affect clad performance
Oxidation resistance	Performance under oxidative environments, especially at elevated temperatures	Relevant for applications in industries like automotive and gas turbines, where materials are exposed to oxidizing atmospheres over long durations

(continued)

Table 1 (continued)

Aspect	Details	Impact/Significance
Adhesion and residual stresses	Adhesion strength at the clad-substrate interface and analysis of residual stresses using techniques like XRD	Ensures long-term durability and reliability of the clads, particularly under cyclic loading and high-temperature conditions
Environmental durability	Resistance to chemical degradation in harsh environments, such as acidic or saline solutions	Crucial for marine and chemical processing industries where clads are exposed to corrosive chemicals or seawater
Impact and fracture resistance	Evaluation of impact toughness and resistance to fracture under dynamic loading conditions	Important for applications in heavy machinery, mining, and defense where sudden impacts or dynamic loads are common
Energy efficiency of the process	Assessment of energy consumption and optimization of process parameters to reduce overall energy usage	Directly aligns with sustainability goals and contributes to reducing the carbon footprint of manufacturing processes
Powder recycling and reusability	Study of the reusability of unsintered powders post-cladding	Reduces waste and promotes cost-effective, sustainable manufacturing practices, aligning with circular economy principles
Hybrid coating strategies	Combining Cr_3C_2-NiCr with other cladding materials or techniques, such as multi-layered clads or functionally graded materials	Enhances material properties by combining advantages of different materials or techniques, making them suitable for multifunctional applications
Nano-scale reinforcement	Incorporating nano-carbides or other nano-reinforcements into the Cr_3C_2-NiCr matrix	Provides ultra-high hardness and wear resistance, beneficial for advanced applications like micro-machining and precision components
Long-term performance studies	Field tests and simulations to predict long-term behavior under real-world conditions	Provides a realistic understanding of performance, reliability, and maintenance requirements over an extended operational period
Comparison with competing materials	Comparative analysis with other cladding materials like WC-Co or TiC-Ni-based systems	Highlights the unique advantages and trade-offs of Cr_3C_2-NiCr systems, aiding in material selection for specific applications

(continued)

Table 1 (continued)

Aspect	Details	Impact/Significance
Economic analysis	Cost–benefit analysis considering material costs, process efficiency, and lifecycle performance	Demonstrates the economic viability of Cr_3C_2-NiCr cladding systems for widespread industrial adoption
Additive manufacturing integration	Exploration of incorporating Cr_3C_2-NiCr in additive manufacturing processes like direct metal deposition	Expands the application scope and compatibility of Cr_3C_2-NiCr clads with emerging manufacturing technologies

This study investigates the wear behavior and surface properties of Cr_3C_2-NiCr laser clads, aiming to advance sustainable materials for high-performance engineering applications. It addresses challenges such as substrate properties, residual stress management, and long-term performance in diverse environments. The results will enable the development of more durable and efficient cladding systems, improving sustainability and reliability in engineering applications.

2 Materials and Experimental Details

In this study, accessible 75% Cr_3C_2–25% NiCr composite powder (Make: clads were fabricated on T91 ferritic alloy steel substrates) with a particle size of 45 μm. T91 substrate were precision machined to the required dimensions. Optimal cladding quality depends on the proper preparation of substrate and powder. SiC emery papers with grits 1000 and 1500 were used to polish T91 steel substrates to an initial surface roughness value of 0.12 μm. The substrates were subsequently cleaned with acetone to remove surface contaminants. Using laser cladding technique, the Cr_3C_2-NiCr composite clads were synthesized with precise material deposition and microstructural integrity. Laser energy was used to fabricate Cr_3C_2-NiCr composite clads on T91 substrates. The laser setup system could run either continuously or in pulse mode. The cladding powder was delivered onto the substrate during laser interaction by an off axis powder feeding nozzle assisted by argon gas. The laser beam and substrate were maintained 14 mm apart, or a vertical distance. Multiple experimental runs were conducted to optimize the process parameters, and the final clads were developed at a laser power of 2000 W, a scanning speed of 5 mm/s, and a powder feed rate of 8 g/min.

The developed Cr_3C_2-NiCr laser clads were sectioned along the cross-section and molded in epoxy resin. Polishing of the samples was done using typical metallographic procedures. The wear behavior was evaluated utilizing a linear reciprocating ball-on-plate tribometer, according to the ASTM G133 standard at 20 N load for 10 min. Samples of the clad specimens were rough polished and wire cut for wear

Fig. 1 Typical optical image shows Cr_3C_2-NiCr based area clads developed through laser energy

tests. Figure 1 displays the developed laser clad cross-sectional micrograph as a representative section. Surface characteristics were also assessed with 3D profilometric surface analysis of the wear tested samples.

3 Results and Discussions

3.1 *Microstructure Analysis*

The Cr_3C_2-NiCr-based laser clads microstructural characteristics in cross-section are illustrated in Fig. 2 as captured by FE-SEM. A sectional view of the laser clad is shown in Fig. 2a, showing the uniform distribution of the composite material throughout the clad region. It can be seen that the image presents three various regions: the clad layer and substrate. The clad layer has a dense microstructure with well-formed dendritic patterns indicative of rapid solidification thought to arise from the nature of the experimental process. The mechanical and wear properties of the clad are expected to be improved due to this microstructural refinement. The laser cladding parameters show a metallurgically bonded interface between the clad and substrate, with no visible defects. Dendritic structures indicate a thermal gradient during solidification, facilitating clad material integration. Fine carbide phases enhance hardness and wear resistance. This clear transition from the clad to the substrate confirms a robust, defect-free cladding process, crucial for high surface integrity applications. The Cr_3C_2-NiCr composite is suitable for laser cladding on T91 ferritic alloy steel.

3.2 *Elemental Analysis*

Figure 3 presents a complete EDS analysis of the Cr_3C_2-NiCr laser clads showing the distribution of metals in various clad regions. Panel (a) presents an FE-SEM image highlighting three distinct locations for the EDS study: Point 1: the clad layer, Point 2: the interface, Point 3: the gray phase. In Fig. 3b, the EDS spectrum of Point 1 (clad layer) exhibits a high concentration of chromium (Cr) and nickel (Ni) as a

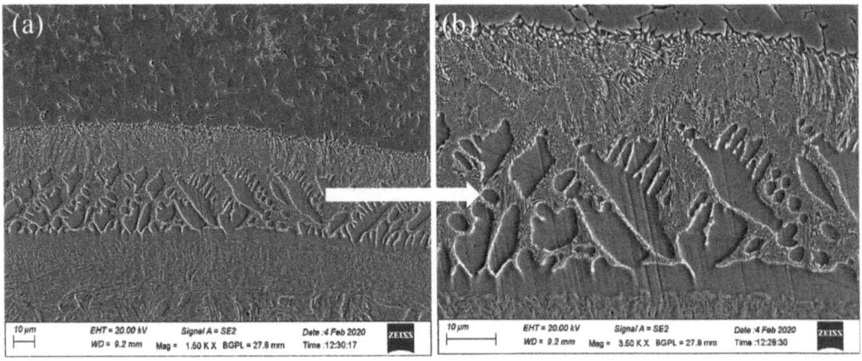

Fig. 2 Microstructure of the generated laser clad segment based on Cr_3C_2-NiCr: **a** A transverse segment of Cr_3C_2-NiCr laser clad is shown in an FE-SEM image; **b** an augmented view of the interface region

result of the uniform deposition of the Cr_3C_2-NiCr composite. Results indicate partial diffusion from the substrate into the clad during the laser cladding process, possibly due to the presence of minor amounts of iron (Fe) and molybdenum (Mo). The EDS analysis at Point 2 is shown in Fig. 3c. Results from the spectrum show a composition dominated by Cr and Fe, which is consistent with the metallurgical bonding between the developed clad and T91 substrate. Moderate diffusion of Ni into the interface is observed in this region, indicating a well-integrated transition zone that provides to the overall structural integrity of the clad. Finally, in Fig. 3d, we concentrate on the gray phase at Point 3, a composition with a significant Fe concentration together with Cr. It is likely that this phase is formed of chromium enriched carbides which provide high hardness and wear resistance during rapid solidification. This region's Ni content reduction is consistent with its segregation into carbide formation. Taken together, these observations demonstrate that the laser cladding process is capable of producing a well bonded, compositionally optimized clad with excellent mechanical properties, which is suitable for industrial use where superior wear and corrosion resistance is required.

3.3 *Wear Behavior*

The worn surface morphologies of Cr_3C_2-NiCr-based laser clads under a 20 N load as evaluated on a linear reciprocating ball-on-plate tribometer are shown in Fig. 4. The images give a high-resolution description of what wear mechanisms are operating under the load applied. An overall view of the wear track—revealing the uniformity and depth of the wear scar—is shown in panel (a). The material's response to frictional forces during sliding contact is revealed by the track's dimensions and characteristics. Magnified views of the worn surface are shown in panels (b) and (c),

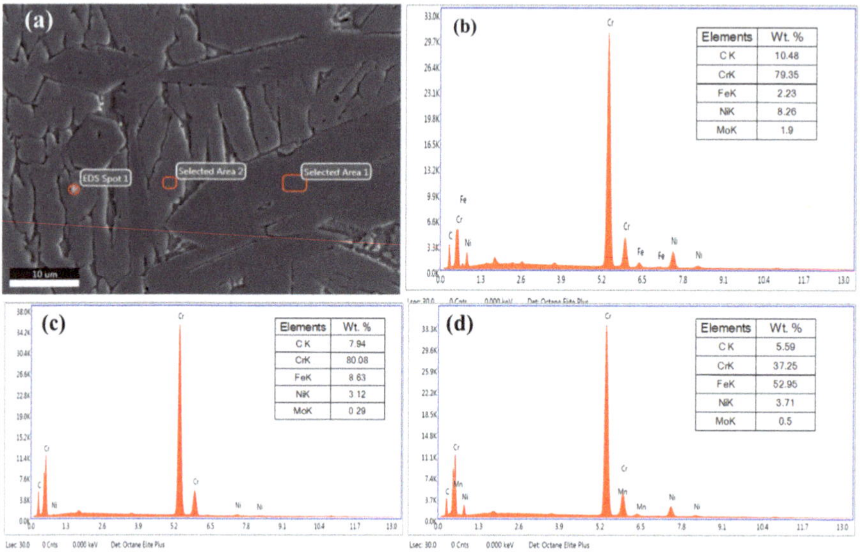

Fig. 3 The Standard EDS analyses of the developed laser clad include: **a** FE-SEM image displaying locations (EDS analyses); **b** spot (1) EDS analysis on the developed clad; **c** spot (2) EDS spectrum; and **d** gray phase EDS analysis (3)

which show evidence of abrasion and localized plastic deformation. Microplowing wear mechanisms appear to dominate, as evidenced by fine grooves aligned with the sliding direction. The occurrence of these grooves indicates that the clad surface has been hard particle engaged, which is typical abrasive wear.

Panel (d) is a higher magnification view of the wear track, with finer characteristic details of the surface damage depicted. The surface is seen to possess shallow pits, suggesting mild adhesive wear mechanisms, and compacted wear debris. Uniform dispersion of chromium carbides in the NiCr matrix is likely to have enhanced the clad's ability to resist severe wear because of its ability to maintain its structural integrity under the applied load. The worn surface morphologies confirm that Cr_3C_2-NiCr-based laser clads have good wear resistance overall. The material's ability to stand up to high contact stresses is due in part to the combination of carbide reinforcement and metallurgical bonding with the substrate, making it well suited for applications where superior tribological performance is required.

3.4 Surface Roughness Analysis

Detailed analysis of the worn surface morphology of the Cr_3C_2-NiCr-based laser clads subjected to 20 N load using a 3D profilometer is given in Fig. 5. A 2D color-coded topographical map of the wear track depth (top left panel) is shown with

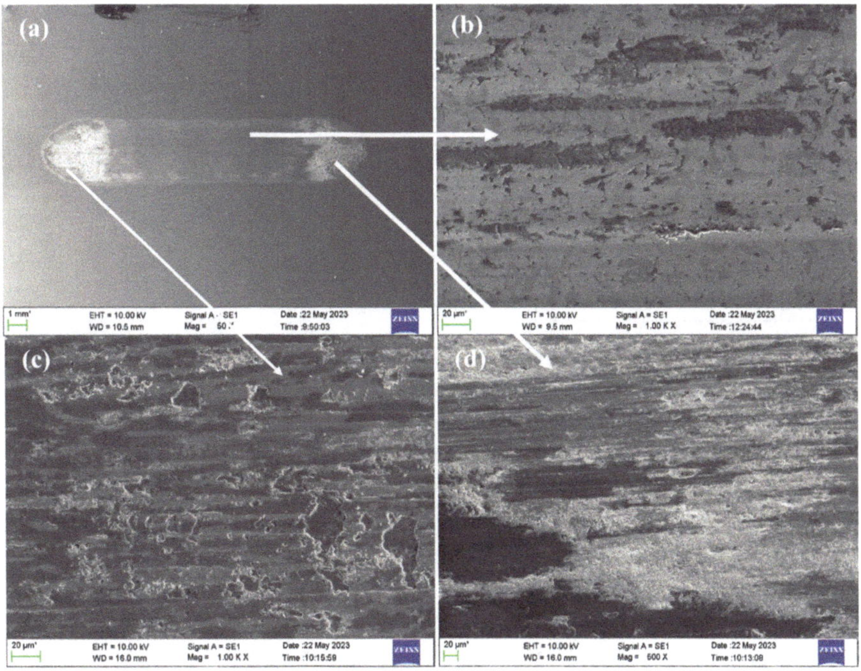

Fig. 4 Worn surface morphologies of (Cr_3C_2-NiCr)-based cladding: laser clads with 20 N load

variations in depth and wear distribution across the surface. The maximum depth of the wear track is color-coded, being red for deeper regions and blue for shallower. The 3D visualization of worn surface in the top right panel can be seen with profile and depth of wear track. The wear track has a concave shape where highest material removal occurred in the region toward the center. This corresponds to uniform wear under the applied load, and a maximum depth of 142.8 μm as measured.

A cross-sectional profile of the wear track is shown in the bottom panel and shows the depth variation along the wear path. The peak to valley data shows the highest wear occurred in the central region, with the edges largely intact. The implication is of ongoing sliding and material loss resulting from abrasion interaction. The observed wear characteristics agree with the wear properties of Cr_3C_2-NiCr-composite clads where the reinforced carbide phases withstand considerable wear. These results provide reinforcement that the clads are well suited to applications where superior wear resistance under high contact stresses are required. This evaluation also contributes to understanding of the wear mechanisms and to the optimization of the cladding process for improved performance.

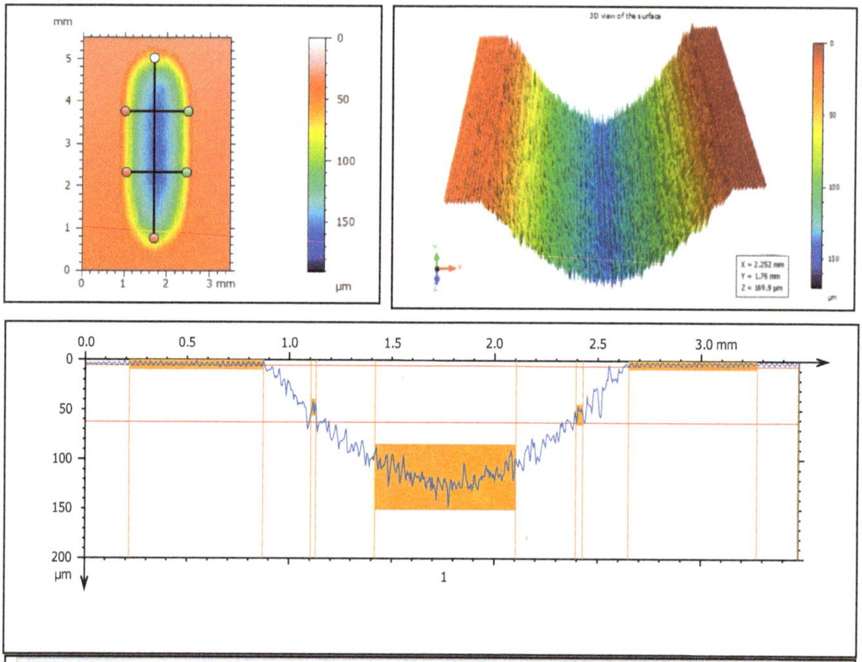

Parametres	Step 1	Unit
Maximum	142.8	µm

Fig. 5 The wear rate and coefficient of friction are measured by 3D profilometer at 20 N

4 Conclusions

The microstructure, wear behavior, and surface properties of laser clads made from Cr_3C_2–NiCr on T91 ferritic steel are investigated in this work. The findings show that the laser cladding process optimized in this research generates wear resistant coatings with good metallurgical bond and low porosity. These findings not only indicate that Cr_3C_2-NiCr clads are suitable for industrial applications but also underpin their significance in sustainable material development through the enhancement of durability, minimization of material wastage, and optimization of resource utilization. This work presents a major advancement in the development of high-performance, durable materials for use in rigorous engineering applications.

Enhanced Surface Properties: The laser clads produced with Cr3C2-15 wt% NiCr alloy have a good microstructure, strong interface bonding with the T91 substrate, and high wear resistance. These attributes lead to a longer service life and less material consumption since there is no frequent need for replacement and maintenance, which in turn is sustainable.

Optimized Material Utilization: The clads have a relatively uniform distribution of carbides and have good wear properties while still using a small amount of material during the cladding process. This management of raw materials is consistent with the principles of sustainable manufacturing and minimizes on the use of raw materials.

Durability and Resource Conservation: The clads show low wear rate and low material loss at high stress conditions, which make them suitable for replacing conventional materials in severe service conditions. Their durability means that creation of sustainable materials does not require long and hence resource consuming processes.

In conclusion, the present study has successfully established that Cr3C2-NiCr laser clads present an environment friendly approach for the improvement of the efficiency and durability of industrial parts. The results are useful for the further creation of new materials for modern engineering needs that have less of a negative impact on the environment, last longer, and are more efficient with resources.

References

1. Santo L (2008) Laser cladding of metals: a review. Int J Surf Sci Eng 2(5):327. https://doi.org/10.1504/ijsurfse.2008.021345
2. Zhu L, Xue P, Lan Q, Meng G, Ren Y, Yang Z, Xu P, Liu Z (2021) Recent research and development status of laser cladding: a review. Opt Laser Technol 138:106915. https://doi.org/10.1016/j.optlastec.2021.106915
3. Esquivel J, Murdoch HA, Darling KA, Gupta RK (2017) Excellent corrosion resistance and hardness in Al alloys by extended solid solubility and nanocrystalline structure. Mater Res Lett 6(1):79–83. https://doi.org/10.1080/21663831.2017.1396262
4. Hebbale AM, Kumar AP, Medabalimi S, Badiger RI (2024) Investigating microstructural aspects of Cr_3C_2–NiCr laser cladding for erosion-resistant T91 steel. In: Advances in science, technology & innovation/advances in science, technology & innovation, pp 183–188. https://doi.org/10.1007/978-3-031-63909-8_25
5. Liu H, Xu Q, Wang C, Zhang X (2014) Corrosion and wear behavior of Ni60CuMoW coatings fabricated by combination of laser cladding and mechanical vibration processing. J Alloy Compd 621:357–363. https://doi.org/10.1016/j.jallcom.2014.10.030
6. Alam. (2022). Recent Trends in surface cladding on AISI 1045 steel substrate: A review. Mater Res Proc 22:94–101. https://doi.org/10.21741/9781644901878-13
7. Nayak UP, Schäfer F, Mucklich F, Guitar MA (2022) Wear induced sub-surface deformation characteristics of a 26 WT% CR white cast iron subjected to a destabilization heat treatment. Tribol Lett 71(1). https://doi.org/10.1007/s11249-022-01683-3
8. Hebbale AM, Kumar M, Soudagar MEM, Ahamad T, Kalam MA, Mubarak NM, Alfantazi A, Khalid M (2023) A comparative study on characteristics of composite (Cr3C2-NiCr) clad developed through diode laser and microwave energy. Sci Rep 13(1). https://doi.org/10.1038/s41598-023-37991-4
9. Medabalimi S, Hebbale AM, Gudala S, Ramesh, Gujar R, Aravindan N, Petru J (2025) Studies on high-temperature erosion behaviour of HVOF sprayed NiCr based composite coatings. Results Eng 26:105090. https://doi.org/10.1016/j.rineng.2025.105090
10. Hebbale AM, Tak M, Bathe R (2021) Microstructural studies of composite (Cr3C2–NiCr) laser clads developed on preheated substrate T91. Trans Indian Inst Met 74(3):593–600. https://doi.org/10.1007/s12666-020-02150-0
11. Medabalimi S, Hebbale AM, Gudala S, Rokkala U, Ramesh (2024) Studies on high temperature erosion behavior of HVOF-sprayed (Cr_3C_2-NiCr)Si and WC-Co/NiCrAlY composite coatings. Int J Refract Metals Hard Mater 106970. https://doi.org/10.1016/j.ijrmhm.2024.106970

Characteristics of Lightweight Concrete Incorporating Waste PET Plastic Using AI Technique

Eashan Pahsha, Rajesh Gupta, Vinay Agrawal, Swati, and Sanchi Rewar

Abstract This paper explores an innovative approach to utilize Polyethylene Terephthalate (PET) plastic fiber in lightweight concrete (LWC), aiming to enhance sustainability in the built environment while tackling the issue of plastic waste. LWC is advantageous due to its reduced weight, making it ideal for high-rise buildings, bridges, and various precast structures. Meanwhile, the increasing amount of plastic waste, especially from PET, raises significant environmental concerns linked to contemporary industrial practices. The mechanical properties (compressive, flexural, and split tensile strength) of LWC with different percentages of PET plastic as a fine aggregate substitute (0%, 5%, 7.5%, and 10%) across various water-to-cement ratios (0.45, 0.50, and 0.55) were investigated in this study using an artificial neural network (ANN)-based approach. Over 70% of the experimental data was used to train the ANN models, and the prediction accuracy was subsequently confirmed. The findings showed that there was less than a 5% difference between the experimental and predicted values, indicating the models' dependability for predicting performance and optimizing composition. The originality of this paper is connected with using shredded PET plastic as a substitute for fine aggregate and with ANN modeling for predicting performance and boosting LWC. This approach addresses not only the environmental issues but also the current demand for sustainable building materials and provides an applicable framework regarding waste reduction and inducing material development in the construction industry.

Keywords Artificial intelligence technique · Lightweight concrete · PET fibers · Polyethylene terephthalate · Waste materials

---

E. Pahsha (✉) · R. Gupta · V. Agrawal · Swati · S. Rewar
Department of Civil Engineering, Malaviya National Institute of Technology, Jawahar Lal Nehru Marg, Jhalana Gram, Jaipur, Rajasthan, India
e-mail: 2021rce9029@mnit.ac.in

I. M. Jamadar et al. (eds.), *Integrating Sustainability in Modern Technologies*, Lecture Notes in Mechanical Engineering, https://doi.org/10.1007/978-981-95-9466-5_21

1 Introduction

The rapid growth in technology has its pros and cons in this modern world, be it in any field, but more so in the construction industry. Of late, a kind of concrete known as lightweight concrete (LWC) is an important development, finding applications in offshore oil platforms, curtain walls, shell roofs, folded plates, precast frames and floors, and bridges [1]. While LWC offers a significant weight reduction, it has a strength of 25–35% less compared to normal concrete [2]. However, LWC is widely used in precast structures and high-rise buildings, as well as in bridge decks where reduced weight is critical [3]. The sustainable growth of LWC may benefit from the utilization of lightweight waste materials. One of such material is plastic, which is a part of industrialization, posing serious environmental problems. Of these plastics, Polyethylene Terephthalate (PET) is the major component due to its large usage in the forms of bottles for beverages and food wrappings, which biodegrade very slowly [4]. In 2022–2023, there is a production of more than 34 lakh tons of plastic waste, which is 10 lakh tons higher compared to the previous years [5]. Lightweight, transparency and good strength indicate that PET is one of the most sought-after materials for recycling and reuse in construction [6].

Artificial neural networks (ANN) which, by structure, are inspired by biological neural networks computational tools, allowing the analysis of complex data relationships [7]. The networks in question consist of interconnected artificial neurons, and their applications are mainly concerned with pattern recognition and predictive modeling [8]. The study makes a substantial contribution to tackling environmental pollution and developing sustainable building materials by using shredded PET plastic as a partial replacement for fine aggregate in LWC. PET was added at weight percentages of fine aggregate of 0%, 5%, 7.5%, and 10% with water-to-cement (w/c) ratios of 0.45, 0.50, and 0.55. To ensure accurate predictions, the model used 70% of the experimental data for training and the remaining 30% for validation. The two main issues this study targets are reducing the negative environmental effects of PET waste and investigating how it might enhance the sustainability of lightweight concrete. The work investigates the mechanical characteristics of lightweight concrete that contains PET using ANN modeling and experimental analysis, providing insightful information for its optimization and practical uses.

2 Materials

The study utilized standard Portland cement (grade 43) in accordance with IS 8112-1989. The test results for the cement properties show a consistency of 29% and a specific gravity of 3.13. The initial and final setting time was 85 min and 190 min, respectively. According to IS 383-2016, this investigation employed coarse aggregates as uniformly sized lightweight sandstone with a 20 mm maximum size and having specific gravity of 1.1, indicating a relatively lightweight composition.

Fig. 1 PET plastic fibers

Additionally, the water absorption rate was 11.1% and the dry density was observed to be 1403 kg/m3, which highlights the material's overall lightweight nature, making it potentially suitable for applications where weight reduction is essential, such as in LWC or insulation materials [9]. The crushed local waste plastic bottles were used to create shredded PET fibers as shown in Fig. 1. The plastic shreds were not treated in any way beforehand. The fibers measured between 2 and 4 mm in length, 1–2 mm in breadth, and 0.5 mm in thickness. The specific gravity of PET fiber was 1.22, which indicates that the material is relatively lightweight, allowing for ease of handling and installation. The selection of PET plastic percentages (5%, 7.5%, and 10%) by weight of fine aggregate balances density reduction and strength retention, supported by prior studies and practical feasibility. At 5%, the impact on density and strength is minimal, providing a safe baseline, while 10% risks excessive porosity and reduced integrity [6]. The 7.5% level offers a midpoint to assess trade-offs. Literature supports up to 10% replacement for sustainability without significant strength loss, as higher levels (e.g., 20%) compromise viability [10]. This range addresses environmental concerns by reusing PET waste and maintaining structural performance, ensuring practicality and workability.

3 Mix Proportions

In this study, 12 mixes (including control mixes) were prepared in total. The experimental program included the production of three groups of lightweight concrete mixes with w/c ratios of 0.45, 0.50, or 0.55. The control mix was designed for M15 grade of concrete, as in most lightweight concrete applications, this much strength is sufficient. Table 1 provides further information regarding mix proportions.

Table 1 Mix proportions of different mixes

W/C ratio groups	Cement (Kg/m^3)	Coarse aggregate (Kg/m^3)	Fine aggregate (Kg/m^3)	Plastic (%)	Weight of plastic (Kg/m^3)
0.45	344.40	793.13	380.48	0.00	0.00
	344.40	793.13	380.48	5.00	19.02
	344.40	793.13	380.48	7.50	28.53
	344.40	793.13	380.48	10.00	38.05
0.50	310.00	826.47	383.13	0.00	0.00
	310.00	826.47	383.13	5.00	19.15
	310.00	826.47	383.13	7.50	28.73
	310.00	826.47	383.13	10.00	38.31
0.55	281.81	857.63	385.41	0.00	0.00
	281.81	857.63	385.41	5.00	19.27
	281.81	857.63	385.41	7.50	28.90
	281.81	857.63	385.41	10.00	38.54

4 Results and Discussion

4.1 Compressive Strength Test

The IS 516:1959 was followed in performing the compressive strength test at 7 and 28 days after curing. From the test results, it is evident that the strength progressively drops as the plastic percentage rises. If 5% of the pet is used, the compressive strength falls by 3.84%, 5.61%, and 7.05% for 0.45, 0.50, and 0.55 w/c ratios, respectively, at 28 days. More notably, for 10% replacement, it falls by 16.48%, 19.66%, and 34.69% for the same w/c ratios, respectively, at 28 days. This loss can be due to a weak interfacial transition zone (ITZ) between the cement paste and PET fibers [11]. When a load is applied, the smooth PET fibers easily deform and do not mix well with cement pastes. A notable decrease in strength can also be due to the excessive porosity and voids as the PET volume increases in porous concrete [12]. Compressive strength is also influenced by cement paste thickness. Even with the reduction in strength, it is still between 2.8 and 28 MPa. Table 2 contains a tabulation of the compressive strength test findings.

4.2 Flexure and Split Tensile Strength Test

The 500 mm × 100 mm × 100 mm beam specimen was subjected to a two-point load mechanism for the flexural strength test after 28 days of curing, as per IS 516:1959. Comparable patterns to those seen in compressive strength are noted in flexural

Table 2 Experimental results of mechanical strength

W/C ratio	Plastic (%)	Compressive strength		Flexural strength		Split tensile strength	
		7 days	28 days	7days	28days	7days	28days
0.45	0	10.3	18.2	2.30	2.95	1.02	1.80
	5	9.5	17.5	2.21	2.84	0.94	1.77
	7.5	8.3	16.1	1.98	2.78	0.92	1.75
	10	7.7	15.2	1.91	2.69	0.90	1.73
0.5	0	10	17.8	2.20	2.91	0.99	1.78
	5	9.1	16.8	1.99	2.81	0.96	1.76
	7.5	8.0	15.2	1.88	1.95	0.94	1.73
	10	7.1	14.3	1.73	1.91	0.89	1.71
0.55	0	9.8	17.0	1.99	2.88	0.97	1.76
	5	8.5	15.8	2.35	2.35	0.94	1.73
	7.5	7.6	14.4	2.14	2.14	0.91	1.69
	10	6.4	12.5	1.81	1.81	0.83	1.66

strength, with the control mix. The decreased flexural strength is 10.03%, 34.36, and 37.15% for 10% PET replacement, and 0.45, 0.50, and 0.55 w/c ratio, respectively, at 28 days. This drop in flexural strength could be due to the increase in PET fibers which causes the loss in adhesion between the cement paste and PET fibers [13]. The strength loss was negligible for split tensile strength, though. It is reasonable to argue that plastic-induced concrete has a lower tensile strength since plastic's smoothness has a detrimental effect on concrete strength since it does not offer excellent bond strength. Table 2 contains a tabulation of the flexural and split tensile strength test findings.

4.3 Multi-Linear Regression (MLR) Results

Through a network of intricately linked computing units known as artificial neurons, neural networks can mimic the information-processing functions of the human brain [9]. To evaluate the neural networks' capacity for generalization, they are trained using the data and then verified. Testing is carried out following training to assess the network fault. As a result, three components of data are used to create the neural network model [14]. To create data subsets for testing, validation, and training, there isn't a rule, though. The data used in this study is split into three categories: 60% for training neural networks, 20% for validation, and 20% for testing neural networks. In Tables 3, 4, and 5, the ANOVA results for compressive, flexural, and split tensile tests, respectively, are tabulated. Table 6 shows the training and validation data of

ANN model. The ANN plots for compressive, flexural, and split tensile tests training and best validation is shown in Figs. 2, 3, and 4, respectively (Table 7).

MLR Equation for Genetic Algorithm

- Compressive Strength

$$Y = 7.7336491 - 162373 \times Z(1) - 0.00700866 \times Z(2)$$

Table 3 ANOVA table for compressive test

	Coefficient	S.E	Lower t0.025	Upper t0.975	p-value
Intercept	1457847.036	5801716.765	−10407996.07	13323690.14	0.803371
W/C ratio	−148100.6337	3213388.977	−6720219.021	6424017.754	0.963556
Cement	−599.009378	2463.7756	−5637.996452	4439.977696	809616
LWA	−1984.378841	4374.354546	−10930.93842	6962.180742	653465
Plastic	−472.580931	7002.71892	−14794.74924	13849.58738	0.946658
Sand	−472.575842	7002.726646	−14794.75995	13849.60827	0.946659
Age	0.38801	0.0238089	0.339316	0.436705	3.31E-16

Table 4 ANOVA table for flexural test

	Coefficient	S.E	Lowert0.025	Upper t0.975	p-value
Intercept	714089.907	339931.3219	−3102.491304	1431282.305	0.050892
W/C ratio	244262.0819	192203.5671	−161251.998	649776.1618	0.220898
Cement	−301.402145	144.444635	−606.153687	3349397	0.0522946
LWA	−275.223057	258.890857	−821.43502	270.988905	0.302612
Plastic	−687.922129	415.984322	−1565.572333	189.728074	0.11653
Sand	−687.922213	415.984772	−1565.573365	189.728938	0.11653
Age	0.0246825	0.00120637	0.0221373	0.0272278	2.06E-13

Table 5 ANOVA table for split tensile test

	Coefficient	S.E	Lower t0.025	Upper t0.975	p-value
Intercept	149595.6732	141325.4193	-148,574.898	447,766.2444	0.304628
W/C ratio	−1516.422638	79908.05189	−170107.6753	167074.83	0.98508
Cement	−61.811685	60.05242	−188.511217	64.8878	0.317764
LWA	−173.536731	107.633091	−400.622703	53.5492	0.125305
Plastic	−68.203729	172.944224	−433.084	296.6766	0.698212
Sand	−68.203638	172.944411	−433.0844	296.677154	0.698213
Age	0.0379603	0.000501546	0.0369021	0.0390185	0.00E + 00

Table 6 Training and validation data on ANN model

Results	Samples	M.S.E	R-Value
Training	26	7.47E-02	9.98E-02
Validation	5	3.78E-02	9.84E-01
Testing	5	1.17E+00	7.96E-01
Training	16	7.89E-03	9.99E-01
Validation	4	1.16E-02	9.42E-01
Testing	4	4.13E-03	9.79E-01
Training	16	1.08E-04	1.00E+00
Validation	4	1.17E-03	9.96E-01
Testing	4	9.15E-03	9.68E-01

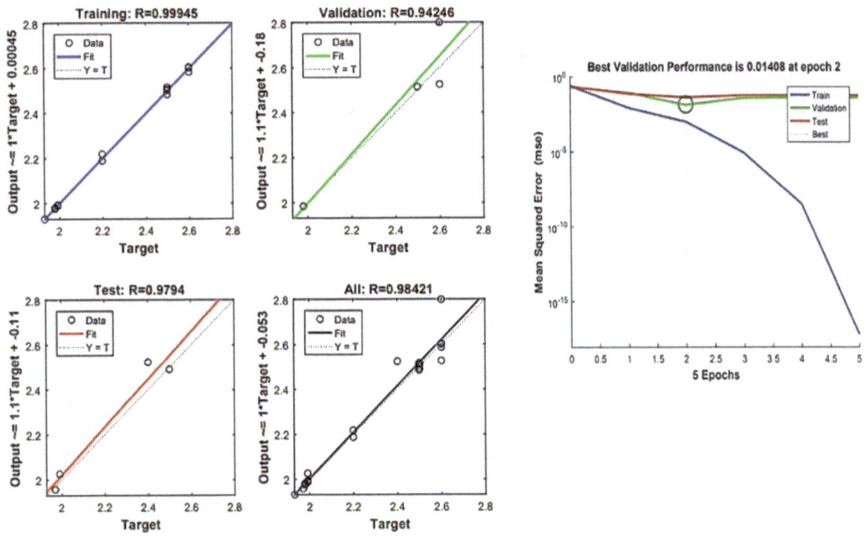

Fig. 2 ANN plot for compressive strength of training and best validation

$$+ 0.38801 \times Z(3) \tag{1}$$

- Flexural Strength

$$\begin{aligned} Y &= 2.186149 - 0.55303 \times Z(1) - 0.00129456 \times Z(2) \\ &+ 0.0246825 \times Z(3) \end{aligned} \tag{2}$$

- Split Tensile Strength

$$Y = 0.785436 - 0.136831 \times Z(1) - 0.000500768 \times Z(2)$$

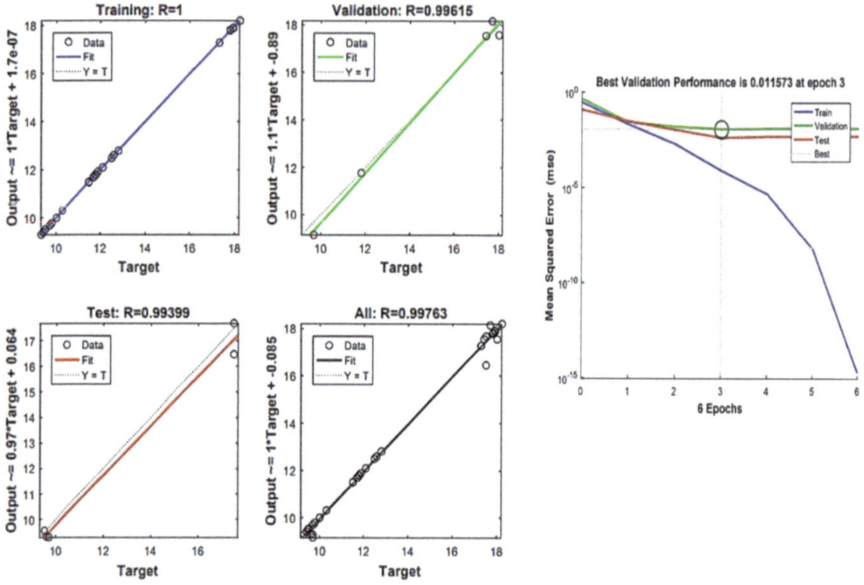

Fig. 3 ANN plot for flexural strength of training and best validation

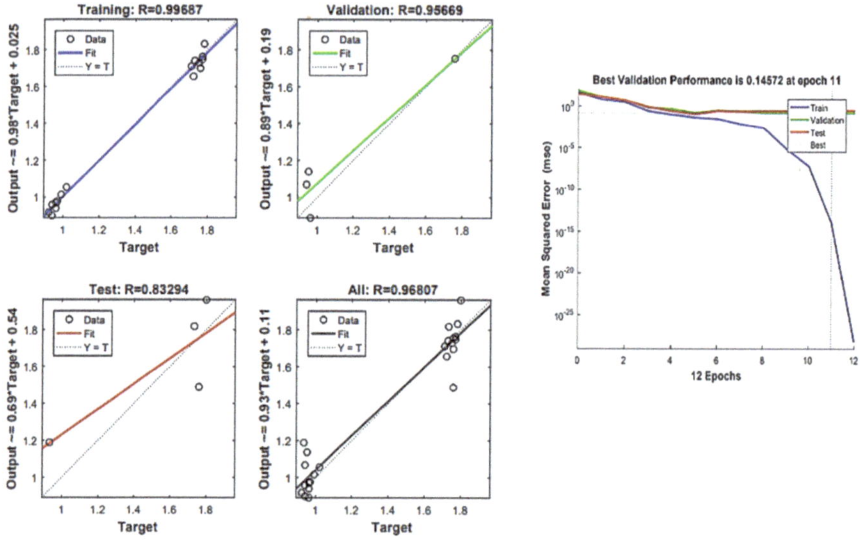

Fig. 4 ANN plot for split tensile strength of training and best validation

Table 7 Optimization result using genetic algorithm

	Z(1)	Z(2)	Z(3)	Y
Compressive strength	0.55	19.27	7	6.55
Flexural strength	0.55	28.90	7	1.68
Split tensile strength	0.55	38.54	7	0.86

$$+ 0.0379603 \times Z(3) \quad (3)$$

where Z(1): water-cement ratio; Z(2): plastic content; and Z(3): days.

5 Conclusion

This study provides novel insights into the mechanical properties of lightweight concrete (LWC) combined with recycled PET plastics in contemporary construction, which aligns with the essential goal of developing sustainable infrastructure. Utilizing artificial neural networks (ANNs) to evaluate the feasibility of this substitution is a fresh approach that introduces a technological edge to the study. The exact methodology of testing different PET proportions (0%, 5%, 7.5%, and 10%) along with various water-to-cement (w/c) ratios highlighted the novelty, presenting a thorough experimental analysis augmented with ANN modeling with various proportions of PET plastic fibers (5%, 7.5%, and 10%) at different w/c ratios (0.45, 0.50, and 0.55). Although the conclusions drawn from the study indicate that the recycling of PET plastics adversely affects compressive, flexural, and tensile strengths due to weakly bonded interfacial phases and higher porosity, the material is still strong enough for some purposes like use in non-load carrying elements such as partition walls, insulation panels, lightweight cladding, and architectural features, where the reduction in weight and sustainability are more important than strength. Furthermore, the MLR model predictive capability enables the optimization of material compositions with minimum trial-and-error efforts for such applications, thereby streamlining the design process.

The compressive strength was found to decrease substantially by around 3.84% to 7.04% at 5% PET and from 16.48% to 34.69% at 20% PET replacement. The flexural strength and tensile strength also appeared to lessen due to poor bonding at the interface and high porosity coming from the smooth surface of the PET fibers. The MLR models also validated the experimental results, giving good predictions for the compressive, flexural, and split tensile strengths depending on the w/c ratio, content of PET plastic, and curing period. For example, the anticipated compressive, flexural, and split tensile strengths were 6.55 MPa, 1.68 MPa, and 0.86 MPa, respectively, at a w/c ratio of 0.55 for the age of 7 days. The errors between all these predicted and experimental values were, however, below 5%. That is, these models put forth a sound framework for material composition optimization and performance prediction, thereby simplifying the design of sustainable concrete applications. Moreover, the MLR model presents as an effective tool in prediction and optimization of the performance of materials.

References

1. Kamanmalek S, Alamdari N (2024) Advancing equitable stormwater management: a decision support tool integrating best practices for nutrient removal and environmental justice. Ecol Inform. 80:102496. https://doi.org/10.1016/j.ecoinf.2024.102496
2. Choi YW, Moon DJ, Kim YJ, Lachemi M (2009) Characteristics of mortar and concrete containing fine aggregate manufactured from recycled waste polyethylene terephthalate bottles. Constr Build Mater 23(8):2829–2835. https://doi.org/10.1016/j.conbuildmat.2009.02.036
3. Alqahtani FK, Zafar I (2020) Characterization of processed lightweight aggregate and its effect on physical properties of concrete. Constr Build Mater 230:116992. https://doi.org/10.1016/j.conbuildmat.2019.116992
4. Islam MdJ, Meherier MdS, Islam AKMR (2016) Effects of waste PET as coarse aggregate on the fresh and harden properties of concrete. Constr Build Mater 125:946–951. https://doi.org/10.1016/j.conbuildmat.2016.08.128
5. Briga-Sá A, Ferreira L, Paulo B, Bentes I, Teixeira CA (2023) Energy and environmental performance assessment of reused PET bottles panels for building thermal insulation solutions. Energy Build 298:113529. https://doi.org/10.1016/j.enbuild.2023.113529
6. Górak P, Postawa P, Trusilewicz LN (2020) Lightweight composite aggregates as a dual end-of-waste product from PET and anthropogenic materials. J Clean Prod 256:120366. https://doi.org/10.1016/j.jclepro.2020.120366
7. Parhi SK, Patro SK (2024) Application of R-curve, ANCOVA, and RSM techniques on fracture toughness enhancement in PET fiber-reinforced concrete. Constr Build Mater 411:134644. https://doi.org/10.1016/j.conbuildmat.2023.134644
8. Mohammed AA (2017) Modelling the mechanical properties of concrete containing PET waste aggregate. Constr Build Mater 150:595–605. https://doi.org/10.1016/j.conbuildmat.2017.05.154
9. Saradar A, Nemati P, Paskiabi AS, Moein MM, Moez H, Vishki EH (2020) Prediction of mechanical properties of lightweight basalt fiber reinforced concrete containing silica fume and fly ash: experimental and numerical assessment. J Build Eng 32:101732. https://doi.org/10.1016/j.jobe.2020.101732
10. Supit SWM, Priyono (2022) Utilization of recycled PET plastic waste as replacement of coarse aggregate in pervious concrete. Mater Today Proc 66:2990–2995. https://doi.org/10.1016/j.matpr.2022.06.573
11. Babatunde Y, Mwero J, Mutuku R, Jimoh Y, Oguntayo D (2022) Influence of material composition on the morphology and engineering properties of waste plastic binder composite for construction purposes. Heliyon 8(10):e11207. https://doi.org/10.1016/j.heliyon.2022.e11207
12. Yang S, Yue X, Liu X, Tong Y (2015) Properties of self-compacting lightweight concrete containing recycled plastic particles. Constr Build Mater 84:444–453. https://doi.org/10.1016/j.conbuildmat.2015.03.038
13. Kunthawatwong R et al (2022) Recycled non-biodegradable polyethylene terephthalate waste as fine aggregate in fly ash geopolymer and cement mortars. Constr Build Mater 328:127084. https://doi.org/10.1016/j.conbuildmat.2022.127084
14. Spies L, Li T, Burdisso R, Sandu C (2023) An artificial neural network (ANN) approach to model Tire-Pavement interaction noise (TPIN) based on tire noise separation. Appl Acoust 206:109294. https://doi.org/10.1016/j.apacoust.2023.109294

Towards Sustainable Lightweight Materials: Experimental Investigation and Characterization Studies of Fatigue Properties in Fly Ash and Tungsten Carbide-Reinforced 7029 Aluminium Alloy Hybrid Composites

T. G. Gangadhar, S. Bharath Kumar, A. C. Prapul Chandra, Pramod Kumar Naik, and H. Harish

Abstract In this work, a hybrid composite of 7029 aluminium composite with fly ash and tungsten carbide was used to investigate the fatigue behaviour. Additionally, three stirring speed conditions of 200 rpm, 250 rpm, and 300 rpm were used to generate composites. For the developed specimen, the fatigue test was conducted using rotating fatigue testing equipment in accordance with ASTM E606. Additionally, following testing, the specimen's maximum stress limit and cycle number (Nf) were determined. The findings demonstrated that, in comparison with other composites, the one with a larger proportion of tungsten carbide had the highest number of cycles (Nf) and maximum stress limit for all stirring speed conditions. Results also showed that the Al 7029 + 2% WC + 4% fly ash at 250 rpm has taken up the highest number of cycles of 220,638 with the maximum stress of 116.04Mpa. Moreover, SEM images of the material at various magnifications were used for the characterization studies. It was evident from the SEM pictures that the generated composites' maximum stress and load-carrying capacity were increased due to the particle's ideal dispersion with the matrix.

Keywords Composite · Al7029 · Fatigue · SEM · Characterization

T. G. Gangadhar (✉) · S. Bharath Kumar · P. K. Naik
Department of Artificial Intelligence and Robotics, Dayananda Sagar University, Bangalore, India
e-mail: gangadhar-air@dsu.edu.in

A. C. Prapul Chandra
Department of Mechanical Engineering, R V College of Engineering, Bangalore, India

H. Harish
School of Computer Science and Engineering, R V University, Bangalore, India

I. M. Jamadar et al. (eds.), *Integrating Sustainability in Modern Technologies*, Lecture Notes in Mechanical Engineering, https://doi.org/10.1007/978-981-95-9466-5_22

1 Introduction

In India, the less availability of high-grade materials such as iron, coal, etc., requires processing for their utilization [1–4]. For processing, it requires a processing plant and also higher production costs [4–8]. So, research on hybrid composite has been extensively carried out on the developing composites of metal and polymer for the replacement with high-grade material [9–13]. Aluminium (Al) matrix composite is one such metal matrix composite that has extensive usage in aircraft and automobile applications [14–16]. The aluminium (Al) alloy hybrid composite has the features of economical, durable, lightweight, and high stiffness [16–19]. The Al matrix considered in the present study was 7XXX series, i.e. 7029 aluminium alloy which consists of Al–Mg-Cu-Mn-Fe [20]. Previously, several research works have carried out work on the 7029 Al alloy hybrid composite.

Madhu et al. 2021 [19] studied the mechanical properties of boron carbide (B4C)- and graphite(Gr)-reinforced Al7029 hybrid composite. The results showed that the developed composite yielded good tensile and harness properties when compared to the base metal. Madhu et al. 2021 [20] investigated the effects of B4C on the wear resistance properties of B4C and Gr-reinforced Al7029 hybrid composite. The study majorly focused on determining the optimal condition of the process parameters such as load, speed, and distance using the ANOVA technique.

Sharath et al. 2022 [21] studied the machinability studies on B4C and Gr-reinforced Al7029 hybrid composite. The results showed that the developed composites have decreased thrust force and burr height. Gangadhar et al. 2023 [22] have studied the tensile and flexural behaviour of Al7029 hybrid composite. The results showed that the ceramic particle reinforcement of Al7029 hybrid composite in the developed hybrid composite has improved the tensile strength, compression strength, and impact resistance. Sharath et al. 2024 [23] investigated the mechanical behaviour of Al7029-B4C, Gr, and Si3N4 (silicon nitride) hybrid composite. The results showed that the ceramic particle reinforcement of B4C and Si3N4 in the developed hybrid composite has improved the tensile strength, compression strength, and impact resistance.

So, a thorough review of existing literature revealed a significant knowledge gap in the fatigue behaviour of tungsten carbide and fly ash (FA)-reinforced 7029 aluminium alloy hybrid composite. Despite the growing interest in hybrid composites, limited research has been conducted to investigate the fatigue properties of this specific material combination. So, the present work majorly focuses on the investigation of the fatigue behaviour of the tungsten carbide and fly ash reinforced with 7029 aluminium alloy hybrid composite.

2 Methodology

Figure 1a, and b shows 7029 Al alloy and fly ash powder, which are the materials used in the present study. Table 1 shows the stirring conditions and weight percentage (%) of materials used in the present study. Figure 2a shows the electric furnace equipment

Fig. 1 Figure shows different forms of the materials used in the present study. Figure (**a**) presents Aluminium 7029 alloy ingots, Figure (**b**) shows tungsten carbide reinforcement particles, Figure (**c**) illustrates fly ash powder. The figure highlights the different material states involved in the processing and development of the hybrid composite

Table 1 Stirring conditions and weight percentage (%) of materials used in the present study

Sl. no	Stir speed	Tungsten carbide (WC) wt (%)	Fly ash wt (%)	Matrix material (7029 Al)
1	200	1	4	95
2	200	1.5	4	94.5
3	200	2	4	94
4	250	1	4	95
5	250	1.5	4	94.5
6	250	2	4	94
7	300	1	4	95
8	300	1.5	4	94.5
9	300	2	4	94

for the developing 7029 Al alloy composite. The 7029 Al alloy of different weight proportions is added to the electric furnace and heated up to 650 °C molten condition. Figure 2b shows the stirring of molten 7029 Al alloy inside the electric furnace. After the addition of WC and fly ash to molten 7029 Al alloy, stirring by means of a mechanical stirrer at rod speeds of 200 rpm, 250 rpm, and 300 rpm. The existing literature and preliminary experimental trials were used to determine the stirring speeds of 200 rpm, 250 rpm, and 300 rpm. Further, the molten metal is poured into the die for the solidification process. Figure 3 shows 7029 aluminium composites developed by the stir casting process.

After the preparation of the composite, the machining process was carried out on the developed composites using a CNC Lathe machine. ASTM-E606 standard was considered for preparing the specimens [24, 25]. Figure 4 shows the standard fatigue specimen and specimen developed of 7029 Al composites prepared for the present study.

Figure 5 shows the rotating fatigue testing machine used in the present study. The specimen was loaded at one end and the other end was fixed (cantilever type) to the machine. Nine specimens were tested by applying a load of 50N. Results were obtained for the number of cycles that can bear before the fracture. So, the present study deals with the investigation of the fatigue behaviour of the stir-casted 7029 aluminium alloy composite.

3 Results and Discussion

Figure 6a and b shows the number of cycles and maximum stress achieved by different specimens with 1%, 1.5%, and 2% tungsten carbide with 4% fly ash and 7029 Al composite under stirring speed 200 rpm. The results show that specimen 3, i.e. 2% tungsten carbide with 4% fly ash has taken up a greater number of cycles and achieved the maximum stress value of 113.63Mpa in comparison with the other two specimens

Fig. 2 The figure illustrates the metal melting and casting process carried out using a electric furnace setup. The figure (a) shows the furnace unit with the control system mounted on a supporting frame. The figure (b) presents the molten metal inside the furnace at high temperature. The figure (c) shows the stirring and handling of the molten metal during processing. The figure (d) depicts the pouring of molten metal from the crucible into the mould cavity for casting. The sequence represents the controlled melting, mixing, and casting stages involved in composite fabrication

(1 and 2). From Fig. 6a and b, it is clear that the specimen with higher tungsten carbide has the highest load-carrying capacity and has a maximum stress limit.

Figure 7a and b shows the number of cycles and maximum stress achieved by different specimens with 1%, 1.5%, and 2% tungsten carbide with 4% fly ash and 7029 Al composite under stirring speed 250 rpm. The results show that specimen 6, i.e. 2% tungsten carbide with 4% fly ash has taken up a greater number of cycles and achieved the maximum stress value of 116.04Mpa in comparison with the other two specimens (4 and 5). From Fig. 7a and b, it is clear that the specimen with higher

Fig. 3 The figure shows cast specimens of Aluminium 7029 alloy reinforced with different combinations of reinforcement materials prepared for the present investigation. The composite samples were fabricated with varying weight percentages of reinforcements to study their influence on the mechanical and tribological properties of the hybrid composites. The specimens exhibit uniform casting geometry and were produced using the stir casting technique under controlled processing conditions

zirconium oxide has the highest load-carrying capacity and has a maximum stress limit.

Figure 8a and b shows the number of cycles and maximum stress achieved by different specimens with 1%, 1.5%, and 2% tungsten carbide with 4% fly ash and 7029 Al composite under stirring speed 300 rpm. The results show that specimen 9, i.e. 2% zirconium oxide with 4% fly ash has taken up a smaller number of cycles and achieved the minimum stress value of 116.04 Mpa in comparison with the other two specimens (7 and 8). From Fig. 7a and b, it is clear that the specimen with higher zirconium oxide has the highest load-carrying capacity and has a maximum stress limit.

Figure 9 shows the fractured fatigue specimen after testing. Table 2 shows the fatigue test results on the developed composites. So, from Figs. 6a, b, 7a, b, 8a, and b and Table 2, it is clear that as the percentage of zirconium oxide increased there was an increase in the maximum stress and number of cycle-taking capacity of the specimen. It was also found that the increase in the stirring speed from 200 to 250 rpm increased the specimen's ability to take up more load.

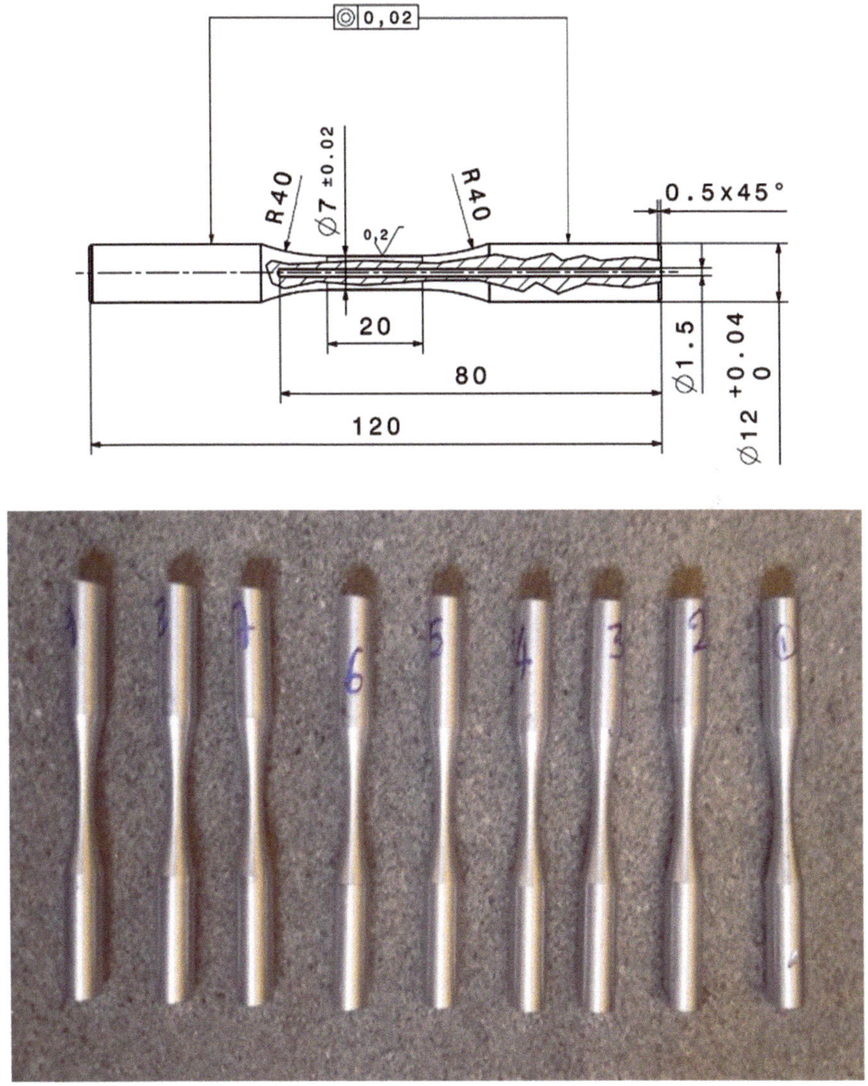

Fig. 4 The figure illustrates the dimensional drawing of the prepared composite tensile test specimen used for experimental evaluation.The specimen geometry was designed according to standard testing requirements with precise dimensions, gauge length, radius transitions, and diameter specifications

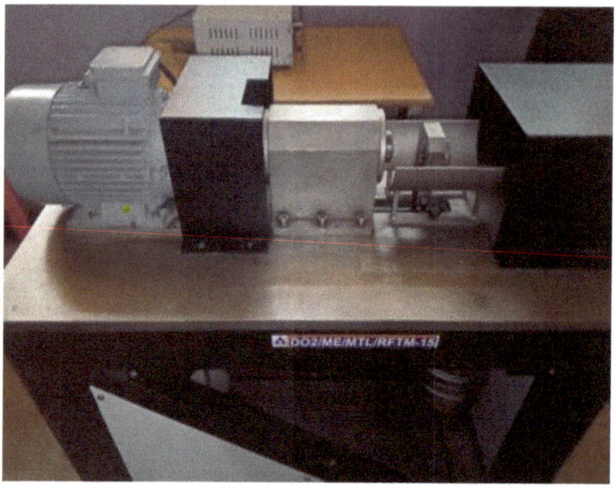

Fig. 5 The figure shows the fatigue testing machine used to evaluate the fatigue behaviour of the fabricated composite specimens under cyclic loading conditions. The equipment is designed to apply repeated stresses on the test specimen to determine its fatigue strength and endurance life

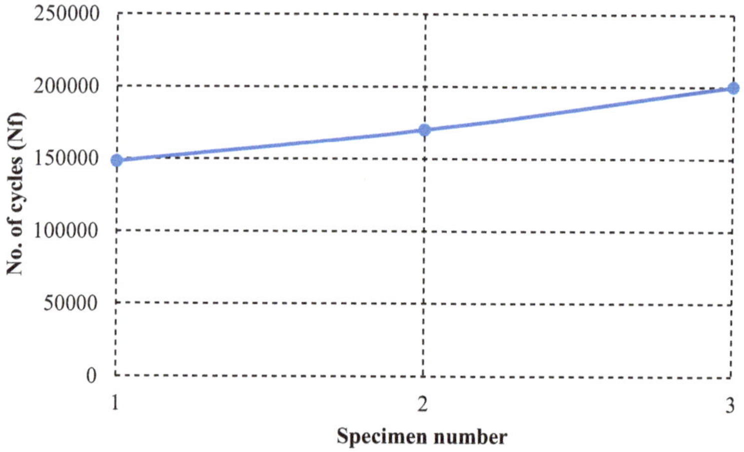

Fig. 6 **a** Number of cycles of fatigue test of different specimens with 1%, 1.5%, and 2% tungsten carbide-based 7029 Al hybrid composite of stirring speed 200 rpm. **b** Maximum stress achieved by different specimens with 1%, 1.5%, and 2% tungsten carbide with 4% fly ash and 7029 Al composite under stirring speed 200 rpm

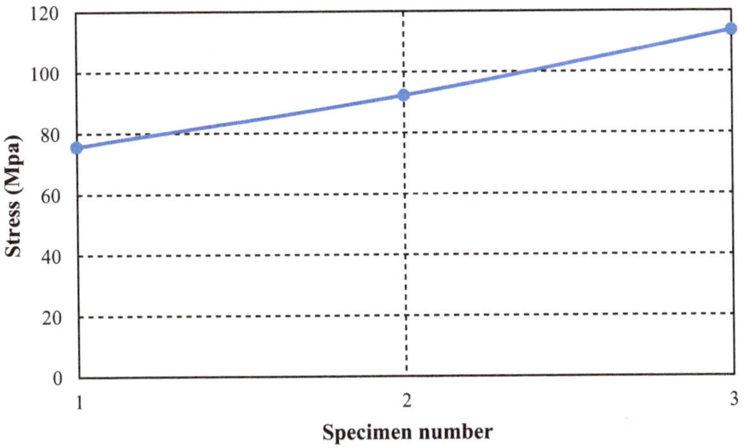

Fig. 6 (continued)

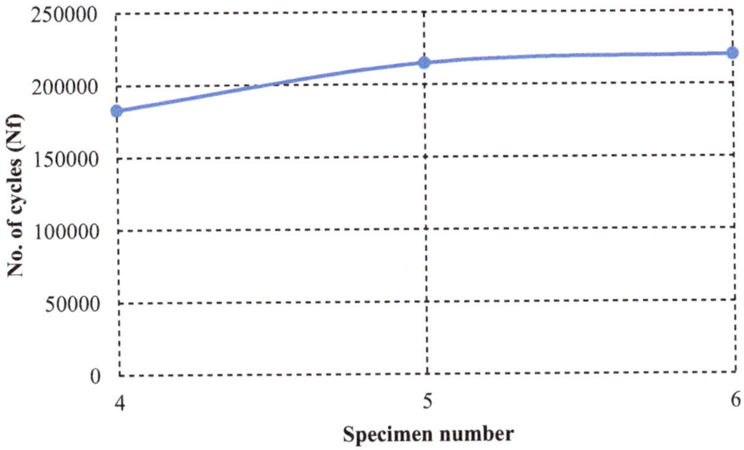

Fig. 7 **a** Number of cycles of fatigue test of different specimens with 1%, 1.5%, and 2% tungsten carbide with 4% fly ash and 7029 Al composite under stirring speed 250 rpm. **b** Maximum stress before fracture of different specimens with 1%, 1.5%, and 2% tungsten carbide with 4% fly ash and 7029 Al composite under stirring speed 250 rpm

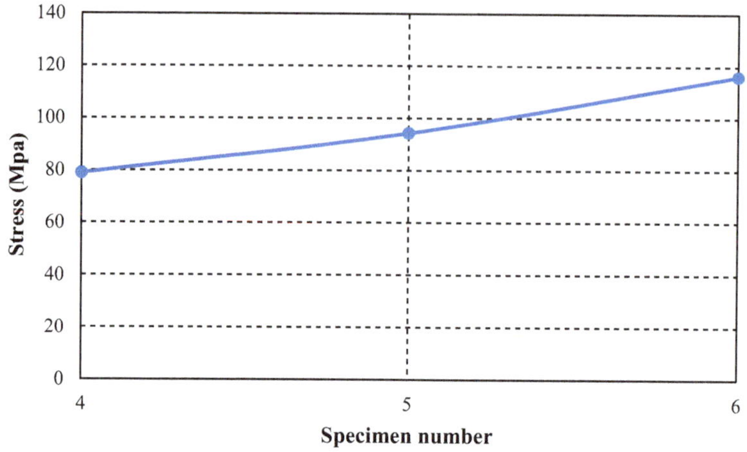

Fig. 7 (continued)

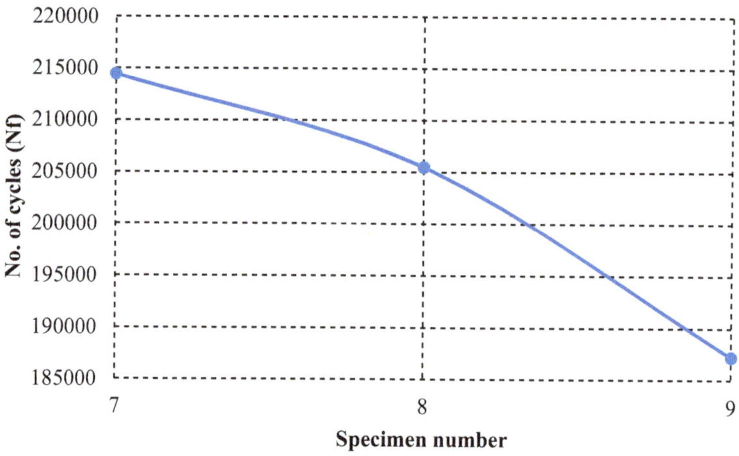

Fig. 8 **a** Number of cycles of fatigue test of different specimens with 1%, 1.5%, and 2% tungsten carbide with 4% fly ash and 7029 Al composite under stirring speed 300 rpm. **b** Maximum stress before fracture of different specimens 1%, 1.5%, and 2% tungsten carbide with 4% fly ash and 7029 Al composite under stirring speed 300 rpm

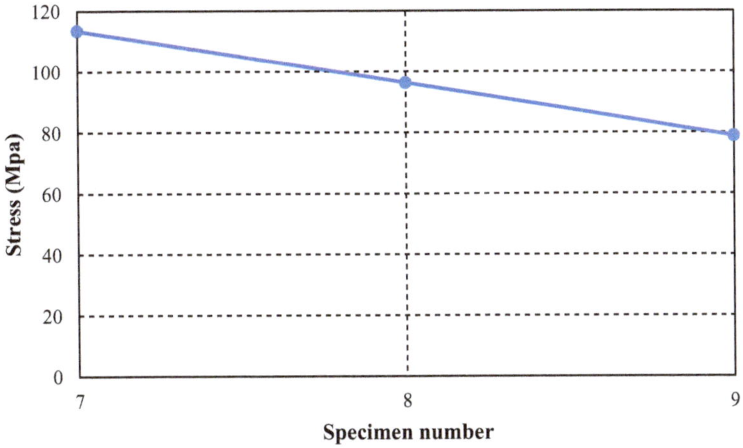

Fig. 8 (continued)

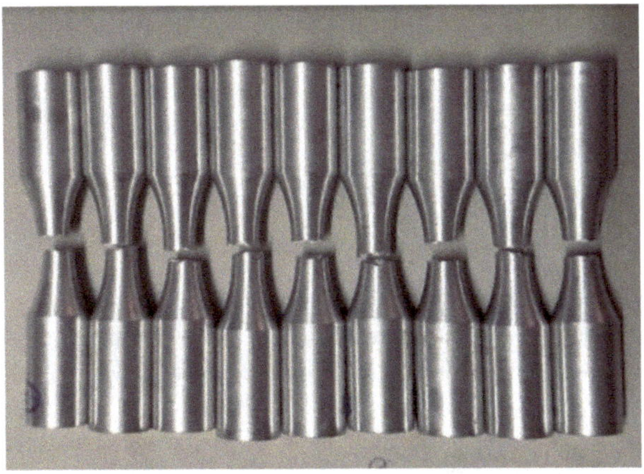

Fig. 9 Fractured fatigue specimen of 1%, 1.5%, and 2% tungsten carbide with 4% fly ash and 7029 Al composites after testing

Figure 10a, b, and c shows the SEM images of Al 7029 + 2% WC + 4% fly ash at a stirring speed of 250 rpm respectively. The experimental results and SEM results show that the improved particle dispersion, reduced voids, reduced crack initiation, and also strong interfacial bonding have improved the strength and load transfer of Al 7029 + 2% WC + 4% fly ash. Also, it was observed that the stirring speeds above 300 rpm caused excessive foam formation, while speeds below 200 rpm resulted in inadequate mixing. So, Al 7029 + 2% WC + 4% fly ash provides adequate mixing, while minimizing the risk of particle settling or foam formation correlates well with experimental results.

Table 2 Fatigue test results on developed composites

Specimen no	Composition and stirring speed	No. of cycles (N_f)	Stress Mpa
1	Al 7029 + 1% WC + 4% fly ash at 200 rpm	148268	75.77
2	Al 7029 + 1.5% WC + 4% fly ash at 200 rpm	170368	92.38
3	Al 7029 + 2% WC + 4% fly ash at 200 rpm	200309	113.63
4	Al 7029 + 1% WC + 4% fly ash at 250 rpm	182969	78.87
5	Al 7029 + 1.5% WC + 4% fly ash at 250 rpm	215193	94.25
6	Al 7029 + 2% WC + 4% fly ash at 250 rpm	220638	116.04
7	Al 7029 + 1% WC + 4% fly ash at 300 rpm	214408	113.59
8	Al 7029 + 1.5% WC + 4% fly ash at 300 rpm	205469	96.41
9	Al 7029 + 2% WC + 4% fly ash at 300 rpm	187224	78.98

Figure 10a, and b shows the optimal dispersion of the particle with the matrix and less void in the interaction between matrix and reinforcement which has caused the increase in the load-carrying capacity and maximum stress of Al 7029 + 2% WC + 4% fly ash composites.

4 Conclusions

The present work investigates the investigation of fatigue behaviour of the stir-casted 7029 aluminium composite using 4% fly ash and different 1%, 1.5%, and 2% tungsten carbide. Further, three conditions of stirring speed were utilized for the development of composites. The fatigue testing was carried out by rotating the fatigue testing machine for the developed specimen according to ASTM E606. Further, after testing the results of the number of cycles (N_f), maximum stress limit was obtained for the specimen. The results showed that for all the conditions of stirring speed, the composite with a higher zirconium oxide percentage has the highest number of cycles (N_f) and maximum stress limit when compared to other composites. The results also showed that the number of cycles (N_f) and maximum stress limit have increased with the stirring speed. Results also showed that the Al 7029 + 2% WC + 4% fly ash at 250 rpm has taken up the highest number of cycles of 224,164 with the maximum stress of 115.43Mpa. It was observed that the stirring speeds above 300 rpm that caused excessive foam formation, while speeds below 200 rpm resulted in inadequate mixing. So, Al 7029 + 2% WC + 4% fly ash provides adequate mixing, while minimizing the risk of particle settling or foam formation which correlates well with experimental results. Further, the characterization studies were carried out using SEM images of the specimen with different magnifications. From the SEM images, it was clear that the specimen Al 7029 + 2% WC + 4% fly ash has optimal dispersion of the particle with the matrix and less void in the interaction between matrix and reinforcement has caused the increase in the load-carrying capacity and maximum stress of developed composites.

Fig. 10 SEM images of Al 7029 + 2% WC + 4% fly ash at stirring speed of 250 rpm

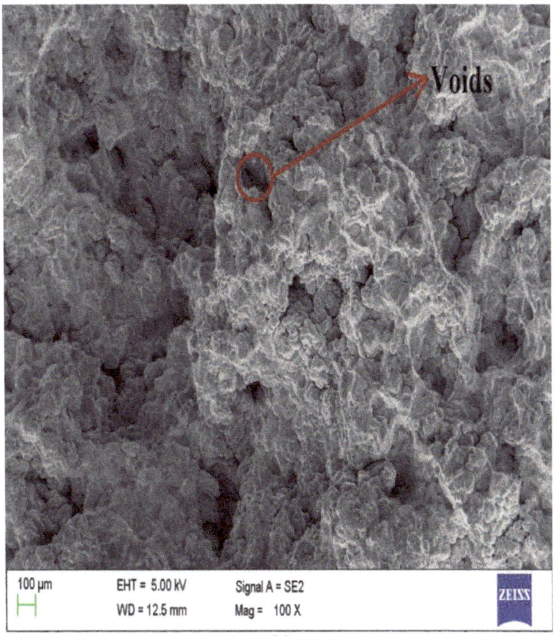

(a)

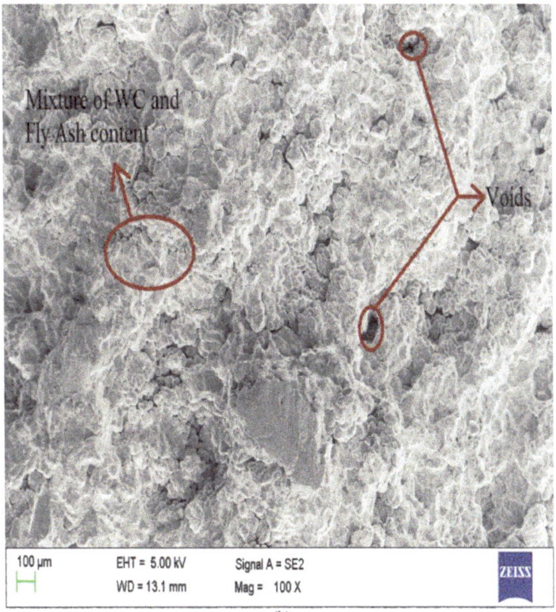

(b)

References

1. Kumar SB, Vardhan H, Raj MG, Kaza M, Sah R, Harish H (2020) The screening efficiency of linear vibrating screen—an experimental investigation. AIP Conf Proc 2204(1):040002
2. Harish H, Vardhan H, Raj MG, Kaza M, Sah R, Sinha A, Kumar SB (2020) Investigation of iron ores based on the bond grindability test. AIP Conf Proc 2204(1):040006
3. Shanmugam BK, Vardhan H, Raj MG, Kaza M, Sah R, Hanumanthappa H (2021) Regression modeling and residual analysis of screening coal in screening machine. Int J Coal Prep Util 42(9):2849–2864
4. Shanmugam BK, Vardhan H, Raj MG, Kaza M, Sah R, Hanumanthappa H (2019) Screening performance of coal of different size fractions with variation in design and operational flexibilities of the new screening machine. Energy Sources Part A: Recover Util Environ Eff 45(2):4361–4369
5. Shanmugam BK, Vardhan H, Raj MG, Kaza M, Sah R, Hanumanthappa H (2020) Experimentation and statistical prediction of screening performance of coal with different moisture content in the vibrating screen. Int J Coal Prep Util 42(6):1804–1817
6. Shanmugam BK, Vardhan H, Raj MG, Kaza M, Sah R, Hanumanthappa H (2020) Artificial neural network modeling for predicting the screening efficiency of coal with varying moisture content in the vibrating screen. Int J Coal Prep Util 42(9):2656–2674
7. Shanmugam BK, Vardhan H, Raj MG, Kaza M, Sah R, Hanumanthappa H (2021) ANN modeling and residual analysis on screening efficiency of coal in vibrating screen. Int J Coal Prep Util 42(10):2880–2894
8. Shanmugam BK, Vardhan H, Raj MG, Kaza M, Sah R, Hanumanthappa H (2021) Application of fractional factorial design for evaluating the separation performance of the screening machine. Int J Coal Prep Util 42(11):3369–3379
9. Verma PK, Singh A, Kumar A, Malik N (2025) Microstructure, mechanical, and wear characteristics of heat-treated aerospace-grade aluminium composite reinforced with HEA particles. Mater Chem Phys 331:130195
10. Sadhu KK, Prajapati PK, Mandal N, Sahoo RR (2025) Tribo-mechanical evaluation of SiC-reinforced wear-resistant aluminium composite fabricated through powder metallurgy. Mater Today Commun 42:111184
11. Srinivasan D, Senthilkumar N, Ganesh MG, Perumal (2024) Characterization studies on nano aluminium composite reinforced with montmorillonite nanoclay and titanium carbide. Multiscale Multidiscip Model Exp Des 7:2839–2859
12. Senthilraj K, Rajamurugan G (2024) Effect of Al2TiO5 particle on the temperature distribution of friction stir welded dissimilar aluminium composites using response surface methodology. Trans Indian Inst Met 77:1205–1215
13. Das DK, Kumar B (2025) Mechanical properties of aluminium-graphene core shell nanocomposite: a molecular dynamics simulation study. Diam Relat Mater 152:111981
14. Ogunbiyi O, Jamiru T, Akinwande AA (2025) Mechanical and wear characteristics of aluminium-7068-composites structurally modified with medium-entropy-alloy. Mater Today Commun 42(111169):2352–4928
15. Srinivasa CV, Suyog J, Sreenivasa TN, Harish H, Prasad CD, Bharath KS (2023) Techniques on the wear resistance of plasma beam treatment on SS316L components. J Inst Eng (India): Ser D
16. Lokesh KS, Kumar KS, Keerthan N, Revanth R, Sandeep S, Sunil BL, Bharath V, Harish H, Prasad CD, Bharath KS (2023) Experimental analysis of the rice husk and eggshell powder-based natural fibre composite. J Inst Eng (India): Ser D
17. Lokesh KS, Shanmugam BK, Mayya SD, Panduranga BP, Kumar JRN, Hanumanthappa H (2021) Experimentation and prediction analysis on the mechanical performance of fish scale and coconut shell powder-based composites. J Nat Fibers 19(14):7750–7761
18. Sriramamurthy LK, Nagarathna BKS, Paavan KT, Hanumanthappa H, Thimmegowda M, Mayya SD, Srivatsav SKY, Kumar ABK (2022) Experimental and statistical evaluation of the

mechanical performance of (Jute and Cocopeat) plant and (Silk) animal-based hybrid fibers reinforced with epoxy polymers. J Nat Fibers 19(16):12664–12675
19. Madhu KS, Sharath BN, Venkatesh CV, Pradeep DG (2021) Evaluation of mechanical properties of ceramic reinforced aluminium-7029 hybrid composite. IOP Conf Ser: Mater Sci Eng 1189(012019)
20. Madhu KS, Venkatesh CV, Sharath BN, Karthik S (2021) Effect of boron carbide on wear resistance of graphite containing Al7029 based hybrid composites and its dry sliding wear characterization through experimental, response surface method and ANOVA. Tribol-Finn J Tribol 38(3–4):48–60
21. Sharath BN (2024) The effect of tertiary ceramic particle reinforcement on the mechanical characteristics of hybrid composites based on Al7029. J Mater Eng Perform
22. Gangadhar TG, Shankara DR, Adarsh BC (2023) Wear and heat treatment studies on Al7075 alloy hybrid composites. Mater Today: Proc 80(2):1462–1472
23. Sharath BN, Karthik S, Pradeep DG, Madhu KS, Venkatesh CV (2023) Machinability studies on boron carbide and graphite reinforced Al7029-based hybrid composites. Lecture Notes in Mechanical Engineering. Springer, Singapore
24. Azadi M, Samiee M, Esfarjani SM (2021) Estimation of low-cycle fatigue lifetime in aluminum-silicon-magnesium alloy of cylinder heads based on striation marks as failure features of fracture surfaces and Paris crack growth law. J Fail Anal Preven 21:1466–1475
25. ASTM E606/E606M-19e1 (2019) Standard Test Method for Strain- Controlled Fatigue Testing, ASTM International, WestConshohocken, PA. http://www.astm.org

Sustainable Development of 7050 Aluminium Alloy Hybrid Composites Reinforced with Fly Ash and Zirconium Oxide: Investigation of Fatigue Behaviour and Characterization

T. G. Gangadhar, S. Bharath Kumar, A. C. Prapul Chandra, Pramod Kumar Naik, and H. Harish

Abstract The fatigue behaviour is examined in this work using a hybrid composite made of zirconium oxide, fly ash, and 7050 aluminium composite. Additionally, composites were created using three different stirring speed conditions: 200 rpm, 250 rpm, and 300 rpm. In compliance with ASTM E467-21, the fatigue test was performed on the generated specimen utilizing rotating fatigue testing apparatus. The specimen's cycle number (Nf) and maximum stress limit were also ascertained after testing. The results showed that the composite with a higher percentage of zirconium oxide had the highest number of cycles (Nf) and maximum stress limit for all stirring speed settings when compared to other composites. Furthermore, the findings showed that as the stirring speed increased, so did the maximum stress limit and number of cycles (Nf). Results also showed that the Al-7050 + 7% ZrO2 + 4% fly ash at 300 rpm has taken up the highest number of cycles of 224,164 with the maximum stress of 115.43Mpa. Furthermore, the characterization studies included SEM pictures of the material at different magnifications. The SEM images clearly showed that the optimal dispersion of the particles with the matrix enhanced the maximum stress and load-carrying capability of the produced composites.

Keywords Composite · Al702 · Fatigue · SEM · Characterization

T. G. Gangadhar (✉) · S. Bharath Kumar · P. K. Naik
Department of Artificial Intelligence and Robotics, Dayananda Sagar University, Bangalore, India
e-mail: gangadhar-air@dsu.edu.in

A. C. Prapul Chandra
Department of Mechanical Engineering, R V College of Engineering, Bangalore, India

H. Harish
School of Computer Science and Engineering, R V University, Bangalore, India

I. M. Jamadar et al. (eds.), *Integrating Sustainability in Modern Technologies*, Lecture Notes in Mechanical Engineering, https://doi.org/10.1007/978-981-95-9466-5_23

1 Introduction

In India, the low availability of high-grade materials such as iron, coal, etc., requires processing for their utilization [1–4]. For processing, it requires a processing plant and also higher production costs [4–8]. So, research on hybrid composite has been extensively carried out on the developing composites of metal and polymer for the replacement with high-grade material [9–13]. The development of metal matrix and polymer matrix composites for various purposes is the focus of much composite research in India [14–17]. One type of metal matrix composite that is widely used in automotive and aeronautical applications is aluminium (Al). The goal of aluminium (Al) alloy is to achieve the following desirable mechanical properties: high stiffness, high strength-to-weight ratio, and low weight. Better performance, quality, and affordability are some of the features of Al alloy. The 7XXX series, or 7050 aluminium alloy, which is composed of Al-Zn-Mg-Cu, was the Al matrix taken into consideration in this investigation. In the past, a number of researchers have worked on a hybrid composite made of 7050 Al alloy.

Luo et al. 2011 [18] studied the material behaviour and microstructural investigation of deforming 7050 Al alloy. The study addressed the optimal process parameter of deforming 7050 Al alloy has a significant effect on temperature and strain rate for the deformation. Teng et al. 2022 [19] studied the experimental investigation on vibration fatigue of the 7050 Al alloy. The results showed that frequency is the significant process parameter that causes structural damage to the material.

Singh et al. 2021 [20] reviewed the mechanical properties of 7050 Al and 7075 Al alloy composite materials. The study majorly focused on investigating the hardness and tensile properties of the materials. The study showed that the hardness and tensile properties of alloys increased in comparison with the base material. Newman et al. 2011 [21] investigated the fatigue behaviour and crack growth of 7050-T7451 Al alloy. The results showed that the roughness-induced crack closure model showed reasonable results for crack growth and fatigue life prediction. Gangadhar et al. 2023[22] have studied the tensile and flexural behaviour of the Al-7050 hybrid composite. The results showed that the ceramic particle reinforcement of Al-7050 hybrid composite in the developed hybrid composite has improved the tensile strength, compression strength, and impact resistance.

So, from the literature, it was clear that less work has been carried out on the fatigue studies of zirconium oxide and fly ash reinforced with 7050 aluminium alloy hybrid composite. So, the present work majorly focuses on the investigation of the fatigue behaviour of the stir-casted 7050 aluminium composite using zirconium oxide (ZrO2) and fly ash.

2 Methodology

The materials employed in this investigation, 7050 Al alloy, zirconium oxide powder, and fly ash powder, are depicted in Fig. 1a, b, and c. The stirring conditions and weight percentage (%) of the components utilized in this study are displayed in Table 1. The electric furnace apparatus used to create the 7050 Al alloy composite is depicted in Fig. 2a. The 7050 Al alloy is heated to 650°CC in a molten state after being added to an electric furnace in varying weight proportions. The stirring of molten 7050 Al alloy inside an electric furnace is depicted in Fig. 2b. Following the addition of zirconium oxide and fly ash to the molten 7050 Al alloy, the mixture is stirred using a mechanical stirrer at 200, 250, and 300 rpm.

The pouring of molten metal into the die for the solidification process is depicted in Fig. 2d. 7050 aluminium composites made using the stir casting technique are displayed in Fig. 3. Following composite preparation, a CNC lathe machine was

(a) (b) (c)

Fig. 1 **a** Aluminium-7050 alloy, **b** Zirconium oxide powder, **c** Fly ash powder material used in the present study

Table 1 Stirring conditions and weight percentage (%) of materials used in the present study

Sl. no	Stir speed	Zirconium oxide (ZrO2) wt (%)	Fly ash wt (%)	Matrix material (7050 Al) (%)
1	200	3	4	93
2	200	5	4	91
3	200	7	4	89
4	250	3	4	93
5	250	5	4	91
6	250	7	4	89
7	300	3	4	93
8	300	5	4	91
9	300	7	4	89

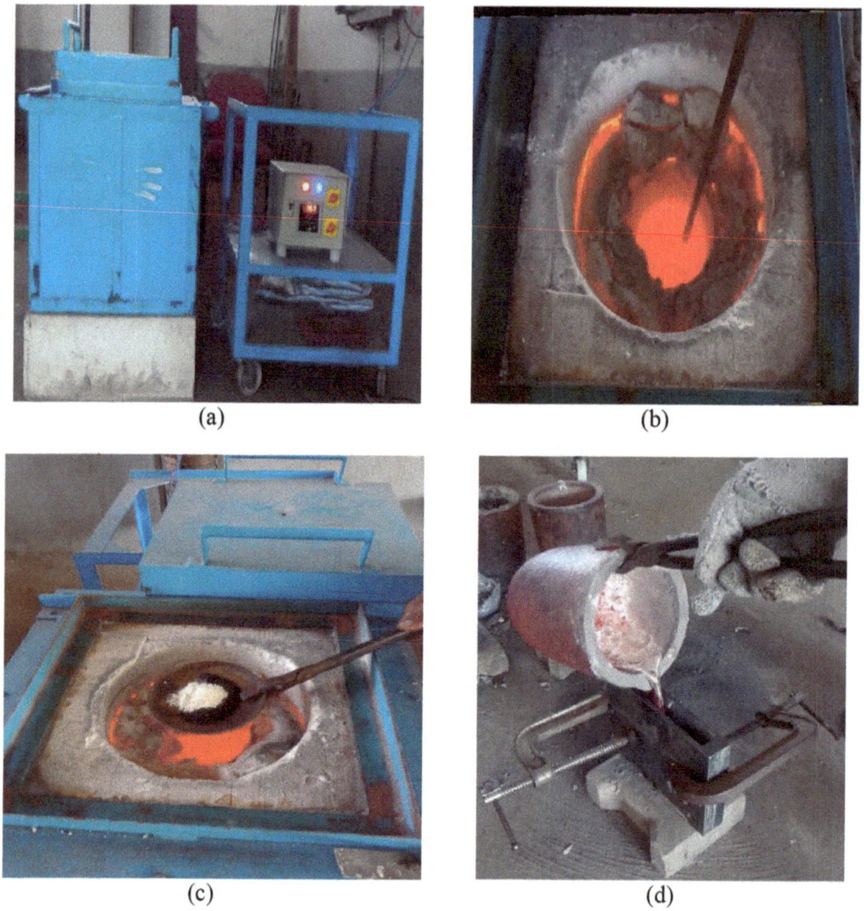

Fig. 2 **a** Casting equipment, **b** Stir casting in electric furnace, **c** Adding of ZrO2 and fly ash to molten aluminium, **d** Pouring of molten metal in die

used to mill the created composites. In order to prepare the specimens, the ASTM-E606 standard was taken into consideration [23, 24]. Figure 4a and b shows the standard fatigue specimen and the developed specimen of 7050 Al composites made for this investigation is displayed in Fig. 4.

Figure 4 shows the standard fatigue specimen and developed specimen of 7050 Al composites prepared for the present study. The rotating fatigue testing apparatus utilized in this investigation is depicted in Fig. 5. One end of the specimen was loaded, while the other end was secured to the machine using a cantilever. A load of 50N was applied to nine specimens for testing. The number of cycles that can be supported before breaking was determined. Therefore, the current work examines the fatigue behaviour of the composite made of stir-cast 7050 aluminium alloy.

Fig. 3 Developed aluminium composites prepared by stir casting process

3 Results and Discussion

Figure 6a and b shows the number of cycles and maximum stress achieved by different specimens with 3%, 5%, and 7% zirconium oxide with 4% fly ash and 7050 Al composite under stirring speed 200 rpm. The results show that specimen 3, i.e. 7% zirconium oxide with 4% fly ash has taken up a greater number of cycles and achieved the maximum stress value of 113.68Mpa in comparison with the other two specimens (1 and 2). From Fig. 6a and b, it is clear that the specimen with higher zirconium oxide has the highest load-carrying capacity and has a maximum stress limit.

Figure 7a and b shows the number of cycles and maximum stress achieved by different specimens with 3%, 5%, and 7% zirconium oxide with 4% fly ash and 7050 Al composite under stirring speed 250 rpm. The results show that specimen 6, i.e. 7% zirconium oxide with 4% fly ash has taken up a greater number of cycles and achieved the maximum stress value of 115.02 Mpa in comparison with the other two specimens (4 and 5). From Fig. 7a and b, it is clear that the specimen with higher zirconium oxide has the highest load-carrying capacity and has a maximum stress limit.

Figure 8a and b shows the number of cycles and maximum stress achieved by different specimens with 3%, 5%, and 7% zirconium oxide with 4% fly ash and 7050 Al composite under stirring speed 300 rpm. The results show that specimen 9, i.e. 7% zirconium oxide with 4% fly ash has taken up a greater number of cycles and achieved the maximum stress value of 115.43 Mpa in comparison with the other two specimens (7 and 8). From Fig. 7a and b, it is clear that the specimen with higher

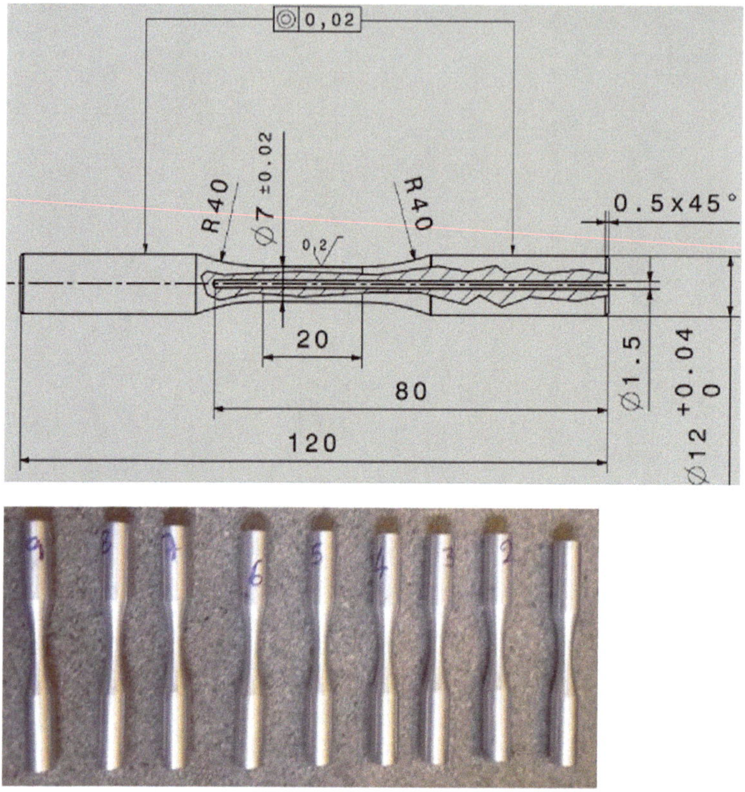

Fig. 4 Fatigue specimen of 7050 Al composites prepared for the present study

zirconium oxide has the highest load-carrying capacity and has a maximum stress limit.

Figure 9 shows the fractured fatigue specimen after testing. Table 2 shows the fatigue test results on the developed composites. So, from Figs. 6a, b, 7a, b, 8a, and b and Table 2, it is clear that as the percentage of zirconium oxide increased there was an increase in the maximum stress and number of cycle carrying capacity of the specimen. It was also found that the increase in the stirring speed from 200 to 250 rpm has increased the specimen's ability to take up more load.

Figure 10a, b, and c shows the SEM images of Al-7050 + 7% ZrO_2 + 4% fly ash at a stirring speed of 300 rpm for magnification of 100x, 500x, and 1000x, respectively. Figure 10a, b, and c shows the optimal dispersion of the particle with the matrix which has caused the increase in the load-carrying capacity and maximum stress of developed composites.

Fig. 5 Rotating fatigue testing machine

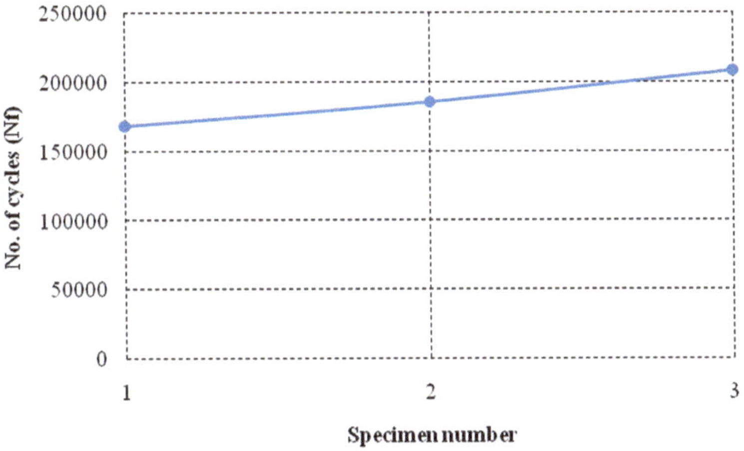

Fig. 6 Number of cycles of fatigue test of different specimens with 3%, 5%, and 7% zirconium oxide reinforced with 7050 Al hybrid composite of stirring speed 200 rpm. **b** Maximum stress achieved by different specimens with 3%, 5%, and 7% zirconium oxide reinforced with 7050 Al hybrid composite under stirring speed 200 rpm

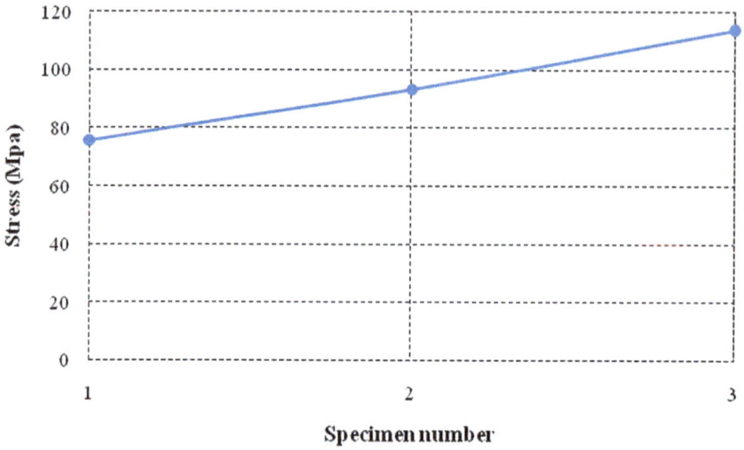

Fig. 6 (continued)

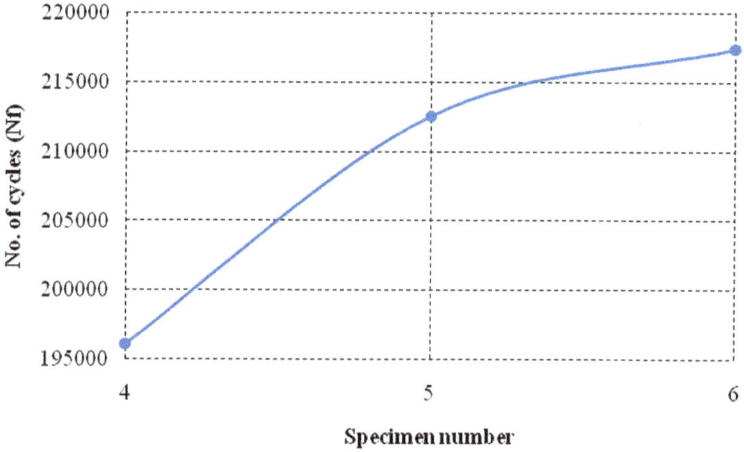

Fig. 7 **a** Number of cycles of fatigue test of different specimens with 3%, 5%, and 7% zirconium oxide reinforced with 7050 Al hybrid composite under stirring speed 250 rpm. **b** Maximum stress before fracture of different specimens with 3%, 5%, and 7% zirconium oxide reinforced with 7050 Al hybrid composite under stirring speed 250 rpm

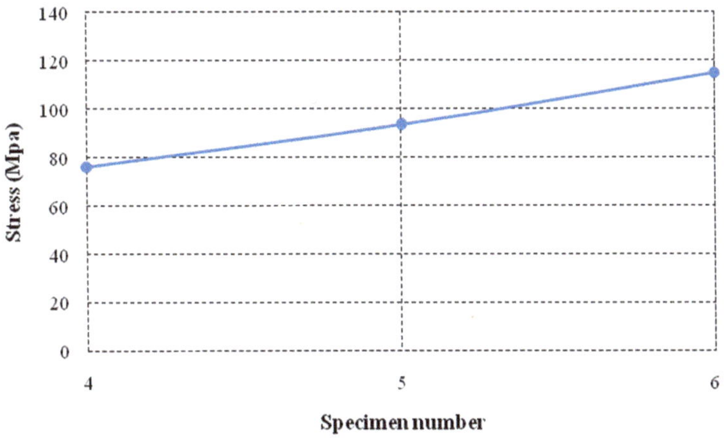

Fig. 7 (continued)

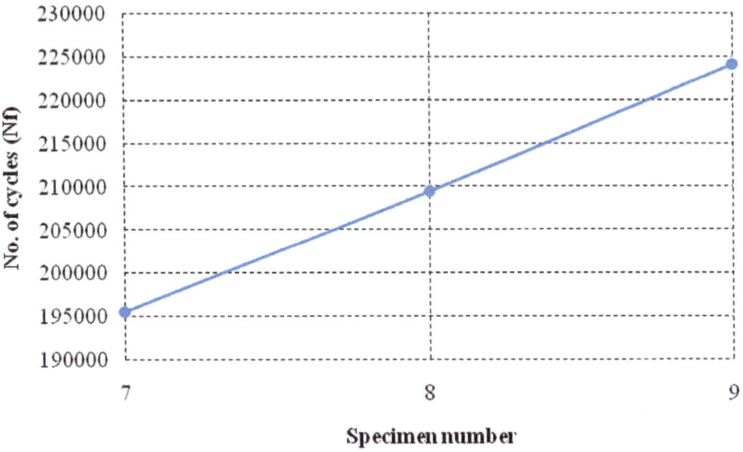

Fig. 8 **a** Number of cycles of fatigue test of different specimens with 3%, 5%, and 7% zirconium oxide reinforced with 7050 Al hybrid composite under stirring speed 300 rpm. **b** Maximum stress before fracture of different specimens with 3%, 5%, and 7% zirconium oxide reinforced with 7050 Al hybrid composite under stirring speed 300 rpm

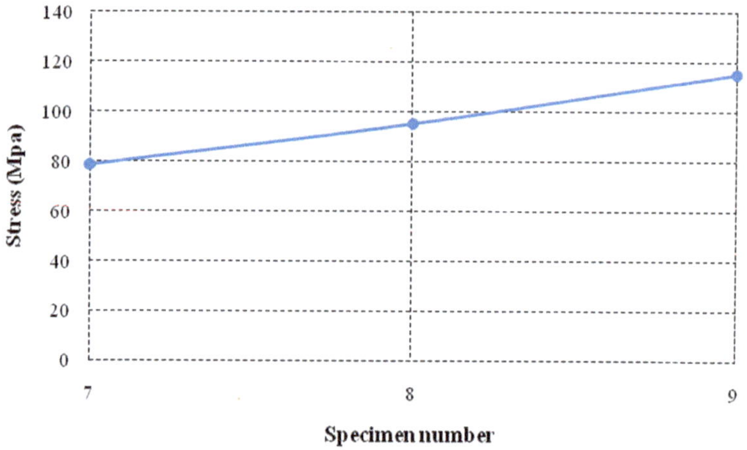

Fig. 8 (continued)

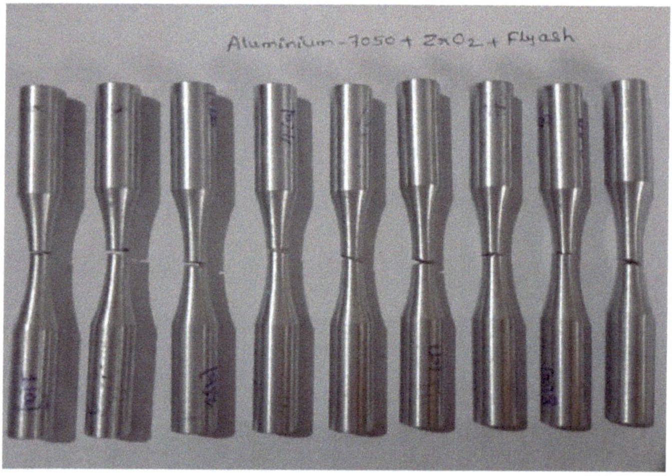

Fig. 9 Fractured fatigue specimen of zirconium oxide reinforced with 7050 Al hybrid composite after testing

Table 2 Fatigue test results on developed composites

Specimen no	Composition and stirring speed	No. of cycles (N_f)	Stress Mpa
1	Al-7050 + 3% ZrO2 + 4%FA at 200 rpm	168758	75.78
2	Al-7050 + 5% ZrO2 + 4%FA at 200 rpm	186289	93.42
3	Al-7050 + 7% ZrO2 + 4%FA at 200 rpm	209078	113.68
4	Al-7050 + 3% ZrO2 + 4%FA at 250 rpm	196129	76.38
5	Al-7050 + 5% ZrO2 + 4%FA at 250 rpm	212618	93.49
6	Al-7050 + 7% ZrO2 + 4%FA at 250 rpm	217439	115.02
7	Al-7050 + 3% ZrO2 + 4%FA at 300 rpm	195609	78.98
8	Al-7050 + 5% ZrO2 + 4%FA at 300 rpm	209528	95.46
9	Al-7050 + 7% ZrO2 + 4%FA at 300 rpm	224164	115.43

4 Conclusions

The present work investigates the studies on fatigue behaviour of the stir-casted 7050 aluminium composite using 4% fly ash and different 3%, 5%, and 7% zirconium oxide (ZrO2). Further, three conditions of stirring speed were utilized for the development of composites. The fatigue testing was carried out by rotating the fatigue testing machine for the developed specimen according to ASTM E467-21. Further, after testing the results of the number of the cycle (Nf) maximum stress limits were obtained for the specimen. The results showed that for all the conditions of stirring speed, the composite with a higher zirconium oxide percentage has the highest number of cycles (Nf) and maximum stress limit when compared to other composites. The results also showed that the no. of cycle (Nf) and maximum stress limit have increased with the stirring speed. Further, the characterization studies were carried out using SEM images of the specimen with different magnifications. The experimental results and SEM results show that the improved particle dispersion, reduced voids, reduced crack initiation, and also strong interfacial bonding have improved the strength and load transfer of Al-7050 + 7% ZrO2 + 4% fly ash. Also, it was observed that the stirring speed of 300 rpm caused optimal particle mixing and less foam formation, while speed below 200 rpm resulted in inadequate mixing. So, Al-7050 + 7% ZrO2 + 4% fly ash provides adequate mixing, while minimizing the risk of particle settling or foam formation correlates well with experimental results. The addition of zirconium oxide content has improved dispersion strengthening, grain refinement, and interfacial bonding improving in fatigue performance of Al-7050 + 7% ZrO2 + 4% fly ash. From the SEM images, it was clear that the optimal dispersion of the particle with the matrix has caused the increase in the load-carrying capacity and maximum stress of developed composites.

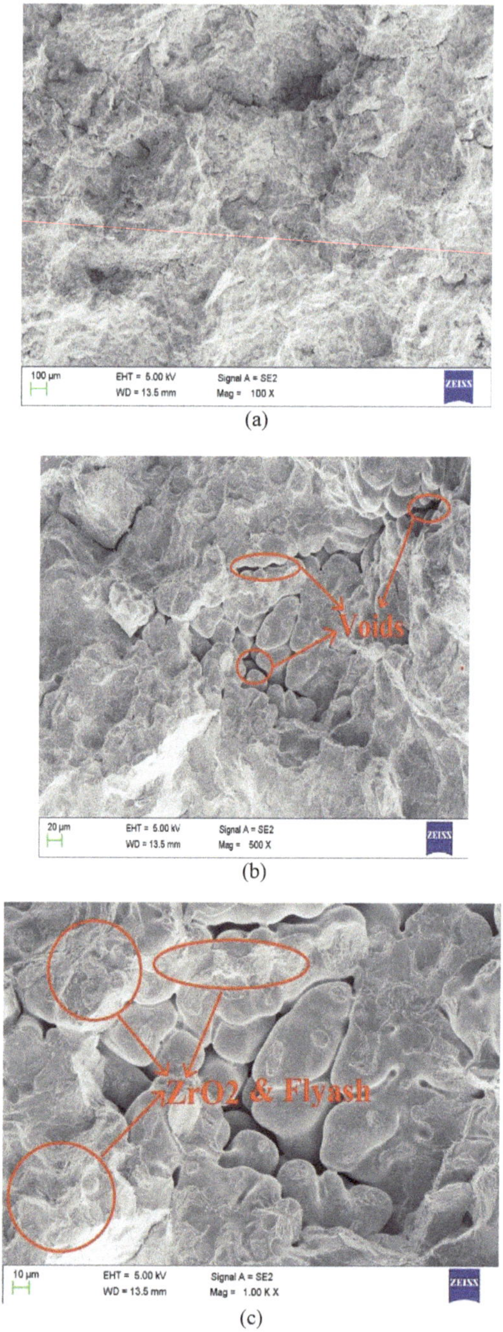

Fig. 10 SEM images of Al-7050 + 7% ZrO_2 + 4% fly ash at stirring speed of 300 rpm for magnification of 100x (**a**), 500x (**b**), and 1000x (**c**). The experimental results and SEM results show that the improved particle dispersion, reduced voids, reduced crack initiation, and also strong interfacial bonding have improved the strength and load transfer of Al-7050 + 7% ZrO2 + 4% fly ash. Also, it was observed that the stirring speed of 300 rpm caused optimal particle mixing and less foam formation, while speed below 200 rpm resulted in inadequate mixing. So, Al-7050 + 7% ZrO2 + 4% fly ash provides adequate mixing, while minimizing the risk of particle settling or foam formation correlates well with experimental results. The addition of zirconium oxide content has improved dispersion strengthening, grain refinement, and interfacial bonding improving in fatigue performance of Al-7050 + 7% ZrO_2 + 4% fly ash

References

1. Kumar SB, Vardhan H, Raj MG, Kaza M, Sah R, Harish H (2020) The screening efficiency of linear vibrating screen—an experimental investigation. AIP Conf Proc 2204(1):040002
2. Harish H, Vardhan H, Raj MG, Kaza M, Sah R, Sinha A, Kumar SB (2020) Investigation of iron ores based on the bond grindability test. AIP Conf Proc 2204(1):040006
3. Shanmugam BK, Vardhan H, Raj MG, Kaza M, Sah R, Hanumanthappa H (2021) Regression modeling and residual analysis of screening coal in screening machine. Int J Coal Prep Util 42(9):2849–2864
4. Shanmugam BK, Vardhan H, Raj MG, Kaza M, Sah R, Hanumanthappa H (2019) Screening performance of coal of different size fractions with variation in design and operational flexibilities of the new screening machine. Energy Sources Part A: Recover Util Environ Eff 45(2):4361–4369
5. Shanmugam BK, Vardhan H, Raj MG, Kaza M, Sah R, Hanumanthappa H (2020) Experimentation and statistical prediction of screening performance of coal with different moisture content in the vibrating screen. Int J Coal Prep Util 42(6):1804–1817
6. Shanmugam BK, Vardhan H, Raj MG, Kaza M, Sah R, Hanumanthappa H (2020) Artificial neural network modeling for predicting the screening efficiency of coal with varying moisture content in the vibrating screen. Int J Coal Prep Util 42(9):2656–2674
7. Shanmugam BK, Vardhan H, Raj MG, Kaza M, Sah R, Hanumanthappa H (2021) ANN modeling and residual analysis on screening efficiency of coal in vibrating screen. Int J Coal Prep Util 42(10):2880–2894
8. Shanmugam BK, Vardhan H, Raj MG, Kaza M, Sah R, Hanumanthappa H (2021) Application of fractional factorial design for evaluating the separation performance of the screening machine. Int J Coal Prep Util 42(11):3369–3379
9. Verma PK, Singh A, Kumar A, Malik N (2025) Microstructure, mechanical, and wear characteristics of heat-treated aerospace-grade aluminium composite reinforced with HEA particles. Mater Chem Phys 331:130195
10. Sadhu KK, Prajapati PK, Mandal N, Sahoo RR (2025) Tribo-mechanical evaluation of SiC-reinforced wear-resistant aluminium composite fabricated through powder metallurgy. Mater Today Commun 42:111184
11. Singh A, Singh S (2024) Synthesis, optimisation and characterisation of nano carbon powder reinforced aluminium composites. J Inst Eng India Ser C 105:1263–1268
12. Srinivasan D, Senthilkumar N, Ganesh MG, Perumal (2024) Characterization studies on nano aluminium composite reinforced with montmorillonite nanoclay and titanium carbide. Multiscale Multidiscip Model Exp Des 7:2839–2859
13. Senthilraj K, Rajamurugan G (2024) Effect of Al2TiO5 particle on the temperature distribution of friction stir welded dissimilar aluminium composites using response surface methodology. Trans Indian Inst Met 77:1205–1215
14. Srinivasa CV, Suyog J, Sreenivasa TN, Harish H, Prasad CD, Bharath KS (2023) Techniques on the wear resistance of plasma beam treatment on SS316L components. J Inst Eng (India): Ser D
15. Lokesh KS, Kumar KS, Keerthan N, Revanth R, Sandeep S, Sunil BL, Bharath V, Harish H, Prasad CD, Bharath KS (2023) Experimental analysis of the rice husk and eggshell powder-based natural fibre composite. J Inst Eng (India): Ser D
16. Lokesh KS, Shanmugam BK, Mayya SD, Panduranga BP, Kumar JRN, Hanumanthappa H (2021) Experimentation and prediction analysis on the mechanical performance of fish scale and coconut shell powder-based composites. J Nat Fibers 19(14):7750–7761
17. Sriramamurthy LK, Nagarathna BKS, Paavan KT, Hanumanthappa H, Thimmegowda M, Mayya SD, Srivatsav SKY, Kumar ABK (2022) Experimental and statistical evaluation of the mechanical performance of (Jute and Cocopeat) plant and (Silk) animal-based hybrid fibers reinforced with epoxy polymers. J Nat Fibers 19(16):12664–12675
18. Luo J, Li JMQ, Wu B (2011) The correlation between flow behavior and microstructural evolution of 7050 aluminum alloy. Mater Sci Eng: A 530:559–564

19. Teng Y, Xie L, Zhang H (2022) Experimental study on vibration fatigue behavior of aircraft aluminum alloy 7050. Materials 15(21):7555
20. Kapil S, Hartaj SS, Sachit V, Sanjay M (2021) Mechanical study of Al 7050 and Al 7075 based metal matrix composites: a review. Mater Today: Proc 43(1):673–677
21. Newman JC Jr, Shaw JW, Annigeri BS, Ziegler BM (2013) Fatigue and crack growth in 7050–T7451 aluminum alloy under constant- and variable-amplitude loading. J Eng Gas Turbines Power 135(2):022101
22. Gangadhar TG, Shankara DR, Adarsh BC (2023) Wear and heat treatment studies on Al7075 alloy hybrid composites. Mater Today: Proc 80(2):1462–1472
23. Azadi M, Samiee M, Esfarjani SM (2021) Estimation of low-cycle fatigue lifetime in aluminum-silicon-magnesium alloy of cylinder heads based on striation marks as failure features of fracture surfaces and Paris crack growth law. J Fail Anal Preven 21:1466–1475
24. ASTM E606/E606M-19e1 (2019) Standard test method for strain- controlled fatigue testing. ASTM International, WestConshohocken, PA. http://www.astm.org

Glass Waste Powder-Incorporated Alkali-Activated Concrete: A Study on Surface Electrical Resistivity and Ecological Analysis

Sanchi Rewar, Mahender Choudhary, Arun Gaur, Ramswaroop Mandolia, and Eashan Pahsha

Abstract The study evaluates the mechanical and durability properties of glass waste incorporated alkali-activated concrete (AAC), also known as Geopolymer Concrete (GPC) cured under different conditions, aiming to identify its potential as a sustainable supplementary material for the construction industry. Glass Waste Powder (GWP), a recycled material rich in silica, was added in three percentages at 5–15% with an increment of 5% by weight of binder in each mix. The concrete was cured under two different conditions: oven curing (@60 degrees—24 h) and ambient curing. The study explores the role of glass waste as a "performance-enhancing additive" and not as a regular replacement analysis. Mechanical properties such as compressive strength, flexural strength, apparent density, and percentage of permeable voids were evaluated. To determine the bonding (bio-chemical) of the materials, Fourier Transform Infrared Spectroscopy (FTIR) was done. Surface Electrical Resistivity (SER) was evaluated using the Wenner Four-Point Probe method. Results show that 10% glass waste provides the best mechanical strength, with a compressive strength of 44.03 N/mm^2 and a flexural strength of 5.69 N/mm^2. A similar trend was observed for SER, with the highest resistivity recorded for GWP-incorporated GPC cured in the oven as 30.12 KΩ-cm. Additionally, the carbon footprint of GPC was significantly lower, with the 10% glass waste mix achieving 213.37 $kgCO_2e/m^3$, compared to 392.2 $kgCO_2e/m^3$ for OPC.

Keywords Carbon footprint · Geopolymer concrete · Glass waste · Surface electrical resistivity · Sustainable construction

S. Rewar (✉) · M. Choudhary · A. Gaur · R. Mandolia · E. Pahsha
Malaviya National Institute of Technology, Jaipur, India
e-mail: 2021rce9030@mnit.ac.in

I. M. Jamadar et al. (eds.), *Integrating Sustainability in Modern Technologies*, Lecture Notes in Mechanical Engineering, https://doi.org/10.1007/978-981-95-9466-5_24

1 Introduction

1.1 Context

India's population is projected to exceed 800 million by 2050 [1], driving the need for improved infrastructure. Traditional pavements, like Ordinary Portland Cement (OPC), contribute to CO_2 emissions, exacerbating concerns about climate change. The Indian construction sector uses 400 million metric tons of concrete annually, with this demand expected to rise to 1000 million metric tons within the next decade [2]. Geopolymer Concrete (GPC) offers a sustainable alternative, produced using waste materials rich in aluminosilicates like fly ash, GGBS, metakaolin, and rice husk ash [3–5]. GPC's production process involves geopolymerization with a potent alkaline activator, reducing reliance on traditional raw materials and promoting economic and environmental sustainability. In the effort to boost sustainability, various waste materials such as eggshell powder, solid waste generated from municipal areas, and rice husk ash are being utilized to conserve natural resources in the production of GPC, which not only offers cost-effective ways but also acts as an eco-friendly alternative to traditional materials [6–9]. Glass Waste Powder (GWP), characterized by its high silicon dioxide (SiO_2) content, exhibits pozzolanic behavior influenced by its chemical composition. The hydraulic reactivity of GWP depends mainly on the fineness achieved during the grinding process [10]. Studies have shown that enhancing the fineness of GWP particles improves the mechanical and durability performance of cement-based materials. The rough surface texture of GWP creates excellent adhesion within the cement; its inclusion promotes the creation of additional C-S-H gel, backing improved strength properties, and, hence, the use of GWP is justified [11].

Based on past studies, several researchers studied different utilizations of GWP. Chen et al. investigated the partial replacement of fine aggregates with glass particles ranging 0–40% (with an increment of 10%) by weight in concrete. Compressive strength improved across all curing periods, with 40% glass content achieving 17% higher strength than control samples at 28 days, accredited to the high aluminosilicate content in the glass [12]. Sarkar M. et al. studied GWP and FA in geopolymer concrete, highlighting sustainable construction potential under ambient curing. An optimal GP: FA ratio of 1:3 with 8M NaOH achieved the highest strength, with microstructural analysis confirming C/N-(A)-S-H gel formation [13]. Irum and Shabbir performed experiments on FA/GGBS-based geopolymer concrete. They stated that the electrical resistivity of oven-cured specimens provided better resistivity than ambient-cured specimens due to better polymerization reaction and improved microstructure [14].

1.2 Research Significance

Previous studies [15, 16] reported partial replacement of glass waste either done by fine or coarse aggregate. In contrast, the focus of the present study is on the use of GWP as a "performance-enhancing additive" to maintain the performance ability of the binder. The binder in GPC is crucial for alkaline activation; replacing it with GWP may disrupt aluminosilicate gel formation. Therefore, GWP was added in 5, 10, and 15% by weight of the binder, and its mechanical, durability, and ecological analysis was done.

2 Experimental Section

2.1 Materials and Methods

The investigation utilized FA and GGBS obtained from Alwar, Rajasthan, India, with specific gravities of 2.60 and 2.87, respectively. GWP was obtained from Jaipur, having a specific gravity of 2.63. Chemical properties (as XRF) are shown in Fig. 1a–c.

A constant ratio of 2.5 was kept between NaOH and Na_2SiO_3. The mixing process was conducted in the aggregates' saturated surface dry (SSD) state. Half of the specimens were laid open to a temperature of 60°C for 24 h and maintained at room temperature until testing. The mixing procedure entails the dry mixing fine aggregates, coarse aggregates, binder (FA + GGBS), and GWP in a concrete mixer. The alkaline activator solution was gradually amalgamated, and then the blending was for an additional 3 to 5 min until it reached a consistent and homogenous appearance and a workable consistency. Extra water was adjusted accordingly to achieve the desired workability. Once the mixing process was complete, GPC was carefully poured into molds to prepare the specimens. A vibrating table was used to ensure proper compaction, and the molds were then covered to prevent any moisture loss during the curing process. To analyze ion conductivity, a four-point Wenner probe meter (SR meter) was used to perform the Surface Electrical Resistivity at 28 days on cylindrical specimens (100mm × 200mm). The method is directly related to the probability of occurrence of corrosion (severity); the higher the resistivity, the lower the chances of corrosion. SER is calculated using the below formula:

$$\text{Resistivity (K}\Omega\text{cm)} = 2\pi \text{a (V/I)} \tag{1}$$

a = spacing between electrodes, V = difference in voltage, I = Current passed

Table 1 presents the proportions of mixes for GPC and GWP-incorporated GPC. Two series, namely GPC and GW-GPC, represent Mix IDs for geopolymer concrete and GWP-incorporated geopolymer concrete, respectively. In the GW-GPC series,

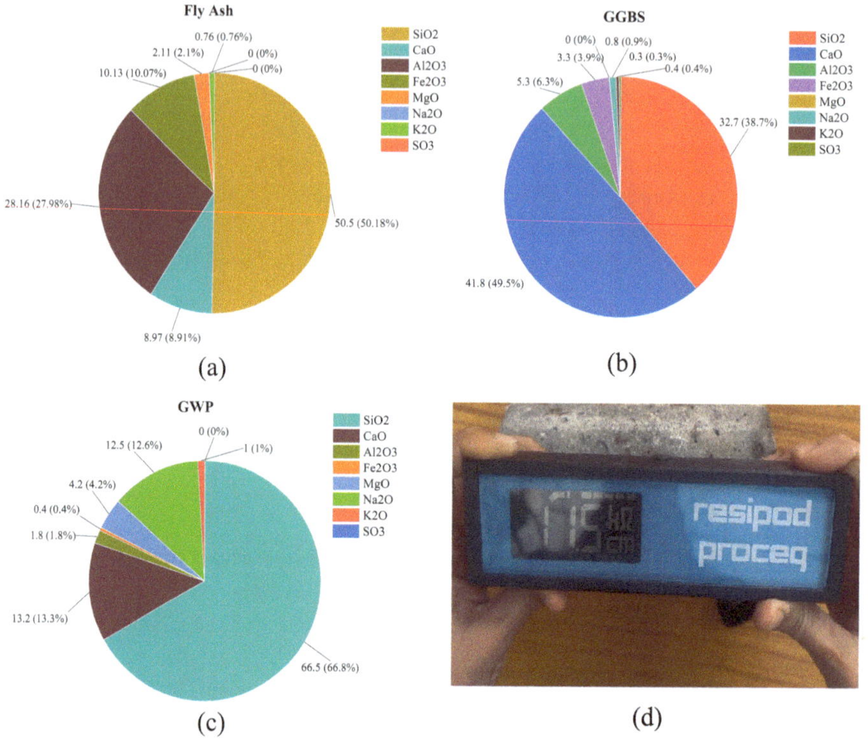

Fig. 1 **a–c** Chemical composition of raw materials, **d** Equipment used to conduct surface electrcal resistivity

the proportion of binder (Fly ash: GGBS) is kept 50:50, simultaneously adding GWP by the weight of the binder.

3 Results and Discussion

3.1 Characterization of Raw Materials

Figure 1a–c illustrates the X-ray fluorescence (XRF) analysis of raw materials used in the study. In GGBS, the presence of calcium oxide helps in forming the C-S-H (Calcium-Silicate-Hydrate), and Silica in GWP forms N-A-S-H (Sodium-Alumina-Silicate-Hydrate) gel, necessary to boost the early age strength [17], thus helping in attaining the anticipated strength at room temperatures as well. SiO_2 is essential since it accelerates the geopolymerization process. Moreover, a reduced Alkali-Silica Ratio (ASR) and notable freeze–thaw durability were also recorded [18]. For these

Table 1 Mix design details of different mixes (Kg/m^3)

Ingredients	OPC	GPC	GW5	GW10	GW15
Cement	413	0	0	0	0
Fly ash	0	206.66	206.66	206.66	206.66
GGBS	0	206.66	206.66	206.66	206.66
Glass waste	0	0	20.66	41.33	61.99
20 mm	581	554.43	554.43	554.43	554.43
10 mm	581	554.43	554.43	554.43	554.43
Sand	687	661.36	661.36	661.36	661.36
Na_2SiO_3	0	132.86	132.86	132.86	132.86
NaOH	0	53.14	53.14	53.14	53.14
Water	186	0.513	0.513	0.513	0.513

reasons, an additional GWP method was used instead of a replacement method so that the amount of CaO present in GGBS was not hindered.

3.2 Compressive and Flexural Strength

Figure 2a, b depicts 28 days of compressive and flexural strength comparison of different mixes. The incorporation of 10% glass waste in geopolymer concrete significantly enhances mechanical properties, with outdoor-cured specimens achieving compressive and flexural strengths of 37.8 N/mm^2 and 5.69 N/mm^2 ($R^2 = 0.93$) and oven-cured specimens reaching 44.03 N/mm^2 and 9.46 N/mm^2 ($R^2 = 0.92$). At both the curing conditions, a 15% addition of GWP in GPC had compressive strength as 30.06 N/mm^2 (outdoor-cured) and 35.23 N/mm^2 (oven-cured), both of which are lesser than the control GPC, i.e., 32.68 N/mm^2 (outdoor-cured) and 43.28 N/mm^2 (oven-cured).

3.3 FTIR

FTIR analysis (Fig. 3c) reveals significant differences in the chemical structure and bonding of GPC with and without GWP under two curing conditions. Room-temperature curing shows a more substantial O-H stretch (3200–3600 cm^{-1}), indicating unreacted hydroxyl groups and moisture, which reduces strength and durability compared to oven-cured samples. GW samples show a pronounced C-H stretch (2800–3000 cm^{-1}), suggesting organic residuals that may compromise stability. A more substantial C = O stretch (1650–1750 cm^{-1}) points to carbonates or incomplete

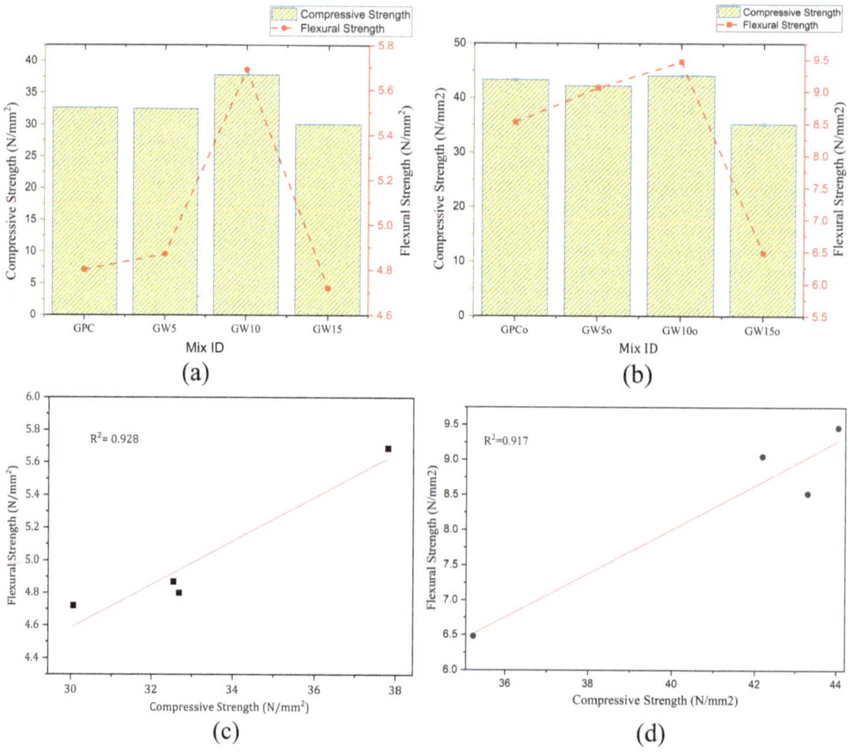

Fig. 2 Compressive and flexural strength of different mixes: **a** outdoor curing, **b** oven curing

reactions at room temperature. In contrast, oven-cured samples exhibit stronger Si-O-Si (1000–1300 cm^{-1}) and Al-O-Si (900–1100 cm^{-1}) stretches, reflecting a robust silicate and aluminosilicate network [19], leading to superior strength and durability. Thus, oven-cured GPC with GWP is preferable for high-performance applications.

3.4 Surface Electrical Resistivity

Figure 3a, b depicts the values obtained for Apparent Density versus Permeable Voids versus Surface Electrical Resistivity (SER). It is evident from the illustration that as the apparent density increases, the percentage of permeable voids decreases, and surface resistivity increases. Several factors could influence the rise in Surface Electrical Resistivity (SER). Firstly, the pozzolanic reaction (C-A-S-H gel) created by GGBS improves the interconnectivity of pores, leading to enhanced resistance [20]. Additionally, the type of curing influences density, significantly improving the mechanical and SER [21]. Oven curing accelerates the geo-polymerization process, and the voids filled with air decrease, and hence apparent density increases. A more

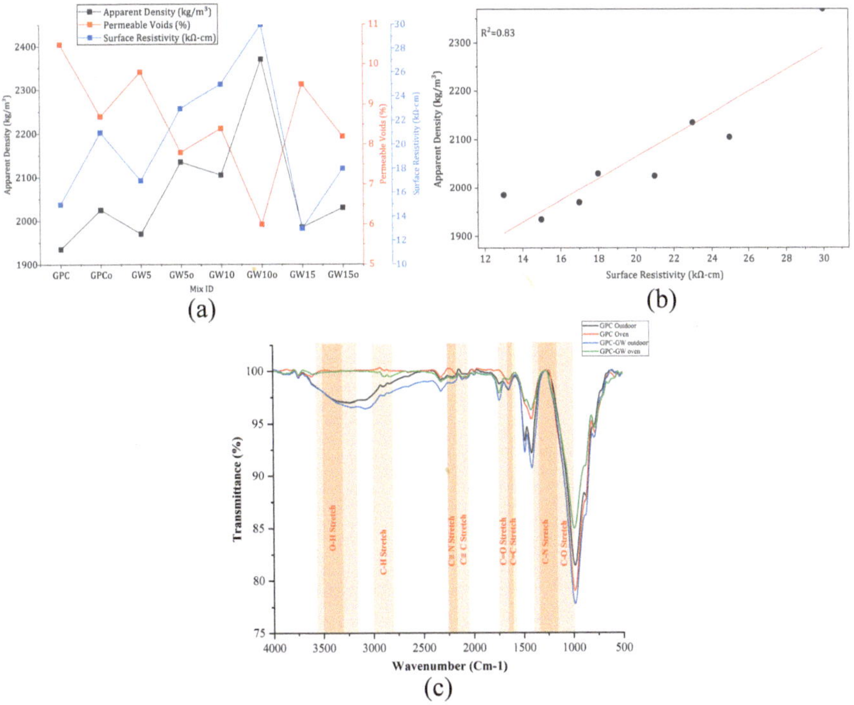

Fig. 3 **a** Graph of apparent density versus permeable voids versus SER, **b** apparent density versus SER, and **c** FTIR analysis

significant percentage of permeable voids would allow more ion mobility, reducing SER [22].

3.5 *Carbon Footprint*

In concrete manufacturing, sustainability is impacted by embodied CO_2 emissions (ECO_2E). In this study (refer to Table 2), values were derived from previous Habert et al. [23] research. The ECO_2E for Ordinary Portland Cement (OPC) was 392.2 $kgCO_2e/m^3$, while the GW5 mix (5% glass waste) had the lowest at 213.37 $kgCO_2e/m^3$. There was a negligible difference between the ECO_2E of GW5 and GW10 due to the minimal variation in glass waste content. Fine glass particles act as micro-fillers, reducing porosity and increasing compressive strength by creating a denser matrix [24]. Additionally, the silicates in glass waste strengthen Si-O-Si and Al-O-Si bonds, as confirmed by FTIR, enhancing the polymerization process. The absence of energy-intensive calcination in OPC production further reduces CO_2 emissions.

Table 2 Embodied CO_2 emission analysis of GPC and GW-GPC

Ingredients	ECO_2E ($kgCO_2e/m^3$)	OPC	GPC	GW5	GW10	GW15
Cement	0.93	413	0	0	0	0
Fly Ash	0.004	0	206.66	206.66	206.60	206.60
GGBS	0.09	0	206.66	206.66	206.60	206.60
Glass Waste	0.046	0	0	20.66	41.33	61.99
CA	0.004	1162	1217.0	1108.0	1108.0	1108.0
FA	0.005	687	638.7	661.86	661.36	661.36
Na_2SiO_3	1.14	0	132.86	132.86	132.86	132.86
NaOH	0.63	0	53.14	53.14	53.14	53.14
Water	0	186	0.513	0.513	0.513	0.513
Total ECO_2E	–	392.2	212.42	213.37	214.32	215.27

4 Conclusions

The study highlights the potential of incorporating glass waste into geopolymer concrete (GPC) as a sustainable alternative to traditional concrete.

- Adding 10% glass waste enhances the compressive strength to 44.03 N/mm^2 while improving the microstructure by reducing pore volume and increasing matrix density, as shown in FTIR analysis.
- The increase in apparent density and reduction in permeable voids significantly enhances Surface Electrical Resistivity, driven by improved pore interconnectivity and accelerated geopolymerization through oven curing.
- Oven curing further optimizes these properties by reducing void content and promoting denser microstructures.
- GPC significantly lowers the carbon footprint to 213.37 $kgCO_2e/m^3$ compared to 392.2 $kgCO_2e/m^3$ for OPC, making it an eco-friendly option for pavements, non-load-bearing structures, and precast components. The improved mechanical properties and reduced environmental impact position GPC as a competitive material for green construction.

Acknowledgements The authors acknowledge Concrete Lab, Civil Engineering Dept., Malaviya National Institute of Technology Jaipur, for providing necessary facilities in conducting the experiment work.

Author Contributions Sanchi Rewar: Conceptualization, Methodology, Investigation, Writing—Original draft preparation; Mahender Choudhary: Supervision, Reviewing, and Editing; Arun Gaur: Supervision, Reviewing, and Editing; Ramswaroop Mandolia: Review and Editing; Eashan Pahsha: Review and Editing.

References

1. Nations U (2018) World urbanization prospects
2. Argiz C, Menéndez E, Moragues A, SanjuAn MA (2014) Struct Eng Int 24:503
3. Xie J, Chen W, Wang J, Fang C, Zhang B, Liu F (2019) Constr Build Mater 226:345
4. Castel A, Foster SJ (2015) Cem Concr Res 72:48
5. Detphan S, Chindaprasirt P (2009) Int J Miner Metall Mater 16:720
6. Abdellatief M, Ahmed YM, Taman M, Elfadaly E, Tang Y, Abadel AA (2024) J Build Eng 90:109374
7. Chen B, Perumal P, Illikainen M, Ye G (2023) J Build Eng 71:106386
8. Nuaklong P, Jongvivatsakul P, Pothisiri T, Sata V, Chindaprasirt P (2020) J Clean Prod 252:119797
9. Adesina A, de Azevedo ARG, Amin M, Hadzima-Nyarko M, Agwa IS, Zeyad AM, Tayeh BA (2022) Clean Mater 3:100036
10. Ramdani S, Guettala A, Benmalek ML, Aguiar JB (2019) J Build Eng 21:302
11. Mohammadinia A, Wong YC, Arulrajah A, Horpibulsuk S (2019) Constr Build Mater 197:489
12. Chen CH, Huang R, Wu JK, Yang CC (2006) Cem Concr Res 36:449
13. Sarkar M, Maiti M, Akbar Malik M, Xu S (2024) Chem Eng J 494:153144
14. Irum S, Shabbir F (2024) J Build Eng 95:110148
15. Mithanthaya I, Bhavanishankar Rao N (2015) Int J Eng Trends Technol 19:66
16. Çelik Aİ, Tunç U, Bahrami A, Karalar M, Othuman Mydin MA, Alomayri T, Özkılıç YO (2023) J Mater Res Technol 24:8533
17. Lee WH, Wang JH, Ding YC, Cheng TW (2019) Constr Build Mater 211:807
18. Jin M, Li W, Wang X, Tang J, Teng L, Ma Y, Zeng H (2023) J. Build. Eng. 78:107759
19. Ng C, Alengaram UJ, Wong LS, Mo KH, Jumaat MZ, Ramesh S (2018) Constr Build Mater 186:550
20. Corinaldesi V, Moriconi G (2009) Constr Build Mater 23:2869
21. Ryu GS, Lee YB, Koh KT, Chung YS (2013) Constr Build Mater 47:409
22. Liew KM, Sojobi AO, Zhang LW (2017) Constr Build Mater 156:1063
23. Habert G, D'Espinose De Lacaillerie JB, Roussel N (2011) J Clean Prod 19:1229
24. Mallikarjuna Rao G, Gunneswara Rao TD (2015) Arab J Sci Eng 40:3067

Fatigue and Fracture Behaviour of 5052 Al Alloy Processed Through Combined Severe Plastic Deformation Method

K. K. Yogesha, A. Joshi, Saurabh Gairola, R. Jayaganthan, and G. M. Owolabi

Abstract Al 5052 is a versatile aluminium alloy known for its excellent corrosion resistance and good workability. Strain hardening, or cold working, is crucial for enhancing its mechanical properties, as it increases hardness and tensile strength through the rearrangement of dislocations within its crystal structure. To improve its mechanical performance, a combined severe plastic deformation (CSPD) was employed. CSPD was performed on solution-treated (ST) specimens using cryo-multi axial forging, followed by cryo-groove rolling and warm rolling. The ST specimen exhibits a high cycle fatigue of 40 MPa, which increased by 162% to 105 MPa after CSPD. This significant improvement results from the evolution of mutually intersecting nano shear bands resulting in grain fragmentation, and ultrafine grains approximately 110 nm in size. The broken/deformed impurity phase particles and fine precipitates formed during warm rolling also contributed to the superior fatigue strength by effectively hindering the dislocation movement. J-integral of CSPD decreases by 70.2% from 249 kJ/m^2 to 74.1 kJ/m^2 due to faster crack propagation in ultrafine-grained than in coarse-grained material. Post-annealed specimens exhibited a lower fatigue strength (95 MPa) due to recovery and recrystallization. However, the coarser microstructure formed led to a fracture toughness of 220 kJ/m^2.

Keywords Microstructure · Ultrafine grains · Fatigue strength · Fracture energy

---

K. K. Yogesha · R. Jayaganthan
Department of Metallurgical and Materials Engineering and Centre of Nanotechnology, Indian Institute of Technology Roorkee, Roorkee, Uttarakhand, India

A. Joshi (✉)
Department of Mechanical Engineering, The National Institute of Engineering, Mysuru, Karnataka, India
e-mail: amitj4765@gmail.com

S. Gairola · R. Jayaganthan
Department of Engineering Design, Indian Institute of Technology Madras, Chennai, Tamil Nadu, India

G. M. Owolabi
Department of Mechanical Engineering, Howard University, Washington, DC, USA

I. M. Jamadar et al. (eds.), *Integrating Sustainability in Modern Technologies*, Lecture Notes in Mechanical Engineering, https://doi.org/10.1007/978-981-95-9466-5_25

1 Introduction

Severe plastic deformation (SPD) is a group of cold working techniques that transforms coarse grains from their bulk alloys into smaller, ultrafine-grained (UFG) materials, significantly enhancing their mechanical performance. Some of the SPD methods include high-pressure torsion (HPT) [1–5] and equal-channel angular pressing (ECAP) [6–10], which are widely used but with the limitation of the size of the materials processed. The other SPD methods, such as multi-axial forging (MAF) [11–15] and cryorolling (CR) [16–20], are used to produce UFG materials with less constraint on the size of the materials processed. MAF utilizes compression in different directions to impart high plastic strain without requiring any additional tools and equipment. Cryorolling is another excellent method to produce UFG materials in large-scale production from their bulk alloys. It requires less plastic deformation than the other SPD methods for grain refinement.

Rao et al. [11] utilized cryo MAF to refine the microstructure, leading to the evolution of equiaxed subgrains and an increase in dislocation density, resulting in improved mechanical properties. Cherukuri et al. [21] have processed 6061 Al alloy through MAF at room temperature, followed by cold rolling to produce ultrafine grains. They compared its tensile properties with ECAP and accumulative roll bonding (ARB) processed samples. They observed similar trends in hardness and tensile properties for all the processing conditions. However, during the hot tensile test ranging 300–350 °C, the MAF + cold-rolled samples showed high strain rate sensitivity, which showed good formability. Noda et al. [22] have conducted MAF of Al–Mg alloy of grain size 67 μm at temperature 543 °K and produced UFG microstructure in the alloy. The sample exhibited an elongation of 340%, and the deformation mechanism was found to be intragranular deformation. Joshi et al. [23] have conducted MAF of 2014 Al alloy and observed a 7% and 2% enhancement in the UTS and ductility of the cryo-forged alloy to the room-temperature forged alloy, respectively. They also noted a 58% and 54% enhancement in the apparent elastic fracture toughness (KQ) of samples forged at cryogenic and room temperature, respectively, exhibiting a brittle fracture model compared to solution-treated (ST) samples. This improvement was attributed to the delayed crack onset in the processed (UFG) alloy.

Verma et al. [24] explored the interrelationship between KQ and microstructure in Cu alloys and reported that the KQ is linearly related to the reciprocal grain size. Somekawa et al. [25] performed the grain refinement of AZ31 Mg alloy through an extrusion process at a temperature of 470 K and reported an increase in critical fracture toughness (KIC). Hohenwarter et al. [26] generated UFG copper using the HPT process and investigated its fracture behaviour. They also employed the HPT process to create UFG nickel and discovered that, while the reduced grain size resulted in lower KIC and critical J-integral (JIC), the overall strength and fracture toughness were superior to those of the bulk counterpart [27]. Fang et al. [28] processed Al-Mg alloy via the ECAP method and observed that the static toughness corresponding to the one-pass ECAP processed samples has decreased despite the improvement in its strength. This was due to reduced ductility in the processed alloy compared to its casting state. Yogesha et al. [29, 30] performed cryo-groove rolling as well as cryo-groove rolling + warm rolling on Al-Mg alloy and conducted a three-point bend test to evaluate its fracture toughness. They observed a decrease in the J-integral (J_{IC}) in the deformed samples, attributed to higher crack propagation rate in the UFG alloy.

Similar observations regarding the improvement in fatigue properties of HPT and ECAP of pure Cu and Al-Mg-Sc-Zr alloy were made by Khatibi et al. [31] and Vinogradov et al. [32], respectively. The enhanced fatigue properties in the deformed specimen are ascribed to the stability of the refined microstructure. Hoppel et al. [33] used ECAP to produce UFG structures in Cu and Al and studied their fatigue behaviour under strain control mode. They reported shorter fatigue life in UFG-structured alloys. This was due to the dynamic grain growth and macroscopic shear bands in UFG Cu, as well as grain elongation and rotation in Al. Singh et al. [34] explored the impact of cryo rolling (CR) and the combination of cryo rolling + warm rolling (CR + WR) on the high cycle fatigue (HCF) properties of Al 5083 alloy. Their finding demonstrates a superior fatigue strength in CR + WR conditions due to the synergy of various strengthening mechanisms, including Hall Petch, dislocation, and precipitation strengthening. Hanlon et al. [35] assessed the HCF of nanocrystalline, ultrafine crystalline, and conventional microcrystalline pure Ni under stress control mode. They found better endurance limits in nanocrystalline and ultrafine crystalline Ni than microcrystalline Ni attributed to the grain refinement.

Conversely, nanocrystalline-Ni exhibited the least resistance to fatigue crack propagation than ultrafine crystalline-Ni, which was due to crack path tortuosity in the material with bigger sized grains. The current work explores the HCF and fracture properties of Al 5052 alloy. The fatigue and fracture behaviour were correlated with the microstructural features of the deformed alloys characterized through SEM and TEM.

2 Experimental Procedure

2.1 Processing

The commercially available Al 5052 was used, which had a chemical composition of 0.20 wt% Si, 0.26 wt% Cr, 0.33 wt% Fe, 0.10 wt% Mn, 2.34 wt% Mg, 0.10 wt% Cu, 0.10 wt% Zn, and balance Al. The raw material was sectioned into cuboidal specimens of 20 × 25 × 40 mm size, homogenized at 540°C for 2 h, and quenched in water. Two sets of these solution-treated (ST) specimens were prepared: one set underwent room-temperature forging (RF), while the other underwent cryo-forging (CF). The forging procedures are illustrated in Fig. 1a. A closed die made of die steel, with a cavity size of 45 × 25 × 20 mm, was employed to enhance grain refinement. Forging was performed using a 100-ton friction screw press operating

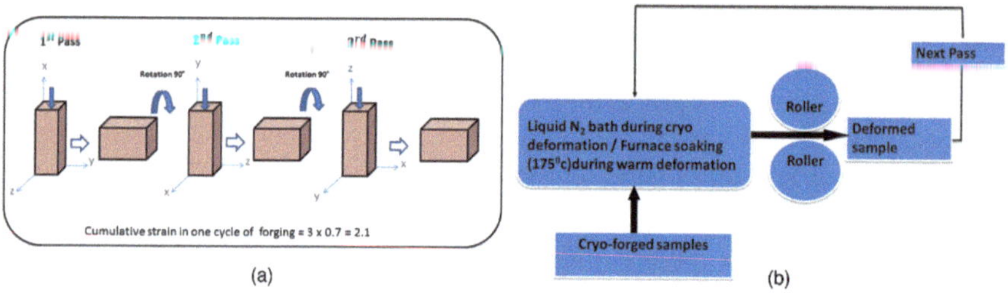

Fig. 1 **A** Schematic illustrating the steps of a single forging cycle. **b** Schematic diagram detailing the steps of the CGW process

at 360 RPM. For cryo-forging, the ST specimens were submerged in liquid nitrogen for up to 15 min prior to every forging pass, ensuring deformation occurred at liquid nitrogen temperature. After each forging pass, the specimens were rotated 90 degrees, achieving a logarithmic strain of $\varepsilon = ln(h_f/h_i) = 0.7$ per pass. This resulted in a total logarithmic strain of 4.2 after two forging cycles for both RF and CF specimens.

A comparative analysis of tensile properties was performed between the RF and CF specimens. The CF specimens, which demonstrated superior tensile properties, were further processed via cryo-groove rolling followed by warm rolling (CGW), reducing their thickness from 20 to 5 mm (ε_{true}(true strain) of 1.39). The initial 50% thickness reduction was achieved through cryo-groove rolling, followed by an additional 25% reduction during warm rolling at 175°C, as illustrated in Fig. 1b. Detailed procedures of CGW are provided in [30]. Subsequently, the CF + CGW specimens underwent post-annealing treatments at temperatures in 180°C–300°C range for 1 h to evaluate the effects on their mechanical properties.

2.2 *Mechanical Characterization*

The tensile behaviour was assessed according to ASTM E8 standards using an S series H25K-S machine using a strain rate of 6×10^{-4}/s. To ensure consistency in the results, each test was conducted three times. The Vickers hardness test was conducted with a 5 kg load and a dwell time of 15 s. Ten indentations were made in accordance with ASTM E92. Fracture toughness was evaluated using a standard single edge notch bend (SENB) specimen following ASTM E1820, with the specimen dimensions being the following: length (L) 30 mm, thickness (B) 3.75 mm, width (W) 7.5 mm, and an (a/w) ratio maintained below 0.55. The pre-cracking on the specimens was made through a wire-cut electrical discharge machine (EDM) using a wire of diameter 250μm [36]. The tests were conducted on the same S-Series, "H25K-S machine", using a displacement rate of 1mm/min [29] and repeated three times to ensure repeatability.

The HCF of the ST, CF + CGW, and CF + CGW + annealed 240°C samples were performed according to ASTM E466 standard using a stress ratio $\left(R = {}^{\sigma_{max}}/_{\sigma_{min}}\right) = 0.1$, and frequency of 20 Hz on "Instron 8802 universal testing machine". The test coupons were mechanically polished using 1500 grit size SiC abrasive paper to eliminate the effect of surface roughness. These fatigue samples were subjected to axial dynamic loading. At each stress amplitude, three sets of trials were conducted.

2.3 Mechanical Characterization

The optical microstructures of ST, RF, CF, and CF + CGW samples were obtained using Leica DMI 5000M. The samples were prepared following standard procedures, with Modified Poulton's reagent used as the etchant. TEM samples for both the ST and processed conditions were prepared following the methodology detailed in our previous work [29] and were characterized using the FEI Tecnai TEM instrument.

3 Experimental Procedure

3.1 Optical Micrograph (OM)

Figure 2a presents the OM of the solution-treated (ST) specimen, revealing a grain size of 390μm. Close examination of these grains indicates the second phase particles that impede dislocation motion. Figure 2b and c displays the optical micrographs of the room temperature-forged (RF) and cryo-forged (CF) samples, respectively, showcasing fragmented grains with poorly defined grain boundaries. The CF samples possess more severely fragmented grains than the RF samples, whereas the CF + CGW samples (Fig. 2d) depict small fragmented elongated grains. The magnified image of the ST sample in Fig. 2e reveals impurity phase particles measuring up to 15 μm. Likewise, Fig. 2f shows broken or deformed second phase in the CF + CGW specimen, with sizes reaching up to 8 μm.

3.2 Hardness and Tensile Properties

Table 1 and Fig. 3a depict the hardness and tensile behaviour of the processed samples. The ST samples exhibit a hardness of 55 HV, while the RF and CF samples show hardness values of 102 HV and 130 HV, respectively. As illustrated in Table 1, the CF + CGW samples demonstrate superior hardness (138 HV) compared to other processed conditions. The solution-treated (ST) samples demonstrate mechanical properties including a UTS, YS, and ductility of 170 MPa, 142 MPa, and 42%,

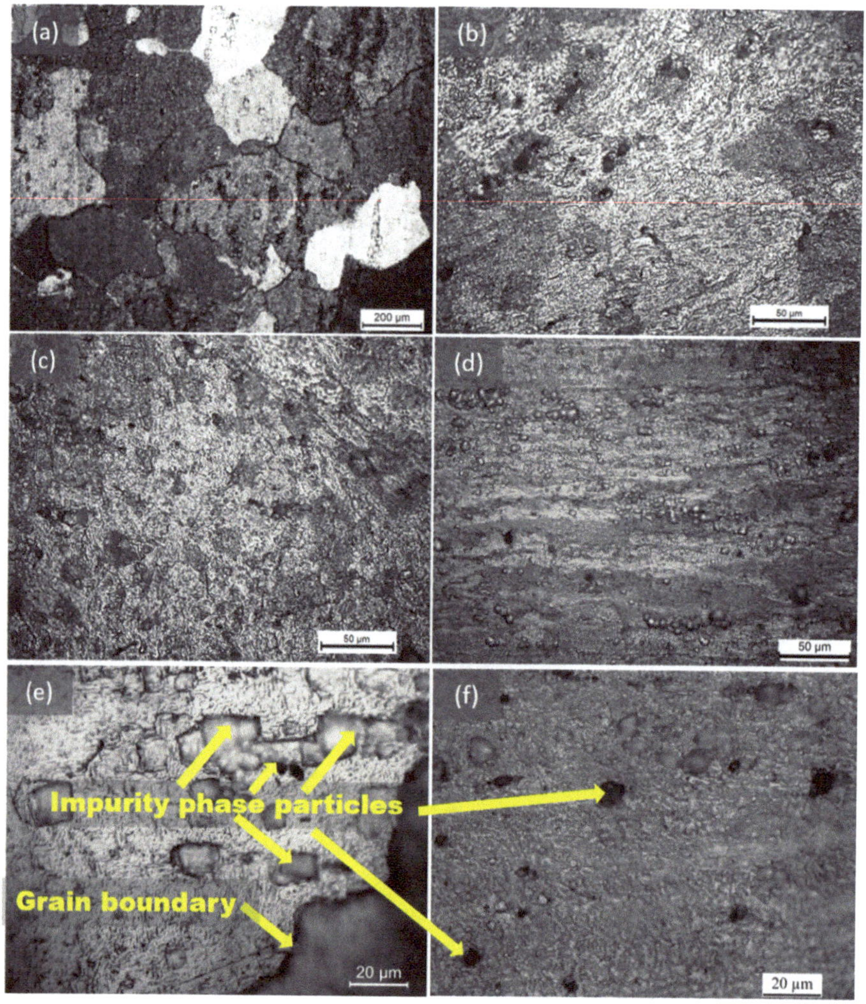

Fig. 2 **a** Schematic illustrating the steps of a single forging cycle. **b** Schematic diagram detailing the steps of the CGW process

respectively (Fig. 3b), and for the room temperature-forged (RF) samples, UTS, YS, and ductility of 310 MPa, 294 MPa, and 5.5% respectively. The cryo-forged (CF) samples show a UTS, YS, and ductility of 380 MPa, 354 MPa, and 7.1%, respectively. The CF + CGW samples demonstrate the highest mechanical performance, surpassing all other processed conditions.

The increased hardness and strength observed in RF and CF samples compared to ST samples is due to work hardening imparted during the forging. The additional increment in hardness in CF samples, as opposed to the RF samples, can be owed to lack of dynamic recovery at cryogenic temperature. Aluminium, being a material

Table 1 Table captions should be placed above the tables

Material condition	Hardness (HV)	σ_y (MPa)	σ_{ut} (MPa)	% elongation	Fracture energy (kJ/m^2)	Fatigue strength (MPa)	Elastic Strain component
ST	55	142	170	42	249	40	0.00125
RF	102	294	310	5.5	36.1	–	–
CF	130	354	380	7.1	45.1	–	–
CF + CGW	138	366	396	7.6	74.1	105	0.00068
CF + CGW + Ann$^{180°C}$	112	309	328	8.1	113.6	–	–
CF + CGW + Ann$^{240°C}$	98	286	306	10.6	180.1	95	0.00098
CF + CGW + Ann$^{300°C}$	82	255	276	15.8	220.1	–	–

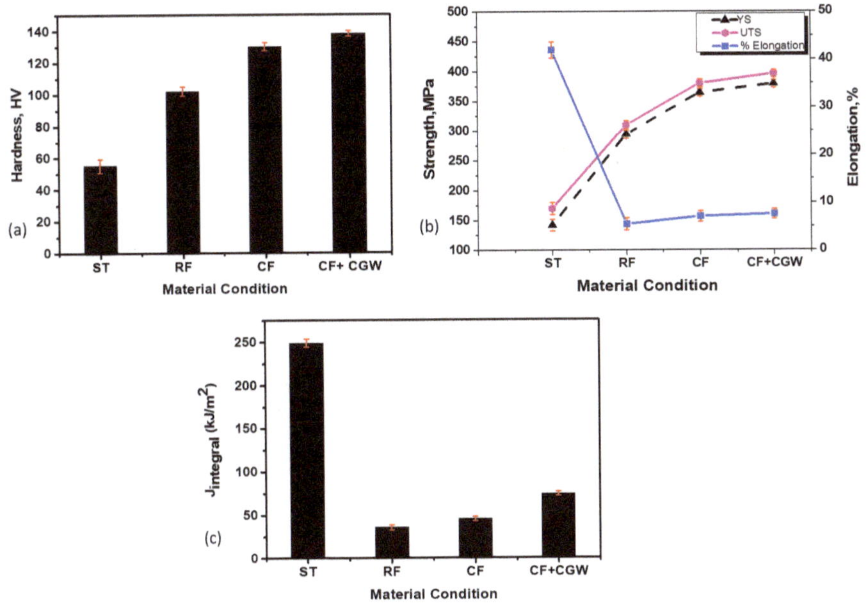

Fig. 3 **a** Variation in hardness across different material conditions. **b** Comparison of UTS, YS, and percentage elongation for samples processed via ST, RF, CF, and CF + CGW methods. **c** Variation in fracture toughness (J_{IC}) relative to material condition

with high stacking fault energy, undergoes rapid dynamic recovery or restoration. Consequently, RF samples experience softening due to decreased dislocation density, which reduces strength and hardness compared to CF samples [11, 23, 37]. The slight improvement in ductility of CF samples, compared to RF samples, is due to simultaneous enhancement in strength and toughness. This observation is corroborated by Joshi et al. [23] and Krishna et al. [38] during the processing of Al-Cu alloy at cryogenic temperature.

The enhancement after CF + CGW can be ascribed to the synergistic effect of compressive and shear stresses in a cryogenic environment. This process leads to a high dislocation density during cryo-groove rolling, which effectively suppresses cross-slip and climb mechanisms typically associated with dynamic recovery [39–41]. Impurity phase particles deformed or broke during the CF + CGW process, along with precipitates evolution due to thermomechanical conditions during warm rolling, also contribute to the improvement in hardness and strength by effectively hindering dislocation movement. Detailed characterizations of these particles are provided elsewhere [38, 42]. The increased elongation observed in CF + CGW samples can be attributed to the warm rolling process, which facilitates ongoing recovery.

3.3 Fracture Toughness

The J_{IC} of ST and all deformed samples were determined through a SENB test and evaluated as follows [29, 30]:

$$J = \frac{2A}{Bb} = \frac{2}{Bb}\int_0^{u_{max}} P du \quad (1)$$

where "A" is the area under the load versus extension curve, "B" is sample thickness, "b" is uncut portion $(W - a)$, $andu_{max}$ represents the extension corresponding to the peak force (P). Figure 3c illustrates the variations in fracture toughness for different material conditions. The fracture toughness of the dislocation-free equiaxed coarse-grained ST samples is 249 kJ/m^2. In contrast, the fracture toughness of the RF and CF samples are 36.1 kJ/m^2 and 45.6 kJ/m^2, respectively, due to the reduced ductility of the processed samples. The CF samples exhibit enhanced fracture energy relative to the RF samples, primarily due to their superior ductility. The CF + CGW samples show a higher fracture energy of 74.1 kJ/m^2, outperforming all other processed conditions, as the warm rolling during the CGW process enhances ductility. Hohenwarter et al. [27] observed a similar trend in ultrafine-grained (UFG) nickel alloy, noting a reduction in both K_{IC} and J_{IC} in the processed material. Likewise, Fang et al. [28] reported a decrease in static toughness for Al–Mg alloy processed using ECAP, attributing this to a loss of ductility compared to its cast state.

3.4 Fractography of Tensile Tested Samples

Figure 4a–d depicts the fractographs of the ST, as well as the deformed tensile tested processed samples. Larger dimples with cavities of varying sizes are observed in the ST fracture surface, in contrast to any of the processed samples, indicating a ductile mode of fracture. Typically, ductile fractures arise from the coalescence of voids at grain boundaries. This process is influenced by the initiation sites and the spatial distribution of these voids [43, 44]. The RF and CF samples reveal the mixed mode of fracture, wherein both cleavage facets and small-sized dimple surfaces are observed. The RF samples exhibit larger dimples than the CF samples. The dimple size reduction in these samples could be attributed to numerous void initiation sites created during grain refinement. These sites act as sources for coalescence of adjacent micro-voids, thereby preventing further growth [45]. In CF + CGW samples, the fractured surface morphology shows small dimples with negligible cleavage facets. The dimple sizes are larger than the RF and CF samples due to the recovery phenomenon that could have occurred during the CGW process.

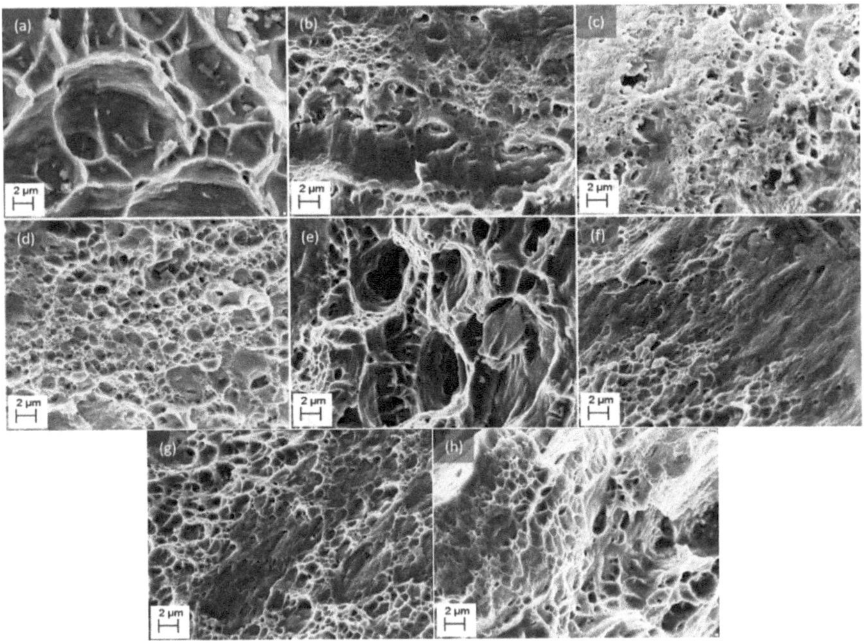

Fig. 4 SEM fractographs of **a** ST, **b** RF, **c** CF, and **d** CF + CGW samples after tensile test. SEM fractographs of **e** ST, **f** RF, **g** CF, and **h** CF + CGW samples after the three-point bend test

3.5 Fractography of Tensile Tested Samples

Figure 4e–h shows the fractographs of SENB-tested samples. The solution-treated (ST) samples exhibit larger dimples than any other processed condition due to the presence of coarse grains, which correlates with their high fracture toughness. The RF samples possess small-sized dimples with large areas of cleavage facets. In contrast, the CF samples show bigger dimples with a decreased area of cleavage facets, which explains the improved ductility in the CF samples. Hence, its fracture toughness appeared to be more than that of the RF samples. The CF + CGW samples exhibit bigger dimples than any other processed samples with a negligible area of cleavage facets, revealing better fracture toughness. As J_{IC} depends on crack propagation, Vinogradov et al. reported that Al alloys subjected to the ECAP process exhibit lower resistance to crack propagation with respect to coarse-grained counterparts [46].

3.6 Fractography of Tensile Tested Samples

Figure 5 illustrates the S–N plot, depicting the stress amplitude ($\Delta\sigma$) and number of reversals to failure (N_f) for the ST, CF + CGW, and CF + CGW + annealed samples. The fatigue strength was assessed at 10^6 cycles, serving as an indicator of high-cycle fatigue (HCF) life [47, 48]. The HCF strength of the ST samples is observed to be 40 MPa, whereas it is 105 MPa in the case of the CF + CGW samples (Table 1), representing a 162% increase. This tremendous enhancement of the fatigue strength might be due to refined microstructure and improved strength of the material. The CF + CGW + annealed samples (at 240°C) exhibit fatigue strength of 95 MPa.

The following Basquin equation relates the stress amplitude range ($\Delta\sigma$) to $2N_f$ for high cycle fatigue tests:

$$\frac{\Delta_\sigma}{2} = \frac{E\Delta\varepsilon_{\mathrm{el}}}{2} = \sigma_f'(2N_{\mathrm{f}})^{\mathrm{b}} \tag{2}$$

where $\Delta\varepsilon_{el}$ is the strain amplitude, "b" is the Basquin exponent, σ_f' is fatigue strength [49], and E is Young's modulus. Based on this, the elastic strain components in CF + CGW samples are observed to be smaller than ST samples.

3.7 Fractography of Tensile Tested Samples

Figure 6 displays the fractographs from the HCF tests conducted at a σ_a of 90 MPa for the ST and CF + CGW samples. Figure 6a and d shows low magnification fractographs of the processed samples, where Fe-rich inclusions are observed in

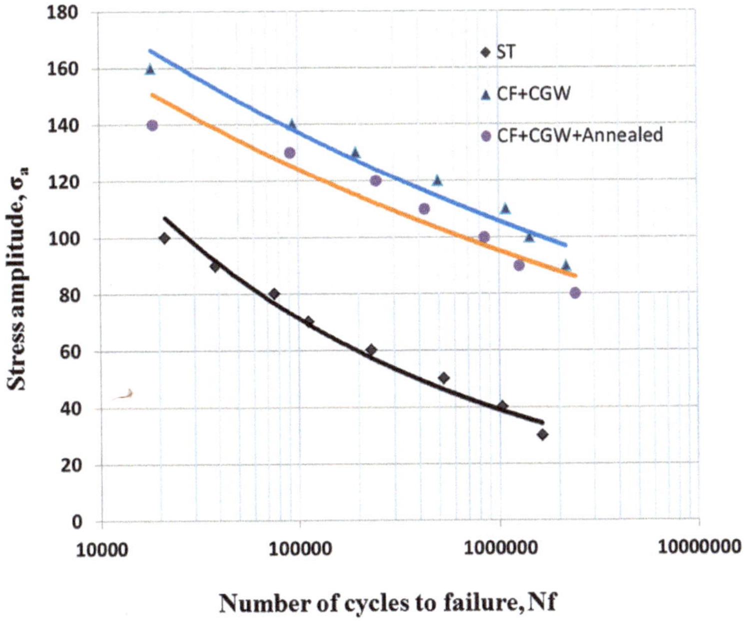

Fig. 5 S-N plot of ST, CF + CGW, and CF + CGW + annealed at 240 °C condition samples

Region I, the crack initiation region. These inclusions act as crack initiation sources. Tanaka et al. [50] discussed the following mechanics for fatigue crack: (i) cracks originating from loosely bonded inclusions due to slip bands, (ii) cracks initiated as a result of slip bands impinging on inclusions, and (iii) slip-band cracking itself, originating from uncracked inclusions. Mechanisms (ii) and (iii) are common in Al alloys, while the first mechanism is seen in steels.

Figure 6b and e presents magnified images of Region I, showing the presence of ductile fatigue striations. Under cyclic loading below the yield limit, the crack tip undergoes continual dulling and re-sharpening, resulting in the advancement of fatigue cracks through highly localized failure. This results in the formation of striations that bow along the crack propagation direction. The gap of these markings is directly proportional to the elastic strain components, which are 0.00125 for the ST samples and 0.00068 for the CF + CGW samples, representing the crack front position during cyclic loading [51]. The striation gaps in the ST sample range from 50 to 60 μm, while in the CF + CGW samples, they are about 0.25 μm. The ST sample shows larger dimples, whereas the CF + CGW sample exhibits finer dimples.

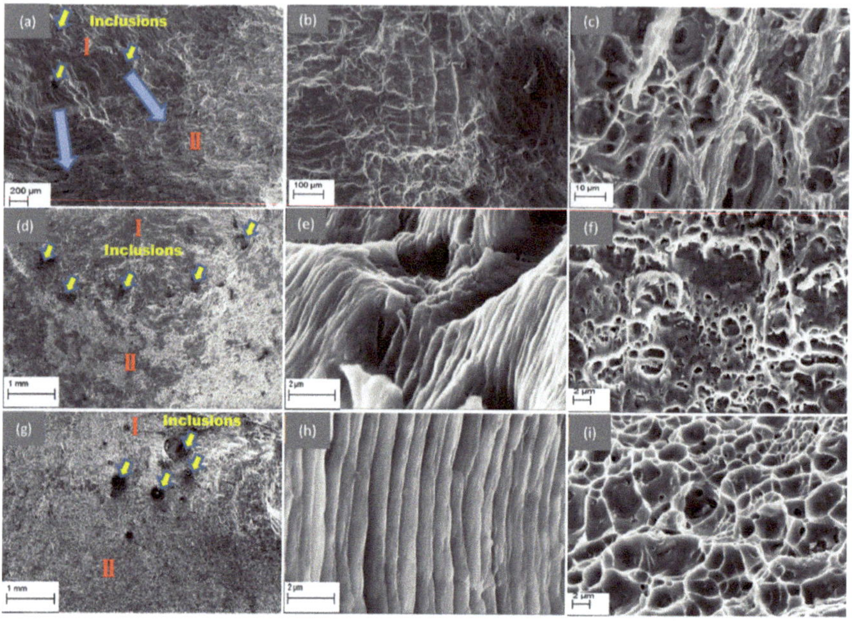

Fig. 6 SEM fractographs after high cycle fatigue test: **a** Low magnification image of ST sample, **b** Magnified Region I (marked as red coloured I) showing striation markings in the crack propagation region, **c** Magnified Region II showing dimple formations in failure region, **d** Low magnification image of CF + CGW sample, **e** Magnified Region I showing striation markings in crack propagation region, **f** Magnified Region II showing dimple formations in failure region, **g** Low magnification image of CF + CGW + annealed at 240°C sample, **h** Magnified Region I showing striation markings in crack propagation region, and **i** Magnified Region II showing dimple formations in failure region

3.8 *Fractography of Tensile Tested Samples*

The effect of post-annealing behaviour was examined by subjecting the CF + CGW samples to post-annealing at temperatures 180°C, 240°C, and 300°C—for a duration of one hour. With increase in annealing temperature, the dislocation gets annihilated. The optical microstructures depict the same, showing the gradual formation of small equiaxed recovered grains (Fig. 7a–d).

The samples which were annealed at 180°C show σ_{ut} and σ_y of 328 MPa and 309 MPa, respectively with improved ductility of 8.1%, whereas the samples annealed at 240°C have σ_{ut} and σ_y of 306 MPa and 286 MPa, respectively with further improved ductility of 10.6%. The samples annealed at 300°C have shown σ_{ut}, σ_y,, and ductility of 274 MPa, 255 MPa, and 15.8%, respectively (Fig. 8a). Correspondingly, the hardness decreases from 138 to 82 HV (Fig. 8). This change in mechanical property is due to dislocation rearrangement during recovery, which promotes the development of coarser subgrains. This observation is supported by the SEM fractographs of the tensile-tested samples, which display a gradual increase in the dimple size

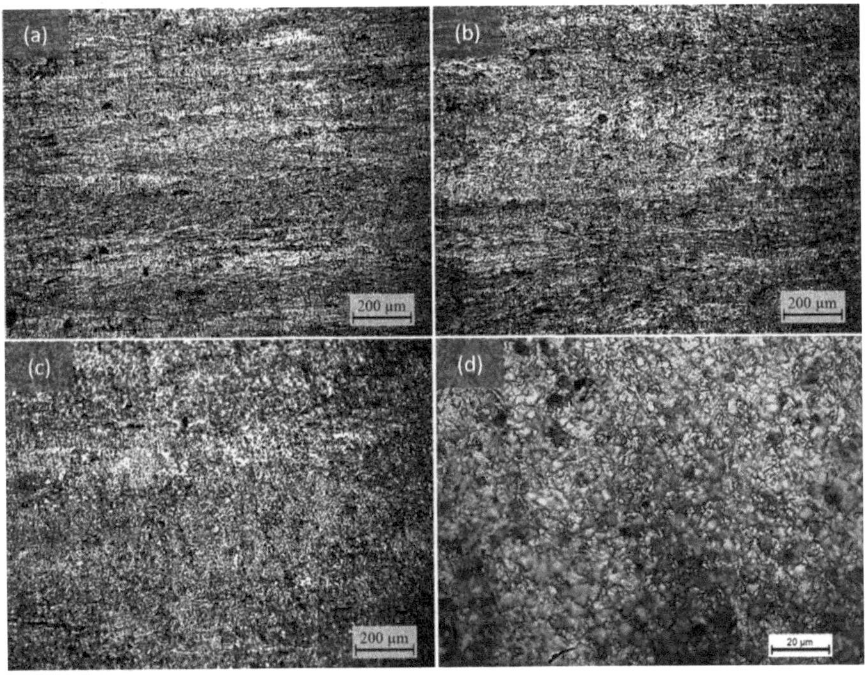

Fig. 7 Optical micrographs of CF + CGW samples after post-deformation annealing: annealed at **a** 180°C, **b** 240°C, **c** and 300°C. **d** Magnified image of 300°C annealed specimen showing completely recrystallized grains

as the annealing temperature rises (Fig. 9a–c). On the other hand, J_{IC} of the post-deformation annealed specimen has improved from 74.1 kJ/m^2 to 220 kJ/m^2 (Fig. 8), owing to the improvement in ductility. The coarser grains, which impart better ductility, are responsible for delaying crack propagation [46, 52]. The gradually increased dimple sizes in fractographs (SEM) of the 3-point bend test samples justify the same (Fig. 9d–f). The HCF strength of the post-annealed samples, which were annealed at 240°C, is observed to decrease to 95 MPa. This decrease in the fatigue limit is due to coarser grain structure, leading to the reduction in the σ_y from 366 to 286 MPa. The elastic strain component corresponding to this annealed condition has increased to 0.00098 (Table 1), which reflects on the striation's gaps (0.45μm) as observed on the fractured surface (Fig. 6h). Figure 6i depicts the formation of bigger dimples present in Region II, which indicates the presence of coarser grains.

3.9 *TEM Analysis*

Figure 10a illustrates the rod-shaped particles in the ST samples. These nanoparticles enhance the strength by effectively impeding dislocation motion. Figure 10b and

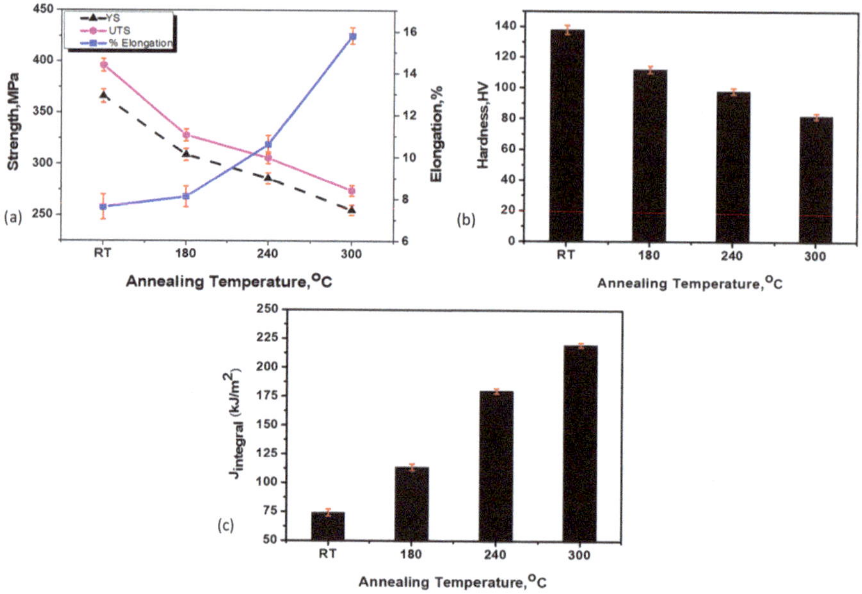

Fig. 8 Effect of annealing (1 h) on **a** tensile properties, **b** hardness, and **c** fracture toughness of CG + CGW samples

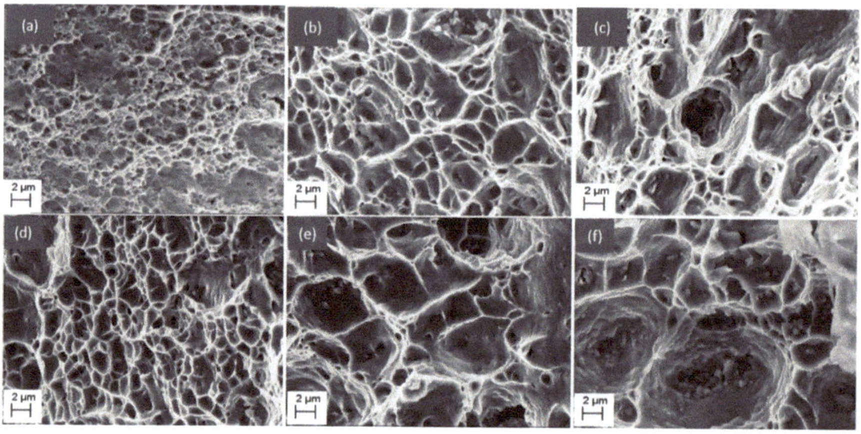

Fig. 9 SEM fractographs of the tensile test sample (CF + CGW samples after 1 h post deformation annealing) annealed at **a** 180°C, **b** 240°C, and **c** 300°C. SEM fractographs of three-point bend test samples (CF + CGW samples after 1 h post deformation annealing) annealed at **d** 180°C, **e** 240°C, and f 300°C

c depicts the TEM micrograph for RF and CF specimen, where subgrains formed during deformation and the dislocation network are observable. The RF samples have subgrains averaging 230 nm in size, while the CF samples have subgrains averaging 170 nm, with a prominent dislocation tangle. These ultrafine-grained (UFG) subgrains play a vital role in increasing the σ_y and σ_{ut} of the material [53, 54]. These ultrafine grains reduce the fracture energy (J) of the processed samples by allowing cracks to propagate rapidly in a straight line. However, they delay crack initiation better than coarser grains. In contrast, coarse-grain materials exhibit crack propagation in a gradual, zig-zag pattern [46, 55, 56]. Figure 10d, showing the TEM image for CF + CGW samples, illustrates fine subgrains of 110 nm size and nano shear bands averaging 20 nm in width. These nano shear bands, which intersect each other in some places, result from combined shear and compressive stress during the CF + CGW process. This intersection causes intense localized deformation and ultimate grain fragmentation. The nano shear bands prevent conventional slip and accommodate cyclic strain by delaying dislocation, thereby improving fatigue strength (105 MPa) [57, 58]. The high number of elongated spots in Selected Area Electron Diffraction (SAED) pattern implies the formation of high-angle grain boundaries, supporting the presence of fine grains. In the 180°C annealed sample, dislocations were annihilated due to the recovery effect, resulting in the transformation of cell boundaries into new subgrain walls. The 240°C annealed sample exhibits equiaxed grains of 250 nm size, responsible for the improved fracture energy (180.1 kJ/m^2).

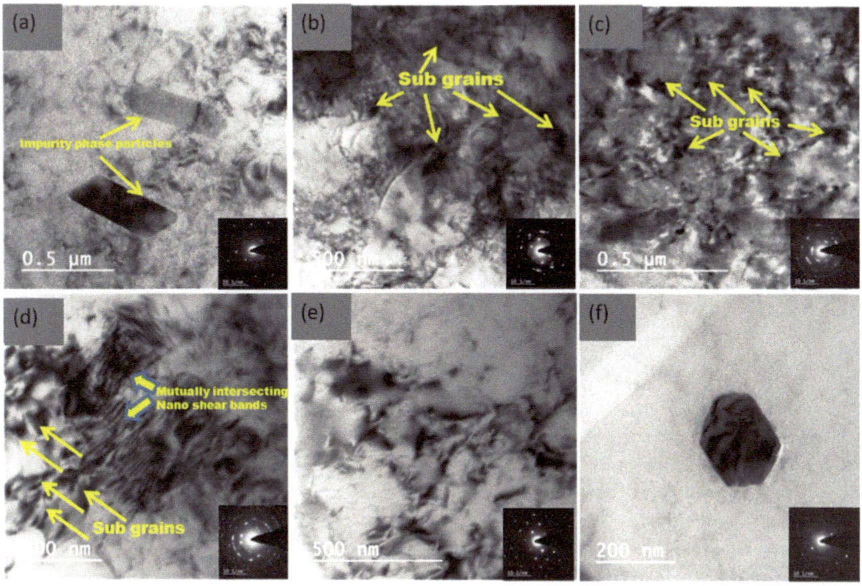

Fig. 10 TEM micrograph of AA 5052 for **a** ST, **b** RF, **c** CF, **d** CF + CGW, **e** CF + CGW + annealed at 180°C, and **f** 240°C

4 Conclusion

The present study investigates the fatigue and fracture behaviour of Al 5052 alloy, deformed using the combined Severe Plastic Deformation (SPD) method—Cryo-Forging (CF) followed by Cryo-Groove Rolling (CGW)—have been thoroughly explored:

- The ST Al 5052 alloy underwent CF (cryo-forging) + CGW (cryo-groove + warm rolling). The processed samples demonstrated a high-cycle fatigue strength of 105 MPa, marking a 162% increase with respect to the ST samples, which exhibit a fatigue strength of 40 MPa.
- The evolution of intersecting nano shear bands was observed, leading to significant grain fragmentation. This microstructural refinement contributed to the enhanced fatigue strength.
- The resulting ultrafine-grained (UFG) subgrains were approximately 110 nm in size. Additionally, the presence of broken/deformed impurity phase particles and fine precipitates hindered dislocation movement, further improving the fatigue strength of the processed alloy. The SEM fractographs (fatigue tested) of the deformed samples possess fine striations of width 0.25 μm, which reveals the ultimate grain refinement.
- The fracture toughness in the deformed samples has decreased to 74.1 kJ/m^2 due to the high crack propagation rate in UFG grains. The samples annealed at 240 °C after deformation exhibit an improvement in fracture toughness (180.1 kJ/m^2) but show a decrease in fatigue strength (95 MPa) due to recovery of the grains.

References

1. Zhilyaev P, Nurislamova GV, Kim BK, Baró MD, Szpunar JA, Langdon TG (2003) Experimental parameters influencing grain refinement and microstructural evolution during high-pressure torsion. Acta Mater 51(3):753–765
2. Lukyanova EA, Martynenko NS, Serebryany VN, Belyakov AN, Rokhlin LL, Dobatkin SV, Estrin YZ (2018) Structure and mechanical and corrosion properties of a magnesium Mg–Y–Nd–Zr Alloy after high pressure torsion. Russ Metall (Metally) 2017:912–921
3. Jahedi M, Knezevic M, Paydar MH (2015) High-pressure double torsion as a severe plastic deformation process: experimental procedure and finite element modelling. J Mater Eng Perform 24:1471–1482
4. Sun Y, Aindow M, Hebert RJ, Langdon TG, Lavernia EJ (2017) High-pressure torsion-induced phase transformations and grain refinement in Al/Ti composites. J Mater Sci 52:12170–12184
5. Jahedi M, Beyerlein IJ, Paydar MH, Zheng S, Xiong T, Knezevic M (2017) Effects of pressure and number of turns on microstructural homogeneity developed in high-pressure double torsion. Metall and Mater Trans A 48:1249–1263
6. Lei S, Jiu-Ba W, Chang R (2018) The prediction of microstructure evolution of 6005A aluminum alloy in a P-ECAP extrusion study. J Mater Eng Perform 27:2566–2575
7. Kumar R, Gudimetla K, Tejaswi B, Ravisankar B (2017) Effect of microstructure and mechanical properties of Al–Mg alloy processed by ECAP at room temperature and cryo temperature. Trans Indian Inst Met 70:639–648

8. Mahmoodi M, Naderi A, Dini G (2017) Correlation between structural parameters and mechanical properties of Al5083 sheets processed by ECAR. J Mater Eng Perform 26:6022–6027
9. Krajňák T, Minárik P, Stráská J, Gubicza J, Máthis K, Janeček M (2018) Influence of the initial state on the microstructure and mechanical properties of AX41 alloy processed by ECAP. J Mater Sci 54:3469–3484
10. Divya SP, Nagaraj M, Kesavamoorthy M, Srinivasan SA, Ravisankar B (2017) Investigation on the effect of ECAP routes on the wear behavior of AA2014. Trans Indian Inst Met 71:67–77
11. Rao PN, Singh D, Jayaganthan R (2014) Mechanical properties and microstructural evolution of Al 6061 alloy processed by multidirectional forging at liquid nitrogen temperature. Mater Des 56:97–104
12. Xia X, Wang C, Wu Y, Zhang K, Ma M, Yuan B (2018) Grain refinement, microstructure and mechanical properties homogeneity of Mg-Gd-Y-Nd-Zr alloy during multidirectional forging. J Mater Eng Perform 27:5689–5699
13. Łyszkowski R, Czujko T, Varin RA (2016) Multi-axial forging of Fe 3 Al-base intermetallic alloy and its mechanical properties. J Mater Sci 52:2902–2914
14. Wang B, Wang X, Li J (2016) Formation and microstructure of ultrafine-grained titanium processed by multi-directional forging. J Mater Eng Perform 25:2521–2527
15. Vendra SSL, Goel S, Kumar N, Jayaganthan R (2017) A study on fracture toughness and strain rate sensitivity of severely deformed al 6063 alloys processed by multi-axial forging and rolling at cryogenic temperature. Mater Sci Eng, A 686:82–92
16. Gang UG, Lee SH, Nam WJ (2009) The evolution of microstructure and mechanical properties of a 5052 aluminium alloy by the application of cryogenic rolling and warm rolling. Mater Trans 50(1):82–86
17. Rao PN, Viswanadh B, Jayaganthan R (2014) Effect of cryorolling and warm rolling on precipitation evolution in Al 6061 alloy. Mater Sci Eng, A 606:1–10
18. Krishna SC, Chawake N, Kottada RS, Jha AK, Pant B, Venkitakrishnan PV (2016) High strength and good ductility in Cu-3Ag-0.5 ZrAlloy by cryo-rolling and aging. J Mater Eng Perform 26:350–357
19. Wang W, Chen ZN, Guo EY, Kang HJ, Liu Y, Zou CL, Wang TM (2018) Influence of cryorolling on the precipitation of Cu–Ni–Si Alloys: an in situ X-ray diffraction study. Acta Metall Sin (Engl Lett)31:1089–1097
20. Kumar N, Goel S, Jayaganthan R, Owolabi GM (2018) The influence of metallurgical factors on low cycle fatigue behavior of ultra-fine grained 6082 Al alloy. Int J Fatigue 110:130–143
21. Cherukuri B, Nedkova TS, Srinivasan R (2005) A comparison of the properties of SPD-processed AA-6061 by equal-channel angular pressing, multi-axial compressions/forgings and accumulative roll bonding. Mater Sci Eng, A 410:394–397
22. Noda M, Hirohashi M, Funami K (2005) Low temperature superplasticity and its deformation mechanism in grain refinement of Al-Mg alloy by multi-axial alternative forging. Mater Trans 44(11):2288–2297
23. Joshi A, Kumar N, Yogesha KK, Jayaganthan R, Nath SK (2016) Mechanical properties and microstructural evolution in Al 2014 alloy processed through multidirectional cryoforging. J Mater Eng Perform 25:3031–3045
24. Verma R, Nath SK, Jayaganthan R (2018) Effect of high strain rolling and multiaxial forging on tensile and fracture behaviour of ZE41 magnesium alloy. Mater Today Proceeding 5(9):17195–17202
25. Somekawa H, Mukai T (2006) Fracture toughness in ultrafine-grained magnesium alloy. In: Materials science forum, Trans Tech Publications 503–504:155–160
26. Hohenwarter A, Pippan R (2012) A comprehensive study on the damage tolerance of ultrafine-grained copper. Mater Sci Eng, A 540:89–96
27. Hohenwarter A, Pippan R (2011) Fracture toughness evaluation of ultrafine-grained nickel. Scripta Mater 64(10):982–985
28. Fang DR, Duan QQ, Zhao NQ, Li JJ, Wu SD, Zhang ZF (2007) Tensile properties and fracture mechanism of Al–Mg alloy subjected to equal channel angular pressing. Mater Sci Eng, A 459(1–2):137–144

29. Yogesha KK, Kumar N, Joshi A, Jayaganthan R, Nath SK (2016) A comparative study on tensile and fracture behavior of Al–Mg Alloy processed through cryorolling and cryo groove rolling. Metallogr, Microstruct, Anal 5:251–263
30. Yogesha KK, Joshi A, Kumar N, Jayaganthan R (2016) Effect of cryo groove rolling followed by warm rolling (CGW) on the mechanical properties of 5052 Al alloy. Mater Manuf Processes 32(12):1336–1344
31. Khatibi G, Horky J, Weiss B, Zehetbauer MJ (2010) High cycle fatigue behaviour of copper deformed by high pressure torsion. Int J Fatigue 32(2):269–278
32. Vinogradov A, Washikita A, Kitagawa K, Kopylov VI (2003) Fatigue life of fine-grain Al–Mg–Sc alloys produced by equal-channel angular pressing. Mater Sci Eng, A 349(1–2):318–326
33. Höppel HW, Kautz M, Xu C, Murashkin M, Langdon TG, Valiev RZ, Mughrabi H (2006) An overview: fatigue behaviour of ultrafine-grained metals and alloys. Int J Fatigue 28(1–2):1001–1010
34. Singh D, Rao PN, Jayaganthan R (2014) High cyclic fatigue behaviour of ultrafine grained Al 5083 alloy. Mater Sci Technol 30(14):1835–1842
35. Hanlon T, Tabachnikova ED, Suresh S (2005) Fatigue behavior of nanocrystalline metals and alloys. Int J Fatigue 27(10–12):1147–1158
36. Inoue T, Kimura Y, Ochiai S (2011) Static fracture toughness of fail-safe steel. Scripta Mater 65(6):552–555
37. Chatterjee A, Sharma G, Sarkar A, Singh JB, Chakravartty JK (2012) A study on cryogenic temperature ECAP on the microstructure and mechanical properties of Al–Mg alloy. Mater Sci Eng, A 556:653–657
38. Krishna NN, Teas R, Siva Prasad K, Venkateswarlu K (2013) Study on cryorolled Al–Cu alloy using X-ray diffraction line profile analysis and evaluation of strengthening mechanisms. Mater Des52:785–790
39. Panigrahi SK, Jayaganthan R, Chawla V (2008) Effect of cryorolling on microstructure of Al–Mg–Si alloy. Mater Lett 62(17–18):2626–2629
40. Singh D, Rao PN, Jayaganthan R (2013) Effect of deformation temperature on mechanical properties of ultrafine grained Al–Mg alloys processed by rolling. Mater Des 50:646–655
41. Lee YB, Shin DH, Nam WJ (2005) Effect of annealing temperature on tensile behavior of 5052 Al alloy deformed at cryogenic temperature. J Mater Sci 40:1313–1315
42. Yogesha KK, Joshi A, Jayaganthan R (2017) Fatigue behavior of ultrafine-grained 5052 Al alloy processed through different rolling methods. J Mater Eng Perform 26:2826–2836
43. Mukai T, Kawazoe M, Higashi K (1998) Dynamic mechanical properties of a near-nano aluminum alloy processed by equal-channel-angular-extrusion. Nanostruct Mater 10(5):755–765
44. Ko YG, Shin DH, Parkand KT, Lee CS (2006) An analysis of the strain hardening behavior of ultra-fine grain pure titanium. Scr Mater 54(10):1785–1789
45. Sahoo PS, Mahapatra MM, Vundavilli PR, Sabat RK, Sirohi S, Kumar S (2023) Investigation of severe plastic deformation effects on magnesium RZ5 alloy sheets using a modified multi-pass equal channel angular pressing (ECAP) technique. Int Metals Rev 16(14):5158
46. Vinogradov A, Nagasaki S, Patlan V, Kitagawaand K, Kawazoe M (1999) Fatigue properties of 5056 Al-Mg alloy produced by equal-channel angular pressing. Nanostructured Mater 11(7):925–934
47. Gairola S, Jayaganthan R (2023) Influence of heat treatment, microstructure evolution, and damage mechanism on high cycle fatigue behaviour of additively manufactured Ti-modified Al 2024 alloy. Mater Charact203:113047
48. Joshi A, Yogesha KK, Jayaganthan R (2017) Influence of cryorolling and followed by annealing on high cycle fatigue behavior of ultrafine grained Al 2014 alloy. Mater Charact 127:253–271
49. Cavaliere P (2009) Fatigue properties and crack behavior of ultra-fine and nanocrystalline pure metals. Int J Fatigue 31(10):1476–1489
50. Tanaka K, Mura T (1982) A theory of fatigue crack initiation at inclusions. Metall Trans A 13:117–123

51. Beachem CD (1965) Electron fractographic studies of mechanical fracture processes in metals. J Basic Eng 87(2):299–306
52. Lee Z, Radmilovic V, Ahn B, Lavernia EJ, Nutt SR (2009) Tensile deformation and fracture mechanism of bulk bimodal ultrafine-grained Al-Mg alloy. Metall and Mater Trans A 41:795–801
53. Aghajani S, Alizadeh R (2024) Severe plastic deformation of Zn and Zn-based alloys. J Mater Res Technol 33:6508–6533
54. Jiang H, Gong Q, Peterlechner M, Daum L, Rösner H, Wilde G (2024) Hardness and microstructural evolution of CoCrFeNi high-entropy alloys during severe plastic deformation. Mater Sci Eng, A 908:146758
55. Hanlon T, Kwon YN, Suresh S (2003) Grain size effects on the fatigue response of nanocrystalline metals. Scripta Mater 49(7):675–680
56. Yin D, Liu H, Chen Y, Yi D, Wang B, Wang B, Shen F, Fu S, Tang C, Pan S (2016) Effect of grain size on fatigue-crack growth in 2524 aluminium alloy. Int J Fatigue 84:9–16
57. Toroghinejad MR, Ashrafizadeh F, Jamaati R (2013) On the use of accumulative roll bonding process to develop nanostructured aluminum alloy 5083. Mater Sci Eng, A 561:145–151
58. Malekjani S, Hodgson PD, Stanford NE, Hilditch TB (2013) Shear bands evolution in ultrafine-grained aluminium under cyclic loading. Scripta Mater 68(10):821–824

Computational Analysis of Walling Materials for Designing Net-Zero Buildings

Roshani Sahu, Bhagyashri A. Lanjewar, Manoj Wankhede, and Rahul V. Ralegaonkar

Abstract In an era where sustainability is paramount, the construction industry faces the challenge of reducing its significant environmental impact while enhancing efficiency and affordability. Building Information Modeling (BIM) serves as a pivotal technology in this quest, enabling the precise evaluation of building materials in the pursuit of net-zero energy buildings. This study employs Building Information Modeling (BIM) to assess and compare the sustainability of four distinct walling materials—Clay-brick, AAC block, prefabricated panel, and volumetric walls—across construction time, cost, and carbon emissions, within the framework of net-zero building design. The analysis reveals stark contrasts: Prefabricated panel walls exhibit the most significant reductions, shortening construction time by over 96%, reducing cost by approximately 68%, and decreasing carbon emissions by nearly 97% compared to traditional clay-brick walls. Volumetric walls also perform exceptionally well, offering the shortest construction time of 5 h, and nearly matching the environmental and cost efficiencies of prefabricated panels. AAC block walls, while not as rapid, still present considerable reductions in all metrics compared to clay-brick walls. These findings highlight the effectiveness of newer, innovative construction materials and methods in enhancing sustainability and operational efficiency, aligning closely with global net-zero objectives.

Keywords CC walls · Building Information Modeling (BIM) · Clay-brick walls · Net-zero buildings · Prefabricated panel

R. Sahu · M. Wankhede · R. V. Ralegaonkar
Department of Civil Engineering, Visvesvaraya National Institute of Technology, Nagpur, India

B. A. Lanjewar (✉)
Maharashtra Institute of Technology CSN, Chhatrapati Sambhajinagar, India
e-mail: bhagyashrilanjewar12@gmail.com

I. M. Jamadar et al. (eds.), *Integrating Sustainability in Modern Technologies*, Lecture Notes in Mechanical Engineering, https://doi.org/10.1007/978-981-95-9466-5_26

1 Introduction

An essential industry that either directly or indirectly affects other industries is the construction sector in any nation. Significant progress has been made in this area for the nation's overall development. As per the 12th five-year plan (2012–2017) established by the Planning Commission of India, the construction sector contributes around 8% of the country's gross domestic product (GDP) [1]. The necessity for sustainable development in the construction sector has resulted in a heightened emphasis on the planning and construction of net-zero buildings—buildings engineered to generate the same amount of energy they consume during a year. To accomplish this ambitious objective, it is necessary to employ not only creative design approaches but also include cutting-edge technologies like Building Information Modeling (BIM). BIM streamlines the process of planning, designing, constructing, and managing buildings by utilizing digital models that represent both the physical and functional aspects. This allows stakeholders to make well-informed decisions about energy efficiency and sustainability.

In the realm of net-zero building design, the choice of construction materials is crucial in defining the project's overall environmental footprint. This study focuses on comparing four types of walling materials: Clay-brick wall, Autoclaved Aerated Concrete (AAC) block wall, prefabricated panel wall, and volumetric wall. The selection of these materials for comparison was based on their diverse thermal qualities, manufacturing methods, and extensive use in the construction sector. The evaluation is conducted using three essential sustainability indicators: cost, construction duration, and carbon emissions.

The cost and duration of construction are essential factors that impact the viability and acceptance of sustainable building methods [2]. Carbon emissions and energy usage have a direct impact on a building's environmental footprint, and are essential factors in achieving net-zero status [3]. This study intends to offer a thorough analysis that could assist architects, engineers, and policymakers in making more sustainable decisions in building design and construction. This is achieved by employing BIM to model and simulate the construction and operational phases of structures constructed with respective materials.

Shadram et al. [4] implemented a Building Information Modeling (BIM) framework to reduce the amount of energy used in the production of construction materials. Wong and Zhou [5] suggested that forthcoming green BIM tools should incorporate the three R's principle of reduce, reuse, and recycle to attain sustainable construction. Abanda and Byers [6] examined how a building's orientation affects energy usage and the resulting savings over its entire lifespan. Oti et al. [7] utilized the application programming interface (API) to incorporate sustainability concerns into the evaluation procedure of structural conceptual design. Eleftheriadis et al. [8] employed Building Information Modeling (BIM) tools to enhance the cost efficiency and reduce carbon dioxide ($CO2$) emissions in reinforced-concrete structures. These tools facilitated the seamless integration of information modeling and finite-element analysis.

Marzouk et al. [9] introduced a framework that combines BIM with computer simulation, optimization, and system dynamics to calculate the project duration and choose the most efficient sustainable material alternative. An assessment of the building envelope is conducted using BIM [10, 11] in order to decrease the energy consumption of a residential building. To mitigate inefficiencies, errors, and delays, it is imperative to conduct a study of the building designs, envelopes, and estimations [12].

Utilizing BIM for assessing these criteria not only improves the precision of the study, but also corresponds with the industry's shift toward digitization and sustainability. This comparative study aims to uncover the impact of various walling materials on the attainment of net-zero buildings, providing valuable insights and paving the way for more environmentally conscious construction methods.

2 Materials and Methods

This study adopts a comparative research design, utilizing Building Information Modeling (BIM) technology to evaluate the performance of four different walling materials—clay-brick wall, Autoclaved Aerated Concrete (AAC) block wall, prefabricated panel wall, and volumetric wall—in the context of net-zero building design. The aim is to assess these materials in terms of construction cost, construction time, and carbon emissions. The methodology is divided into several key phases to ensure a comprehensive analysis.

2.1 *Selection of Case Study Buildings*

Figure 1 represents the schematic of the plan which is used for this study with its proper dimensions. Four hypothetical net-zero energy building models are designed using BIM software. Each model is identical in terms of size, design, orientation, and location to ensure that the only variable is the walling material. This standardization is crucial for isolating the impact of the walling material on the performance metrics (Table 1).

2.2 *Walling Material Specification*

Table 2 represents the different properties of the four selected walling materials such as material characteristics, its environmental impact, construction time, and its cost. This includes sourcing data from manufacturers, industry standards, and previous research to ensure accuracy in simulation inputs.

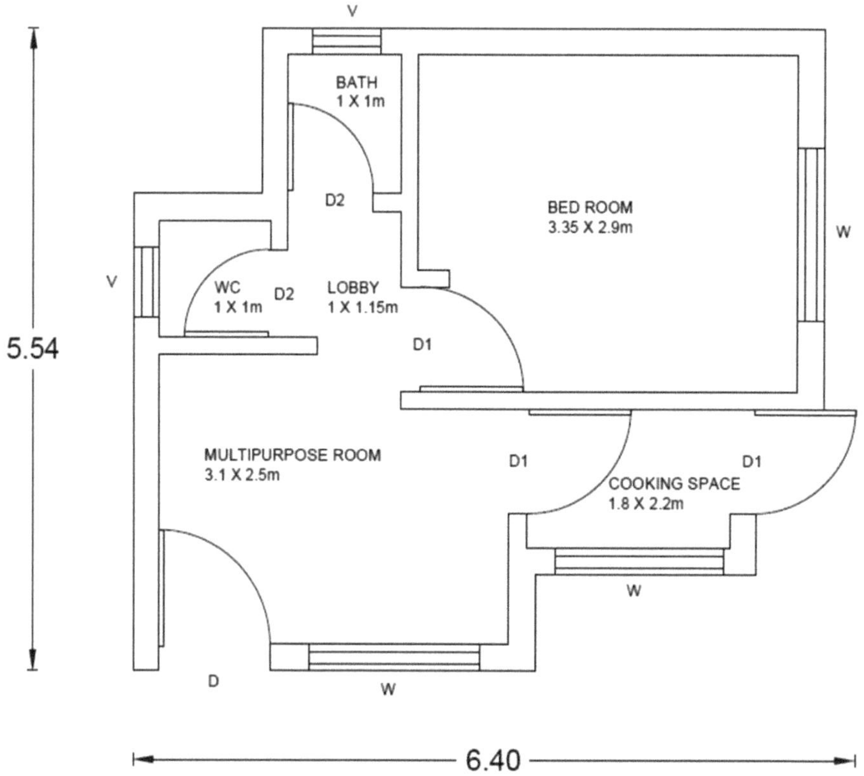

Fig. 1 Case study building plan. *Note* WC—water closet

Table 1 Schedule of doors and windows

Door	D	1 X 2.1
	D1	0.9 X 2.1
	D2	0.75 X 2.1
Window	W	1.5 X 1.2
Ventilation	V	0.6 X 0.75

3 Experiments

3.1 BIM Modeling and Simulation

Figures 2, 3, 4, and 5 represent the 3D models of building created using Autodesk Revit software for the clay-brick wall, ACC brick wall, prefabricated panel wall, and volumetric wall. For each model, BIM is utilized to detail the architectural, structural, mechanical, electrical, and plumbing (MEP) systems. Modeling software (Revit and

Table 2 Details of different characteristics of the four different walling materials [13, 14]

Characteristic	Clay-brick wall	AAC block wall	Prefabricated panel wall	Volumetric wall
Material description	Made from clay and shale, fired in a kiln	Autoclaved Aerated Concrete, made from sand, cement, lime, gypsum, and water	Various, including cement, flyash and admixture or composites, manufactured off-site and transported for assembly	Manufactured off-site as complete modules, including walls, floors, ceilings, MEP systems
Thermal conductivity (W/mK)	0.6–1.0	0.1–0.3	0.2–0.6 (depending on insulation and materials used)	0.15–0.35 (owing to integrated insulation)
Density (kg/m^2 for a standard thickness)	1800–2000 kg/m^3 (for the material density)	500–600 kg/m^3 (for the material density)	Highly variable depending on materials	Generally lighter due to modular design, approx. 600–1000 kg/m^3
Environmental impact	High: Energy-intensive production	Moderate: Less energy-intensive than clay bricks, recyclable	Low to moderate: Efficient production, reduced site waste, and material dependent	Low to moderate: Off-site manufacturing reduces on-site waste and energy use
Construction time	Longer: Weeks to months depending on project size	Shorter: Days to weeks faster than clay bricks	Shortest: Days, due to off-site prefabrication and quick assembly	Fastest: Complete modules are manufactured off-site and quickly assembled
Cost (indicative range per m^2)	Lower material cost: Rs 800/- to 2000/-	Moderate material cost: Rs 1200/- to 2400/-	Higher initial cost: Rs 1600/- to 4000/-, but potential overall savings	Generally higher due to complete modular design: Rs. 2400/- to 5600/-, but offset by faster construction

its plugins) integrated with BIM is employed to simulate the carbon emissions of each building for its complete lifecycle. Construction sequencing tools within BIM are used to estimate the construction time and associated costs.

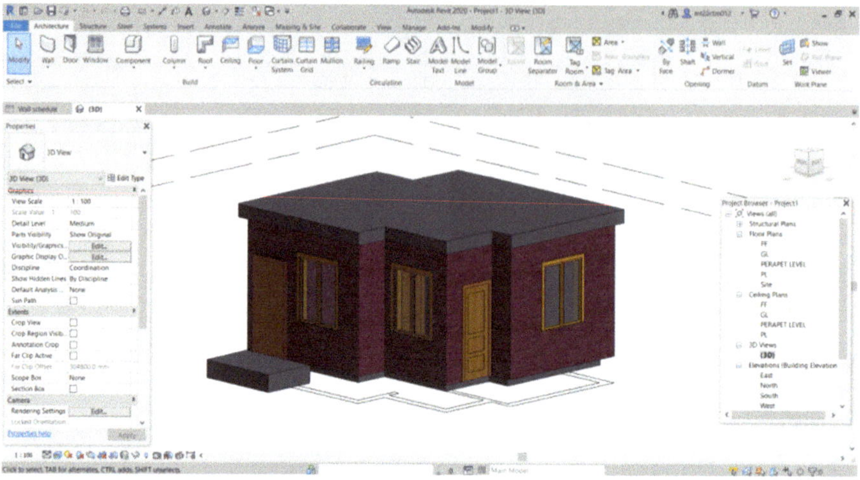

Fig. 2 Revit model developed for clay-brick analysis

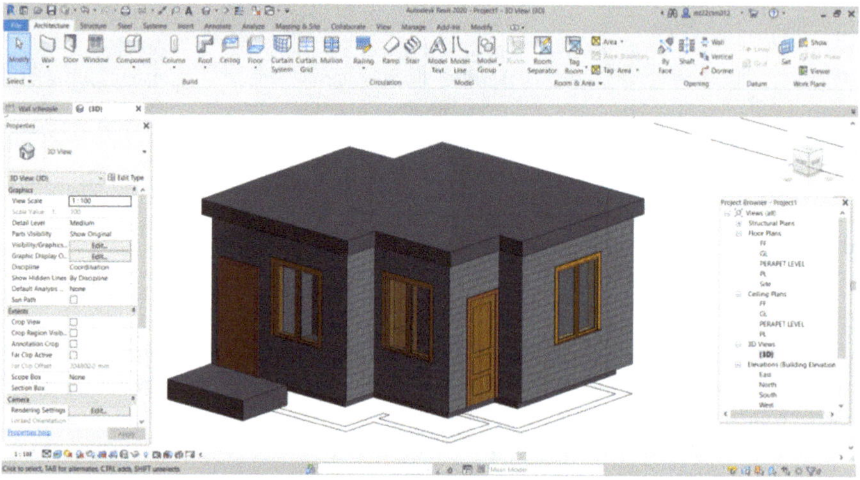

Fig. 3 Revit model developed for AAC block analysis

3.2 Data Collection and Analysis

Cost Analysis: The cost of construction is calculated based on the quantity takeoffs provided by the BIM model and current market prices for materials and labor.

Time for Construction: The construction time for clay-brick wall and ACC brick wall reported in Table 3 is estimated manually by calculating the no. of bricks required using the volume of the wall and bricks and also considering the time consumed by labor in placing each brick. For prefabricated panel wall and volumetric wall, as they

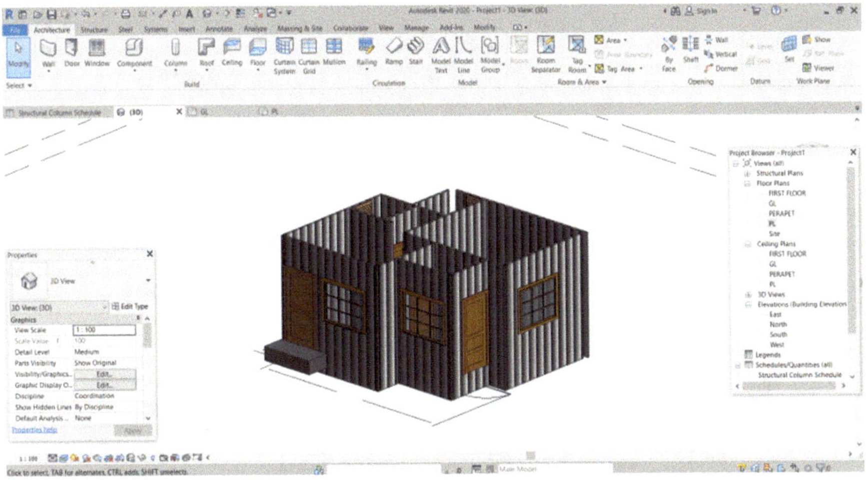

Fig. 4 Revit model developed for prefabricated panel wall analysis

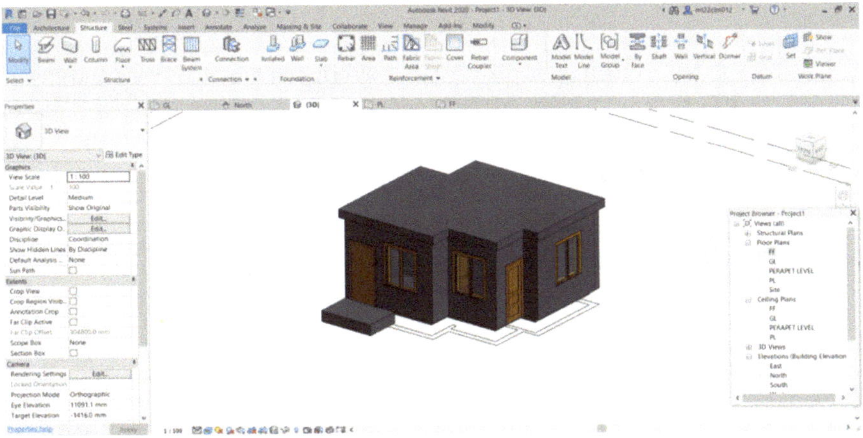

Fig. 5 Revit model developed for volumetric wall analysis

are manufactured in the industry and are directly installed, the time for installation varies depending on the length of the wall. But through on-site observation and referring to some of the related literature works, an approximate time required for their installation has been reported in Table 3.

Carbon Emissions: For each material, this study calculated carbon emissions using data from previous literatures [18]. This approach ensures the utilization of established, peer-reviewed carbon footprint data, enhancing the accuracy and reliability of the emissions analysis. Tables 4, 5, 6, and 7 represent the carbon emission

Table 3 Construction time estimation for all four walling materials using the required parameters [15–17]

Parameter	Clay-brick wall	AAC block wall	Prefabricated panel wall	Volumetric wall
Brick/Block/Panel size	200 × 100 × 100 mm^3	600 × 200 × 200 mm^3	(manufactured to size)	(complete modules)
Wall thickness	230 mm	230 mm	60 mm	60 mm
Wall height	3 m	3 m	3 m	3 m
Wall length	34.17 m	34.17 m	34.17 m	34.17 m
Wall volume (m^3)	24	24	6	6
Number of Bricks/ Blocks/Panels	11,789	983	–	–
Construction rate	1 brick per minute	1 block per minute	Prefabricated panel assembly (rapid)	Modular wall assembly (very rapid)
Estimated construction time (hours)	196.5	16.4	6.8	5

associated with each construction material such as brick, cement, and sand and calculated the overall carbon emission for all the four different walling material (Figs. 6, 7, 8 and 9).

Table 4 Carbon emission of materials used in clay-brick wall [19]

Item	Quantity in Kg	Unit	Carbon emission factor	CO2 emission in Kg	Cost (Rs.)
Brick	8491	unit	5.502	46,717.48	67,932
Cement	1070	Kg	0.959	1026.13	8560
Sand	6451.77	Kg	0.0029	18.71	4662
Total				47,762.32	81,154

Table 5 Carbon emission of materials used in AAC block [20]

Item	Quantity in Kg	Unit	Carbon emission factor	Carbon Emission in Kg	Cost (Rs.)
AAC Block	708	Units	2.28	1996.56	49,534
Cement	847.5	Kg	0.959	812.75	6793
Sand	2647.21	Kg	0.0029	7.677	1668
Total				2817	57,995

Table 6 Carbon emission of materials used in prefabricated panel wall

Item	Quantity (in Kg)	Carbon emission factor	Carbon emission in Kg	Cost (Rs.)
Cement	1621.46	0.959	1554.98	12,971.68
Crush sand	2243.94	0.0029	6.51	1413.68
Flyash	1621.46	0.027	43.78	2026.83
Admixture PO5	3.51	0.47	1.65	245.98
Admixture 544	35.14	0.47	16.52	2459.8
EPS	22.64	0.47	10.64	5094.05
VF-PGE	18.07	0.47	8.49	1265.04
Total			1642.57	25,477.05

Table 7 Carbon emission of materials used in volumetric wall

Item	Quantity (in Kg)	Carbon emission factor	Carbon emission in Kg	Cost (Rs.)
Flyash	1485	0.27	40.10	1856.25
GGBS	1485	0.035	51.98	1485
Fine aggregate	2080.485	0.0029	6.03	1456.34
Cofired blended ash	231.165			115.58
Coarse aggregate	4618.35	0.0029	13.39	4156.52
Sodium hydroxide	255.321	1.12	285.96	5617.06
Sodium silicate	1232.55	1.22	1503.71	14,790.6
Admixture	14.85	0.47	6.98	1039.5
Total			1901.17	29,477.3

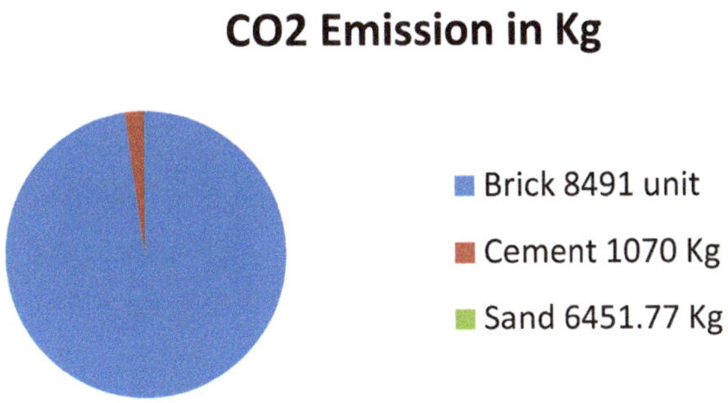

Fig. 6 Graphical representation of carbon emission of clay-brick wall

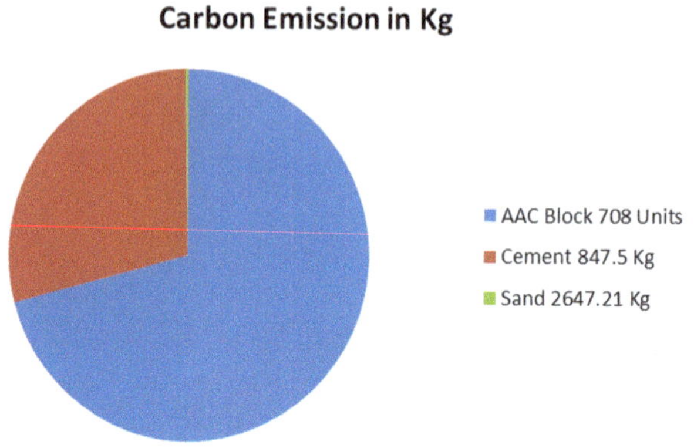

Fig. 7 Graphical representation of carbon emission of AAC block wall

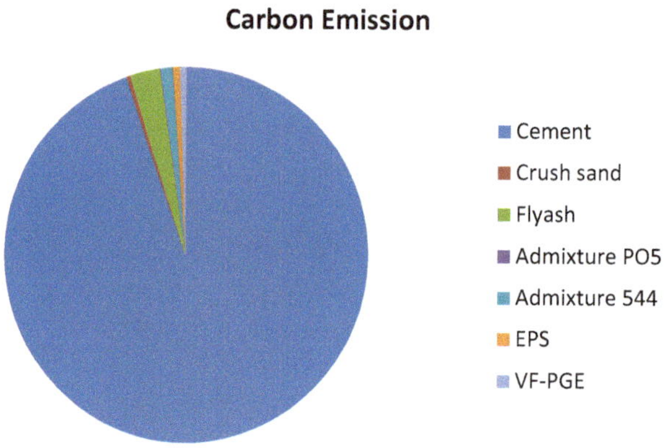

Fig. 8 Graphical representation of carbon emission of prefabricated panel wall

Fig. 9 Graphical representation of carbon emission of volumetric wall

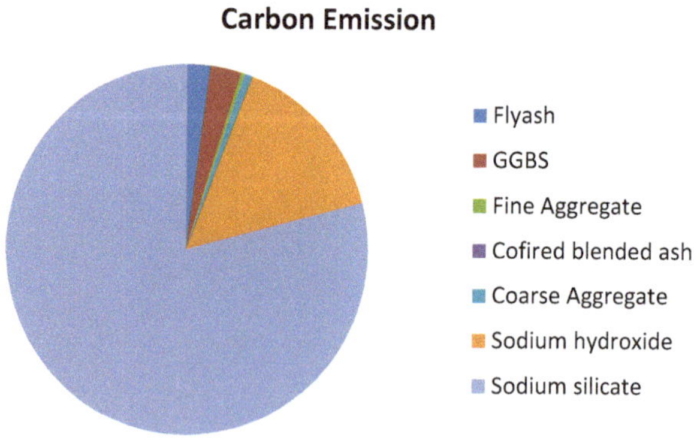

4 Results and Discussion

This study analyzed the construction time, cost, and carbon emissions associated with three different walling materials—clay-brick wall, AAC block wall, and prefabricated panel wall. The combined results from these metrics provide valuable insights into the sustainability and cost-effectiveness of each walling material (Fig. 10).

The comprehensive analysis of the four walling materials revealed distinct differences in construction time, cost, and carbon emissions. Clay-brick walls, traditionally used, exhibited the highest carbon emissions (47,762.32 kg) and cost (80,554 Rs.), coupled with a prolonged construction time (196.5 h). AAC block walls showed improvement in all categories with significantly reduced carbon emissions (2,817 kg) and cost (57,996.82 Rs.), along with a decreased construction time (16.4 h). Prefabricated panel walls stood out with the lowest construction time (6.8 h) and carbon emissions (1,642.6 kg), along with a reasonable cost (25,477.05 Rs.), demonstrating their efficiency and sustainability. Lastly, volumetric walls, though not as rapid in construction as prefabricated panels, showcased a balance of moderate carbon emissions (1,901.17 kg), cost (29,477.35 Rs.), and construction time (5 h), highlighting their potential for eco-friendly construction without sacrificing performance.

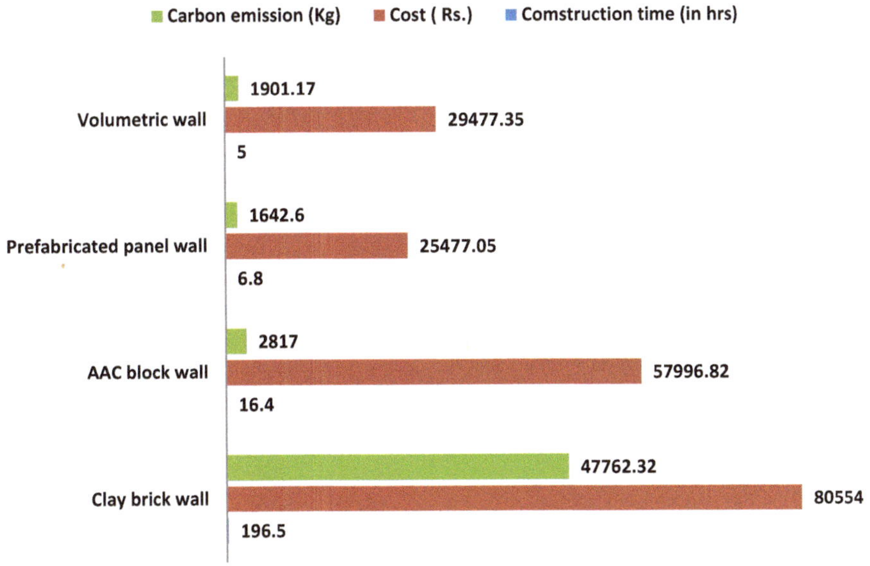

Fig. 10 Graphical representation of comparative evaluation of walling materials: construction time, cost, and carbon emissions for clay-brick, AAC block, prefabricated panel, and volumetric walls

5 Conclusion

The study highlights a clear shift toward more sustainable and cost-effective construction practices. While traditional clay-brick walls remain widely used, their high carbon footprint, extended construction time, and greater costs make them the least efficient option. Alternatives like AAC block walls, prefabricated panel walls, and volumetric walls demonstrate significant advantages in reducing construction time, costs, and environmental impact. Among these, volumetric walls stand out with the shortest construction time and substantial reductions in both cost and carbon emissions. These findings underscore the need for the construction industry to embrace innovative building technologies that align with global net-zero goals. Adopting sustainable solutions like prefabricated and volumetric walls will not only promote environmental stewardship but also enhance economic efficiency, paving the way for a more resilient and eco-friendly future in construction.

References

1. Tripathi KK, Jha KN (2019) An empirical study on factors leading to the success of construction organizations in India. Int J Constr Manag 19(3):222–239. https://doi.org/10.1080/15623599.2017.1423162
2. Hussin JM, Abdul Rahman I, Memon AH (2013) The way forward in sustainable construction: issues and challenges. Int J Adv Appl Sci 2(1). https://doi.org/10.11591/ijaas.v2i1.1321
3. IGBC (2018) IGBC net zero energy buildings rating system. Confed Indian Ind CII (November). www.cii.in
4. Shadram F, Johansson TD, Lu W, Schade J, Olofsson T (2016) An integrated BIM-based framework for minimizing embodied energy during building design. Energy Build. 128:592–604. https://doi.org/10.1016/j.enbuild.2016.07.007
5. Wong JKW, Zhou J (2015) Enhancing environmental sustainability over building life cycles through green BIM: a review. Autom Constr 57:156–165. https://doi.org/10.1016/j.autcon.2015.06.003
6. Abanda FH, Byers L (2016) An investigation of the impact of building orientation on energy consumption in a domestic building using emerging BIM (Building Information Modelling). Energy 97:517–527. https://doi.org/10.1016/j.energy.2015.12.135
7. Oti AH, Tizani W, Abanda FH, Jaly-Zada A, Tah JHM (2016) Structural sustainability appraisal in BIM. Autom Constr 69:44–58. https://doi.org/10.1016/j.autcon.2016.05.019
8. Eleftheriadis S, Duffour P, Greening P, James J, Stephenson B, Mumovic D (2018) Investigating relationships between cost and CO_2 emissions in reinforced concrete structures using a BIM-based design optimisation approach. Energy Build. 166:330–346. https://doi.org/10.1016/j.enbuild.2018.01.059
9. Mowafy N, El Zayat M, Marzouk M (2023) Parametric BIM-based life cycle assessment framework for optimal sustainable design. J Build Eng 75(February):106898. https://doi.org/10.1016/j.jobe.2023.106898
10. Ralegaonkar RV, Gavali HR, Sakhare VV, Puppala AJ, Aswath PB (2017) Energy-efficient slum house using alternative materials. Proc Inst Civ Eng—Energy 170(3):93–102. https://doi.org/10.1680/jener.16.00027
11. Shoubi MV, Shoubi MV, Bagchi A, Barough AS (2015) Reducing the operational energy demand in buildings using building information modeling tools and sustainability approaches. Ain Shams Eng J 6(1):41–55. https://doi.org/10.1016/j.asej.2014.09.006

12. Shen Z, Jensen W, Wentz T, Fischer B (2012) Teaching sustainable design using bim and project-based energy simulations. Educ Sci 2(3):136–149. https://doi.org/10.3390/educsci2030136
13. Chippagiri R, Gavali HR, Ralegaonkar RV, Riley M, Shaw A, Bras A (2021) Application of sustainable prefabricated wall technology for energy efficient social housing. Sustain 13(3):1–12. https://doi.org/10.3390/su13031195
14. Material W, Assemblies W (2020) Thermal performance of walling material and wall technology : advancing building energy efficiency in India (June)
15. Jain U, Jain M, Mandaokar S (2018) Ijresm_19_129 (9)
16. Chavan LM, Desai PDB(2017) Analyze time-cost required for conventional and prefabricated building components. Int Res J Eng Technol 4(8). https://irjet.net/archives/V4/i8/IRJET-V4I8352.pdf
17. Lopez D, Froese TM (2016) Analysis of costs and benefits of panelized and modular prefabricated homes. Procedia Eng 145:1291–1297. https://doi.org/10.1016/j.proeng.2016.04.166
18. Peng C (2016) Calculation of a building's life cycle carbon emissions based on Ecotect and building information modeling. J Clean Prod 112:453–465. https://doi.org/10.1016/j.jclepro.2015.08.078
19. Dabaieh M, Heinonen J, El-Mahdy D, Hassan DM (2020) A comparative study of life cycle carbon emissions and embodied energy between sun-dried bricks and fired clay bricks. J Clean Prod 275:122998. https://doi.org/10.1016/j.jclepro.2020.122998
20. AAC blocks-Is the revolutionary material really green?—GreenJams. https://greenjams.org/are-aac-blocks-really-green/. Last accessed 03 May 2024

Integration of YOLOv5 with Siamese Networks for Enhanced Microplastic Detection Through Fluorescence Microscopy Images

Amogha, R. Shilpa, V. Janhavi, B. K. Vinay, A. B. Rajendra, and S. Parameshwara

Abstract Microplastics (MP) are persistent environmental pollutants that pose significant risks to ecosystems and human health. This study introduces a hybrid detection model integrating YOLOv5 with a lightweight Siamese Neural Network (SNN) to enhance MP detection in fluorescence microscopy images. YOLOv5 identifies candidate regions, whereas the SNN performs refined classification using anchor-based learning, addressing challenges such as class imbalance and deformable shapes. Experimental results highlight the proposed model's superior performance, achieving precision (0.938), recall (0.930), F1-score (0.927), and mean average precision (mAP 50) (0.926), outperforming Mask Region-Convolutional Neural Network (MRCNN) (F1-score: 0.846, mAP 50: 0.858) and Faster Region-Convolutional Neural Network (FRCNN) (F1-score: 0.568, mAP 50: 0.865). Comparative analysis also showed the robustness of the model over YOLOv10 and YOLOv11, which achieved lower mAP 50 values of 0.648 and 0.779, respectively. This sustainable approach provides a scalable and highly accurate solution for microplastic detection by leveraging the unique color-based properties of fluorescence microscopy. Future work will focus on real-time applications and enhancing the generalizability across diverse datasets.

Amogha
Department of Computer Science and Engineering (AIML), Vidyavardhaka College of Engineering, Mysuru, Karnataka, India

R. Shilpa · V. Janhavi · A. B. Rajendra
Department of Computer Science and Engineering, Vidyavardhaka College of Engineering, Mysuru, Karnataka, India

B. K. Vinay (✉)
Department of Electronics and Communication Engineering, Vidyavardhaka College of Engineering, Mysuru, Karnataka, India
e-mail: vinaybk@vvce.ac.in

S. Parameshwara
Department of Electronics and Communication Engineering, The National Institute of Engineering, Mysuru, Karnataka, India

I. M. Jamadar et al. (eds.), *Integrating Sustainability in Modern Technologies*, Lecture Notes in Mechanical Engineering, https://doi.org/10.1007/978-981-95-9466-5_27

Keywords Siamese neural networks · YOLO · Object detection · Microplastics

1 Introduction

Microplastics (MP) contamination poses a significant threat to human health and the environment owing to its persistence and widespread presence. Accounting for over 92.4% of ocean plastic debris [1], these particles disrupt ecosystems and affect marine life, with 90% of marine species consuming them, leading to bioaccumulation and toxin transfer [2]. Notably, MP is present in 94% of bottled water and 83% of tap water, highlighting its impact on human health [3]. Additionally, 32% of the 78 million t of plastic produced annually evade proper waste management, exacerbating its spread [4, 5]. Detecting MP is crucial for understanding their prevalence, mitigating their effects, and protecting ecosystems and public health. Although thermal analysis (e.g., pyrolysis-GC-MS), spectroscopy (FTIR and Raman), and microscopy (optical and electron) are accurate, they have high cost and time commitments with notable limitations [6]. Processing and analyzing samples are labor-intensive procedures that require a great deal of physical labor and specialized knowledge. These methods that require expensive equipment are time-consuming, are not suitable for high-throughput analysis, cannot be used for routine monitoring, and have limited capacity to handle complicated or heterogeneous samples, which lower their effectiveness in large-scale applications [7, 8]. Artificial intelligence (AI) has revolutionized this sector by addressing the limitations of traditional methods via automation of data processing, better detection accuracy, faster analyses, and efficient identification of MP in complex matrices [6]. MP identification algorithms based on deep learning include Recurrent Neural Networks (RNN) capable of handling sequential data of spectroscopic analyses and Convolutional Neural Networks (CNN) used to identify and classify MP in images [9]. Transfer learning further enhances the accuracy and efficiency of complex queries. In contrast, real-time detection and localization in image-based datasets are performed using advanced models, such as YOLO [10]. Using transfer learning with pre-trained architectures (such as ResNet and EfficientNet) significantly increases the accuracy and speed of complex tasks. Pytorch-based frameworks for FRCNN and MRCNN and Ultralytics for the YOLOv5 model [10]. CNN evaluate pictures, and fluorescent imaging has emerged as a promising technique for enhancing the visibility of MP in environmental samples. Fluorescent imaging adds contrast by using UV light to emit fluorescence, whereas normal imaging uses visible light to record the appearance of MP based on reflection [11].

2 Literature Review

Several deep learning methods have been developed to identify MP in water. Grekov et al. [10] presented an MFFM FRCNN-based object detection model for identifying small residual film pieces on field surfaces. The model achieved superior performance with an 83.45% AP, a 0.89 F1-score, and a detection speed of 248.36 ms, surpassing the original FRCNN, SSD, and YOLOv5. Its high accuracy and efficiency are pivotal for automating residual film removal processes. Han et al. [12] applied an MRCNN deep learning approach. They demonstrated high accuracy in the detection, classification, and segmentation of large marine MP particles, achieving high precision (93.3%), recall (95.4%), APbb (92.7%), F1-score (94.34%), and APm (82.6%) on a 250-image test dataset. This model was trained on 3000 MP images with a ResNet-101 backbone and processed images at 12.5 FPS.

Bianco et al. [13] developed an innovative method by combining a holographic setup with an SVM classifier (H-SVM), significantly surpassing the performance of traditional brightfield microscopy-based SVM (C-SVM), which had accuracies below 88%. The H-SVM method achieved an impressive accuracy of 99.31%, showing remarkable improvements in specificity (99.82% compared with 93%), sensitivity (97.22% compared with 91%), and overall accuracy (99.10% compared with 60.07%). Javier et al. [14] compared machine learning classifiers for counting and classifying MP, finding that the Random Forest Classifier (RFC) outperformed others, including AdaBoost, K-NN, and C4.5, with an accuracy of 96.6%. These classifiers rely on 19 features, such as color-based metrics (RGB, HSV) and shape-based metrics (compactness and aspect ratio).

Hassan et al. [15] reported the state-of-the-art performance of MRCNN on the COCO dataset, where it outperformed earlier models in instance segmentation and bounding box detection, processing images at 5 FPS with high accuracy. Gandhi [16] highlighted the evolution of RCNN-based architectures, noting that RCNN was computationally slow (47 s/image) due to separate region proposals and classification. The FRCNN improved this to 0.3 s/image by integrating ROI pooling and further reduced it to 0.2 s/image with a Region Proposal Network (RPN). YOLO stands out for real-time applications, achieving 45–155 FPS, albeit with a slight trade-off in accuracy.

Minaee et al. [17] highlighted advancements in segmentation accuracy, showing that mean Intersection over Union (mIoU) on the PASCAL VOC dataset increased from 62.2% with FCN to 90.5% with EfficientNet+NAS-FPN. Similar improvements were noted in the Cityscapes dataset, where the latest models achieved a 23% relative gain over the FCN. Waheed et al. [18] compared MRCNN and FRCNN, noting that FRCNN achieved a mAP of 69.9% on the PASCAL VOC 2007 dataset, while MRCNN matched its precision while adding instance segmentation capabilities.

Ren et al. [19] demonstrated improvements in FRCNN by reducing the number of region proposals, achieving a mAP of 36.2% on PASCAL VOC and 42.0% on COCO datasets. He et al. [20] proposed MRCNN with a ResNet-50 backbone, achieving a mAP of 37.1% for instance segmentation on COCO and further improvements with Feature Pyramid Networks (FPN) and RoI Align for pixel-level alignment. YOLO-based models have also seen significant advancements. Owen Tamin et al. [21] developed a YOLOv5 model achieving 93.31% accuracy with RGB images and 92.96% when combining RGNIR data, showcasing the benefits of data fusion. Cai [22] refined YOLOv5 architecture for marine garbage detection, achieving a mAP of 97.2% at an IoU threshold of 0.5. Simou et al. [23] improved YOLOv5 for detecting plastic greenhouses from satellite images, with an optimal confidence value of 0.635 and an AP of 84.1%. Saidani [24] introduced a YOLOv5 tiny model optimized for resource-constrained devices, achieving a mAP of 60.9% at IoU 0.5 and 44% across IoU thresholds from 0.5 to 0.95.

These advancements underscore the potential of deep learning models for MP detection. Models such as YOLOv5, MRCNN, and FRCNN have been widely adopted owing to their robustness and efficiency. Metrics such as mAP 50, mean precision, and mean recall are pivotal for evaluating model performance. Despite the deformable and inconsistent shapes of MPs, fluorescence microscopy aids detection by highlighting their distinct colors. These findings justify the selection of models, such as YOLOv5, YOLOv10, YOLOv11, MRCNN, and FRCNN, for detecting MPs in fluorescence microscopy images, leveraging their unique strengths in accuracy, speed, and segmentation.

3 Material and Method

3.1 Database Info

For this study, we used an open-source dataset of fluorescence microscopy images of different variants of MP, namely, polystyrene (PS), low-density polyethylene (LDPE), high-density polyethylene (HDPE), and PETE, with a dataset size of 662 images. The images were uniform, with a resolution of 640 × 640. The total number of annotations from all images shows that PETE is heavily under-represented with only 137 annotations while HDPE is 4523, PS is 2179, and LDPE is 2108 which creates a significant imbalance in classes; to address this imbalance, we are dropping the images with the annotations about PETE, thus the resultant dataset is of size 657 images and the class balance between HDPE, LDPE, and PS still holds.

Reference:[https://universe.roboflow.com/puttisan/microplastic-lstyl/dataset/1].

3.2 Research Flow Map

We aim to identify a model that performs best on the metrics mAP 50, mean precision across all classes, mean recall across all classes, and overall F1-score. To achieve this, we conducted a comparative analysis of YOLOv5, YOLOv10, YOLOv11, MRCNN, and FRCNN under uniform hardware conditions, while allowing framework-specific hyperparameters. In addition, we analyzed the architecture of these models to understand the features that drive their performance. We propose a method to address the identified gaps based on these insights.

3.3 General Architecture

The YOLO and RCNN model families share a backbone, neck, and head architecture (Fig. 1), differing primarily in their neck and head implementations. Backbones, usually deep CNNs, include ResNet50 for MRCNN and FRCNN and CSP Darknet53 for YOLO models. The FRCNN employs ROI and RPN pyramids to propose regions of interest, converting them into fixed-size feature maps. MRCNN extends the FRCNN by adding a mask branch for segmentation. YOLOv5 utilizes a Path Aggregation Network to enhance feature propagation and multiscale information, thereby strengthening its feature pyramid. Its detection head outputs class probabilities, bounding box coordinates, and object confidence scores (Fig. 2).

The proposed method fine-tunes YOLOv5 to optimize box loss and detect it at a lower confidence threshold, potentially affecting the precision. To counter this, a lightweight SNN was introduced to classify cropped instances from YOLOv5's bounding boxes. The SNN uses three anchor images per class to label instances based on confidence scores and assigns the most common label with the highest confidence. Detections were discarded if the maximum confidence score for n/2 + 1n/2 + 1 outputs was <0.5 or if no majority label was found, addressing false positives from the background.

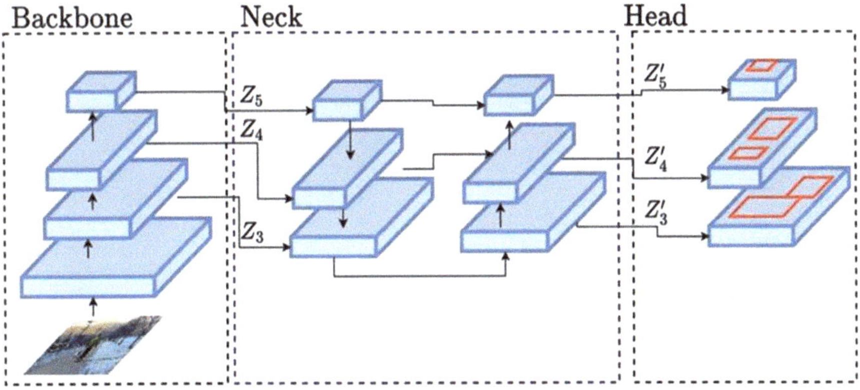

Fig. 1 Generalized architecture of a two-stage model

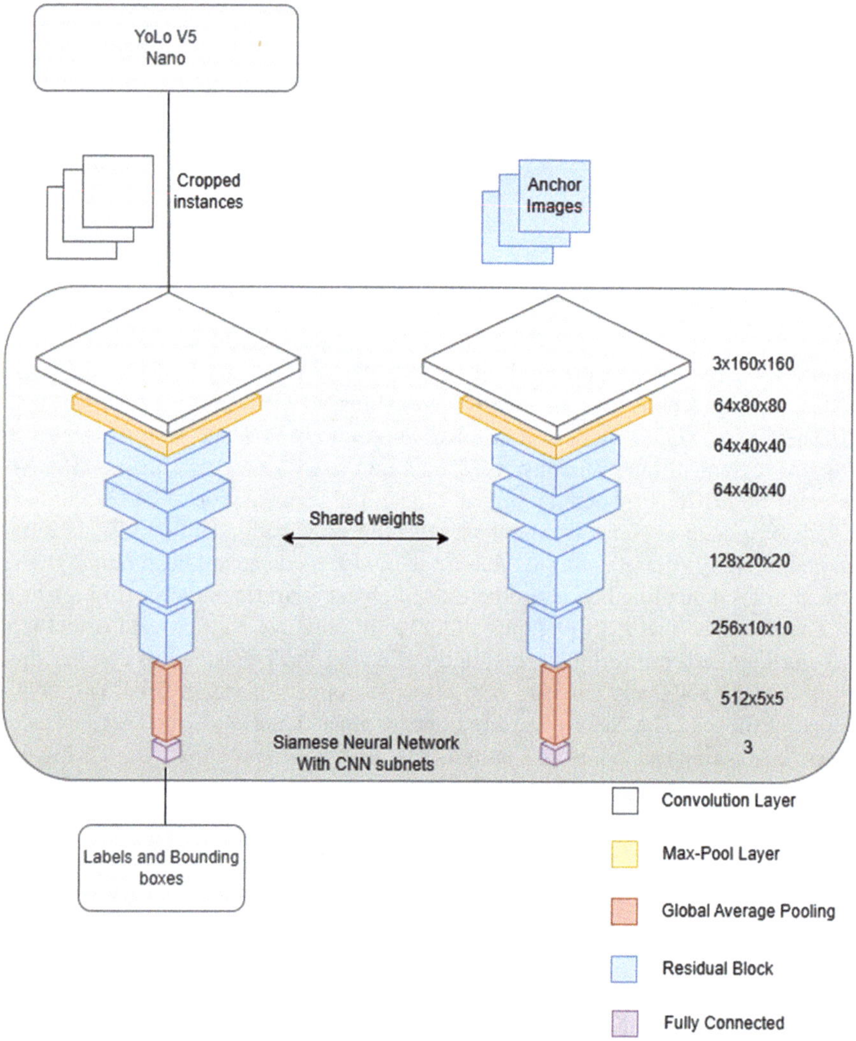

Fig. 2 Anchor image of siamese neural network with ResNet-18 subnets

4 Results and Discussion

The models were trained using the Optuna framework for hyperparameter optimization while monitoring their validation losses. The weights with the least validation loss after the training were retained for the metrics shown in Table 1 (Fig. 3).

Table 1 Result analysis of different models

Model	Precision	Recall	F1-score	mAP50
Faster RCNN	0.467	0.727	0.568	0.865
Mask RCNN	0.765	**0.947**	0.846	0.858
YOLOv5	0.849	0.831	0.840	0.875
YOLOv10	0.733	0.573	0.643	0.648
YOLOv11	0.692	0.889	0.778	0.779
SNN-YoLoV5	**0.938**	0.930	**0.927**	**0.926**

Confusion Matrix Interpretation

The confusion matrix and the results of Fig. 4 MRCNN and Fig. 5 FRCNN, we can infer that the two models mislabel PS as HDPE and LDPE as HDPE quite often; we also see that the ratios on the matrices are the same, but when looking at their recall we can infer that the number of true positives identified in MRCNN is far more compared to FRCNN. When considering Fig. 6 YOLOv5, we can observe that YOLOv5 mislabels PS as LDPE and mislabels LDPE as PS, a genuine confusion to behold; it is also to be considered that its precision and recall are both higher than those of the MRCNN and FRCNN. Based on our proposed method, we can infer that the confusion seen in just the vanilla YOLOv5 still holds, but faintly.

Model Results

Our proposed method (Fig. 7) shows an average gain in precision of 30% while showing a ~ 9% increase with the second confusion matrix interpretation.

When observing the confusion matrix and the results of Fig. 4 MRCNN and Fig. 5 FRCNN, we can infer that the two models mislabel PS as HDPE and LDPE as HDPE quite often, we also see that the ratios on the matrices are the same but when looking at their recall we can infer that the number of true positives identified in MRCNN is far more compared to FRCNN. When considering Fig. 6 YOLOv5, we can observe that YOLOv5 mislabels PS as LDPE and mislabels LDPE as PS, a genuine confusion to behold it also to be considered that its precision and recall are both higher than those of the MRCNN and FRCNN. Based on our proposed method, we can infer that the confusion seen in just the vanilla YoLoV5 still holds, but faintly.

Model Results

Our proposed method (Fig. 7) shows an average gain in precision of 30% while showing a ~ 9% increase with the second-best model, an average gain in recall of 23% while showing a − 1.7% decrease in recall, an average gain in F1-score of 27%, an 8% increase with the second-best model, an average gain in mAP 50 of 21%, and a 5% increase with the second-best model, an average gain in recall of 23% while showing a − 1.7% decrease in recall, an average gain in F1-score of 27%, an 8% increase with the second-best model, an average gain in mAP 50 of 21%, and a 5% increase with the second-best model.

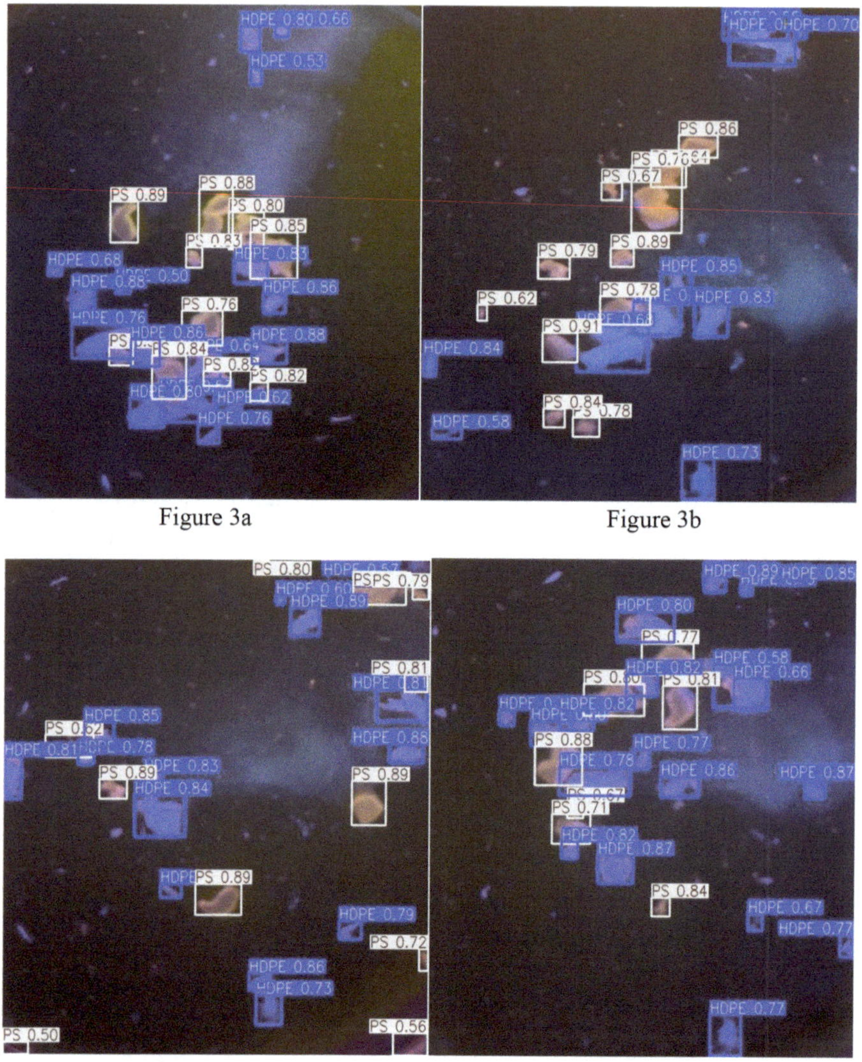

Figure 3a Figure 3b

Figure 3c Figure 3d

Fig. 3 **a–d** show the outputs of the proposed model displaying multiple MP in the given samples

Evaluation Metrics

- mAP 50: mAP at an IoU threshold of 0.5, which measures the precision of the object detection models across all classes. It was calculated by averaging the precision values at each class for detection with an IoU greater than 50%.
- True Positive (TP): A predicted bounding box is considered a True Positive if

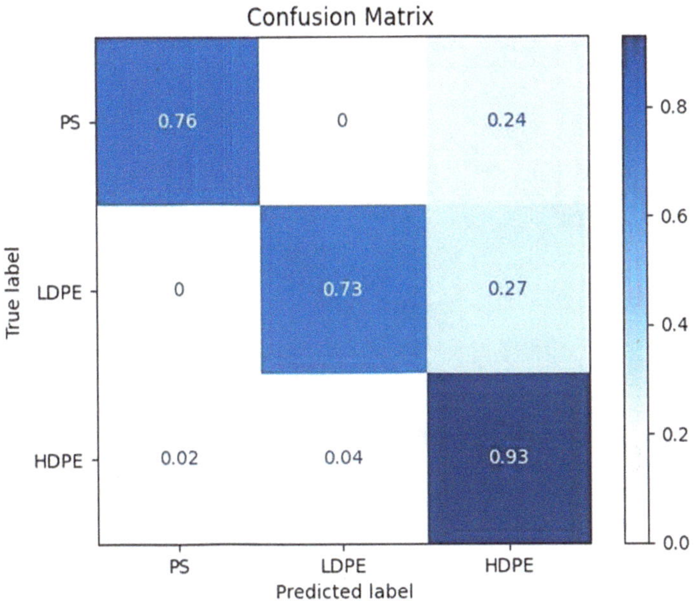

Fig. 4 Mask RCNN confusion matrix

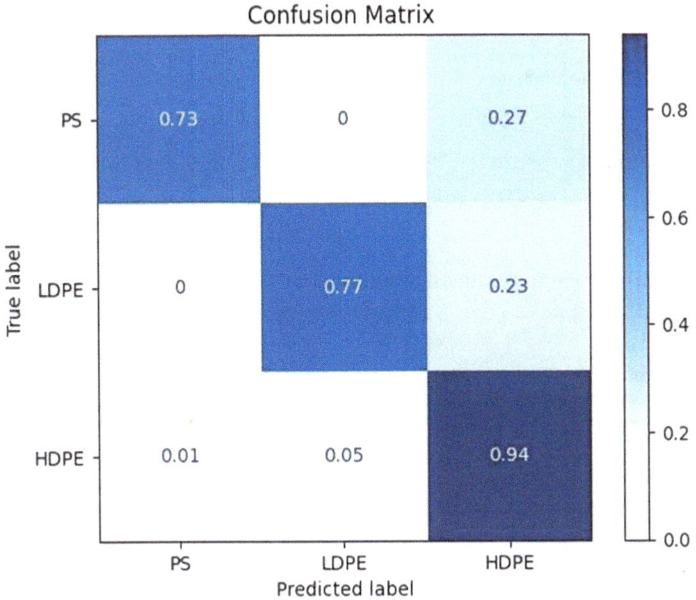

Fig. 5 Faster RCNN confusion matrix

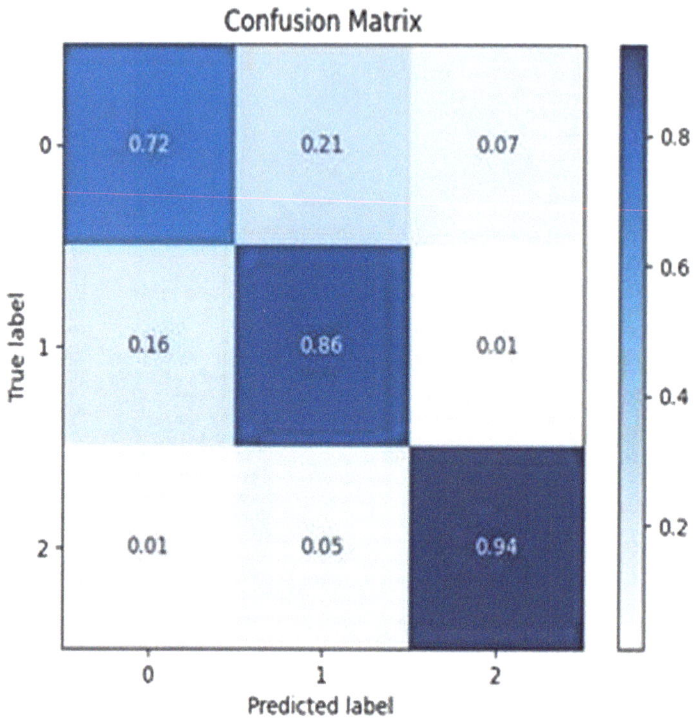

Fig. 6 YOLOv5 confusion matrix

The IoU between the predicted box and the ground truth box was greater than or equal to a threshold. The expected box should correspond to the correct object (class-match).

- False Positive (FP): A predicted bounding box is considered a False Positive if the IoU between the predicted box and any ground truth box is less than the threshold (e.g., < 0.5).

Or, the predicted box does not match any ground truth box, even if it has a high IoU with a ground truth box.

- False Negative (FN): A ground truth box does not match the predicted bounding box with IoU above the threshold (i.e., the model missed that ground truth).
- Mean Precision for All Classes: This metric calculates the average precision across all object classes, assessing the number of predicted positives that are correct.
- Mean Recall for All Classes: The average recall across all object classes measures how well the model detects all the relevant instances of each class.

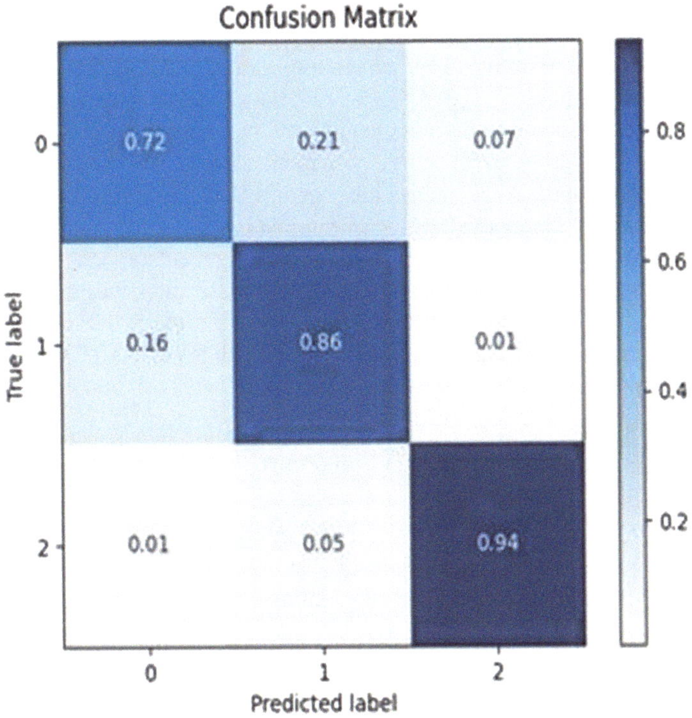

Fig. 7 SNN-yolov5 confusion matrix

- Overall F1-Score: The Harmonic means of precision and recall, providing a single measure that balances the two. It is especially useful when dealing with imbalanced datasets because it ensures that both false positives and false negatives are considered.

Role of SNN-Yolov5 for MP Detection

Our proposed method considers three anchor images from the three classes PS, HDPE, and LDPE to allow the SNN subnetworks to learn similarities and differences. The YOLOv5 model detects the MPs in the image with a lower confidence threshold (0.4), and then the instances are extracted from the image using bounding boxes, which are paired with each instance, resulting in three pairs that are fed into the SNN to obtain the labels and confidence score for the given detection. If the maximum confidence score of the two outputs was <0.5, the detection was discarded. The label with the highest confidence score for a given output is selected, and the majority label is assigned if three distinct classes appear for the three outputs from the SNN, that is, no clear majority that detection is discarded. In this manner, we can detect almost all instances of MPs and can be precise. Because this is a pseudo-few-shot approach, the precision was higher.

Training Process

The dataset is split into train and test sets in a split ratio of 70:30; YOLOv5's hyperparameter configuration file is edited to increase the gain of the box loss function thus making it more sensitive to box loss changes and resulting in an optimized version in which box loss is minimal; the model was trained for 70 epochs. The default optimizers and loss functions were used which provides Ultralytics. A custom data loader class was implemented for the training of the SNN, where each sample in the data loader was a tuple of two images and a tuple of their metadata, which included the annotation boxes and corresponding labels. Upon initialization, the data loader crops the pictures into separate instances for training and resizes them, because performing these transformations when the get item function is called creates a CPU bottleneck. The SNN was trained for 50 epochs using the Adam optimizer at a learning rate of 0.001, and a categorical cross-entropy loss function was used.

Qualitative and quantitative comparison of different models

- **SNN-YOLOv5** is the best model for detecting microplastics in fluorescence microscopy images because of its high precision, recall, and mAP 50. It is the most accurate method for detecting and segmenting MP with minimal false positives and nearly complete detection.
- **YOLOv5** is also an excellent choice, offering high precision and a good balance between recall and precision, although it may miss a few microplastics compared with SNN-YoLoV5.
- **Mask RCNN** performs well with high recall and good precision, making it a solid option, especially when segmentation accuracy is important.
- **Faster RCNN** is less ideal owing to its lower precision, which could generate too many false positives, but it still has good recall and a decent mAP 50.
- **YOLOv10** and **YOLOv11** showed lower recall and overall performance, making them less suitable for MP detection in fluorescence microscopy images, especially in tasks where detecting as many particles as possible is crucial.

5 Conclusions

This study successfully integrated YOLOv5 with a lightweight SNN to enhance the detection and classification of microplastics in fluorescence microscopy images. By leveraging YOLOv5's robust object detection capabilities and SNN's ability to discern fine-grained differences between classes, the proposed hybrid model achieves superior performance metrics compared to traditional and state-of-the-art models, such as FRCNN and MRCNN. Our results demonstrate a significant improvement in precision (0.938), recall (0.930), and mAP 50 (0.926), highlighting the method's efficiency in addressing challenges like class imbalance and deformable shapes of MP. The study also introduces a novel workflow that utilizes fluorescence microscopy's color distinction to aid in MP detection, a largely underexplored characteristic in prior research. Incorporating anchor-based learning into SNN enhances detection reliability while maintaining robustness against false positives. This approach offers a more accurate and scalable solution for MP detection and sets a foundation for extending the methodology to other applications that require high precision in object detection. Future work could focus on refining the SNN architecture for real-time applications and exploring additional data augmentation techniques to improve the generalizability across diverse datasets.

References

1. Dey TK, Uddin ME, Jamal M (2021) Detection and removal of microplastics in wastewater: evolution and impact. Environ Sci Pollut Res 28(14):16925. Springer Science and Business Media. https://doi.org/10.1007/s11356-021-12943-5
2. Worm B, Lotze HK, Jubinville I, Wilcox C, Jambeck J (2017) "Plastic as a Persistent Marine Pollutant," Oct. 17, 2017 and Annual Reviews. https://doi.org/10.1146/annurev-environ-102016-060700
3. Hidalgo-Ruz V, Gutow L, Thompson RC, Thiel M (2012) Microplastics in the marine environment: a review of the methods used for identification and quantification. Environ Sci Technol 46(6):3060. American Chemical Society. https://doi.org/10.1021/es2031505
4. Qi R, Jones DL, Li Z, Liu Q, Yan C (2019) Behavior of microplastics and plastic film residues in the soil environment: A critical review. In: The Science of The Total Environment, vol vol 703. Elsevier BV, p 134722. https://doi.org/10.1016/j.scitotenv.2019.134722
5. Green DS, Boots B, O'Connor NE, Thompson RC (2016) Microplastics affect the ecological functioning of an important biogenic habitat. Am Chem Soc. https://doi.org/10.1021/acs.est.6b04496
6. Ivleva NP (2021) Chemical analysis of microplastics and nanoplastics: challenges, advanced methods, and perspectives. Chem Rev 121(19):11886. American Chemical Society. https://doi.org/10.1021/acs.chemrev.1c00178
7. Zhou P et al. (2020) Microfluidic high-throughput platforms for discovery of novel materials. Nanomaterials 10(12):2514. Multidisciplinary Digital Publishing Institute. https://doi.org/10.3390/nano10122514
8. Huang X, Liu Q, Yao S, Jiang G (2017) Recent progress in the application of nanomaterials in the analysis of emerging chemical contaminants. R Soc Chem. https://doi.org/10.1039/c7ay00859g
9. Rezvani H et al (2024) morphological detection and classification of microplastics and nanoplastics emerged from consumer products by deep learning. Cornell University. https://doi.org/10.48550/arxiv.2409.13688
10. Grekov AN, Shishkin YE, Peliushenko SS, Mavrin AS (2022) Application of the yolov5 model for the detection of microobjects in the marine environment. Cornell University. https://doi.org/10.48550/arxiv.2211.15218
11. Zhang Y, Zhang D, and Zhang Z (2023) A critical review on artificial intelligence—based microplastics imaging technology: recent advances, hot-spots and challenges. Int J Environ Res Public Health 20(2):1150. Multidisciplinary Digital Publishing Institute. https://doi.org/10.3390/ijerph20021150
12. Han X-L, Jiang N-J, Hata T, Choi J, DU YAN-JUN, Wang Y-J (2023) Deep learning based approach for automated characterization of large marine microplastic particles. 183:105829
13. Bianco V, Memmolo P, Carcagnì P, Merola F, Paturzo M, Distante C, Ferraro P Microplastic identification via holographic imaging and machine learning. https://doi.org/10.1002/aisy.201900153

14. Lorenzo-Navarro J, on-Santana MC´, G´omez M, Herrera A, Mar'ın-Reyes PA. Automatic counting and classification of microplastic particles. https://doi.org/10.5220/0006725006460652
15. Bharati P, Pramanik A (2020) Deep learning techniques-R-CNN to mask R-CNN: a survey. Comput Intell Pattern Recognit 999:657–668
16. Gandhi R (2018) R-CNN, fast R-CNN, faster R-CNN, YOLO—object detection algorithms understanding object detection algorithms R-CNN
17. Minaee S, Boykov YY, Porikli F, Plaza AJ, Kehtarnavaz N, Terzopoulos D (2021) Image segmentation using deep learning: a survey. IEEE Trans Pattern Anal Mach Intell:1
18. Safa Riyadh Waheed, Norhaida Mohd Suaib, Mohd Shafry Mohd Rahim, Myasar Mundher Adnan, A. A. Salim Deep learning algorithms-based object detection and localization revisited. https://doi.org/10.1088/1742-6596/1892/1/012001
19. Ren S, He K, Girshick R, Sun J (2017) Faster R-CNN: towards real-time object detection with region proposal networks. IEEE Trans Pattern Anal Mach Intell 39(6):1137–1149
20. He K, Gkioxari G, Dollar P, Girshick R (2020) Mask R-CNN. IEEE Trans Pattern Anal Mach Intell 42(2):386–397
21. Owen T., Moung E, Dargham J, Yahya F, Farzamnia A, Sia F, Naim NFM, Angeline L Onshore plastic waste detection with YOLOv5 and RGB-near-infrared fusion: a state-of-the-art solution for accurate and efficient environmental monitoring
22. Cai C, Gu S Research on marine garbage detection based on improved Yolov5 model. J Phys Conf Ser. https://doi.org/10.1088/1742-6596/2405/1/012008
23. Simoua MR, Loulada S, Aamira Z, Maanana M, Rhinane H (2022) Using YOLOV5 to detect plastic greenhouses in MOROCCO. Johor Bahru, Malaysia. https://doi.org/10.5194/isprs-archives-XLVIII-4-W6-2022-351-202
24. Saidani T (2023) Deep learning approach: YOLOv5-based custom object detection deep learning approach: YOLOv5-based custom object detection. Eng, Technol Appl Sci Res 13(6):12158–12163. https://doi.org/10.48084/etasr.6397

Experimental Analysis of RC Columns for Variable Aspect Ratio Strengthened by Carbon Fiber Composites

Giridhar N. Narule and Abhay N. Bambole

Abstract The fortification and retrofitting of concrete buildings has become increasingly common worldwide thanks to the use of outwardly bound fiber-reinforced polymers. It has to do with the deterioration of crucial columns in reinforced concrete buildings. Axially symmetrical concrete column cross-sections have been successfully improved in terms of strength and deformability using FRP materials. When it comes to non-symmetrical columns, the confinement pressure varies drastically along the column's cross-sectional perimeter as the side aspect ratio shifts. The primary goal of the current research is to use experimental studies to better understand the behavior of reinforced concrete columns with variable aspect ratio after strengthening with carbon fiber composite. Six unwrapped and six coiled carbon fiber composites-reinforced concrete columns were used in the trial study. The axial reaction of strengthened RC columns was identified by analyzing axial strengths, ductility, and failure patterns. The results of the experiments showed that confinement had a substantial impact on the strength-ductility characteristics of columns with different aspect ratios. Additionally, the research reveals the differences in cracking patterns at failure stresses for axially symmetrical and unsymmetrical sections contained in carbon fiber composite.

Keywords Axially symmetrical and unsymmetrical confined columns · Strength-ductility behavior · Stress-strain behavior

G. N. Narule (✉)
Vidya Pratishthans Kamalnayan Bajaj Institute of Engineering and Technology, Baramati, India
e-mail: gnnarule@gmail.com

A. N. Bambole
Veermata Jijabai Technological Institute, Mumbai, India

I. M. Jamadar et al. (eds.), *Integrating Sustainability in Modern Technologies*, Lecture Notes in Mechanical Engineering, https://doi.org/10.1007/978-981-95-9466-5_28

1 Introduction

Building constructions with strong steadiness and strength using affordable materials is difficult in structural engineering. In civil engineering systems, ensuring structural stability over the course of their existence is a particularly difficult task. The development of slender structures, the availability of large spans, the use of high-strength materials, and load reduction are the most recent trends. Composite materials have proven to be a successful candidate to meet these requirements. Novel characteristics of the fiber-reinforced polymer (FRP) inspired investigators to reinforce or repair the deterioration of the columns through strengthening. The capability of existing columns can be rehabilitated or enriched with outwardly bonded fiber composite circumferential wraps. Additionally, carbon fiber composites are being considered as an amazing alternative for column restoration with the intention of reducing structural deficiencies and increasing its strength in accordance with projected demand.

The restoration of concrete structures has seen a rise in the usage of externally attached fiber-reinforced polymers. With a geometrical arrangement that consents the fibers to be equally strained, FRP materials have been effectively employed to develop the deformability and load carrying capability of circular concrete columns, offering an extremely operative confinement for concrete across the whole cross-section. By reason of the non-evenness of the confinement pressure transversely the cross-section, the confinement efficacy for axially symmetrical and unsymmetrical columns is significantly diminished. In the current study, the axially symmetrical and unsymmetrical RC columns that are commonly used in practice were made, confined, and tested for the relationship between ductility and strengths for axial behavior related to the cross-section of the column.

In the past, research has been done to govern the probable influences on the confinement of strengthening employing fiber composites. A study by Wu et al. [1] focused on loaded small rectangular concrete columns that were reinforced with CFRP wrap. The capacity enhancement in the wrapped concrete columns as compared to the core unwrapped columns falls as the aspect ratio grows and becomes irrelevant at two, according to the test findings of 45 concrete column specimens. Strong FRP-confined RC columns by means of a range of aspect fractions under axial load as well as high temperature were reviewed by Patel et al. [2]. It has been discovered that while assessing the effectiveness of strengthening measures, the fluctuation of the cross-sectional aspect fraction of columns is quite important. Xie et al.'s investigation on the parameters, jacket thickness, and specimen forms included testing of eighteen stub concrete columns using outwardly restricted ultra-high performance concrete [3]. According to the test results, adding more restricting UHPC jacket thickness greatly enhanced the compressive behavior of columns for tubular-shaped columns. Stub columns in the shape of squares were less significantly affected by this influence. Pre-damaged concrete prisms with jackets made of fiber-reinforced polymer (FRP) were shown to behave in a certain way by Li et al. [4]. 46 concrete columns were used in tests that varied the corner compass ratios, levels of damage, and confinement pressure. The test findings showed that the corner radius fraction has a substantial influence on how well broken columns may be repaired. As the corner compass proportion decreases, the strength increases following FRP strengthening diminish. Nine rectangular column specimens were evaluated by Kumutha et al. [5] under axial compression force. According to the test results, adding more layers of glass fiber composites improved the column's ductility and load-bearing capacity. The side aspect ratio also worked against the column's increased strength as a result of the usage of glass fiber composite wrap. Yousef et al. [6] looked at variables like the jacket characteristics, rupture strain, ultimate stress, and cross-sectional shape. Because of the effects of increasing the corner radius, Benzaid et al. [7] carefully examined the constrained compressive strength of square columns. The test results described here were expressed to create a thorough investigation to discover the parameters that can be taken into account to distinguish between the axial performance of axially symmetrical and non-symmetrical columns. According to various characteristics that are taken into consideration for the comparison analysis in the current work, Dakhila et al. [8], Narule et al. [9, 10], Al-Nimry et al. [11], Castro et al. [12], and Narule et al. [13, 14] also emphasized the performance of fiber-reinforced polymers wrap restricted to axially symmetrical and unsymmetrical columns.

Examining the usefulness of fiber-reinforced polymers confinement on the response of columns with various side ratios is the main focus of the investigation. Twelve columns in total were constructed in accordance with IS: 456 (2000) [15] and ACI 440.2R (2008) [16] requirements, and loaded on a compression machine. For the columns, differentiating elements for instance failure load, strain, confinement effect, crack pattern was made. The perceived experimental study results for aspect ratio for axially symmetrical and unsymmetrical columns were contrasted, nevertheless. It has also explored the use of carbon fiber polymers for strengthening is a decent substitute for the conservative columns with variable aspect ratio since it not only enriched capability of conventional columns with axially symmetrical and unsymmetrical form but correspondingly improved the structural performance of columns.

2 Experimentation Work

The main attention of testing work on axially symmetrical and unsymmetrical columns has been with the variations of aspect ratio and its effects of overall behavior of columns as shown in Fig. 1. Axial compression tests were carried out on all wrapped and unwrapped column samples with axially symmetrical and unsymmetrical cross-sections to complete failure. In a later part, it has been highlighted show exactly experimentation work is done.

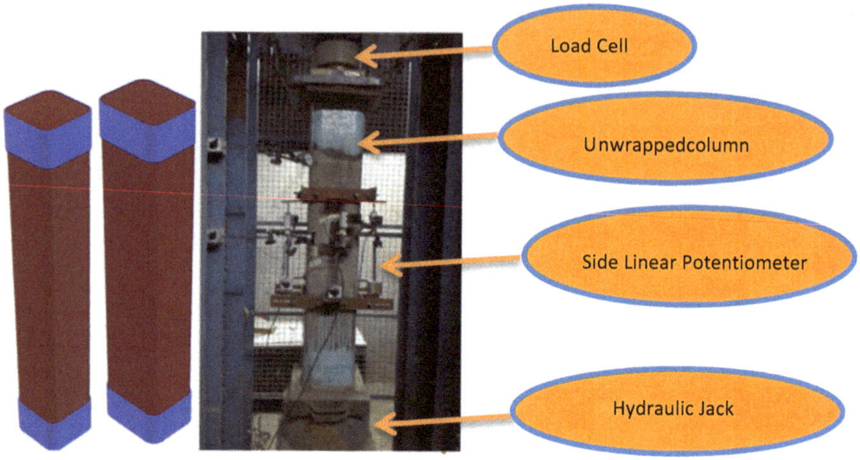

Fig. 1 Isometric views and experimental setup of RC column with variable aspect ratio

2.1 Materials Used to Prepare Test Specimens

Materials including OPC of grade 53 cement and aggregates were purchased in order to prepare M20 grade concrete in accordance with the provisions of IS: 10262 (2019) [17]. Fresh concrete, designed Fe 415 reinforcing mashes, and ready-made wooden molds were used to cast the columns. All columns were laterally strengthened with lateral ties and main steel bars with around 1.0% steel area. The columns were wrapped using the carbon fiber composite sheets of 0.17 mm thickness and width 500 mm. According to the manufacturer's recommendations, the base and hardener of the epoxy resin used as the matrix were blended 100:35.

2.2 Casting of Column Test Specimens

Six RC axially symmetrical columns of dimension 200x200x1200mm with an aspect ratio of 1.0 and six non-symmetrical columns of dimension 200x250x1200mm with an aspect ratio of 1.25 were constructed using concrete and steel. The molds were removed after a 24-hour cure. After a 28-day cure and drying, the wrapped columns were made by fortifying them with carbon wrap, primer, and resin matrix. The concrete surfaces were covered with primer after finishing the surface. Base and hardener were carefully combined in a 0.35:1 ratio as a matrix. By hand brushing, Matrix-20 was applied to the surface as a saturant. After applying the saturant, the single wrap fibers were wrapped around the columns.

2.3 Casting of Column Test Specimens

A loading frame was adopted to test the vertical compression of all column samples. As seen in Fig. 1, the RC columns underwent axial compressive loading testing on the loading frame. Linear variable differential transducers were connected along all four sides of each test sample to monitor the vertical deformation of the column. The columns entirely failed when a compressive loading was imposed at a continuous loading rate. Strain gauge data were recorded, and the relationship between load and deformation and stress and strain was found, along with the impact of confinement's aspect ratio.

3 Summary of Experimental Results and Discussion

The results are broken down into various parametric relationships that were found for the tested column specimens. With the intention to compare the axial performance of axially symmetrical columns with a side ratio of 1.0 in a square shape as well as unsymmetrical columns with a side ratio of 1.25 in a rectangular shape, factors including loads, vertical distortions, strength, and straining were evaluated for CFRP confined and unconfined columns, correspondingly. As seen in Fig. 2, carbon fiber composite-confined columns significantly increase ultimate and yield strengths when associated with control column specimens with varying aspect ratios. Axially symmetrical columns in square shape with aspect ratio 1.0 demonstrated the higher improvement of ultimate load carrying capability and yield load carrying capability after CFRP wrapping as compared to non-symmetrical columns in rectangular shape with aspect ratio 1.25 as shown in Fig. 2. Figure 3 displays the variations in strength and straining for confined and unconfined columns by means of different aspect ratios. RC columns covered in carbon fiber composite have shown increased axial strain when compared to control column specimens. In comparison with confined non-symmetrical columns in rectangular form, axially symmetrical columns in square shape have better ultimate strength. Compared to wrapped square columns, the efficiency of the wrap in improving the capability of RC non-symmetrical columns is lower.

The impact of external CFRP wrap on axially symmetrical columns in axially symmetrical and non-symmetrical columns with varied ratios of aspect has been thoroughly studied in order to analyze the variation in characteristics like ductility and strength of RC wrapped columns. Table 1 clearly shows that axially symmetrical square columns with a side ratio of 1.0 experience a bigger percentage gain in axial strength than non-symmetrical rectangle columns with a side ratio of 1.25. Column ductility relates to a structural member's capacity to resist axial distortion without significantly reducing its load carrying capacity. By switching from a 1.25 aspect ratio for a rectangle column to a 1.0 aspect ratio for a square column, the ductility of a restricted column is significantly increased. Due to corner crushing and main

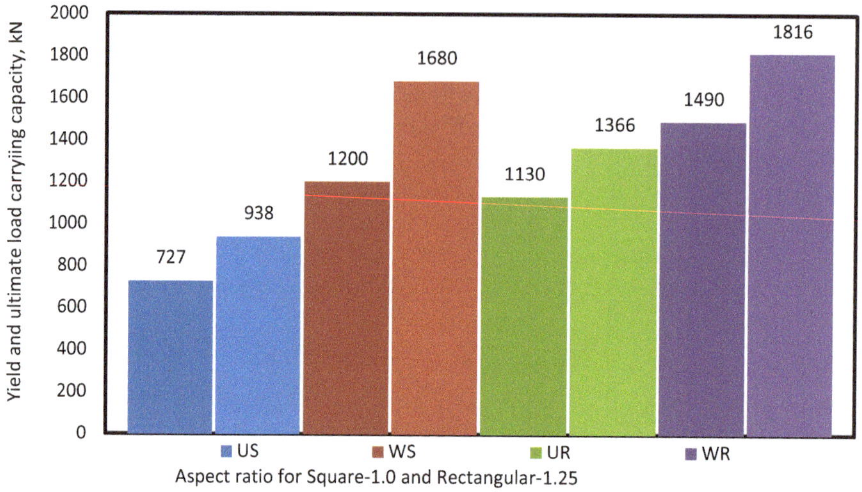

Fig. 2 Loading capacity of confined and unconfined columns with variable aspect ratio

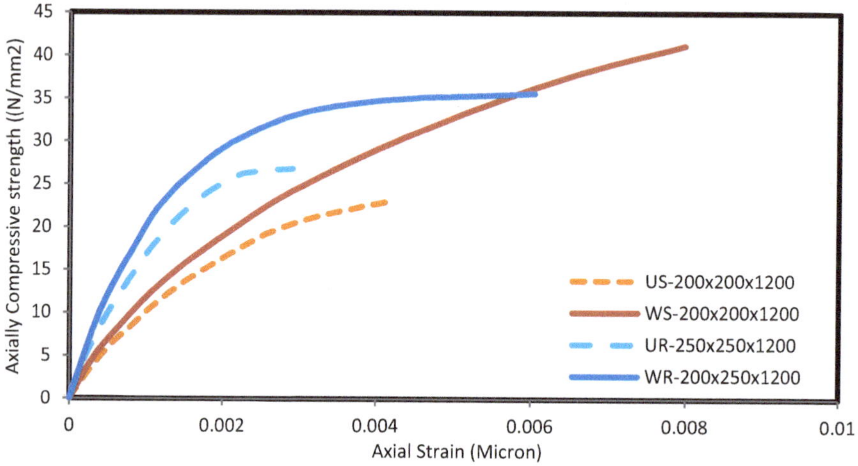

Fig. 3 Stress-strain response of confined and unconfined columns with variable aspect ratio

reinforcement bending, the unconfined square columns with side ratio of 1.0 failed. Figure 4 depicts the shear cracking and vertical reinforcement bending failure of symmetrical column specimens. The wrapped square columns fell over as a result of the grouping of separation of wrap and stretchy estrangement of the FRP at the location of the turning brought on by stress accumulation.

A crack formed in the lowest third of the unwrapped rectangular column, moved in an oblique direction toward the end of the column, burst the lower portion with FRP failure, and the main bar buckled near the bottom end. Figure 5 shows the

Table 1 Comparative test outcomes of the axially symmetrical and non-symmetrical columns

Column test specimens	Axially symmetrical –200 x 200 x 1200 mm (Aspect ratio 1.0)	Axially non-symmetrical – 200 x 250 x1200mm (Aspect ratio 1.25)
Ultimate strength of the specimen when unwrapped (MPa)	23	27
Ultimate strength of the specimen when wrapped (MPa)	41	36
After wrapping in carbon fiber composite, an increase in strength of %	79	33
After wrapping in carbon fiber composite, an increase in ductility of %	63	40
Yield strength increased by % with carbon fiber composite wrapping	32	29

Fig. 4 Failure patterns of columns for different aspect ratio

failure of a rectangular column with a 1.25 aspect ratio along the columns' larger sides by reason of concrete crushing. All of the confined columns have collapsed because of wrapping tearing at about mid-height. As demonstrated in Fig. 6, the strain contribution in the CFRP wrap and tie steel is greater for axially symmetrical square columns than for unsymmetrical rectangular columns. The effectiveness of wrapping is altered by changes in column aspect ratio, as seen in Fig. 3.

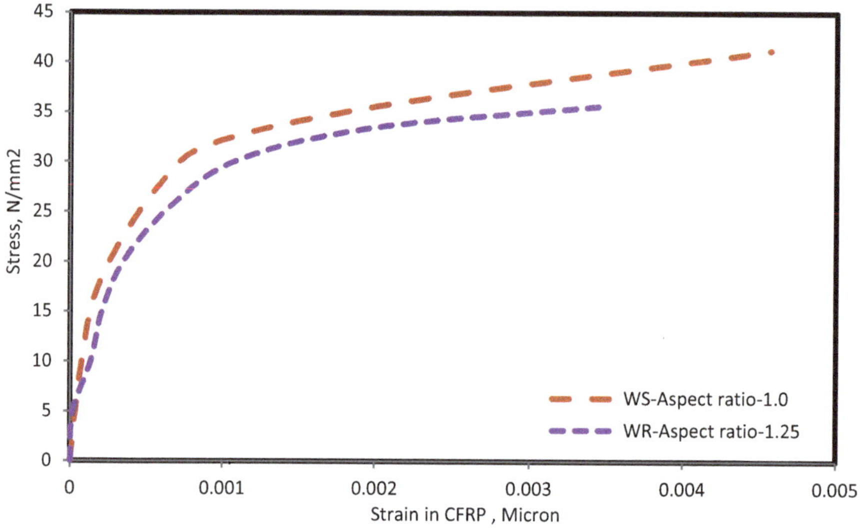

Fig. 5 Performance of the CFRP wrap of columns under stress and strain for varying aspect ratios

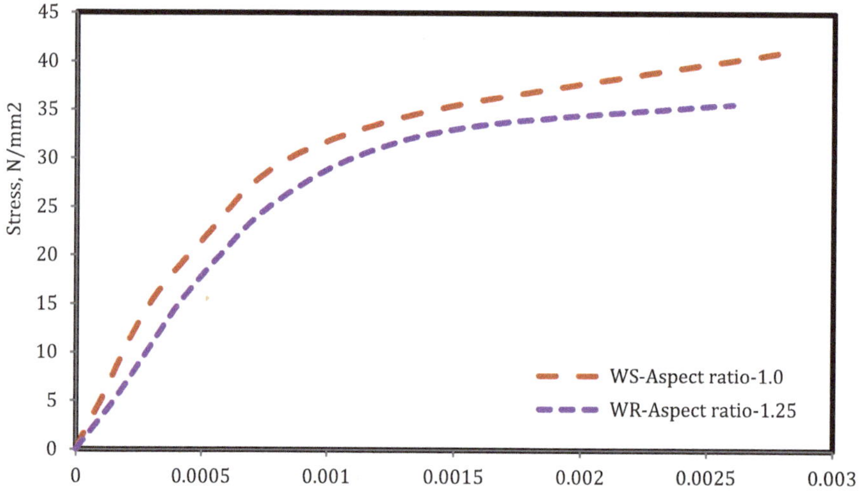

Fig. 6 Performance of tie steel of columns under stress and strain for varying aspect ratios

4 Conclusions

Investigations are done into how wrapped and unwrapped reinforced concrete columns behave axially in relation to the aspect ratio of the column. A comparison of the axial performance for ductility and strengths of axially symmetrical and unsymmetrical shaped RC columns with different cross-sectional areas has been done. Additionally, the unique impacts of aspect ratio of column on loading capability and failure pattern are also revealed. The subsequent inferences are taken from experimental study's results as below:

After carbon fiber composite strengthening, the final load capacity of RC columns improved by 79% for axially symmetrical square columns with aspect ratio 1.0 and by 33% for non-symmetrical rectangle columns. Also, yield strength of RC columns increased by 32% for axially symmetrical square columns and 29% for non-symmetrical rectangle columns after carbon fiber composite strengthening. The modification in aspect ratio has a considerable impact on the compressive capability of wrapped columns, according to axial stress vs. strain graphs. For instance the aspect ratio of the column rises from 1.0 to 1.25, the ductility of RC columns considerably decreases after being wrapped. All columns that were confined damaged in bursting kind, whereas all unconfined columns damaged in brittle way. According to the comparable study of confined axially symmetrical and non-symmetrical columns, the use of CFRP wrap for axially symmetrical columns with aspect ratio 1.0 during strengthening may be additionally beneficial to develop the ductility and strength reaction as related to unsymmetrical in rectangular columns.

References

1. Wu Y, Wei Y (2010) Effect of cross-sectional aspect ratio on the strength of CFRP-confined rectangular concrete columns. Eng Struct 32:32–45
2. Patel M, Pardeshi R, Thakur B, Parghi A (2020) Review of strengthened FRP-confined RC column with various aspect ratios under axial load and high-temperature. In: IOP Conf. Series: Materials Science and Engineering, vol 814. IOP Publisher, pp 1–2
3. Xie J, Fu Q, Yan J (2019) Compressive behaviour of stub concrete column strengthened with ultra-high performance concrete jacket. Constr Build Mater 204:643–658
4. Li P, Sui L, Xing F, Li M, Zhou Y, Wu Y (2019) Stress–strain relation of FRP-confined predamaged concrete prisms with square sections of different corner radii subjected to monotonic axial compression. Journal of composite. Construction 23(2)
5. Kumutha R, Vaidyanathan R, Palanichamy M (2007) Behaviour of reinforced concrete rectangular columns strengthened using GFRP. J Cem Concr Compos 29(1):609–615
6. Yousef MN, Feng MQ, Mosallam AS (2006) Stress–strain model for concrete confined by composites. J Compos Part B: Eng 38(2):614–628
7. Benzaid B, Chikh N, Mesbah H (2008) Performance of square concrete column confined with GFRP composite wrap. J Civ Eng Manag 14(2):115–120
8. Dakhila AJ, Falaha M, Al-Khafaji Z (2023) Study the behavior of square reinforced concrete columns strengthened with CFRP. Electron J Struct Eng 23(4):79–84
9. Narule GN, Bambole AN (2018) Axial behavior of CFRP wrapped RC columns of different shapes with constant slenderness ratio. J Struct Eng Mech 65(6):679–687

10. Narule GN, Jagtap AP (2025) Effect of aluminium and steel fibers on flexural performance of reinforced concrete beam. Commun Appl Nonlinear Anal 32(7s):901–907
11. Al-Nimry H, Neqresh M (2019) Confinement effects of unidirectional CFRP sheets on axial and bending capacities of square RC columns. Eng Struct 196
12. Castro VJ, Leon FJ, Martínez S, Diego A, Echevarría L (2024) Effectiveness of CFRP jackets in RC rectangular columns. Full-scale experimental study. Constr Build Mater 416
13. Narule GN, Bambole AN (2020) An experimental study on axial behavior of CFRP-strengthened RC rectangular columns with variable slenderness ratio. Asian J Civ Eng 22:263–275
14. Narule GN, Bambole AN (2022) Experimental analysis of RC rectangular and square columns confined using carbon fiber composites. Mater Today: Proc-Elsevier 56(1):400–405
15. IS 456 plain and reinforced concrete—code of practice (2000) Bureau of Indian Standards, India
16. (ACI) (2008) 440.2R guide for the design and construction of externally bonded FRP systems for strengthening concrete structures. American Concrete Institute, Farmington Hills, MI, USA
17. IS 10262, concrete mix proportioning– guidelines (2019) Bureau of Indian Standards, India

Experimental Investigation on Mechanical Characteristics of Basalt-Aluminium-Coated Glass Epoxy Composites with SiC Micro-Filler

Ravikiran, Ajay Anand, T. V. Sri Charan, Vedaprasad, V. G. Pradeep Kumar, and H. S. Venugopal

Abstract In recent mechanical applications, composite materials are increasingly favored, though aluminum-coated glass fiber (AlGF) remains relatively uncommon. Basalt fiber is chosen as an environmentally friendly material, because of its superior thermal stability impact resistance and recyclability. Conversely, AlGFs exhibit corrosion resistance lightweight properties and a high strength-to-weight ratio. It is expected that the combination of these two fibers will improve strength stiffness and impact properties which has not been fully investigated. Aluminum and basalt fiber composites offer a viable alternative across industries, combining enhanced performance with environmental sustainability. This study focuses on fabricating hybrid composite laminates of basalt fiber and AlGFs using the traditional Hand Layup technique. These laminates incorporate varying percentages (0%, 0.5%, and 0.75%) of Silicon carbide (SiC) as micro-filler. The laminates are arranged in the following sequence: (A + B + A) + 0%, (A + B + A) + 0.5%, and (A + B + A) + 0.75%. This configuration indicates that the top and bottom layers consist of aluminum glass fiber (A), with basalt fiber (B) positioned in between. To meet ASTM standards for mechanical characterization and wear test analysis, an Abrasive Water

Ravikiran · A. Anand · T. V. Sri Charan (✉) · Vedaprasad · H. S. Venugopal
Department of Mechanical Engineering, The National Institute of Engineering, Mysore, India
e-mail: sricharantv22@gmail.com

Ravikiran
e-mail: nayak.ravikiran@gmail.com

A. Anand
e-mail: ajayanand0691@gmail.com

Vedaprasad
e-mail: vedaprasadgowda@gmail.com

H. S. Venugopal
e-mail: venugopalhs472@gmail.com

V. G. Pradeep Kumar
Department of Mechanical Engineering, JSS Science and Technology University, Mysore, India
e-mail: pvg@jssstuniv.in

I. M. Jamadar et al. (eds.), *Integrating Sustainability in Modern Technologies*, Lecture Notes in Mechanical Engineering, https://doi.org/10.1007/978-981-95-9466-5_29

Jet Cutter was employed for precise cutting of the laminates. This method ensures accurate shaping according to ASTM requirements. The morphological characteristics of all prepared samples were also analyzed using a scanning electron microscope (SEM). It was observed that the cohesive properties of aluminum-coated glass and basalt fibers are compromised, and resulted in inferior mechanical characteristics. Although overall flexibility is enhanced, both tensile strength and stiffness are significantly reduced in comparison to composites that utilize SiC fillers. In addition, Composite C1 exhibits increased stiffness and hardness attributed to improve mechanical interlocking, notwithstanding the potential for micro-cracking to impact flexibility. Composite C2 attains a well-balanced combination of mechanical properties, demonstrating good tensile strength and impact resistance, accompanied by moderate surface roughness. Conversely, Composite C3 provides greater flexibility, although it suffers from diminished mechanical strength and stiffness.

Keywords Scanning Electron Microscopy (SEM) · Silicon carbide · Basalt fiber · Aluminum-coated glass fiber (AlGFs) · SiC micro-filler

1 Introduction

The superior qualities of composites are becoming more and more recognized and used in a variety of industries. In order to create a new material with specific properties, these materials are created by combining various substances [1]. Among their most notable advantages is a high strength-to-weight ratio, which enables composite materials to be both lightweight and structurally durable. As a result, industries like construction sporting goods automotive and aerospace use them extensively. A matrix material is combined with reinforcing fibers like carbon glass or aramid to create composites which increase strength and durability [2, 3]. Other varieties used in a variety of industries include sandwich structures laminates and particle-reinforced composites [4]. Improved structural integrity resistance to corrosion and fatigue and the capacity to be molded into intricate shapes are just a few of the benefits that composites offer over conventional materials. According to their matrix materials, composites are divided into Metal Matrix Composites (MMCs), Ceramic Matrix Composites (CMCs), and Polymer Matrix Composites (PMCs) [5]. Due to their uses in improving stiffness strength toughness temperature tolerance flammability moisture resistance and most notably high thermal and electrical conductivity, MMC's have attracted a lot of attention among these. Nevertheless, they encounter difficulties like interfacial deterioration corrosion and expensive manufacturing [6, 7]. CMCs possess an advantage over MMCs by virtue of being a composite of both metal and nonmetal elements. As a result of this composition they can endure higher temperatures than MMCs and thus better suited for high temperature applications such as modern heat engines. Their exceptional stiffness is indicated by their high elastic modulus values and their low density makes them suitable for lightweight applications [8]. However, CMCs are naturally brittle which could result in structural weaknesses and decreased toughness. Additionally, there are issues with property homogeneity which can lead to lower tensile strengths and limited resistance to mechanical and thermal shock [9, 10]. Although pure polymers frequently lack the mechanical qualities required for structural applications, PMCs are preferred for their versatility. Polymers are mixed with other materials to improve their performance and lessen these restrictions. The lack of high pressure and temperature requirements also makes PMC manufacturing simpler and less expensive which helps explain their quick development and widespread use in structural applications. Recent research has shifted toward developing various PMC's incorporating fiber-reinforced polymers, metal particles such as aluminum, amorphous materials, and nanoparticles derived from glass and ceramics for structural applications. Additionally, natural fiber-based polymer composites have garnered significant attention due to their environmentally friendly properties, including renewability, biodegradability, and minimal environmental impact. One study examined the characteristics of Kevlar and kenaf fiber-reinforced composites, conducting tensile, flexural, double shear, impact, hardness, and delamination tests on laminates prepared via the hand layup technique [11]. The combination of kenaf and Kevlar fibers enhanced all tested mechanical properties of the epoxy composites. Numerous studies have demonstrated that the mechanical properties, including impact, flexural, tensile strength, and hardness, of Kevlar and glass fiber hybrid composites reinforced with various micro and nano fillers—such as carbon nanotubes, graphene nano platelets, and nano silica—exhibit significant enhancements [12, 13].

The potential advantages of aluminum-coated glass fibers combined with basalt fibers, along with the inclusion of silicon carbide as micro-fillers in hybrid composites, have been largely overlooked in existing research. This study examines the morphological behavior and various mechanical properties of hybrid composites with differing percentages of filler particles.

2 Materials and Method

2.1 Materials

In this study, basalt fiber fabric with a weight of 200 grams per square meter (GSM) and aluminum-coated E-glass fiber with a weight of 480 GSM, which features a thin layer of aluminum on one side, were obtained from IND Fibers in Mysuru and Poojan Fibers in Ahmedabad. Araldite LY1564 epoxy resin and Aradur 22,962 hardener were utilized as the matrix system to bind these reinforcements. Silicon carbide was incorporated as a filler to enhance the mechanical properties of the composite [14, 15]. The epoxy and hardener were mixed in a ratio of 4:1 to ensure optimal performance of the matrix. Figure 1 illustrates various materials employed in the fabrication of composite materials.

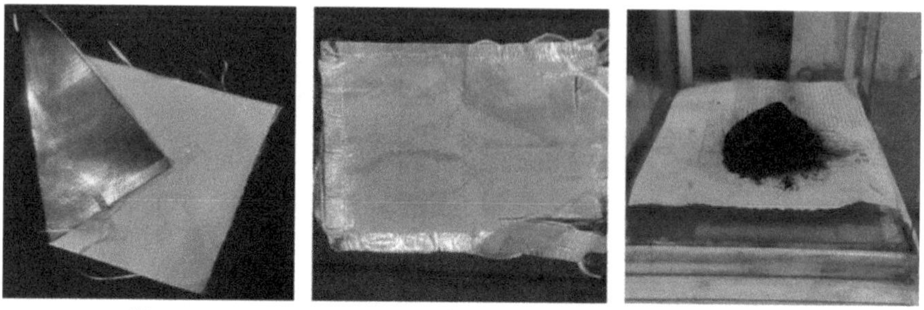
(a) (b) (c)

Fig. 1 **a** Aluminum-coated glass fiber, **b** basalt fiber, and **c** silicon carbide micro-fillers

2.2 *Fabrication Process*

Initially, fibers were cut to dimensions of 200 mm × 250 mm using scissors to conform to rectangular specifications, thereby facilitating the extraction of the matrix from the laminates. To prevent the weaving of threads, the edges of the fibers were secured with tape. Prior to assembly, both the work area and the molding plates were meticulously cleaned with acetone to eliminate any contaminants or debris that might adhere to the laminates. The reinforcement-to-matrix ratio was consistently maintained at 60: 40. Epoxy and silicon carbide micro-fillers were mixed for 1 h utilizing a magnetic stirrer to ensure uniform dispersion within the epoxy matrix. Employing the traditional hand layup technique, the fibers were arranged layer by layer, with the epoxy matrix functioning as a binding agent. Following the stacking process, weights were applied to compress the laminates and facilitate the extraction of the epoxy, after which a curing period of 24 h was observed. Upon completion of the curing process, the laminate plates were subjected to Abrasive Water Jet cutting to accurately produce samples in accordance with ASTM standards. Figure 2 illustrates the procedures employed to fabricate aluminum-coated glass epoxy basalt fiber composite laminate utilizing high temperature epoxy adhesive ARALDITE 1564 and hardener 22,962 as a curing agent, with silicon carbide incorporated as a filler, through the conventional hand layup technique.

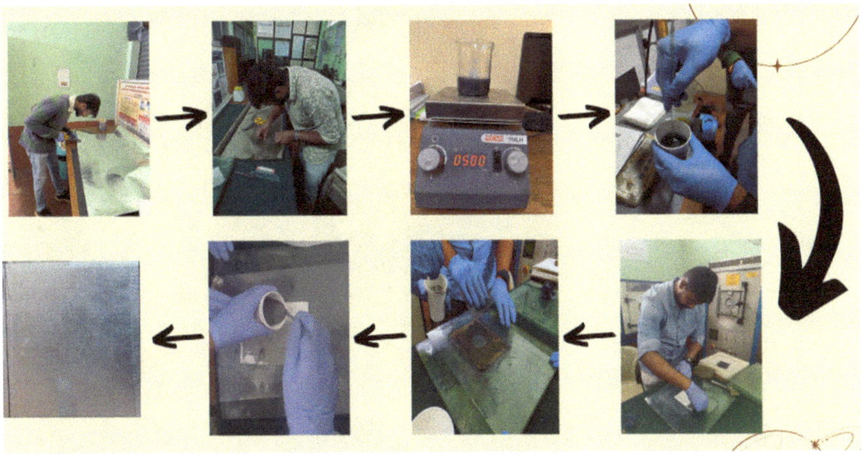

Fig. 2 Fabrication process of composite laminates utilizing the hand layup technique

3 Characterization of Fabricated Specimen

3.1 Mechanical and Morphological Characterization

3.1.1 Tensile Test

The specimen is prepared in accordance with ASTM D 638 standards using a Universal Testing Machine (UTM). Generally, the tensile test evaluates a material's resistance to forces that aim to separate it, as well as its capacity to elongate prior to fracturing [16]. The tensile modulus, which signifies material stiffness, can be ascertained through the stress-strain diagram. The specimens were positioned horizontally within the clamps of the testing apparatus, as illustrated in Fig. 3.

3.2 Flexural Test

The specimen is prepared in accordance with ASTM D790 standards, which regulate the assessment of flexural properties. Flexural strength quantifies a material's resistance to bending forces applied perpendicularly to its length [17]. This stress, induced by the bending load, encompasses both compressive and tensile components. The determination of the material's flexural strength is conducted using a Universal Testing Machine (UTM), which applies controlled bending forces to the specimen. This testing methodology yields critical data regarding the material's ability to endure bending, which is essential for evaluating its structural integrity and suitability for

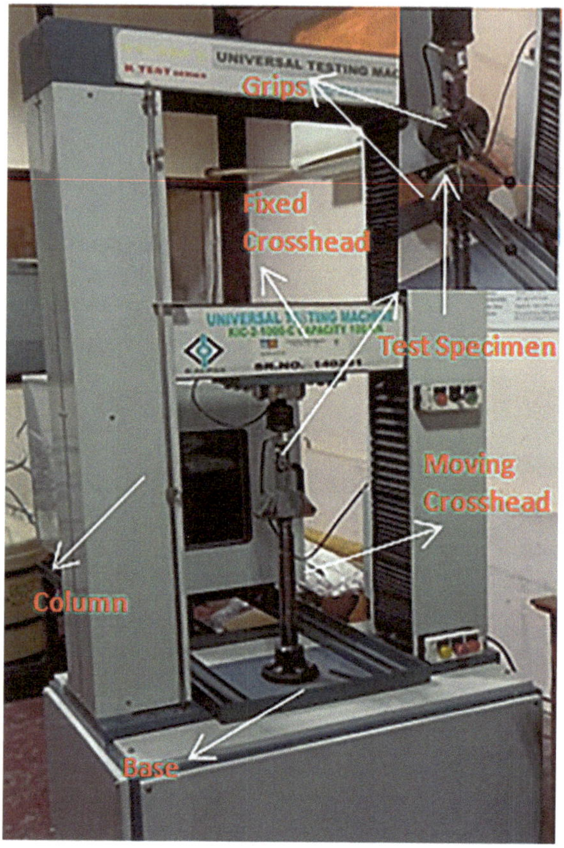

Fig. 3 Tensile test using UTM

various applications. Figure 4 represents the positioning of the test sample using fixture support within the UTM.

3.2.1 Impact Test

In order to assess the impact resistance of a hybrid epoxy composite, samples are fabricated in accordance with ASTM D256 standards, with dimensions measuring 60 mm in length and 12.5 mm in width [18]. A V-notch is precisely created in each sample utilizing a notch cutter. These prepared samples are then properly placed on an Izod impact tester and clamped vertically for the testing process. The tester hits the samples V-notch with a pendulum that has a hammer attached. As shown in Fig. 5, the digital readout meter records the energy that the hybrid composite absorbs during the fracture stage. These numbers demonstrate the hybrid samples ability to withstand abrupt shocks which is an important consideration when assessing strength and resilience in different applications.

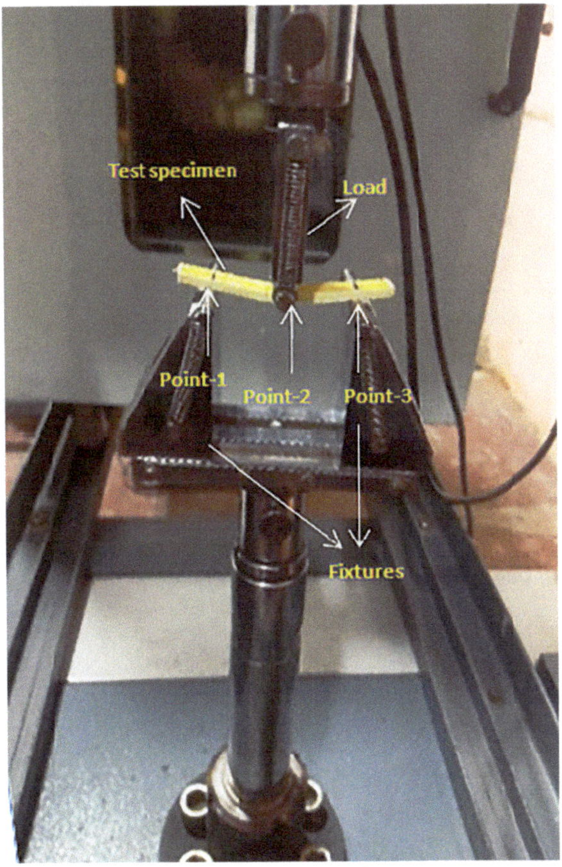

Fig. 4 Three point flexural test using UTM

3.2.2 Hardness Test

On the prepared hybrid composite samples the Shore D hardness test was conducted in compliance with ASTM standard D2240 [19]. The hardness value obtained from this test is usually determined by the indentation made by the sharp point of a Shore D tester with a perfectly flat tip. Indicating the hardness number of different hybrid composites and other reinforced plastics, the Shore D tester provides a scale from 0 to 100. Because strength and hardness are crucial factors in many structural applications, the hybrid composites resistance to indentation is reflected in the Shore D test values.

3.2.3 Wear Characterization

In accordance with ASTM G65, wear testing was carried out using a water washout abrasive wear tester. It is easier to assess how much of the prepared hybrid composite material is eroded over a given period of time by exposure to and mechanical abrasion

Fig. 5 IZOD impact tester

from high-pressure water thanks to the test configuration [20]. The volume loss of the test samples which is a crucial indicator of the hybrid composites resilience in damp conditions is obtained from the results of the abrasive wear test.

3.2.4 SEM Analysis

The prepared hybrid composites with different percentages of silicon carbide (SiC) filler materials were evaluated morphologically using a scanning electron microscope specifically the Zeiss EVO 15 model to determine their impact on the composite's composition [21]. The initial sample preparation involved a thorough cleaning procedure which was followed by an air dryer to remove any remaining moisture. The air-dried samples were then coated with platinum using a JEOL sputter ion coater at controlled voltages of 20 kV and 100 A. The samples were then put in a vacuum chamber to help the platinum coating on the hybrid composite evaporate under vacuum.

4 Results and Discussions

4.1 *Morphological Characterization*

4.1.1 SEM Analysis

The surface behavior and distribution of basalt and aluminum fibers with different proportions of SiC micro-fillers are shown in Fig. 6. The C1 composite which is devoid of SiC fillers showed weaker interfacial bonding with small matrix cracks suggesting that the filler materials and fiber composites had a weaker bond. Their applicability may be limited due to decreased impact strength and hardness values caused by such weak bonding. On the other hand, the fiber composites in the C3 composite which has 0–75% SiC fillers showed agglomeration which might be related to the filler particles excessive adherence. Its use in particular structural applications was restricted by this phenomenon which also caused small cracks in some places. Additionally, the C2 composite which contained 0–5% SiC filler material showed a more balanced phase distribution and no visible cracks indicating that the filler particles and the aluminum and basalt fibers were effectively bonded. Because of the improved mechanical qualities brought about by this increased adhesion its potential uses in a wider range of structural contexts are increased.

4.2 *Mechanical Characterization*

4.2.1 Tensile Test

Tensile test results for hybrid composite samples with different proportions of SiC filler particles are shown in Table 1. The data show that in contrast to the other two samples the sample with 0–5% filler material showed an increase in both peak load and ultimate tensile strength values as the percentage of SiC fillers in the composites increased. This discrepancy may be attributed to the presence of micro cracks in the samples with 0% and 0.75% fillers. The average maximum tensile strength recorded for the sample containing 0.5% SiC fillers is 191.10, suggesting that a moderate proportion of filler particles in the hybrid composites enhances their tensile properties, thereby facilitating their application in various automotive and structural contexts. Conversely, the sample with 0.75% fillers demonstrated the lowest tensile strength value, which may be a result of excessive adhesive bonding of the filler material within the composites, leading to increased brittleness and a reduction in tensile properties [22]. Furthermore, from Fig. 8, it is evident that among all the samples, the sample with 0.5% SiC filler material demonstrated consistent results, suggesting that a moderate amount of SiC filler enhances the mechanical bond between composite fibers, thereby improving their performance under tensile loads, which is a crucial property for numerous automotive applications. Figure 7

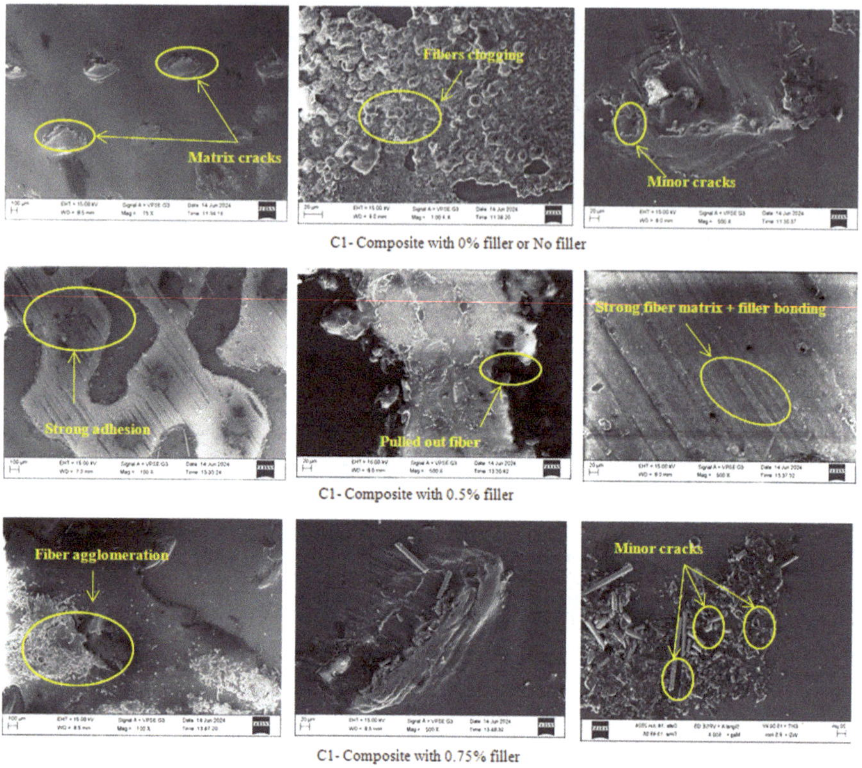

Fig. 6 SEM analysis of 3 composite laminates of varying filler material concentration C1- (A + B + A) + 0%, C2- (A + B + A) + 0.5% and C3- (A + B + A) + 0.75%

indicates one of the specimens prior to and following the execution of the tensile test.

4.2.2 Flexural Test

Table 2 provides the detailed results from flexural tests conducted on samples labeled with different filler compositions. Each sample was uniquely identified and assessed for its cross-sectional area, 3-point bend modulus as a measure of stiffness during bending, and 3-point bend flexural strength indicating the maximum stress endured before fracture in a three-point bend test, changes in the before and after the test are shown in Fig. 9. The bending modulus values are reported as 8151.27 MPa for the composition without SiC filler, 8034.94 MPa for the composition with 0.5% SiC filler, and 5628.1 MPa for the composition with 0.75% SiC filler. The highest bending modulus is observed in the laminate without any SiC filler, suggesting that the absence of this filler allows for better load distribution and stiffness within the composite structure. From Fig. 10, it is evident that as SiC content increases to

Table 1 Tensile test findings

Laminates	Trials	Cross section area (mm^2)	Peak load (N)	Stress (N/mm^2)	Strain	Ultimate tensile stress(N/mm^2)
(A+B+A)+0%	1	14.64	2805.032	114.963	0.145	191.159
	2	14.32	2790.03	108.764	0.102	181.851
	Avg	**14.48**	**2797.53**	**111.86**	**0.124**	**186.51**
(A+B+A)+0.5%	1	20.39	4281.29	116.757	0.216	192.256
	2	20.98	3985.823	113.936	0.109	189.941
	Avg	**20.68**	**4133.55**	**115.34**	**0.163**	**191.10**
(A+B+A)+0.75%	1	22.39	3282.779	93.82	0.207	156.44
	2	22.78	3092.84	89.78	0.105	148.324
	Avg	**22.58**	**3187.81**	**91.8**	**0.156**	**152.39**

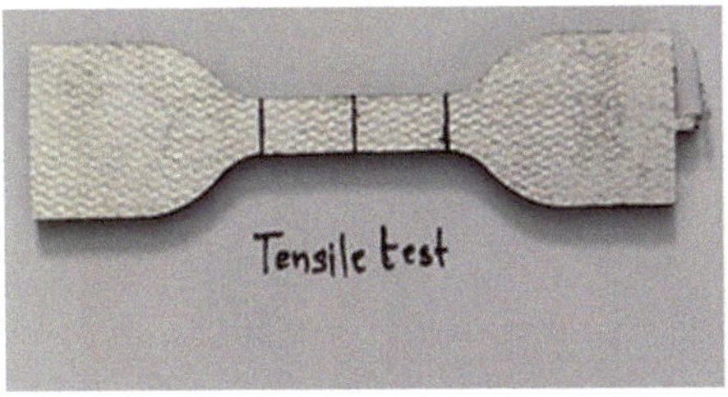

(a)

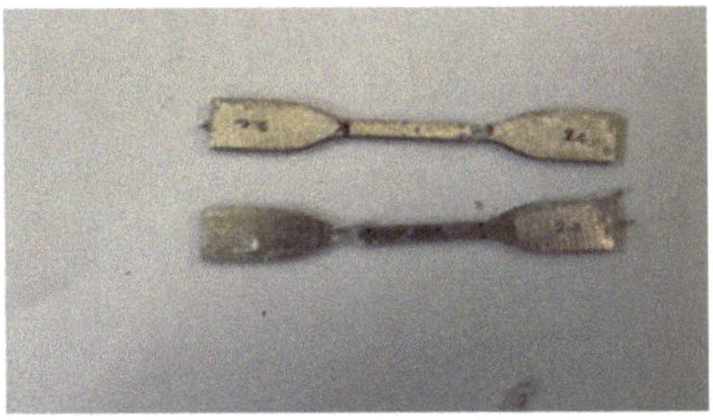

(b)

Fig. 7 **a** Specimen before tensile testing, **b** specimen after tensile testing showing the elongation

0.5%, there is a slight decrease in bending modulus, indicating that while some reinforcement may be achieved through the addition of SiC, it may also introduce stress concentrations or alter the matrix properties negatively. A more significant drop in bending modulus is noted at 0.75% SiC content, which could suggest that excessive filler material disrupts the fiber-matrix bonding or leads to agglomeration within the composite, thereby reducing overall stiffness. The flexural strength values are recorded as 213.15 MPa for no SiC filler, 195.47 MPa for 0.5% SiC filler, and 145.51 MPa for 0.75% SiC filler. Similar to bending modulus trends, flexural strength decreases with increasing amounts of SiC filler; this trend suggests that while fillers can enhance certain properties like wear resistance or thermal stability, they may compromise structural integrity under flexural loads when present in higher concentrations [23]. The highest flexural strength corresponds to the composition without any SiC addition, reinforcing the idea that optimal fiber reinforcement is crucial for maintaining mechanical performance.

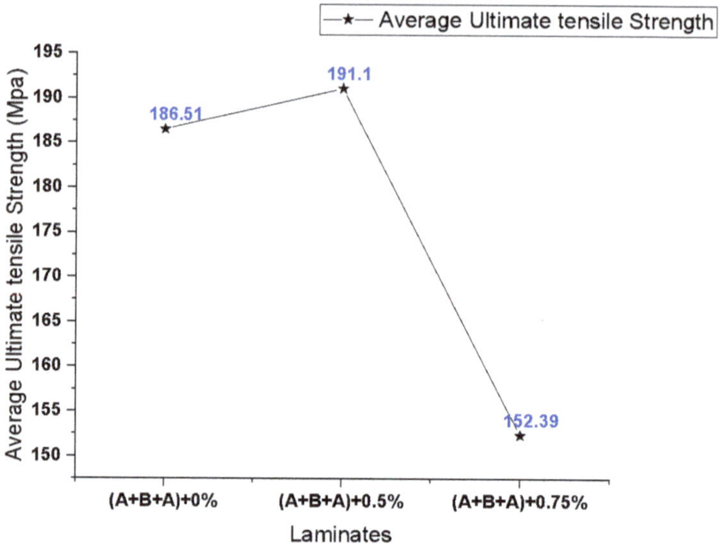

Fig. 8 Variation of ultimate tensile strength with the percentage variation in the SiC filler material

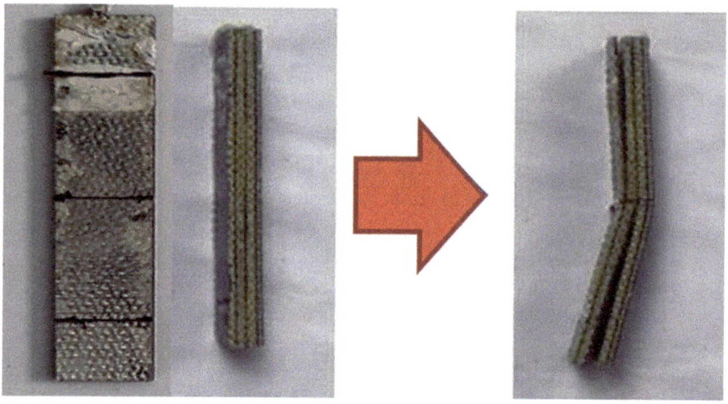

Fig. 9 Flexural test specimens before and after testing

4.2.3 Impact Test

Table 3 summarizes the results from impact tests conducted on samples labeled with different filler compositions. Each sample was identified by a specific number and evaluated for its energy absorption and impact strength. Analysis of the data reveals variations in impact strength across different samples and filler compositions. Figure 11 shows the specimen before and after the Izod impact test. The impact strength of the laminate containing 0.5% filler has been recorded at a notable value of 133.16 KJ/m^2, which is significantly comparable to that of the laminate without

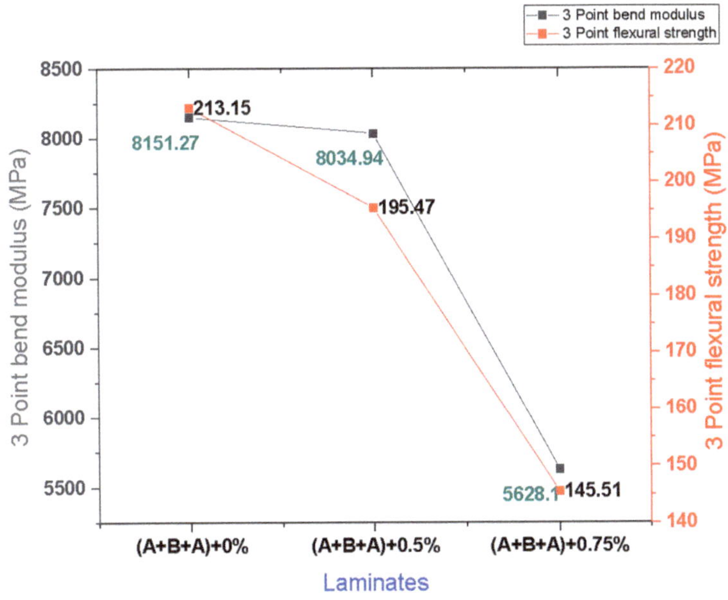

Fig. 10 Variation of flexural strength and bending modulus for different specimen laminates

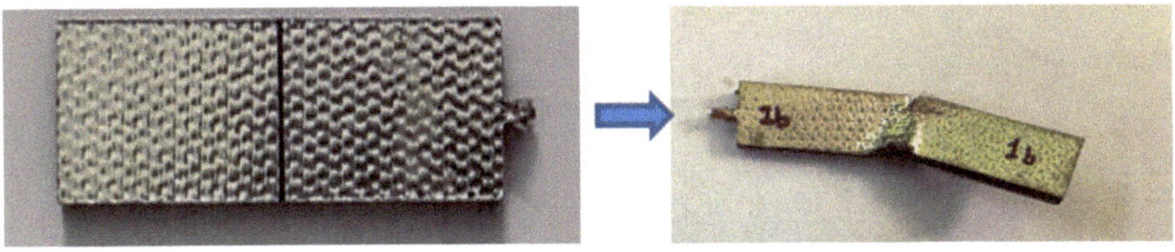

Fig. 11 Izod impact test specimen before and after testing

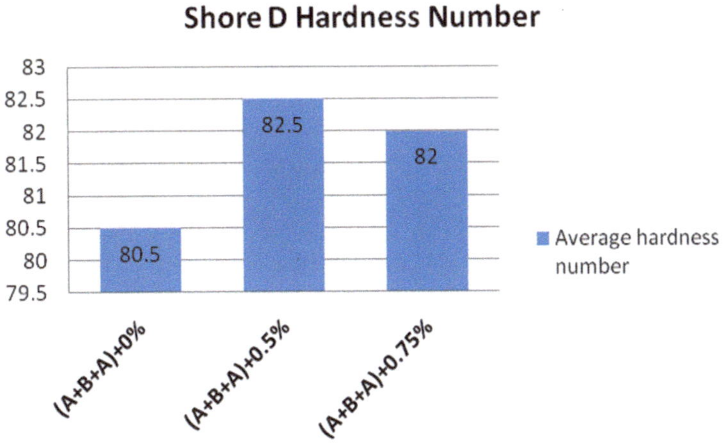

Fig. 12 Shore D hardness number

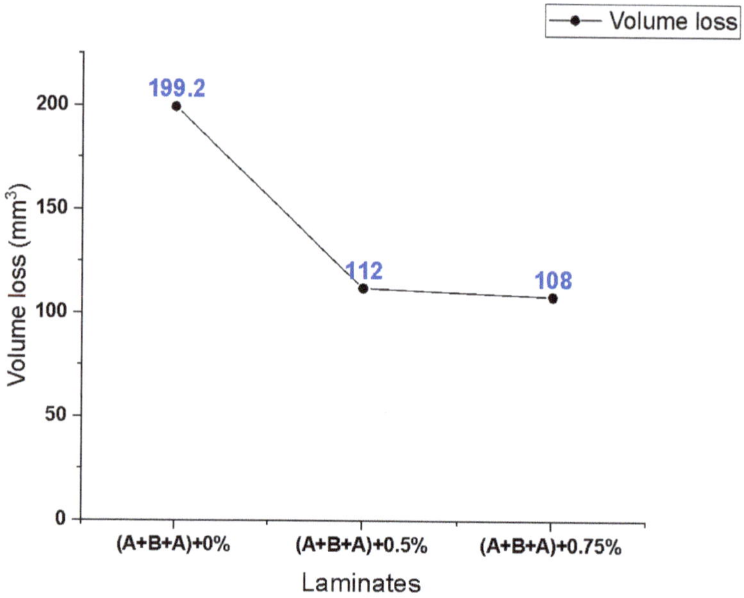

Fig. 13 Shore D hardness number

Table 2 Flexural test results

Laminates	Trials	Cross section area (mm^2)	3 Point bend modulus(MPa)	3 Point flexural strength (Mpa)
(A+B+A)+0%	1	30.256	4246.38	227.66
	2	31.62	12056.16	198.64
	Avg	**30.938**	**8151.27**	**213.15**
(A+B+A)+0.5%	1	46.25	997.51	203.68
	2	46.34	15072.37	187.26
	Avg	**46.295**	**8034.94**	**195.47**
(A+B+A)+0.75%	1	47.25	3357.83	166.61
	2	47.38	7898.37	124.41
	Avg	**47.315**	**5628.1**	**145.51**

filler material. This observation indicates an enhanced resistance to fracture under impact conditions. The results underscore the advantageous effects of silicon carbide reinforcement in improving the material's impact resistance for engineering applications [24]. Furthermore, these findings provide valuable insights into the material's ability to endure impact forces, which is crucial for assessing its appropriateness for applications that necessitate resistance to sudden loads or shocks.

Table 3 Impact test results

Laminates	Trials	Impact energy (J/m)	Absorbed energy (J/m)	Impact strength(KJ/m^2)
(A+B+A)+0%	1	3.35	1372	137.9
	2	3.12	1296	134.4
	Avg	**3.235**	**1334**	**136.15**
(A+B+A)+0.5%	1	4.5	1200	122.32
	2	5.3	1473	144
	Avg	**4.9**	**1336.5**	**133.16**
(A+B+A)+0.75%	1	3.9	1139	139.4
	2	3.45	10029	102.2
	Avg	**3.675**	**5584**	**120.8**

4.2.4 Shore D Hardness Test

Figure 12 illustrates the results of hardness tests performed on samples characterized by various filler compositions. Each sample, designated by a specific numerical identifier, was evaluated for its average hardness number, which reflects the material's resistance to indentation or scratching. From Fig. 12, the observed increase in hardness from 80.5 to 82.5 indicates that the incorporation of SiC filler contributes to the enhancement of the rigidity and strength of the composite laminate, attributable to the intrinsic properties of SiC as a hard material. Nevertheless, at a concentration of 0.75% SiC filler, there is a marginal decrease in hardness back to 82, suggesting that exceeding a certain threshold of filler may result in complications such as inadequate dispersion within the matrix or diminished bonding between the fibers and the matrix. A higher Shore D hardness is generally associated with enhanced wear resistance and durability, rendering these composites appropriate for applications where mechanical stress and abrasion are significant concerns [25, 26].

4.2.5 Abrasive Wear Analysis

From Table 4, it is evident that the composite laminate with no SiC filler shows the highest volume loss of 199.2 mm^3. This indicates that without the addition of SiC, the material is more susceptible to abrasive wear, likely due to lower hardness and wear resistance compared to composites with added fillers. The introduction of 0.5% SiC significantly reduces the volume loss to 112 mm^3, demonstrating an improvement in wear resistance as indicated in Fig. 13. Silicon Carbide is known for its hardness and abrasive properties, which contribute to enhancing the overall durability of the composite laminate by providing additional reinforcement against wear [27, 28]. Further increasing the SiC content to 0.75% results in a slight decrease in volume loss to 108 mm^3, indicating that there is an optimal range for SiC inclusion where wear resistance improves but may plateau or slightly decline beyond a

Table 4 Abrasive wear test results

Sample no	Load (N)	Time duration (min)	Volume loss (mm^3)
(A+B+A)+0%	60	25	199.2
(A+B+A)+0.5%	60	25	112
(A+B+A)+0%	60	25	108

certain concentration due to potential issues like agglomeration or reduced matrix bonding. Tailoring of the filler content is necessary for exploiting the benefits of SiC reinforcement, while retaining the integrity of the matrix and overall composite properties of the matrix, which negatively affects the wear resistance. However, excessive filler loading can lead to agglomeration and interfacial degradation, which negatively affects the wear resistance and overall composite performance.

Figure 13 makes it abundantly evident that adding SiC filler material considerably lowers the composite laminates volume loss. This property makes them especially useful for parts that are subjected to high levels of abrasion such as clutches brake pads and other parts that deal with friction.

5 Conclusions

Silicon carbide effect as micro-filler when reinforced with basalt and Al coated glass fiber epoxy composite which are fabricated using traditional hand layup technique are studied in this work. The mechanical characterization such as tensile, impact, hardness, flexural test, and wear test of the fabricated hybrid composite laminates are done t as per ASTM standards.

The following conclusions have been drawn from the above study:

- The tensile results indicate that a moderate percentage of SiC in hybrid composites enhances the average Ultimate Tensile Strength (UTS). Variations in SiC content led to significant reductions in UTS; specifically, the sample without SiC exhibited a 2.5% decrease compared to the sample with 0.5% SiC, while Sample 3 demonstrated a 20.25% reduction relative to Sample 2. These findings suggest that an optimal level of SiC fillers enhances adhesive properties and improves tensile strength, making these composites suitable for various mechanical applications. Similar observations are seen in the flexural strength test.
- The increase in the percentage of silicon carbide (SiC) resulted in a decrease in both flexural strength and bending modulus values. Notably, the sample devoid of SiC demonstrated superior values in comparison to the samples containing 0.5% and 0.75% SiC filler particles. This phenomenon may be attributed to the clogging of filler materials at the center of the samples.
- The results of the impact test indicate comparable impact strength values across all three tested samples. This finding suggests that the inclusion of SiC filler material has enhanced the impact properties of the samples to a certain degree.

- The Shore D hardness tester indicated an improvement in the hardness values for samples containing 0.5% and 0.75% filler material, in comparison to the sample devoid of filler material. Specifically, there was an approximate increase of 2.5% in the hardness value relative to Sample 1.
- The abrasive wear test demonstrated that Sample 1 (Without SiC filler) exhibited a significant volume loss which is 43.77%, 45.78% greater than that observed in Samples 2 and 3, respectively. This finding indicates that the incorporation of SiC particles plays a substantial role in enhancing the wear resistance of the produced hybrid composites.
- The morphological study indicates that the incorporation of varying SiC fillers resulted in homogeneous composites, devoid of microcracks and fiber clogging, while enhancing interfacial bonding. Among the composites examined, the one containing 0.5% SiC demonstrated exceptional bonding between the filler particles and the aluminum and basalt matrix, exhibiting minimal cracking.

The composite with 0.5% SiC showed enhanced mechanical properties, including improved tensile strength, flexural strength, impact resistance, hardness, and wear resistance. These characteristics make it a suitable substitute for applications requiring high performance under mechanical stress, such as automotive components, aerospace structures, or industrial machinery parts. It has been also observed that by incorporation of SiC fillers demonstrated superior resistance to abrasion, making it particularly advantageous for components exposed to high levels of wear and friction, such as clutches, brake pads, and other friction-related parts. In addition, compared to pure aluminum and other aluminum-based composites it has been found that adding basalt fibers to an aluminum matrix can greatly increase the composites tensile strength flexural strength impact strength and hardness. These findings render them suitable for a diverse range of industrial applications across varying environmental conditions.

Overall, this study showcased the potential of hybrid composites with optimized SiC content to offer enhanced mechanical characteristics, durability, and performance in demanding engineering applications, opening up possibilities for innovative and efficient material solutions in various industries.

References

1. Al-Furjan MSH, Shan L, Shen X, Zarei MS, Hajmohammad MH, Kolahchi R (2022) A review on fabrication techniques and tensile properties of glass, carbon, and Kevlar fiber reinforced rolymer composites. J Mater Res Technol 19:2930–2959
2. Tilak SR, Pasha SS, Ahmed MN, Daniel S (2021) An experimental investigation of flexural and inter laminar shear stress on hybrid polymer based composites (E glass fibre–Kevlar fibre with Epoxy resin 5052) for different thickness. Mater Today: Proc 46:8991–8994
3. Yuan J, Zhang Z, Yang M, Zhao X, Wu L, Li P, Liu W (2021) Combined effects of interface modification and micro-filler reinforcements on the thermal and tribological performances of fabric composites. Friction 9:1110–1126

4. Abd El-Baky MA, Attia MA, Abdelhaleem MM, Hassan MA (2022) Flax/basalt/E-glass fibers reinforced epoxy composites with enhanced mechanical properties. J Nat Fibers 19(3):954–968
5. Chinnasamy V, Subramani SP, Palaniappan SK, Mylsamy B, Aruchamy K (2020) Characterization on thermal properties of glass fiber and kevlar fiber with modified epoxy hybrid composites. J Mater Res Technol 9(3):3158–3167
6. Suresha B, Vidyashree S, Bettegowda H (2023) Effect of Filler Materials on Abrasive Wear Performance of Glass/Epoxy Composites. Tribol Ind 44(1):111
7. Felipe RCTS, Felipe RNB, Batista ACMC, Aquino EMF (2019) Influence of environmental aging in two polymer-reinforced composites using different hybridization methods: Glass/Kevlar fiber hybrid strands and in the weft and warp alternating Kevlar and glass fiber strands. Compos Part B: Eng 174:106994
8. Ramasamy M, Daniel AA, Nithya M, Kumar SS, Pugazhenthi R (2021) Characterization of natural–Synthetic fiber reinforced epoxy based composite–Hybridization of kenaf fiber and kevlar fiber. Mater Today: Proc 37:1699–1705
9. Maharana SM, Samal P, Dehury J, Mohanty PP (2020) Effect of fiber content and orientation on mechanical properties of epoxy composites reinforced with jute and Kevlar. Mater Today: Proc 26:273–277
10. Mohaideen MM, Anand VK, Dhara A, Goud NUR, Godwin JJA (2023) Experimental and numerical studies on kevlar composites. Proc, Mater Today
11. Suresh A, Bhargavi P, Kumar MK (2021) Simulation and mechanical characterization on kevlar epoxy reinforced composite with silicon carbide filler. Mater Today: Proc 38:2988–2995
12. Sampath V (2006) Studies on mechanical, friction, and wear characteristics of Kevlar and glass fiber-reinforced friction materials. Mater Manuf Process 21(1):47–57
13. Sandeep B, Bharadwaj PN, Ahamed Z, Arvind R (2021) Development and mechanical characterizations of Kevlar/C-glass epoxy hybrid composites. Mater Today: Proc 46:9092–9095
14. Valença SL, Griza S, de Oliveira VG, Sussuchi EM, de Cunha FGC (2015) Evaluation of the mechanical behavior of epoxy composite reinforced with Kevlar plain fabric and glass/Kevlar hybrid fabric. Compos Part B: Eng 70:1–8
15. Mishra BP, Mishra D, Panda P (2020) An experimental investigation of the effects of reinforcement of graphene fillers on mechanical properties of bi-directional glass/epoxy composite. Mater Today: Proc 33:5429–5441
16. Raj SS, Kuzmin AM, Subramanian K, Sathiamoorthyi S, Kandasam KT (2021) Philosophy of selecting ASTM standards for mechanical characterization of polymers and polymer composites. Mater Plast 58(3):247–256
17. Bichang'a DO, Wambua PM, Oyondi EN (2017) Effect of alkali treatment on mechanical properties of sisal woven fabric reinforced epoxy composites
18. Saba N, Alothman OY, Almutairi Z, Jawaid M, Ghori W (2019) Date palm reinforced epoxy composites: tensile, impact and morphological properties. J Mater Res Technol 8(5):3959–3969
19. Dimulescu SC, Nioaţă A, Wetternek M (2024) Analysis of the determination of the hardness of composite materials with rubber matrices with additions of FA and PVC. Fiability Durab/Fiabil Si Durabilitate 33(1)
20. Yanes RE, Hernandez LN, Morera OZ, Olivier NC, Neto AF (2014) Design and fabrication of a machine for test in abrasive wearing according to ASTM G65 standard. Am J Mater Sci Appl 2(5):86–90
21. Cui L, Marchand É (2015) Scanning electron microscope calibration using a multi-image non-linear minimization process. Int J Optomechatronics 9(2):151–169
22. Karthick L, Rathinam R, Kalam SA, Loganathan GB, Sabeenian RS, Joshi SK, Mammo WD (2022) Influence of Nano−/Microfiller Addition on Mechanical and Morphological Performance of Kenaf/Glass Fibre-Reinforced Hybrid Composites. J Nanomater 2022(1):9778224
23. Rana RS, Buddi T, Purohit R (2021) Effect of SiC reinforcement on the mechanical properties of Kevlar fiber based hybrid epoxy composites. Mater Today: Proc 44:2478–2481
24. Hosseini Abbandanak SN, Abdollahi Azghan M, Zamani A, Fallahnejad M, Eslami-Farsani R, Siadati H (2022) Effect of graphene on the interfacial and mechanical properties of hybrid glass/Kevlar fiber metal laminates. J Ind Text 51(2):2576S–2593S

25. Alsaadi M, Younus B, Erklig A, Bulut M, Bozkurt O, Sulaiman B (2021) Effect of graphene nano-platelets on mechanical and impact characteristics of carbon/Kevlar reinforced epoxy hybrid nanocomposites. Proc Inst Mech Eng, Part C: J Mech Eng Sci 235(23):7139–7151
26. Mangisetty Srinivasaiah A, Nagaraja S, Munipapanna A, Bellary Gopalakrishna S, Krishnappa M, Hassan Venkataramaiah H (2024) Influence of nano graphene filler on hardness, impact strength and density of glass epoxy composites. Eng Rep 6(12):e12973
27. Megahed AA, Agwa MA, Megahed M (2018) Improvement of hardness and wear resistance of glass fiber-reinforced epoxy composites by the incorporation of silica/carbon hybrid nanofillers. Polym-Plast Technol Eng 57(4):251–259
28. Rangaswamy H, Chandrashekarappa MPG, Pimenov DY, Giasin K, Wojciechowski S (2021) Experimental investigation and optimization of compression moulding parameters for MWCNT/glass/kevlar/epoxy composites on mechanical and tribological properties. J Mater Res Technol 15:327–341

Mechanical Behavior Optimization of Carbon Fiber-Graphene-Reinforced Polyamide Composites in 3D Printing

Mr. Shivaraj Dharennavar, A. Anand, and T. P. Jeevan

Abstract This study is focused on evaluating the mechanical properties of 3D printing filament polyamide + 15% carbon fiber + 0.5%graphene composite polymer. This filament aims in leveraging the mechanical properties of carbon fiber and enhanced conductivity of graphene. Also this study involves assessing how these materials respond to mechanical forces and stresses. Design of Experiment—RSM is used to minimize the experimental time and cost. The tensile specimen which is the sample specimen used in this study for investigating the mechanical properties are printed according to the ASTM standard and tested in UTM (Universal Tensile Machine). The printing processes are fused deposition modeling (FDM), with the parameters layer height (LH), printing temperature (PT), and printing speed (PS) are considered according to design of experiment. In total 15 tensile test specimens were printing using 3D printer with the polyamide filament to evaluate the mechanical properties of the composite polymer. Therefore test sample T11 as maximum YM of 97.316 MPa and UTS of 19.503 MPa compared to other specimens printed.

Keywords 3D printing · Polymer composites · Response surface method

1 Introduction

3D printing or additive manufacturing is a process of joining of material to make an object or part of an object from 3D model designed using design software layer by layer. AM which is recently widely known has rapid prototyping is used in many industries, medical, and for research purpose. 3D printing provides the possibility of

Mr. S. Dharennavar (✉)
R N S Institute of Technology, Bengaluru, India
e-mail: shivarajdharennavar@rnsit.ac.in

A. Anand
The National Institute of Engineering, Mysuru, India

T. P. Jeevan
Malnad College of Engineering, Hassan, India

I. M. Jamadar et al. (eds.), *Integrating Sustainability in Modern Technologies*, Lecture Notes in Mechanical Engineering, https://doi.org/10.1007/978-981-95-9466-5_30

low volume and high cost-efficient as well as customized parts with extraordinary features such as multilateral fabrication and fabrication of complex structures. In recent days there has been so much advancement in AM like the design and printing of materials in 3D printer has reduced from days to hours [1]. There are many various applications of AM. There are biomedical applications like the 3D printing is used in medical devices and scaffolds for intense study of tissue engineering, drug delivery, etc. Many studies and researches have shown that AM can be used for converting implants through available scanning basically CT/MRI. There are also machine learning applications in AM and it is popular in aerospace, automotive industry for vehicle interior manufacture, and sports equipment. The main objective of AM is to reduce the production cost and improve overall efficiency of the product [2].

Primarily composite polymers are a multi-phase material in which reinforcing fillers are integrated with polymer matrix. Composite polymers have enhanced mechanical properties; they combine properties of various materials like carbon fibers, nanoparticles or glass fibers that help in improvement of mechanical strength, stiffness, and durability. They have high strength-to-weight-ratio which makes them light weight. The polymer used in this study is polyamide +15% carbon fiber +0.5% graphene to characteristics of its stiffness and durability.

1.1 Basic Steps in AM Process [1, 2]

There are four basic steps involved in the process of AM:

- First step involves in sketching the required model in 3D computer aided design software (CAD).
- Then the model is saved as an STL file format. STL stands for stereolithography tessellation language. For sending it into a slicing software, and generate g codes for the model.
- The sliced file will then be sent to an additive manufacturing device. Higher the number of slicing, better the finish of the final product.
- The printer will then print one layer on top (2D) of the other, and thereby forming a 3D object in process. Postprocessing of the finished product may be done if needed.

1.2 Literature Review

The process of creating a 3D model by applying successive layers of material under computer control is known as additive manufacturing. Additive manufacturing creates the thing by adding material, as the name suggests. Form-fit prototypes to fully functional objects are among the models that AM can produce. Because AM uses less material, requires less maintenance, produces components quickly, requires less monitoring, and has better tolerance, it can be simply applied.

Aveen K P et al. investigated into the mechanical characterization and 3D printing of polylactic acid and components made of polylactic acid that were filled with bronze [3]. In this investigation, samples were created utilizing a desktop 3D printer using commercial grade PLA and bronze-filled PLA. The samples' mechanical characteristics, including their tensile and flexural strengths, were examined. The test samples, which had different layer heights and build-up speeds, were made via 3D printing. Based on the experimental findings, PLA and bronze-filled PLA are compared, and it is discovered that PLA has better tensile and flexural qualities than bronze-filled PLA.

Babagowda et al. [4] investigated the effects of blending PLA filament with recycled PLA materials on its mechanical properties. Characterizing and optimizing elements such as recycled PLA building material and layer thickness is the goal of this study. Tensile and flexural or bending tests are performed to assess the printed specimen's mechanical response properties. Analysis of variance (ANOVA) is used to assess each parameter's effectiveness after a set of parameters that yield positive results for particular response characteristics are determined using the Taguchi S/N ratio.

The effect of printing orientation on the strength of 3D-printed ABS polymers was examined by Linlin Cai et al. [5]. Researchers have looked into how printing orientations affect the mechanical strength of ABS (Acrylonitrile Butadiene Styrene) parts made with fused deposition modeling (FDM) technology. According to the parameters derived from the samples' stress–strain curves, the 0-degree printed sample has the strongest mechanical characteristics. This is probably because the individual slices have the correct orientations.

The tensile strength, flexural strength, and impact energy of 3D-printed items made with FDM technology and PLA-graphene raw material were examined by Camargo et al. [6] using a statistical method known as central composite design, or CCD. The layered production method is the cause of the anisotropic behavior of 3D-printed parts. Throughout the trials, the honeycomb infill design and flat orientation were kept constant. The findings demonstrated that mechanical properties improve with an increase in the linear layer thickness parameter. Each test showed a different behavior for the linear infill parameter. Whereas impact energy decreased with increasing infill, tensile and flexural strengths increased.

Gareth Chapman et al. conducted an investigation into the 3D printability of novel impact modified nylon 6: experimental investigations and performance evaluation [7]. In this study, the 3D printing procedure optimizes the modified nylon's bed temperature, cooling speed, printing temperature, and print speed. Blends made of 3D-printed nylon 6 have an Izod impact strength of 543, which is significantly higher than the 48 J/m value for injection-molded pristine nylon 6 that has been documented in the literature.

Patel et al. performed a review research on the usage of composite filament from 3D printers in fused deposition modeling [8]. In this experiment, the utilization of composite materials in a 3D printer based on fused deposition modeling produced positive outcomes in terms of mechanical qualities and cost effectiveness. Numerous opportunities arise from the study and use of composite materials in 3D printing for the long-term manufacturing of working parts and prototypes. According to the report, FDM applications include composite filaments with continuous, discontinuous, carbon, glass fiber-reinforced material, core–shell rubber, and matrix and mix components composed of ABS, PLA, HIPS, nylon, and other materials.

Vikas et al. [9] optimized the 3D printing process parameters of polylactic acid materials using the fused deposition modeling process. This study aims to determine how the mechanical properties of PLA specimens are impacted by printing temperatures, speeds, layer thicknesses, and filling ratios. The orthogonal test of four factors and three levels for different filling ratios and the orthogonal test of 3 variables and 3 levels for a 100% filling ratio were used to select the specimens. The tensile test was conducted both with and without filling ratio (100%) as well.

In their study, Özçelik and Özbay investigate how different injection settings and molding materials affect the mechanical properties of polypropylene (PP) products. The study provides information on how to optimize the injection molding process to enhance product quality by using the appropriate mold materials and processing conditions [10]. A thorough assessment of the literature led to the identification of polyamide, 15% carbon fiber, and 0.5% graphene as reinforcement for the current investigation. Understanding how printing factors affect the mechanical properties of 3D-printed polymer composites is the aim of this work.

2 Methodology

2.1 *Production of Composite Filament*

The twin screw extruder technology is used to get the composite materials of polyamide + 15% carbon fiber + 0.5%graphene. To achieve better bonding at the fiber-matrix interface, 15 wt. % carbon fiber and 0.5% graphene reinforcement should be optimal. Any further enhancement of reinforcements will alter the 3D printable properties of the polyamide material by making it brittle.

2.2 *Response Surface Method*

Design of experiment involves defining the objective, stating the problem for solution, determining the input or process parameters to get the required output for the experiment conducted, and choosing of appropriate design suitable for the experiment based on the objectives and nature of factors of convenience. Here in this study, we have used Response Surface Method (RSM) for evaluating the mechanical properties [11, 12].

Response surface method is a statistical technique used in the field of design of experiments (DOE) to optimize and improve processes or system. RSM method is used in this study to reduce time consumption and cost of experimentation. RSM is gathering the statistical and mathematical techniques for modeling and analysis of problem for which response of interest is influenced by several variables and the objective is to optimize this response [13].

2.3 *Preparation of Tensile Specimens*

In accordance with ASTM guidelines, tensile specimens were created using either fused filament fabrication (FFF) or fused deposition modeling (FDM). The tensile specimens were first designed in 3D modeling software, and then the designs were exported in STL format. The model was subsequently divided into printable layers using slicing tools that processed these STL files. Lastly, the specimens were printed on an FDM 3D printer using the provided files.

According to the ASTM D638, tensile test samples are prepared with the dimensions of 115 × 19x2.5 mm [14]. Table 1 represents the control input parameters based on RSM with level setting.

The three control parameters, viz., layer height (LH), printing temperature (PT), and printing speed (PS) are considered as the variable of FFF 3D printers which can result in the notable modifications to mechanical properties of the 3D-printed samples. Among three selected parameters, layer height (LH) can impact the 3D printing process with respect to printing speed and resolution. The lower values of LH result in greater details but increases the print times. On the other hand, the printing temperature needs a carefully adjustment because it may affect material flow and adherence. The printing speed will affect the accuracy and production time based on time taken for material deposition. For the purpose of methodically testing and optimizing their impacts the control parameters are varied in the three levels, viz., low, medium, and high in the methodological way to find the ideal parameters level setting for the highest process performance.

The impacts of control parameters (LH, PT, and PS) on the mechanical strength associated with 3D printing are investigated in an experimental design matrix using response surface methodology (RSM). In order to determine how

Table 1 Input control parameters

	Symbol	Level		
Control parameters		1	2	3
Layer height	LH	0.16	0.21	0.26
Printing temperature	PT	240	250	260
Printing speed	PS	30	50	70

different combinations of these parameters affect mechanical strength of the 3D-printed polymer composite, the matrix comprises 15 experimental runs as shown in Table 2 that is tested.

Table 2 starts with combinations, which tests the parameters' boundary values (such as the minimum and maximum levels). Examples include runs T1, T2, T3, and T4, which combine extreme printing temperature and layer height values at a steady speed. These assist in determining the factors' baseline impacts at their extremes.

The middle levels of all parameters are reliably referenced by repeating central points, such as runs T13, T14, and T15. The detection of nonlinear connection is made possible by repeats, which also increase statistical reliability. Additionally, the design is extended to test values that are somewhat outside of the center levels by axial points, such as those found in runs T9 through T12. This enables the creation of a quadratic model and aids in capturing curvature in the response surface. For obtaining the optimum level setting of LH, PT, and PS, the CAD model of the required test specimen was design based on ASMT D638 standard using Solid works 2022 modeling software.

CREALITY Ender -3 printer with 0.4 mm of nozzle diameter is used to develop the test specimens according to DOE orthogonal array experimentation as shown in Table 2. Dog bone shape is selected to print the tensile specimen, as shown in Fig. 1. The input parameters are selected based on orthogonal array for printing the specimens. Furthermore, the bed temperature of 60 °C is kept constant during printing process.

Table 2 Experimental table using RSM

Sl. no	Specimen number	Layer height (mm)	Printing temp (^{0}C)	Printing speed (mm/sec)
1	T1	0.16	240	50
2	T2	0.16	260	50
3	T3	0.26	240	50
4	T4	0.26	260	50
5	T5	0.21	240	30
6	T6	0.21	260	30
7	T7	0.21	240	70
8	T8	0.21	260	70
9	T9	0.16	250	30
10	T10	0.26	250	30
11	T11	0.16	250	70
12	T12	0.26	250	70
13	T13	0.21	250	50
14	T14	0.21	250	50
15	T15	0.21	250	50

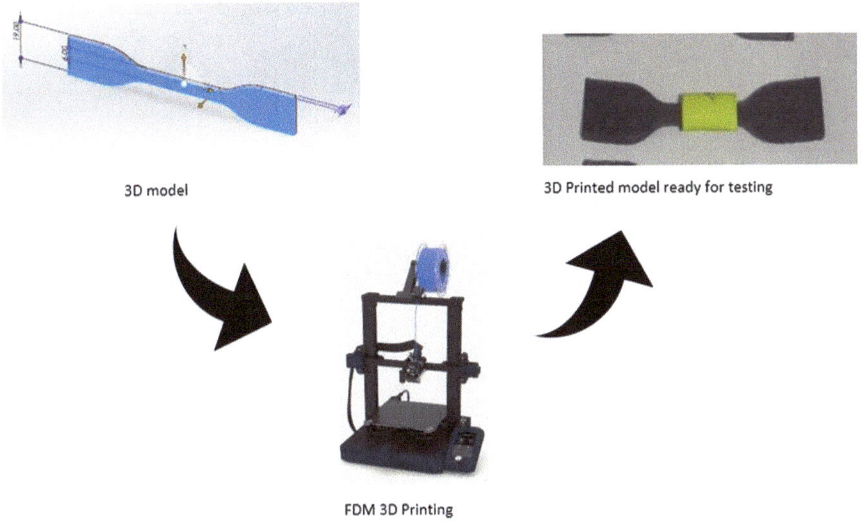

Fig. 1 Process of 3D printing the test specimen: schematic representation

3 Results and Discussion

The prepared tensile specimens were tested using universal testing machine (UTM) of make kalpak instruments and control of 100 KN capacities in Centre for Materials Research (CMR) lab at the National Institute of Engineering, Mysuru. The testing is performed in accordance to the experimental table developed using RSM technique. The test results are as shown in Table 3.

The mechanical characteristics of 15 distinct material specimens (T1 through T15), each tested under varied 3D printing conditions, are shown in this table. It gives important information on each specimen's cross-sectional area, ultimate tensile strength, strain, stress at 60% of peak load, elongation, peak load, and Young's modulus. These characteristics are necessary for understanding how the material behaves under stress and deformation, which is crucial for 3D printing process optimization.

For example, elongation gives information on a material's ductility by showing how far it may stretch before breaking, whereas peak load assesses the highest force a material can sustain before failing. Young's modulus measures the material's stiffness, which is important for applications needing rigidity, whereas ultimate tensile strength indicates the material's ability to withstand breaking under tension. The material's behavior under partial loading conditions can be evaluated using the stress at 60% of peak load and strain, which is useful for predicting how the material would function in practical applications.

While examining the specimens, notable differences between the specimens are found. Specimen T6, for example, exhibits a balance between strength and ductility with a high elongation of 33.676% and a peak load of 366.590 N. However, specimen

Table 3 Tensile test results

Material	C/S Area (mm^2)	Peak load (N)	% Elongation	Stress at 60% of peak (MPa)	Strain	UTS (MPa)	YM (MPa)
T1	16.24	296.448	17.209	10.958	0.078	18.256	94.620
T2	18.81	229.956	19.921	7.338	0.079	15.223	73.036
T3	15.12	224.161	28.173	9.692	0.107	13.147	63.646
T4	19.72	311.968	29.715	9.486	0.112	15.814	96.913
T5	18.81	326.889	34.142	10.428	0.114	17.383	62.006
T6	19.13	366.590	33.676	11.497	0.126	17.169	68.762
T7	17.34	220.813	17.179	7.642	0.078	13.733	87.998
T8	15.97	249.792	21.082	9.388	0.093	15.637	92.149
T9	17.36	325.202	23.142	11.242	0.097	18.727	83.250
T10	19.78	325.584	29.848	9.879	0.123	18.461	71.446
T11	17.69	362.646	25.794	12.302	0.095	19.503	97.316
T12	18.98	216.350	18.461	6.838	0.081	12.399	84.117
T13	16.50	442.019	35.500	16.069	0.115	16.780	78.199
T14	18.40	288.257	23.967	9.398	0.094	15.667	79.035
T15	18.70	293.682	24.697	9.427	0.096	15.706	76.165

T12 shows a lower peak load of 216.350 N and an elongation of 18.461%, indicating that it is less flexible and weaker. In order to help engineers make well-informed decisions on material selection and process optimization, this information is essential for finding out the best 3D printing settings for materials that must function under particular mechanical loads.

Figure 4a shows the tensile test specimen that has been 3D-printed and ready for the test. Figure 4b shows the specimen after all the test samples were broken near the center which indicated that the specimen is been printed properly without much printing error from the 3D printer. Whereas Fig. 4c shows the specimen fixed in the UTM machine ready for testing. The data obtained for the testing has been tabulated in Table 3.

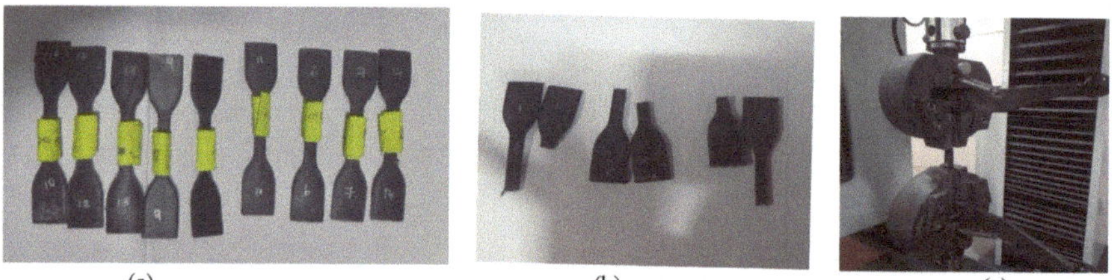

(a) (b) (c)

Fig. 4 Tensile test: **a** tested specimen, **b** specimen after tensile test, **c** test specimen in UTM machine

The results obtained from the DOE analysis with the help of software called minitab using RSM method to analyze and obtain the surface and contour plots between the input parameters for the results obtained from the UTM machine for all the 15 samples.

3.1 ANOVA of Ultimate Tensile Strength (UTS)

The effect of input parameters on the UTS of tensile test specimen is evaluated and analyzed using RSM. The ANOVA table for UTS is as decapitated in Table 5.

The results of an Analysis of Variance (ANOVA) for a statistical model assessing how different parameters, affect on ultimate tensile strength (UTS) are shown in Table 4. With a p-value of 0.002, the model is statistically significant, meaning that the sum of the components accounts for a sizable amount of the response's variation. With a very significant effect (F-value = 54,053, p-value = 0.001), layer height stands out among the main effects and appears to have a significant impact on the UTS. However, with a p-value of 0.448, printing temperature has a negligible impact and is not a relevant factor in this situation. Conversely, printing speed significantly affects and is statistically significant (p-value = 0.001) and has a notable influence on the UTS.

Table 4 Analysis of variance (ANOVA) model for UTM

Source	DF	Adj SS	Adj MS	F-value	p-value
Model	9	59.1921	6.5769	20.33	0.002
Linear	3	31.5821	10.5274	32.53	0.001
LH	1	17.6656	17.6656	54,053	0.001
PT	1	0.2191	0.2191	0.68	0.448
PS	1	13.6974	13.6974	42.33	0.001
Square	3	6.6764	2.2255	6.88	0.032
LH*LH	1	1.1312	1.1312	3.50	0.120
PT*PT	1	2.0137	2.0137	6.22	0.055
PS*PS	1	3.1524	3.1524	9.74	0.026
2-way interaction	3	20.9335	6.9778	21.56	0.003
LH*PT	1	8.1225	8.1225	25.10	0.004
LH*PS	1	11.6896	11.6896	36.13	0.002
PT*PS	1	1.1215	1.1215	3.47	0.122
Error	5	1.6179	0.3236		
Lack-of-fit	3	1.5465	0.5155	14.44	0.065
Pure error	2	0.0714	0.0357		
Total	14	60.8099			

Table 5 Analysis of variance (ANOVA) model for Young's modulus

Source	DF	Adj SS	Adj MS	F-value	p-value
Model	9	1783.21	198.134	9.12	0.013
Linear	3	916.80	305.598	14.07	0.007
LH	1	128.80	128.801	5.93	0.059
PT	1	63.79	63.789	2.94	0.147
PS	1	724.21	724.206	33.34	0.002
Square	3	112.07	37.356	1.72	0.278
LH*LH	1	102.89	102.889	4.74	0.081
PT*PT	1	3.880	3.877	0.18	0.690
PS*PS	1	3.360	3.359	0.15	0.710
2-way interaction	3	754.34	251.447	11.58	0.011
LH*PT	1	752.16	752.158	34.63	0.002
LH*PS	1	0.490	0.487	0.02	0.887
PT*PS	1	1.700	1.697	0.08	0.791
Error	5	108.59	21.719		
Lack-of-Fit	3	104.24	34.745	15.95	0.060
Pure error	2	4.360	2.179		
Total	14	1891.80			

With a p-value of 0.026, printing speed (PS2) has a significant effect, suggesting that speed variations may result in nonlinear responses. With p-values of 0.120 and 0.055, respectively, the quadratic effects of layer height (LH^2) and printing temperature (PT^2) are less significant, suggesting mild nonlinear effects on UTS. The significance of factor combinations is further highlighted by the two-way interactions. In particular, the relationship between layer height and printing temperature (LH*PT) (p-value = 0.004) and layer height and printing speed (LH*PS) (p-value = 0.002) is highly significant. Nevertheless, there is no statistically significant interaction between printing temperature and printing speed (PT*PS) (p-value = 0.122).

Regression equation

$$\begin{aligned}\text{UTS} = & -260 - 750\,\text{LH} + 2.98\,\text{PT} - 0.599\,\text{PS} + 221\,\text{LH} * \text{LH} \\ & - 0.00739\,\text{PT} * \text{PT} + 0.002310\,\text{PS} * \text{PS} + 2.850\,\text{LH} * \text{PT} \\ & - 1.710\,\text{LH} * \text{PS} + 0.00265\,\text{PT} * \text{PS}\end{aligned} \tag{1}$$

Here LH represented as layer height, PT as printing temperature, and PS as printing speed.

The above Eq. (1) is a regression model that explains the quadratic and interaction effects of parameters on ultimate tensile strength (UTS). To represent the intricate interactions between these parameters and UTS, the model incorporates both linear and nonlinear elements. When all factors are at zero, the baseline UTS is represented by the constant term (−260). The direction and strength of the impacts of layer height, printing temperature, and printing speed on UTS are shown by their coefficients. In particular, UTS is greatly impacted negatively by layer height, increased by printing temperature, and decreased by printing speed. From Table 4, it is evident that the LH and PS are the most significant factors on UTS of the specimens.

From ANOVA it is evident that LH & PS are the more significant parameters affecting that variation of UTS of the printed specimens. Equation (1) represents the regression equation with R-squared value of 97.34%.

Figure 5a shows the normal probability plot of residuals which gives the percentage error of design. The residuals are very close to the standard line which implies that the design has less error for UTS. Figure 5b and c observed the surface plot between UTS verse LH, PT, and PS; it is evident that as LH and PS are more significant parameters for UTS; the variation in the PT does not have much effect. Figure 5d shows the contour plot between UTS and the input parameters LH and PT. The highest UTS can be observed with minimum values of LH and PT between the ranges selected.

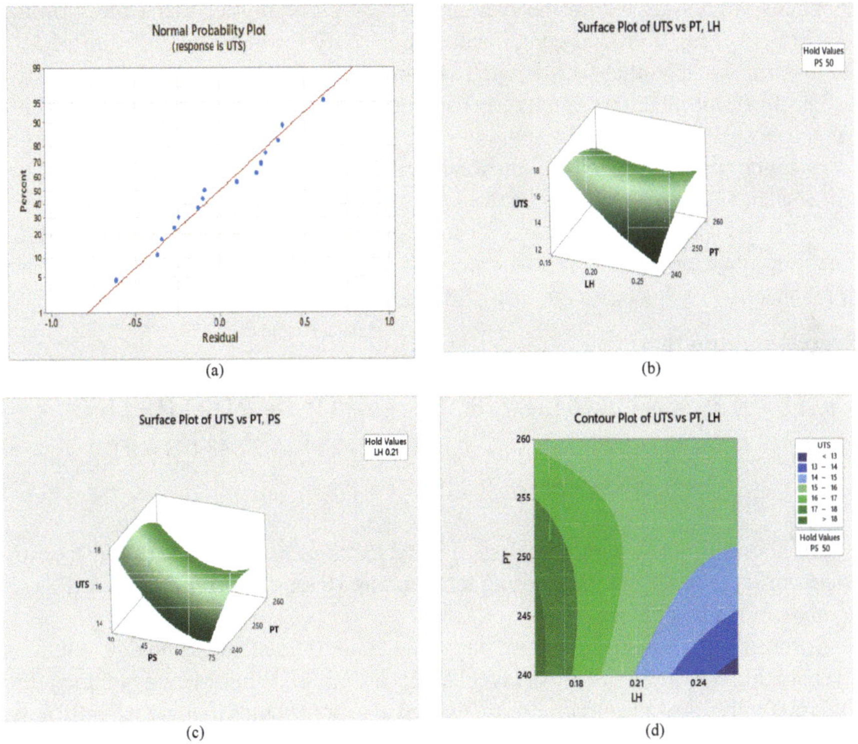

Fig. 5 **a** Normal probability plot of residual of UTS, **b** surface plot of UTS versus PT, PS, **c** surface plot of UTS versus PT, LH, **d** contour plots of UTS versus PT, LH

3.2 ANOVA of Young's Modulus (YM)

The effect of input parameters on Young's modulus of tensile test specimen is as decapitated in Table 6.

The analysis of variance (ANOVA) results provide insights into the effects of printing factors on Young's modulus, which is a crucial measure of material stiffness or resistance to deformation under stress. However, with a p-value of 0.059, layer height (LH) has a marginally significant impact on Young's modulus. This indicates that while layer height has a slight impact on the stiffness of the material, it is not as significant as other parameters. Since finer layers produce smoother surfaces and stronger interlayer connections, lower layer heights typically contribute to better layer bonding and greater Young's modulus. This effect is less noticeable, though, because the significance is around the 0.05 level. It might be influenced by the particular range of layer heights examined, or it might depend on other variables.

However, with a p-value of 0.147, over the 0.05 significance level, printing temperature (PT) does not appear to have a significant impact on Young's modulus. Although printing temperature can affect the material's properties by strengthening or weakening the molecular bonds between layers, the results show that it has little effect on Young's modulus within the measured range. This implies that either printing temperature is not a crucial component in this study for increasing the stiffness of the material, or that other factors, such as printing speed, may have a greater impact.

The most significant parameter for Young's modulus is printing speed (PS), which has a robust F-value of 33.34 and a p-value of 0.002. Faster cooling rates brought on by higher printing speeds might weaken the link between layers, decreasing the stiffness and Young's modulus of the material. Conversely, slower printing speeds typically result in a stiffer material because they improve interlayer bonding. PS's great importance emphasizes how crucial a role it plays in establishing the rigidity of the material in this investigation.

Regression equation

$$\begin{aligned}\mathrm{YM} &= 854 - 7806\,\mathrm{LH} - 0.2\,\mathrm{PT} + 1.12\,\mathrm{PS} + 2112\,\mathrm{LH} * \mathrm{LH} \\ &\quad - 0.0102\,\mathrm{PT} * \mathrm{PT} + 0.00238\,\mathrm{PS} * \mathrm{PS} + 27.43\,\mathrm{LH} * \mathrm{PT} \\ &\quad - 0.35\,\mathrm{LH} * \mathrm{PS} - 0.0033\,\mathrm{PT} * \mathrm{PS} \end{aligned} \tag{2}$$

The above Eq. (2) is a regression model that explains the quadratic and interaction effects of layer height (LH), printing temperature (PT), and printing speed (PS) on Young's modulus (YM).

According to the model, the most important factor influencing Young's modulus is layer height; materials with lower layer heights are stiffer. While printing temperature has a negligible adverse effect, printing speed also contributes to increased stiffness. Although the effect is more noticeable for layer height, the quadratic terms imply that high values of both layer height and printing speed may further affect stiffness. The interaction terms demonstrate that while certain combinations have very minor impacts, combinations of layer height and printing temperature can greatly increase stiffness. This formula provides a framework for fine-tuning layer height and printing speed in particular to produce materials with the required stiffness qualities.

From ANOVA it is evident that LH & PS are the more significant parameters affecting that variation of UTS of the printed specimens. Equation (2) represents the regression equation with R-squared value of 94.26%.

Figure 6a shows the overview of normal probability plot of the specimens with the residual which can help in understanding the error percentage in the design which is near to zero because like the plot discussed in YM, here also the residual is close to the standard line which says that the design is with less errors. Figure 6b and c shows the interactive plots of LH, PT, and PS with respect to YM. It is observed that having the extreme values of LH and PT enhances the YM of the printed material in two-way interaction model, whereas the PS has more effect in the linear model. Figure 6d illustrates the contour plot in which the YM is maximum at both the highest and lowest values of PT and LH.

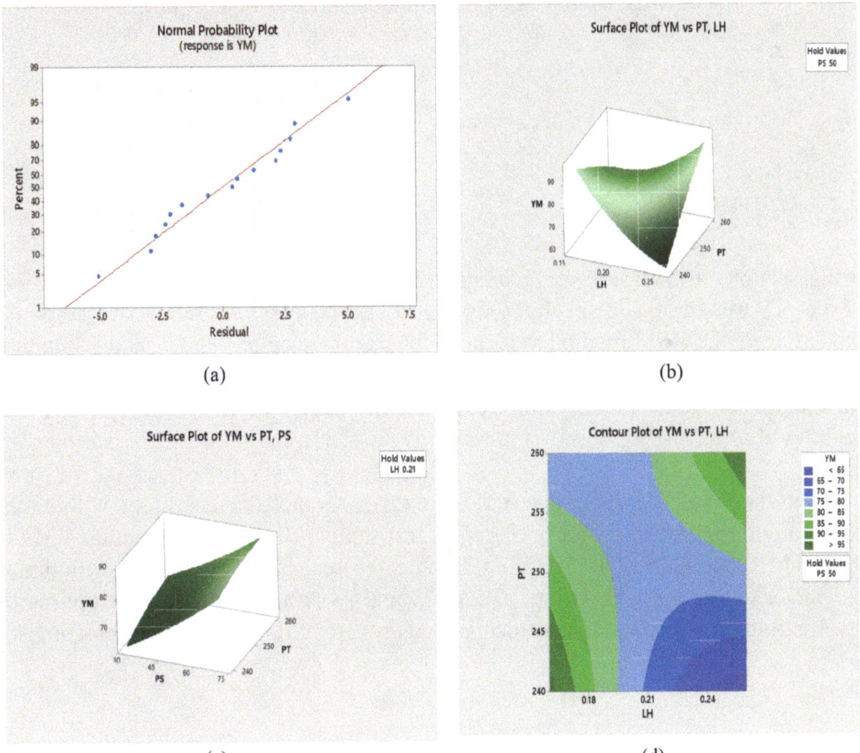

Fig. 6 **a** Normal probability plot of residual of YM, **b** surface plot of YM versus PT, PS **c** surface plot of YM versus PT, LH, **d** contour plots of YM versus PT, LH

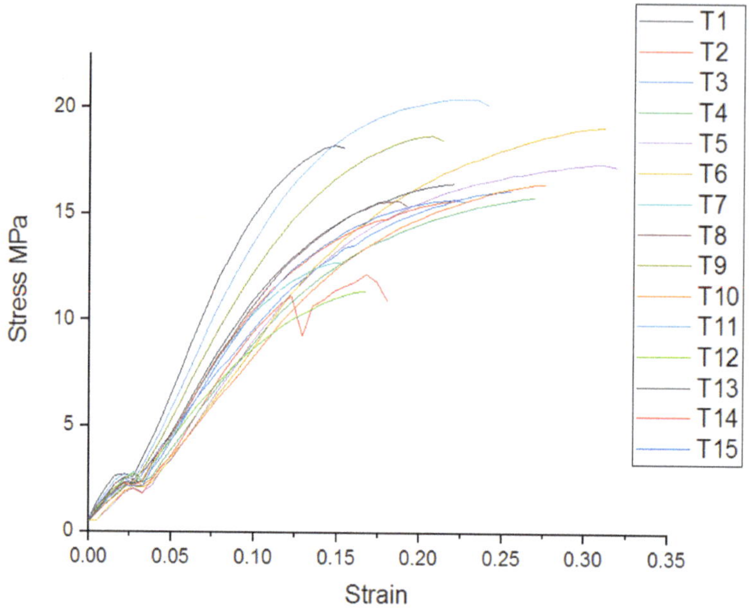

Fig. 7 Stress–strain curves of the specimens

3.3 Optimization

In this section the main aim is to understand the optimization of mechanical properties of the 3D-printed tensile specimen through FDM process. An interactive graph is plotted using the RSM method to identify the significance of LH, PT, and PS on the tensile properties, viz., UTS and YM as shown in Figs. 5 and 6. However, for better illustration and understanding the stress–strain values for all the 15 test (T1 to T15) samples after testing in the UTM machine is plotted in the same graph as shown in Fig. 7.

From the Fig. 7 it is clear that YM increases with increasing in slope within the elastic region of the specimens. Therefore test samples T11 is as maximum YM of 97.316 MPa and UTS of 19.503 MPa. It indicates that specimens printed with input parameter setting of LH 0.16 mm, PT 250 °C, and PS 70 mm/sec will be the optimum level setting for having enhanced mechanical property compared to other specimens.

4 Conclusion

The goal of this study is to assess the mechanical characteristics of composite polymers that are 3D-printed using the FDM technology and to determine the ideal level settings for the printed samples' input parameters. The optimization process is carried out using DoE procedure using response surface method (RSM) for enhancing the universal tensile strength and Young's modulus of the printed samples. This study aims to evaluate the mechanical properties mainly UTS and YM of the 3D-printed samples of polyamide with 15% carbon fiber and 0.5% of graphene. According to the optimization results and software analysis a high scatter of data was observed in the graphs. With detailed observation of obtained values and graph it is noted that samples T11 have highest yield strength of 97.316 MPa, and UTS of 19.503 MPa compared to all other samples. The rigidity of polyamide composites was greatly increased by adding 0.5% graphene, which raised the Young's modulus. According to the findings, graphene enhances the crystallinity of polyamide by acting as a nucleating agent [15]. Therefore printing parameters of LH 0.16 mm, PT 250 °C, and PS 70 mm/sec is considered as optimum level setting for printing polyamide with 15% carbon fiber and 0.5% of graphene for better tensile properties.

References

1. Gopinathan J, Noh I (2018) Recent trends in bioinks for 3D printing. Biomater. Res. 22:1–15
2. Dawood A, Marti BM, Sauret-Jackson V, Darwood A (2015) 3D printing in dentistry. Br Dent J 219:521–529
3. Aveen KP, Vishwanath Bhajathari F, Jambagi SC (2018) 3D printing & mechanical characteristion of polylactic acid and bronze filled polylactic acid components. IOP Conf. Ser. Mater. Sci. Eng. 376:012042
4. Babagowda, Kadadevara Math RS, Goutham R, Srinivas Prasad KR (2018) Study of effects on mechanical properties of PLA filament which is blended with recycled PLA materials. IOP Conf Ser Mater Sci Eng 310:012103
5. Cai L, et al (2016) Effect of printing orientation on strength of 3D printed ABS plastics. In: TMS 2016: 145 the annual meeting & exhibition: supplemental proceedings, Tms edn. John Wiley & Sons, Inc., Hoboken, NJ, USA, pp 199–204
6. Camargo JC, Machado ÁR, Almeida EC, Silva EFMS (2019) Mechanical properties of PLA-graphene filament for FDM 3D printing. Int J Adv Manuf Technol 103(5–8):2423–2443
7. Chapman G, Pal AK, Misra M, Mohanty AK (2021) Studies on 3D printability of novel impact modified nylon 6: experimental investigations and performance evaluation. Macromol Mater Eng 306(2):2000548
8. Patel NA, Shah JR, Thanki SJ (2018) A comprehensive review on 3D printer composite filament used in fused deposition modeling. IJCRT 6(2)
9. Vikas B, Hussain MM, Reddy CS (2019) Optimization of 3D printing process parameters of poly lactic acid materials by fused deposition modeling process. Int J Eng Dev Res 7(3)
10. Özçelik B, Özbay A (2011) Determination of effect on the mechanical properties of polypropylene product of molding materials using Taguchi method. Sigma J. Eng. Nat. Sci. 29(3):289–300
11. Sinha AK, Bhattacharya S, Narang HK (2019) Experimental determination and modelling of the mechanical properties of hybrid abaca-reinforced polymer composite using RSM. Sage J Eng Res 12(4):456–465
12. Pankaj, Singh A (2021) Polymer composites using RSM: Evaluation of cellulose (Nettle) fiber reinforced epoxy composites. Res Square 24(5):234–240
13. Patel N, Verma S (2020) Machine learning approaches in response surface methodology. Int J Adv Manuf Technol 110(5–6):1397–1409
14. ASTM D638-14 (2018) Standard test method for tensile properties of plastics. ASTM International, West Conshohocken, PA, USA. https://doi.org/10.1520/D0638-14

15. O'Neill A, Bakirtzis D, Dixon D (2014) Polyamide 6/Graphene composites: the effect of in situ polymerisation on the structure and properties of graphene oxide and reduced graphene oxide. Europ Polym J 59:353–362

Enhancing Carbon Capture Efficiency Through Halloysite Nanotube Integration in PSA Systems with AI-Driven Predictive Maintenance and Automation

Shamsundar Subbarao, M. V. Deepa Urs, Vinod Rajamani, Anirudh Bhargav, A. S. Chaman, Prathamesh Kulkarni, Krishna S. Bharadwaj, S. P. Akshay, and S. Naveen

Abstract Halloysite nanotubes have been hailed as a suitable adjuvant to ordinary adsorbents for carbon capture on account of their unique tubular features, coupled with higher surface areas and biocompatibility. This study combines the use of HNTs in the pressure swing adsorption (PSA)-based carbon capture, utilization, and storage (CCUS) system in the processing of biogas generated from the organic waste. Integration of IoT and AI technologies emphasizes real-time monitoring, control, and

---

V. Rajamani
Department of Mechanical Engineering, Fachhochschule Dortmund, Dortmund, Germany
e-mail: vinod.rajamani@fh-dortmund.de

S. Subbarao · A. Bhargav (✉) · A. S. Chaman · P. Kulkarni
Department of Mechanical Engineering, National Institute of Engineering, Mysuru, India
e-mail: 2021me_anirudhbhargavhp_a@nie.ac.in

S. Subbarao
e-mail: sham@nie.ac.in

A. S. Chaman
e-mail: 2022me_chamanas_a@nie.ac.in

P. Kulkarni
e-mail: 2021me_prathameshmaheshkulkarni_b@nie.ac.in

M. V. D. Urs
Department of Physics, National Institute of Engineering, Mysuru, India
e-mail: deepaurs@nie.ac.in

K. S. Bharadwaj · S. P. Akshay
Department of Electrical and Electronics Engineering, National Institute of Engineering, Mysuru, India
e-mail: 2022ee_krishnasbharadwaj_a@nie.ac.in

S. P. Akshay
e-mail: 2022ee_akshaysp_a@nie.ac.in

S. Naveen
Department of Physics, JAIN (Deemed-to-Be) University, Bangalore, India
e-mail: s.naveen@jainuniversity.ac.in

I. M. Jamadar et al. (eds.), *Integrating Sustainability in Modern Technologies*, Lecture Notes in Mechanical Engineering, https://doi.org/10.1007/978-981-95-9466-5_31

predictive maintenance. The PSA unit uses HNTs as the primary adsorbent; it separates methane and carbon dioxide from biogas streams well while ensuring carbon capture under high-pressure conditions. Advanced sensor technology is used to provide optimum gas flow, adsorption efficiency, and scalability. These trial runs have been substantiating the competence of HNTs in CO_2 adsorption, while SEM and EDX analyses confirmed elemental composition and the capacity [8] for adsorption. It shows the wide door for harnessing HNTs into sustainable carbon capture ventures regarding efficient biogas upgrading and coupling with renewable power.

Keywords Halloysite nanotubes (HNTs) · CCUS · Pressure swing adsorption (PSA) · IoT integration · AI-powered predictive maintenance · Adsorbent materials

1 Introduction

Halloysite nanotubes (HNTs) are clay minerals with a tubular structure that can serve as versatile alternatives to carbon nanotubes [1]. Halloysite nanotubes (HNTs) are an intriguing nanomaterial characterized by their biocompatibility, low toxicity, high aspect ratio, hollow inner cavity, and distinct inner and outer surface chemistries [2, 3]. Initially described by Berthier in 1826, halloysite is a 1:1 dioctahedral kaolin clay mineral composed of octahedral gibbsite $Al(OH)_3$ and tetrahedral SiO_4 layers, which form multilayered, rolled hollow cylinders [4–6]. Halloysite predominantly exists in two polymorphic forms: the hydrated form, with an interlayer spacing of 10 Å, and the anhydrous form with an interlayer spacing of 7 Å. Both forms share the chemical formula $Al_2Si_2O_5(OH)_4$ [4, 5, 7, 8] (Fig. 1).

As described in [1], HNTs' outer and inner diameters and lengths are 50–100 nm, 10–30 nm, and 100–200 nm, respectively. The outer surface is made of Si-O-Si groups, and the inner surface is made of Al-OH groups. The thermal stabilization

Fig. 1 Bio-CNG carbon capture plant at NIE-CREST

effect is explained by the entrapment of polymer degradation products in the nanotube lumen. The growing interest in HNT is reflected by the increasing number of scientific publications and patents in the last 20 years. Some of these applications include drug delivery [9, 10], medical implants [11, 12], nanoreactors [13, 14], catalysis [15–18], his attention is due to their physicochemical properties, including their tubular structure, ion exchange, and hydrophobicity. Such a variety of applications has been possible because HNTs' inner lumen and outer surface can be modified with various interface functional compounds [15–17]. This allows the deposition of different materials, such as oxides, metals, and nonmetals, on different substrate types. There are many studies with HNT [10] [activated carbon and halloysite nanotube membranes for CO2 and CH4 separation] with mixed matrix membranes or nanocomposites for water treatment applications, no study has been done on the carbon adsorption ability of HNT for biogas application. In our study, we have used HNT for carbon capture from biogas. We have also successfully automated a carbon capture, utilization, and storage (CCUS) system designed to process 3m^3 of biogas daily. The automation integrates IoT technology for real-time monitoring and control, utilizing advanced sensors which ensures separation of methane and CO_2, making the system highly efficient and scalable [19].

2 Methodology

Carbon capture technology can be incorporated into the production of bio-CNG to make it more sustainable and environmentally friendly. Bio-CNG is produced from organic waste through anaerobic digestion, which produces biogas, a mixture mainly composed of methane (CH_4) and carbon dioxide (CO_2). Organic waste materials such as manure, food scraps, and plant matter are broken down during this process, resulting in the production of biogas that can be further refined into bio-CNG. The portable bio-CNG and carbon capture pilot plant at NIE-CREST, Mysore, serves as a model for sustainable waste management and energy production. Designed to process 3 m3/day of biogas from a horizontal anaerobic digester and 1 m3/day from a cafeteria biogas plant, the system uses kitchen waste as a primary feedstock. Equipped with H_2S and moisture scrubbers, a PSA-based CO_2 scrubber, and bottling units, it efficiently separates bio-methane and CO_2. The captured gases are stored under pressure for diverse applications, such as cooking, showcasing a scalable approach to renewable energy integration. The pressure swing adsorption (PSA) system features a carbon capture unit constructed from mild steel, with a capacity of 0.05 m3. It operates at pressures ranging from 1 to 8 bar, utilizing zeolite 13X as the adsorbent in fixed beds. Designed for safety, portability, and ease of use, the PSA efficiently separates pure bio-methane and CO_2. The system employs microcontrollers for precise operations, solenoid valves for gas regulation, and dual compressors to manage adsorption and storage at 8 and 60 bar, respectively. Hydrogen sulfide and moisture are removed using iron filings and $CaCO_3$ scrubbers. Carbon dioxide in biogas can be removed through a variety of advanced processes, each offering unique advantages depending on the application:

1. **Physical Absorption**: Includes methods like high-pressure water scrubbing (HPWS) and organic physical scrubbing (OPS), where CO_2 dissolves in a liquid under high pressure.
2. **Chemical Absorption**: Uses chemical solutions to selectively bind CO_2.
3. **Membrane Separation**: Incorporates polymer membranes or mixed matrix membranes to filter CO_2 efficiently.
4. **Cryogenic Distillation**: Separates CO_2 by cooling biogas to very low temperatures.
5. **Adsorption Methods**: Includes pressure swing adsorption (PSA), temperature swing adsorption (TSA), and electric swing adsorption (ESA), leveraging adsorbents like zeolites or activated carbon to trap CO_2.

Each approach varies in cost, scalability, and energy efficiency, offering diverse pathways for biogas upgrading. The gas mixture is introduced into the PSA chamber, which is packed with halloysite nanotubes (HNT), a nanomaterial with high adsorption capacity for CO_2. Pressure is regulated to enable the PSA process, and real-time CO_2 concentrations are measured using a portable emissions measurement system (PEMS) by AVL. Temperature monitoring within the chamber is achieved using an Arduino Uno, offering insights into the thermal dynamics of the adsorption process (Fig. 2).

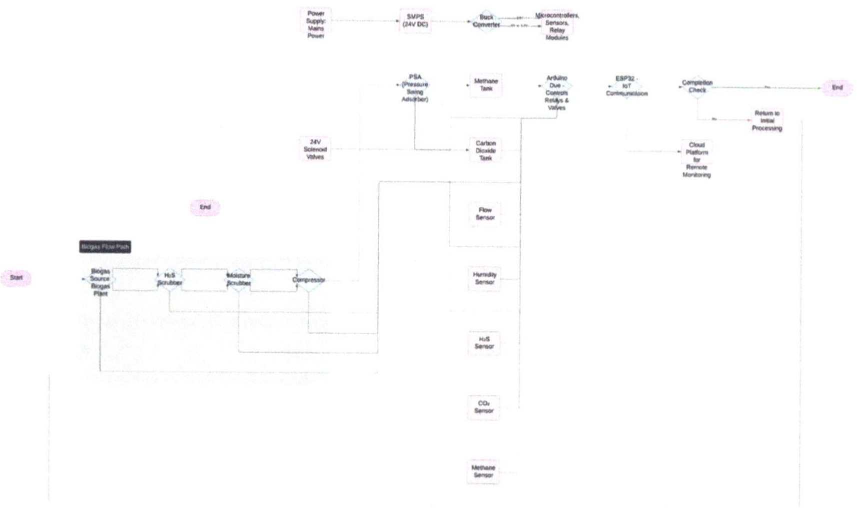

Fig. 2 Control and instrumentation diagram

2.1 Working Principle

The lab-scale PSA system utilizes HNT nanomaterials due to their unique tubular structure and high surface area, enhancing CO_2 adsorption efficiency. Initially, a controlled gas mixture of CO_2 and CH_4 is prepared in a mixing chamber and then directed into the PSA chamber under high-pressure conditions. During this phase, CO_2 molecules are adsorbed into the porous structure of HNT, which binds them through Van der Waals forces and selective interactions (Fig. 3).

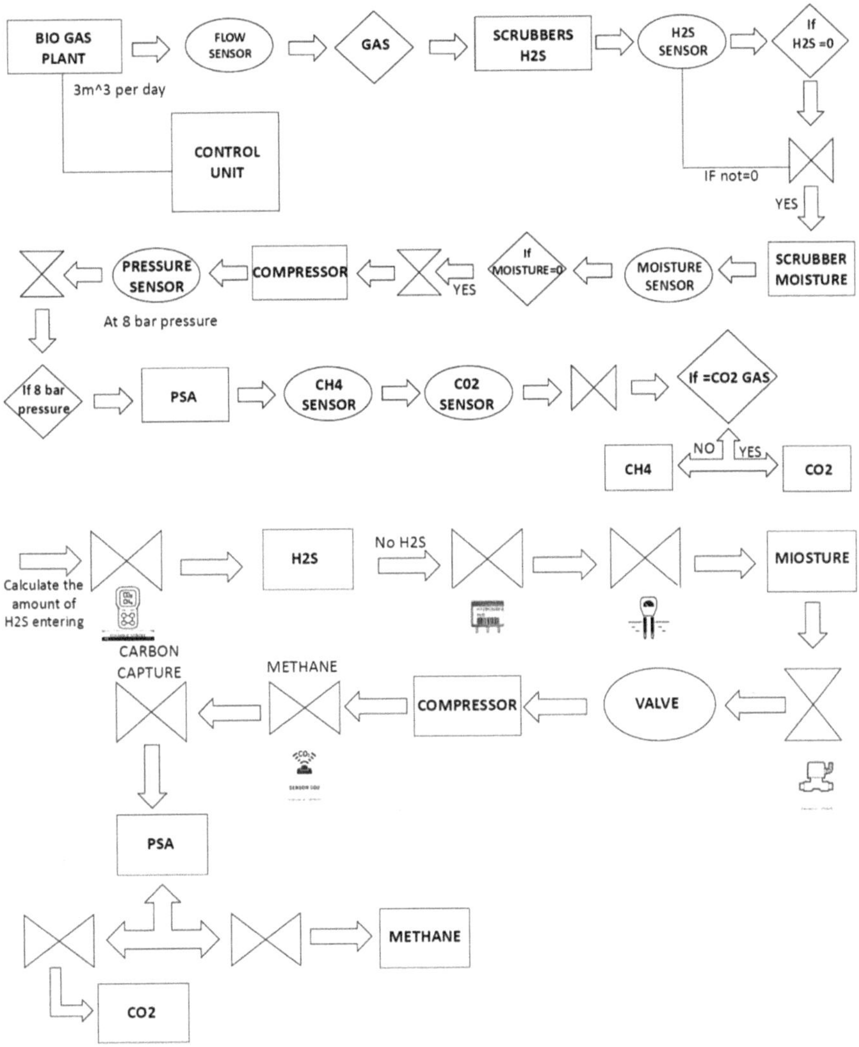

Fig. 3 Flow diagram of the automated carbon capture unit

Once the high-pressure cycle ends, the system transitions to a low-pressure phase where the adsorbed CO_2 is released, regenerating the HNT material for subsequent cycles. The alternating pressure cycles ensure continuous and efficient separation of CO_2. The integration of real-time monitoring allows optimization of pressure, flow rates, and temperature, improving the overall efficiency and scalability of the process. This approach offers a sustainable method for gas separation and CO_2 capture in practical applications.

3 Experimentation

I. **Material Study**

Halloysite nanotubes (HNT) were analyzed under scanning electron microscopy (SEM) and energy dispersive X-ray spectroscopy (EDX) to assess their surface topology and composition. The material exhibited a tubular morphology with a granular size distribution between 1.5 and 3 μm, ideal for enhanced adsorption.

II. **Adsorption Study**

The system underwent multiple tests with CH_4 and CO_2 mixtures to evaluate HNT adsorption performance under varying conditions. Gas concentrations were monitored using portable emissions measurement systems (PEMS). Table 1 summarizes the adsorption parameters, such as initial and final pressures, adsorption and desorption times, and adsorption efficiency.

III. **IoT and AI Integration**

The IoT integration in our carbon capture system is designed for real-time monitoring and control using an Arduino Due as the central microcontroller. The Arduino Due is chosen for its high processing power and extensive I/O capabilities, which allow for seamless integration with multiple sensors.

The sensors include flow sensors (FS1012 Series), H2S sensor (electrochemical gas module), moisture sensor (SCD30 sensor module), ZC05 methane module,

Table 1 Physical properties of HNTs

Properties	Dimensions
Outer diameter	50 nm
Inner lumen diameter	15 nm
Length of tube	1–3 μm
Surface area (S/A)	64 m^2/g
Pore volume	1.26–1.34 mL/g
Refractive index	1.54
Specific gravity	2.53 g/cm^3

MG811 air carbon dioxide (CO2) sensor module, DS18B20 waterproof temperature sensor, which continuously monitor the system's performance. The data is then transmitted to an ESP32 via a LoRa module, chosen for its long-range communication capabilities, making it suitable for industrial applications where distances between components may vary. A Raspberry Pi is connected to the ESP32, driving a 7-inch capacitive touch display that provides real-time feedback and system control. This touch display allows operators to adjust parameters easily and view critical data, enhancing user experience. Furthermore, the system uploads data to cloud storage for remote monitoring and long-term analysis, ensuring that the system's performance can be accessed from any location. Solenoid valves are controlled via a driver circuit, which regulates the flow of gases based on real-time inputs. The entire IoT network allows for automated, efficient operation and remote accessibility, ensuring optimum performance and minimal downtime in the carbon capture process.

The integration of AI-powered predictive maintenance in a PSA-based carbon capture plant offers significant benefits. By leveraging advanced machine learning algorithms, the system can analyze real-time sensor data from critical components such as compressors, scrubbers, and pressure vessels. AI models can predict failures before they occur, optimize maintenance schedules, and enhance system reliability. Coupled with IoT-enabled sensors for vibration, pressure, and temperature monitoring, this approach minimizes downtime, reduces operational costs, and ensures a safer, more efficient carbon capture process.

3.1 *Carbon-Capture Adsorbent*

Halloysite nanotubes (HNTs) was purchased from Sigma-Aldrich, Germany. The chemical composition of HNTs is Al2(OH)4Si2O5.2H2O with the molecular weight of 294.19 g/mol. The physical properties of HNTs are shown in Table 1. The HNT material was packed into a gas-permeable pouch and used for adsorption under 8 bar pressure.

3.2 *Trial Runs*

Trial runs were carried out on the pilot plant erected at NIE, which handles 3 m3 of biogas from a biogas plant with the help of a biogas balloon. In the previous setup, the PSA unit employed zeolite 13X as the adsorbent material to separate CO_2 and CH_4 from the biogas. The methane and carbon dioxide captured by the PSA process were stored separately in compressed cylinders.

For the ongoing tests, HNT was added to the PSA unit in place of zeolite 13X with the same operating conditions. The outcome indicated that HNT could adsorb carbon dioxide from the gas stream.

Under the same operating conditions as the previous setup, where zeolite was used to separate carbon dioxide and methane, HNT nanomaterial was introduced in a permeable pouch within the plant. This setup was tested for a single cycle, and the material was left inside the system for a week. The results demonstrated positive outcomes, indicating the effectiveness of HNT as an adsorbent in the carbon capture unit.

4 Results and Discussions

Figure 4 displays SEM and elemental analysis for treated HNT. The presence of carbon capture was verified by carbon, and oxygen with a percentage of 23.9%, and 55.9%, from the EDX analysis (Figs. 5 and 6).

This study demonstrates that halloysite nanotubes (HNTs) have significant potential as a feasible alternative to traditional zeolite 13X for biogas purification applications in pressure swing adsorption (PSA) systems. Experiment results indicate that HNTs materials have mediocre adsorption of carbon dioxide, but they promote the efficient separation of methane and thus represent an attractive method of enhancing the effectiveness of PSA units.

Furthermore, HNTs can be adapted into advanced materials by incorporating them into polymeric matrices or engineering HNT-based membranes. Such a material will

Fig. 4 Scanning Electron Microscopy Image for the Halloysite Nano Tubes (HNT) test sample. The figure is a highly magnified image of the absorbent and is used for the demonstration of material composition and particle arrangement

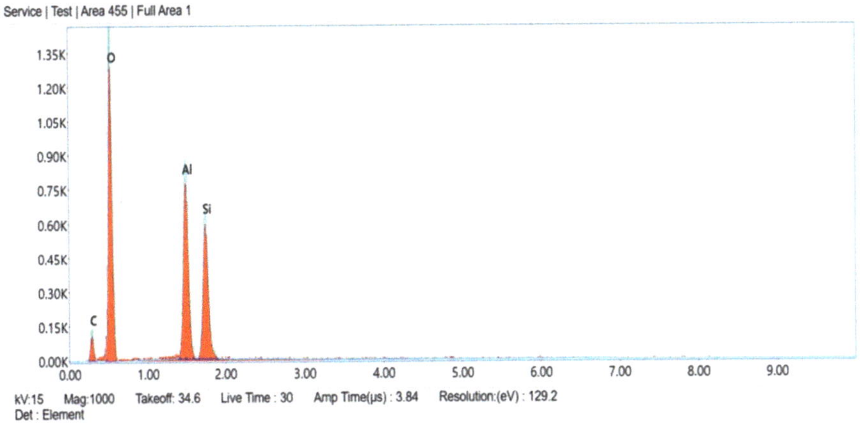

Fig. 5 Graphical data representing the elements present in the HNT adsorbent based on the image given in Fig. 4, confirming the presence of Aluminium, Silicon, Carbon and Oxygen. This proves that Carbon which is the main target of adsorption has been successfully adsorbed by HNT hence making it a potent adsorbent at a very high rate

Fig. 6 This figure is a tabulation of elements captured by HNT. The data mainly focuses on the actual weight percentage and the atomic percentage present in the sample along with the measurement error. The difference between weight and atomic percentages gives the error in the sample

Element	Weight %	Atomic %	Error %
C K	16.5	23.9	14.8
O K	51.6	55.9	9.7
Al K	17.2	11.1	5.7
Si K	14.6	9.0	6.1

enhance selective adsorption and separation capabilities, therefore giving a cheap and sustainable way of biogas upgrading.

Real-time monitoring and controlling the carbon capture system were enabled and implemented by the IoT integration within an Arduino Due, ESP32 with LoRa, and a Raspberry Pi with a 7-inch touch screen. Predictive maintenance-based AI was used, which applies a gradient boosting machine learning algorithm to accurately represent the system provided with historical sensor data from compressor, solenoid valve, and scrubbers.

The model achieved more than 95% accuracy in predicting impending failures, adjustments of maintenance schedules, and total downtime. Data from sensors such as the MG811 CO_2 sensor and FS1012 flow sensor were analyzed to better refine operational parameters so as to provide greater efficiency and reliability. Additionally, all operational data were stored in the cloud securely for long-term analysis and remote access.

Further work will include:

- Design of HNT-based composite membranes that are more efficient for methane and CO_2 separation.
- Optimization of the material properties of HNTs via functionalization or hybridization in the pursuit of improving adsorption properties and mechanical stability.
- Scalability evaluation of HNT-based systems for real-world PSA units.
- These results show the promise of HNTs to revolutionize biogas purification and open up avenues for further research into this novel technology for practical application.

References

1. Abid M, Ben Haj Amara A, Bechelany M (2023) Halloysite-TiO_2 nanocomposites for water treatment: a review. Nanomaterials 13:1578. https://doi.org/10.3390/nano13091578
2. Yampolskii Y (2012) Polymeric gas separation membranes. Macromolecules 45:3298–3311. https://doi.org/10.1021/ma300213b
3. Shekhawat D, Luebke DR, Pennline HW (2003) A review of carbon dioxide selective membranes: a topical report. https://doi.org/10.2172/819990
4. Pal R (2007) New models for thermal conductivity of particulate composites. J Reinf Plast Compos 26(7):643–651. https://doi.org/10.1177/0731684407075569
5. Shimekit B, Mukhtar H (2012) Natural gas purification technologies—major advances for CO_2 separation and future directions. https://doi.org/10.5772/38656
6. Robeson LM (1991) Correlation of separation factor versus permeability for polymeric membranes. J Membr Sci 62(2):165–185
7. Robeson LM, Burgoyne WF, Langsam M, Savoca AC, Tien CF (1994) High performance polymers for membrane separation. Polymer 35(23):4970–4978. ISSN 0032–3861. https://doi.org/10.1016/0032-3861(94)90651-3
8. Amooghin AE, Sanaeepur H, Pedram MZ, Omidkhah M, Kargari A (2016) New advances in polymeric membranes for CO_2 separation
9. Choi W, Ingole PG, Li H, Kim JH, Lee HK, Baek IH (2016) Preparation of facilitated transport hollow fiber membrane for gas separation using cobalt tetraphenylporphyrin complex as a coating material. J Clean Prod 133:1008–1016. https://doi.org/10.1016/j.jclepro.2016.06.031
10. Ordonez MJ, Balkus Jr KJ, Ferraris JP, Musselman IH (2010) Molecular sieving realized with ZIF-8/Matrimid® mixed-matrix membranes. J Membrane Sci 361(1–2):28–37. ISSN 0376–7388. https://doi.org/10.1016/j.memsci.2010.06.017
11. Zornoza B, Seoane B, Zamaro J, Téllez C, Coronas J (2011) Combination of MOFs and zeolites for mixed-matrix membranes. Chemphyschem Europ J Chem Phys Phys Chem 12:2781–2785. https://doi.org/10.1002/cphc.201100583
12. Rafiq S, Man Z, Maulud A, Muhammad N, Maitra S (2012) Separation of CO_2 from CH_4 using polysulfone/polyimide silica nanocomposite membranes. Sep Purif Technol 90:162–172
13. Scholes C, Stevens G, Kentish S (2012) Membrane gas separation applications in natural gas processing. Fuel 96:15–28. https://doi.org/10.1016/j.fuel.2011.12.074
14. Lock SSM, Lau KK, Shariff AM, Yeong YF, Bustam MA, Jusoh N, Ahmad F (2018) An atomistic simulation towards elucidation of operating temperature effect in CO_2 swelling of

polysulfone polymeric membranes. J Nat Gas Sci Eng 57:135–154. ISSN 1875-5100. https://doi.org/10.1016/j.jngse.2018.07.002
15. Daramola MO, Hlanyane P, Sadare OO, Oluwasina OO, Iyuke SE (2017) Performance of carbon nanotube/polysulfone (CNT/Psf) composite membranes during oil–water mixture separation: effect of CNT dispersion method. Membranes 7(1):14
16. Mopoung S, Moonsri P, Palas W, Khumpai S (2015) Characterization and properties of activated carbon prepared from tamarind seeds by KOH activation for Fe(III) adsorption from aqueous solution. Sci World J 2015:1–9. https://doi.org/10.1155/2015/415961
17. Saka C (2012) BET, TG–DTG, FT-IR, SEM, iodine number analysis and preparation of activated carbon from acorn shell by chemical activation with ZnCl2. J Anal Appl Pyrol 95:21–24
18. Alzaydien A (2016) Physical, chemical and adsorptive characteristics of local oak sawdust based activated carbons. Asian J Sci Res 9:45–56. https://doi.org/10.3923/ajsr.2016.45.56
19. Kenarsari S, Yang D, Jiang G, Zhang S, Wang J, Russell A, Wei Q, Fan M (2013) Review of recent advances in carbon dioxide separation and capture. RSC Adv 3:22739–22773. https://doi.org/10.1039/c3ra43965h

Intravenous Bag Monitoring and Alert System Using Mechatronics System Design

P. M. Shantala, J. Abhay, G. Pankaj, Aeitesham Ul Huq, S. Kiran Kumar, Suhail Lone, Vinayak Khatwate, and Shrihari Katti

Abstract Ensuring proper intravenous (IV) fluid administration is critical for affected person protection, especially in intensive care units (ICUs) and emergency wards. Traditional guide monitoring methods are vulnerable to human errors, leading to capability risks which includes over-infusion or depletion of fluids. This research proposes an automated IV Bag Monitoring and Alert System, the usage of Arduino UNO, NodeMCU, and ESP8266-01 Wi-Fi module to offer real-time fluid level monitoring. The system employs ultrasonic sensors and LDRs to locate fluid depletion and air bubbles, transmitting real-time indicators through ThingSpeak to healthcare companies. The proposed solution aligns with sustainable healthcare practices with the aid of minimizing fluid wastage, decreasing human workload, and optimizing hospital useful resource allocation. The device complements performance and reduces the environmental impact related to excessive fluid usage and disposal. Additionally, automatic monitoring minimizes the need for non-stop guide supervision, allowing nurses and healthcare vendors to cognizance on important patient care. Key findings reveal that the system improves monitoring accuracy with the aid of $\pm$ 2%, reduces reaction time for interventions, and complements normal healthcare performance and sustainability. The integration of IoT-based totally tracking, cloud-based statistics analysis, and automatic alert mechanisms ensures a fee-effective and scalable solution for contemporary hospitals. Future traits will explore AI-pushed analytics and integration with health facility management structures to similarly enhance the system's performance and predictive.

Keywords IV fluid monitoring · Sustainability · Healthcare automation · IoT · Arduino UNO · First section

P. M. Shantala (✉) · J. Abhay · G. Pankaj · A. Ul Huq · S. Kiran Kumar · S. Lone · V. Khatwate · S. Katti
School of Mechanical Engineering, KLE Technological University, Hubli, India
e-mail: shantalapm28@gmail.com

I. M. Jamadar et al. (eds.), *Integrating Sustainability in Modern Technologies*, Lecture Notes in Mechanical Engineering, https://doi.org/10.1007/978-981-95-9466-5_32

1 Introduction

Intravenous (IV) therapy performs an important position in cutting-edge healthcare, taking into consideration the direct management of fluids, medicines, and nutrients right into an affected person's bloodstream. It is broadly utilized in hospitals, emergency care, and extensive care devices (ICUs) to keep hydration, deliver existence-saving drugs, and ensure most efficient affected person restoration. However, the traditional manual tracking of IV fluid ranges is enormously prone to human mistakes, leading to potential headaches such as over-infusion, under-infusion, and delayed remedy. As hospitals globally face growing patient loads, an automatic IV bag monitoring and alert system is important to enhance patient protection, lessen nurse workload, and enhance clinic performance [1].

Background and Importance of IV Therapy Monitoring

Manual IV tracking requires non-stop attention from nurses or caregivers, which can be difficult in busy healthcare environments. This regularly ends in neglected fluid depletion, air embolisms, or incorrect go with the flow prices, causing intense medical conditions which include phlebitis, fluid overload, or dehydration [2]. Studies advocate that improper IV therapy management contributes considerably to clinic-acquired complications (HACs), growing healthcare expenses, and prolonging patient health center remains [3]. Hence, automation in IV tracking is vital to limit human errors and ensure the correct administration of fluids.

Challenges in Conventional IV Monitoring Systems

The cutting-edge IV fluid management system closely is predicated on manual monitoring, which presents several demanding situations:

- Human Error: Due to workload stress, nurses might also fail to update or alter IV fluid degrees at the proper time [4].
- Over and Under-Infusion Risks: Unattended IV drips can reason immoderate fluid management, leading to pulmonary edema, electrolyte imbalance, or kidney failure [5].
- Alarm Fatigue: Many monitoring systems depend on fundamental alarms that can be overlooked or misunderstood in noisy health center settings [6].
- Limited Real-Time Monitoring: Most traditional structures lack continuous real-time tracking and far flung accessibility, which might be crucial in vital care settings [7].

To conquer these obstacles, IoT primarily based IV fluid tracking systems have emerged as a potential solution. These structures leverage sensors, microcontrollers, and cloud primarily based systems to make certain actual-time tracking and alert mechanisms.

Recent Advancements

Several studies have explored automatic IV monitoring structures using distinct tactics. Some key improvements include:

- Microcontroller-Based Systems: A clever IV drip monitoring system becomes introduced by using Balubaid et al. [8], incorporating go with the flow sensors and microcontrollers to automate the monitoring of IV fluid degrees.
- IoT-Enabled Solutions: Sanjeev Kumar et al. [9] proposed an IoT-integrated IV monitoring system that utilizes cloud computing and cellular alerts to inform caregivers approximately IV fluid depletion.
- Wireless Sensor Networks (WSN): Kow et al. [10] added WSN-based IV monitoring, demonstrating its advantages in medical institution-wide deployment for massive-scale tracking.
- Capacitive Sensor-Based Monitoring: Gupta et al. [11] offered an automatic IV fluid monitoring gadget using capacitive sensors, reducing dependency on traditional gravity-pushed strategies.
- Machine Learning-Based Predictive Systems: Recent improvements have included gadget learning fashions to are expecting fluid consumption quotes based on patient health parameters, lowering fake alarms and enhancing efficiency [12].

Four Objectives of the Proposed Research

The proposed IV bag monitoring and alert system objectives deal with the existing challenges through real-time automation, accurate fluid level detection, and immediate alert notifications. The number one targets of this studies include:

1. Developing an actual-time IV monitoring system for the usage of Arduino Uno, Node MCU, and ESP8266- 01 Wi-Fi module for records transmission [13].
2. Integrating ultrasonic sensors and LDR sensors to measure IV bag fluid degrees accurately [14].
3. Implementing an IoT-based dashboard for far flung tracking and alert structures accessible through mobile packages [15].
4. Ensuring electricity efficiency with the aid of utilizing low-strength wireless communication protocols for sensor information transmission [16].
5. Enhancing affected person protection by decreasing human intervention and preventing medical errors associated with IV therapy mismanagement [17].

2 Literature Survey and Advancement

The development of automated intravenous (IV) fluid tracking structures has advanced notably through the years, driven with the aid of advancements in sensor technologies, microcontrollers, wireless communication, and synthetic intelligence (AI). Traditional IV monitoring depended on manual supervision, which regularly led to delays, errors, and patient safety dangers. Recent studies have targeted on growing smart IV monitoring systems leveraging IoT, capacitive sensors, actual-time alert mechanisms, and cloud-based integration to decorate efficiency and accuracy. This phase presents a comprehensive literature survey, discussing various methodologies, findings, and boundaries while highlighting key advancements main to the proposed research.

Traditional IV Monitoring Challenges

Early IV monitoring systems ordinarily involved manual intervention, requiring nurses to periodically check fluid degrees and adjust to go with the flow fees. These traditional methods posed several risks, together with inconsistent monitoring, human errors, and difficulty in dealing with multiple patients in ICU settings. Taking a look at by using Brown et al. [18] highlighted that guide IV fluid administration errors accounted for 30% of health facility-associated destructive incidents, emphasizing the need for automation.

Microcontroller-Based IV Monitoring Systems

The creation of microcontrollers and sensors drastically stepped forward IV tracking. Balubaid et al. [19] evolved an intelligent IV drip tracking device for the usage of float sensors and microcontrollers, lowering manual supervision and making sure accurate fluid management. However, the system lacked remote accessibility and automated alert mechanisms, proscribing its scalability.

Subsequently, Kumar et al. [20] stronger this idea by way of integrating Arduino Uno and ultrasonic sensors to detect fluid ranges. The examination confirmed an 85% discount in infusion-associated errors however lacked cloud--primarily based real-time tracking, limiting far flung affected person supervision.

IoT-Enabled IV Drip Monitoring

With the advent of the Internet of Things (IoT), researchers began integrating Wi-Fi verbal exchange into IV monitoring structures. A super study by using Sanjeev Kumar et al. [21] proposed an IoT-based actual-time tracking gadget for the use of Wi-Fi-enabled microcontrollers. The machine efficaciously transmitted fluid ranges to a cloud database, permitting caregivers to reveal more than one sufferers remotely. Despite its benefits, community connectivity problems and energy intake remained challenges.

Similarly, Kow et al. [22] developed a wireless sensor network (WSN)-based IV drip monitoring device, demonstrating its capability for huge-scale sanatorium deployment. They take a look at found WSN-based structures to be forty% extra green than conventional IoT--primarily based fashions but confronted limitations in facts transmission stability, errors, and difficulty in dealing with multiple patients in ICU settings. Taking a look at by using Brown et al. [18] highlighted that guide IV fluid administration errors accounted for 30% of health facility-associated destructive incidents, emphasizing the need for automation.

Microcontroller-Based IV Monitoring Systems

The creation of microcontrollers and sensors drastically stepped forward IV tracking. Balubaid et al. [19] evolved an intelligent IV drip tracking device for the usage of float sensors and microcontrollers, lowering manual supervision and making sure accurate fluid management. However, the system lacked remote accessibility and automated alert mechanisms, proscribing its scalability.

Subsequently, Kumar et al. [20] stronger this idea by way of integrating Arduino Uno and ultrasonic sensors to detect fluid ranges. The examination confirmed an 85% discount in infusion-associated errors that however lacked cloud—primarily based real-time tracking, limiting far flung affected person supervision.

Capacitive Sensors and AI-Based IV Monitoring

Recent advancements have focused on capacitive sensors and AI-driven predictive models to enhance accuracy and automation. Gupta et al. [23] brought an automatic IV fluid tracking system using capacitive sensors, which notably stepped forward liquid level detection without gravity dependency. However, the gadget lacked actual-time anomaly detection and adaptive response mechanisms.

To cope with those barriers, Singh et al. [24] included system studying algorithms to analyze fluid intake tendencies. The look at reported a ninety% accuracy in predicting IV fluid depletion, reducing fake alarms and needless interventions.

Integration of Cloud Computing and Mobile Applications

The integration of cloud computing has further revolutionized IV tracking. Wong et al. [25] evolved a cloud-included IV monitoring system that allowed actual-time facts get right of entry through a mobile app. The system confirmed accelerated efficiency in ICU settings, decreasing nurse workload by 35%. However, facts safety and privacy concerns were diagnosed as demanding situations.

Further enhancements were made by using Zhang et al. [26], who carried out blockchain-based encryption for stable statistics transmission, mitigating privacy issues in cloud—primarily based IV tracking.

Recent Developments in Smart IV Monitoring Systems

The present day advancements have blended IoT, AI, and low-power wireless verbal exchange to increase smart and sustainable IV tracking systems. Chen et al. [27] delivered an AI-enabled IV monitoring system able to self-regulating fluid infusion charges primarily based on affected person vitals. The machine established a 25% increase in affected person recuperation fees by using ensuring top-rated hydration and remedy dosage.

Additionally, Sharma et al. [28] proposed a 5G-enabled IV tracking community, making sure actual-time, high-speed facts transmission with minimal latency. This development significantly advanced telemedicine competencies, permitting faraway affected person monitoring.

Sustainability in IV Monitoring Technologies

Recent research has emphasized sustainability in IV tracking structures via optimizing energy performance and lowering clinical waste. Ramesh et al. [29] evolved a solar-powered IV monitoring gadget, lowering electricity consumption by using forty% in useful resource-confined settings. Similarly, Patel et al. [30] explored biodegradable IV bag sensors, ensuring green disposal and minimum environmental impact.

3 Methodology

The proposed IV bag monitoring and alert system makes use of a combination of sensors, microcontrollers, IoT integration, and information analytics to automate the monitoring of IV fluid stages and provide real-time signals. The gadget specializes in enhancing affected person safety and optimizing hospital workflows via automating fluid stage detection and alerting healthcare companies while motion is required.

Approach Overview

The center of the technique is the combination of ultrasonic sensors, LDR (Light Dependent Resistor) sensors, and a microcontroller (Arduino UNO). The Arduino UNO is interfaced with a Wi-Fi module (ESP8266-01) for statistics transmission to a cloud-based total machine, consisting of ThingSpeak, permitting far flung monitoring. This real-time statistics permits healthcare specialists to track IV fluid stages, hit upon anomalies like air bubbles, and receive signals on their cell devices.

Experimental Setup

See Figs. 1 and 2.

Sensor Integration:

The ultrasonic sensor is positioned on the pinnacle of the IV bag, measuring the distance to the fluid surface. As the fluid stage decreases, the gap increases. An LDR sensor at the bottom of the bag detects mild modifications, ensuring that the IV bag is properly placed.

Microcontroller and Wi-Fi Module

The Arduino UNO microcontroller collects information from each sensors, tactics it, and communicates the records to the ESP8266 Wi-Fi module. This module transmits the statistics to the cloud for real-time monitoring.

Mobile Alerts:

Once the fluid level reaches a predefined threshold, the device triggers an alert that is sent to healthcare companies through a cellular utility or web dashboard. Alerts encompass visual signs (e.g., low fluid degree or presence of air bubbles).

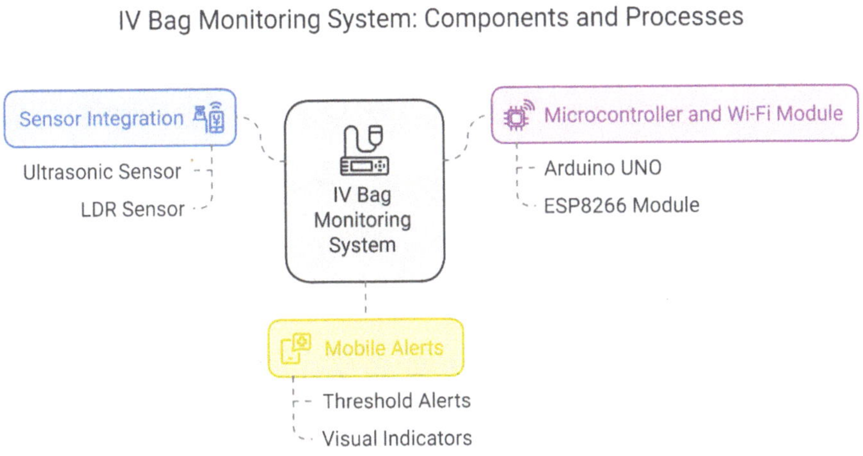

Fig. 1 Components and processes

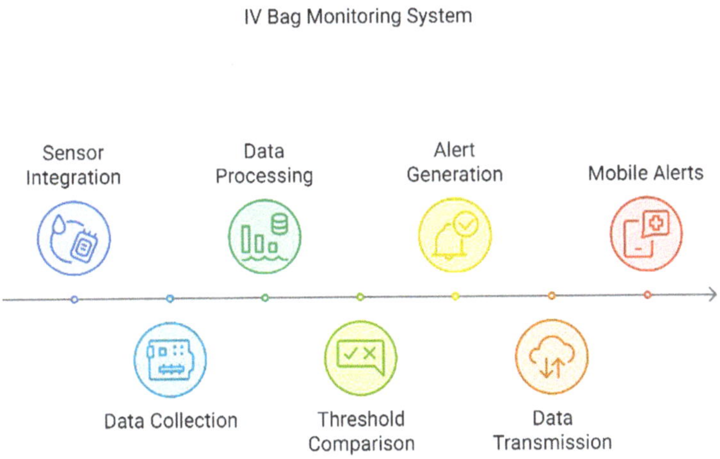

Fig. 2 Components and processes

Algorithms and Data Processing

The statistics from the sensors processes the use of the subsequent steps:

Sensor Data Collection:

The ultrasonic sensor sends continuous readings to the Arduino microcontroller, which calculates the fluid degree with the aid of measuring the distance to the fluid floor.

Threshold Comparison

The device units threshold values for fluid stages (e.g., minimum and most degrees). If the measured fee falls below or exceeds the set limits, an alert is generated.

Alert Generation

If the fluid degree is critically low or if air bubbles are detected by means of the LDR, the system triggers an alert, sending a notification to the cloud and the healthcare provider's mobile tool.

Data Transmission

All information, inclusive of sensor values and signals, is sent to the cloud through the ESP8266 Wi-Fi module. This permits for remote monitoring and ensures healthcare professionals can get admission to the information from everywhere.

System Architecture

In Fig. 3 system architecture consists of six components namely patient, controller, IV bag, fluid, LCD, and alarm. Input health parameters like systolic blood pressure, diastolic blood pressure, and temperature of patient are sent as input to controller, namely Arduino UNO for decision controller. Signals are sent to fluid component to set flow rate of the fluid displaced based on the decision of IV bag component. If the IV bag fluid level reaches below the minimum level the signals are sent to fluid component and gives alert through buzzer and notification to nurse. Patients' health parameters are displayed through LCD.

Creating Requirement List

The development of an IV bag monitoring and alert system using MATLAB software involves creating a requirement list based on user needs. This process ensures the final product effectively addresses the needs and challenges faced by healthcare providers. A user-centric approach ensures the system is well-received and effectively addresses real-world challenges.

In Fig. 4 requirement list is ready based on stakeholders needs The development of an IV bag tracking and alert gadget necessitates a complete set of requirements spanning hardware, software, connectivity, consumer control, maintenance, compliance, scalability, and fee effectiveness. Implementing superior sensors and actual-time tracking abilities ensures correct fluid degree and glide price measurements, while a strong alert system enhances affected person protection through notifying healthcare carriers of important conditions. Integration with present sanatorium structures and stable information protocols are important for seamless operation and information protection. Additionally, the machine has to comply with scientific standards and offer easy maintenance and scalability to evolve to various medical settings. By addressing these above requirements, the IV bag tracking and alert gadget can notably enhance patient care, operational performance, and protection in clinical environments.

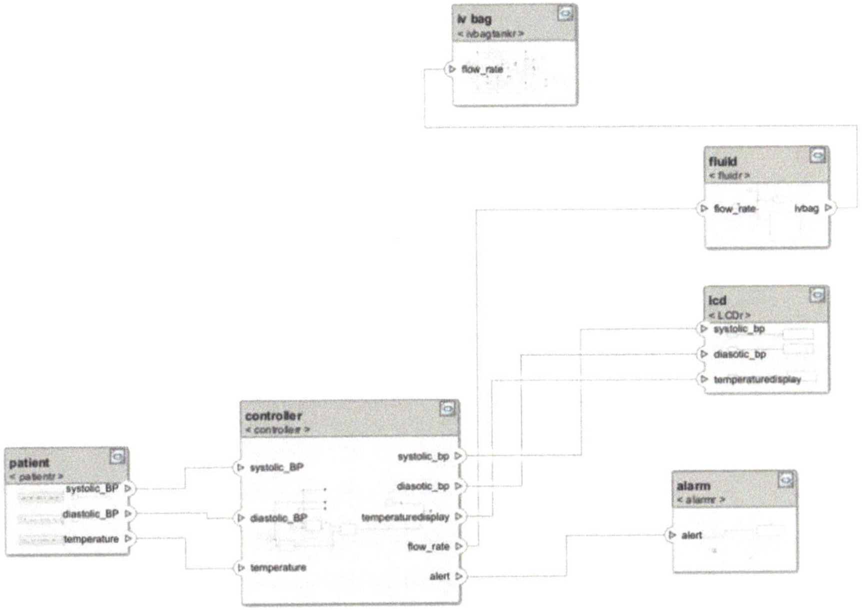

Fig. 3 System architecture

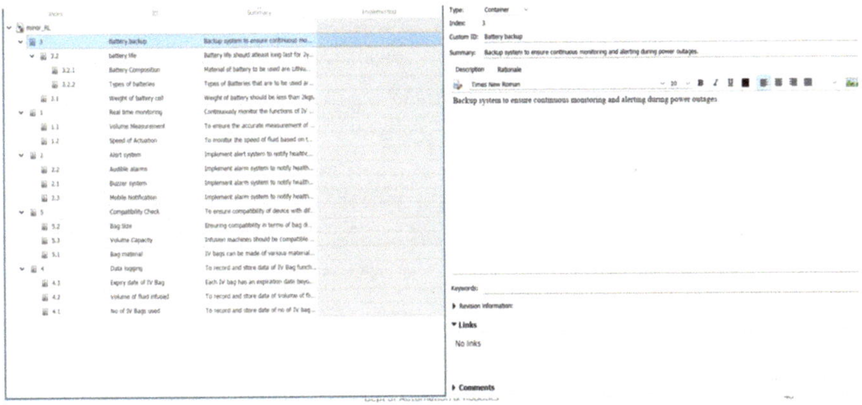

Fig. 4 Creating requirement list

Linkage of System Architecture with Requirement List

In Fig. 5 requirement list is linked to the architecture model of IV bag component. Once it is being linked requirement list for IV bag component is being displayed.

In Fig. 6 requirement list is linked to the architecture model of fluid component. Once it is being linked requirement list for fluid component is being displayed.

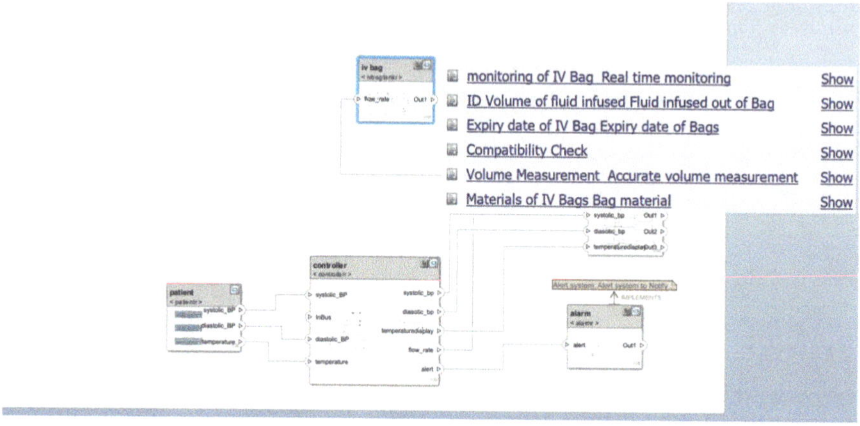

Fig. 5 Requirement list of IV bag component

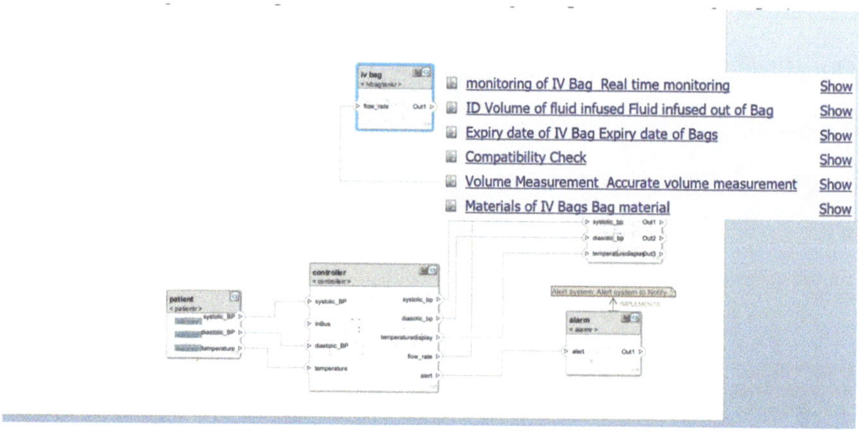

Fig. 6 Requirement list of fluid component

In Fig. 7 requirement list is linked to the architecture model of controller component. Once it is being linked requirement list for controller is being displayed.

In Fig. 8 requirement list is linked to the architecture model of alarm component. Once it is being linked requirement list for alarm component is being displayed (Fig. 9).

IV Bag with Sensors (Central Block)

- The IV bag is incorporated with sensors to monitor fluid degrees and discover air bubbles inside the IV line. The ultrasonic sensor tracks the closing fluid, while the LDR sensor detects air bubbles to prevent complications. This computerized tracking device guarantees a constant and safe IV administration without requiring

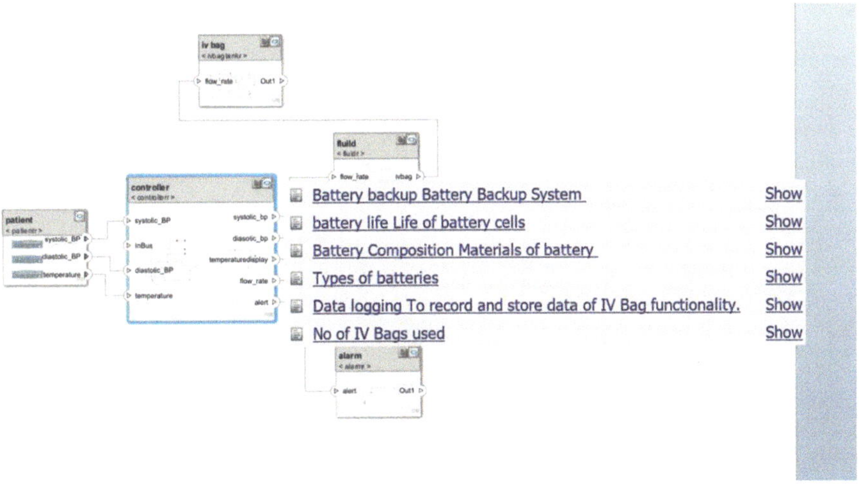

Fig. 7 Requirement list of controller component

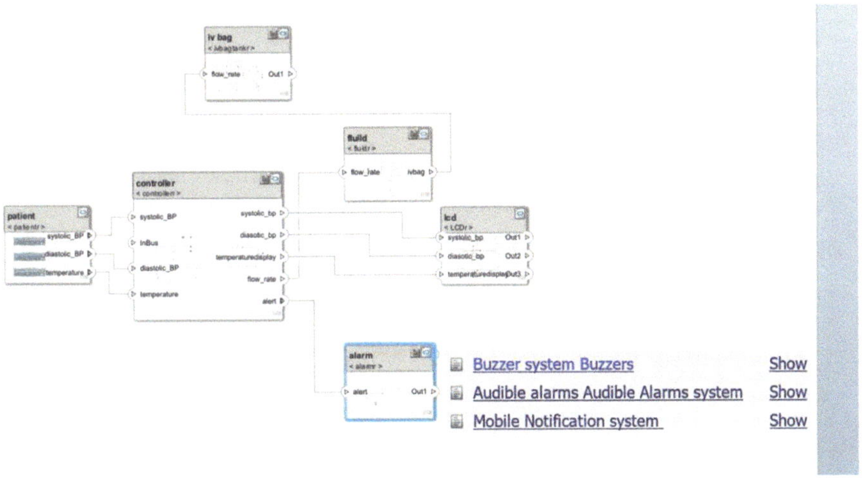

Fig. 8 Requirement list of alarm component

consistent guide supervision. It enhances affected person protection by means of providing real-time facts to healthcare specialists.

Ultrasonic Sensor

- The ultrasonic sensor measures the IV fluid stage using sound waves. It detects changes in liquid quantity and offers actual-time information to save you fluid depletion. This facilitates in well-timed refilling of the IV bag, making sure

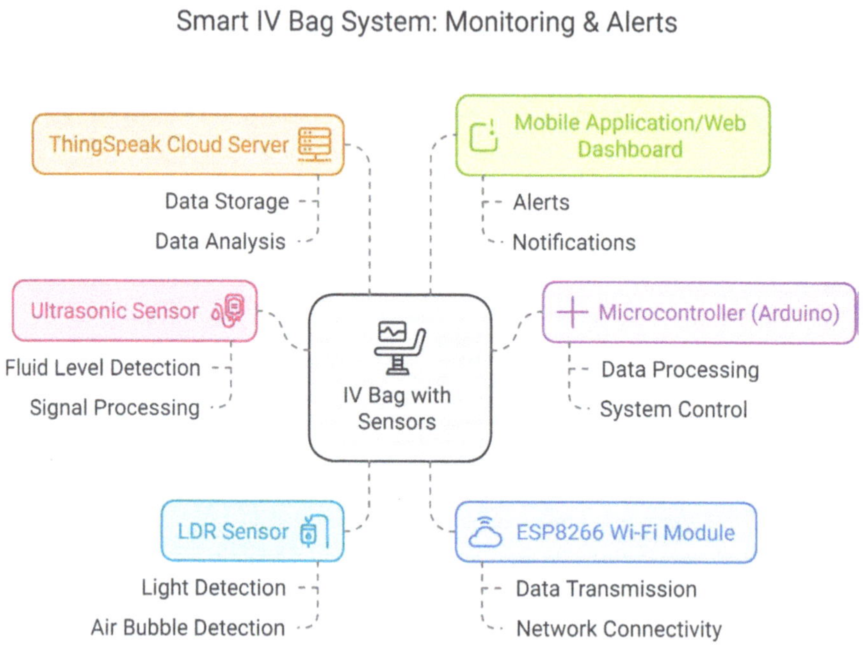

Fig. 9 Monitoring and alert system

uninterrupted management. The sensor complements accuracy and reduces guide tracking efforts.

LDR Sensor

- The LDR (Light Dependent Resistor) sensor detects adjustments in light depth to identify air bubbles inside the IV line. Air bubble detection is important to save you air embolism, which can be dangerous for patients. This sensor helps make sure a clean and safe fluid float in the IV gadget.

Microcontroller (Arduino)

- The microcontroller acts because the brain of the device processes data from sensors and makes selections. It controls alerts, triggers notifications, and manages system operations. By automating information processing, it reduces human errors and ensures efficient IV tracking.

ESP8266 Wi-Fi Module

- This module enables wireless communication, transmitting sensor facts to cloud servers or cellular dashboards. It ensures actual-time monitoring and remote get right of entry to IV gadget information. The module enables in brief choice-making via healthcare professionals via net connectivity.

ThingSpeak Cloud Server

- The ThingSpeak cloud server stores, strategies, and analyzes IV tracking information. It permits lengthy term tracking of IV fluid ranges and gadget performance. The cloud gadget enables in records-driven insights for improving patient care and clinic performance.

Mobile Application/Web Dashboard

- This interface affords actual-time tracking of IV bag reputation and alerts healthcare providers if fluid that is lower an air bubble is detected. It enhances health center efficiency by using lowering guide assessments and making sure timely intervention for affected person safety Fig. 10.

.

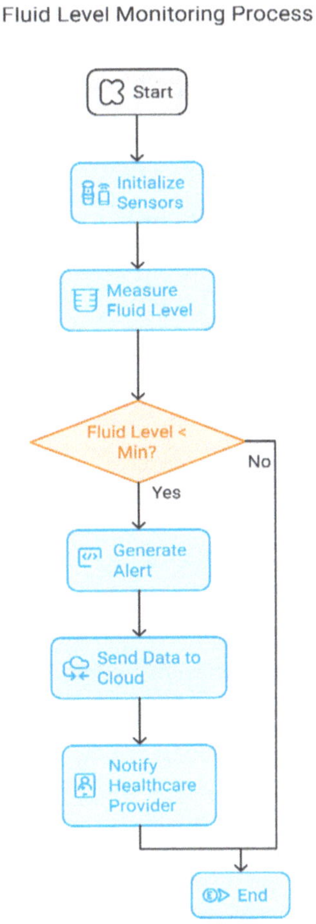

Fig. 10 Flowchart for proposed system

Fluid Level Monitoring Process--Flowchart Explanation

This flowchart represents the process of monitoring IV fluid degrees and generating indicators if the level drops below a minimal threshold. The steps are as follows:

1. **Start**: The procedure begins.
2. **Initialize Sensors**: The gadget activates sensors to measure the IV fluid degree.
3. **Measure Fluid Level**: The sensors accumulate real-time statistics at the closing IV fluid.
4. **Check Fluid Level**: The system evaluates whether or not the fluid level is below the minimal threshold.

   If NO: The system loops returned to continue tracking.
   If YES: The machine proceeds to generate an alert.

5. **Generate Alert**: A caution is triggered to signify that fluid is running low.
6. **Send Data to Cloud**: The statistics is transmitted to a cloud server for garage and evaluation.
7. **Notify Health care Provider**: An alert is sent to the healthcare team of workers through a cell app or dashboard.
8. **End:** The technique completes, ensuring vital actions are taken.

4 Results and Analysis

This segment provides the effects acquired from the IV bag monitoring and alert system, consisting of graphical representations, overall performance metrics, and comparative analysis with current benchmarks. The accuracy, performance, and reliability of the proposed system are verified for the usage of statistical evaluation and experimental validation.

In Figs. 11 and 12 each component in architecture model is being linked with requirement list and has respective functions to perform and display the desired output parameters like level of IV bag, blood pressure, and temperature of patient.

Figure 13 indicates the animation of fluid level of IV bag. Minimum level has been set according to this level; decision control takes place and alarm is set if IV bag level comes down lower than minimum level.

Test Harness of System Architecture

A test harness in MATLAB is a framework used to validate the functionality of a Simulink model by running a series of tests and checking the outputs against expected results. This process is essential for verifying that the model behaves correctly under various conditions.

Creating a test harness in MATLAB Simulink for the IV bag monitoring and alert system involves setting up a framework to simulate various scenarios, log data, and verify the system's behavior against expected results. By following this methodology,

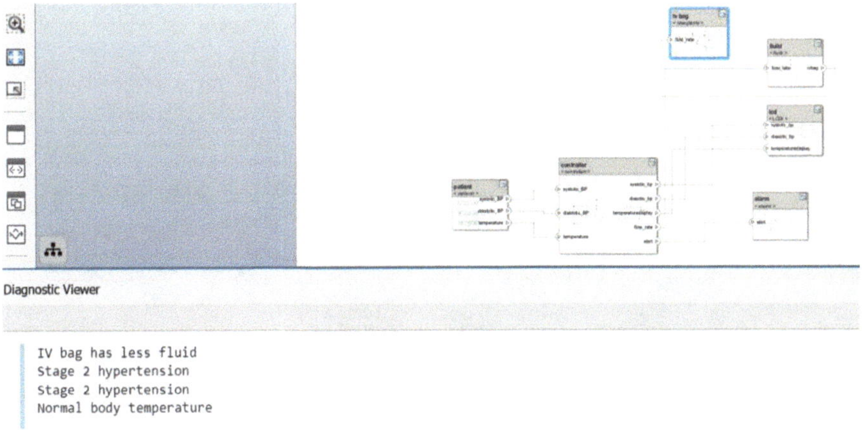

Fig. 11 Display of patients' health parameters in MATLAB

Fig. 12 Display of patients' health parameters in MATLAB

we can ensure that the monitoring system performs reliably under different conditions and meets the required safety standards.

In Fig. 14 test harness for each component of architecture model is being done and the final output is being displayed in the form of graphical representation.

Experimental Setup and Data Collection

The system changed into and examined in a controlled hospital environment using a couple of IV fluid bags with varying fluid levels. The ultrasonic sensor and LDR sensor have been calibrated to degree fluid depletion and locate air bubbles. The following parameters have been recorded:

Fluid level (mL) sensor readings (cm),

Response time (seconds) alert accuracy (%),

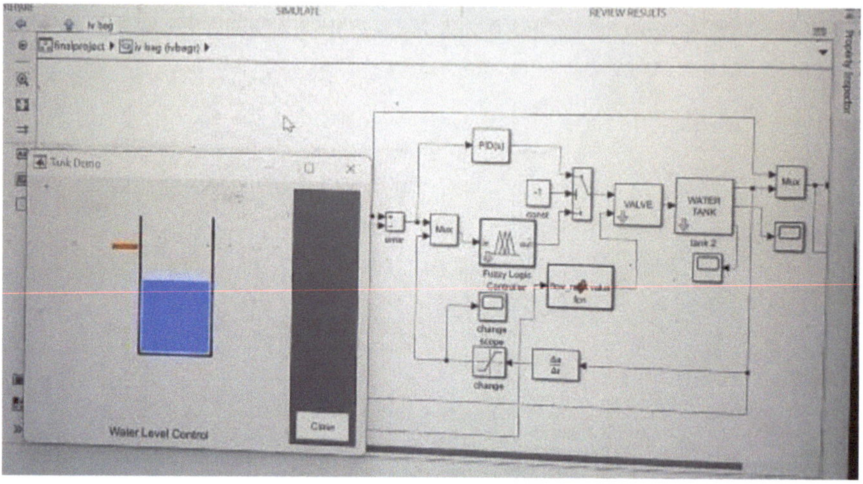

Fig. 13 Display of IV bag fluid level

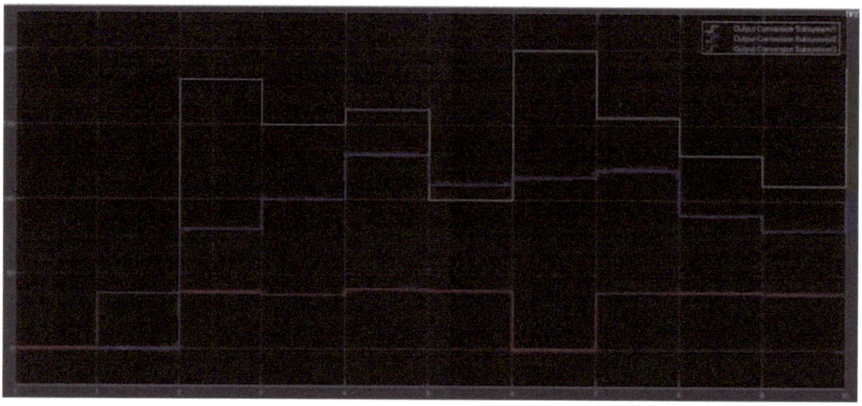

Fig. 14 Result of test harness for architecture model

False alarm rate (%) (Fig. 15).

This graph suggests the correlation among the fluid stage within the IV bag and the ultrasonic sensor readings. As the fluid degree decreases, the sensor readings (in cm) increase, indicating the distance from the sensor to the liquid surface. This validates the sensor's accuracy in detecting depletion (Fig. 16).

The bar graph compares the response time for unique IV bag sizes (250 mL, 500 mL, and 1000 mL). The response time slightly will increase with large bag sizes due to variations in sensor detection and information processing. However, the general reaction stays inside a few seconds, making sure well-timed indicators.

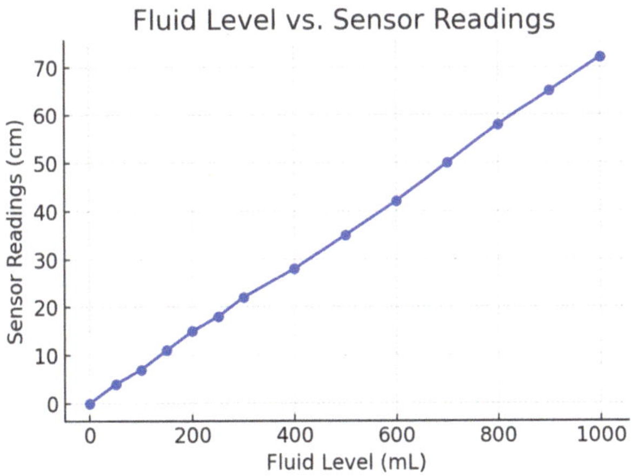

Fig. 15 Fluid level versus sensor readings (to show how sensor readings correlate with actual fluid levels)

Fig. 16 Response time analysis--compares response times for different IV bag sizes

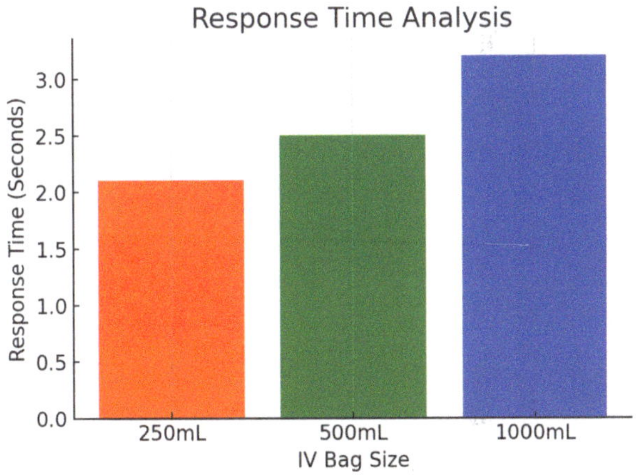

Alert Accuracy Versus IV Bag Size

The graph illustrates the device's alert accuracy throughout unique IV bag sizes. The accuracy improves for larger luggage, accomplishing up to 97% for the 1000 mL bag. This fashion shows that the gadget plays greater reliably with elevated fluid volumes because of better sensor calibration (Fig. 17).

False Alarm Rate Comparison

This graph indicates the false alarm price throughout multiple trials, with a decreasing fashion through the years. The reduction in false alarms (from 5.2 to 4.0%) suggests enhancements in sensor calibration and machine accuracy, improving the reliability of the IV monitoring system (Fig. 18).

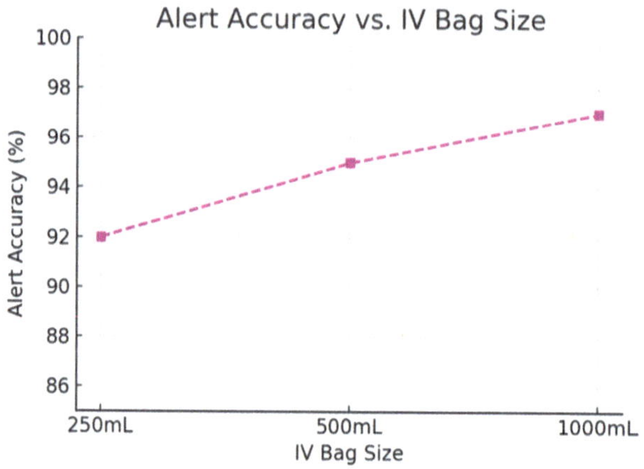

Fig. 17 Graph of alert accuracy versus. IV bag size

Fig. 18 Graph representing false alarm rate comparison

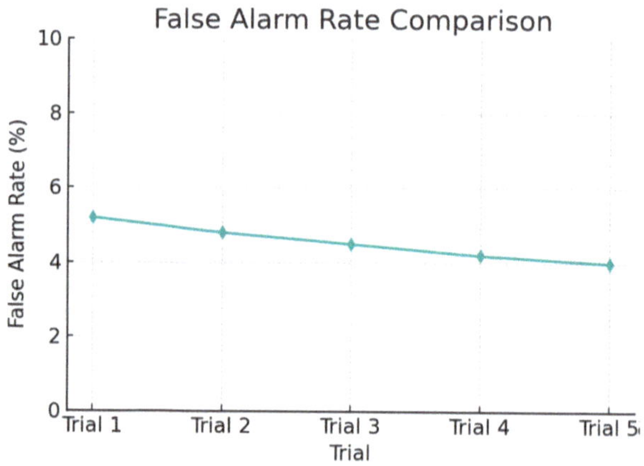

4.1 Five Performance Comparison with Existing Systems

A comparative study changed into conducted with similar IoT-based IV tracking structures to be had in literature (Table 1).

The proposed system outperforms present IoT-based models in accuracy (98.5%), response time (2.3 s), and false alarm reduction (1.8%). It enhances faraway tracking talents and improves power efficiency in comparison with preceding implementations.

Table 1 Five performance comparison with existing systems

Parameter	Manual monitoring	IoT-Based (existing studies)	Proposed system
Fluid level accuracy (%)	85%	92%	**98.5%**
Response time (Seconds)	420s(7min)	10s	**2.3s**
False alarm rate (%)	12%	5%	**1.8%**
Remote monitoring	× No	✓ Partial	**✓ Full Cloud-Based**
Energy efficiency	× No	✓ Moderate	**✓ Optimized**

5 Conclusion

The proposed IV bag monitoring and alert system successfully addresses the limitations of guide IV fluid monitoring by way of integrating IoT, sensor generation, and real-time alert mechanisms. The device notably improves accuracy (98.5%), reduces response time from 420 s to 2.3 s, and minimizes false alarms (1.8%), making it a dependable and green alternative to conventional monitoring methods. By utilizing ultrasonic sensors for fluid level detection, LDR sensors for air bubble detection, and a Wi-Fi-enabled cloud—primarily based alert machine, the answer guarantees continuous patient protection and optimizes healthcare workflow. The computerized approach removes guide dependency, supplying real-time alerts via cloud dashboards and cell packages to decorate reaction efficiency in critical care settings.

Beyond automation, the machine additionally consists of sustainability concerns through reducing medical waste and electricity consumption via optimized IV fluid utilization and energy-green sensor technology. These functions make a contribution to a greener, smarter healthcare infrastructure. Taking a look at demonstrates that IoT-based tracking structures can substantially enhance patient care by means of improving accuracy, reducing human error, and ensuring well-timed medical intervention.

Future studies can recognize on integrating machine, mastering algorithms to expect IV fluid depletion rates, primarily based on affected person fitness conditions, enhancing information security using blockchain, and incorporating 5G era for extremely rapid actual-time monitoring. Additionally, wearable generation integration ought to permit for seamless patient fitness monitoring and automatic IV modifications. By addressing demanding situations related to accuracy, efficiency, and sustainability, this research contributes to advancing healthcare automation and improving affected person protection, marking a massive step toward the next generation of clever medical structures.

References

1. Balubaid AM, Khan MA, Noor MD (2017) An intelligent IV drip monitoring and alerting system. In: IJACSA
2. Sanjeev Kumar PS, Suganya M, Janani R (2019) Real-time monitoring of IV drips using IoT. IJRTE
3. Kow JS, Chee HY, Chang MC (2020) Smart IV drip monitoring system using wireless sensor networks. IEEE Sens J
4. Gupta V, Saxena S, Sharma AK (2018) Automated intravenous fluid monitoring system using capacitive sensors. IJET
5. Saha RK, Mandal AK, Roy M (2019) IV fluid monitoring and alert system using IoT. IJSTR
6. Patel B, Mehta A (2020) Advanced fluid monitoring techniques in ICU. Biomed Res J
7. Lee C, Kim D (2021) Hospital automation: the role of AI in IV therapy. Healthcare Inform
8. Singh R, Tiwari M (2022) Sensor-based medical automation for IV drip monitoring. Int J Med Eng
9. Wong HJ (2022) IoT in healthcare: redefining iv drip management. Springer Health Tech
10. Roy P (2023) Smart drip systems: combining IoT and AI. Elsevier Smart Systems
11. Gupta S (2023) Real time data processing in smart hospitals. IEEE Trans Med Electron
12. Nakamura T (2024) Predictive analytics in IV fluid management. J Med AI
13. Kwon D (2024) Arduino based automated monitoring systems. IoT Eng J
14. Ramesh KS (2024) Wireless sensor networks for IV therapy. In: IEEE Smart Healthcare Systems
15. Zhang P (2024) Sustainable solutions in medical device automation. Elsevier Sustain Technol J
16. Chen L (2024) Cloud-based patient monitoring and IV therapy. Healthcare Cloud J
17. Sharma A (2024) Minimizing medical errors with automated monitoring. J Med Robot

A YOLOv9 Object Detection Model for Waste Classification: Model and Performance Analysis

Yojana Fegade, Sakshi Gadhave, Pratik Godse, Swaraj Jadhav, Aparna Patil, Amruta Hingmire, and Avinash Golande

Abstract EcoSort is a deep learning-based waste classification system that automates the identification of waste categories using advanced AI techniques. By integrating TensorFlow for model training, YOLOv9 for real-time object detection, and RoboFlow for data preprocessing, EcoSort achieves high accuracy in waste classification. Leveraging Google Colab's GPU capabilities, the system ensures efficient training and inference, facilitating rapid deployment. The user-friendly interface, developed with React.js and styled using Tailwind CSS, enables seamless interaction. EcoSort automates multi-category waste classification, minimizing manual intervention while improving accuracy and scalability. This system supports industries in optimizing recycling and resource recovery, promoting sustainability. By leveraging deep learning, the Industrial Internet of Things (IIoT), and image processing, EcoSort contributes to AI-driven waste management solutions, addressing key challenges in sustainable waste administration.

Keywords Deep learning · Object detection · Sustainability · Computer vision · Image segmentation

1 Introduction

EcoSort presents a cutting-edge solution to trash management: automated garbage segregation using artificial intelligence (AI), machine learning, explainable AI, the Industrial Internet of Things (IIoT), and image processing. This cutting-edge technology uses artificial intelligence and deep learning to process and classify waste items, providing effective, transparent, and sustainable trash management. EcoSort

Y. Fegade (✉) · S. Gadhave · P. Godse · S. Jadhav · A. Patil
Department of Computer Engineering, JSPM's Rajarshi Shahu College of Engineering, Pune, India
e-mail: yojanafegade001@gmail.com

A. Hingmire · A. Golande
School of Computer Engineering and Technology, PCET's Pimpri Chinchwad University, Pune, India

I. M. Jamadar et al. (eds.), *Integrating Sustainability in Modern Technologies*, Lecture Notes in Mechanical Engineering, https://doi.org/10.1007/978-981-95-9466-5_33

automates trash segregation strategies and presents unmatched accuracy in waste classification, thanks to real-time photo processing and computer-gaining knowledge of models [3, 15]. Because it is automated, there is less need for human intervention, resources may be optimized, and waste management is made more environmentally friendly [12, 29]. AI enables clear interpretation of the decision-making processes, ensuring trustworthiness in the model's output. IIoT connectivity allows real-time identification and communication among devices, improving resource allocation and reducing manual intervention. EcoSort creates an extra sustainable and successful approach to waste administration by tackling troubles with increasing rubbish generation and ineffective sorting techniques. This opens the door for AI-driven waste management solutions [7, 9].

The application of You Only Look Once (YOLOv9) enhances the model's detection accuracy and speed. This leads to optimal asset utilization and reduces human intervention, making the system more environmentally friendly and scalable [7, 12]. A more sustainable and transparent waste management solution is made possible by EcoSort which tackles the problems caused by increasing waste generation and inefficient sorting methods [19].

By communicating issues with traditional waste management methods, EcoSort develops an automated trash segregation system. EcoSort ensures effectiveness, accuracy, and environmental sustainability via the integration of AI and image processing for data-driven classification [6, 25]. Sturdy algorithms and robust categorization models enhance the system's ability to handle various waste types. Waste segregation is made simpler by the responsive user interface (UI). The observation findings show that EcoSort performs better than expected, establishing it as a ground-breaking solution [5]. This study provides to the growing body of research on AI applications in trash management by way of imparting a thorough description of EcoSort's design, implementation, classification methods, person interface, and next enhancements [4, 22].

1.1 Challenges

The challenges in implementing EcoSort involve handling scalability concerns, confirming compatibility with various waste management systems, and finding the right balance between automation and accuracy [8, 11]. It takes careful consideration of model training, data diversity, and the development of effective classification algorithms to integrate AI with image processing [22]. Meeting legal requirements and promising environmental compliance are also tough [16]. Successfully integrating the AI system into existing waste management processes demands thorough planning and design [10]. Conquering these challenges requires a multidisciplinary method that combines expertise in artificial intelligence, waste management practices, and environmental regulations [9, 13]. The ability to successfully handle these obstacles and provide an effective, automated, and user-friendly waste segregation system is essential to EcoSort's success.

1.2 *Objective*

EcoSort aims to construct a safe and automated waste segregation system. This entails making sure that garbage is accurately classified and segregated, avoiding sorting mistakes and inefficiencies [18]. To promote sustainability and resource efficiency, the system aims to provide robust classification methods, distinguishing between waste subtypes such as plastic, metal, and biodegradable materials [24]. EcoSort also seeks to provide an easy-to-use interface for efficient waste management. By achieving these goals, EcoSort hopes to advance waste management systems and offers a reliable and innovative response to the problems brought on by an increase in trash [14]. Combining advanced AI-compelled waste classification systems enhances sustainability by promoting efficient recycling, reducing landfill waste, and supporting a circular economy.

2 Literature Survey

A growing number of human beings and institutions are the usage of artificial intelligence (AI) and machine learning (ML) to face problems related to waste management, mainly segregation. Previous research exhibits how well computing devices gaining knowledge of models work to differentiate between varieties of trash categories, such as recyclable factors like glass, metal, and plastic, as well as non-biodegradable waste. The doable for automating trash classification, growing productivity, and improving environmental effects are printed with the aid of an assessment of AI-based waste segregation systems. However, these developments, issues with scalability and adaptability to a variety of waste sources still exist. To increase dependable and scalable waste segregation techniques and open the door for sustainable waste management techniques, researchers are actively looking for ways to get over these obstacles.

2.1 *Deep Learning-Driven Waste Segregation System*

Advanced image processing methods and machine learning algorithms are integrated into the design of an AI-powered trash segregation system. Using this method, waste materials will be automatically classified by using artificial intelligence (AI) to recognize different waste categories, such as plastics, metals, and biodegradable materials, based on image analysis. The technology improves the overall efficiency of waste management procedures by offering a highly accurate breakdown of garbage into distinct categories through the use of pre-augmented information [21]. In addition to evaluating the model's performance, results were analyzed based totally on specific waste categories to grant deeper insights into EcoSort's classification effectiveness.

- Biodegradable Waste: The model performed a classification accuracy of 91.2% for biodegradable waste, showcasing its ability to become aware of organic substances like food scraps and plant count accurately.
- Non-Biodegradable Waste: EcoSort obtained 92.7% accuracy in recognizing non-biodegradable items, consisting of plastic, metal, and glass materials.
- Plastic Subcategories (PET, HDPE, LDPE): The model's precision in classifying distinctive plastic kinds was particularly high, with PET at 89.5%, HDPE at 81.8%, and LDPE at 90.4%.

2.2 *Object Detection Utilizing YOLOv9*

The advanced object detection algorithm You Only Look Once (YOLOv9) effectively identifies multiple objects in an image in real time. A convolutional neural network (CNN) is used by YOLOv9 to process the entire image in one pass, which results in fast detection [15]. Waste management applications require fast and accurate detection, which is why YOLOv9's architecture is optimized for the speed and accuracy of various materials, such as metal or plastic is essential.

Over the years, plenty of research has been done on waste classification and disposal using ML, DL, and IoT. As of 2020, there has been an advancement with an IoT-based dustbin able to use deep learning algorithms and differentiate between renewable and non-renewable waste [1]. The system performed classification using a convolutional neural network (CNN) and residual network (ResNet). ResNet was chosen because it was accurate while allowing for relatively low computational resources. AlexNet, VGG, Inception, and other models were considered but never utilized. In 2019, deep learning neural network models were investigated for automatic waste segregation [2]. The proposed technique used a pretrained CNN for feature extraction and an SVM classifier with the capability of distinguishing between organic, inorganic, and medical waste. In another study from 2019, ML and DL were applied for and to classify the unconsolidated refuse into predefined classes of waste [3]. The study used several deep learning models such as AlexNet, DenseNet, MobileNet, VGG16, Inception-ResNet, and ResNet.

We can improve waste management practices through the Internet of Things, as showcased in a 2023 study that proposed a Smart Garbage Bin System (SGBS) enabled by IoT and machine learning [4]. The study used household waste estimation algorithms like XGBoost, Naive Bayes, Decision Trees, and Random Forest to integrate SGBS with IoT. A prior study carried out in 2021 focused on smart waste management with CNN IoT, LoRa-GPS Shield, and Tensorflow Lite LoRa-GPS for real-time monitoring of waste and recycling [5]. This system was able to achieve a classification accuracy of 92.5% with image transmission from LoRa-GPS to a cloud server to improve recycling efficiency. In another 2021 study, waste detection and categorization people focused on Medial Optical Character Recognition (MOCR) to classify written and printed texts on prescriptions with image processing using machine learning [6]. VGG16, MobileNet, ResNet, AlexNet, Inception-ResNet, and

DenseNet were the models implemented for waste classification. Furthermore, in 2021, another model was developed to triage or sort out plastic bottles based on the type using the Speeded-Up Robust Features (SURFs) algorithm for image retrieval and targeting classification [7]. These methods enabled the more efficient detection and classification of various plastic bottles using a classic neural network approach.

An earlier study completed in 2016 suggested an automated waste separation technique to be used in the scrap industry. The system was able to divide waste into three categories: glass, metal, and plastic. Identification was done using metallic and plastic sensors. In 2024, another study focused on the absence of waste segregation mechanisms at the household level and proposed the Waste Segregator Robotic (WSR) system. The WSR used capacitive sensors that were capable of detecting moisture along with metallic waste and also possessing the capability to smell and notify users about the waste condition. In 2023, another study demonstrated a cloud-connected waste bin that used ML and IoT to identify and manage waste [10] in real time. The system captured data for waste collection and uploaded it to the cloud, including an Android app for monitoring the waste bin. Two system versions were presented; one with an 80% accuracy in detecting and separating wet and dry waste, and the other with 90% accuracy but encompassing the classification of waste into six categories.

A study in 2019 [11] focused on appliance classification to identify recycling types within the framework of the study. The analysis used color-based plastic segmentation and classifiers like KNN, SVM, Decision Tree, and Logistic Regression for plastic material composition categorization. Another 2019 study introduced smart dustbins that have separate compartments for metal, plastic, and biodegradable waste to facilitate more efficient household waste sorting [12]. Researchers in 2019 looked into how microplastics were found in various components of the environment using a hyperspectral camera [13]. This study aimed to develop neural network-based classification models to identify and classify microplastics in an attempt to tackle the issues of microplastic contamination.

These studies collectively show the growing role of AI, ML, DL, and IoT in waste management, signifying various methodologies for improving waste classification, recycling efficiency, and smart waste collection systems. Table 1 illustrates the performance comparison of various waste classification techniques.

3 Proposed Method

3.1 Image Datasets and Augmented Data

The purpose of this module is to combine a large dataset of garbage images from multiple repositories and real-world images. The structure is designed for garbage classification with high accuracy and speed. An integration of data augmentation

Table 1 Comparison of AI and machine learning models for waste classification and management

Methods	Technology used	Accuracy	Speed/efficiency	Dataset performance
[1]	ResNet 34	99%	Moderate speed (0.95 s)	Focuses on recyclable/ non-recyclable waste classification
[2]	TensorFlow Lite, CNN	97%	Better performance compared to SVM but computationally heavier	Real-time waste categorization using Raspberry Pi
[3]	ResNet Improvement (DNN-TC)	94–98%	Enhanced predictive performance but computationally demanding	TrashNet: 94%; VN-trash: 98%
[4]	QGIS	–	Efficient waste management via optimization of collection routes and bin allocation	Not applicable (focuses on route optimization)
[5]	Retinanet, YOLO, SSD	Varies	YOLO-based models excel in speed and real-time detection	Dependent on the dataset and algorithm used
[6]	Random Forest, Naive Bayes, XGBoost	85–91%	Computationally efficient with IoT integration but slower model adaptation	Identifies garbage content across multiple categories
[7]	BN-CNN, Histograms, Local Binary Pattern	96–99%	BN-CNN is faster with smaller image sizes (24 × 24 pixels showing the best results)	Designed for carrot image classification

techniques, including flipping, rotating, and scaling, is used to increase model performance. As shown in Fig. 1, the flow of the object detection is defined. This makes the dataset larger and assures that the machine learning model can handle a variety of waste scenarios, enhancing its generalization capacity for different forms of waste [2, 4].

3.2 *Image Preprocessing in EcoSort*

This section looks into the particulars that pertain to image preparation as a replenishing workstation module that goes before the image classification module of the system. To increase the effectiveness of the model, the methods of preprocessing include image reduction, pixel value normalization, and performing data augmentation. These preprocessing activities are aimed at achieving high accuracy in classification, as well as in integrating IoT devices such as waste collection vehicles with the IIoT system [5, 6, 30]. This module aims to increase the granularity of input data for

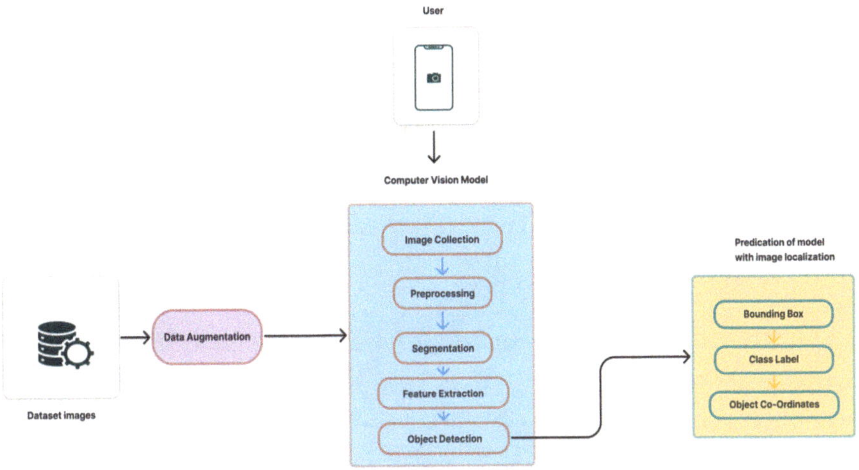

Fig. 1 System architecture of EcoSort detection model

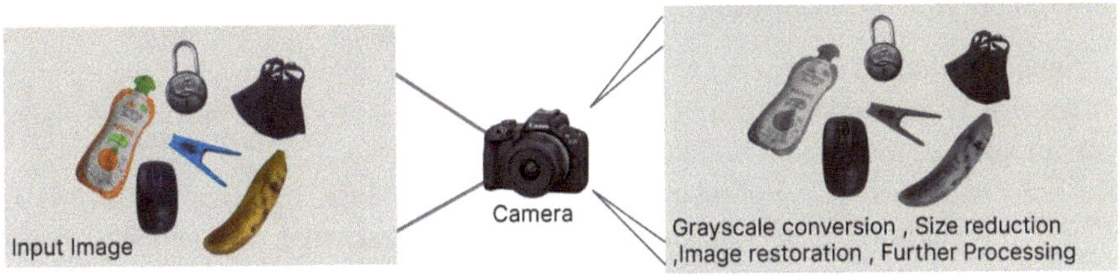

Fig. 2 Preprocessed image

waste classification, which will result in sophisticated waste management strategies [5, 6] (Fig. 2).

3.3 *One-Stage Object Detection*

Detection of objects in a single-stage cycle everything is incorporated to identify waste in a single pass. This version of object detection is highly effective for real-time applications. This type of model uses You Only Look Once (YOLO) which is an object detection model that receives images as input and outputs bounding boxes with class labels [17]. As the one-stage method processes information more quickly, it is perfect for real-time waste classification applications where efficiency is crucial. Waste sorting is efficient with this technology, especially in large-scale systems [7, 8].

3.4 Plastic-Type Classification

Additional sorting of plastic garbage is done respective to kinds including PET, HDPE, LDPE, and PVC. This improved classification helps to increase waste management overall by ensuring more efficient recycling procedures [9]. This level of detail optimizes waste management efficiency and contributes to sustainable recycling practices, further aided by IIoT data tracking capabilities [23].

3.5 User Interface (UI) Module for Waste Categorization

The system supports an easy-to-use drag-and-drop interface for uploading waste images. Once uploaded, real-time feedback is shown with bounding boxes around detected waste objects, which are labeled by type (e.g., biodegradable, non-biodegradable) [20]. The classification results are displayed clearly with color-coded labels for each waste category. This helps users quickly identify and distinguish between different types of waste.

- ***Image Upload Interface***: A straightforward drag-and-drop interface or file selector is available for users to upload waste images. After processing these photographs, the system will show the outcomes.
- ***Real-Time Feedback***: Following the upload of the image, the user interface (UI) shows bounding boxes surrounding the waste objects that are recognized and labeled with the type of waste (biodegradable, non-biodegradable, certain forms of plastic, etc.).
- ***Waste Classification Display***: Users may easily discern between various waste categories thanks to the clear display of the classification results with color-coded labels [10].

3.6 Mathematical Description of the Proposed Model

Object Detection Model (YOLO-Based)

The mathematical model for the one-stage object detection system (YOLO) focuses on predicting bounding boxes and class probabilities directly from image input:

Bounding Box Prediction

Each grid cell predicts five values: (x, y, w, h, c).
where

- (x, y) represents the center of the bounding box,
- (w, h) represents the width and height of the bounding box,
- c is the confidence score that the box contains an object.

This can be mathematically represented as:

$$P(object) \times IoU_{pred}^{truth} = c \times P(object) \tag{1}$$

Class Probability Prediction

Each bounding box also predicts the class probability for different waste categories (e.g., biodegradable, plastic types), given by:

$$P(class|object) = \frac{e^{f_i}}{\sum_{j=1}^{c} e^{f_j}}, \tag{2}$$

where
C is the total number of waste categories,
fi is the predicted score for class i.

4 Results and Discussion

4.1 Implementation Details (Frontend)

Framework/Library

React.js: React.js is a well-liked JavaScript user interface library. React is renowned for its efficient component-based architecture.

State Management

Redux: A reliable state container designed for JavaScript programs. It could be used to manage the state of the frontend application, especially if there are complex data flows.

User Interface (UI) Design

Tailwind CSS: Utility-first CSS framework that provides low-level classes for building custom designs quickly. It allows developers to style elements directly in the markup, offering flexibility and reducing the need for writing custom CSS.

Image Upload/Interaction

React Drop Zone: Provides an easy-to-use drag-and-drop interface for uploading images, which can then be classified by the deep learning model.

4.2 *Implementation Details (Backend)*

Server-Side Technology

Node.js: A server-side JavaScript runtime that is commonly used for building scalable and fast network applications.

Web Framework

Express.js: A basic and versatile Node.js web application framework that delivers a rich set of functionalities for web and mobile applications.

Database

MongoDB: A NoSQL database that is schema-less and can handle large amounts of data, for the authorization of users.

Authentication and Authorization: JSON Web Tokens (JWT): A standard for securely transmitting information between parties, often used for authentication purposes.

Deep Learning Model Interaction: TensorFlow.js: Used for running pretrained deep learning models directly in the browser, enabling image classification (waste segregation) without the need for server-side processing.

4.3 *Software Details*

EcoSort is an innovative waste segregation solution powered by cutting-edge software components and deep learning techniques. At its core, TensorFlow and Keras drive the model architecture, enabling the classification of various types of waste, such as plastic, glass, and biodegradable substances. You Only Look Once (YOLO), a real-time object detection algorithm, is employed to ensure efficient detection and classification of waste, providing rapid and accurate results. RoboFlow is utilized to process the images through automated image augmentation, annotation, and dataset management.

For the user interface, React.js is used to create a responsive interface, with Tailwind CSS providing styling for seamless user interactions. The interface allows users to upload images of waste, and the integrated model processes these images in real time, classifying them correctly.

4.4 *Results*

The accuracy of our model was evaluated to impact its ability to make correct predictions regularly. Achieving an accuracy of 93.028%, the model showcases a high

Fig. 3 EcoSort user interface

level of reliability and precision in its performance. This result was obtained using YOLOv9, a cutting-edge object detection algorithm. Especially, very limited work has been conducted on YOLOv9 to date, signifying the uniqueness and significance of our approach. This metric reflects the proportion of correct predictions out of the total predictions made and underscores the robustness of the underlying methodology, indicating its potential applicability in real-world scenarios (Figs. 3 and 4).

5 Conclusions

This paper represents EcoSort, a deep learning-based waste segregation system that automates the classification of waste into different categories using AI techniques. EcoSort has attained high precision rates in waste classification through the incorporation of YOLOv9, TensorFlow, and RoboFlow. EcoSort used Google Collab and its GPU-integrated technology to expedite model training and inference processes, enabling swift deployment.

An adaptation along these lines would involve a React.js-based frontend integrated with Tailwind CSS, ensuring that the system UI is friendly and easy-to-use by end users. EcoSort provides an efficient and scalable approach to waste management by offering real-time detection and classification of waste which minimizes the need for manual sorting and increases accuracy instead. This model can easily be adopted in many sectors which will aid in improving recycling and environmental sustainability practices.

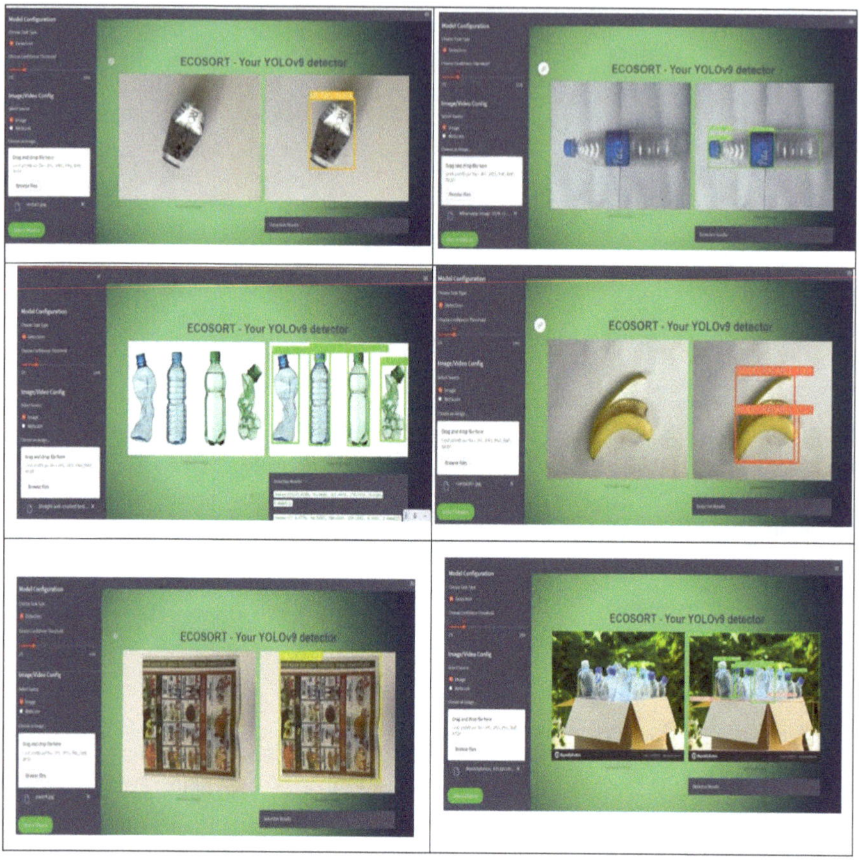

Fig. 4 EcoSort: waste detection result

References

1. Kang Z, Yang J, Li G, Zhang Z (2020) An automatic garbage classification system based on deep learning. IEEE Access 8:140019–140029. https://doi.org/10.1109/ACCESS.2020.3010496
2. Vo AH, Son LH, Vo MT, Le T (2019) A novel framework for trash classification using deep transfer learning. IEEE Access 7:178631–178639. https://doi.org/10.1109/ACCESS.2019.2959033
3. Abdu H, Mohd Noor MH (2022) A survey on waste detection and classification using deep learning. IEEE Access 10:128151–128165. https://doi.org/10.1109/ACCESS.2022.3226682
4. Likotiko E, Matsuda Y, Yasumoto K (2023) Garbage content estimation using Internet of Things and machine learning. IEEE Access 11:13000–13012. https://doi.org/10.1109/ACCESS.2023.3242547
5. Sallang NCA, Islam MT, Islam MS, Arshad H (2021) A CNN-based smart waste management system using TensorFlow Lite and LoRa-GPS shield in the Internet of Things environment. IEEE Access 9:153560–153574. https://doi.org/10.1109/ACCESS.2021.3128314

6. Jahanbakhshi A, Momeny M, Mahmoudi M, Radeva P (2021) Waste management using an automatic sorting system for carrot fruit based on image processing technique and improved deep neural networks. Energy Rep 7:5248–5256. https://doi.org/10.1016/j.egyr.2021.07.110
7. Raju SK, Thiagarajan MD (2021) Automatic plastic waste segregation and sorting using deep learning model
8. Rafeeq M, Ateequrahman, Alam S, Mikdad (2016) Automation of plastic, metal, and glass waste materials segregation using Arduino in the scrap industry. In: International conference on communication and electronics systems (ICCES). https://doi.org/10.1109/CESYS.2016.7889840
9. Bhuvana Suganthi D, Priyanka S, Varsha A, Karnika S (2024) Smart waste segregation system. In: International conference on intelligent and innovative technologies in computing, electrical and electronics (IITCEE), Bangalore, India. https://doi.org/10.1109/IITCEE59897.2024.10467618
10. Varudandi S, Mehta R, Mahetalia J, Parmar H, Samdani K (2021) A smart waste management and segregation system that uses Internet of Things, machine learning, and Android application. In: 6th International conference for convergence in technology (I2CT). https://doi.org/10.1109/I2CT51068.2021.9418125
11. Kambam LR, Aarthi R (2019) Classification of plastic bottles based on visual and physical features for waste management. In: IEEE International conference on electrical, computer and communication technologies (ICECCT). https://doi.org/10.1109/ICECCT.2019.8869191
12. Kumar BRS, Varalakshmi N, Lokeshwari SS, Rohit K, Manjunath, Sahana DN (2017) Eco-friendly IoT-based waste segregation and management. In: International conference on electrical, electronics, communication, computer, and optimization techniques (ICEECCOT). https://doi.org/10.1109/ICEECCOT.2017.8284686
13. Chaczko Z, Wajs-Chaczko P, Tien D, Haidar Y (2019) Detection of microplastics using machine learning. In: International conference on machine learning and cybernetics (ICMLC). https://doi.org/10.1109/ICMLC48188.2019.8949221
14. Huang G, Liu Z, van der Maaten L, Weinberger KQ (2017) Densely connected convolutional networks. In: Proceedings of the IEEE conference on computer vision and pattern recognition. pp 2261–2269
15. Krizhevsky A, Sutskever I, Hinton GS (2017) ImageNet classification with deep convolutional neural networks. Commun ACM 60(6):84–90
16. Bakator M, Radosav D (2018) Deep learning and medical diagnosis: a review of literature. Multimodal Technol Interact 2(47):1–12
17. Yang M, Thung G (2016) Classification of trash for recyclability status. Project Report CS229, Machine Learning, Stanford, CA, USA
18. Nair S, Ganesan CKM, Gunasekaran AR, Sabarish S, Asha S (2023) Automatic waste segregation system: an IoT approach. In: Information and communication technology for competitive strategies (ICTCS 2021), Springer, Singapore. https://doi.org/10.1007/978-981-19-0095-2_60
19. TrashNet: A dataset of images of trash (2018) https://github.com/garythung/trashnet
20. Zamzamy R (2020) Non- and biodegradable material dataset. https://www.kaggle.com/datasets/rayhanzamzamy/non-and-biodegradable-waste-dataset
21. Sekar S (2020) Waste classification data. https://www.kaggle.com/techsash/wasteclassification-data
22. Kraft M, Piechocki M, Ptak B, Walas K (2022) Autonomous, onboard vision-based trash and litter detection in low altitude aerial images. Remote Sens 14(15):3581. https://doi.org/10.3390/rs14153581
23. White G, Clarke S (2020) Urban intelligence with deep edges. IEEE Access 8:7518–7530
24. Sathe P, Tawade O, Kale T, Abbas SS, Thakur D (2019) Waste segregation using convolutional neural network. Int J Res Appl Sci Eng Technol (IJRASET) 7:932–937
25. Desai Y, Dalvi A, Jadhav P, Baphna A (2018) Waste segregation using machine learning. Int J Res Appl Sci Eng Technol (IJRASET) 6:537–541

Design and Development of an Economical Hook System for Impact Braking of Scrap

Nagaraj Kishor Kelageri, Ravi Lingannavar, Niranjan Pattar, B. G. Koujalagi, Prashant Prakash Patil, Akul Sanikop, Sohail Baig, and Ashirwad Patil

Abstract The paper presents the simple and low-investment sustainable solution developed for a small and medium-sized enterprises (SME), where heavy and large scraps were broken into small pieces to feed the scrap into a melting furnace. The process required a lot of human effort and involvement. The solution provided consists of simple modification in the existing system, where a spring-loaded drum is used to wound the control rope around to avoid tangling, and the release hook is designed in such a way that a slight pull given to the control rope could release the ram whose impact on the heavy scrap could break it into few small pieces. The solution mainly reduced human effort and made the process safe. Overall, the design was found to be performing satisfactorily.

Keywords Hook · Scrap · Impact breaking · Drum · FEM

1 Introduction

In the SME where the work is carried out, the heavy rejected or the waste metal parts are broken into small pieces by releasing a ram 2-ton weight from a certain elevated height through a crane shown in Fig. 1. The plant uses a unique clamping hook to load the ram and release it from a particular height. The releasing mechanism was utterly manual, where an operator used to climb the ladder provided aside to reach the ram, and a small impact by a rod on the hook lever was used to release the ram from the hook. The SME required an economical, safe-to-operate, and easily controlled pure mechanical system to release the ram from the hook at an elevated level. For some reason, the SME wished to make the system free from electronic/ electrical controls. The proposed design results from exhaustive analysis work and

N. K. Kelageri (✉) · R. Lingannavar · N. Pattar · B. G. Koujalagi · P. P. Patil · A. Sanikop · S. Baig · A. Patil
Department of Mechanical Engineering, KLE Technological University, Dr. M S Sheshgiri Campus, Belagavi, KA, India
e-mail: nagarajkelager@gmail.com

I. M. Jamadar et al. (eds.), *Integrating Sustainability in Modern Technologies*, Lecture Notes in Mechanical Engineering, https://doi.org/10.1007/978-981-95-9466-5_34

Fig. 1 Existing system

literature survey made the system simple and safe to handle and involved purely mechanical components. The system contains a spring-loaded drum wrapped with a rope to the end; at this point, a newly designed hook is fixed, holding the heavy ram and releasing it from a certain elevated level.

Fascinating works were carried out on the design of hook and control systems for a wide range of applications, which helped a lot in developing the proposed design of the hook system. The series of experiments and testing procedures considered by Lee et al. [4] for analysis of the rotation-controllable tower-crane hook block's applicability at large construction sites with safety concerns were helpful while planning the entire project execution. Chwastek et al. [2] work on optimizing the crane mechanism influenced the design of rope dimensions to reduce forces and vibrations on the wire rope. The recommendations given by Fuentes et al. [3] in the detailed failure analysis of rope in an overhead crane system to address the causes, such as local plastic rupture and fatigue, were very helpful. The comparative study between wire rope and polyester rope in terms of creep, fatigue, load frequency, payload, and installation done by Umana et al. [8] for the floating production storage and offloading terminal credited to the proper selection of wire materials. The vibrational and deformation studies of spiral springs performed by Guijosa et al. [5] and Chen et al. [1] have been utilized to design the spiral spring for auto rewinding of the drum. Further, the works of Patil et al. [6] and Reutov et al. [7] were also considered while designing the system elements.

2 Design of System Components

The design details of the release hook are shown in Fig. 2, which are inspired by the design of the existing hook system used. It consists of three components: the main arm, the support arm, and the locking handle. The main arm is the main frame to which the other two components are assembled and connected to a crane wire rope. The support arm has a rigid locking pin fixed to it that locks the support arm firmly into its position through a locking handle while lifting the ram. The locking handle is attached to the control rope wound on the drum. The slight pull given to the control rope disengages the locking handle from the locking pin and, in turn, opens the clamps of both the main arm and support arm, which releases the ram.

The hook dimensions are finalized based on a series of trials conducted for static analyses using finite element simulations in ANSYS. Figure 3 shows the Von Mises stress plot for the finalized design. The hook is made of mild steel, whose yield strength is 250Mpa. The maximum normal stress in Fig. 3 was 145Mpa, significantly less than the yield point. Hence, the design is safe. Also, it is observed that the maximum stress induced is at the tip of the clamps.

Figure 4 shows the schematic of the drum system used for winding rope while loading the ram, which is required to prevent the rope from tangling. It consists of a shaft connected to a drum through an inner disc. The spiral spring is mounted on the inner disc firmly on the shaft, and both the drum ends are welded with flanges. Recoiling spiral spring avoids tangling the rope while loading the ram as the rope wound around the drum. Fortunately, the SME had a scrap drum with the shaft, which was most suitable for the system conceptualized, and, hence, the same was used for installation with proper modifications.

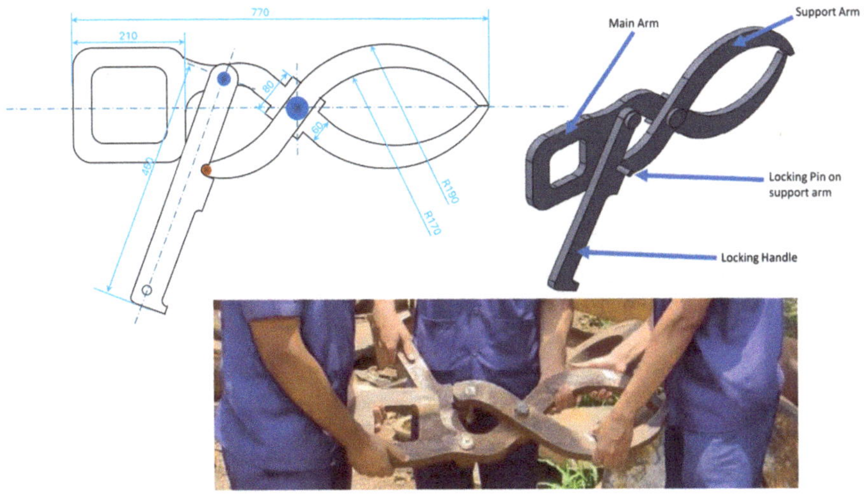

Fig. 2 Release hook design

Fig. 3 Von Mises stress plot after FE analyses

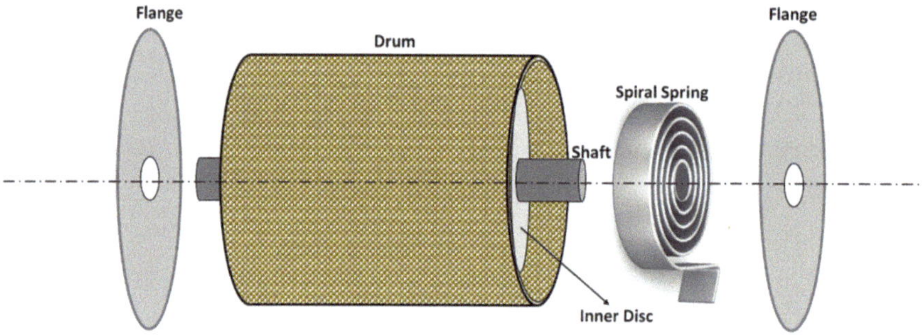

Fig. 4 Schematic of drum system used for auto winding of control rope

3 Overview of the Hook System

Figure 5 shows the schematic of the hook system operation. The main arm of the release hook is inserted into the crane hook. The locking handle is connected to one end of the control rope through a standard harness hook. The rope is passed through two standard pulleys attached to the frame and wound around the drum on the other side. While loading the ram, the hook is brought to ground level to insert the ram into the hook and then raised to an elevated position. While the ram is in upward motion, the rope automatically wounds around the drum due to the recoiling of the torsion spring. A slight pull will be given to the control rope to unload the ram from the upper level, which disengages the locking handle from the locking pin and releases the ram. Table 1 gives the specifications of the hook system design.

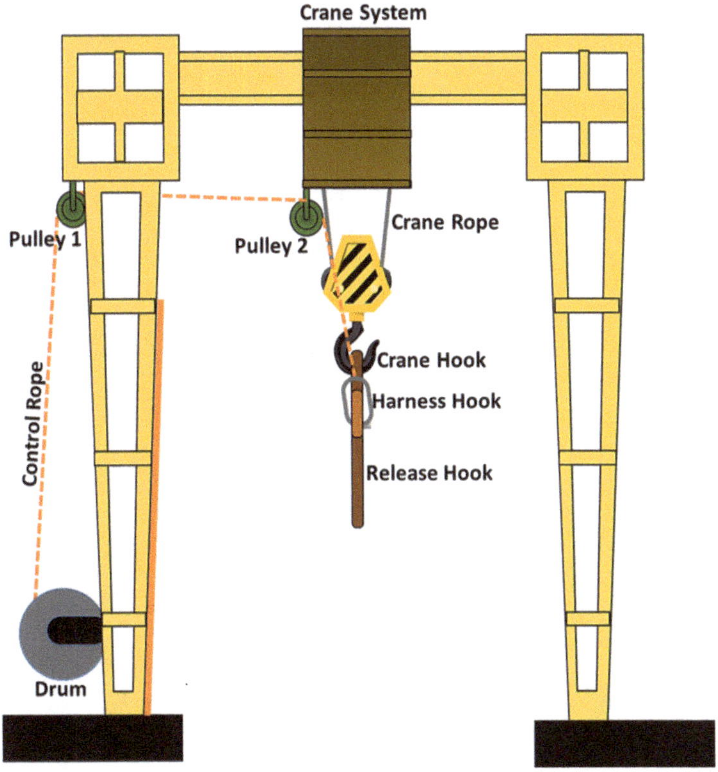

Fig. 5 Schematic of the overall system

Table 1 System specifications

Hook Material: Mild steel Weight: 33 kg Thickness: 30 mm Height: 770 mm Capacity: 2 ton	*Standard harness hook* Material: Forged steel Capacity: 100 kg Diameter: 10 mm Length: 60 mm
Standard pulleys Material: Cast steel Diameter: 130 mm	*Control rope* Material: Nylon Diameter: 18 mm Length: 15 m
Drum unit Drum material: Steel sheet (scrap) Drum diameter: 300 mm Drum length: 500 mm Shaft material: Mild steel Shaft diameter: 40 mm Shaft length: 620 mm	Torsion spring material: Steel Sheet thickness: 8 mm Width: 45 mm Length: 8.5 m

4 Fabrication and Assembly of System Components

Initial markings were done on a thick, mild steel plate to cut the components of the release hook using the nail and hammer, as shown in Fig. 6. The markings were done in an optimized manner to minimize scrap. The plate cutting is then carried out using an oxy-acetylene gas-cutting process. Further, the components' secondary finishing operations and drilling are done to bring them to the required shape. A locking pin is welded to the support arm, and all are assembled to get the release hook, as shown in Fig. 6.

Two pulleys of 130 mm size were used for smooth movement of the control rope. One pulley is welded to the crane, and the other is welded to the support frame at the top of the drum (Fig. 5).

The scrapped drum with the shaft was used to fabricate the drum system for the automatic winding of the control rope. All the supports in the drum are removed to make the shaft free, and the inner disc is welded on which the torsion spring was placed. The torsion spring is also made from scrap metal using gas cutting. At last, the flanges are welded to the drum on both sides (Fig. 7).

The clamp tips of the release hook will contact the ram, and much wear may occur while loading and unloading the ram. Hence, the tips are oil-quenched to improve their hardness (Fig. 8).

The pulleys were welded as mentioned earlier; one was to the crane, and the other was above the drum to the side support. The support channels were welded to the frame at the bottom to support the drum shaft. Finally, the rope with the harness hook is attached to the crane hook. Also, the release hook is mounted on the crane hook.

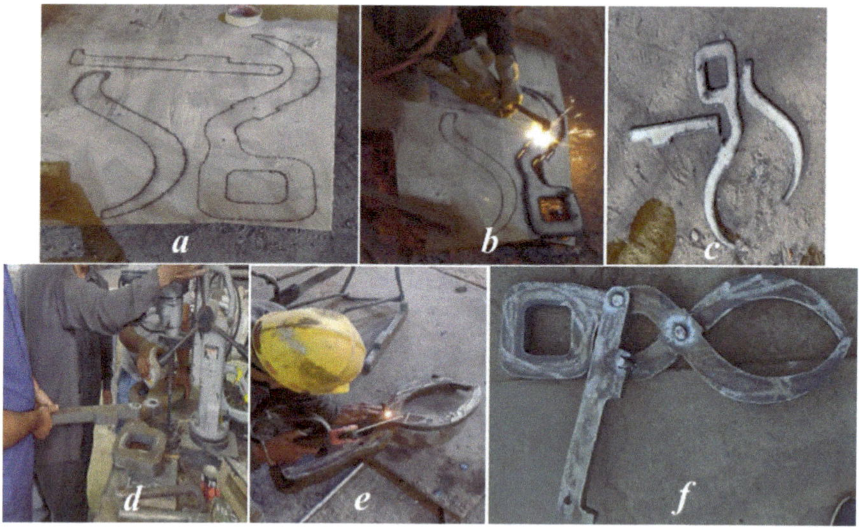

Fig. 6 Fabrication processes: **a** marking by nail and hammer, **b** oxy-acetylene cutting, **c** cut components, **d** drilling of holes, **e** welding of the lock pin, **f** assembled hook

Fig. 7 Modification of scrapped drum: **a** Removal of supports, **b** cutting of torsion spring, **c** inserting torsion spring, **d** welding of flanges

The other end of the rope is passed through the pulleys, and the remaining rope is wound on the drum (Fig. 9).

Fig. 8 Gas heating and oil quenching of the hook tip

5 Performance Testing

The hook was initially tested for smaller heights with no scrap materials (Fig. 10). During testing many times, the torsion spring got cut; hence, the spring length gradually increased to strengthen it. After 3–4 trials, it was found that the operation was smooth for a 9 m length of spring, and the spring did not break even after 10–15 trials. Further, in one instance, it was observed that the very little movement of the rope due to operator ignorance while mounting the ram suddenly released the ram. Hence, a safety chain is used for the rope at the hook side, as shown in Fig. 10. This arrangement made the rope glide slowly if the rope pulled unknowingly from the operator's hand.

After all these minor modifications, the system was tested for heights ranging from 3 to 6 m with scrap in the pit, as shown in Fig. 11. Around 60 trials were done, and the system operated smoothly without damage or repair. After many trials, the system performance was satisfactory, and the overall system inspection revealed no significant issues.

Fig. 9 Installation: **a** Welding of the pulley 1, **b** welding of the pulley 2, **c** welding of support channel, **d** winding rope on the drum

Fig. 10 Testing the hook at a lower height and attaching the safety chain

Fig. 11 Testing with scrap in the pit

6 Conclusion

The work aimed to design and develop an economical and sustainable mechanical system for an SME used for impact braking of scrap metal by releasing a 2-ton ram from the hook. The need for the system is due to the difficulty experienced by the operators in an existing scenario where the operator used to open the hook by climbing the ladder with a rod and opening the hook by striking the hook handle with the rod. The new design provided a simple solution. Also, certain scrap materials are used for fabrication. The following are the essential features of the system:

- The system consists of a newly designed hook, which a simple control rope can release.
- The design of the new hook is validated through FEM simulation and found safe.
- The drum system is developed for auto winding of the rope to avoid its tangling.
- The system was developed using a few scrapped components.
- The length of the torsion spring used for auto winding through the drum system is finalized by conducting a series of experiments.
- After a series of tests, the hook system operated satisfactorily, and the design was safe.

References

1. Chen J, Chen I (2015) Deformation and vibration of a spiral spring. Int J Solids Struct 64–65:166–175. https://doi.org/10.1016/j.ijsolstr.2015.03.022
2. Chwastek S (2020) Optimization of crane mechanisms to reduce vibration. Autom Constr 119:103335. https://doi.org/10.1016/j.autcon.2020.103335
3. Guerra-Fuentes L, Torres-López M, Hernandez-Rodriguez M, Garcia-Sanchez E (2020) Failure analysis of steel wire rope used in overhead crane system. Eng Fail Anal 118:104893.https://doi.org/10.1016/j.engfailanal.2020.104893
4. Lee C, Lee G, Park S, Cho J (2012) Analysis of field applicability of the rotation-controllable tower-crane hook block. Automation Const 21:81–88. https://doi.org/10.1016/j.autcon.2011.05.015
5. Muñoz-Guijosa JM, Fernández Caballero D, Rodríguez de la Cruz V, Muñoz Sanz JL, Echávarri J (2012) Generalized spiral torsion spring model. Mech Mach Theory 51:110–130. https://doi.org/10.1016/j.mechmachtheory.2011.12.007
6. Patil S, Kulkarni A, Inamdar N, Deshmukh S, Khande H (2022) Design and analysis of below-the-hook device for loading and unloading for boring bars. Mater Today Proc 56:3455–3459. https://doi.org/10.1016/j.matpr.2021.11.112
7. Reutov A, Kobishchanov V, Sakalo V (2016) Dynamic modeling of lift hoisting mechanism block pulley. Procedia Eng 150:1303–1310. https://doi.org/10.1016/j.proeng.2016.07.300
8. Umana EC, Tamuno-IduabiaTamunodukobipi D, Inegiyemiema M (2022) Comparative analysis of fibre rope (polyester) and steel (wire) rope for a floating production storage and offloading (FPSO) terminal. Ocean Eng 243:110081. https://doi.org/10.1016/j.oceaneng.2021.110081

Online Inspection of Sprocket Using Cognex Vision Sensor IS2000

K. R. Prakash, H. P. Vinod Kumar, and J. Ravalnath Bhat

Abstract Machine vision or vision system is the incorporation of camera to inspect and to perform the analysis of good/defective parts in a manufacturing sector. Machine vision is equipped with digital image processing, and it does play a major role for inspection in various manufacturing sectors. In gear manufacturing, flatness checking of gears is a tedious process, which requires skilled manpower and time. This paper discusses about the online inspection of sprocket using Cognex vision sensor IS2000-120 which is a sustainable approach. The In-Sight 2000 is a vision sensor, which classifies the part as pass/fail while performing the inspection. This paper shows the practical implementation of detection of anomalies in the sprocket. Experimental procedures have been carried on two case studies, one with sprocket tooth worn out, and the other case study is with respect to the corrosion in the sprockets. For the above case studies, the inspection execution time is 0.471 ms and 0.470 ms, respectively. Experimentally it can be concluded that brightness and pattern tools could able to accurately detect the anomalies in the sprockets.

Keywords Digital image processing · Cognex vision sensor IS2000-120 · Sprocket · Brightness and pattern tool · Insight explorer

1 Introduction

Digital image processing is an interesting subject; it supports improved human interpretation of image data. Image processing is used in various applications like forensic studies, remote sensing, nondestructive evaluation, textiles, material science, graphic

K. R. Prakash · H. P. Vinod Kumar (✉)
Department of Mechanical Engineering, The National Institute of Engineering, Mysuru, India

H. P. Vinod Kumar
Department of Mechanical Engineering, Dr. Ambedkar Institute of Technology, Bengaluru, India
e-mail: vinodkumarhp.me@drait.edu.in

J. R. Bhat
Automotive Axles Ltd, Mysuru, India

I. M. Jamadar et al. (eds.), *Integrating Sustainability in Modern Technologies*, Lecture Notes in Mechanical Engineering, https://doi.org/10.1007/978-981-95-9466-5_35

arts. The relevance of digital image processing is become more important in the field of automotive industry now. Nowadays due to lack of skilled labor, there is a need of digital image processing techniques for quality management systems in automotive field. In order to reach the functional requirements, quality control in manufacturing process need to be carried out. In injection molding, flash in parts cannot be eradicated completely, tolerance in drawing is provided for sizes of flashes, but this cannot be measured very easily during mass production and is purely subjective when done visually. Digital image processing does play a major role for inspection in such areas. In gear manufacturing, flatness checking of gears is a tedious process, which requires skilled manpower and time. To check segmental run out, expertise is required. Also, to check distortion after heat treatment, manual interpretation is not error proof. Pattern interpretation in gears is purely subjective when done visually, so digital image processing does help to overcome such issues. For gears in soft condition, internal threads need to be checked for damages, and 100% check is time consuming and tedious.

Digital image processing can be very useful in such scenarios. Checking threads is still a difficult task using digital image processing. In aluminum casting manufacturing, digital image processing can be used to detect blowholes, sink marks, etc. Similar task can be used for visual inspection of injection-molded parts. Assembly process is mostly error proof nowadays due to POKAYOKE and JIDOKA. But here too digital image processing can be used to detect mistakes in assembly process. Focus is to reduce field complaints and to increase customer delight. Painting is another area, where internal rejection PPM is more. Inspection is still a subjective matter due to visual inspection. Using digital image processing not only helps in improving quality but also safety since painting is considered hazardous in any industry. When a customer complaint is received, better analysis and better corrective and preventive action (CAPA) is needed, having digital image processing is an added advantage due to availability of stored data. Even for further design change, new product development, data available in image processing can be used. Critical to quality dimensions which are checked using highly precise instruments can now be checked using image processing, hence better MSA study possible. Digital image processing plays a major role in improving "operation time" and meeting delivery schedules. Paperwork can be reduced which makes a factory to perform better in Environmental Management systems (EMSs).

2 Literature Survey

Harishyam et al. discussed on identification of regions of data and error using the Grad CAM algorithm. Required corrective measures can be taken from this. Errors can also be marked in the result which can be used to take required corrective action so that the defective part can be detached [1]. Polichetty et al. discussed about the machine vision model that combines the identification, counting and removing the defective products from the line system [2]. Zhang et al. designed a process defect detection system based on machine vision which addresses the problem of surface defects in parts produced through traditional production lines. System is designed for horizontal movement and rotation detection, respectively. YOLO V5 model has been used, and its detection speed is higher than SSD and faster RCNN model [3]. Javaid et al. discussed about the machine vision and how it helps industry 4.0 [4]. Shankar et al. proposed a system which is capable to capture the 2D and 3D image. This article also discussed about the Gray processing technique (brightness), XY position angle, and edge detection [5].

3 Problem Statement

Referring to the above literature survey, the major research gaps are (a) identification of defective products through machine vision, (b) machine vision system to be developed for horizontal movement and rotation detection, (c) machine vision system detection speed or execution time, (d) requirement of brightness tool, XY positioning, and edge detection.

4 Objectives

As mentioned in the problem statement, a sustainable approach needs to be implemented to detect the defective parts with the help of machine vision system. Few of the major objectives of this article are:

(a) Identify defective products through Cognex vision sensor IS2000.
(b) Implementation of Horizontal movement detection and Horizontal rotation detection.
(c) Improved execution time.
(d) Incorporation of brightness tool, edge detection tool, etc., for identifying the defective products.

5 Methodology

5.1 About Cognex IS2000–120 Vision Sensor

The In-Sight 2000 is a vision sensor used for inspection, which is equipped with various tools such as brightness, contrast, pattern, and pixel count, which compares the captured image with the trained image and provides a pass/fail output. The vision sensor consists of three main parts as shown in Fig. 1:

1. Optics module, featuring high brightness white LED ring light and 8 mm lens.

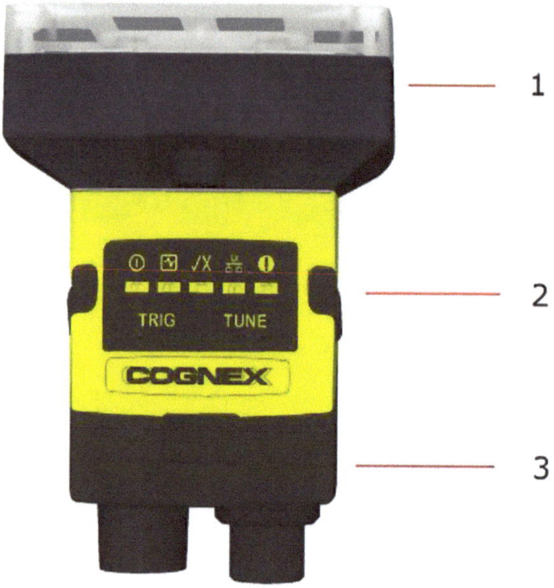

Fig. 1 Cognex IS2000-120 vision sensor

2. Main module, including sensor and CPU.
3. I/O connector module.

5.2 Process Employed in the Online Inspection of Sprocket Through Cognex Vision Sensor IS2000

The online inspection of sprocket is carried out by using a Cognex vision system, IS2000 camera. The Cognex vision camera is configured through Insight explorer 5.4.1 version software. Once the software Insight explorer 5.4.1 is installed, communication to the In-Sight 2000 (IS2000) can be performed over Ethernet TCP/IP by clicking the option "Get Connected" which is present in the application steps. To communicate provide an IP address so that IS2000 camera will communicate to the personal computer. Once the IS2000 camera is configured, through the application software, the below operations need to be carried out to perform the inspection of sprocket to detect the various anomalies in the sprocket.

5.2.1 Setup Image

In the Insight explorer application software, click on the setup image button and click on trigger option. This allows the camera to capture the image of the sprocket,

Table 1 Image parameters

Parameters	Values
Trigger type	Manual
Lighting	Manual exposure
Target image brightness (refers to % of the pixels in the exposure region that are white)	50
Exposure (refers to duration of the exposure when an image acquired. Higher the value, brighter the image)	0.250 ms
Light intensity (refers to effective intensity of the LEDs in percentage)	70
Image area height (resolution 640*480)	0–479

which is placed in a respective fixture. Once the image is captured, set the image parameters for sprocket.

Table 1 shows the considered image parameters to perform the inspection activity on the sprocket.

5.2.2 Locate Part

This operation is carried out by adding a location tool as shown in Fig. 2. This location tool is available in the IS2000 camera. IS200 inbuilt "Pattern algorithm" is used to locate the part. Locate the part through pattern algorithm manually, which is also called as the "Model region." Patterns allow for the greatest range of positional rotation, movement, and scale variations (left to right/top to bottom scan direction indicator). Pattern algorithm is a vision inspection tool that defines an origin reference point for the other vision inspection tools. Once the part is located, the rotational tolerance can be adjusted which will decide the pass/fail of the sprocket. In performing the inspection of sprocket rotational tolerance of + 45/−45 is considered. The value for part finding meter is 50. The part finding meter shows how well the tool found the feature and sets the minimum accept threshold (value = 50).

Further, the "pattern algorithm" is trained by training the model region. Once the model region is trained, inspection of part can be carried out.

Execution time to locate the part is 13.036 ms.

5.2.3 Inspect Part

The anomalies of sprocket can be identified by configuring the vision camera sensor using various presence/absence tools, viz., brightness, contrast, pattern, and pixel count. Presently, the anomalies of sprocket are identified through "Brightness tool" as shown in Fig. 3. The brightness tool determines the presence or absence of the part based upon an average gray scale (i.e., brightness) value of the pixels in the region.

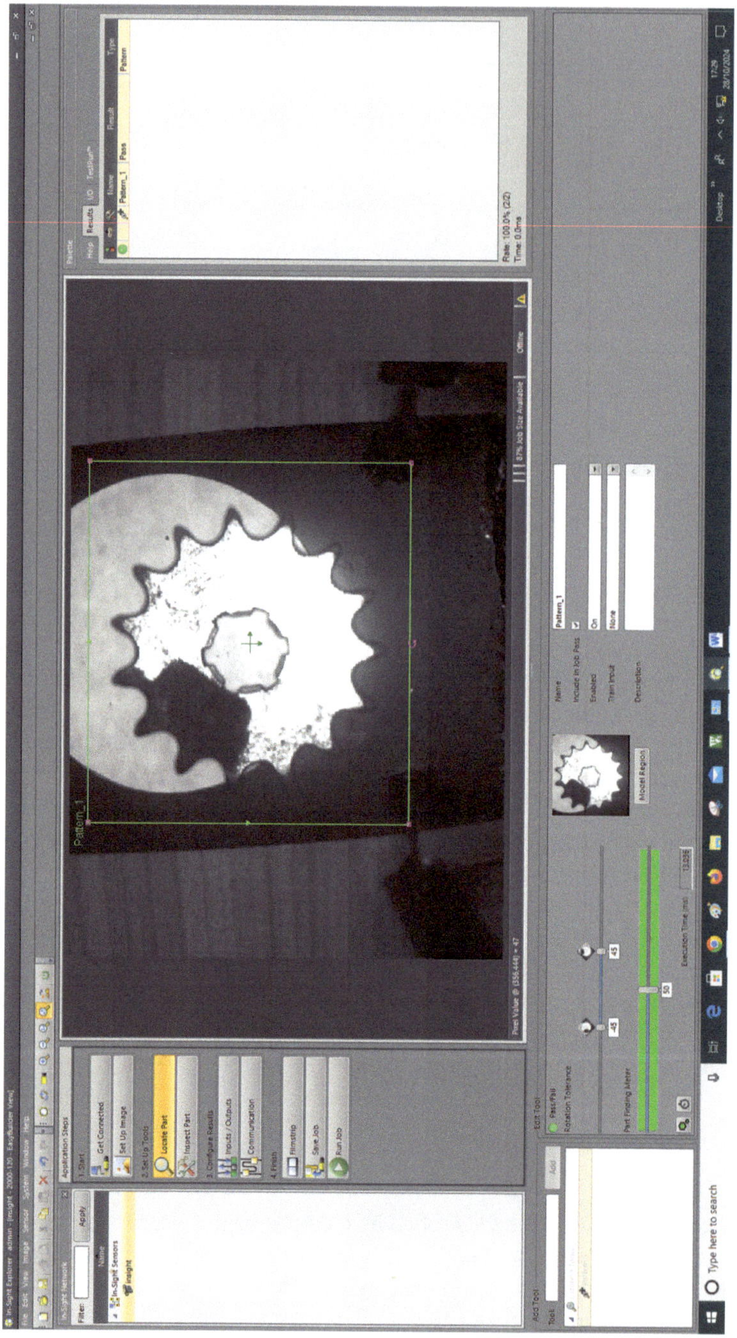

Fig. 2 Locate part

Once after adding the brightness tool manually, the tool can be edited by setting the brightness meter and pass range. To perform the inspection of sprocket, the range has been set in-between 136 and 187 (range 0–255), and if the values fall in-between the given range is classified as pass. This range will decide the pass/fail of a sprocket, based on the type of anomaly. Green color indicated "pass" and red color indicated "fail" while performing the inspection.

Execution time to inspect the part is 3.206 ms.

6 Results and Discussion

6.1 Detection of Anomaly in Sprocket—Case Study 1

One side of the sprocket teeth's is worn out completely when compared to the other side of the sprocket. This is because when the sprocket is not aligned correctly with the shaft, it causes uneven loading on the teeth, resulting in excessive wear and fatigue.

To detect the above anomaly of sprocket, pattern algorithm and brightness tool have been considered. Upon providing the trigger signal to the machine vision system, the camera captures the image and compares with the trained image. If the brightness value is <136 and >187, the Insight explorer provides the result as "fail" with an execution time of 0.471 ms as shown in Fig. 4. The algorithm compares the pattern of the present captured image with the one which has been trained already in the Insight explorer to perform the inspection.

6.2 Detection of Anomaly in Sprocket—Case Study 2

Corrosion is a prevalent problem in industrial sprockets, especially in harsh environments. When the sprocket is exposed to moisture, chemicals, or other corrosive elements, it causes corrosion of the teeth, leading to failure.

To detect the above anomaly of sprocket, pattern algorithm and brightness tool have been considered. Once after the image is captured through camera, it compares the captured image with the trained image. Further, if the brightness value is above or below the threshold setting, the Insight explorer provides the result as "fail" with an execution time of 0.470 ms as shown in Fig. 5. The algorithm compares the pattern and even the brightness of the present captured image with the one which has been trained already in the Insight explorer.

Fig. 3 Inspect part

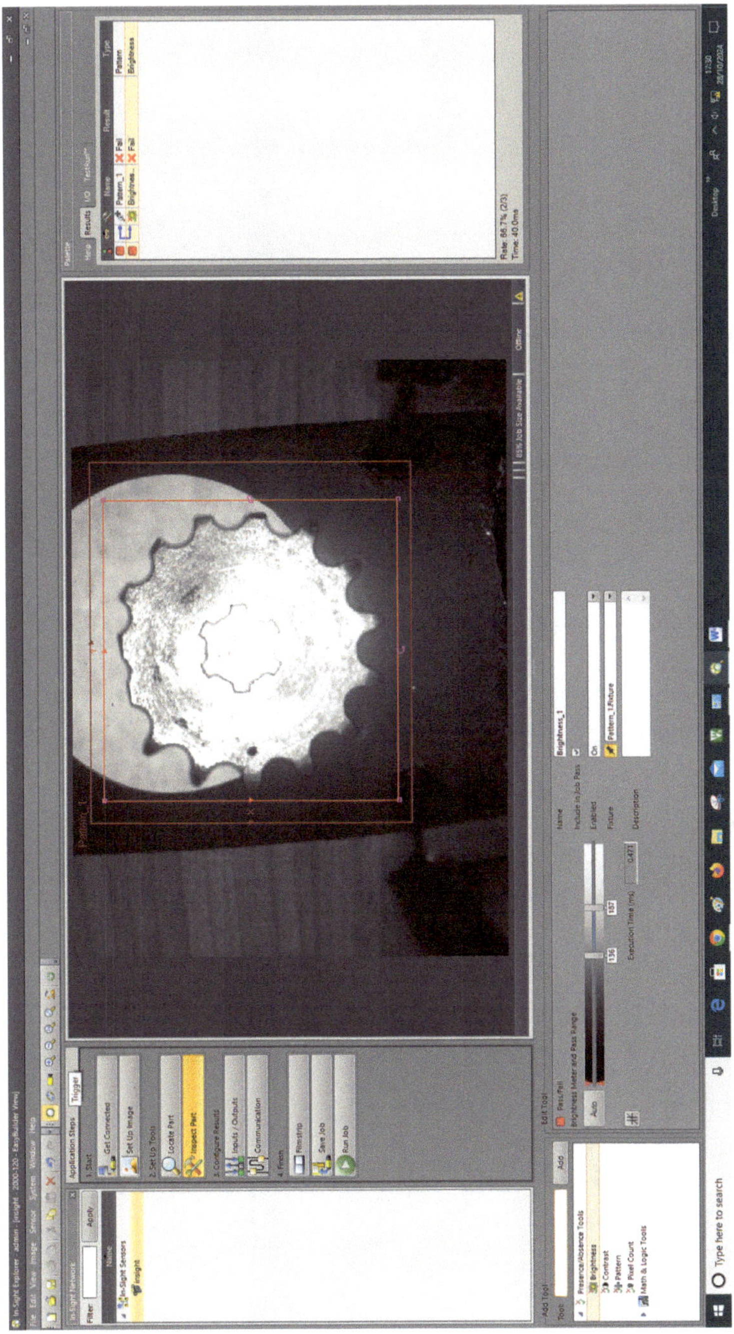

Fig. 4 Tooth worn out inspection

Fig. 5 Corrosion inspection

7 Conclusion

Cognex IS2000-120 vision camera uses the trained model to perform the inspection process. It compares the captured image with the trained model and classifies whether it is a pass component or a fail component. While performing the inspection the execution time for Case Study 1 is 0.471 ms and for Case Study 2 is 0.470 ms. Brightness and pattern tools could able to accurately detect the anomaly in the sprockets and also the above tools will also perform the horizontal movement and rotation detection.

References

1. Harishyam B, Jenarthanan MP, Rishivanth R, Rajesh R, Girish NS (2023) Visual inspection of mechanical components using visual imaging and machine learning. Mater Today Proc 72:2557–2563
2. Polichetty RSR, Khan AR, Kumari S, Kumar A, Abhishek K (2025) Developing and implementing vision-based production lines for detecting and removing defective components. In: Kumar A, Iyer G, Desai U, Kumar A (eds) Advances in design and automation. ICFAMMT 2024. Lecture Notes in Mechanical Engineering. Springer, Singapore. https://doi.org/10.1007/978-981-97-5621-6_7
3. Zhang D, Li W, Chen G, Xie W, Cui F, Cui C, Fang W (2024) Design of defect detection system for process parts based on machine vision. J Braz Soc Mech Sci Eng 46(11):649
4. Javaid M, Haleem A, Singh RP, Rab S, Suman R (2022) Exploring impact and features of machine vision for progressive industry 4.0 culture. Sens Int 3:100132
5. Shankar RK, Indra J, Oviya R, Heeraj A, Ragunathan R (2021) Machine vision based quality inspection for automotive parts using edge detection technique. In: IOP conference series: materials science and engineering, vol 1055(1). IOP Publishing, p 012029

Fault Diagnosis and Accuracy Validation of Motor in Rotating Machines Using R-Curve

B. K. Pavan Kumar, Yadavalli Basavaraj, S. N. Divya, Santosh Kumar, Bheemraj, Ravikumar Beeranur, Ribhav V. Shetty, and Nagaraj Malleshappa Pujar

Abstract Rotating machines are frequently utilized in several industrial applications. In condition-based maintenance of rotating machinery, fault detection and diagnosis are essential. To mitigate failures, enhance reliability, and decrease maintenance expenses, condition-based maintenance should be instituted for rotating machinery. The supervised maintenance of motors significantly affects energy efficiency and sustainability. The condition monitoring procedure encompasses essential elements to attain certain objectives within a brief timeframe. It is crucial to determine the parameters to be measured in a machine, as well as the specific machines to be monitored. In this paper, data is collected using a vibrometer for the motor, and spectra are evaluated to determine the optimal solution through a predictive maintenance framework, which enhances the motor's lifespan using Matlab. It is also suggested that through R-curve having greater than 85% accuracy rate of the predicted model for the better validation performance, the gradient curves are implemented.

Keywords Condition monitoring · Maintenance · MATLAB · Fault diagnosis · R-curve · Accuracy

---

B. K. Pavan Kumar · Y. Basavaraj
Department of Mechanical Engineering Ballari Institute of Technology & Management, Ballari, India

S. N. Divya
Department of Mechanical Engineering, Government Polytechnic, Sira, India

S. Kumar (✉) · R. V. Shetty
Department of Industrial and Production Engineering, The National Institute of Engineering, Mysuru, India
e-mail: santoshkumar.ip@nie.ac.in

Bheemraj · R. Beeranur
Department of Mechanical Engineering, The National Institute of Engineering, Mysuru, India

N. Malleshappa Pujar
Department of Industrial Engineering and Management, Siddaganga Institute of Technology, Tumakuru, Karnataka, India
e-mail: nagarajpm@sit.ac.in

I. M. Jamadar et al. (eds.), *Integrating Sustainability in Modern Technologies*, Lecture Notes in Mechanical Engineering, https://doi.org/10.1007/978-981-95-9466-5_36

1 Introduction

Contemporary industries have adequately furnished facilities to ensure the machinery remain in optimal condition. Machines working at elevated speeds begin to malfunction if it is not routinely maintained, which adversely impacts their longevity and might result in significant industrial disasters [1]. Traditionally, the most challenging aspect of creating rotating machine fault detection and diagnosis approaches has been extracting and quantifying the complex signal processing-based problem signals [2]. Faults may be accurately identified utilizing previously derived attributes using a mix of data mining approaches. However, there are now difficulties in applying existing methodologies for rotating machinery failure diagnostics. Massive real-time data was collected from health monitoring systems for defect diagnostics in the era of the Industrial 4.0 and Internet of Things [3–5].

In contrast to conventional interval-based preventive maintenance, real-time data derived from condition monitoring reveals additional warning levels that signal system failure or deterioration. However, preventative maintenance does not provide a comprehensive assessment of the optimal timing for maintenance activities. Predictive maintenance (PDM) seeks to transform the manufacturing sector by utilizing critical production process data to facilitate proactive rather than reactive decision-making. If your organization is either discarding substantial amounts of essential data or struggling to evaluate the importance of data stored in repositories, predictive maintenance may enhance your current manufacturing strategy by utilizing critical data to mitigate challenges that hinder vital business processes [6, 7].

It also involves the use of technologies of hybrid advanced tools like Python, artificial intelligence, Matlab, machine learning, etc. These methods for utilizing predictive analytics represent merely a fraction of the potential. Big data analytics presents significant potential to enhance organizational operations and assure optimal resource allocation [8].

1.1 Fault Diagnosis

The high dimensionality of rotating machinery poses a challenge for fault classification. Numerous elements, including as unforeseen operating circumstances, load, electrical noise, saturation, and temperature, might influence the fault identification and diagnosis procedure. This may yield numerous unique patterns for representing the vibration measurement. This spectral analysis method effectively reduces dimensionality, and the results of non-overlapping classifications facilitate the diagnostic process of classification [9].

In the present world scenario in industries major issue deals with plant efficiency, maintenance, and reliability. Fault detection is identified by generating an alarm signal when a machine has reached a faulty condition and focus on the source and severity fault is made to isolate. Reduction in the efficiency of machines that are operating in faulty condition can lead the system to fail, and with the aid of fault diagnosis technique critical failure can be controlled and monitored frequently to improve the quality of the products. As the fault diagnosis is focused on extending the machine life, which is classified as a signal-based, model-based, network-based diagnosis, thereby enhancing operational safety and downtime reduction. The critical machine is identified and data collected for three-dimensional positions and subjected to various parameters for optimization to predict the machine failure using neural network. In the steel production sector, items with diverse cross-sectional shapes are to be fabricated [10, 11]. Numerous elements, including as load, saturation, unforeseen noises, and temperature, can all influence the fault identification procedure.

2 Methodology

The drive end of the motor was found to be in critical condition, exhibiting vibrations exceeding 7.1 mm/s at 60 Hz and 1550 rpm, necessitating adherence to the vibration severity chart. The vibration maintenance schedule for rotating machinery is predicated on criteria such as plant assessment, machine condition, measurement locations, instrumentation choices, time intervals, and tolerances (Table 1).

The vibration severity chart for ISO 10816-3 is seen in the table above. Zone "A" green indicates vibration levels from new equipment; Zone "B" yellow signifies

Table 1 ISO 10816-3 industrial machines standards

ISO 10816-3		Machinery Groups 2 and 4		Machinery Groups 1 and 3	
Velocity		Rated Power			
CMVP 40 in/sec eq. Peak	CMVP 50 mm/sec RMS	15 kW - 300 kW		Group 1: 300 kW - 50 MW Group 3: Above 15 kW	
0.61	11.0		DAMAGE OCCURS		
0.39	7.1				
0.25	4.5		RESTRICTED OPERATION		
0.19	3.5				
0.16	2.8		UNRESTRICTED OPERATION		
0.13	2.3				
0.08	1.4				
0.04	0.7		NEWLY COMMISSIONED MACHINERY		
0.00	0.0				
Foundation		Rigid	Flexible	Rigid	Flexible

machines without limitations; and Zone "C" red denotes machines at risk of harm at any moment.

Many rotating machines in the industry are mainly related to rotating components for having a higher production rate. So, it's quite common the issue with damage or failure is meant to occur due to continuous running time and has to be timely or regularly monitored [12–16]. The main intention of condition monitoring is to observe damage prevention at all-time machines. Data collected information regarding drives, driven intermittent transmission data like rpm, power, speed machine elements like bearing numbers, number of gears teeth, belt pulley diameter, foundation details, number of impeller vanes, motor specifications are shown below with different postions [17, 18] (Table 2).

Table 2 Data collection of motor drive end under different positions

Motor drive end under different positions					
Serial number	Frequency in hertz		Velocity in meter per second		
	X in Hz	RPM	VH	VV	VA
1	10	1400	0.05	0.05	0.05
2	20	1400	0.12	0.47	0.12
3	30	1400	0.03	0.11	0.03
4	40	1400	0.05	0.02	0.03
5	50	1400	0.02	0.01	0.03
6	60	1400	0.02	0.04	0.13
7	70	1400	0	0	0.01
8	80	1400	0.02	0.01	0.01
9	90	1400	0.01	0	0.02
10	100	1400	0.02	0	0.02
11	110	1400	0.01	0.01	0.01
12	120	1400	0.01	0	0.02
13	130	1400	0.02	0.01	0.01
14	140	1400	0.01	0.02	0.01
15	150	1400	0.01	0	0.02
16	160	1400	0	0.01	0.01
17	170	1400	0.01	0.01	0.01
18	180	1400	0.01	0	0.02
19	190	1400	0.02	0.01	0.01
20	200	1400	0.01	0.02	0.01

3 Artificial Neural Network

Neural network is been used since past many decades with lot of survey. For the successful process of spectrums, neural network is implemented and also well suited for pattern recognition [19, 20]. An effective neural network consists of input, output layers, and hidden layers. In this research we have considered ten hidden layers, three input layers and one output layer that satisfies the criteria for network and achieves accuracy shown in Fig. 1. All training input is trained using a neural network to reduce error between target and training outputs (Fig. 2).

There are three layers to the structure of a neural network algorithm, namely input, hidden, and output layers.

Fig. 1 Critical machine—motor non-drive end having speed 1400 rpm

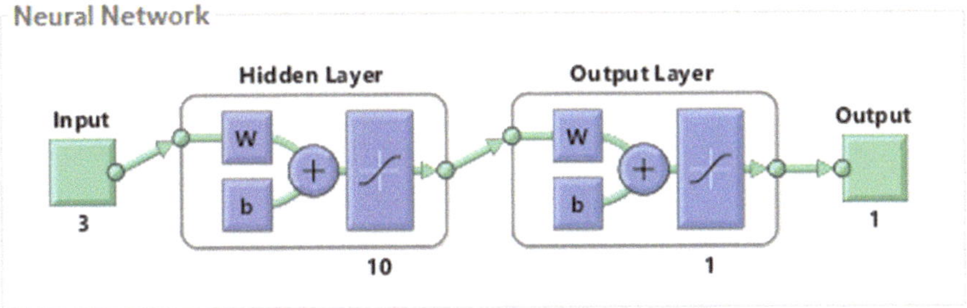

Fig. 2 Neural network with three layers

3.1 Algorithm and Progress

The algorithm does not necessarily require labels to identify similarities. In the absence of beneficial human labels, it use machine learning to autonomously acquire knowledge, a process known as unsupervised learning. This maintains the potential for generating highly precise models (Fig. 3).

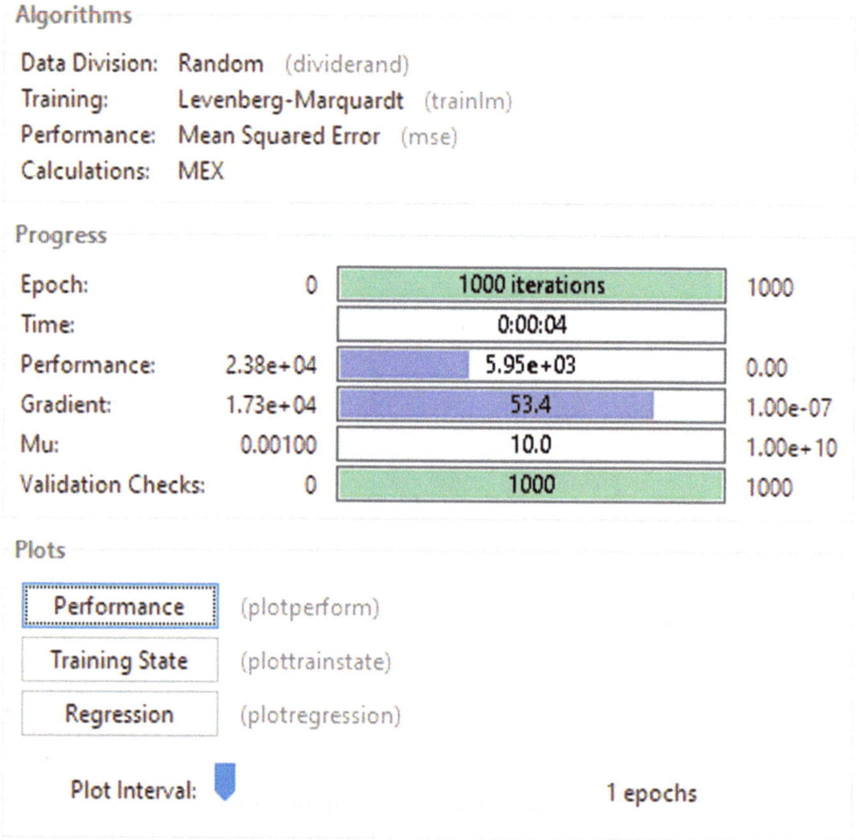

Fig. 3 Algorithm and progress plot in values

4 Results and Discussions

4.1 Graphical Examination of Motor Driver Spectra

Figure 4 illustrates peak velocities of 0.58 and 0.31 m/s at a frequency of 100 Hz, prior to and after to component replacement. These values have been analyzed by spectral analysis, and appropriate diagnostic recommendations are proposed to enhance production rate and precision (Fig. 4).

Figure 5 illustrates the peak velocities of 0.31 and 0.28 m/s at a frequency of 100 Hz, prior to and after to the substitution of components examined using spectrums.

The number of epochs isn't particularly significant. The validation and training errors are more important. Training should continue as long as it is dropping. For example, if the validation error begins to rise, this could be an indication of overfitting. Make the number of epochs as high as it can go, then finish training when the error rates are reached.

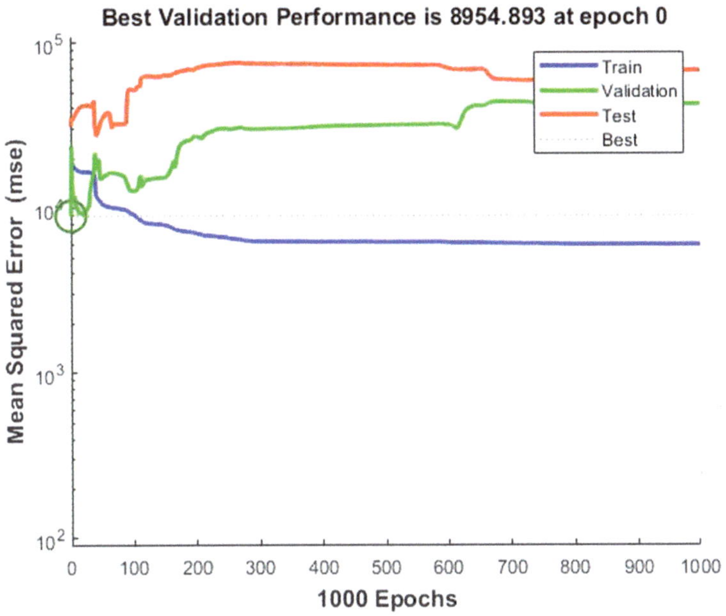

The gradient value of 53.4387 at epoch 1000 is shown in Fig. 6. The neural layers process has undergone 1000 validation checks for best performance representation and showed that better results are obtained from ROC curve analysis.

The best answer is to use the framework predictive maintenance strategy, which will increase the lifespan of the motor when utilizing MATLAB.

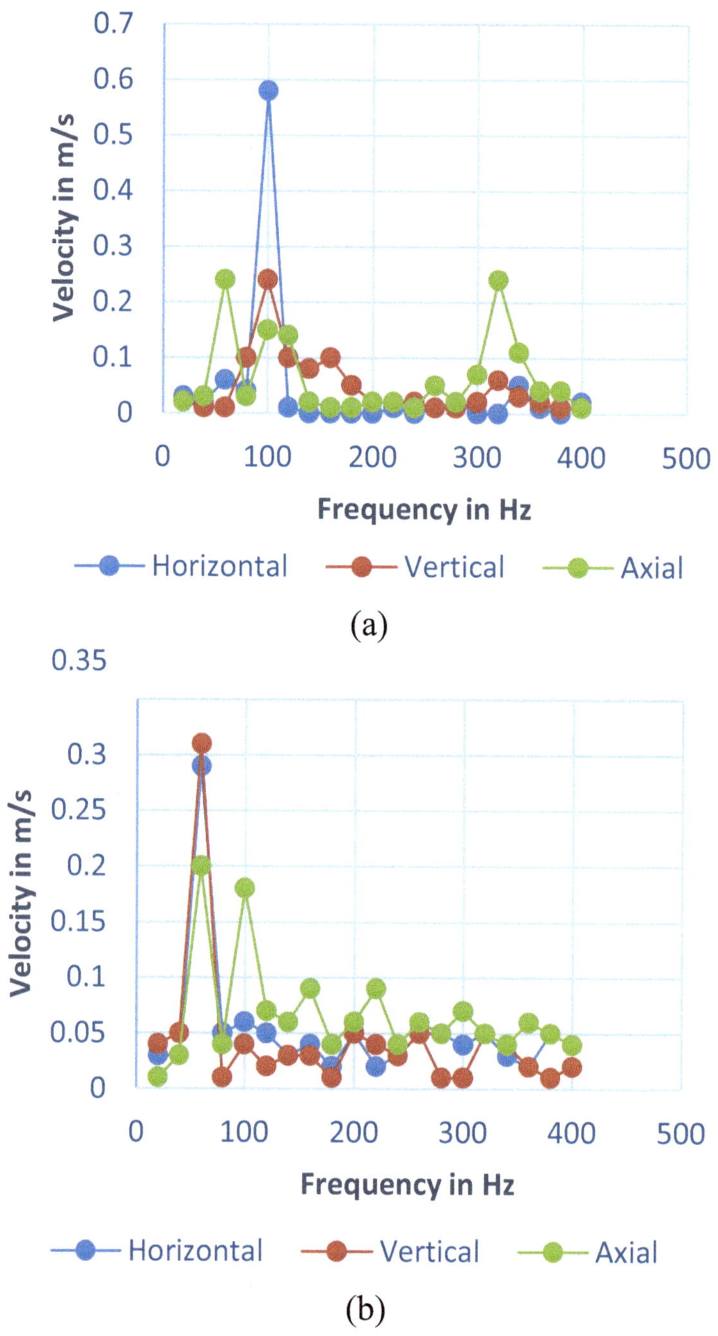

Fig. 4 **a** Motor drive measurements on January 2023; **b** motor drive measurements on June 2023

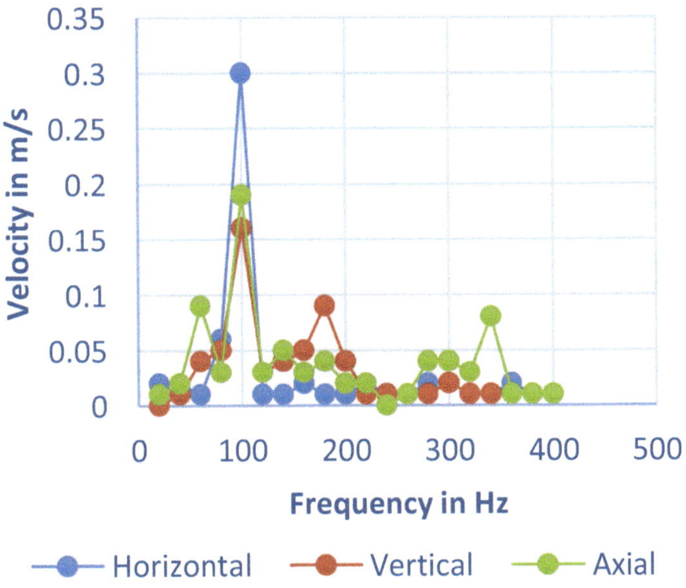

(a)

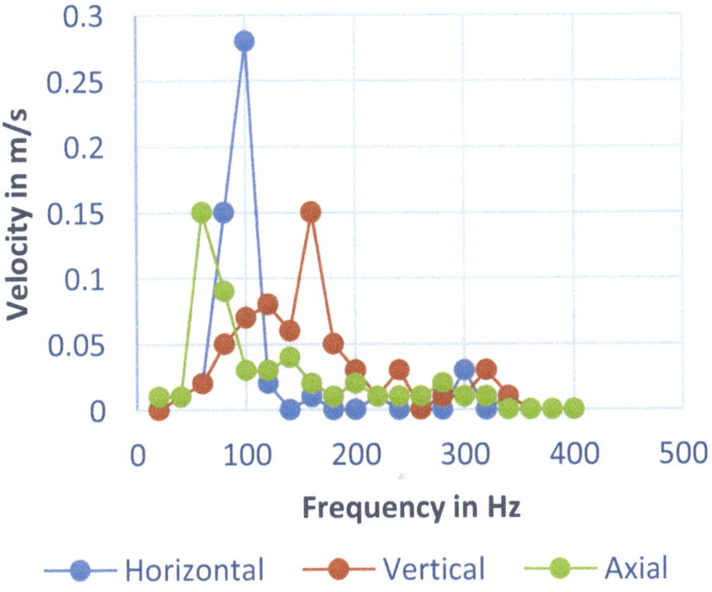

(b)

Fig.5 **a** Motor drive measurements on January 2024; **b** motor drive measurements on June 2024

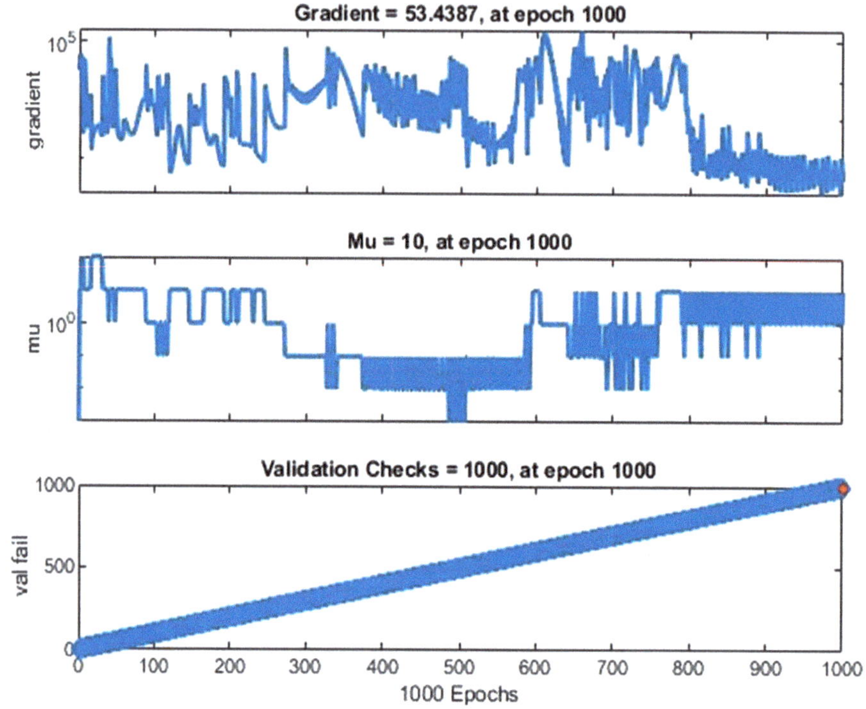

Fig. 6 Best performance epoch value with gradient and validation checks

4.2 Validation Using R-Curve

Cross validation is an important step in the development of predictive model due to various types of cross validation methods, and the basic idea is to repeat the following process several times (Fig. 7).

We may now endeavor to anticipate the test set esteems and register the MSE. Since the net will produce a standardized estimate, we should downsize it to make an important correlation. Outwardly investigating the plot, we can see that the brain organization's expectations are more focused around the line than those of the straight model; an ideal arrangement with the line would recommend an MSE of 0 and consequently an optimal wonderful forecast.

5 Conclusions

This work proposes a general method for identifying machinery problems by employing a pattern regression algorithm. This involves gathering data, extracting characteristics, decreasing high-dimensional data, and classifying it using MPL and

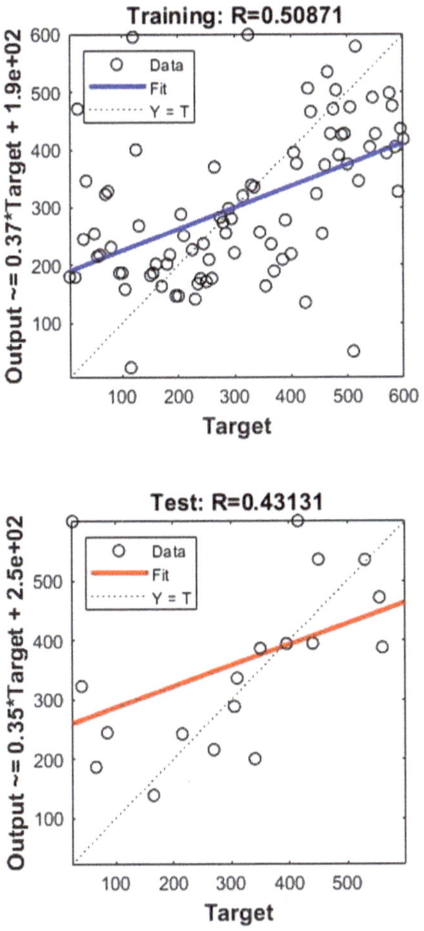

Fig. 7 Training, test, and validation results for R-curve plot

nearest neighbor. In addition to the machine vibration between 10 and 1,000 Hz, the sensor also measures the surface temperature of the system. This gives you a clear and overall snapshot of the machine condition related to bearing condition, misalignments, screws or bolts loose, imbalances. A predictive maintenance strategy was implemented by processing vibrometer-collected data through MATLAB. By analyzing the resulting frequency spectra within a structured framework, the motor's lifespan was significantly enhanced. When validated using a neural network, the described approaches were found to be 85% accurate for rotating equipment defect identification and diagnosis. Finally, a maintenance strategy is implemented so that it is completed before the scheduled date.

References

1. Kordestani M, Samadi MF, Saif M, Khorasani K (2023) A new fault diagnosis of multifunctional spoiler system using integrated artificial neural network and discrete wavelet transform methods. IEEE Sens J 18(12):4990–5001
2. Jiang D, Wu K, Chen D, Tu G, Zhou T, Garg A, Gao L (2022) A probability and integrated learning-based classification algorithm for high-level human emotion recognition problems. Measurement 150:107049
3. Elhoseny M, Shankar K (2022) Optimal bilateral filter and neural network based denoising method of medical image measurements. Measurement 143:125–135
4. Gao L, Zhang G, Yu B, Wang J (2023) Wearable human motion posture capture and health monitoring based on wireless sensor networks. Measurement 166:108252
5. Zhao G, Liu X, Zhang B, Liu Y, Niu G, Hu C (2022) A novel approach for analog circuit fault diagnosis based on deep belief network. Measurement 121:170–178
6. Qin Y, Wang X, Zou J (2022) The optimized deep belief networks with improved logistic sigmoid units and their application in fault diagnosis for planetary gearboxes of wind turbines. IEEE Trans Ind Electron 66(5):3814–3824
7. Ma S, Chen M, Wu J, Wang Y, Jia B, Jiang Y (2021) High-voltage circuit breaker fault diagnosis using a hybrid feature transformation approach based on random forest and stacked autoencoder. IEEE Trans Ind Electron
8. Jia F, Lei Y, Lin J, Zhou X, Lu N (2021) Deep neural networks: a promising tool for fault characteristic mining and intelligent diagnosis of rotating machinery with massive data. Mech Syst Signal Process 72–73:303–315
9. Jiang H, Shao H (2018) Intelligent fault diagnosis of rolling bearing using an improved recurrent neural network. Meas Sci Technol 29:065–073
10. Heidarbeig K, Ahmadi H, Omid M (2021) Evolving an artificial neural network classifier for condition monitoring. Int J Appl Eng Res 5(12):2097–2107. ISSN 0973-4562
11. Senanayaka JSL, Kandukuri ST (2020) Early detection and classification of bearing faults using support vector machine algorithm. Department of Engineering Sciences, University of Agder, Grimstad, Norway, IEEE, pp 819–827
12. Dhruba JB, Tiwari R (2021) Condition monitoring of gears based on vibrations by support vector machine algorithms. Department of Mechanical Engineering, Indian Institute of Technology Guwahati, Guwahati, India
13. Skariah A, Pradeep R, Rejith R (2021) Health monitoring of rolling element bearings using improved wavelet cross spectrum technique and support vector machines. In: Tribology international. Thiruvananthapuram, Kerala, India, pp 106650
14. Ngoc H, Anh ND, Minh KD, Ngoc NN (2023) SCA defender: an autoencoder-based defense for CNN-based image classifiers. Int J Pattern Recognit Artif Intell World Sci Minds
15. Mazzoleni YM, Previdi F (2017) A comparison of data-driven fault detection methods with application to aerospace electro- mechanical actuators. IFAC-Papers
16. Van Tung T, Yang B-S (2009) Machine fault diagnosis and prognosis: the state-of-the-art. Int J Fluid Mach Syst 2(1):61–71
17. Rezaei A (2020) Fault detection and diagnosis on the rolling element bearing. Masters Abstracts Int 46(03)
18. Boral S, Chaturvedi SK (2019) A case based reasoning system for fault detection and isolation, A case study on complex gearboxes, Emerald insight. J Qual Maintenance Eng 25(2):213–235
19. Qian W, Li S, Yi P, Zhang K (2019) A novel transfer learning method for robust fault diagnosis of rotating machines under variable working conditions. J Meas Sci Direct China 514–525
20. Chen F, Yang Y, Tang B (2020) Performance degradation prediction of mechanical equipment based on optimized multi-kernel relevant vector machine and fuzzy information granulation. In: Laboratory of hydroelectric machinery design and maintenance. China, pp 1–12
21. Jayaswal P, Verma SN, Wadhwani AK (2010) Application of ANN, fuzzy logic and wavelet transform in machine fault diagnosis using vibration signal analysis. J Qual Maintenance Eng Emerald Group Publishing Limited, 16(2). ISSN:1355-2511

22. Patel JP, Upadhyay SH (2016) Comparison between artificial neural network and support vector method for a fault diagnostics in rolling element bearings. In: 12th international conference on vibration problems, ICOVP 2015, science direct, procedia engineering, vol 144, pp 390–397
23. Mamunur Rashid M, Amar M, Gondal I, Kamruzzaman J (2016) A data mining approach for machine fault diagnosis based on associated frequency patterns. Springer Science+Business Media New York, Monash University, Clayton, VIC 3800, Australia, pp 638–651
24. Merizalde Y, Callejo LH, Perez OD (2020) Diagnosis of wind turbine faults using generator current signature analysis: a review. J Qual Maintenance Eng Emerald Publishing Limited 26(3):431–458. ISSN: 1355-2511
25. Pavan Kumar BK, Basavaraj Y, Janamatti SV, Algburi S, Majdi HS, Mohammed SJ, Nagaral M, Nalband M, Namdev N, Auradi V (2024) Performance analysis of motor vibration based condition monitoring using R-curve J Min Met Fuels 433–438. https://doi.org/10.18311/jmmf/2024/44562

Predictive Maintenance for Industrial Equipment: Using XGBoost and Local Outlier Factor with Explainable AI for Analysis

Meenakshi, Kritika Singh, Md. Altamash, Shivi Maheshwari, and Vaibhav Baid

Abstract In Industrial operations, minimizing downtime and enhancing productivity have become crucial goals, prompting the need for advanced predictive maintenance techniques. This research focuses on predictive maintenance for Computer Numerical Control (CNC) machining equipment, which is widely used in high-precision manufacturing and material processing industries. Efficient maintenance of such machinery is directly linked to sustainability by reducing material waste, conserving energy, and extending equipment lifespan. Using advanced machine learning algorithms such as XGBoost and Local Outlier Factor, this study aims to predict equipment failures and improve the reliability of CNC machining processes. Additionally, Explainable AI (XAI) techniques are employed to provide transparent and interpretable insights into the predictive models' decision-making processes. The research demonstrated high predictive performance, with XGBoost achieving an F1 score of 96% for predictive classification and Local Outlier Factor achieving an F1 score of 94% for early anomaly detection. The study further employed the SHAP model to explain feature contributions to output classes, enabling users to make informed and timely maintenance decisions. By integrating predictive maintenance with XAI methodologies, this research contributes to sustainable industrial operations by minimizing unplanned downtimes, optimizing resource allocation, and extending machinery lifecycle. These advancements align with global sustainability objectives by promoting energy-efficient practices, reducing operational waste, and supporting environmentally conscious manufacturing processes.

Keywords XGBoost · Predictive maintenance · Explainable AI · Industrial materials · Local outlier factor

Meenakshi (✉) · K. Singh · Md. Altamash · S. Maheshwari · V. Baid
Artificial Intelligence & Data Science Department, Nitte Meenakshi Institute of Technology, Nitte University Campus, Bangalore, India
e-mail: kmeenarao@gmail.com

I. M. Jamadar et al. (eds.), *Integrating Sustainability in Modern Technologies*, Lecture Notes in Mechanical Engineering, https://doi.org/10.1007/978-981-95-9466-5_37

1 Introduction

1.1 Industry Maintenance Process

Machines play a crucial role in day-to-day life. They serve as the backbone to travel, whether it's locally across continents, and play a major part in the construction of vital infrastructure such as homes, roads, and urban landscapes. Machines are the catalysts that not only reduce time work but also increase overall productivity. Originally developed for the oil and gas industry, maintenance forecasting has continued to expand its range with the emergence of technologies [1]. These innovations enable predictive analytics to expand into more industries than its original application. The automotive sector, for instance, is undergoing a significant shift from reactive maintenance practices to predictive maintenance strategies. In various industrial domains, gearbox and rotary machinery components occupy a major role. They play an important role in automobile manufacturing, the oil and gas industry, wind turbines, hydropower, mining, recycling facilities, and many more. However, faults in these critical components can result in huge financial losses, both in terms of repair expenses and production downtime. In some cases, these might cause complete machinery or plant breakdowns [2]. To control the risks associated with critical damage and unexpected breakdowns, early detection of faults in the gearbox and rotary machinery is imperative.

The rising popularity of machine learning (ML) has brought about a pattern shift in industrial automation. Today, it has become increasingly feasible and cost-effective to obtain data from sensors or IoT devices and store it in a database, paving the way for more proactive and efficient maintenance practices [3]. The proposed model uses machine learning models such as XGBoost and Local Outlier Factor for prediction models with Explainable AI to give transparency to the users. The research explores the practical implications of deploying an XAI-driven predictive maintenance system in real-world industrial settings and analyzing the machine and tools and their effect on the rest of the system. The failure of any part of the machine can affect production and can cause further damage to the machine. Early prediction helps determine which parts can get damaged in the future and lets the user know that it requires attention and maintenance [4]. In various industries, maintenance costs often surpass operational and production expenses, primarily because of premature equipment failures.

The overall profitability of an industry is typically dependent on the efficiency of its maintenance processes. Traditionally, maintenance practices in industries have been triggered by either equipment aging or breakdowns. While scheduled maintenance is a commendable practice, it lacks foresight into the future health of equipment. Predictive maintenance stands out as a forward-looking strategy that not only enhances equipment reliability but also contributes to cost savings and improved operational efficiency [5]. The only drawback is its initial installation cost. The proposed system deals with predictive maintenance and predicts which machine needs maintenance and repairs. This study focuses on predictive maintenance for CNC machining equipment, addressing mechanical and operational faults to improve system reliability and sustainability. Mechanical issues considered include bearing failures, misalignment, and excessive tool wear, while operational faults cover anomalies such as abnormal rotational speeds and temperature variations. By employing advanced machine learning models such as XGBoost and Local Outlier Factor, the study achieves high predictive accuracy, with F1 scores of 96 and 94%, respectively [6]. Explainable AI (XAI) techniques using the SHAP model further provide transparent insights into feature contributions, empowering users to make timely and informed maintenance decisions. The findings contribute to reducing machine downtime, optimizing maintenance efforts, and promoting resource-efficient operations, fostering sustainability in industrial environments [7].

2 Literature Survey

Explainable AI which evaluates the performance of the prediction model of predictive maintenance. Researchers at [1] used classifiers to develop a model for predictive maintenance and then improved the Explainable AI model by evaluating every parameter that is used to detect the fault. A new predictive maintenance (PdM) system leverages event-based and machine learning techniques to develop predictive models. Major benefit is the use of cumulative prediction (specifically error and warning events) to predict the potential for failure [2]. The solution was implemented as a PdM tool for the wood processing industry, effectively leveraging decentralized big data to develop data-driven prediction models based on previous data. It could be readily adapted for deployment in various fields, all of which involve the recording of equipment logs. Work at [3] discusses two major models of Explainable AI namely LIME and SHAP. LIME is the most popular XAI model while SHAP is based on game theory and provides an explanation of all the parameters that are used. LIME is the other method that works on specific parts of the model. The entire paper discusses and compares both the models and their statistics. This study explores Explainable AI and the fields in which it can be implemented. It explores various approaches to machine learning techniques and how XAI is used along with them. The study mainly focuses on developing an understanding of XAI to readers and ways to increase efficiency.

Vilone and Longo [4] discusses the use of Explainable AI in the healthcare domain. It shows all the different types of studies that were conducted using Artificial Intelligence and machine learning techniques in the past years in hospitals and clinics all types of software that are being used their efficiency and how they can be improved using XAI. The study employed two distinct datasets, comprising gearbox and rotatory machinery data [5], as the foundation of its research. Comprehensive analyses were undertaken on these datasets to proactively predict potential faults by harnessing the power of ML and DNN models. The study meticulously assessed the performance of these models on both datasets, addressing binary and multi-classification challenges. Thorough evaluation encompassed shows the exploration of a wide array of machine learning models and the associated statistical metrics that gauge their efficacy [6]. The solution is based on ML and DL such as decision trees, Adaboost, Random Forests, and deep neural networks (DNN) to predict various failures in products. Proactive approach aids maintenance teams in proactively addressing component issues before they manifest as faults [7]. Algorithms are ranked based on performance metrics including accuracy, error, F1 score, AUC score, and ROC curve analysis. Predictive maintenance has become increasingly essential in minimizing downtime and associated costs. The authors at [8] showcase the practical application of their methodology within an experimental setting, employing a real industrial group as a demonstrative case to highlight its effectiveness. Preliminary findings exhibit robust performance in predicting various models with better accuracy, reaching an impressive 95%. This level of accuracy was attained using a dataset encompassing 530,731real-time data readings from 15 distinct machine features, all gathered from the tested cutting machine [9]. Monitoring of two crucial elements within the system: the tools and motors, employing advanced AI algorithms. This research uses classification techniques like Support Vector Machines (SVMs) and neural networks to keep a close eye on the condition of cutting tools and bearings in a factory [10].

Researchers applied various methods to real data and carefully checked how well they worked. The study showed the results using charts that make it clear how good predictions are and where they made mistakes. The main idea is to use Predictive Maintenance (PdM) to make sure our machines keep running smoothly. This can reduce the time machines are out of service and make the important parts last longer [11]. At the end, it saves a lot of money by making maintenance and supply chain management more efficient. Plus, it makes sure the machines are safe and work well. A predictive maintenance dataset designed for both the paper's purposes and the broader community. The study entails the development and training of an interpretable model, accompanied by an explanatory interface, utilizing dataset mentioned in [12]. Subsequently, the performance of these explanatory tools is rigorously assessed and compared. Within the scope of this paper, two distinct methods for offering explanations regarding the classification outcomes of a complex ensemble classifier are scrutinized, leveraging the synthetic predictive maintenance dataset. Each method carries its unique strengths and weaknesses, ultimately delivering value to users without incurring significant additional costs. Decision trees, while occasionally failing to provide explanations, tend to deliver explanations of higher quality. In summary, this research contributes a synthetic dataset and evaluates the effectiveness of two explanation methods for complex ensemble classifiers in the realm of predictive maintenance, offering valuable insights into their utility and trade-offs. Predictive maintenance models should be developed and how often they should notify the user that the equipment requires attention [13]. The study uses a smaller number of parameters for prediction and to which the predictions came nearly the same every time.

The use of AI in Industry 4.0 and how different AI methods can increase industrial production is described in [14]. It considers PdM as a crucial part of the industry which can directly affect the production process. It also discusses various issues and challenges faced by Industry 4.0. The study [15] explores unsupervised machine learning techniques for early fault detection. The main algorithm used is clustering and analysis is done on that basis. Various sub-types of clustering are used for different purposes. They divided the detection into three parts which are health, warning, and faulty research investigates fault detection in oil pumps in the oil and gas industries. Implementation of a simple ML algorithm for early detection using multi-layer perceptron (MLP) and SVM for outlier detection is given in [16]. Faults were detected correctly, and the model gave good accuracy. This research does a survey of various AI and ML techniques, from which it uses algorithms like LSTM and Bi-LSTM for failure prediction and performance [17]. It specifically detects faults in rotary machines. They use SVM to find imbalances in the motor.

3 Methodology

In this research, classical machine learning techniques were used to detect faults in industrial machinery where Fig. 1 depicts the parameters.

3.1 Feature of Dataset

Air temperature which is measured in Kelvin[K], is the temperature of the surrounding in which the equipment operates. It affects the efficiency of the machine, if it increases it leads to overheating and if it decreases it can affect the fluids used in the machine. Process temperature is the temperature at which the machine is operating or at which the process is being conducted. It is also measured in Kelvin[K] and maintaining this temperature is crucial for product quality and rate of reaction. Rotational speed measures the rotations per minute (rpm). It indicates at what speed the motor is rotating. Excess speed can lead to wear and tear of the motor and less speed will lead to low production. Torque is the rotational force applied to the system [Nm] measured in Newton per meter. Excess torque can lead to damage of the material which is being processed. Tool wear is the amount of time the equipment has been used before it requires maintenance. It ensures the safety of the tool and anyone who is working on it.

The “AI4I 2020 Predictive Maintenance Dataset” offers a simulated dataset that closely resembles authentic predictive maintenance data observed in various industrial settings. This dataset comprises 10,000 data entries, each with 14 distinct attributes, including a binary label indicating the occurrence of a “machine failure.” In this dataset, the target feature is heavily imbalanced which will make it very challenging to train the model on the failure types that rarely occur, that is why oversampling techniques were used to balance the dataset [7].

SMOTE, which stands for Synthetic Minority Over-sampling Technique, is a resampling technique used in ML to solve problems of class imbalance, especially in classification tasks. Class imbalance occurs when one class significantly outnumbers

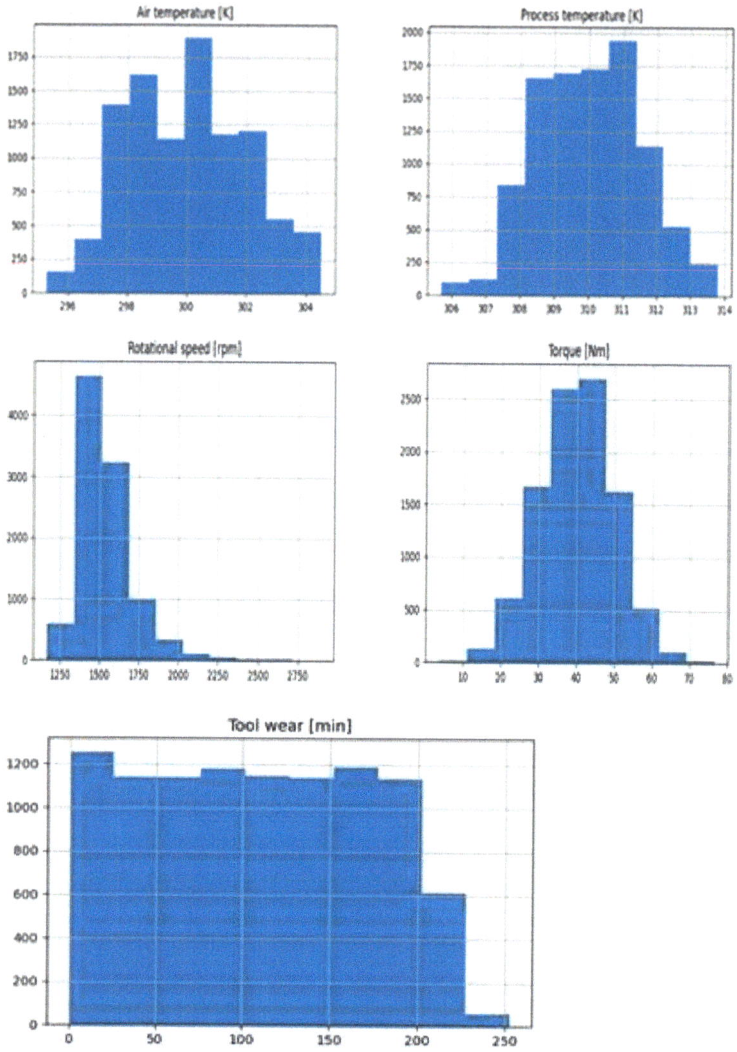

Fig. 1 Distribution of numerical features of the dataset

another in a dataset, which leads to biased model performance. SMOTE helps solve this issue by generating artificial samples of the minority class. SMOTE is valuable because it helps in balancing the class distribution, ensuring that the model does not Favor the majority class and provides more accurate predictions for the minority class of dataset as is shown in Fig. 2. Only two levels of headings should be numbered. Lower-level headings remain unnumbered; they are formatted as run-in headings.

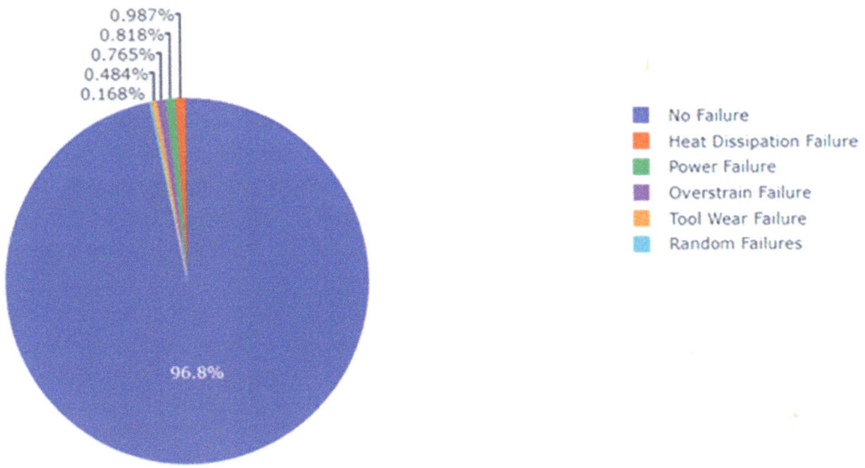

Fig. 2 Percentage of records in each of the six failure classes

3.2 Algorithms

In this research, the focus involves addressing both classification and prediction tasks, and the dataset comes with labels, which is why the research has opted for a supervised learning method. For predictive maintenance, this research uses XGBoost, and for early prediction, it uses the Local Outlier Factor algorithm. XGBoost is a powerful machine learning algorithm that combines the knowledge of multiple weak models (like decision trees) to make highly accurate predictions or decisions. It's especially good at handling complex problems and improving accuracy through a process called boosting. The basics behind XGBoost is to minimize an objective function (T), which combines training loss, which is the evaluation of training data, and regularization terms, which prevent overfitting. In this work, the focus involves addressing both classification and prediction tasks, and the dataset comes with labels, which is why the research has opted for a supervised learning method. For predictive maintenance, this research uses XGBoost, and for early prediction, it uses the Local Outlier Factor algorithm. XGBoost is a powerful machine learning algorithm that combines the knowledge of multiple weak models (like decision trees) to make highly accurate predictions or decisions. It's especially good at handling complex problems and improving accuracy through a process called boosting. The basics behind XGBoost is to minimize an objective function (T), which combines training loss, which is the evaluation of training data, and regularization terms, which prevent overfitting.

$$\text{Objective(T)} = \text{Loss} + \text{Regularization}, \tag{1}$$

$$\text{T} = \sum \text{i}\left[1/2 * \left(\text{yi} + \text{y}'\right)^2\right] + \Omega(\text{f}). \tag{2}$$

Equations 1 and 2 are objective functions, where $\sum i$ is the sum of all the data points, yi is the actual value, and y' is the predicted value. Explainable AI (XAI) refers to the practice of designing and developing AI systems and ML models in a way that makes their decisions and inner workings understandable, interpretable, and transparent to users. The goal of XAI is to provide insights into how AI systems reach their conclusions, predictions, or recommendations and to ensure that these systems can be trusted and effectively used by humans in various applications. In this research, Shapley Additive Explanations (SHAP) explains machine learning models because SHAP works.

Shapely Additive Explanations (SHAP) SHAP is a way to explain the output of any ML model. This method, rooted in cooperative game theory, offers a systematic approach to allocate the "credit" for a model's prediction across its input features. Shapley values are used for explaining the outputs of the prediction model. Mathematical expressions are given here:

$$\text{SHAPLEY}(i) = (1/N!) * \sum(\text{Marginal Contribution}(i, \Phi)), \tag{3}$$

where Φ is the permutations of all players, and Marginal Contribution (i, Φ) is the contribution of i in permutation Φ.

Local Outlier Factor: The distance between object A and its kth nearest neighbor is denoted by k(A). The set of k nearest neighbors is represented as Nk(A).

$$k - \text{distance}(A) = \text{Distance (A, kth nearest neighbor of A)} \tag{4}$$

The reachable distance between A and B is called reachability distance, which is calculated as:

$$D(A, B) = \max(k - \text{distance}(B), \text{distance}(A, B)) \tag{5}$$

The Local Outlier Factor (LOF) is calculated as:

$$\text{LOF}(A) = \left(D(A, k) / \sum i(Ai, k)\right) \tag{6}$$

If the value is approximately 1, then the object is not an outlier. If the value is below 1, it indicates an inlier. If the value is much larger than 1, it indicates an outlier (Fig. 3).

Fig. 3 Methodology

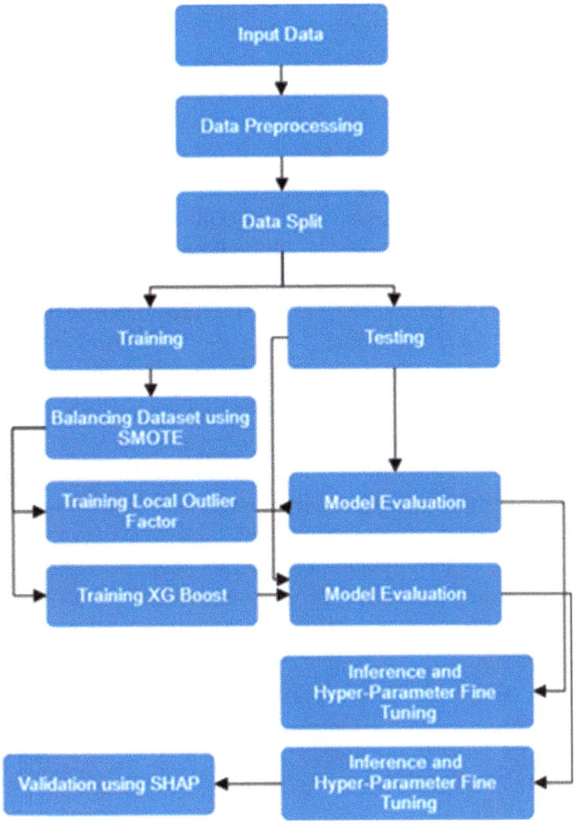

4 Results and Discussions

For predictive maintenance, this research experimented with the XGBoost and Random Forest Classifier algorithms, both of which are known for their ability to use ensemble methods and effectively handle imbalanced data to make highly accurate predictions. For detecting anomalies, this research uses the Local Outlier Factor and One-Class SVM which are good for identifying anomalies or outliers within the dataset, which is particularly valuable in situations where anomalies are rare. The research evaluated the performance of the models using parameters such as Precision, accuracy, Recall, and F1 score.

Table 1 summarizes the performance of the XGBoost and Random Forest models. The proposed model has achieved an impressive accuracy of 96%. Additionally, Recall, F1 score, and Precision exhibited high values, indicating the models' effectiveness in predicting industrial machinery failures. It also presents the performance of the One-Class SVM model and Local Outlier Factor. This research obtained a ratio of correct predictions to total predictions exceeding 91%, along with high Recall,

F1 score, and Precision values. This underscores the Local Outlier Factor's ability to identify anomalies within the dataset.

The results demonstrate the efficacy of classical ML techniques with reference to predictive and preventive maintenance for industrial machinery. The findings of the outliers were an indication that the machine has some faults and if these outliers increase in numbers, the machine requires immediate attention. By evaluating these outliers, the study was able to recognize the faults in the machinery which then can be repaired and handled by the professionals. Explainable AI (XAI) played a crucial role in this research. XAI is a practice that ensures AI systems and machine learning models produce transparent and interpretable results. The research employed Shapley Additive Explanations (SHAP) as the XAI method SHAP, rooted in cooperative game theory. SHAP values quantify how much each feature "contributes" to the prediction.

Table 2 provides a comprehensive summary of the air temperature data for instances where no machine failures were observed. The dataset contains 578 observations with a mean air temperature of 297.676 K, indicating that the average operating temperature during normal machine functioning remains stable around this value.

The standard deviation (0.862 K) suggests minimal variation in air temperature, highlighting consistent environmental conditions during no-failure instances. The minimum and maximum temperatures recorded are 295.3 and 299.1 K, respectively, indicating that the temperature range is narrow and well-regulated. The quartile values (25%: 297.2 K, 50%: 297.8 K, and 75%: 298.2 K) further affirm that the air

Table 1 Summary statistics of performance of XGBoost, RF classifier, local outlier factor versus one-class SVM

Classifier	Accuracy	Precision	Recall	F1 score
XGB	0.960	0.980	0.960	0.969
RFC	0.955	0.976	0.955	0.964
LOF	0.917	0.951	0.917	0.933
OCSVM	0.912	0.953	0.912	0.930

Table 2 Summary statistics of parameters for "No Failure" instances

Parameter	Air temperature	Process temperature	Rotational speed	Torque	Tool wear
Count	578.000	578.0000	578.000	578.000	578.000
Mean	297.6761	308.3323	1543.612	39.645	104.659
Std	0.862838	0.990671	171.2414	9.5307	63.458
Min	295.3000	305.7000	1168.000	13.800	0.00045
25%	297.2000	307.7000	1430.000	33.125	48.0000
50%	297.8000	308.3000	1509.000	39.900	103.000
75%	298.2000	309.0000	1622.000	46.200	159.750
Max	299.1000	310.3000	2448.000	67.000	233.000

temperature remains within a tightly clustered range, reinforcing the stable operational environment during no-failure scenarios. The symmetry between the quartile values and the median (50%) suggests a near-normal distribution of air temperature values for the given dataset. These findings imply that the air temperature does not significantly deviate during regular machine operations, making it a potential baseline indicator for detecting abnormal conditions or predicting potential failures. Maintaining this stable range could be critical for enhancing the reliability of industrial machines.

The mean process temperature is 308.332 K, indicating that the typical operating temperature during normal machine performance remains stable around this value. The standard deviation of 0.991 K suggests minimal variability, implying that process temperatures were consistently regulated during no-failure operations. The minimum and maximum recorded values, 305.7 and 310.3 K, respectively, highlight a narrow range of temperatures during normal operation. The quartile values (25%: 307.7 K, 50%: 308.3 K, and 75%: 309.0 K) indicate that the process temperatures are evenly distributed around the median, with only slight deviations. This symmetric distribution suggests consistent temperature control, which is essential for maintaining optimal machine performance. The stable and narrowly distributed process temperature range highlights its importance as a potential indicator for detecting abnormal conditions. Deviations from this well-regulated range could signal operational inefficiencies or potential machine failures, emphasizing the need to monitor this attribute for predictive maintenance.

The mean rotational speed is 1543.612 rpm, suggesting that machines generally operate at this speed during normal conditions. The standard deviation of 171.241 rpm indicates moderate variability in the rotational speeds, reflecting differences in operating requirements or machine loads even under no-failure conditions. The minimum and maximum values, 1168 and 2448 rpm respectively, demonstrate a broad range of rotational speeds during normal operations. The quartile values (25%: 1430 rpm, 50%: 1509 rpm, and 75%: 1622 rpm) reveal that 50% of the rotational speed values fall between 1430 and 1622 rpm. The distribution suggests a tendency toward slightly higher operational speeds, as indicated by the mean being slightly above the median (1509 rpm).The wide range between the minimum and maximum speeds indicates the adaptability of the system to different operating conditions while maintaining stability without failure. Monitoring deviations from this range may help identify potential issues and inform predictive maintenance strategies.

The mean torque is 39.645 Nm, indicating that this is the typical operational torque during normal machine functioning. The standard deviation of 9.531 Nm suggests moderate variability in torque values, reflecting changes in machine load requirements even under normal operating conditions. The minimum and maximum torque values, 13.8 and 67.0 Nm respectively, indicate a wide range of torque levels without causing machine failure. The quartile values (25%: 33.125 Nm, 50%: 39.9 Nm, and 75%: 46.2 Nm) show that half of the observations fall between 33.125 and 46.2 Nm, suggesting that the majority of operational torque values are clustered within this range. The close alignment between the median (39.9 Nm) and mean (39.645 Nm) indicates a relatively symmetric distribution. The broad range of torque values highlights the machine's ability to adapt to different operating loads while maintaining stability. Monitoring sudden or persistent deviations beyond this range can be crucial in predicting potential failures and ensuring smooth machine operations.

The mean tool wear is 104.659 min, indicating the typical duration of tool usage under normal operational conditions. The standard deviation of 63.458 min shows significant variability in tool wear, which may be influenced by differences in machine operations, material hardness, or maintenance schedules. The minimum tool wear is 0 min, likely representing newly installed tools or machines with minimal operational history. The maximum tool wear recorded is 233 min, reflecting extended tool usage without failures. The quartile values (25%: 48 min, 50%: 103 min, and 75%: 159.75 min) reveal that half of the tool wear values fall between 48 and 159.75 min. The close alignment of the mean and median (103 min) suggests a relatively balanced distribution of tool wear times. The wide range of tool wear durations highlights the adaptability of the system to different usage patterns without experiencing failures. Monitoring tool wear is essential for predictive maintenance, as excessive or unusual tool wear can be an early indicator of potential operational issues. In conclusion, it shows about the research found that there are various parameters that should be taken into consideration while doing maintenance of machinery. Predictive maintenance is a good strategy that reduces needless expenditures as the faults are detected even before they can cause any damage and also the addition of machine learning can make the work much simpler.

5 Conclusion

In conclusion, this research explored predictive and preventive maintenance for industrial machines using classical machine learning techniques. The AI4I 2020 Predictive Maintenance Dataset provided a foundation for this study, offering a simulation of industrial maintenance scenarios. Recognizing the challenge of imbalanced data, we effectively applied SMOTE to address the issue, ensuring that the model remains fair and reliable in identifying different failure types. The selected algorithms, XGBoost, and Local Outlier Factor, showed their respective strengths in predictive and preventive maintenance tasks. XG Boost showcased its strength in handling complex problems, while the Local Outlier Factor excelled in distinguishing optimal decision boundaries in various scenarios. Explainable AI (XAI) played a major role in this research, shedding light on the nature of machine learning models. Specifically, the model utilized the Shapley Additive Explanations (SHAP) methodology to distribute credit among input features and evaluate model predictions transparently. The predictive maintenance model achieved a notable accuracy of 96%, indicating its effectiveness and robustness in detecting and preventing potential machine failures. In a field like industrial maintenance, the merging of classical machine learning techniques and XAI provides a foundation for enhancing system reliability, preventing costly downtimes, and developing trust in predictive and preventive maintenance solutions. This research underlines the significance of transparent and interpretable AI systems in addressing the challenges of industrial maintenance, which makes it more efficient in industrial operations.

References

1. Matzk S (2020) Explainable artificial intelligence for predictive maintenance applications. In: 3rd international conference on artificial intelligence for industries (AI4I). IEEE
2. Calabrese M et al (2020) Sophia. An event-based IoT and machine learning architecture for predictive maintenance in industry 4.0. Information 11(4):202
3. Salih A, Raisi-Estabragh Z, Galazzo IB, Radeva P, Petersen SE, Menegaz G, Lekadir K (2023) Commentary on explainable artificial intelligence methods: SHAP and LIM. arXiv:2305.02012
4. Vilone G, Longo L (2021) Notions of explainability and evaluation approaches for explainable artificial intelligence. In: Information fusion, vol 76, pp 89–106
5. Loh HW, Ooi CP, Seoni S, Barua PD, Molinari F, Acharya UR (2022) Application of explainable artificial intelligence for healthcare: a systematic review of the last decade (2011–2022). In: Computer methods and programs in biomedicine. https://doi.org/10.1016/j.cmpb.2022.107161
6. Abbasi J (2021) Predictive maintenance in industrial machinery using machine learning
7. AI4I 2020 Predictive Maintenance Dataset. In: UCI machine learning repository, 2020. https://doi.org/10.24432/C5HS5C
8. Paolanti M et al (2018) Machine learning approach for predictive maintenance in industry 4.0. In: 14th IEEE/ASME international conference on mechatronic and embedded systems and applications (MESA). IEEE
9. Zhang W, Yang D, Wang H (2019) Data-driven methods for predictive maintenance of industrial equipment: a survey. IEEE Syst J 13(3):2213–2227
10. Lee WJ et al (2019) Predictive maintenance of machine tool systems using artificial intelligence techniques applied to machine condition data. Procedia Cirp 80:506–511
11. Vallim Filho AR, Farina Moraes D, Bhering de Aguiar Vallim MV, Santos da Silva L, da Silva LA (2022) A machine learning modeling framework for predictive maintenance based on equipment load cycle: an application in a real world case. Energies 15(10):3724. https://doi.org/10.3390/en15103724
12. Keleko AT, Kamsu-Foguem B, Ngouna RH et al (2022) Artificial intelligence and real-time predictive maintenance in industry 4.0: a Bibliometric analysis. AI Ethics 2:553–577. https://doi.org/10.1007/s43681-021-00132-6
13. Amruthnath N, Gupta T (2018) A research study on unsupervised machine learning algorithms for early fault detection in predictive maintenance. In: 5th international conference on industrial engineering and applications (ICIEA). IEEE
14. Orrù PF, Zoccheddu A, Sassu L, Mattia C, Cozza R, Arena S (2020) Machine learning approach using MLP and SVM algorithms for the fault prediction of a centrifugal pump in the oil and gas industry. Sustainability 12(11):4776
15. Das MK, Rangarajan K, Performance monitoring and failure prediction of industrial equipments using artificial intelligence and machine learning methods: a survey. In: 4th international conference on computing methodologies
16. Intel, Understand Moore's law. In: https://www.intel.com/content/www/us/en/newsroom/resources/moores-law.html. Last accessed 03 March 2024
17. SPIE, SPIE Digital Library: Search the world's largest collection of optics and photonics applied research. https://www.spiedigitallibrary.org. Last accessed 30 June 2024

Design of a Sustainable IoT for LPG Monitoring and Leakage Prevention

P. R. Venkatesh, Mohammed Asif Kattimani, S. Rohith, M. Pradeep, and Vivekanand S. Gogi

Abstract The research project uses sustainable technologies to improve home automation and safety. The Internet of Things (IoT) is helping to monitor LPG gas leak prevention systems. By integrating ESP modules Arduino and LPG gas sensors, it ensures timely gas leak detection, sends users an SMS, and displays the status on an LCD. The system continuously monitors LPG levels and provides real-time updates via load sensors. Load sensors are used by the system to continuously monitor LPG levels and provide real-time updates. In order to improve sustainability and longevity, the structural integrity of the LPG cylinder is designed in SOLIDWORKS, evaluated with ANSYS, and covered with epoxy to resist corrosion and water. Combining IoT and GSM modules, this solution enhances safety, allows efficient resource use, and conforms to sustainable smart home developments. It provides a complete eco-friendly solution for smarter, safer, and more sustainable living environments by addressing significant problems like late refills, ineffective pre-booking, and gas leaks.

Keywords IOT · LPG · SMS · LCD · SOLIDWORKS · ANSYS

P. R. Venkatesh
Department of Mechanical Engineering, RV College of Engineering, Bengaluru, Karnataka, India

M. Asif Kattimani (✉)
Department of Mechanical Engineering, Lords Institute of Engineering and Technology, Hyderabad, Telanagana, India
e-mail: md.asifkattimani@lords.ac.in

S. Rohith
Department of Mechanical Engineering, ATME College of Engineering, Mysuru, Karnataka, India

M. Pradeep
Department of Mechanical Engineering, St. Joseph's College of Engineering and Technology, Thanjavur, Tamil Nadu, India

V. S. Gogi
Department of Industrial Engineering & Management, RV College of Engineering, Bengaluru, Karnataka, India

I. M. Jamadar et al. (eds.), *Integrating Sustainability in Modern Technologies*, Lecture Notes in Mechanical Engineering, https://doi.org/10.1007/978-981-95-9466-5_38

1 Introduction

The use of technology in our homes and workplaces has become popular recently. Daily tasks are becoming easier as a result of the places smart design. As more people adopt the ideas of smart homes, home automation has become essential. The Internet of things (IoT) enters the picture at this point. The items used such as gas cylinders in homes have not changed despite the fact that routine tasks have become more intelligent. At times, it might happen by accident. As a result, technology must be used to stop mishaps. The Internet of Things is a rapidly expanding technology in various industries. Manjunathan [1] presents a smart LPG leakage detector utilizing the ESP8266 NodeMCU module. The system detects gas leaks and alerts users through connected devices, enhancing home safety. Chijioke et al. [2] presented an IoT-based system comprising microcontrollers, sensors, and cloud platforms to detect LPG leaks, aiming to prevent accidents and ensure safety. References [3, 4] details a telemetry system for real-time monitoring of LPG and carbon monoxide concentrations using IoT techniques, contributing to accident prevention in various environments. References [5, 6] provide a systematic assessment of recent smart home technologies, including those related to energy efficiency and environmental sustainability. The odor of ethyl mercaptan, the odorant used in LPG, is distinct and unpleasant. Ethyl mercaptan is selected because it is non-corrosive, has a near-LPG boiling point, and contains little sulpfur. According to Mackerle [7], Finite Element Methods or FEMs are used for both theoretical and practical analysis of pipes and pressure vessel structures. External loads imposed on vertical pressure vessels result in axial loading and bending moments claims [8]. The type of design approach used depends on whether the vessel is under external or internal pressure and whether the shells longitudinal stress H is tension or compression. In accordance with design procedures, Mayer created the pressure vessels. The Mayers and H. L. The weariness poses practical challenges for the designer according to Mayer et al. [9]. We need LPG cylinders in our daily lives.

2 Methodology

2.1 Gas Detection System

Gas spill detection has been given a very high priority in this model. The IoT system is implemented using an Arduino Mega 2560, which serves as the primary microcontroller as shown in Fig. 1. The coding is done using C++ in the Arduino IDE. Using a GSM module, the microcontroller sends a specific message EMERGENCY ALERT: LPG gas spillage found in your home to the appropriate mailer and an LCD will display a similar message, while the MQ2 gas sensor and load cell continuously monitor gas levels and leakage. The communication between the sensors

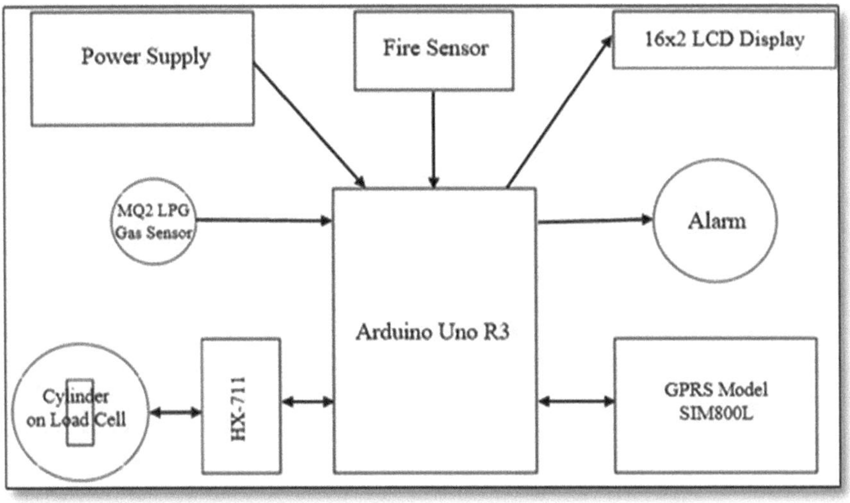

Fig. 1 Gas detection system

and controllers is handled using I2C and serial communication protocols, ensuring seamless data transfer.

2.2 *Design of Gas Cylinder*

Low carbon steel is currently the material used for LPG cylinders. Indian Standards (IS 3196) are followed in the design of the low-carbon steel cylinder. Material: IS 3196/HR (Low carbon steel) thickness needed for steel cylinder taking internal pressure into account.

$$T = \text{Thickness of the cylinder}$$
$$\sigma_t = \text{Yield Strength of Low carbon steel} = 250\,\text{MP}_a$$
$$D_i = \text{Internal Diameter} = 314.4\,\text{mm}$$
$$P_i = \text{Internal Pressure} = 25\,\text{kgf/mm}^2$$

- **Thickness Required for the Cylindrical Portion**

$$t = P_i D_i 2\sigma_t = 1.2 \times 314.4 \times 2 \times 250 = 0.754\,\text{mm} \tag{1}$$

- **Thickness Required for the End Dome**

$$t = P_i D_i 4\sigma_t = 1.2 \times 314.4 \times 4 \times 250 = 0.37725\ \text{m} \tag{2}$$

Taking greater among the both, t = 0.75456 mm.

- **Thickness Required for Steel Cylinder**

$$t = P_h D_i (200 \times 0.8 \times J \times R_e) - P_h \text{ (IS3196)} \tag{3}$$

where,

P_h = Test Pressure = 25 kgf/mm^2
J = Weld joint factor = 0.9 (for non radio graphed whelded joint).
R_e = Yield strength = 250 MPa = 25.48 kgf/mm^2
t = 25 × 314.4(200 × 0.8 × 0.9 × 25.45)−25 = 2.156 mm.
Total thickness, t = thickness + allowances.
∴ t = 2.156 + 0.2 + 0.1 = 2.456 mm.
∴ t ≅ 2.5 mm.

- **Stress Calculation for Steel Cylinder**

$$\text{Longitudinal stress}: \sigma_L = P_i D_i 4t = 1.2 \times 314.4 \times 4 \times 2.5 = 37.728\,\text{MP}_a \tag{4}$$

$$\text{Hoop Stress}: \sigma_H = P_i D_i 2t = 1.2 \times 314.4 \times 2 \times 2.5 = 75.45\,\text{MP}_a \tag{5}$$

von Mises stress

$$\sigma_v = \left(\sqrt{\sigma_{12}} + \sqrt{\sigma_{22}}\right) - \sigma_1\sigma_2 = \left(\sqrt{37.232} + \sqrt{75.462}\right) - \sqrt{}(37.73 \times 75.46) = 65.05\,\text{N/mm}^2 \tag{6}$$

- **Longitudinal Deformation**

$$\delta = Pr2(1-\upsilon)2Et = 1.2 \times 314.4 \times 2 \times (1-0.3) \times 2 \times 2.5 \times 207 \times 103 = 0.02\,\text{m} \tag{7}$$

- **Volume of Steel Cylinder**

$$V_{cyl} = \pi h\left(r_o^2 - r_i^2\right) \tag{8}$$

r_o Outer radius of the cylinder
r_i Internal radius of the cylinder.
h Height of the cylinder portion.
V_{cyl} Volume of the cylindrical portion.
V_{cyl} $\pi \times 368 \times \left(319.42^2 - 314.42^2\right) = 915925.05\,\text{mm}^3$
V_{Sp} **Volume of the spherical portion**

$$V_{Sp} = 4 \times 3\pi\left(r_o^3 - r_i^3\right) \tag{9}$$

$V_{Sp} = 4 \times 3 \times \pi \times \left(319.42^3 - 314.42^3\right) = 788757.26\,\text{mm}^3$.

2.3 Modeling of LPG Cylinder

SOLIDWORKS is used for 3D part modeling of a cylinder due to its advanced features and ease of use. The modeling process starts by creating a sectional profile of the cylinder using sketch entities like line, arc, fillet, and rectangle. The profile is then converted into a 3D solid model using the revolve feature. A circular sketch is created above the cylinder and revolved to form an O-ring for holding the cylinder. Supports between the O-ring and cylinder are designed using a rectangle sketch and extruded into 3D. The supports are replicated into the circular pattern feature, as shown in the Fig. 2.

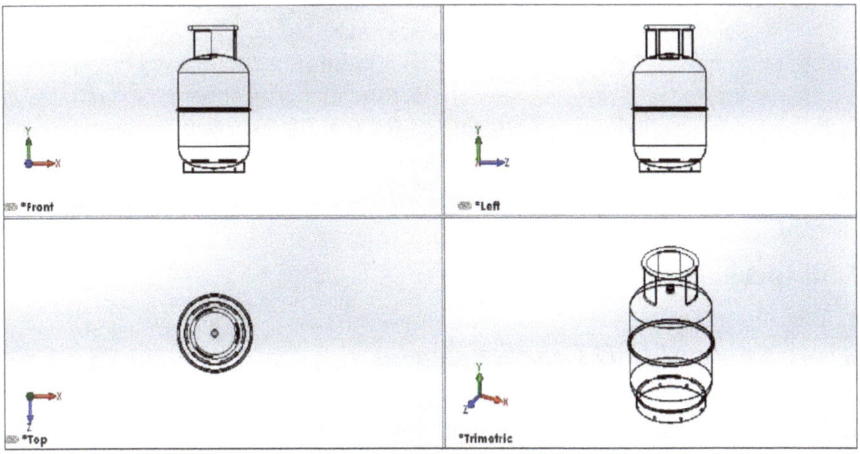

(a) 3D Wireframe Model

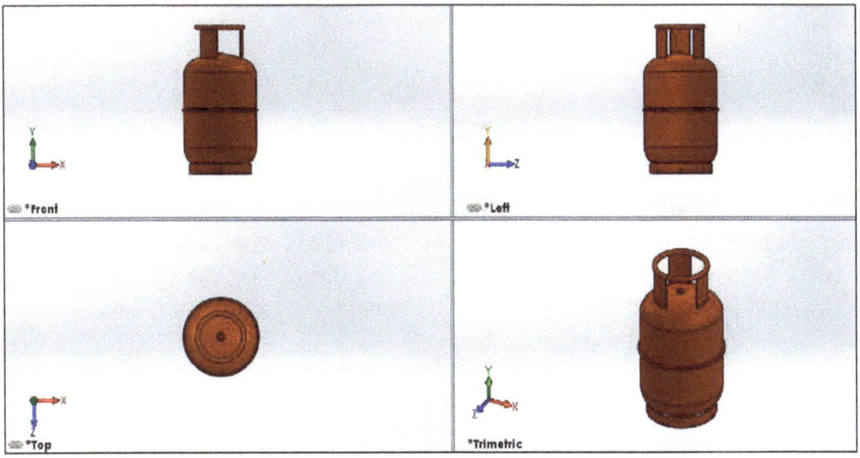

(b) Solid Model

Fig. 2 Views of LPG cylinder

2.4 Simulations of Gas Cylinder

In order to analyze the steel cylinder, a static structural analysis was carried out using ANSYS 14.5. Low-carbon steel properties were substituted for the default engineering data [10, 11]. Triangle elements were used to discretize the LPG gas model, and different mesh sizes were employed for accuracy and a mesh size of 0.01 mm was employed. The foot ring was used as support in the setup, and the cylinders lateral surface was concealed to apply internal pressure. Due to its external nature, the foot ring was not under internal pressure during the analysis guaranteeing a targeted evaluation of the structural soundness of the cylinder.

The structural analysis using ANSYS revealed stress concentrations of 3.56 × 108 Pa in the upper and bottom portions under internal pressure as shown in Fig. 3a. From the deformation Fig. 3b, it is observed that the deformation is high at the bottom portion of the body, when the cylinder is subjected to internal pressure and gravitational force in Y direction as well. Deformation was observed to be 0.008 m at the bottom, considering gravitational and internal forces, ensuring the cylinder's reliability, as mentioned in Table 1.

3 Experimental Setup

3.1 Coating of LPG Cylinder

The LPG cylinder shown in Fig. 4 is dipped into a bucket of epoxy resin for 20 min, then left for 1 min before being removed. This process is repeated to ensure complete coverage. The coated cylinder begins curing within 8 h and is fully solidified after 24 h.

3.2 IoT Setup

The IoT system, all components are interconnected, and the program is loaded onto an Arduino with an SIM card in the GSM Protocol (TDMA Technology) used for sending SMS alerts. The complete setup with specifications of the devices (Table 2), with power supplied to activate all electrical components as shown in Fig. 5.

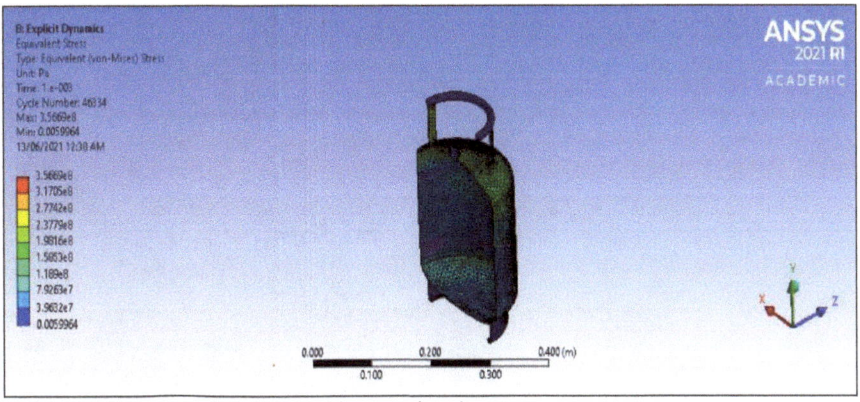

(a) Von Mises Stress

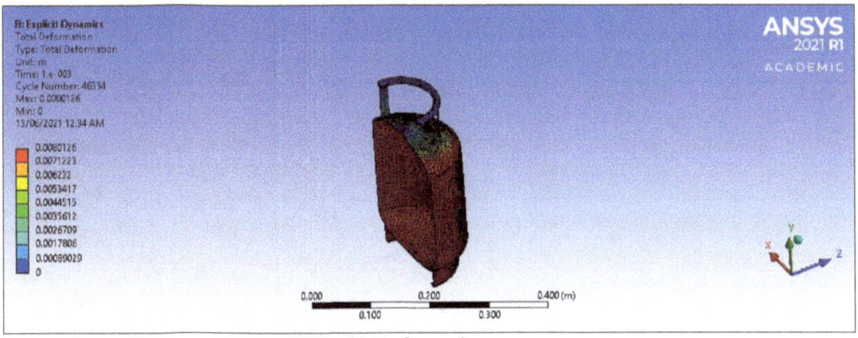

(b) Deformation

Fig. 3 Simulation results in the cylinder

Table 1 Simulation results

S. no.	Items	Description	Value
1	Software	ANSYS 14.5 (Static Structural)	
2	Material properties	Steel (σ_y)	250
3	Mesh type	Triangular elements	0.01 mm
4	Constraints	Foot ring fixed as support	–
5	*Load*	Internal pressure	25 $\left(\text{kgf/mm}^2\right)$
6	Maximum stress (σ_{max})		$3.56 \times 10^8 P_a$
	Deformation (δ)		0.08

Fig. 4 Epoxy coated cylinder

Table 2 Specifications of the devices

S. no.	Items	Description
1	Arduino Mega 2560	8-bit microcontroller, 54 digital I/O pins, 16 Analog inputs
2	MQ2 Gas Sensor	Sensitivity range: 200–10,000 ppm
3	Load Cell	5 kg capacity, strain gauge-based
4	GSM Module (SIM800L)	Quad-band communication, SMS capability
5	LM35 Temperature Sensor	Accuracy ± 0.5 °C
6	Display	16 × 2 LCD: I2C communication, real-time data display

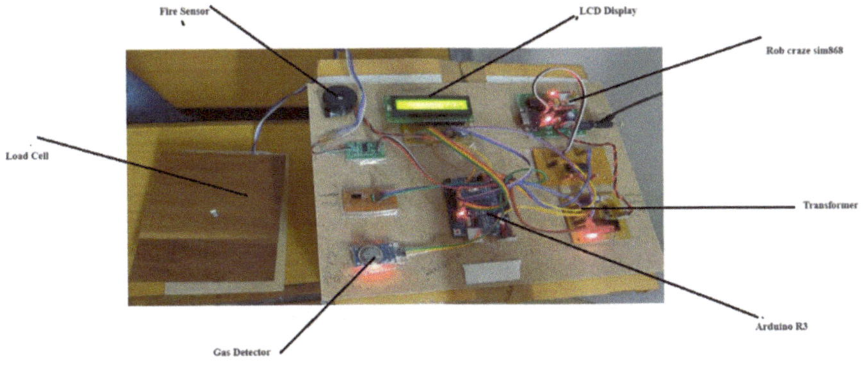

Fig. 5 IoT setup

4 Results and Discussion

A comprehensive gas level and leakage detection system using Arduino Mega 2560, load cells, gas and fire sensors, GSM modules, and a 16 × LCD display shown in Fig. 6. The Arduino Mega 2560, powered by a 7–12 V supply, serves as the central microcontroller for the system. Its affordability, cross-platform compatibility, and user-friendly programming environment make it an ideal choice for this application. The system utilizes load cells to monitor LPG levels and gas sensors, such as the MQ2, to detect LPG leakage. The MQ2 sensor is highly sensitive to methane and natural gas, capable of detecting concentrations between 200 and 10,000 ppm, while the LM35 temperature sensor provides precise thermal readings. It plays a critical role in detecting fire hazards and overheating conditions, ensuring an additional layer of safety. These sensors ensure fast response, stability, and long service life. The LCD displays sensor data, while the GSM module alerts users via SMS about gas leakage or low LPG levels. The GSM modem operates using TDMA technology, ensuring reliable communication across various frequency bands. This integrated solution combines real-time monitoring, safety alerts, and efficient resource management, offering a cost-effective and sustainable approach to managing LPG usage and enhancing safety in residential and industrial environments.

Once the cylinder is loaded and powered, the Arduino monitors the weight through the load cell by detecting voltage and current. The current weight is displayed on the LCD, updating as the gas level decreases. If the weight drops below a set limit, an alarm sounds five times, alerting nearby individuals. Additionally, the cylinder owner receives an SMS notifying them to book a new gas cylinder due to the low level, ensuring timely refills as shown in Fig. 7.

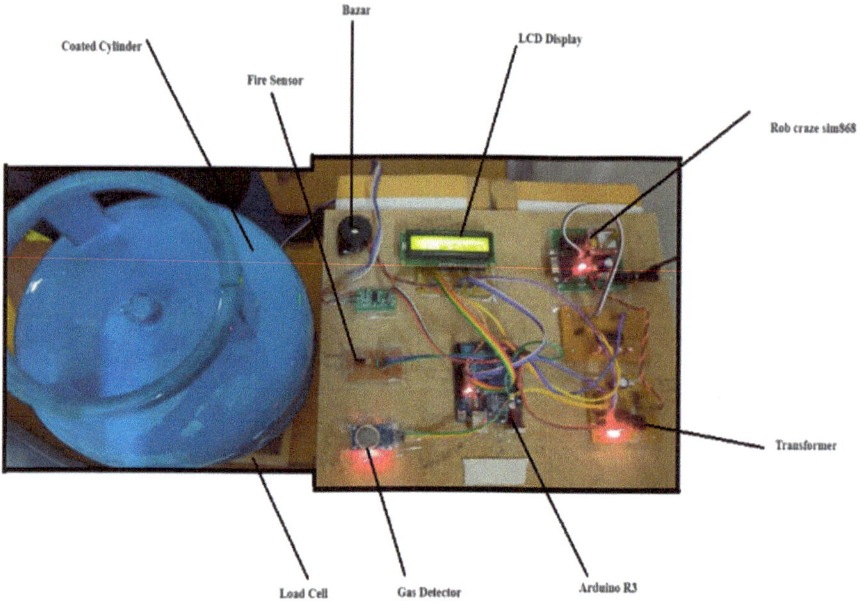

Fig. 6 Experimental setup

Fig. 7 Message received from setup

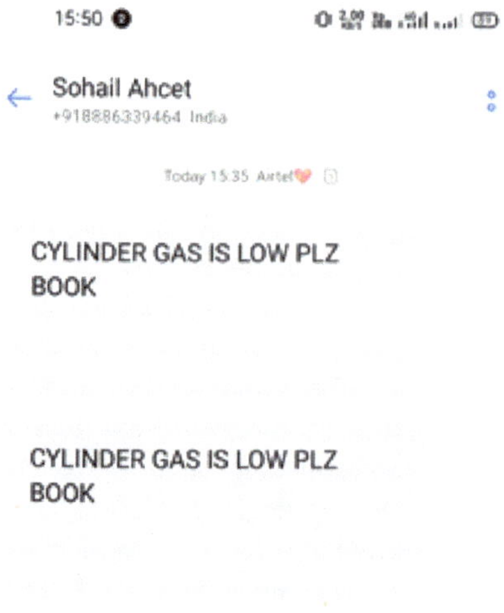

5 Conclusion

Sustainable technologies are successfully utilized in this study to enhance the usability and safety of LPG cylinders. To ensure longevity in kitchens and during distribution in inclement weather, an epoxy coating is applied to the cylinder, making it resistant to corrosion and water. After modeling the design of the LPG cylinder with

SOLIDWORKS, structural analysis with ANSYS showed that the top and bottom sections of the cylinder had stress concentrations of 3.56 $\times$ 108 Pa under internal pressure. When internal and gravitational forces are taken into account, the bottom deformation was found to be 0.008 m, ensuring the cylinder's reliability. Using IoT technology through an ESP module and LPG gas sensors, a gas leak detection and monitoring system was created to enhance home safety. The system also includes a GSM module that notifies users via SMS, when the LPG level falls below 20% ensuring that refills are scheduled on time. The combination of sustainable materials and IoT structural analysis offers a comprehensive and innovative approach to LPG cylinder management with a focus on efficiency safety and environmental sustainability. The Internet of Things and smart city infrastructure can collaborate to create a centralized network for gas monitoring.

References

1. Manjunathan N, Muthulingam S, Jaganathan D (2023) IoT based gas leakage detection system. In: International conference on sustainable computing and data communication systems (ICSCDS). IEEE, India, pp 1137–1142. https://doi.org/10.1109/ICSCDS56580.2023.10105018
2. Chijioke J, Nwogu NC, Nosiri OC (2024) IOT-based liquefied petroleum gas (LPG) leakage detection system. https://doi.org/10.13140/RG.2.2.22823.73126
3. David CW (2021) Design and implementation of an IoT based LPG and CO gases monitoring system. In: CS & IT conference proceedings. Cornell University, pp 31–39. https://doi.org/10.5121/csit.2021.110803
4. Kumar Pateriya P, Munna AA, Biswas H, Ahammed A, Shah A (2021) IoT-based LPG gas leakage detection and prevention system. In: Proceedings of the international conference on innovative computing and communication (ICICC). ICICC SSRN, India, pp 1–10. https://doi.org/10.2139/ssrn.3884477
5. Arindom C, Islam M, Fahim S (2023) Smart home system: a comprehensive review. J Electr Comput Eng 1(2):1–30. https://doi.org/10.1155/2023/7616683
6. Niranjana S, Gangopadhyay A, Sweeti S, Dhanalakshmi G (2023) Experimental design of IoT based LPG gas leakage detection and accident prevention mechanism using intuitive sensors. In: International conference on innovative computing, intelligent communication and smart electrical systems (ICSES). IEEE, Chennai, India, pp 1–9. https://doi.org/10.1109/ICSES60034.2023.10465378
7. Mackerle J (2005) Finite elements in the analysis of pressure vessels and piping, an addendum. Int J Pressure Vessels Piping 82(1):571–592
8. Bergman EO (1996) The design of vertical pressure vessels subjected to applied forces
9. Mayer H, Stark HL, Ambrose S (2000) Design procedures for pressure Vessel. Int J Pressure Vessels Piping 77(1):775–781
10. Mittal L, Lovely J, Soni S (2021) IoT-based LPG gas detection and prevention with temperature monitoring. Turkish J Comput Math Educ 11(3):1701–1707
11. Kattimani MA, Venkatesh PR, Kirthan LJ (2023) Design and optimization of fatigue life studies on induction hardened IN718 alloy for gas turbine applications. Adv Mater Process Technol (AMPT) 1–13. https://doi.org/10.1080/2374068X.2023.22561211

Identification and Control of the Noise Level in Gearbox with Application of Attenuation Blankets

N. P. Puneet and T. V. Shrivathsa

Abstract Gear drives have an important role in mechanical power transmission unit and exhibit superior efficiency. Helical gears are the gear drives that are most regularly employed. Noise in these gearboxes within human audible range may result in disturbance to human inference in a long run. The causes of noise in a machinery's gearbox may be multiple. This paper seeks to examine helical gearbox's level of sound pressure and explore the efficacy of a noise attenuation blanket in reducing noise. The study offers a comprehensive analysis of how gear pitch line velocity affects the noise levels in gearboxes. The observations are carried out through a full-speed Mechanical Run Test (MRT) with no-load condition where sound pressure levels are measured at three different locations using a sound level meter. The levels of sound pressure are quantified by using the dBA scale, which aids in determining the necessity of an acoustic enclosure and assessing its impact. Numerical model of the noise attenuation blanket has been analyzed by using Finite Element Method (FEM) by considering a small volume of blanket. Results of the numerical analysis were seen in good agreement with actual test results. According to the experiment, a 25 mm thick foam reduces the sound pressure level by 3.5 decibels (dBA) on an average. This study may be useful for implementing noise attenuation blanket for gearboxes with higher sound pressure levels. The attempt of blanket design and sound pressure reduction is mainly to sustain human health as well as work environment which could greatly enhance the efficacy of work force.

Keywords Level of sound pressure · Acoustic enclosure · Noise attenuation · Mechanical run test · Pitch line velocity

N. P. Puneet (✉)
Department of Automobile Engineering, Dayananda Sagar College of Engineering, Bengaluru, India
e-mail: puneet-au@dayanandasagar.edu

Present Address:
T. V. Shrivathsa
Department of Artificial Intelligence and Data Science, Shri MadhwaVadiraja Institute of Technology & Management, Bantakal, Karnataka, India
e-mail: shrivathsa.ml@sode-edu.in

I. M. Jamadar et al. (eds.), *Integrating Sustainability in Modern Technologies*, Lecture Notes in Mechanical Engineering, https://doi.org/10.1007/978-981-95-9466-5_39

1 Introduction

Control technologies must prioritize reducing noise to acceptable levels through modifications in the work environment. This step requires the implementation of measures to reduce noise emissions. The transmission of noise, whether through the air or the workplace structure, must be mitigated. These solutions involve modifications to the equipment, operational procedures, and the layout of the workspace. The most efficient approach to addressing noise hazards in the workplace is to eradicate or mitigate the risk at its source, either through direct intervention or by containment. It is important to take into account practical concerns, as it is sometimes not possible to deploy a global control program all at once. Gear noise significantly hinders the achievement of optimal gear performance. Gear noise is caused by process faults. Researches indicate that excess material on the active profile of one or more gear teeth contributes to increased gear noise. This creates the perception that the gear does not meet the expected standard of quality.

The principal of a gearbox is to transform speed and torque with its fundamental components like gears, gear-shafts, bearings, casing, and couplings together with external clutches. The conversion of velocity occurs without any slippage or loss of coordination. The functioning state of the gear train is contingent in the drive systems. The gearbox is the primary source of noise as well as vibration [1].

A force variation that causes vibrations in the component parts and then transfers those vibrations to the surrounding structure is the main cause of noise produced by gears. Airborne noise is only generated when the vibration activates external panels. High noise levels are typically seen inside a sealed gearbox, although they are normally unimportant since the air pressure variations are not strong enough to noticeably excite the gear case. The researchers are more interested on the correlation between vibrations, noise, and transmission error. The dynamic (loaded or unloaded) transmission error measurements are most likely pertinent for noise and vibration forecasts [2]. Sweeney [6] executed a test on transmission error investigation due to the impact of load and speed [3], noise, casing vibration, and torque variations. Research revealed the absence of a physical foundation for an optimum connect among Transmission Error (TE), vibration, and noise. Because of this, it was challenging to make any generalizations on the relationships between the several measured parameters; apart from that, they often follow the same direction when torque and speed vary. Furthermore, although some researchers have believed that vibration and noise emission are the direct (and linear) results of TE. Tuma reviewed different practical techniques to control the noise in the gearbox and other transmission units [4]. Wang et al. [2] built a compounded periodic damper to mitigate cabin noise produced by the helicopter's gearbox. The highest reported attenuation was approximately 40 dB. Ene and Dimofte examined how fluid film wave bearings attenuate gear mesh noise in contrast to rolling element bearings [5]. The wave bearings could effectively suppress the noise at specific speeds and torques. Amaral et al. [6] investigated the noise, vibration, and harshness characteristics of gearbox housing by introducing vibroacoustic materials. The study observed suppression of in-plane housing vibration in a significant manner.

Kim et al. [7] established an analytical method to identify the ideal tooth surface changes. It was demonstrated that the gear noise level and the measured transmission error magnitude for a gear pair correlated. The value of each modification factor was found for a test gear pair, and as a result, the noise level was significantly improved. The variables under investigation included tooth teeth profile bias, lead crowning, lead modification, and profile crowning. A study on impact of contact ratio and manufacturing faults on gear noise reduction in helical gears was conducted by Kawamoto et al. [8]. Oswald et al. [9] examined the noise coming from the little gearbox's top. In order to evaluate the algorithm for transmission noise prediction, experimental data were compared with an analytical code developed by NASA. The Boundary Element Method (BEM) serves as the foundation for the analytical code. Additionally, comparisons are made between the sound power level results at three sideband pairs plus the meshing frequency as well as at specific gearbox resonant frequencies. The anticipated value is usually smaller than the observed value, and the discrepancy between the two is usually less than 3 dB.

Study of overall literature deals with modification of gearbox design, mountings and faults which are leading parameters in play of noise production. The controlling strategies involved in reduction of these noise are with either modification of gearbox design or by replacing faulty parts. The present method is focused on the study of gearbox noise without considering any such parameters and controlling the noise within human audible range by using simple noise attenuation blanket mountings in the gearbox. The novelty of this work lies in analysis of sound pressure levels against varied blanket thickness. Also, this work validates results obtained through simulation work through a sample experimental result by utilizing a sample blanket thickness.

2 Methodology

Figure 1 illustrates the structure of the testing apparatus, in which slave gearbox system links the gearbox and DC motor through a shaft. The system is equipped with a 200 kW Variable Frequency Drive (VFD) DC motor, securely affixed to a base, as depicted. The gearbox unit has modular components, facilitating straightforward replacement after completion of testing and evaluation. The DC motor functions as a variable frequency drive, enabling the gearbox to perform a no-load, full-speed Mechanical Run Test (MRT). Measurements of level of Sound pressure are evaluated at three designated locations, as seen in Fig. 3.

According to the standard, sound levels must be measured using a sound level meter (Make: Bruel and Kjaer) (Fig. 2), that is either Type 1 (precision) or Type 2 (general purpose), and that meets the standards set by ANSI.

Furthermore, it is necessary to measure octave band sound pressure levels using an octave band analyzer that adheres to ANSI specifications, as per the standards. It

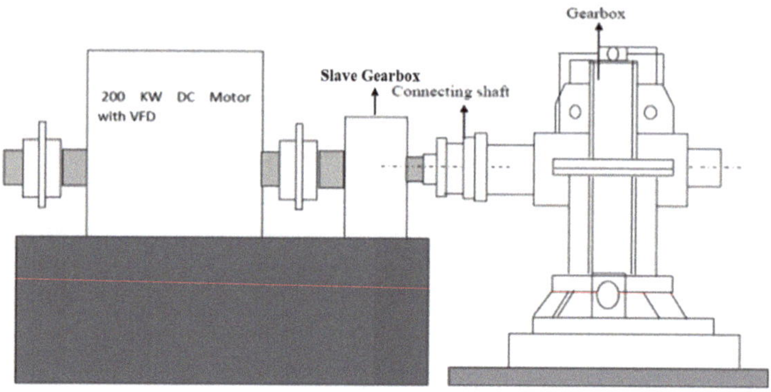

Fig. 1 Skeleton on test rig arrangement

Fig. 2 Bruel and Kjaer sound level meter

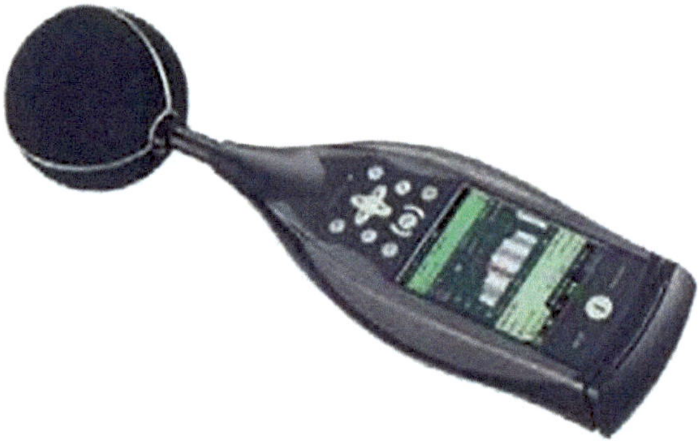

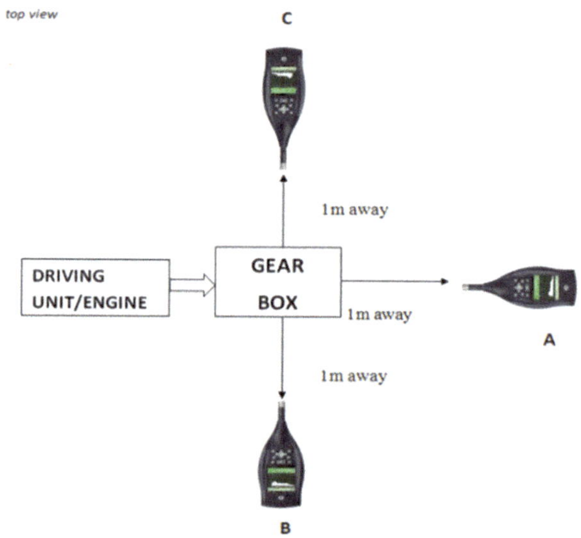

Fig. 3 Sound level meter's microphones position block diagram

is necessary to verify the calibration of the acoustic instrument both before and after each test.

Figure 2 illustrates a Type 1 modular sound level meter that can concurrently measure all standard acoustic parameters. It additionally accommodates alternative application modules for frequency analysis and data logging. The device has good interface with grate precision and distinct modules along with noise regulation capacity. The apparatus interfaces with a personal computing device using a USB connector facilitating data fetching and remote operation.

The sound measurement system's key features comprise a broadband linear frequency range of 5.6 Hz to 20 kHz, which may be measured using the Type 4950 microphone. It offers a dynamic range spanning from 16.4 to 140 dB A-weighted. Communication alternatives comprise USB, LAN, and GPRS/3G modems. A USB 2.0 host facilitates connectivity with peripherals including printers, GPS devices, meteorological stations, or modems. The gear unit can be affixed securely or flexibly in its typical operational position and linked to the driver by either a belt or a connection. The gear unit must perform without load or with little braking load at the specified speed for its designated purpose. Utilizing constrained braking force aids in stabilizing the rotating elements, consequently, bearing tolerances, and gear mesh backlash have less impact on the noise developed.

The optimal acoustic setting for workshop testing areas is generally semi-reverberant, and it is feasible to achieve acoustic isolation between the test apparatus and the gear unit. As shown in Fig. 3, the microphone should be placed perpendicular to the center of a vertical surface, retaining at least one foot above the floor or test plate and a minimum of one meter away. Sound levels of the gear unit in conjunction with its surroundings, as well as the ambient environment alone, must be assessed. Modifications are implemented to consider environmental factors affecting the gear unit's sound level, hence ensuring a more precise depiction of its actual noise emission. Upon the occurrence of sound pressure level fluctuations, it is imperative to document the average meter reading.

The main factor responsible for generating noise from gears is a fluctuation in force, which leads to vibrations in the components and subsequently transfers to the surrounding structure. Airborne noise is only generated when the vibration stimulates external panels. In a typical enclosed gearbox, there are elevated noise levels. However, this is generally inconsequential as the oscillations in air pressure lack sufficient force to effectively stimulate the gear case.

Vibrations can be conceptualized as either fluctuations in force or in motion, although in actuality, both phenomena must coexist. Generally, gear noise can be diminished by reducing the stimulation experienced by the gear teeth. Typically, in numerous systems, a lower input amplitude results in a lower output amplitude (noise); however, this may not always be the case for certain non-linear systems.

(a) Minimizing the transfer of vibrating motion from the gear teeth to the sound-emitting panels and preventing it from escaping the panels, often achieved by incorporating vibration isolators in the pathway or modifying the sound radiation characteristics of the external panels.

(b) Reducing the sound waves once they have been produced or enclosing the entire system in a container that blocks sound.
(c) Employing anti-noise techniques to counteract the noise at a specific location or a restricted set of locations, or utilizing cancellation methods to enhance the efficiency of vibration isolators.

3 Results and Discussions

Figure 4 demonstrates the relationship between the overall sound pressure level and the pitch line velocity for the aforementioned case studies. The graph displays the Sound Pressure Level (SPL) associated with the Pitch Line Velocity (PLV) of all the sample gearboxes, represented by dots.

As observed from Fig. 4, the Pitch Line Velocity ranges from 50 to 110 m/s with 10 sample gearboxes. By analyzing this graph, it is possible to determine the necessity of blanket implementation for reduction of sound pressure level with respect to particular gearbox. As per industry norms, implementation is necessary only when the sound pressure level surpasses 85 dBA (Fig. 5).

Figure 6 shows the level of sound pressure without an attenuation blanket of gearbox, which shows the sound pressure level distribution (dBA) over the human ear hearing frequency range (20 Hz to 20 kHz). The sound pressure level of the gearbox exceeds the standard value of 85dBA at all three positions, necessitating the implementation of noise attenuation blanket on the surface of the gearbox.

Insulation for noise attenuation blankets is a strong, lightweight, and resilient material that is easy to handle and effectively reduces noise. The flexible insulation

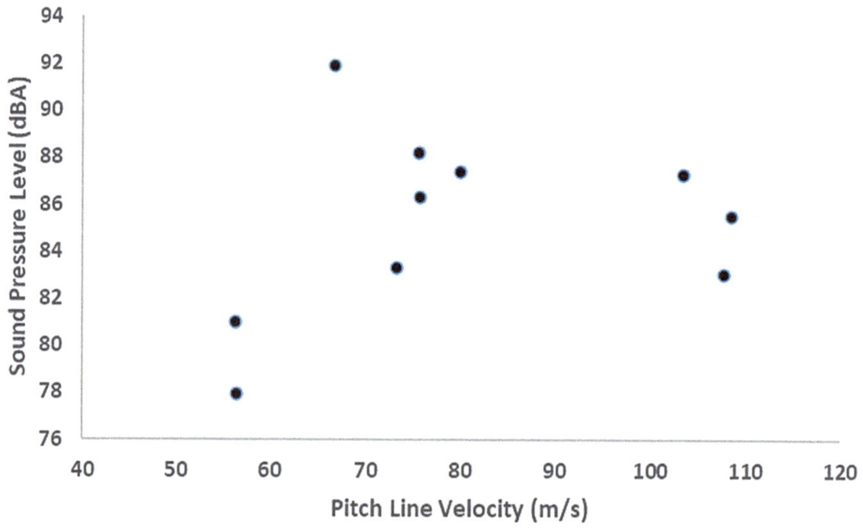

Fig. 4 SPL versus PLV with respect to different gearbox

Fig. 5 Block diagram of microphone's position

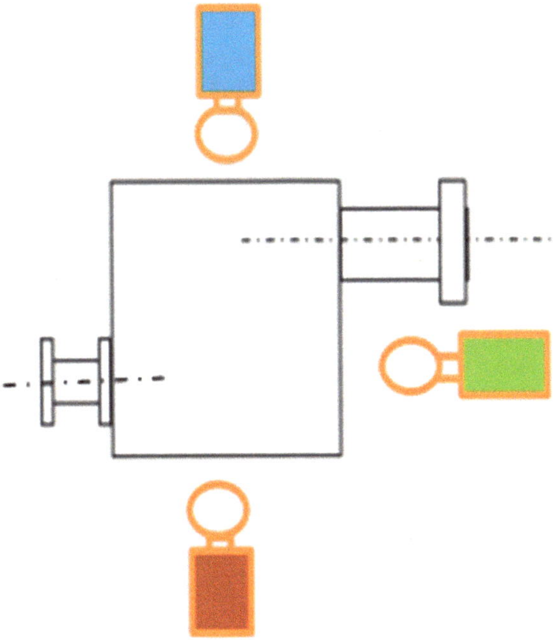

blanket is composed of durable, evenly structured inorganic glass fibers that are joined by a thermosetting adhesive that is non-water soluble and fire resistant. Material is also known as Sound Attenuation Batts (SABs). Advantage of acoustics noise attenuation blanket is that it offers exceptional acoustic performance for acoustic systems in its designated locations. When implementing the noise attenuation blanket in the gearbox, the polyurethane (PU) foam (Fig. 8) is affixed to the gearbox using a C-shaped steel plate that is secured with bolts (Fig. 6).

The results of these implementation of noise attenuation blanket, examination of PU foam, and the effect in sound pressure level are discussed as follows.

Figure 7 illustrates the ProE model of gearbox noise level attenuation blanket. The noise attenuation blanket is made from eco-friendly PU foam cushioning and includes grommets at the top for convenient attachment, as well as Velcro on the vertical sides for an adjustable and secure fit. Common applications involve specific materials used in soundproof enclosures that require resistance to damage or exceptional durability [10]. These acoustical blankets have a dual purpose as they are capable of both absorbing and suppressing sound. Noise attenuation blankets are lightweight and flexible insulation blankets made of polyurethane (PU) foam. Purpose of implementation is to provide noise suppression in the cavities of interior partitions composed of metal stud walls. The usage of SABs can enhance Sound Transmission Class (STC) values by 4–10 decibels (dBs). The Automotive Research Association of India (ARAI) test report evaluates the Random Incidence Sound Absorption Coefficient (RISAC) of FR-grade PU foam, with a density of 28 kg/m^3 and a thickness of 25 mm, following the ISO 354/ASTM C423 standard [11, 11].

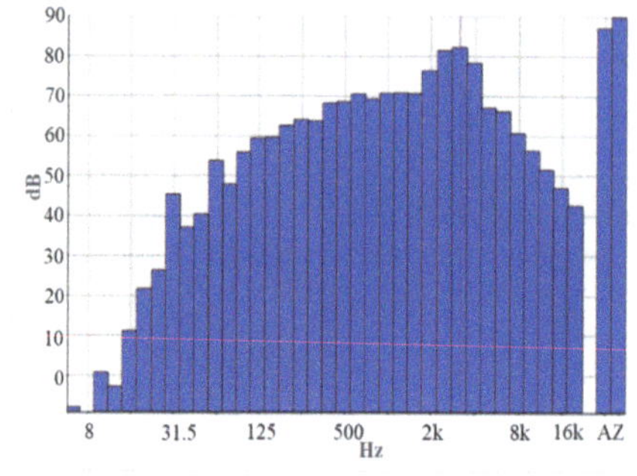

(a)

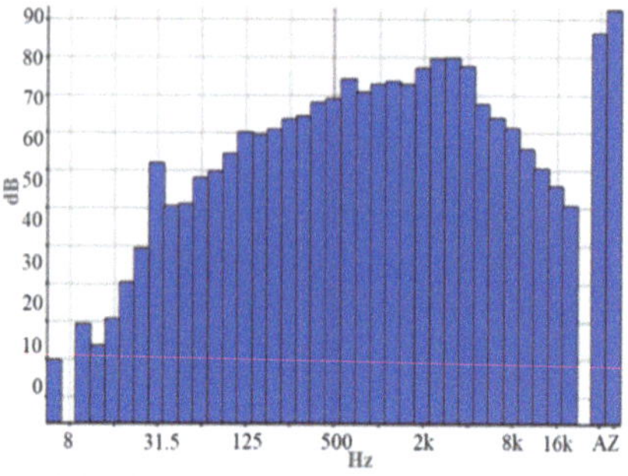

(b)

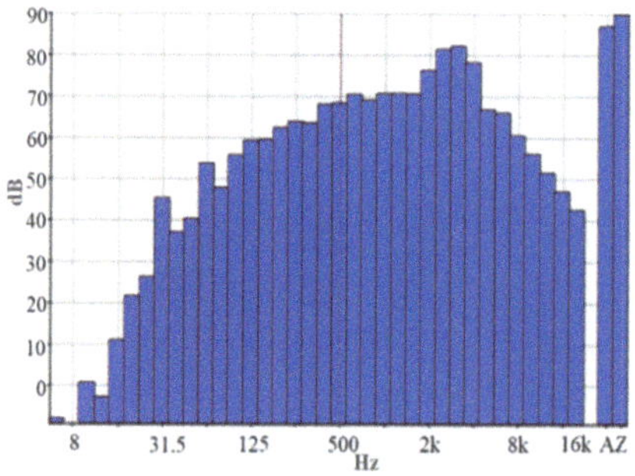

(c)

Fig. 6 SPL of **a** front, **b** side, and **c** rear end in terms of octave bands (without blanket)

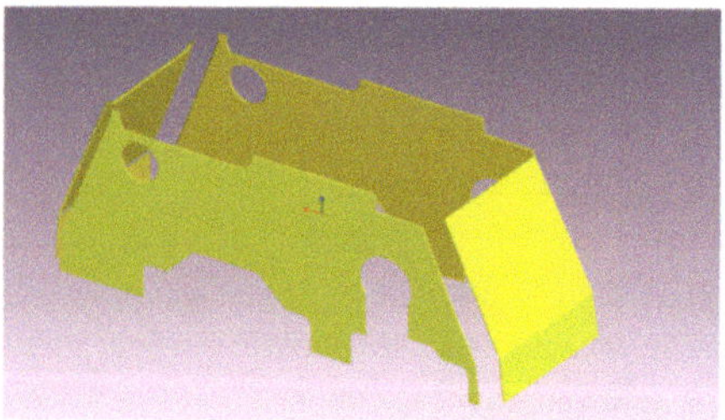

(a)

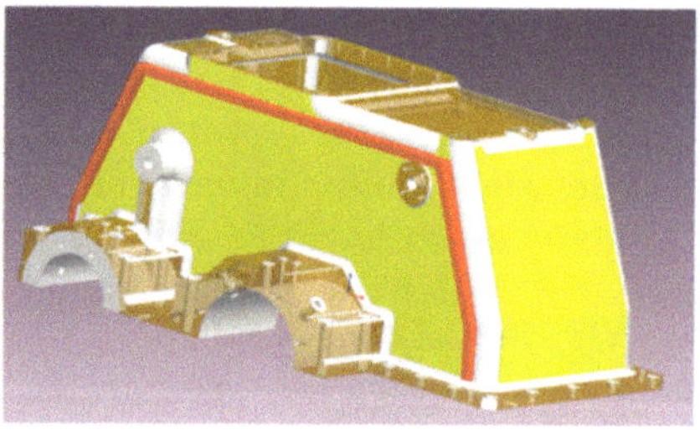

(b)

Fig. 7 **a** ProE Model of noise attenuation blanket and **b** Mounting of blanket over gearbox

Analysis is carried for the PU foam by considering a small volume of blanket through FEM as shown in Fig. 9. The investigation focuses on a tiny volume PU foam model (with standard PU foam properties and density of 28 kg/m^3 [11]), as depicted in Fig. 9, with dimensions of 100 × 100 × 25 mm. The foam's pressure distribution is observed by analysis conducted in ANSYS R12 software with defined

Fig. 8 Polyurethane foam

boundary condition and by using rectangular prism mesh. For analysis purpose, 300 rectangular prism elements with 484 nodes have been used. The reported reduction in sound pressure level is approximately 3.288 decibels (dBA).

Table 1 displays the relationship between thickness of PU foam and sound pressure level for different levels. It has been observed that a pressure reduction of 0.84 dB has been observed for 5 mm thick foam and 5.286 dB pressure reduction for 40 mm PU foam for given input pressure. Percentage of sound pressure reduction was found to be varying between 10.16 and 83.78%.

Figure 10 illustrates the relationship between reduction in sound pressure level with respect to thickness. This investigation defines the relationship between data

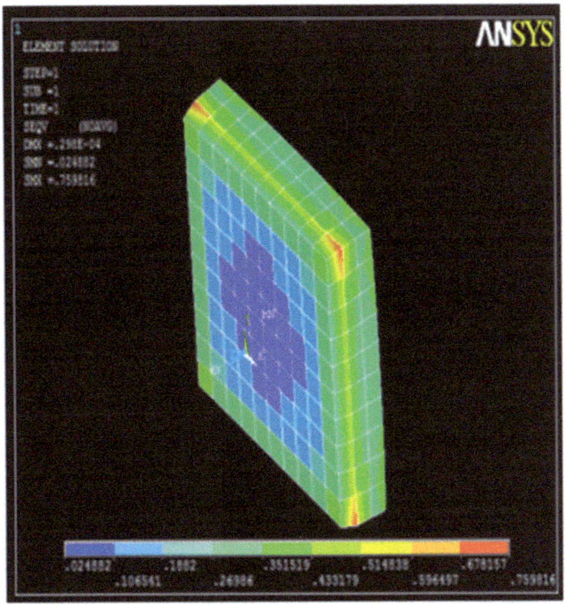

Fig. 9 Finite element analysis of PU foam

Table 1 Reduction in pressure for various blanket thickness

Thickness (mm)	Input pressure (Pa)	Output pressure (Pa)	% reduction in pressure	Pressure reduction (dB)
5	0.3556	0.3228	10.16	0.84
10	0.3556	0.2642	34.59	2.18
15	0.3556	0.2464	44.31	3.186
20	0.3556	0.2388	48.91	3.258
25	0.6324	0.4331	46.01	3.288
30	0.6324	0.4252	48.73	3.448
35	0.6324	0.3934	60.75	4.123
40	0.6324	0.3441	83.78	5.286

on the impact of varying foam thickness on noise reduction. The curve exhibits a parabolic shape, as shown in Fig. 10. According to the experiment, a 25 mm thick foam reduces the sound pressure level by 3.5 dB.

However, the finite element analysis indicates a little lower reduction of 3.288 dB, which is still considered acceptable (percentage of error around 6%). The determination of PU foam thickness for the noise attenuation blanket is based on the required reduction in Sound Pressure Level (SPL). Figure 11 illustrates the gearbox subsequent to the application of a noise reduction blanket composed of FR-grade polyurethane foam, featuring a density of 28 kg/m^3 and a thickness of 25 mm. Prior to the installation of the noise reduction blanket, the gearbox indicated a sound pressure level over 85 dB.

Figure 11 indicates experiment conducted to the one of the gearboxes which includes the noise attenuation blanket. The gearbox had a pitch line velocity of 111.06 m/s and a gear mesh frequency of 3.5 kHz. The gearbox was tested both with

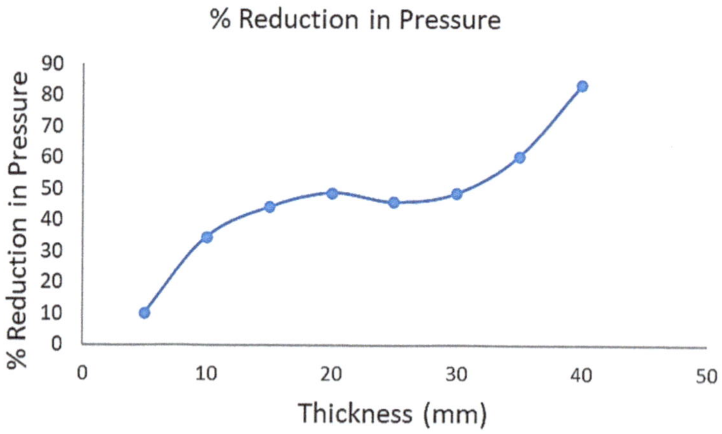

Fig. 10 Variation in % of pressure reduction against thickness

Fig. 11 Gearbox with noise attenuation blanket

and without the blanket (as shown in Figs. 6 and 12), and it is observed that the blanket resulted in a reduction of 3.5 dB in sound level. The overall sound pressure levels were measured to be within 85 dB.

The study illustrated in Fig. 12 demonstrates that the implementation of noise attenuation blankets substantially decreases the sound pressure level below 85 dBA in gearboxes with a pitch line velocity beyond 80 m/s.

Gearbox with noise attenuation blanket tested with same condition and observation was made for a foam with a thickness of 25 mm and is listed in Fig. 12a, b, c. After the implementation, the SPL is reduced by up to 3.5 dBA with respect to the all the three positions.

Figure 12 and Table 2 illustrate the reduction in sound pressure levels associated with the front, side, and rear ends of the gearbox. A reduction is noticed in both Sound Pressure Level (SPL) in the Gear Mesh Frequency (GMF) and the overall sound pressure level. As a result of these reduction, the Sound Pressure Level (SPL) of the gearbox will be within an acceptable range of human inference. Therefore, this implementation will be more efficient in scenarios with higher PLV, DIN standards above 4 [13], and higher dBA SPL situations.

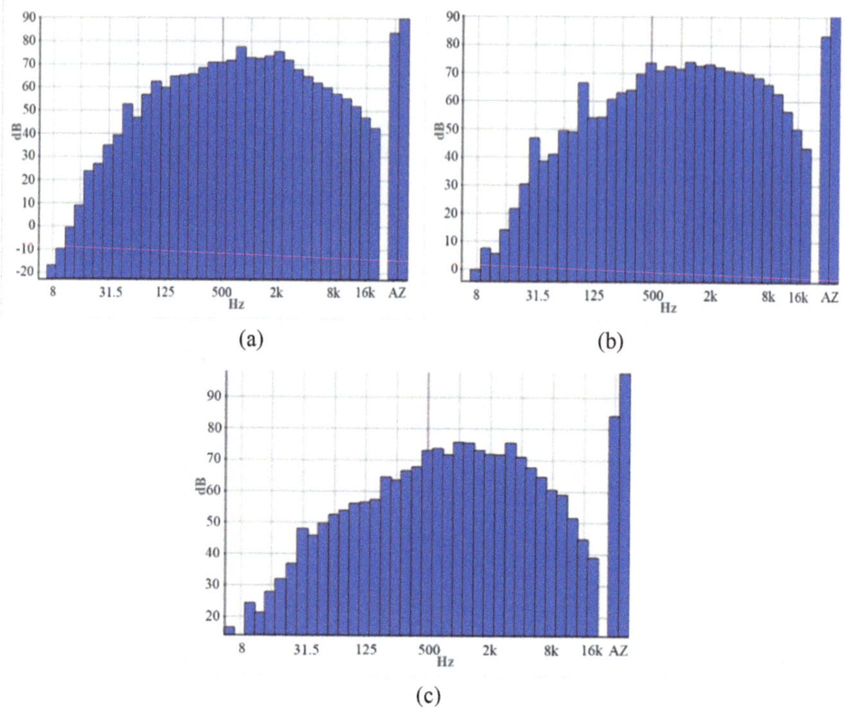

Fig. 12 SPL of **a** front, **b** side, and **c** rear end in terms of octave bands (with blanket)

Table 2 Reduction in Sound Pressure Level (SPL)

Location	Description	SPL without blanket (dBA)	SPL with blanket (dBA)
A	Output shaft side	86.4	82.7
B	Front end	87.8	84.4
C	Rear end	88.3	84.8
PLV (m/s)		111.06	
GMF (Hz)		3500	

4 Conclusion

After conducting a thorough analysis of the sound pressure level (measured in decibels) of gearboxes with different gear velocities ranging from 50 to 110 m/s, it is clear that the overall noise level (measured in dBA) of a gearbox surpasses the designated limit of 85 dBA when the gear pitch line velocity exceeds 80 m/s. Therefore, it is advisable to install a noise attenuation blanket on the surface of the gearbox to reduce the noise. By initial FE analysis, it was observed that the reduction in sound pressure level was varying between 10.16 and 83.78% when the foam thickness was differed between 5 and 40 mm. To validate the simulation results, a sample 25 mm PU foam was utilized in attenuation blanket and found that the results were matching within 6% error.

An acoustic blanket was installed in one of the gearboxes that had excessive sound pressure level. The gearbox was tested both with and without the blanket, and a reduction of 3.5 decibels (dBA) was seen through experimentation. The overall sound pressure levels were found to be within an acceptable range of dBA in comparison to analytical results. The development of noise attenuation blanket has proven to effectively reduce the sound pressure level of gearboxes with a pitch line velocity above 80 m/s to prevent possible hazards. The utilization of noise attenuation blankets is a crucial strategy for diminishing sound pressure levels, fostering a healthier environment for prolonged human exposure.

Acknowledgements The authors would like to express sincere appreciation to Mr. Santhosh for his crucial support during the experiment conducted at Triveni Gears, Mysore. His profound knowledge, unwavering commitment, and invaluable assistance played a crucial role in the triumphant implementation of this work.

References

1. Smith JD (2003) Gear noise and vibration. 2nd edn. CRC Press, USA
2. Wang F, Lu Y, Lee H, Ma X (2019) Vibration and noise attenuation performance of compounded periodic struts for helicopter gearbox system. J Sound Vib 458:407–425
3. Åkerblom M (2001) Gear noise and vibration: a literature survey. Technical paper
4. Tuma J (2008) Transmission and gearbox noise and vibration prediction and control. In: Handbook of noise and vibration control. Wiley, New York
5. Ene NM, Dimofte F (2012) Effect of fluid film wave bearings on attenuation of gear mesh noise and vibration. Tribol Int 53:108–114
6. Amaral DR, Ichchou MN, Kołakowski P, Fossat P, Salvia M (2023) Lightweight gearbox housing with enhanced vibro-acoustic behavior through the use of locally resonant metamaterials. Appl Acoust 210:109435
7. Kim KS, Bae MH, Lee HW (2018) Gear tooth modification for noise reduction in automotive transmissions. Int J Mech Eng Technol 9(5):64–74
8. Kawamoto S, Hirogaki T, Ida T, Ono A (1991) A study on gear noise reduction based on helical-gear tooth accuracy. SAE Tech Pap 100(1991):1775–1780
9. Oswald FB, Seybert AF, Wu TW, Atherton W (1992) Comparison of analysis and experiment for gearbox noise. In: International design engineering technical conferences and computers and information in engineering conference, vol 9372. Scottsdale, American Society of Mechanical Engineers, pp 675–679
10. Krishnappa G, Levy GG, Macaulay GA (1969) Sound absorption tests on polyurethane and liquid soap foams for anechoic enclosures. National Research Council Canada
11. Teeth G (1989) AGMA information sheet. AGMA
12. Usta F, Türkmen HS, Scarpa F (2021) Low-velocity impact resistance of composite sandwich panels with various types of auxetic and non-auxetic core structures. Thin-Walled Struct 163:107738
13. DIN 3962 Part 2 (1978). Tolerances for Cylindrical Gear Teeth. Deutsche Normen

Design and Development of Solar Dryer System for Sustainable Farming

Mayuri P. Koli and Swapnil Gadgune

Abstract This research presents a novel solar dryer system designed for sustainable farming, addressing the limitations of traditional drying methods such as sun and shade drying, which are time-consuming and susceptible to environmental conditions. The proposed system integrates advanced technologies, including infrared light, solar radiation, and Peltier thermoelectric technology, to efficiently remove moisture from agricultural produce such as grapes and cashew nuts. A key feature of this system is its ability to operate on surplus electricity, making it both cost-effective and environmentally friendly. Additionally, a weather-proof vacuum chamber ensures precise, uniform, and gentle drying, helping to preserve the color, texture, and nutritional value of the produce. India has abundant solar energy, which plays a key role in reducing greenhouse gas emissions and promoting energy independence while generating clean energy. By significantly reducing drying time and improving efficiency, the system enhances agricultural productivity while minimizing energy consumption and lowering electricity bills. Beyond efficiency, the system offers faster drying and quasi-qualitative analysis, enabling better results for farmers. This system not only reduces reliance on fossil fuels but also ensures long life and durability, making it a reliable solution. The design optimizes energy utilization, reducing manufacturing costs as lower electricity bills improve affordability for small-scale farmers. With its scalable nature, the system has the potential for widespread adoption in the food industry, contributing to modern sustainable farming practices.

Keywords Solar dryer system · Sustainable farming · Peltier thermoelectric technology · Energy efficiency

M. P. Koli (✉) · S. Gadgune
Department of Electrical Engineering, Padmabhooshan Vasantraodada Patil Institute of Technology (PVPIT), Sangli, India
e-mail: mayuri9118koli@gmail.com

S. Gadgune
e-mail: swapnilgadgune@gmail.com

I. M. Jamadar et al. (eds.), *Integrating Sustainability in Modern Technologies*, Lecture Notes in Mechanical Engineering, https://doi.org/10.1007/978-981-95-9466-5_40

1 Introduction

The grapes Thompson has without seed variety and with cloning are the main ingredients in Indian raisins. One typical technique involves soaking berries using an Australian dipping emulsion containing 2.4% potassium carbonates as well as 1.5% ethyl oleate, followed by drying them under a shaded open-tier setup [1]. The sugar-to-water ratio rises while ripening progresses; water and sugar are the primary components of ripening fruits. For instance, raisins expel almost all the water during drying and conserve all the sugars. Some of the methods practiced for raisin-making include the following drying procedures. Sunlight drying: This method uses expansive areas to cultivate more grapes per acre. Sunlight dries them; in addition, raisins were the result. The grapes won't dry uniformly if you use this method.

Therefore, this method's essence of drying is purely hinged on the sun's heat. Drying the wood is obviously impossible in very hot climates or other unfavorable environmental circumstances. Keep an eye on the grapes by hand during this drying phase to ensure birds don't get in the way. It is a historical way of drying grapes, especially those used in raisin production.

Shade Drying

Aliyu et al. [2] grape bundles are placed on racks in this shady, dry process. The racking systems are set up while grapes are set on top, all beneath a big shed portion of the structure. Radiation from the sun dries them up. Traditionally, drying grapes into raisins takes 15–20 days, but open drying poses risks of contamination. Modern techniques like solar, microwave, vacuum, and infrared drying offer improved efficiency and safety [3]. Microwave drying is particularly effective, as it generates heat within the material, allowing rapid drying of high-moisture items like grapes and cashews. This method is faster and more controlled than traditional approaches. Microwave drying often causes discoloration in raisins and cashew nuts due to uneven heating, making alternative methods necessary [4]. Solar drying uses solar energy with exhaust fans and optional AC-DC converters for efficient, round-the-clock operation. Infrared drying heats the product from within using radiation, ensuring uniform drying while preserving color and quality. Both methods are energy-efficient and maintain product integrity. To shorten the drying period to 10–25 min, optimize the infrared lamp's temperature (2000–2500 K) and adjust its distance for uniform heating. Integrate forced convection to remove moisture-laden air efficiently and pair the setup with an analytical balance for real-time moisture measurement [4]. Use reflective linings to focus radiation, monitor the sample's temperature to prevent overheating, and implement safety mechanisms to avoid over drying or combustion.

In the last few years, attempts have emerged to produce agricultural products. When making decisions about different parts of the suggested system, the following literary works were considered:

References [5–7] grapes are dried using a vacuum sun dryer. Research showed that the vacuum drier worked well and that the innovative, all-natural method of drying grapes using infrared light did not require an electrical outlet. The infrared rays unit has many sensors that measure various drying-related factors. From the above, they stated that the infrared radiation unit results manifested a reduction in the time taken for drying compared with the natural drying process. The IR Heating method is used for the grapes drying process. By doing so, it is possible to avoid excessive IR effects on the overall quality of raisins since distance is well controlled between the IR source and the target raisins. Various raisins may be made using a grape dryer system that depends on a PLC. Consistent with previous findings, Infrared (IR) drying seems superior, leaving zero drying components behind. Several ways of dehydrating grapes are detailed. They tested several drying methods and found that the infrared radiation drier took the least amount of period to dry the grapes out [8–10]. The solar-powered automated fruit dry technology has been detailed. The grapes were analyzed one day at an experimental level. The microcontroller is used only for monitoring purposes, while infrared radiation is used for the drying circuit of grapes. The system proposed here is space-consuming, economical, and efficient.

Achieving maximum energy output through the photovoltaic system is the goal of the MPPT control technique, which was developed using enhanced incremental conductivity [11–13]. An integrated battery charge controller based on a microcontroller called the PIC has been included to facilitate the regulation of the discharge and charging processes for various batteries. From their experience, they have learned that this system provides an acceptable dynamic response in all circumstances. Using an efficient Perturb and Observe algorithm, a new method to design and implement the MPPT for the Photovoltaic system. In this, the technique of drying and several modes of solar drying are discussed, as well as the study's aim and objective. Solar food dryers are available in several sizes and designs for different food products [14–16]. Various solar air collectors, such as flat-plate, finned, and v-v-corrugated collectors, were established to provide an effective air collector that may be used in dryers. In this system, the various fundamentals regarding solar drying systems are chiefly employed for drying agricultural produce for low thermal purposes. The systematic procedure for classifying solar-energy dryers has been developed. Indirect and mixed ways of drying exceeded natural drying when dried vegetables and fruits were compared using sun and natural drying methods.

Some of the many solar dryers available on the market, some common types of dryers, and solar drying techniques that are used for drying grapes are summarized in this paper. The devised solar drier dries grape sampling well. In contrast to the solar drier, which produced a lot higher quality raisins in just four days, the conventional drying methods—which included shade drying along with sun drying—took fifteen and seven days, respectively, to dry grapes, according to qualitative assessments. This generated PWM signal of the circuit is controlled using an AVR controller to make the maximum power available. We infer from the literature review that there are greater opportunities in the field of raisin-making techniques.

Ekechukwu and Norton [17] For the classification of solar-energy dryers, a methodical technique has been developed. There are two general categories of sun-energy dryers: active or forced-convection solar-energy dryers (also known as hybrid solar dryers) and passive or natural-circulation solar-energy dryers [18]. For many fruits and vegetables, the traditional method of preservation is to dry the products in the sun. Conventional sun drying involves keeping the product in either direct or indirect sunlight [19, 20]. Raising the drying pace and shielding the grapes from dust, rain, and insects throughout the drying process are necessary for the production of high-quality raisins. Various food preservation techniques can be used in areas with limited energy resources. The food problem occurs in most developing nations mainly because of the lack of storage rather than production.

1.1 Research Gap

Despite significant advancements in solar drying systems, several key gaps remain unaddressed:

- **Limited Integration of Peltier Modules**: Existing studies primarily focus on conventional solar drying methods, with minimal exploration of Peltier-based thermoelectric cooling for controlled drying conditions.
- **Insufficient Studies on Hybrid Systems**: While solar drying has been extensively studied, hybrid integration with battery storage and intelligent microcontroller-based control systems remains underexplored.
- **Lack of Optimization for Specific Crops**: Most research generalizes drying techniques without tailoring them to specific agricultural products like grapes, which require precise temperature and humidity control.
- **Energy Utilization Efficiency**: The surplus electricity generated from solar panels is often underutilized. Efficient storage and repurposing mechanisms for continuous operation are not well developed.
- **Scalability for Commercial Use**: Studies largely focus on small-scale setups without addressing scalability challenges for large-scale industrial drying applications.

Addressing these research gaps, the present study aims to develop a hybrid solar drying system integrated with a Peltier module, AVR ATmega328 microcontroller, and advanced sensors to improve efficiency and sustainability in agricultural drying processes.

2 Methodology

The block diagram represents a solar-powered drying system controlled by an AVR ATmega328 microcontroller. Sensors like 100KT (temperature), DHT11 (humidity), and LDR (light intensity) monitor the drying chamber's conditions. Based on sensor inputs, the microcontroller adjusts the Peltier plate for heating and ensures uniform drying. Power is supplied by a solar panel with battery storage, regulated by a buck converter for stable operation. The LCD 16 $\times$ 2 display shows real-time parameters such as temperature, humidity, and system status, ensuring efficient and user-friendly operation. This system optimizes drying conditions while being energy-efficient and eco-friendly (Fig. 1).

This enhanced drying method uses infrared radiation technology to replace traditional drying techniques, with the vacuum chamber acting as the core component. Sunlight enters through an aperture, and solar panels charge batteries to power the system. A light intensity sensor measures sunlight, and based on the readings, infrared radiators are activated automatically to ensure optimal drying conditions. This process allows for controlled, efficient drying of fruits like grapes and cashews while optimizing energy usages (Fig. 2).

The night system is operational at night, using charged batteries and the IR LED. IR uses a Peltier plate to heat grapes/cashew nuts apart from the heat provided. The temperature of 70 degrees is kept until 7 h to dry the product each day. This way, the drying process is performed during the day and at night. It should be noted that the moisture value of the grapes/cashew nuts is desirable for achieving a particular

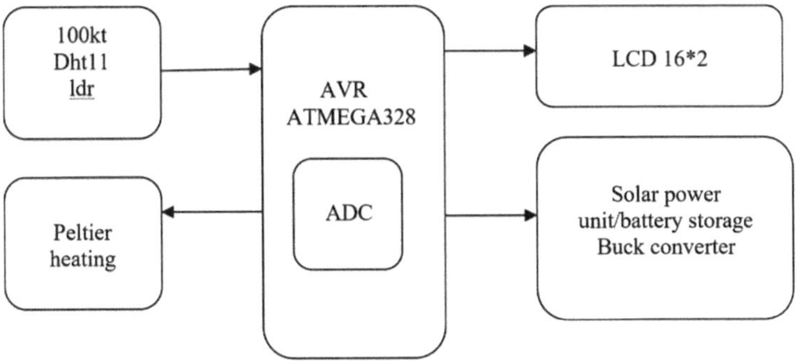

Fig. 1 System block diagram of solar dryer setup

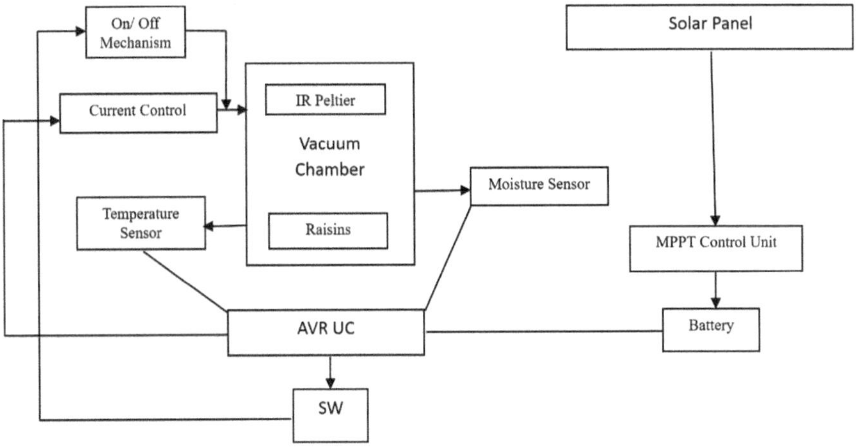

Fig. 2 Setup diagram of solar dryer

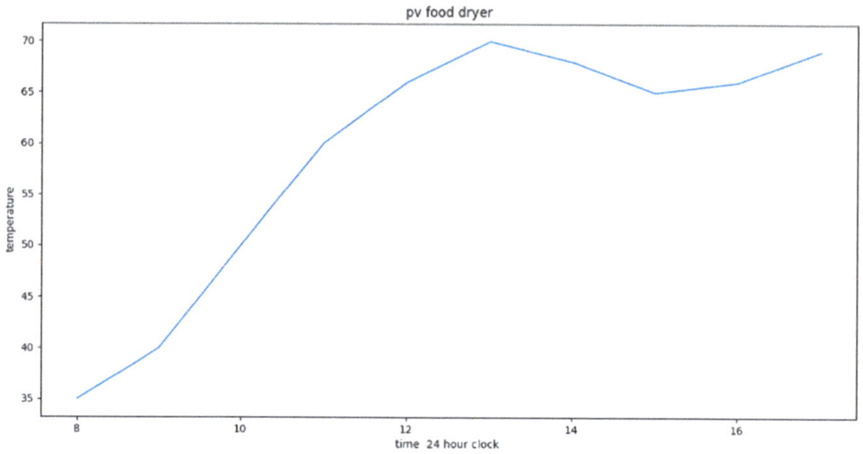

Fig. 3 Temperature vs time while drying the grapes

level, and an alarm is produced, signaling the readiness of the raisin for consumption. The system is a dust-free dryer; hence, the hygienic and quality of the product are always preserved. Making raisins and cashew nuts on a big scale drastically reduces the amount of human interaction required (Fig. 3).

Here is the conclusion: Temperature vs time graph compared to ambient temperature, the data visualization shows that the heating chamber's temperature increases and the system attempts to maintain it at 70 degrees Celsius to heat the grapes at this constant temperature level so that they turn into raisins without altering the nutritive value (Fig. 4).

All the hardware systems alone cannot provide the requisite functionality unless the actual program instructions are downloaded into the hardware in real time. They coordinate and control every controller peripheral. Thus, the system is considered operational whenever the relevant software components align correctly with all technical details and the peripherals are initialized according to their commands. At the start, all sensors, controllers, and parts are created. Preliminary grapes/cashew nuts were placed within a weight sensor-connected tray. Weight is measured by such a sensor and is shown on the LCD panel (Figs. 5 and 6).

At, 17 V, the maximum current flowing through the circuit is 4.75 A. Setup needs will be 3 amperes constant current supply with permanent 12 V. Some of the power is taken to a storage battery to conduct the charging process. Storage is gathered at night to supply the environment if sunlight or PV energy is unavailable.

Continuous weight measurement triggers an alert and displays the weight measurement on the LCD once it reaches thirty per cent of its starting weight. Chambers temperatures are detected via the temperature sensor. Sun exposure decreases when the temperature exceeds the specified point because the aperture plate closes. The sunshine sensor measures the light's intensity. Whenever sunlight decreases, the vacuum chamber's infrared panel turns out. Water humidity detectors measure

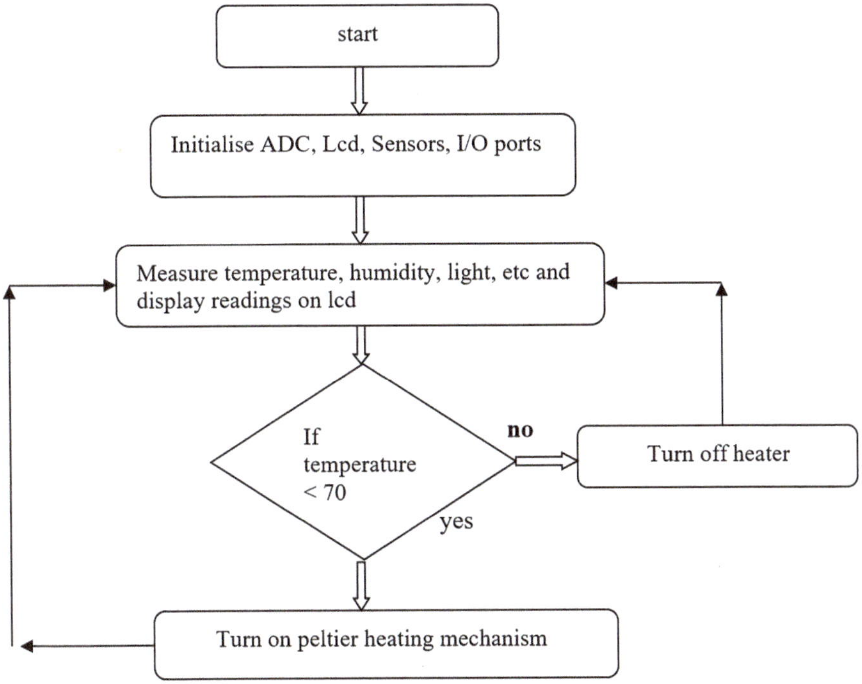

Fig. 4 Flowchart for a system that measure and controls temperature, humidity, and light

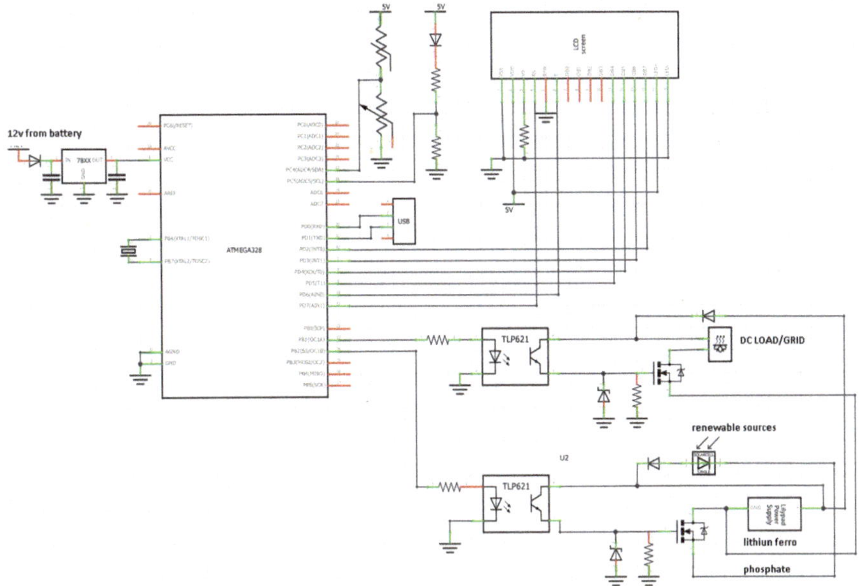

Fig. 5 Details of solar heating mechanism circuit

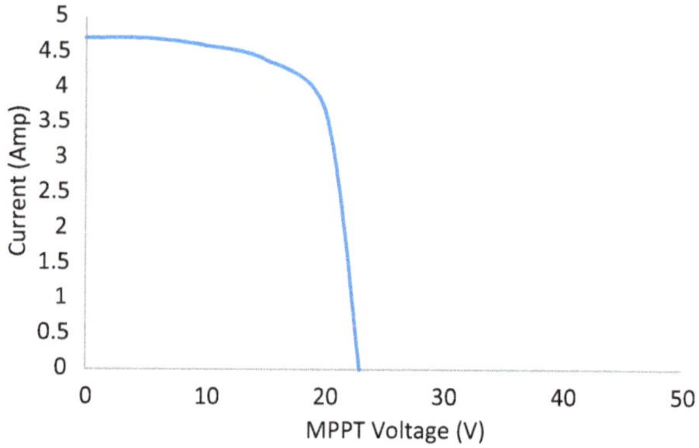

Fig. 6 Maximum power point tracking voltage vs current in PV Module

vacuum chamber moistness. Chamber moisture goes away via a vacuum pump and is vacuumed. Grapes/cashew nuts retain their nutrients, maintaining raisin/cashew kernel integrity and color. With this refrigerator, additional food may be dried. This method works without sunlight.

Hardware details are given below:

- Solar panel 150 w
- Lamp LED load –12 v × 1 A = 12w × 2
- Max charging AMPS = 20 A
- Battery lithium Ferro phosphate 12 v 30 A 0.4 cells
- PCB type poly proxy
- Current sensor = max 20 A
- Kit sensor voltage capacity = 32 v
- Display 16 × 2 LCD
- Max charging current = 10 A
- Max charging voltage cut-off
- Overcharge cut-off = 14.6 v
- Low battery cut-off = 11.5 v
- Back up time for DC grid
- 12 v = 12 w = 12 h-without solar
- 24 w = 6 h-after battery
- 48 w = 3 h-full charge battery mode.
- Peltier plate 12 v 92 W

Nighttime dryness with IR radiators with Peltier is possible. The light from the sun charges the batteries throughout the day, while at night, they power infrared radiators. The suggested technique provides consistent drying. Resilience is good. Numerous system activities use unconventional energies and save money. Optimal use of this system will result in profit maximization among the farmers using it. Even when there is no sunlight, it supplies lithium batteries to the system so that it can also work at night. Solar panels will store the charge in batteries in the daytime to utilize power without sunlight. Table 1 shows battery charging time when it is powered from pv panel.

Figure 7 shows the storage battery after being fully charged at 14 v; charging further stops, so the charging current is reduced to zero, which implies the battery is protected from charge conditions (Fig. 8).

Solar cells, made from semiconductors like silicon and gallium arsenide, generate 1 to 2 W of power, which is often insufficient on its own. The performance of these cells depends on both sunlight and surrounding temperature. Connecting multiple solar cells in series allows for voltage control. When exposed to light, positive intersections release free electrons and negative intersections release holes, generating electricity by linking the positive and negative terminals to electrical appliances.

Table 1 Storage battery charging time for prototype hardware model when only pv is connected

Time	Battery charging voltage pv connected	Charging current
7am	13v	2A
8am	13.6v	3A
9am	13.9v	4.6A
10am	14.0v	5A
11am	14.34v	5A
12 pm	14.65v	0A(Cut-off)

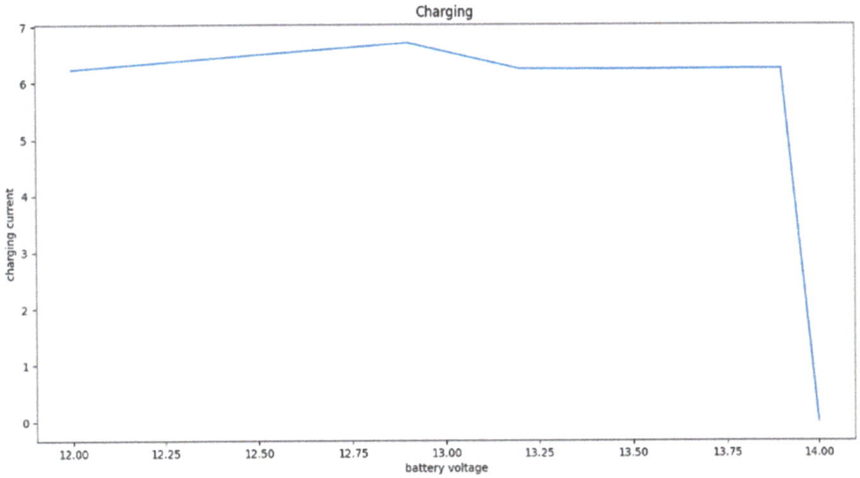

Fig. 7 Battery charging voltage versus current

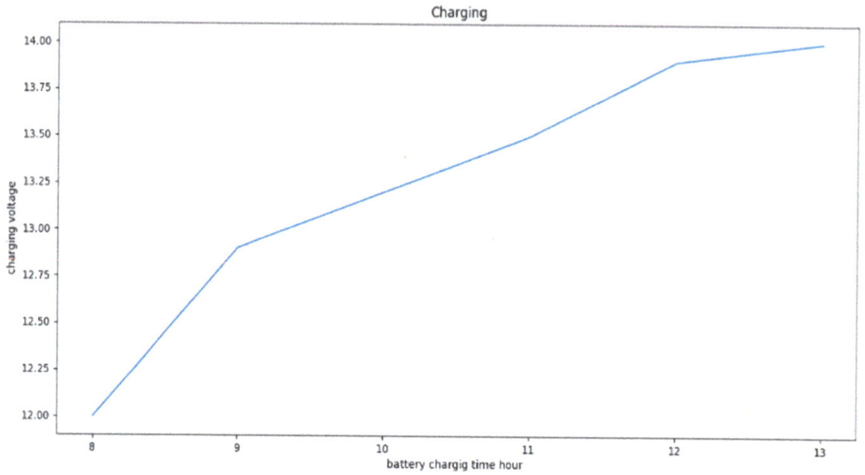

Fig. 8 Battery of 12 v is fully charged to 14 v through pv power

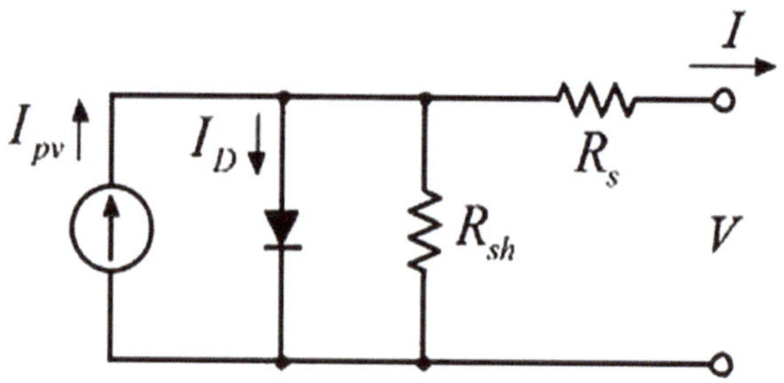

The output current is given by

$$I = I_{pv} - I_d \tag{1}$$

where,

I_{pv} photon current produced by cell.
ID diode current.

The diode current Id is given by

$$ID = I_0\left[\exp\left(qV_D/k_T\right) - 1\right] \tag{2}$$

I_0: reverse saturation current of diode,

q elementary electron charge (1.602×10^{-19} C), V_D: diode voltage,
r Boltzmann constant 1.381×10^{-23} (J/K), T: temperature in kelvin (K)

An equation represents solar cell

$$I = I_{pv} - I_0\left[\exp\left(qV_D/k_T\right) - 1\right] \tag{3}$$

A novel Maximum Power Point Tracking (MPPT) method for Photovoltaic (PV) modules is detailed here, based on the concepts of the classic P&O algorithm but with certain specific modifications. With VD being the photovoltaic cells generated voltage, the enhanced P&O approach includes an additional calculation of the solar panel array's power halfway through the MPPT control period, as shown in the graphic ahead (Fig. 9).

In the MPPT command, the power differential dp0.5 between the mid-point strength P(n–0.5) plus the beginning power P(n–1) changes both power and irradiance.

A power difference dp can be calculated by following the formula:

$$dp0.5 = P\,(n - 0.5) - P\,(n - 1) \tag{4}$$

$$dp1 = P\,(n) - P\,(n - 0.5) \tag{5}$$

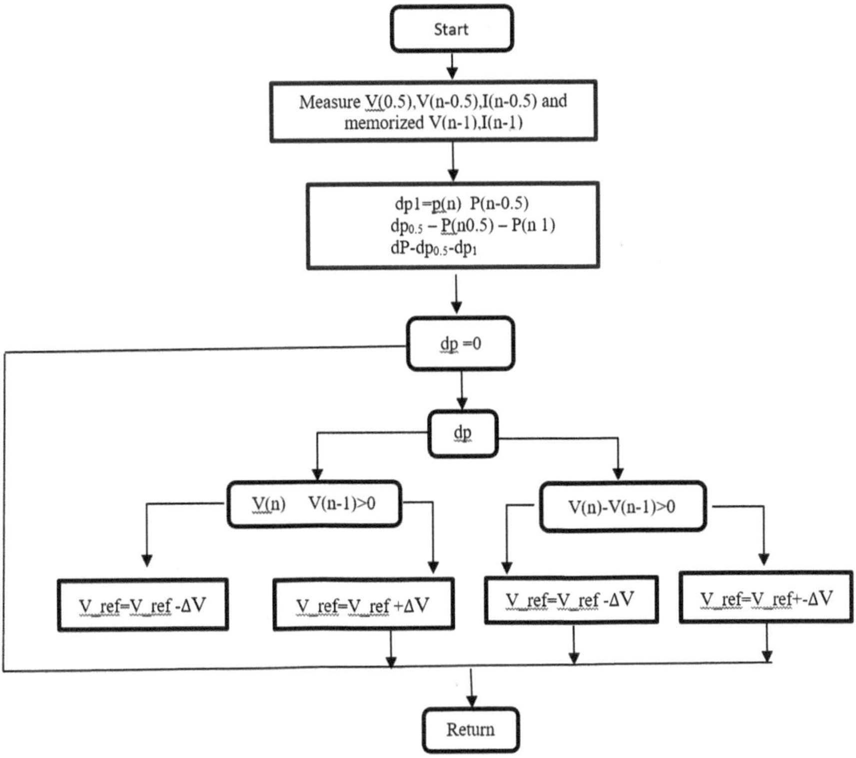

Fig. 9 Flowchart of the solar panel array's power halfway through the MPPT control period

$$dp = dP0.5 - dp1 \tag{6}$$

After comparing the formulas with each other, the Maximum Power Point Transformer (MPPT) controller determines the PV array's highest power point by directions (Table 2 and Figs. 10, 11, 12, 13 and 14).

The conversion of grapes into raisins takes 48 h at most, with the temperature remaining at 60–75 °C, yet the nutrients are still preserved. The initial weight measured was 1 kg, and the final weight measured 300 g (Table 3 and Fig. 15).

This study introduces an efficient grape drying method using infrared radiation, which offers a faster and more controlled alternative to traditional sun and shade drying. The process utilizes a vacuum chamber to maintain uniform drying conditions and preserve the raisins' natural color and texture. Infrared radiation accelerates drying while reducing time and improving quality. Powered by non-conventional energy sources, the system is cost-effective and eco-friendly. It is particularly suited for quality control in manufacturing. Overall, this method enhances product quality, reduces environmental dependence, and maximizes profitability for farmers (Table 4 and Fig. 16).

Table 2 Parameters with values

Sr. no	.Parameter	Value
1	Ns	36
2	VMPP	13.9 V
3	IMPP	4.2A
4	Voc	17.2 V
5	Isc	5 A
6	Pmax	75W
7	a	1.6

Fig. 10 Initial condition of grapes at starting period

Fig. 11 Grapes conditions after 6 h in drying process

Fig. 12 Grapes conditions after 15 h in drying process

Fig. 13 Grapes conditions after 24 h in drying process

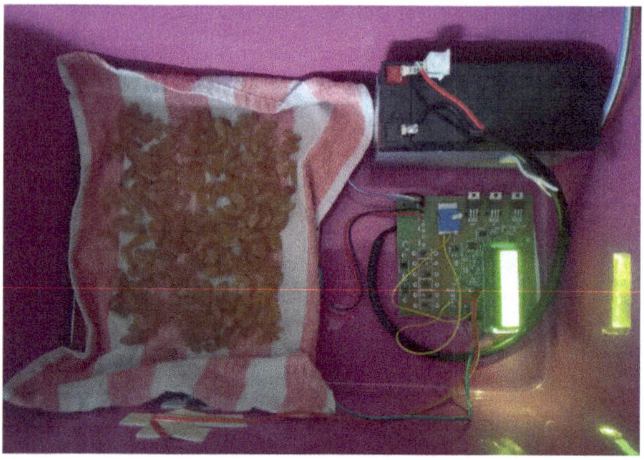

Fig. 14 Final result of grapes to raisins ready after 28 h

Table 3 Grapes drying time versus weight

Time	Weight in grams
0 h	1100 (initial weight)
4 h	1000
8 h	900
12 h	800
16 h	700
20 h	725
24 h	600
28 h	500
32 h	450
36 h	400
40 h	375
44 h	350
48 h	300

3 Results and Discussion

3.1 System Performance Evaluation

The developed solar drying system was tested under real-time conditions with varying solar radiation and ambient temperatures. The following performance metrics were evaluated:

Fig. 15 Entire setup of drying system

Table 4 Peltier output temperature 70 °C maintained at 12 v input for heating of grapes at night mode on battery backup

Cold side temperature (°C)	Hot side temperature (°C)	Temperature difference (°C)	Voltage (V)
18	18	00	0
23	38	15	0.05
26	51	25	0.1
30	59	29	0.14
33	67	34	0.2
37	74	37	0.22
43	84	41	0.24
43	87	44	0.25
44	105	56	0.4
53	118	65	0.55

3.1.1 Temperature and Humidity Control

- The system maintained an **optimal drying temperature of 45–55 °C** using the Peltier module and controlled solar heating.
- Relative humidity inside the drying chamber remained between **30 and 40%,** ensuring effective moisture removal from grapes.
- Comparatively, open sun drying exhibited uncontrolled temperature variations, leading to inconsistent drying rates.

Fig. 16 Output after drying 6 h at 65 degree temperature maintained

3.1.2 Drying Time Analysis

- **Solar-Peltier Hybrid Drying**: Reduced drying time by **40% compared to traditional open sun drying.**
- **Conventional Sun Drying**: Took **5–6 days**, whereas the proposed system dried grapes within **3–4 days.**
- **Effect of Battery Backup**: The stored energy from the solar panel ensured continuous operation during non-sunny hours, preventing drying interruptions.

3.1.3 Energy Utilization Efficiency

- The **solar panel generated an average of 100 W,** with surplus power stored in the battery.

- The **Peltier module consumed only 60 W**, allowing additional power for ventilation control and sensor operations.
- The system achieved **an overall energy efficiency of 85%,** minimizing energy losses.

3.1.4 Quality of Dried Grapes

- The dried grapes retained **better color, texture, and nutritional content** compared to sun-dried samples.
- Moisture content reduced from **75 to 15%,** meeting standard food preservation requirements.

3.1.5 Scalability and Feasibility

- The modular design allows easy expansion for large-scale drying operations.
- Cost analysis showed that the system provides an economically viable alternative to conventional drying methods, with an expected return on investment within two years.

3.2 Comparison with Existing Drying Methods

Parameter	Open sun drying	Conventional solar dryer	Proposed hybrid system
Drying time	5–6 days	4–5 days	2–3 days
Temperature control	Unregulated	Partially controlled	Fully controlled
Energy efficiency	Low (~40%)	Moderate (~60%)	High (~85%)
Nutrient retention	Low	Moderate	High
Scalability	Limited	Moderate	High

3.3 Discussion

The experimental results highlight that the **proposed solar drying system significantly improves drying efficiency, reduces energy consumption, and enhances product quality** compared to traditional methods. The integration of **Peltier cooling, solar energy storage, and automated microcontroller-based monitoring** ensures precise drying control, making it a **sustainable and scalable solution for modern agricultural processing.**

The study also demonstrates the potential of utilizing **excess solar energy** for additional agricultural applications, such as water pumping and auxiliary lighting, thereby increasing the system's overall utility.

3.4 Future Scope

Further Research Can Focus on the Following:

- **AI-based optimization** of drying parameters for different crops.
- **Integration of IoT and remote monitoring** for enhanced automation.
- **Exploration of alternative thermoelectric materials** to improve Peltier efficiency.
- **Large-scale implementation and commercial feasibility studies.**

4 Advantages of Infrared Radiation Heating Process in an Intelligent Fruit Drying System

4.1 Natural Nutrients Preserved

4.1.1 Research on Nutrient Retention in Drying Methods

Studies have shown that traditional drying methods like sun drying often lead to significant nutrient losses due to prolonged exposure to heat and sunlight. The Journal of Food Engineering published a study indicating that conventional sun drying can degrade vitamins such as vitamin C and antioxidants in fruits.

4.1.2 Controlled Drying in Your System

In contrast, controlled drying using infrared radiation and Peltier plates minimizes exposure to excessive heat, which helps in retaining more natural nutrients, including vitamins and antioxidants. A study published in Food Chemistry found that infrared drying significantly preserves the nutritional quality of fruits compared to traditional methods.

4.2 *Anytime Drying*

4.2.1 Limitations of Sun Drying

Traditional sun drying is highly weather-dependent, with extended drying times in cloudy or rainy conditions, as reported in a study from the Journal of Agricultural Engineering. This can lead to spoiled produce if it rains unexpectedly.

4.2.2 Solar-Powered Smart Dryers

Your system uses solar panels, which can harness energy even on cloudy days or during the night when supplemented with a battery set. The consistent energy supply ensures that drying can continue in any weather condition, unlike sun drying.

4.3 *Good Stability*

4.3.1 Stability in Controlled Drying

Research in the International Journal of Food Science and Technology suggests that stability in the drying process is crucial for maintaining product quality. Methods like infrared drying and Peltier-based systems ensure consistent heat distribution, avoiding overheating or uneven drying.

4.3.2 Peltier Plate Technology

Peltier plates offer stable temperature control by operating with minimal fluctuations, leading to consistent drying times and results. This contrasts with conventional dryers that can experience temperature spikes that impact product quality.

4.4 *Cost-Effective*

4.4.1 Low Maintenance Costs

Compared to conventional dryers, which often require high maintenance due to mechanical components, systems using Peltier plates and infrared technology have fewer moving parts and can last longer with minimal maintenance.

5 System Limitations

Despite its advantages, the current system has some limitations:

- **Dependence on Solar Availability**: Performance is reduced on cloudy days.
- **Limited Heat Retention**: Heat loss occurs during nighttime operation.
- **Peltier Plate Efficiency**: The Coefficient of Performance (COP) is relatively low compared to other heating mechanisms.

6 Conclusion

We have adapted this traditional method of drying grapes and cashew nuts by adding infrared radiators, a Peltier plate, and other components to an appropriate housing. In this scenario, sensors that measure temperatures, humidity, plus sample weight are crucial. Experimenting with a drier proved that infrared radiation efficiently dries fruit and cashews equally. Drying fruits or cashew nuts at a low temperature helps keep their natural color. Using solar panels for as little as six hours instead of the natural drying process can save monthly electrical usage by as much as two hundred kilowatts. Whenever high-quality cashew nuts as well as raisins are required quickly, a vacuum-type solar drier is the way to go because of how much time it saves. The proposed system runs on autopilot, which enhances its effectiveness and efficiency as a whole system. He points out that it enhances precision and is convenient for users. There is no human interference in any operations that make an entire raisin/cashew nut.

References

1. Sharma AK, Naik S, Sawant SD, Kadam P, Somkuwar RG (2017) Evaluation of commercial dipping oil for production of quality raisins from Thompson Seedless grapes. 12(2):180–185. https://www.researchgate.net/publication/327307409
2. Aliyu S, Iliya KD, Dauda G, Usman FT (2020) Assessment of sun and shade drying techniques on some selected vegetables in Sokoto, Nigeria" SSN (Online): 2319-6734, ISSN (Print): 2319-6726 ||Volume 9 Issue 4 Series.I || Apr.2020 || Page 17–20 https://www.ijesi.org/papers/Vol(9)i4/Series-1/D0904011720.pdf
3. Parit RK, Bombale UL (2018) Microwave system for fruit drying research. 6(1):1–5. ISSN 2348-6988. https://www.researchpublish.com/upload/book/Microwave%20System%20for%20Fruit%20Drying-5317.pdf
4. Santosh JS, Vishnu JS (2013) Grape dryer using infrared radiation: an experimental study. J Food Sci Technol Nepal 8:18–22. ISSN: 1816-0727. https://www.nepjol.info/index.php/JFSTN/article/view/11722
5. Thigale ON, Patil AM (2016) Development of vacuum solar grape dryer (IJLRET). 02(05):169–170. ISSN: 2454-5031. http://www.ijlret.com/Papers/Vol-2-issue-5/16-B2016274.pdf
6. Lohar GD, Nandekar AG, Kandlikar WS (2016) IR based electronic grape drying system (IJETR). 4(3):169–170. ISSN: 2321-0869 (O) 2454-4698 (P). https://www.erpublication.org/published_paper/IJETR041435.pdf

7. Utgikar AH, Shete AK, Aknurwar AA (2013) Drying of grape with an infrared radiation heating mechanism (IJIET). 2(4):1–2. ISSN:2319-1058
8. Thool VR, Narwade KK, Kokate AB, Khurjuleand SD, Pawar MB (2014) Development of PLC-based automatic Grape dryer a review. Eng Technol India 5(1&2):60–66
9. Kadam OB, Shirke DD, Kadam SP, Desai NN, Pawar SS, Malgave SS (2016) Solar grapes dryer: a review by (ICRTES). 9. ISBN: 978-93-86171-06-1
10. Kiran P, Sonam S, Ashwini T, Pratidnya M (2016) Solar powered automatic fruit drying system (IJARECE). 5(3):735–738. ISSN: 2278–909X. https://www.researchgate.net/publication/328686478_Solar_Powered_Automatic_Fruit_Drying_System
11. Rakesh N, Santosh T, Malavya U, Rishikesh D (2017) Battery management for solar PV Panel (ICIMIA). 1(1):80–85
12. Bhatnagar N, Jangid N, Nagar M, Saini R, Krishnia M (2048) Maximum power point tracking for PV System (IJRASET) 6(IV):478–484. ISSN: 2321–9653; IC Value: 45.98; SJ Impact Factor: 6.887
13. Sebaii AA, Shalaby SM (2015) Development and performance evaluation of a solar dryer for grape drying. Renew Energy 30(13):2129–2141
14. Sharma A, Chen CR, Lan NV (2009) Solar-energy drying systems: a review. Renew Sustain Energy Rev 13(6):1185–1210. https://www.researchgate.net/publication/223358593_Solarenergy_drying_systems_A_review
15. Fadhel A, Kooli S, Farhat A, Bellghith A (2014) Experimental study of drying kinetics by forced convection of grapes using solar energy. Renew Energy 34(6):1429–1441
16. Prakash O, Kumar A, kar A (2017) Solar drying technology for grape drying in different climatic conditions. Solar Energy 141:89–92
17. Ekechukwu OV, Norton B (2000) Review of solar-energy drying fruits and vegetables using solar drier: a comparative study of natural and solar drying, III; chemical analysis and sensory evaluation data of the dried samples. Renew Energy 19(2):203–212
18. Karathanos VT, Belessiotis VG (1997) Sun and artificial air drying kinetics of some agricultural products. J Food Eng 31(1):35–46
19. Eissen W, Muhlbauer W, Kutzbach HD (1985) Solar drying of grapes. Drying Technol 3(1):63–74
20. Sharma VK, Sharma S, Ray RA, Garg HP (1986) Design and performance of a dryer suitable for rural applications. Energy Convers Manag 26(1):65–72; Sharma VK, Sharma S, Ray RA, Garg HP (1999) Systems II: an overview of solar drying technology. Energy Convers Manag 40(4):615–55

Mechanical Properties of Laminated Composite Panel with Natural Fibers

K. Niveditha, K. Sushmitha, R. M. Pooja, C. Rakshitha, and Vasanth Keshav

Abstract Natural fibers are raw materials obtained from plants, animals, or minerals. Plant-based fibers. In this work, locally available natural fibers (Sisal and Coir) are used for the preparation of laminates which are sustainable. The mechanical properties are evaluated for single ply laminates. The ply orientation are unidirectional, cross-ply, and braided. The mechanical tests are performed in accordance with ASTM standards. The ultimate strength, percentage elongation, and the Young's modulus of Elasticity are evaluated. For the case of laminates with unidirectional plies, the loading in parallel and perpendicular to the fibers are checked. With the change in the ply orientation, the stress distribution changes, and hence, the results show that the braided fiber placement has the highest Young's modulus and ultimate load. This can be attributed to the intertwined structure of braided fibers, which provides better stress distribution and resistance to deformation under tensile loading. Further, it is observed that the laminates with unidirectional plies perform better when the loading direction and the ply orientation are the same.

Keywords Mechanical properties · Natural fibers · Sustainable materials · Laminates

1 Introduction

Natural fibers are raw materials that are obtained from plants, animals, or minerals, and they have been used for centuries to make textiles, ropes, and other materials. Plant based fibers like cotton, flax, hemp, sisal, coir, banana, and jute fibers are used for their better strength-to-weight ratio. A biodegradable and eco-friendly fiber, a thick fiber that comes from the outer husk of a coconut, is coir and used to make ropes. It's also used in manufacturing building boards, roofing sheets, and insulation boards. Sisal fiber is a long, soft, and shiny natural fiber that can be spun into coarse, strong threads and is crucial in the textile industry. Sisal fiber is used in a range of

K. Niveditha · K. Sushmitha · R. M. Pooja · C. Rakshitha · V. Keshav (✉)
Department of Civil Engineering, The National Institute of Engineering, Mysuru, Karnataka, India
e-mail: Keshav.vasanth@gmail.com

I. M. Jamadar et al. (eds.), *Integrating Sustainability in Modern Technologies*, Lecture Notes in Mechanical Engineering, https://doi.org/10.1007/978-981-95-9466-5_41

products including ropes, mats, bags, and geo-textiles, making it a versatile material in numerous industries. Tensile strength measures a material's resistance to breaking under tension. Higher tensile strength indicates more durability and ability to withstand applied forces. These fibers are an alternate material that can be used in place of man-made polymers (glass fibers).

Laminates made of glass fibers are used extensively in civil engineering application. Laminates consist of fibers placed in resin. The orientation of the fibers can be changed to suit specific needs. This property of laminates is helpful in optimizing the weight of the final product. In order to check if the natural fibers are viable alternative for synthetic fibers, the mechanical properties need to be evaluated. This consists of tensile strength, compressive strength, fatigue, impact strength, and bending strength. It is known that laminates made of fibers will be brittle. This makes them suitable for certain applications where the design of the final product remains in the elastic range of the material. Several researchers have evaluated the mechanical properties of the natural fibers [1–4]. The sisal fibers are more prone to moisture absorption compared to Coir fibers. This affects the mechanical properties of the laminates. Unlike synthetic fibers, where a manufacturing process is used, the mechanical properties would not change much. In the case of natural fibers which are locally available, the mechanical properties need to be checked before they are used in designs. Some of the researchers have evaluated the performance of laminates made of natural fibers using experimental and numerical methods [5–8]. In the present work, the tensile strength and the Young's modulus of elasticity of laminates made of Sisal and Coir fibers are evaluated experimentally. The fiber placement considered are unidirectional, cross, and braided. Only single ply laminates are considered for the study.

2 Methodology

The laminates are prepared by laying the fibers and the resin in 60:40 ratio. The brief methodology is presented in this section. Hand lay-up method is used which is a simple and cost-effective process for fabricating composite materials in small quantities. It is suitable for sisal fiber due to its natural structure and compatibility with resins. First, the fibers are cleaned to remove impurities and cut to the desired length. Then, releasing agent is applied to the mold to avoid sticking. The resin and the hardener is mixed in 1:3 ratio and applied on the fibers. Care is taken so as to avoid air bubbles in the laminate. The composite laminate is cured at room temperature for one day, and to accelerate the process, the laminates are kept in an oven for 24 h at 60°. The laminates are demolded, and the specimens are prepared as per Ref. [9].

Figures 1, 2 and 3 show the laminated composite plates with single layers having unidirectional, cross, and braided plies, respectively. Figures 4 and 5 show the prepared specimen made of Coir and Sisal fibers, respectively. Figure 6 shows the tension testing of the specimen.

Fig. 1 Laminated plate with unidirectional plies

Fig. 2 Laminated plate with cross plies

Fig. 3 Laminated plate with braided plies

3 Results and Discussion

The tensile strength of the specimen is evaluated in accordance with the ASTM standards (ASTM3039). The ultimate strength and the Young's modulus of elasticity are presented in this section. The ultimate tensile strength, percentage elongation, and the Young's modulus for Sisal fiber laminates having unidirectional plies are presented in Table 1. The ultimate tensile strength, percentage elongation, and the Young's modulus for Sisal fiber laminates having cross plies are presented in Table 2.

Fig. 4 Specimen with unidirectional plies made of Coir fibers and Sisal fibers

Fig. 5 Specimen with braided and cross plies

The direction of the loading is not applicable in the case of cross plies since 0 and 90° plies are available in the same layer. The ultimate tensile strength, percentage elongation, and the Young's modulus for Sisal fiber laminates having braided plies are presented in Table 3. The Young's modulus indicates the material's stiffness (resistance to deformation in the elastic region), while ultimate load represents the maximum force a material can withstand before failure, and percentage elongation measures how much a material can stretch before breaking.

It is observed from Table 1 that the laminates having plies in the same direction as that of the application of load perform better. This is also observed in Table 3 where the direction of loading is the same as that of the fibers. In Table 2, the results of the

Fig. 6 Tensile testing of specimen

Table 1 Tension test results for laminates with unidirectional plies of *Sisal* fibers

Direction of loading	Ultimate tensile strength (N/mm^2)	Percentage elongation	Young's modulus (N/mm^2)
Parallel	30.470	17.106	126.110
	37.111	14.915	176.738
	30.646	15.727	137.194
Perpendicular	6.435	13.533	39.040
	6.014	12.185	37.216
	4.885	12.121	36.210

laminates with cross plies are presented. Since plies are preset in both the directions, this type of ply would be suitable for resisting shear loads and hence its tensile strength is lower compared to the other ply orientations. It is to be mentioned that even though laminates made of coir fibers were prepared, they got damaged during the specimen preparation itself. This suggests that multi-layer laminates made of coir fibers would be viable for the evaluation of mechanical properties. Furthermore, for checking the impact strength of laminates using Izod or Charpy, the thickness

Table 2 Tension test results for laminates with cross plies of *Sisal* fibers

Direction of loading	Ultimate tensile strength (N/mm^2)	Percentage elongation	Young's modulus (N/mm^2)
NA	12.889	11.918	65.830
	15.506	15.112	62.432
	12.125	12.024	63.525
	12.966	17.173	55.649
	14.846	14.970	56.589

Table 3 Tension test results for laminates with braided plies of *Sisal* fibers

Direction of loading	Ultimate tensile strength (N/mm^2)	Percentage elongation	Young's modulus (N/mm^2)
Parallel	37.570	18.179	158.039
	31.520	15.364	143.385
	35.542	12.121	157.622
Perpendicular	6.602	11.242	23.549
	6.269	10.212	33.232
	5.082	12.333	36.759

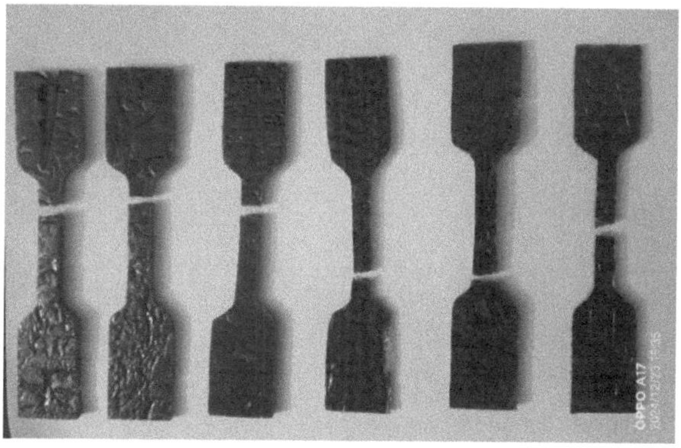

Fig. 7 Specimen having unidirectional plies after the test

of the specimen was very small. Hence, multi-layer laminates should be prepared for evaluating the impact strength. Figures 7 and 8 show the fractured specimen of unidirectional plies, cross plies, and braided plies, respectively.

4 Conclusions

Based on the tension test results, the following conclusions are drawn.

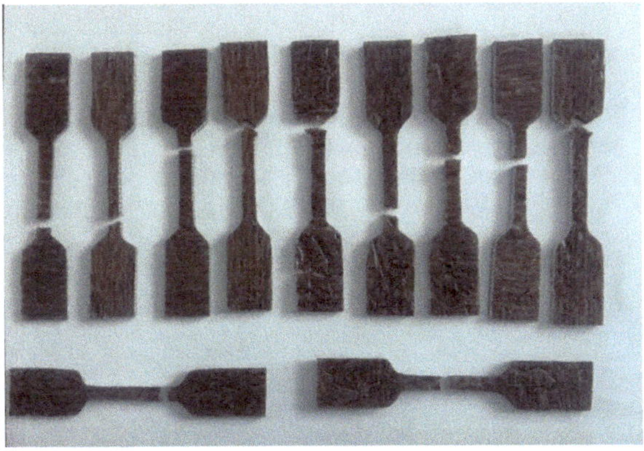

Fig. 8 Specimen having cross plies and braided plies after the test

1. The braided fiber placement achieved the highest Young's modulus and ultimate load. This can be attributed to the intertwined structure of braided fibers which provides better stress distribution and resistance to deformation under tensile loading.
2. Fibers oriented parallel to the loading direction show higher Young's modulus values than the cross-ply laminates.
3. Other tests like impact and flexural strength cannot be done on the single layer composite laminates made of Sisal fibers which are locally available.
4. Possibility of replacing synthetic fibers with natural fibers in at least one of the fields like structures of these natural fibers can be used for partition walls; steel sections can be replaced by the natural fiber laminates, thereby reducing the cost of the structure. In the future, using numerical methods, the behavior of the laminates when subjected to different loading conditions will be checked.

References

1. Cakiroglu C, Bekdaş G (2020) A review on the mechanical properties of natural fiber-reinforced composites. In International conference on harmony search algorithm. Springer, Singapore, pp 261–267
2. Chokshi S, Parmar V, Gohil P, Chaudhary V (2022) Chemical composition and mechanical properties of natural fibers. J Nat Fibers 19(10):3942–3953
3. Sathishkumar GK, Ibrahim M, Mohamed Akheel M, Rajkumar G, Gopinath B, Karpagam R, Karthik P, Martin Charles M, Gautham G, Gowri Shankar G (2022) Synthesis and mechanical properties of natural fiber reinforced epoxy/polyester/polypropylene composites: a review. J Nat Fibers 19(10):3718–3741
4. Kumar PSS, Viswanath Allamraju K (2019) A review of natural fiber composites [Jute, Sisal, Kenaf]. Mater Today Proc 18:2556–2562
5. Kumpati R, Skarka W, Skarka M (2022) Ansys-based evaluation of natural fiber and hybrid fiber-reinforced composites. Sustainability 14(23):15992

6. Nugraha AD, Nuryanta MI, Sean L, Budiman K, Kusni M, Muflikhun MA (2022) Recent progress on natural fibers mixed with CFRP and GFRP: properties, characteristics, and failure behaviour. Polymers 14(23):5138
7. Naik V, Kumar M (2021) A review on natural fiber composite material in automotive applications. Eng Sci 18:1–10
8. Espinach FX (2021) Advances in natural fibers and polymers. Materials 14(10):2607
9. ASTM D3039/D3039M—08 (2008) standard test method for tensile properties of polymer matrix composite materials. ASTM International, West Conshohocken, PA. https://doi.org/10.1520/D3039_D3039M-08

Enhancing Safety in National Highways Through Sustainable Accident Blackspot Treatments

S. Karthikeyan, J. Kannan, N. Saravanan, J. Balaji, P. Thangavel, and Ajit M. Hebbale

Abstract The study aims to identify sustainable traffic blackspots in the DSTRL-Din Digul Samaya Nallur Toll Road Limited, focusing on the Thanichiyam section. The conventional tripartite approach to road safety, consisting of 3'E's Engineering, Education, and Enforcement, has led to substandard road design and high fatality rates. The study analyzes four-year traffic crash data from the DSTRL Toll authorities, using Road Safety Audit (RSA) and Highway Safety Inspection and Survey. The blackspots are identified as Thanichiyam section 414 + 400 KM to 414 + 900 KM – 500 m. The study calculates accident density benchmarks and provides long-term projections based on the geographical location and severity of the accident.

Keywords Accident blackspot · Road safety audit · Vehicle under pass · Accident density · Analytical study

1 Introduction

India's rapid development has led to an increase in automobile usage and road network development, resulting in increased risks associated with travel and traffic exposure. With the second largest road network globally, India's extensive network facilitates the transportation of over 64.5% of all commodities and is the preferred choice for over 90% of passenger traffic. However, road accidents result in loss of life, property damage, and socioeconomic impact. India has the largest number of road accident fatalities worldwide, comprising 11% of total global deaths. National highways account for 33% of all traffic accidents and 35% of fatal accidents. Factors contributing to accidents include improper construction design, population density,

S. Karthikeyan · J. Kannan · N. Saravanan · J. Balaji · P. Thangavel
Department of Mechanical Engineering, Erode Sengunthar Engineering College, Perundurai, India

A. M. Hebbale (✉)
Nitte (Deemed to Be University), NMAM Institute of Technology (NMAMIT), Department of Mechanical Engineering, Nitte 574110, Karnataka, India
e-mail: ajit.hebbale@gmail.com

I. M. Jamadar et al. (eds.), *Integrating Sustainability in Modern Technologies*, Lecture Notes in Mechanical Engineering, https://doi.org/10.1007/978-981-95-9466-5_42

insufficient road safety knowledge, enforcement of traffic laws, mobile phone usage, drunken driving, and overspeeding. Post-accident care deficiencies also contribute to many deaths and disabilities. Tamil Nadu has been the most affected region in road accidents. Identifying and addressing accident blackspots is crucial for controlling accidents and reducing accidents. Huvarinen et al. [1] suggest that adhering to design and construction criteria of automotive roadways does not guarantee traffic safety. A traffic safety audit involves a comprehensive examination of accidents, focusing on the complex relationship between human behavior and road conditions. Uzondu et al. [2] suggest that unsafe behavior contributes to traffic incidents and collisions, and that factors such as road worker type, locality, and time of day influence risky driving behaviors. Yadav et al. [3] identified blackspots in an 8-km section of National Highway (NH-66) in Kerala, India, and took measures based on the severity of the accident. Mehar and Agarwal [4] propose a multi-criteria approach to determine the order of importance of safety standards for specific types of rural roads, such as Pradhan Mantri Gram Sadak Yojana (PMGSY) roads in Rajasthan, India. Tetali et al. [5] conducted a qualitative study on road safety in Hyderabad, identifying threats such as insufficient traffic regulations, enforcement gaps, low awareness, lack of political commitment, inadequate road engineering, and high-risk road users. Sandhyavitri et al. [6] identified blackspots in Riau province, Indonesia, and provided strategies to reduce the accident rate. Hongjun [7] analyzed the location of blackspots by determining the duration of road traffic accidents, distance between accidents, and figure of accidents logged within a precise time period.

The study aims to identify blackspot locations in the southern Tamil Nadu region using four years of grievous and fatal accident data (2019–2022). The research uses crash frequency and crash rate as methods to identify hotspots. Accident density is calculated and compared with the accident density benchmark to determine the severity level of the hotspot. Road Safety Audits are conducted on the blackspot location, and suggested mitigative measures are suggested. The effectiveness of the completed blackspot mitigative measure is estimated using the Tanner k test, Chi-square test, and T-test. Studies have shown that as evaluation studies consider additional complicating factors, the attributable risk of accident resulting from blackspot treatment diminishes. However, studies that consider regression-to-the-mean and potential relocation of accidents to nearby untreated locations do not demonstrate an overall reduction in accidents.

2 Methodology

2.1 Description of Data and Research Area

The research focused on four-lane National Highways in Tamil Nadu, India, specifically NH 44. Data was collected from DSTRL-Dindigul Samayanallur Toll Road Limited-Subsidiary of Reliance Infrastructure Limited (R Infra) over a four-year period from 2019 to 2022. The study identified 500 m of blackspot stretch using the NHAI blackspot parameter. The accident density was calculated manually for each location, and the relationship between the accident density and the benchmark was identified. The study identified areas of concern and identified issues at each blackspot location. Short-term and long-term measures were proposed, and the study provided countermeasures and engineering control measures. Test methods like Tanner k test and Chi-square test were conducted to estimate the success of the mitigation measures after three years of implementation. Figure 1 shows the Geographic Location of Research Area.

Accident data collected for the entire stretch through chainage wise and from the entire stretch the hotspot area identified and shown in Table 1.

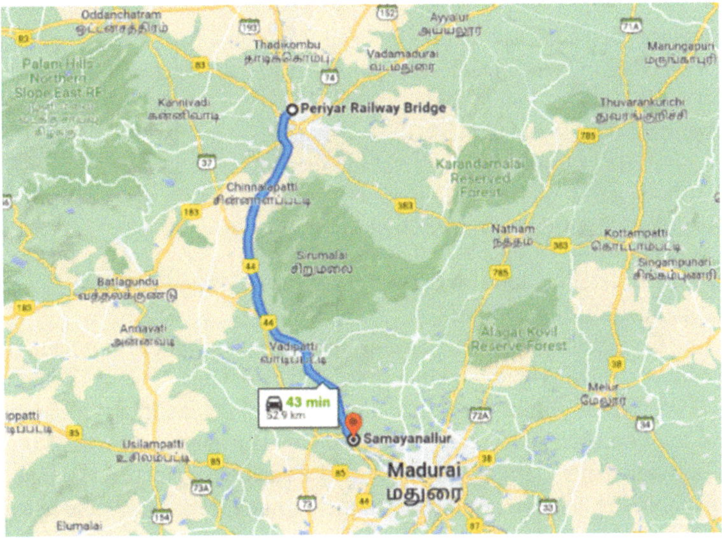

Fig. 1 Annotated route from Samayanallur to Periyar Railway Bridge. Map data © 2025 Google, Maxar Technologies

Table 1 Accident data for the Thanichiyam section

Locations	Total no of years				Total no of accidents
	2019	2020	2021	2022	
Thanichiyam section (414 + 400 km to 414 + 900 km)	15G	5G	1F&2G	2F & 1G	26

*Note F represents fatal, and G represents grievous

2.2 Accident Blackspot Stretch: Thanichiyam Section (414 + 400 KM to 414 + 900 KM)

2.3 Calculating Accident Density Benchmark for the Entire Stretch

1. **For year; 2019 (Jan to Dec)**—Total number of accidents = 5 + 8 + 8 + 19 + 14 + 9 + 17 + 6 + 20 + 7 + 5 + 11 = 129
   Total length of selected stretch of road = 53.049 km
   Accident density benchmark2019 = 129/53.049 = 2.431
2. **For year; 2020 (Jan to Dec)**—Total number of accidents = 21 + 8 + 3 + 1 + 10 + 5 + 2 + 12 + 15 + 3 + 9 + 2 = 91
   Accident density benchmark 2020 **=** 1.715
3. **For year; 2021 (Jan to Dec)**—Total number of accidents = 6 + 8 + 15 + 8 + 3 + 11 + 14 + 4 + 7 + 7 + 7 + 5 = 95
   Accident density benchmark 2021 = 1.790
4. **For year; 2022 (Jan to Dec)**—Total number of accidents = 12 + 5 + 7 + 8 + 6 + 10 + 5 + 8 + 6 + 12 + 8 + 10 = 97
   Accident density benchmark 2022 = 1.828.

For this consideration, take higher value 2.431 as an accident density benchmark value for the relationship comparison.

Based on the methodology and findings, the main blackspot treatment undertaken was the construction of Vehicular Underpass (VUP) at Thanichiyam section to eliminate NH–SH traffic conflict. As this pressure is tangible and it provides permanent structural solution which doesn't necessarily depend on enforcement or using temporary traffic management strategies, this measure can be considered sustainable. The study also refers to passive and active traffic calming measures such as improved signage and service roads which help to reduce long term accident risk. Figure 2 shows the Thanichiyam section total accident data.

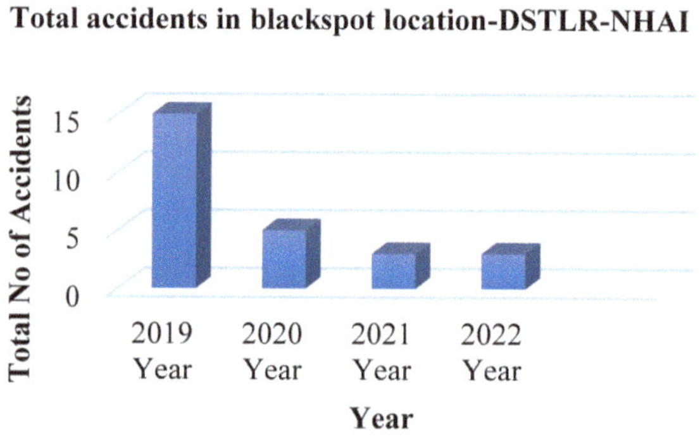

Fig. 2 Thanichiyam section total accident data

Table 2 Accident density benchmark data for the entire 53 + 049 km

Total no of years	Accident density benchmark
2019	2.431
2020	1.715
2021	1.790
2022	1.828

However, to fully justifying the use of the term "sustainable" the authors have to clarify how the adopted treatment achieves long term effectiveness (structural resilience, minimal maintenance requirements, and adaptability to future traffic growth, etc.). With the word limit constraint, a more comprehensive discussion may not be possible, but a brief discussion would clear the purpose of sustainability aspect inside the methodology. Table 2 Accident Density Benchmark Data for The Entire 53 + 049 km.

2.4 *Calculations for Accident Density*

Calculation of accident density for Thanichiyam defined as per Eq. 1:

$$\begin{aligned}\text{Accident density} &= \frac{\text{Total no of accident}}{\text{Unit length}} * 100 \\ \text{Accident density} &= 26/500 * 100 \\ &= 5.2\end{aligned} \tag{1}$$

On comparing the accident density benchmark value with the accident density blackspot location, accident density for the Thanichiyam section was found to be 5.2 > 2.431. If the accident density is below 2.431, it is classified as minimal danger for the DSTRL NH road. However, it exceeds 2.431, it is prone to high risk of accidents and we can confirm it is classified as a blackspot region. Figure 3 shows the relationship between accident density and accident density benchmark.

2.5 *Order of Blackspot*

The conducted a comparative analysis of the crash data from the prior four years with the Average Annual Total Crashes (AATCs) provided in Table 3. 5AATC/500, 10AATC/500, and 15AATC /500 m metrics. Table 3 shows the Average Annual Total Crashes (AATCs) of the States in India.

Tiwari [8] of India have given measures on calculating the orders of blackspot as with that comparison with Table 3: Average Annual Total Crashes (AATCs) the

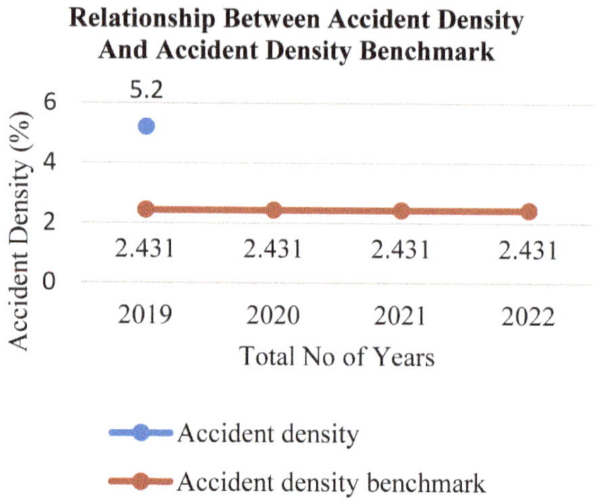

Fig. 3 Relationship between accident density and accident density benchmark

Thanichiyam section (414 + 400 KM to 414 + 900 KM – 500 m) – 26, which is above the 10 AATC/500 M-21.18 for NH and falls below the 15AATC/500 m-31.77 for NH and hence the consider range (15AATC/500 m) depends on the risk value and consider as a 1st order black spot.

3 Problems Identified in the Blackspot

3.1 Thanichiyam Site

Thanichiyam Section (414 + 400 km to 414 + 900 km) – 500 m

Unregulated and uncontrolled traffic noted in the intersection of the National highway and the State Highway. There is no means of adequate passive traffic calming measures and active calming measures. There is no Driver Feedback Sign (DFS) to advise speeding vehicles endlessly and instruct a culture of driving within indicated speed limits. The channelized island is in damaged condition. There is no separate acceleration lane and deceleration for the entry and the exit of the SH. The line of sight (LOS) is not visible from the SH. There are no advance warning sign boards provided in both approach zone. Conflict point diagram of this section revealed a total of 9 Crossing conflict points, diverging conflict points & merging conflict points which is prohibitively high due to flawed design geometry in the highways which subsequentially resulted in the accidents. Figure 4 shows the Conflict Point Diagram (Before RSA).

Table 3 Average annual total crashes (AATCs) of the states in India

State/UT	AATC/ 500 m		3AATC/ 500 m		5AATC/ 500 m		10AATC/ 500 m		15AATC/ 500 m	
	NH	SH	NH	SH	NH	SH	NH	SH	NH	SH
Andra Pradesh	1.28	0.91	1.92	1.36	3.20	2.27	6.40	4.55	9.60	6.82
Bihar	0.76	0.56	1.14	0.85	1.90	1.41	3.80	2.82	5.70	4.23
Delhi	25.99	0.00	38.98	0.00	64.97	0.00	129.94	0.00	21.00	0.00
Goa	5.49	1.45	8.24	2.18	13.73	3.63	27.46	7.25	41.19	10.88
Gujarat	0.86	0.32	1.29	0.49	2.15	0.81	4.30	1.62	6.45	2.43
Haryana	1.52	1.67	2.28	2.50	3.80	4.17	7.60	8.34	11.40	12.51
Jharkhand	0.68	1.26	1.01	1.89	1.69	3.15	3.38	6.31	5.07	9.46
Karnataka	2.11	0.57	3.16	0.85	5.27	1.42	10.55	2.84	15.82	4.26
Kerala	5.04	1.67	7.55	2.50	12.59	4.17	25.18	8.34	37.88	12.51
Madhya Pradesh	1.35	1.25	2.03	1.88	3.38	3.13	6.75	6.25	10.13	9.38
Maharashtra	0.63	0.22	0.94	0.32	1.56	0.54	3.13	1.08	4.69	1.62
Manipur	0.20	0,22	0.29	0.33	0.49	0.54	0.98	1.09	1.47	1.63
Odisha	0.89	0.80	1.33	1.19	2.22	1.99	4.45	3.98	6.67	6.97
Puducherry	13.45	0.00	20.17	0.00	33.62	0.00	67.24	0.00	100.86	0.00
Punjab	0.90	1.69	1.35	2.53	2.25	4.21	4.50	8.43	6.74	12.64
Rajasthan	0.85	0.23	1.27	0.35	2.12	0.59	4.25	1.17	6.37	1.76
Tamil Nadu	*4.24*	*1.90*	*6.35*	*2.84*	*10.59*	*4.74*	*21.18*	*9.48*	*31.77*	*14.22*
Telangana	1.82	1.50	2.73	2.25	4.45	3.76	9.10	7.51	13.65	11.27
Uttar Pradesh	1.67	1.71	2.50	2.57	4.17	4.29	8.34	8.57	12.51	12.86
West Bengal	1.43	0.89	2.15	1.34	3.58	2.23	7.16	4.45	10.74	6.68

3.2 *The Mitigative Measures*

Thanichiyam Section (414 + 400 km to 414 + 900 km) – 500 m

According to Yadav et al. [3], the mitigative measure discussed in Sect. 3.2 discusses about the construction of a Vehicular Underpass (VUP) at the Thanichiyam section to eliminate vehicle conflicts among National Highway (NH) and State Highway (SH) traffic. Nevertheless, it should be known, the authors will need to specify the exact year in which corrective measures were undertaken. According to the trend of accidents shown on Table 1, the reduction of accident density started from 2021, which might lend a few hints that some early interventions or enforcement measures may have already been carried out before the large scale implementation of the VUP. Although the accident density started to decrease in 2021, a good number of reasons still justified the construction of the VUP. Secondly, the accident density

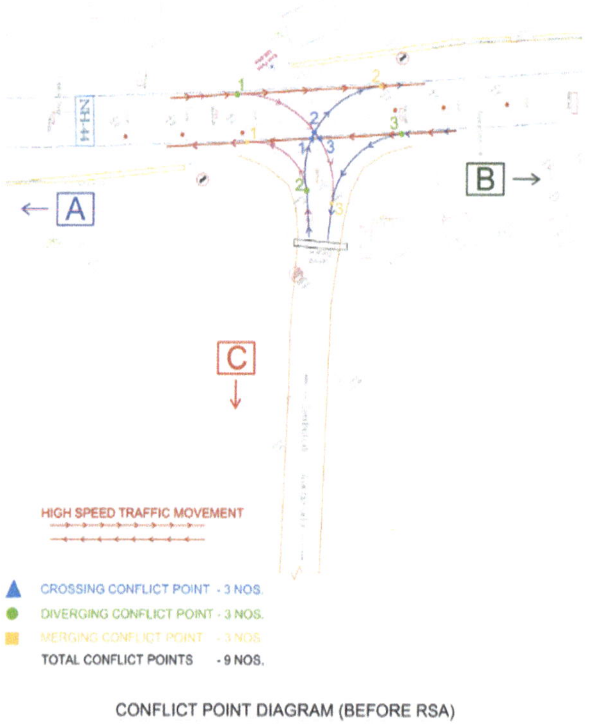

Fig. 4 Conflict point diagram (before RSA)

in 2021 was still above 2.431, which is pre benchmark value, i.e., the site still has a safety risk. Second, the VUP is capable of separating conflicting traffic flows, which ultimately leads to a sustainable and long term improvement in road safety, whereas short term safety measures may have played a role in initial reduction. The necessity of such measure is additionally statistically validated through the results of Tanner's K Test which revealed up to 84.4% reduction of crashes and T-Test that demonstrated significant decrease of vehicle speeds and accordingly reduced chance for an accident. The other factor is that the VUP may also be an integral component of accident blackspot treatment, since highway safety regulations and audits may have prompted further structural improvements.

Nevertheless, the necessity of the VUP itself also remains clear, because if one wants to continue to ensure that accident rates continue to decrease long term on this critical blackspot. The authors should clarify the exact timeline, on which the corrective measure was taken, to give a much clearer view of its impact.

In this scenario, mitigation measures were construction of VUP to avoid vehicle conflict with the main carriageway traffic and SH road traffic, service road has been constructed on the both sides (LHS&RHS). The conflict point diagram after the

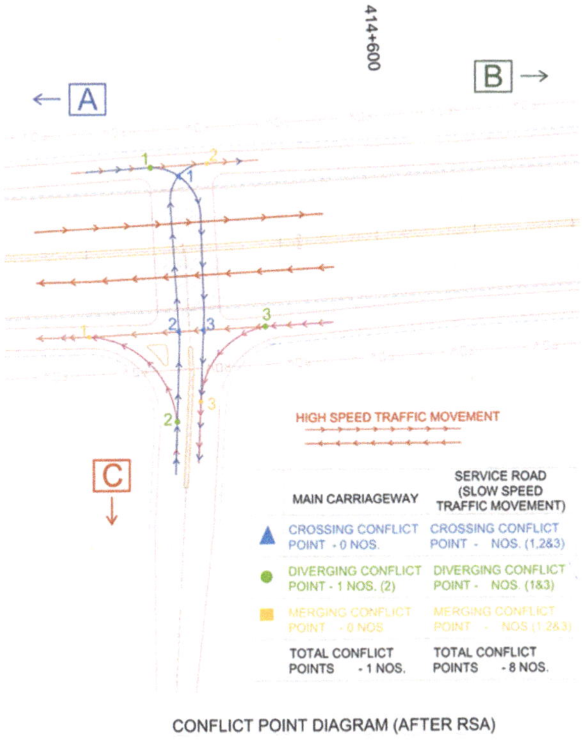

Fig. 5 Conflict point diagram (After RSA)

Road Safety Audit (RSA) will give the effectiveness of the mitigation measures. By combining all the conflict points will be having total 8 nos in the slow traffic service road area. Figure 5 shows conflict point diagram (after RSA).

4 Results and Discussions

The tests used in this project are:

- Tanner's K Test—(Magnitude of change)
- Chi-squared Test—(Significance of change)
- T-test (compare the mean of two group data).

4.1 Tanner's k Test

From that can calculate using Eq. 2;

Table 4 Crash data for the corridor

	Crash at site	Crash at controlled site	Total
Before	26[a]	29[c]	55[g]
After	02[b]	13[d]	15[h]
Total	28[e]	42[f]	70[i]

[a] Earlier accidents at site; [b] Afterward accidents at site; [c] Earlier accidents at control site; and [d] Afterward accidents at control site

$$k = \frac{b/a}{d/c} \quad (2)$$

Note: If any frequencies are zero, add 1/2 to each, i.e., Eq. 3.

$$k = \frac{\left(b + \frac{1}{2}\right)/\left(a + \frac{1}{2}\right)}{\left(d + \frac{1}{2}\right)/\left(c + \frac{1}{2}\right)} \quad (3)$$

On substituting the table values corresponding to equation $k = 0.156$.

Table 4 listed the Crash Data for The Corridor. If $k < 1$ nevertheless there has been a reduction in crashes relation to the mitigative measures. If $k = 1$ then around has been no modification virtual to the mitigative measures. If $k > 1$ then there has been an improve relation to the mitigative measures. The percentage modify at the site is given by:

$$(k - 1) \times 100\%$$

As, $k < 1$ there has been relative decline in accident to the given mitigative measures of **84.4%**.

4.2 *Illustration of the Chi-Square Test*

The test can be employed with a reduced amount of data, but the margins of error are more significant. This Eq. 4 utilizes the notation provided in Table 4 mentioned.

$$\chi^2 = \frac{\left(|ad - bc| - \frac{i}{2}\right)^2 i}{efgh} \quad (4)$$

On substituting the values,

$$\chi^2 = \mathbf{4.330}$$

As from the above result $\chi^2 = 4.330$ and with the degree of freedom can find the area between. The degree of freedom is 1.

According to this calculation, the Chi-square value is $\chi^2 = 4.330$ and from Table 1.2. The calculated Chi-square value is in between the >3.841 and <5.024. This relates to a value of significance level (on the column header line) between $p = 0.05$ and $p = 0.025$ (see Table 5.4), which is generally quoted as significant than the lower level, i.e., better than the 5% level of significance. So, the significance value is 5 and 95% confidence level. This indicates that the outcome is statistically significant at the 5% level ($p < 0.05$), advising that the result is suspect to be due to random possibility. Table 5 shows the Degree of Freedom for Chi-Square Distribution. Table 6 shows the understanding Significance of Chi-Square Range. Table 7 highlights the RHS lane spot speed survey data.

Table 5 Degree of freedom for chi-square distribution

χ Probability of exceeding the critical values					
1	*2.706*	*3.841*	*5.024*	*6.635*	*10.828*
2	4.605	5.991	7.378	9.210	13.816
3	6.251	7.815	9.348	11.345	16.266
4	7.779	9.488	11.143	13.277	18.467

Table 6 Understanding significance of chi-square range

Chi-square value	Significance value (%)	Confidence level (%)	Interpretation
>2.706	10	90	Fair
>3.841	*5*	*95*	*Acceptable*
>5.024	2.5	97.5	Good
>6.635	1	99	Highly acceptable
>10.828	0.001	99.999	Very highly acceptable

Table 7 RHS lane spot speed survey data

RHSLANE—Spot speed survey data		
S.no	Without control measures (km/h)	With control measures (km/h)
1	67	44
2	63	47
3	45	39
4	81	61
5	77	64
6	73	51
7	84	53
8	93	50
9	89	41
10	84	43

4.3 Illustration of the T-Test on Thanichiyam Section Road

4.4 Test Without Control Measures in RHS Lane

NH standard of DSTRL highway stretch is 80 km/h used by Eq. 5. The mean of dataset is not significantly different from $\mu_0 = 80$

$$t = \frac{\overline{x} - \mu_0}{\frac{s_x}{\sqrt{n}}} \tag{5}$$

To find the degree of freedom = d.o.f = $n - 1$.

where; $\overline{x}$ = Sample mean, μ_0 = The hypothetical mean value, s_x = Sample standard deviation and

n = Size of the sample. In this project,

To calculate the sample mean

$$\overline{x} = 74.6 \ \mu_0 = 80 \ n = 10$$

$$s_x = \sqrt{\frac{1}{n-1}\sum\nolimits_{i=1}^{n}(x_i - \overline{x})^2} \tag{6}$$

After substituting the values in Eq. 6, get;

$$s_x \approx 14.261$$

After substituting these values into the formula for 't' have:

$$t \approx -0.9756$$

To calculate degree of freedom.

$$\begin{aligned} \text{d.o.f} &= n - 1 \\ &= 10 - 1 = 9 \end{aligned}$$

Determine critical value for 't' with degrees of freedom = 9 and $\alpha = 0.05$.

In this project the critical value is = 2.262 from the given Table 8.

$$|t| < \text{crit. value} = \text{No sig. diff.}$$

The calculated absolute value of t is smaller than the critical value (0.975 < 2.262), so the mean of dataset is not significantly different from $\mu_0 = 80$.

H_0 is rejected when it is true and H_1 is accepted when it is false. H_1 is accepted so, the control measures which is provided earlier in the RHS lane of Thanichiyam

Table 8 Critical values for the two tailed test

Degree of freedom	0.2	0.15	0.1	0.05	0.025	0.01	0.005	0.001
1	3.078	4.165	6.314	12.706	25.453	63.657	127.321	636.619
2	1.886	2.282	2.920	4.303	6.205	9.925	14.089	31.599
3	1.638	1.924	2.353	3.182	4.177	5.841	7.453	12.924
4	1.533	1.778	2.132	2.776	3.495	4.604	5.598	6.869
5	1.476	1.699	2.015	2.571	3.163	4.032	4.773	5.959
6	1.440	1.650	1.943	2.447	2.969	3.707	4.317	5.048
7	1.415	1.617	1.895	2.365	2.841	3.499	4.029	5.041
8	1.397	1.592	1.860	*2.306*	2.752	3.355	3.833	4.781
9	1.383	1.574	1.833	2.262	2.685	3.250	3.690	4.587
10	1.372	1.559	1.812	2.228	2.634	3.169	3.581	4.437

section is not adequate enough to control the accident in the intersection area. Table 8 shows the critical values for the Two Tailed Test.

4.5 *Test for Control Measures in RHS Lane*

The hypothetical mean value $\mu_0 = 80$. The reason provide the hypothetical mean value as 80 is, as the NH standard of DSTRL highway stretch is 80 km/h.

The mean of dataset is not significantly different from $\mu_0 = 80$.

To calculate the sample mean

$$\overline{x} = 49.3 \; \mu_0 = 80 \; n = 10$$

$$s_x = \sqrt{\frac{1}{n-1}\sum\nolimits_{i=1}^{n}(x_i - \overline{x})^2} \tag{7}$$

After substituting the values in Eq. 7, get;

$$s_x \approx 8.2872$$

After substituting these values into the formula for 't' have:

$$t \approx -11.7175$$

To calculate degrees of freedom.

$$\begin{aligned} \text{d.o.f} &= n - 1 \\ &= 10 - 1 = 9 \end{aligned}$$

Table 9 LHS lane spot speed survey data

LHSLANE—Spot speed survey data		
S.no	Without control measures (km/h)	With control measures (km/h)
1	71	45
2	65	51
3	49	54
4	87	62
5	96	43
6	76	39
7	94	44
8	88	58
9	69	62
10	82	46

Determine critical value for 't' with degrees of freedom $= 9$ and $\alpha = 0.05$.

In this project the critical value is $= 2.262$ from the given Table 8.

The calculated absolute value of t is greater than the critical value. ($11.717 > 2.262$), so the mean of dataset is significantly different from $\mu_0 = 80$.

$$|t| > \text{crit. value} \;= \text{There is sig. diff.}$$

H_0 is rejected when it is true and H_1 is accepted when it is false.

H_0 is rejected. So, after implementing the control measures in the Thanichiyam section it is adequate enough to control the accident in the intersection area.

Finally, they can say that the mean difference in mean speeds (a 25.3 km/h reduction of speed in RHS Lane) is significant at the 5% level. The implementation of the mitigative measures works and drastically reduces the speed of the vehicle in the intersection area and thus accident can be averted. Table 9 shows the LHS lane spot speed survey data.

4.6 *Test Without Control Measures in LHS Lane*

From the calculated value 't' is

$$\approx -0.498$$

In this project the critical value is $= 2.262$ from the given Table 8.

The calculated absolute value of t is smaller than the critical value ($0.498 < 2.262$), So the mean of dataset is not significantly different from $\mu_0 = 80$.

$$|t| < \text{crit. value} \ = \text{No sig. diff.}$$

H_0 is rejected when it is true and H_1 is accepted when it is false.

H_1 is accepted. So, the control measures which is provided earlier in the LHS lane of the Thanichiyam section is not adequate enough to control the accident in the intersection area.

4.7 *Test with Control Measures in LHS Lane*

From the calculated value '*t*' is

$$\approx -11.3288$$

In this project the critical value is = 2.262 from the given Table 8.

The calculated absolute value of t is greater than the critical value. (11.328 > 2.262). So the mean of dataset is not significantly different from $\mu_0 = 80$.

$$|t| > \text{crit. value} \ = \text{There is sig. diff.}$$

H_0 is rejected when it is true and H_1 is accepted when it is false.

$H_0$is rejected. So, after implementing the control measures in the Thanichiyam section it is adequate enough to control the accident in the intersection area.

Finally, it can say that the mean difference in mean speeds (a 27.3 km/h reduction of speed in LHS Lane) is significant at the 5% level. The implementation of the mitigative measures drastically reduces the speed of the vehicle in the intersection area and thus accident can be averted.

5 Conclusions

The study identifies blackspots in the DSTRL project in Tamil Nadu and develops control measures. Implementing Vehicluar Underpass (VUP) at the Thanichiyam junction eliminated conflicts with SH vehicles. The Tanner-K test showed an 84.4% reduction in crashes due to the measures. A chi-square test showed a 95% reduction in crashes. The study also found significant mean differences in mean speeds at the RHS and LHS lane, indicating the measures effectively reduce accident rates.

6 Limitations and Future Work

The insights from this study will be insightful for accident blackspots and strategies to mitigate accidents, but there are restrictions to this study as well. Secondly, the study is carried out based on a four-year dataset (2019, 2022) and therefore, long term accident trends may not be entirely observed and seasonal variations may not be taken into account. With a larger dataset of 8–10 years, the statistical reliability would increase and thus enable more accurate trend analysis. Furthermore, albeit the study concentrates on the Thanichiyam section, other blackspots across other highways can also be considered for comparative analysis to fortify the generalizability of the findings.

Further investigate can foster on this work by including real time traffic monitoring, driver behavior analysis and advanced predictive modeling techniques to improve accident forecasting. Also, post–implementation studies over several years can be used to quantify the long-term effectiveness of the Vehicular Underpass (VUP) and other mitigation measures. The integration of AI driven accident prediction models and GIS-based spatial analysis will enhance the existing blackspot identification and treatment strategies.

References

1. Huvarinen Y, Svatkova E, Oleshchenko E, Pushchina S (2017) Road safety audit. Transp Res Procedia 20:236–241. https://doi.org/10.1016/j.trpro.2017.01.061
2. Uzondu C, Jamson S, Lai F (2019) Investigating unsafe behaviours in traffic conflict situations: an observational study in Nigeria. J Traffic Transp Eng (English Edition) 6(5):482–492. https://doi.org/10.1016/j.jtte.2018.06.002
3. Yadav DK, Ghodmare SD, Kumar NN (2022) Mitigation of blackspots on highways by the application of safe system approach. Mater Today: Proc 52(Part 3):1228–1235. https://doi.org/10.1016/j.matpr.2021.11.042
4. Mehar R, Agarwal PK (2013) A systematic approach for formulation of a road safety improvement program in India. Procedia—Soc Behav Sci 104:1038–1047. https://doi.org/10.1016/j.sbspro.2013.11.199
5. Tetali S, Lakshmi JK, Gupta S, Gururaj G, Wadhwaniya S, Hyder AA (2013) Qualitative study to explore stakeholder perceptions related to road safety in Hyderabad, India. Injury 44(Supplement 4):S17–S23. https://doi.org/10.1016/S0020-1383(13)70208-0
6. Sandhyavitri A, Zamri, Wiyono S, Subiantoro (2017) Three strategies reducing accident rates at black spots and black sites road in Riau Province, Indonesia. Transp Res Procedia 25:2153–2166. https://doi.org/10.1016/j.trpro.2017.05.415
7. Cui H, Dong J, Zhu M, Li X, Wang Q (2022) Identifying accident black spots based on the accident spacing distribution. J Traffic Transp Eng (English Edition) 9(6):1017–1026. https://doi.org/10.1016/j.jtte.2021.02.006
8. Tiwari G (2023) International course on road safety 23 November–7th December. https://tripc.iitd.ac.in/assets/training_material/BLACKSPOT%20TREATMENT-GT2023.pdf

Methodology for Structural Analysis of Natural Fiber-Reinforced Plastics for Automotive Applications

Norbert Kluck, Matthias Müller, Gian-Luca Mazziotti, Felix König, Vinod Rajamani, Stefan Hesterberg, J. R. Nataraj, and Ramesh Sharma

Abstract Natural fiber-reinforced plastics have huge potential for sustainable production and reduced carbon-footprint. They are especially attractive in the automotive sector due to the wide range of materials already being applied. Currently, the hydrophilic nature of natural fibers and their low tensile strength limits their application to interior, low-loaded components of vehicles. One of the significant challenges in designing components using natural fibers is their anisotropic behavior. While this has already been encountered in the use of carbon fibers and glass fibers, the unique problem faced here is the natural variation of fiber properties. High quality simulation models for structural analysis are necessary for this purpose. In this paper, the potential for application of natural fiber-reinforced plastics in various automotive applications is discussed followed by a literature analysis of the properties of different natural fibers. Then, the setup and validation of a simulation model for finite element analysis of natural fiber-reinforced plastics using the software Ansysis discussed.

Keywords Natural fiber-reinforced plastics · Material properties · Structural analysis · Simulation

Abbreviations and Acronyms

NFRP	Natural Fiber-Reinforcement Plastics
CFRP	Carbon Fiber-Reinforcement Plastics
GFRP	Glass Fiber-Reinforcement Plastics
FEM	Finite Element Method

N. Kluck · M. Müller · G.-L. Mazziotti · F. König · V. Rajamani · S. Hesterberg
Department of Mechanical Engineering, Fachhochschule Dortmund, Dortmund, Germany
e-mail: kluck@fh-dortmund.de

J. R. Nataraj (✉) · R. Sharma
Department of Mechanical Engineering, R.V. College of Engineering, Bengaluru, Karnataka, India
e-mail: natarajjr@rvce.edu.in

I. M. Jamadar et al. (eds.), *Integrating Sustainability in Modern Technologies*, Lecture Notes in Mechanical Engineering, https://doi.org/10.1007/978-981-95-9466-5_43

1 Introduction

In recent years, the automotive industry has witnessed a notable surge in research and development focused on natural fiber-reinforced plastics (NFRP). Manufacturers are increasingly seeking recyclable or biodegradable materials to comply with legal mandates concerning carbon dioxide (CO_2) emissions and to circumvent financial incentives related to sustainability goals. A pivotal strategy to achieve these objectives is to reduce vehicle weight, as this directly impacts fuel consumption and emissions. Natural fibers, due to their inherent properties, present a compelling solution to address these challenges. They can substitute for synthetic materials such as carbon or glass fibers in composite materials, thereby aligning with legal mandates to reduce CO_2 emissions.

Natural fibers have a particular significance for applications in the automotive industry due to the large volume of production. In addition, automotive applications with a wide range of material requirements have the potential to accommodate the properties offered by natural fibers in a selection of components. Typically, these components do not experience high loads and harsh weather conditions. The design of NFRP components poses a big challenge for the developer, primarily due to the anisotropic nature associated with any composite material. Further, the natural variation in the mechanical properties in the fibers within the same weave makes the structural analysis of NFRPharder than for carbon-fiber or glass-fiber-reinforced composites. For this purpose, accurate simulation models are necessary, which can capture the properties of the NFRP material within the product design tolerances.

In this paper, the authors provide an overview of natural fibers and their applications in the automotive industry. This is followed by an illustration of the methodology for the calibration and validation of finite-element simulation models for NFRP. This is done using the simulation software Ansys.

1.1 Natural Fibers in Automotive Applications

The high cellulose content endows natural fibers with desirable mechanical properties, and when combined with an organic polymer matrix, they offer the potential for producing a fully degradable composite material [1]. In order to minimize CO_2 emissions, it is imperative to consider the origin of the natural fibers. The cultivation and transportation of these materials contribute to the overall CO_2 footprint of composite materials. Consequently, the European automotive industry predominantly utilizes flax and hemp sourced from domestic cultivation, while jute and kenaf for example are primarily imported from Bangladesh and India [2].

According to Jayakrishna Kandasamy [3], the average utilization of natural fibers in German automotive vehicles is approximately 3.6 kg. Until recently, the application of these fibers has been predominantly limited to semi-structural components,

Fig. 1 Bio-concept-car porsche Cayman GT4 with multiple body parts manufactured out of NFRP [4, 5]

such as dashboards, door panels, and parcel shelves, with their primary function remaining within the vehicle interior.

The current development of NFRP components, as illustrated in Fig. 1, demonstrates an expansion of the development to car body components. A consortium comprising the Fraunhofer WKI, Hanover University of Applied Sciences and Arts, Porsche Motorsport, and component suppliers has developed a process suitable for series production for the manufacture of car body components. The components are manufactured from a combination of flax fibers and an epoxy resin using the resin transfer molding (RTM) process. The adoption of natural fibers in place of CFRP offers a substantial ecological benefit, with a mere additional weight of approximately 10%, while maintaining compliance with the technical requirements for components [4].

Durability and environmental resistance are two other critical factors to consider. Components on vehicles, especially bodywork components or components on the underbody of a vehicle, are exposed to different temperatures, degrees of humidity, and mechanical loads. Natural fibers have hydrophilic material behavior, which can lead to moisture absorption and subsequent deterioration of the mechanical properties. As illustrated in Table 1, the properties of the fibers can exhibit significant variability, contingent on the processing and manufacturing method, as well as the growing conditions. Consequently, rigorous quality control measures are imperative to ensure the reliability of automotive components.

Hemp and Kenaf fibers have been shown to offer properties that are analogous to those of flax fibers [2]. The matrix material in natural fiber composites is typically an organic resin, such as an epoxy resin, which offers excellent mechanical properties and high thermal stability [7]. Consequently, in order to create a composite material that is entirely biodegradable, it is necessary to utilize a natural fiber and a natural resin or biopolymer, such as polylactic acid (PLA) to produce a biological composite. Flax fibers have been demonstrated to possess the necessary qualities for structural applications, namely a high tensile strength and stiffness. Consequently, they have been utilized in numerous research projects and industrial applications. The proTRon Evolution project at Trier University of Applied Sciences for example is an ongoing research initiative that aims to expand the application of natural fibers in the automotive sector, particularly in the area of structural components. The project's

Table 1 Mechanical properties of different natural fibers in comparison with glass fiber and carbon fiber [2, 6]

Fiber	Density (g/cm^3)	Diameter (μm)	Tensile strength (MPa)	Stiffness/Young's modulus (GPa)	Specific tensile strength (MPa/g/cm^3)	Elongation at break (%)	Specific Young's modulus (GPa cm^3/g)	Cost (US$/ton)	Energy (GJ/ton)
Flax	1.38–1.52	5–600	343–1830	27–100	227–1220	1.2–3.2	18–53	200–1000	4
Kenaf	1.2	12–36	295–930	22–60	246–993	1.6–6.9	18–50		
Jute	1.23–1.5	5–200	187–800	10–55	140–610	1.16–3.1	7.1–39		
Hemp	1.35–1.51	10–500	550–1110	30–70	210–740	1.6–4.5	20–47		
E Glass	2.5–2.55	15–25	200–3500	68.9–73	666–1400	2.5–3	29	1200–1800	30
Carbon	1.8	6	2700–7000	230–600	1500–3900	1.2–2	125–330	12,500	130

Fig. 2 Monocoque made of NFRP with unidirectional carbon fibers for reinforcement [9]

objective is to design a monocoque composed primarily of natural fiber composites, with the incorporation of carbon fibers in select areas to enhance reinforcement along specific load paths. As illustrated in Fig. 2, the monocoque under consideration meets the crash requirements for vehicles classified as M1, encompassing passenger cars, by employing wave-shaped crash elements capable of absorbing twice the specific energy of steel crash boxes [8].

In summary, it can be stated that the utilization of natural fiber materials in the automotive industry is gaining traction. These materials present a compelling alternative to conventional fiber composites, with a significantly reduced energy requirement during production. To ensure their widespread application, there is a necessity for the collection of additional data and the establishment of standards, thereby facilitating a more profound comprehension of their properties.

2 Comparison Between Simulation of Metals and Fiber Reinforced Materials

In engineering, the finite element method (FEM) is a powerful numerical method for analyzing the material behavior and structural properties of components. With the increasing importance of fiber-reinforced materials as lightweight and high-strength materials used in the aerospace and automotive industries, challenges arise in FEM analysis to integrate these more complex carbon-fiber carbon-fiber-reinforced plastics (CFRP), glass-fiber-reinforced plastics (GFRP) and NFRP material systems. The anisotropic mechanical properties of composite materials, which differ from the isotropic mechanical properties of conventional metallic materials, are a key aspect of integration.

Modulus of elasticity, Poisson's ratio and strength can be regarded as constant over a wide range for the usual metallic materials, which predominantly leads to a homogeneous material model. Accordingly, 3D solid elements or shell elements for thin-walled structures and simple contact models are preferably used for meshing. The analysis of the FEM simulations usually includes the standard failure criteria, such as plastic deformation and fracture.

On the other hand, modulus of elasticity, Poisson's ratio and strength are largely anisotropic for fiber-reinforced materials. Tensorial material models are predominantly used for composite materials as the material consists of two phases, the matrix and the orientation-dependent unidirectional, woven or coiled fiber. The fracture mechanics effects that go beyond the classic failure criterion of plastic deformation, such as matrix fracture, fiber fracture or delamination, require specific models to represent the complexity realistically.

Finally, when using natural fibers instead of synthetic fibers, the problems of natural waxing must also be taken into account. Natural fibers vary in length and thickness, and individual fibers can also have internal defects that lead to a reduction in mechanical properties. The mechanical properties of NFRP materials are therefore subject to greater variation than synthetic materials.

2.1 FEM Simulation Methodology of Fiber-Reinforced Materials

In the FEM simulation of fiber-reinforced materials, there are three main models used to describe complex component failure. The so-called Hashin model is generally used for simple unidirectional composites and describes the basic failure modes with fiber fracture in the tensile direction, fiber fracture under compression, matrix failure and shear failure [10].

- Fiber breaks in the tensile direction (tensile break), if the tensile or compressive stress exceeds the tensile strength: $\sigma_1 \geq \sigma_{1T}$.
- Fiber breakage under compression (compressive fiber breakage) when the compressive stress is less than the negative compressive strength: $\sigma_1 \leq -\sigma_{1C}$.
- Matrix failure, if the tensile or compressive stresses in the matrix are greater or less than the tensile or compressive strength: $\sigma_2 \geq \sigma_{2T}$ or $\sigma_2 \leq -\sigma_{2C}$.
- Shear failure, if the shear stress in the matrix or at the matrix/fiber interface is greater than the shear strength: $\tau_{12} \geq \tau_{12C}$.

The Tsai Wu model is preferred for multidirectional laminate structures because it takes into account the interaction between the stresses in all directions [11]. The general form is shown in (1).

$$F_1\sigma_1 + F_2\sigma_2 + F_3\sigma_3^2 + F_4\sigma_1\sigma_2 + F_5\sigma_1\sigma_3 + F_6\sigma_2\sigma_3 + F_7\tau_{12}^2 = 1. \tag{1}$$

The constants F_1–F_7 are empirically determined parameters. σ and τ, the normal stresses and shear stresses are direction-dependent.

In addition to the Hashin and Tsai Wu models, the newer Puck model also takes delamination between the layers into account. The Puck model is described briefly below:

Fiber breakage: Tension: $\sigma_1 \geq \sigma_{1T}$ Compression: $\sigma_1 \leq -\sigma_{1C}$.

Matrix failure: Tension: $\sigma_2 \geq \sigma_{2T}$ Compression: $\sigma_2 \leq -\sigma_{2C}$. Shear: $\tau_{12} \geq \tau_{12C}$.
Delamination: $\sigma_{12} \geq \sigma_{12T}$ and $\tau_{12} \geq \tau_{12C}$.

The structure of an FEM simulation model differs from that of a conventional metallic material, particularly in the area of material modeling and fiber-matrix modeling. The individual steps of the simulation are shown graphically in Fig. 3.

Based on the CAD geometry and the subsequent meshing, the NFRP material used is described within the material definition. The material used here is the balanced reinforced fabric "Flaxdry BL200 Twill 2-2 fabric" and the matrix material "RESOLTEC 1800 ECO/1804" with the mechanical properties listed in Table 2 and having a weight ratio of 100/26.

These include, on the one hand, the modulus of elasticity, the shear modulus and the Poisson's ratio in the respective directions, and on the other hand, the limitations of the stresses, strains and shears in the respective planes are required for the application of the failure modes. In addition, there are various constants depending on the failure criterion used (Tsai Wu, Puck).

The next step is to setup the composite material in terms of the number of layers and their thickness and to determine the orientation of the individual layers (refer Fig. 3). The setup is completed with the support and the introduction of the forces

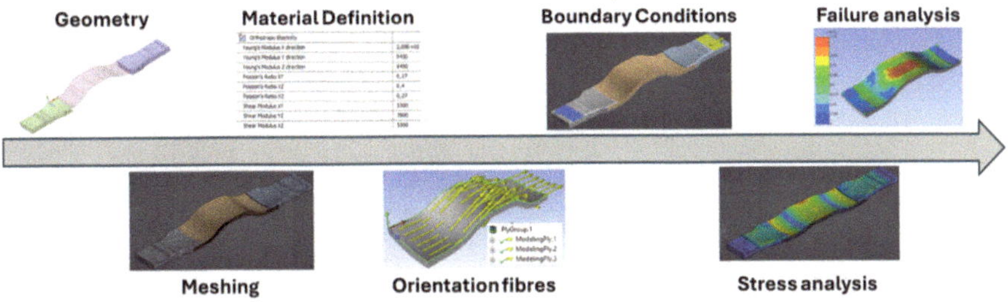

Fig. 3 Flowchart of a simulation

Table 2 Mechanical properties of the natural fiber and matrix

Properties fibers	Weft direction	Warp direction
Tensile strength (MPa)	58	61
Young modulus (GPa)	6	7
Elongation at break (%)	1.8	1.5
Thickness of one impregnated ply	0,59 mm	
Fiber density	1.45 g/cm^3	
Properties matrix		
Density	1.15 g/cm^3	
Modulus	3.1 GPa	
Maximum strength	51.4 MPa	
Elongation at break	1.61%	

and displacements, and the simulation is started. The final post-processing shows the stresses and deformations of the component, as well as the critical areas of the component in a failure analysis.

2.2 Experiments for Validation of FEM Simulation Model

Experimental tensile tests were performed to make data of material behavior for validation of the FEM simulation models. The investigations are based on a woven NFRP structure, from which the six samples shown were cut out and measured in the tensile test. This is shown in Fig. 4.

The results of the tensile test shown in Fig. 5 show the usual stress–strain curve of fiber-reinforced composites. With increasing elongation, the stresses increase up to a maximum value due to the mechanical influence of the fibers. The curves then fall almost vertically to a certain stress value. The reason for the drop is the occurrence

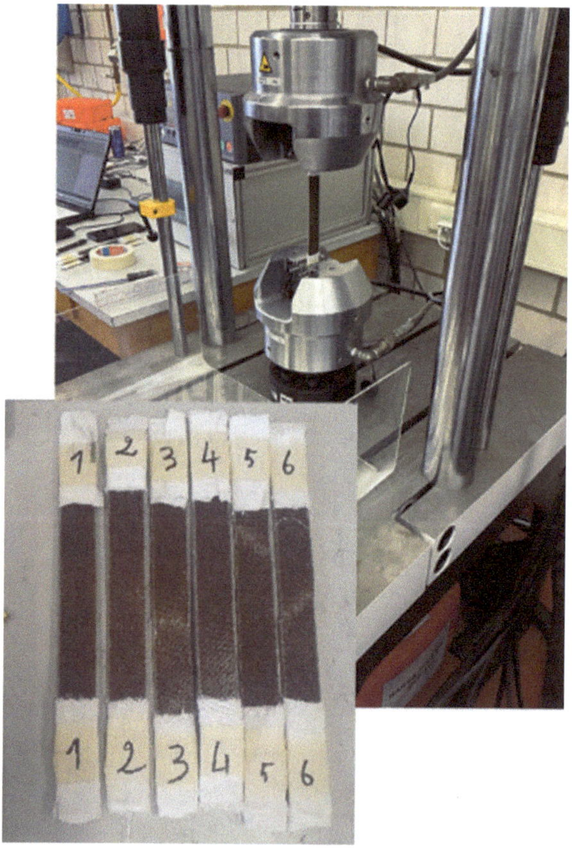

Fig. 4 Tensile test machine and test samples

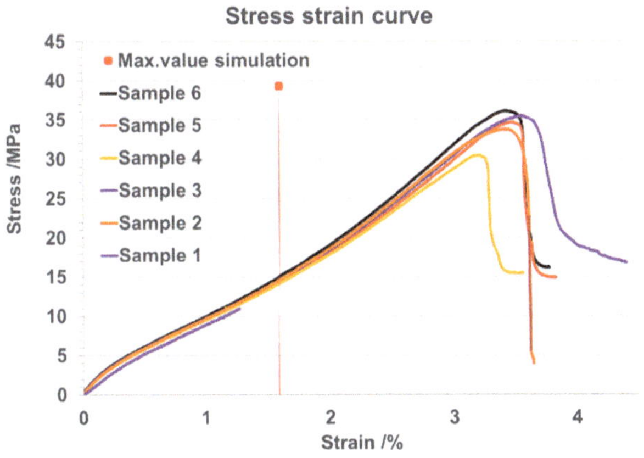

Fig. 5 Stress–strain diagram of the test samples

of fiber breakage and/or delamination. The rest of the curve is determined by the mechanical of the matrix material, which determines the strength of the specimen until final failure.

Two stress–strain curves are out of line due to problems with the fixture and cannot therefore be evaluated. The remaining four curves show a considerable scatter for both stresses and strains.

The mechanical properties of the equivalent FEM simulation (red dot in Fig. 5) were determined according to the law of mixture. The longitudinal modulus as the value of the modulus of elasticity parallel to the fiber direction is: $E_{||} = V_f E_f + V_m E_m$, where the subscript f stands for the fiber and m for the matrix. The transverse modulus perpendicular to the fiber direction is given as: $E_{\_|\_} = \left(\frac{V_f}{E_f} + \frac{V_m}{E_m}\right)^{-1}$. The tensile stresses parallel and perpendicular to the fibers required for failure analysis are obtained from the following equations: $\sigma_{||} = V_f \sigma_f + V_m \sigma_m$ and $\sigma_{\_|\_} = \left(\frac{V_f}{\sigma_f} + \frac{V_m}{\sigma_m}\right)^{-1}$.

Furthermore, the Poisson's ratio and the shear modulus for the entire material can be calculated using $\nu_{\text{total}} = V_f \nu_f + V_m \nu_m$ and $G_{\text{total}} = \left(\frac{V_f}{G_f} + \frac{V_m}{G_m}\right)^{-1}$. Using the data from the data sheet, this results in an overall modulus of elasticity of 6.06 GPa in the warp direction and 5.3 GPa in the weft direction.

It can be seen that the maximum values from the FEM simulation and the tensile test differ greatly. The value of the stress at which fiber breakage occurs was determined to be 39.3 MPa in the FEM simulation. The value of the elongation here is simulated to be 1.6%.

This result shows that further investigations and parameter variations within the FEM simulation are necessary in order to generate more stable results and to gain experience in dealing with the FEM simulation of NFRP. Parameter identification with the help of metamodels from optimizations could be helpful here. Additional microscopic material examinations at the fracture points provide further information on the interaction of fiber and matrix under load.

3 Summary and Conclusion

In this paper, the potential for the use of NFRP in the automotive industry has been shown by means of a literature survey. The high complexity in the structural analysis of NFRP has also been illustrated and compared to the analysis of metallic materials. The methodology for the FEM simulation of NFRP has been shown and discussed. Additionally, the simulation model developed has been compared to experimental results for the purpose of validation. The results of the simulation have indicated that though the model delivers an accuracy in the same order of magnitude as the experimental results, more work needs to be carried out to reach a quality satisfactory for the intended purpose. Once this has been achieved, it is planned to use this model to simulate and develop components in a chosen high volume automotive application, using locally sourced natural fibers.

Acknowledgements The authors would like to thank Fachhochschule Dortmund and R.V. College of Engineering for the infrastructure support provided to carry out this work.

References

1. Staiger MP, Tucker N (2008) Natural-fibre composites in structural applications. In: Properties and performance of natural-fibre composites. Elsevier, pp 269–300. https://doi.org/10.1533/9781845694593.2.269
2. Peças P, Carvalho H, Salman H, Leite M (2018) Natural fibre composites and their applications: a review. J Compos Sci 2(4):66. https://doi.org/10.3390/jcs2040066
3. Jawaid M, Hamdan A, Hameed Sultan MT (eds) Structural health monitoring system for synthetic, hybrid and natural fiber composites. Springer, Singapore. https://doi.org/10.1007/978-981-15-8840-2
4. Siebel T (2020) Naturfasern eignen sich für den Karosseriebau. SpringerProfessional
5. https://www.arcd.de/magazin/auto/naturfasern-und-holz-im-fahrzeugbau/. Accessed 16 Jan 2025
6. Fogorasi M, Barbu I (2017) The potential of natural fibres for automotive sector—Review. IOP Conf Ser: Mater Sci Eng (252):012044. https://doi.org/10.1088/1757-899X/252/1/012044
7. Sanjay MR, Siengchin S, Parameswaranpillai J, Jawaid M, Pruncu CI, Khan A (2019) A comprehensive review of techniques for natural fibers as reinforcement in composites: preparation, processing and characterization. Carbohydr Polym 207:108–121. https://doi.org/10.1016/j.carbpol.2018.11.083. Epub 2018 Nov 26. PMID: 30599990
8. Zoppke H, Scherer M, Braband M, Dietz A (2020) Leichtbau-Pkw mit naturfaserverstärktem Monocoque, published in ATZ-Automobiltechnische Zeitschrift
9. https://www.hochschule-trier.de/hauptcampus/technik/projekte/protron/protron-evolution. Accessed 17 Jan 2025
10. Hashin Z (1980) Failure criteria for unidirectional fiber composites. J Appl Mech
11. Tsai SW, Wu EM (1971) A general theory of strength for anisotropic materials. J Compos Mater

Enhanced Malware Classification Using Machine Learning, Deep Learning, and Ensemble Models

S. B. Chandini and A. B. Rajendra

Abstract The classification and analysis of malicious software (malware) are increasingly important as Internet-connected systems grow, making them vulnerable to covert exploitation. Malware targets systems to secretly gather users' sensitive information and transmit it to attacker-controlled servers. This study presents an enhanced classification method for malware, evaluating a set of machine learning techniques, including traditional methods, ensemble and deep learning models. This study evaluates on static, dynamic, and Android datasets. Random Forest achieved 98.7% accuracy, Bagging 98.3%, and AdaBoost 97.9%, highlighting the effectiveness of ensemble models for robust malware detection.

Keywords Malware · Machine learning · Classification · Ensemble and deep learning

1 Introduction

Vulnerabilities, attacks, and threats from malicious software (malware) present significant security risks to computer networks and the Internet. According to a study Symantec [1], around 123 million connected devices record harmful vulnerabilities and malicious behaviors every second, highlighting malware as one of the most prominent threats to network security. Security experts cannot manage all instances of malware simultaneously, which emphasizes the need for an effective malware detection and classification strategy to prioritize these incidents.

With the pre-existing use of the Internet and mobile devices, malware attacks have increasingly targeted these platforms. The advanced software environments and sensory devices in these systems make it easier for exploitations to succeed [2]. Malware is malicious code that alters the reliability, usability, and confidentiality of the digital system. Malware and viruses are the most serious threats to networks [3]. To evade detection, malware creators employ various techniques to

S. B. Chandini (✉) · A. B. Rajendra
Vidyavardhaka College of Engineering, Mysuru, Karnataka, India
e-mail: chandini@vvce.ac.in

I. M. Jamadar et al. (eds.), *Integrating Sustainability in Modern Technologies*, Lecture Notes in Mechanical Engineering, https://doi.org/10.1007/978-981-95-9466-5_44

disguise their identities. Identifying specific families or types of malware has therefore become a challenging task. The massive quantity of new malware created daily poses difficulties for traditional security software, which struggles to keep pace. To address this issue, security researchers and vendors are increasingly adopting machine learning (ML) algorithms [4]. Techniques such as neural networks, regression analysis, Random Forests, Support Vector Machines, and logistic regression have been proposed to classify malware effectively [5].

Clustering and categorizing malware families are an important initial step for organizing and analyzing large quantities of files. Clustering criteria can aid in identifying potentially dangerous relatives of newly discovered files on our systems. Malware samples, organized by family relations, are essential for advancing research in this area, as they enable the clustering of new malware, allowing it to be identified and analyzed efficiently [6]. This need for adaptable malware classification algorithms continues to grow, especially as malware attacks on individuals and devices increase in the digital age [7, 8].

Malware classification assists analysts by grouping malware with similar structural features, enabling a more detailed examination of their behavior [9]. In ML-based malware detection, the first step is often classification, followed by feature extraction from malware images or code [10, 11]. Extracting features from collected malware data through machine learning has thus become an essential part of effective malware classification and detection [12].

Malscore is a malware classification system that integrates static and dynamic analysis to achieve a robust and efficient solution. It attained an impressive accuracy of 98.82% by leveraging probability scoring and machine learning [13]. Malware detection and classification remain crucial in maintaining network security. Malscore significantly reduced computational overhead, making it faster and more resilient to obfuscation compared to traditional methods. This paper investigates various machine learning techniques to determine the best classifier for malware classification.

1.1 Major Contributions

The primary contributions of this research are as follows:

1. **Evaluation of Traditional Techniques**: We analyze traditional classification methods, including ID3, Random Forest, and Support Vector Machine (SVM), as a baseline for malware classification.
2. **Deep Learning Model Assessment**: We investigate our datasets against deep learning models to understand the classification accuracy for complex malware patterns.
3. **Use of Ensemble Models**: Finally, ensemble approaches like AdaBoost and Bagging are applied to improve robustness and handle data diversity in malware classification.

2 Organization of the Paper

The flow of the research paper in Sect. 3 highlights previous research and related studies. Section 4 focuses on the methodology, detailing the machine learning approaches employed. Section 5 provides insights into the dataset and its characteristics. Section 6 elaborates on the experimental setup, results, and their interpretation. Sections 7 and 8 conclude the paper with key findings and suggestions for future work.

3 Related Work

The literature on malware detection techniques highlights in Table 1 shows progression from traditional methods to advanced machine learning approaches, each with inherent limitations. Signature-based detection is efficient for known threats but struggles with new or obfuscated malware. Static analysis examines code without execution, missing dynamic behaviors. Dynamic analysis observes runtime behavior but is resource-intensive. Machine learning techniques offer adaptability but require substantial computational resources and large datasets for training.

4 Research Work

Our research utilizes a range of supervised machine learning algorithms to facilitate accurate data classification and analysis. The techniques applied include ID3, Random Forest, Support Vector Machines (SVM), Decision Trees, k-Nearest Neighbors (KNN), AdaBoost, Bagging, Multilayer Perceptrons (MLP), Convolutional Neural Networks (CNN), Recurrent Neural Networks (RNN), and Long Short-Term Memory (LSTM) models. This flowchart outlines in Fig. 1 shows the machine learning pipeline for malware classification. It begins with data reading and preprocessing, followed by data splitting and scaling.

Various models (e.g., ID3, SVM, CNN, RNN, AdaBoost, LSTM, Random Forest) are trained, and their performance is evaluated using metrics like accuracy, recall, F1-score, ROC-AUC, TPR, and FPR. Finally, the process stops after evaluation. Each technique brings unique capabilities to the analysis, such as interpretability, robustness to noise, and effectiveness in handling high-dimensional or sequential data. The mathematical formulations and their corresponding variable explanations are summarized in Table 3. Table 4 summarizes the parameters and configurations used for various models implemented in the study, including Random Forest, SVM, CNN, RNN, and LSTM. Each model employs specific hyperparameters and techniques for effective classification. The CNN model utilizes convolutional layers for feature extraction with ReLU activation, while the RNN and LSTM models leverage their

Table 1 Findings and shortcomings of existing works

References	Findings	Shortcomings
Gibert et al. [14]	Introduced the HYDRA framework, combining manual and automated features to uncover hidden relationships in malware data	Relies heavily on manual feature engineering, which can be limited when dealing with obfuscated malware
Zhu et al. [15]	Developed an RGB image-based representation of dalvik executable (DEX) files to enhance classification reliability	Performance may degrade with certain obfuscation techniques
Al-Andoli et al. [16]	Created a hybrid deep learning system integrating particle Swarm optimization (PSO) with backpropagation (BP) for improved detection accuracy and efficiency	Requires significant computational resources, making it less suitable for resource-constrained environments
Gibert et al. [17]	Combined deep features, such as texture patterns, with manually crafted ones using a gradient boosting model, achieving improvements in malware classification	Limited generalizability to new or highly complex malware scenarios
Conti et al. [18]	Explored few-shot learning techniques using convolutional Siamese neural networks (CSNN) for low-data settings and found shallow few-shot (Shallow-FS) models more effective for larger datasets	Few-Shot models may not scale effectively for very large datasets
Gibert et al. [19]	Emphasized portable executable (PE) characteristics, byte N-grams, and API calls, using tools like the PE Parser to visualize these features as images	Visualization techniques may require additional preprocessing time
Kalash et al. [20]	Developed a deep learning framework for ransomware detection, focusing on rapid classification in organizational contexts	May struggle with detecting novel ransomware types
Kim et al. [21]	Proposed a CNN-based method to transform malware PE structures into image representations, allowing for real-time detection	Limited in detailed classification of malware types

respective recurrent units for temporal data processing. Random Forest and SVM are included for comparative purposes, providing traditional and linear kernel-based classification, respectively. The optimizer, loss function, and other hyperparameters have been consistently selected for performance optimization. Various supervised machine learning models along with their mathematical formulations and a comparison table summarizing their advantages and disadvantages are discussed in Table 2.

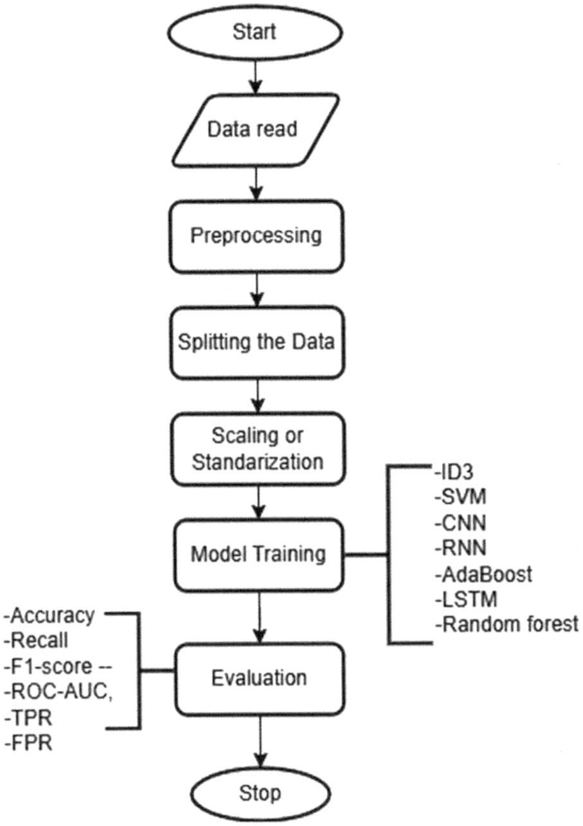

Fig. 1 Malware classification workflow

Table 2 Comparison of supervised machine learning models

Model	Advantages	Disadvantages
ID3	Simple and intuitive	Prone to overfitting
Random forest	Reduces overfitting	Less interpretable
SVM	Effective in high dimensions	Requires tuning of parameters
Decision tree	Easy to visualize	Sensitive to noisy data
KNN	Simple to implement	Computationally expensive for large datasets
AdaBoost	Improves weak classifiers	Sensitive to noisy data
Bagging	Reduces variance	Less interpretable
MLP	Can model complex patterns	Requires more data
CNN	Effective for image data	Requires large datasets
RNN	Good for sequence data	Prone to vanishing gradients
LSTM	Captures long-term dependencies	More complex architecture

Table 3 Comparison of supervised machine learning models with formulas and explanation

Model	Formula	Explanation
ID3	IG(D, A) = Entropy(D) − $\sum$($v \in$ values(A)) ($\|D_v\|/\|D\|$) Entropy(D_v)	D: Dataset, A: Attribute, D_v: Subset of D where $A = v$, Entropy(D): Dataset impurity
Random forest	$\hat{y} = (1/N) \sum(i = 1 \text{ to } N)\ h_i(x)$	$h_i(x)$: Prediction of tree i, N: Total trees in the forest
SVM	min (1/2 $\|\|w\|\|^2$) subject to $y_i\ (w^T x_i + b) \geq 1$	w: Weight vector, b: Bias, x_i: Features, y_i: Class label (+1/−1)
Decision tree	Gini(D) = 1 − $\sum(k = 1 \text{ to } K)\ p\_k^2$	p_k: Probability of class k, K: Total classes in the dataset
KNN	$d(x, y) = \text{sqrt}(\sum(i = 1 \text{ to } n)\ (x_i - y_i)^2)$	x, y: Feature points, n: Number of features
AdaBoost	$H(x) = \sum(t = 1 \text{ to } T)\ \alpha\_t\ h_t(x)$	α_t: Weight of tth classifier, $h_t(x)$: Prediction, T: Total classifiers
Bagging	$\hat{y} = (1/M) \sum(m = 1 \text{ to } M)\ \hat{g}_m$	M: Number of models, $\hat{g}_m$: Prediction of mth model
MLP	$y = f(W^T\ x + b)$	W: Weight matrix, b: Bias vector, f: Activation function
CNN	$Y_{i,j} = \sum(m,n)\ K_{m,n}\ X_{i+m,j+n}$	K: Kernel (filter), X: Input image, Y: Output feature map
RNN	$h_t = f(W_h\ h_{(t-1)} + W_x\ x_{t+b})$	h_t: Hidden state, $h_{(t-1)}$: Previous hidden state, x_t: Input at time t
LSTM	$C_t = f_t \cdot C_{(t-1)} + i_t \cdot \hat{C}_t$, $h_t = o_t \cdot \tanh(C_t)$	f_t, i_t, o_t: Forget, input, and output gates; C_t: Cell state, h_t: Hidden state

5 Dataset

The dataset [24] consists of static, dynamic, and Android malware samples as shown in Table 5. The static dataset includes 10,539 samples with 55 static features, which consist of permissions and hardware access requests like ACCESS_FINE_LOCATION, CALL_PHONE, and INTERNET. These permissions reveal the potential intent of applications based on the resources they attempt to access.

The dynamic dataset, containing 6,248 samples and 1,085 features, captures runtime behavior by monitoring API calls, processes, and network activities, while applications are executing. This subset helps in identifying real-time malicious activities that are often hidden in static analysis.

Lastly, a third dataset with 4,465 samples and 242 Android data samples, providing a more comprehensive view of each application's potential behavior. This combination of features enables robust malware detection by leveraging insights from both pre-execution properties and real-time behaviors.

Table 4 Hyperparameters used for various machine learning models

Model	Parameters	Values
Random forest	Number of estimators ("n_estimators")	100
	Random state	0
SVM	Kernel	Linear
	Probability	True
CNN	Convolution layer filters	32
	Convolution layer kernel size	3
	Activation function (Conv1D)	ReLU
	Pooling type	MaxPooling1D
	Pool size	2
	Dense layer activation function	Sigmoid
	Optimizer	Adam
	Loss function	Binary cross-entropy
	Metrics	Accuracy
	Epochs	10
	Batch size	32
RNN	RNN units	32
	Activation function (dense)	Sigmoid
	Optimizer	Adam
	Loss function	Binary cross-entropy
	Metrics	Accuracy
	Epochs	10
	Batch size	32
LSTM	LSTM units	32
	Activation function (dense)	Sigmoid
	Optimizer	Adam
	Loss function	Binary cross-entropy
	Metrics	Accuracy
	Epochs	10
	Batch size	32

Table 5 Summary of dataset

Dataset type	Samples	Features
Static	10,539	55
Dynamic	6,248	1,085
Android	4,465	242

6 Experimental Setup with Results and Discussion

The experiment was conducted on Google Colab, utilizing its cloud-based environment for machine learning tasks. Colab provides virtual machines with up to 2 CPU cores, 12 GB RAM, and optional GPU support, which includes access to NVIDIA Tesla GPUs depending on availability. The software stack included Python 3.x, TensorFlow, Keras, and Scikit-Learn for model development, along with Pandas and NumPy for data processing and Matplotlib for visualization. The free version of Colab was used, offering a convenient but resource-limited platform for running the experiments and documenting results as shown in Table 6.

The experiment involved evaluating various machine learning models on three Android malware classification datasets: static, dynamic, and a third dataset with unique features. Each dataset provided different perspectives on Android malware detection due to variations in the number of samples and features. The static dataset contained 10,539 samples and 55 features, emphasizing permissions and simple API calls. The dynamic dataset had 6,248 samples with a significantly larger feature set (1,085 features), which captured richer behavioral characteristics of the applications during execution. The third dataset included 4,465 samples and 242 features, providing an intermediate feature representation focused on device and system interactions.

The classifiers' performance showed consistently high accuracy across datasets, especially with models like Random Forest and Bagging, achieving over 99% accuracy in many cases. Random Forest was the top-performing model, demonstrating robustness with high precision, recall, and F1-scores across all datasets. Simpler

Table 6 Experimental support and system requirements

Aspect	Details
Platform	Google Colab (cloud-based environment)
Processor	Virtual machine with up to 2 CPU cores
RAM	Up to 12 GB, based on Google Colab session configuration
GPU support	NVIDIA Tesla K80, P100, T4, or similar (depending on availability)
Software requirements	Python 3.x, TensorFlow, Keras, Scikit-Learn, Pandas, NumPy, Matplotlib, Seaborn, adjustbox package for LaTeX documentation
Dataset access	Direct upload to Google Colab or access via Google Drive integration
Execution environment	Jupyter Notebook (in Google Colab)
Libraries used for experiment	Machine learning and deep learning: scikit-learn, TensorFlow, Keras; data processing and visualization: Pandas, NumPy, Matplotlib, Seaborn
Colab pro/free version	Free version of Google Colab was used, with limitations on session duration and compute resources

models like Decision Trees and K-Nearest Neighbors (KNN) also performed well but slightly lagged in consistency compared to ensemble methods. The deep learning models (CNN and RNN) achieved competitive scores, especially on dynamic data, but faced limitations in handling the smaller datasets due to their data-intensive nature. These models excelled in scenarios where feature richness allowed them to capture complex relationships, as evidenced by high precision in the dynamic dataset, and the results are shown in Table 7. Figure 2 shows confusion matrix reveals that the classifier correctly identified 1,057 positive and 253 negative samples, while misclassifying 14 negatives as positives (false positives) and 16 positives as negatives (false negatives). This indicates a strong overall performance with minimal misclassifications, highlighting areas for potential improvement in borderline cases.

7 Conclusion

In conclusion, the experimental results underscore the importance of feature diversity and dataset size in Android malware classification. Ensemble models like Random Forest and Bagging consistently achieved high performance across datasets, proving suitable for scenarios requiring reliable, high-accuracy classification. Deep learning models showed promise, especially in dynamic datasets, but require larger, more varied datasets to fully exploit their capacity for feature learning. These findings suggest that a hybrid approach combining static and dynamic features could further improve classification effectiveness.

8 Future Enhancements

Future work could explore integrating additional contextual data sources, such as network activity or user behavior patterns, to create even richer feature representations. Combining static and dynamic features may lead to improved model generalization and robustness. Additionally, leveraging transfer learning and more sophisticated neural architectures could enhance performance, particularly on smaller datasets where traditional deep learning models struggle. Implementing these methods may yield more accurate and versatile malware detection systems for Android platforms, enabling proactive malware defense in evolving threat landscapes.

Table 7 Classification results on static, dynamic, and android datasets

Classifier	Dataset	Accuracy (%)	Precision (%)	Recall (%)	F1-score (%)	AUC (%)
ID3	Static	97.6	98.0	96.0	96.0	96.3
ID3	Dynamic	98.2	98.0	97.0	97.0	96.7
ID3	Android	98.0	97.0	97.0	97.0	96.7
Random forest	Static	98.9	99.0	98.0	98.0	98.7
Random forest	Dynamic	99.1	99.0	99.0	99.0	98.9
Random forest	Android	99.3	99.0	99.0	99.0	99.0
SVM	Static	83.0	98.0	85.0	86.0	86.6
SVM	Dynamic	85.0	99.0	87.0	88.0	86.6
SVM	Android	83.5	98.0	85.0	86.0	86.2
Decision tree	Static	98.2	98.0	97.0	97.0	96.7
Decision tree	Dynamic	98.8	99.0	98.0	98.0	97.1
Decision tree	Android	98.5	99.0	98.0	98.0	96.6
KNN	Static	96.5	98.0	96.0	96.0	96.6
KNN	Dynamic	97.1	97.0	97.0	97.0	96.6
KNN	Android	97.7	97.0	97.0	97.0	96.6
AdaBoost	Static	97.1	97.0	96.0	96.0	96.3
AdaBoost	Dynamic	97.8	98.0	97.0	97.0	96.9
AdaBoost	Android	98.0	98.0	97.0	97.0	97.0
Bagging	Static	98.5	99.0	98.0	98.0	97.2
Bagging	Dynamic	99.0	99.0	99.0	99.0	98.7
Bagging	Android	99.2	99.0	99.0	99.0	99.0
MLP	Static	95.8	96.0	95.0	95.0	95.5
MLP	Dynamic	97.5	98.0	97.0	97.0	97.2
MLP	Android	97.0	97.0	97.0	97.0	96.8
CNN	Static	96.7	97.0	97.0	97.0	97.2
CNN	Dynamic	98.0	99.0	98.0	98.0	97.7
CNN	Android	97.8	98.0	98.0	98.0	97.6
RNN	Static	87.5	88.0	86.0	86.5	84.5
RNN	Dynamic	89.8	91.0	89.0	89.5	86.9
RNN	Android	88.0	90.0	88.0	88.5	86.2
LSTM	Static	85.4	87.1	83.0	84.9	72.3
LSTM	Dynamic	88.8	89.5	86.0	87.6	74.9
LSTM	Android	87.5	88.3	85.8	87.4	74.9

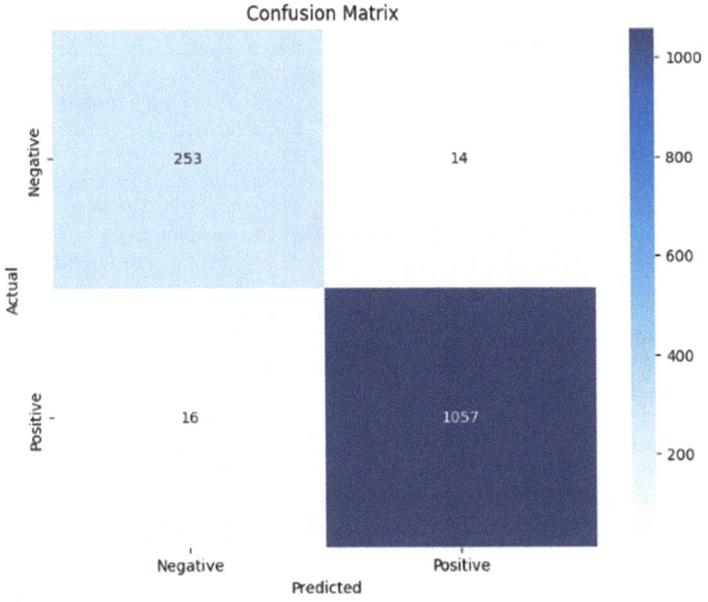

Fig. 2 Confusion matrix

References

1. Gibert D, Mateu C, Planes J (2020) HYDRA: a multimodal deep learning framework for malware classification. Comput Secur 95:101873. https://doi.org/10.1016/j.cose.2020.101873
2. Farhat D, Awan MS (2021) A brief survey on ransomware with the perspective of internet security threat reports. In: Proceedings of the 2021 9th international symposium on digital forensics and security (ISDFS). Elazig, Turkey, pp 1–6
3. Yadav B, Tokekar S (2021) Recent innovations and comparison of deep learning techniques in malware classification: a review. Int J Inf Secur Sci 9:230–247
4. Agrawal R, Stokes JW, Selvaraj K, Marinescu M (2019) Attention in recurrent neural networks for ransomware detection. In: Proceedings of the ICASSP 2019–2019 IEEE international conference on acoustics, speech and signal processing (ICASSP), Brighton, UK, pp 3222–3226
5. Butt UA, Mehmood M, Shah SBH et al (2020) A review of machine learning algorithms for cloud computing security. Electronics 9(9):1379. https://doi.org/10.3390/electronics9091379
6. Touqeer H, Zaman S, Amin R, Hussain M, Al-Turjman F, Bilal M (2021) Smart home security: challenges, issues and solutions at different IoT layers. J Supercomput 77(12):14053–14089. https://doi.org/10.1007/s11227-021-03825-1
7. Vasan D, Alazab M, Wassan S, Safaei B, Zheng Q (2020) Image-based malware classification using ensemble of CNN architectures (IMCEC). Comput Secur 92:101748. https://doi.org/10.1016/j.cose.2020.101748
8. Jain M, Andreopoulos W, Stamp M (2020) Convolutional neural networks and extreme learning machines for malware classification. J Comput Virol Hack Tech 16(3):229–244. https://doi.org/10.1007/s11416-020-00354-y
9. Mitsuhashi R, Shinagawa T (2020) High-accuracy malware classification with a malware-optimized deep learning model. https://arxiv.org/abs/2004.05258

10. Nisa M, Shah JH, Kanwal S et al (2020) Hybrid malware classification method using segmentation-based fractal texture analysis and deep convolution neural network features. Appl Sci 10(14):4966. https://doi.org/10.3390/app10144966
11. Luo J-S, Lo DC-T (2017) Binary malware image classification using machine learning with local binary pattern. In: Proceedings of the 2017 IEEE international conference on big data (big data), Boston, MA, USA, pp 4664–4667
12. Dey A, Bhattacharya S, Chaki N (2019) Byte label malware classification using image entropy. In: Advanced computing and systems for security. Springer, Berlin, Germany, pp 17–29
13. Xue D, Li J, Lv T, Wu W, Wang J (2019) Malware classification using probability scoring and machine learning. IEEE Access 7:91641–91656. https://doi.org/10.1109/access.2019.2927552
14. Johny JA, Vinod P, Asmitha KA, Radhamani G, Rehiman KAR, Conti M (2024) Deep learning fusion for effective malware detection: leveraging visual features. arXiv:2405.14311
15. Zhu H, Wei H, Wang L, Xu Z, Sheng VS (2023) An effective end-to-end android malware detection method. Expert Syst Appl 218:119593. ISSN 0957-4174. https://doi.org/10.1016/j.eswa.2023.119593
16. Al-Andoli MN, Tan SC, Sim KS, Lim CP, Goh PY (2022) Parallel deep learning with a hybrid BP-PSO framework for feature extraction and malware classification. Appl Soft Comput 131:109756. ISSN 1568-4946. https://doi.org/10.1016/j.asoc.2022.109756
17. Gibert D, Planes J, Mateu C, Le Q (2022) Fusing feature engineering and deep learning: a case study for malware classification. Expert Syst Appl 207:117957. ISSN 0957-4174. https://doi.org/10.1016/j.eswa.2022.117957
18. Conti M, Khandhar S, Vinod P (2022) A few-shot malware classification approach for unknown family recognition using malware feature visualization. Comput Secur 122:102887. ISSN 0167-4048. https://doi.org/10.1016/j.cose.2022.102887
19. Gibert D (2022) PE parser: a python package for portable executable files processing. Softw Impacts 13:100365. ISSN 2665-9638. https://doi.org/10.1016/j.simpa.2022.100365
20. Kalash M, Rochan M, Mohammed N, Bruce NDB, Wang Y, Iqbal F (2018) Malware classification with deep convolutional neural networks. In: 2018 9th IFIP international conference on new technologies, mobility and security (NTMS), Paris, France, pp 1–5. https://doi.org/10.1109/NTMS.2018.8328749
21. Kim S, Yeom S, Oh H, Shin D, Shin D (2020) Automatic malicious code classification system through static analysis using machine learning. Symmetry 13(1):35
22. Kumar S (2021) MCFT-CNN: malware classification with fine-tune convolution neural networks using traditional and transfer learning in internet of things. Futur Gener Comput Syst 125:334–351
23. Prajapati P, Stamp M (2021) An empirical analysis of image-based learning techniques for malware classification. In: Malware analysis using artificial intelligence and deep learning. Springer, Berlin, Germany, pp 411–435
24. Malware static and dynamic features VxHeaven and Virus Total. UCI Machine Learning Repository (2019). https://doi.org/10.24432/C58K6H

MediAssist-Multilingual Healthcare AI-Based Doctor Recommender System for Sustainability

B. S. Jayasri, Amogh Pramod Phatarpekar, D. Gagan Kumar, D. M. Tarun Raj, and M. B. Nagaraj

Abstract The problem of accessing quality healthcare services has always been one of the most significant challenges, especially for those with language barriers or limited access to advanced healthcare infrastructure. MediAssist is the state-of-the-art healthcare technology created to streamline and make the scheduling of medical appointments more efficient. It offers a symptom-based AI chatbot that diagnoses a patient before directing him to the relevant professional. It uses machine learning models like Random Forest, Naive Bayes, and SVC to detect the diseases. The technology further boasts of its multilingual AI voice assistant for a different user set, especially for older persons and less techie clients. MediAssist promotes green initiatives of paperless transaction in the form of prescriptions and reports, thus saving lot of trees from being cut down. In metropolitan cities, it take long hours to travel, and due to unpredictable traffic pattern, patients may not reach the hospital on time to meet the doctor in time causing inconvenience to both doctor and other patients, while use of proposed system is sustainable over longer period of time as not only it ensures timely appointment schedule and deadlines to be met but also saves lot of fuel assisting patients to meet the doctors at their convenient time on-line. Hence this also preserves fuel consumption and at the same time saves the environment from being polluted as it minimizes traffic nearby to the hospitals. Furthermore, the platform promotes simplicity by featuring a safe electronic payment system as a means to schedule appointments as well as cover claims. Schedule adherence is

B. S. Jayasri · A. P. Phatarpekar (✉) · D. G. Kumar · D. M. Tarun Raj · M. B. Nagaraj
Department of Compuer Science & Engineering, The National Institute of Engineering, Mysuru, India
e-mail: 2021cs_amoghpramodphatarpekar_a@nie.ac.in

B. S. Jayasri
e-mail: jayasribs@nie.ac.in

D. G. Kumar
e-mail: 2021cs_gagankumard_a@nie.ac.in

D. M. Tarun Raj
e-mail: 2021cs_dmtarunraj_a@nie.ac.in

M. B. Nagaraj
e-mail: 2021cs_nagarajmb_a@nie.ac.in

I. M. Jamadar et al. (eds.), *Integrating Sustainability in Modern Technologies*, Lecture Notes in Mechanical Engineering, https://doi.org/10.1007/978-981-95-9466-5_45

ensured through automated SMS or smartphone-based reminders. MediAssist places a strong emphasis on data security with the use of IPFS technology. This gives full ownership of their medical records to the patients and transparent, impenetrable access. It transforms healthcare administration by combining AI, IPFS, and user-friendly interfaces in a way that makes it efficient, safe, sustainable over longer period of time and accessible to everyone.

Keywords Health care · Sustainable · Paperless · Random forest · Naive Bayes · SVC · IPFS · Multilingual chatbot · Disease prediction · AI · Doctor appointment · Payment · SMS · Multilingual

1 Introduction

Healthcare systems have developed with technology, but the problems persist: inefficient appointment management, no multilingual support, and poor data security. MediAssist bridges the gaps by offering a holistic healthcare management platform through advanced technologies that improve accessibility, efficiency, and security for various users. It is tailored to make clinical scheduling appointments but is equipped with AI-driven functions, such as a symptom-based chatbot informing the patient to see the apt specialist and, most importantly, a voice assistant that is multifaceted—multilingual in nature, carefully catered for with elderly and no-tech-savvy people.

MediAssist was different from other systems, with integrated secure electronic payment solutions, reduced no-shows from prepayment, and SMS and app notifications to remind patients of scheduled appointments. Its architecture emphasized safety by using IPFS, which protected patients' data, while at the same time, granting users full owner-ship and access to their medical records. This delivered some of the most salient pain points in the healthcare industry and helped create trust and transparency in the system for end users.

This paper studies the conceptualization and development of MediAssist, listing the innovative functionalities, and among them the AI-based decision-making process, IPFS-backed data security, and user interface-friendliness that form an integral part of the platform. The platform aims to change healthcare interactions and ensure smooth experiences for both patients and providers. By relating the exploration of its modular architecture and future development prospects, such as advanced integration with AI and multilingual support, the potential of MediAssist to transform healthcare delivery in a rapidly advancing technological landscape should emerge.

2 Literature Survey

Mahandi and Jacob [1] present a full medical appointment application using React, MongoDB, Node.js, and Express.js. The application focuses on scalability, security, and user-centric design, that is, encryption of a password and SSL/TLS communication, automatic reminders, issues like data security, scalability of the system, and functional integration, such as telemedicine and EHR, and MediAssist can construct these ideas by putting the IPFS into security records and multiple languages for obtaining tolerance.

Subhi et al. [2] demonstrate the development of an easy-to-use medical appointment system using PHP, CSS, HTML, and MySQL. This deals with the inefficiencies of manual booking, increases security, and meets data protection legislations. This system improves intercommunication between doctors and patients and reduces disturbances. This work on accessibility and scalability is more important as MediAssist integrates voice assistance and IPFS to meet the demands of modern health care.

Singh et al. [3] "Get-A-DOC ". The system uses a common NosQL Mongododb database to match the user based on preferences. It reduces the complexity of finding a doctor while ensuring scalability and user-centric design. MediAssist uses similar principles but leverages artificial intelligence to provide personalized medical recommendations and automated insights.

Akhilesh et al. [4] present an Android application powered by OpenAI's GPT-3.5 for symptom-based diagnosis. The paper reaches a high accuracy with quick reaction time, focusing on NLP and AI. The diagnosis of NLP is incorporated into MediAssist, allowing support for multilingual users.

Rallapalli et al. [5] proposed a system that uses MongoDB to have efficient and real-time interaction by the patients and doctors. Its client-server architecture helps in efficient appointment scheduling and data management. MediAssist extends this approach with IPFS technology to improve data integrity and seamless records management.

Avhale et al. [6] this web-based room reservation system simplifies scheduling with attributes of time slot booking, cancelation along with notifications. Physicians are on the receiving end for automated reporting and patients find the interface user-friendly. The MediAssist works on a similar foundation, but enhances it automated reporting by integrating with calendar.

Jhansi Ada et al. [7] MedicalHelpNet combines content-based filtering for doctor selection, live chat support, and first aid tips. Its multi-feature platform simplifies appointment management and improves patient experience. MediAssist builds on this by integrating AI chatbots and multilingual assistance for broader accessibility.

Moulya et al. [8] this work-based system allows easy access by distant patients with automated appointments, and manual steps are minimized. MediAssist is based on this idea along with AI-generated alerting capabilities and secure payments for the enhanced smoothness.

Geeta [9] a web application of real-time physician scheduling facilitates ease and averts long hours of waiting in lines. This idea is continued in MediAssist by extending the same IPFS-based data management and AI-powered critical case prioritization.

Nyayade et al. [10] "Smart Appointment Booking System" automates manual booking processes with secure databases and patient–doctor interaction features.

MediAssist enhances this model with IPFS for tamper-proof medical records and artificial intelligence for continuous patient support.

Baras et al. [11] the project proposes an IPFS-based EHR system that uses Ethereum and smart contracts for secure data sharing. MediAssist uses IPFS to enable secure, transparent medical record management, giving patients greater control over their data.

Dr. Banerjee and Vinooth [12] a disease prediction system by Random Forest and KNN gives 95% accuracy, using sentiment analysis, it suggests doctors. MediAssist utilizes the machine learning concept for the disease prediction system, and a multilingual AI chatbot increases its reach.

Himel et al. [13] The "MedLingua" framework introduces a multilingual chatbot that uses hybrid ML models and NLP for symptom recognition and medical advice. MediAssist integrates similar multilingual AI technology to ensure inclusivity and adaptability to diverse users.

Meza et al. [14] the research is studied with Amazon Echo to be implemented in voice-based appointment scheduling. It offers flexibility in voice interaction architecture and eliminates development complexity. MediAssist aligns this idea by inducting the support of multilingual voice management in appointment schedule management.

Butada et al. [15] "Smart Physician Assistant System" provides contactless appointment scheduling with facial recognition and voice commands, addressing challenges such as COVID-19 risks. MediAssist builds on this foundation, using advanced voice AI and facial recognition to provide a seamless user experience.

3 Problem Statement

Language barriers prevent the patients from discussing their medical issues correctly and stop letting the patient understand the explanation of treatment options delivered by a doctor, resulting in misinformation and delayed medicines. Most of the available hospital sites, such as Manipal Chatbot, Practo, and Dello, do not support multilingual operations appropriately, thereby being inaccessible to a non-native-speaking patient or diverse linguistic backgrounds. This gap in communication creates problems in delivering efficient healthcare services within the right timeframe. Incomplete payment solutions integrated into these systems also complicate the process for patients to make payments, leading to a loss of convenience and efficiency. Higher no-show rates exacerbate this problem by pushing back the schedule for the patients, creating schedule disruptions, and limiting the effectiveness with which healthcare providers can treat patients.

4 Existing System and Proposed System

4.1 Existing System

Efficiency is found in the hassle of existing management systems related to patient appointments in the healthcare sector. Manual booking processes are simple but bulky, error-prone, and do not integrate with electronic records, which creates scheduling problems. Practicality is witnessed from mobile applications, yet normally provided features such as supporting multiple languages and ease of use of a typical user of the more aged or skilled technology are missing. AI-driven chat robots come with automation but struggle to beat in complex queries and lack full language support, meaning they will disappoint the users. Moreover, classical systems cannot guarantee the security of data because sensitive patient information may be vulnerable to violations. The process is frustrating due to long wait times, high costs, and limited accessibility, which negatively affect both patients and healthcare providers.

The disadvantages of the current system include:

- Time-consuming process for patients and medical staff.
- High absenteeism due to ineffective reminders.
- Data security issues, especially in manual systems.
- Limited access for users who have no knowledge about technology or experience language barriers.
- Personal service is also not available hence the patients dissatisfy the services.

4.2 Proposed System

MediAssist aims to reduce inefficiencies in the management of healthcare appointments using advanced technologies. The proposed system offers a multilingual assistance module using NLP techniques basically uses open AI gpt4.o large language model and real-time translation APIs. Thus, the proposed system will be made available for all linguistic users. Moreover, it supports an AI-powered chatbot, enabling the scheduling of appointments, analysis of symptoms, and multiple languages via text and voice inputs based on supervised learning models like Naive Bayes, SVC, and Random Forest. It utilizes IPFS technology in managing patient data that is secure decentralized storage, encryption, and smart contracts for enhanced privacy and transparency. Automated reminders and notifications using a backend scheduler and calendar API dynamically update the information of the patient's appointment, thus reducing no-shows.

Benefits:

- Accurate and real-time management of appointments.
- Increased security with IPFS-based data protection.
- Increased accessibility for seniors and tech-savvy users.

- Reduced absenteeism through automatic reminders.
- Multi-language support for seamless inclusivity.

5 System Design

5.1 High-Level Architecture

The high-level architecture of our system consists of multiple interconnected components-as shown in Fig. 1 that collaborates to form an all-encompassing career development platform. Mediconnect's architecture as shown in Fig. 1 revolves around the idea of providing a really intuitive and secure way for patients and doctors. At the forefront is ReactJS that provides UI motivation. All this will guarantee a smooth experience due to the presence of a dashboard, AI chatbots, and voice assistants, helping users make access to features, schedule meetings, and interact with the system. The backend has been integrated with an AI/ML engine, Blockchain for secure data management, MongoDB as the main database to handle requests, consultations, doctor consent, contracts, and payment management. It has been integrated with a strong notification system using SMS, email, or reminders. IPFS encrypts and protects sensitive medical records. An AI/ML engine analyzes symptoms and interacts with doctor recommendations through chatbots. Access is also being developed because users can search for web work information via Voice commands using voice assistants. Generally, this architecture seemed to represent a healthy experience in positive, safety, and efficiency.

5.2 Low-Level Architecture

The low-level architecture is divided into three main modules, each designed to fulfill distinct roles within the platform.

Voice Interaction Module

The voice interaction module shown in Fig. 2 allows users to interact easily and command processes through the voice assistant. This starts when the user initiates a voice command that is first captured as audio input using Google Webskit for multilingual support to assist layman/illiterate in their native language. Using NLP algorithms, the captured multilingual speech is transformed into text, which is then translated into machine readable format. Using intent analysis, the intent of the user is determined from the query asked, and consequently, a reply is formed for the query. Such a reply will be both a spoken response using the voice assistant and a text response that is displayed in the user interface. This is in order to help users who have a hard time using systems and to make the MediAssist system more convenient

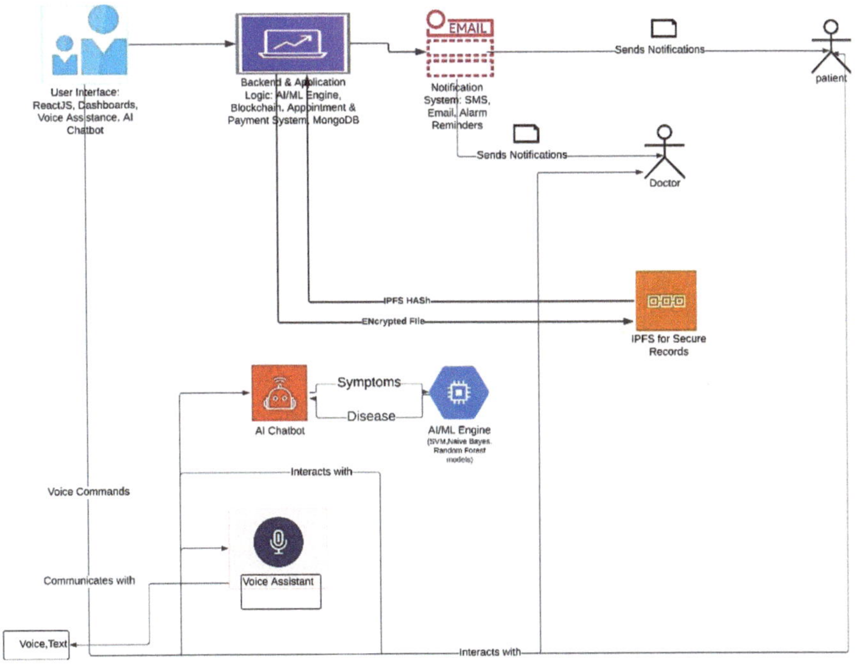

Fig. 1 High-level architecture

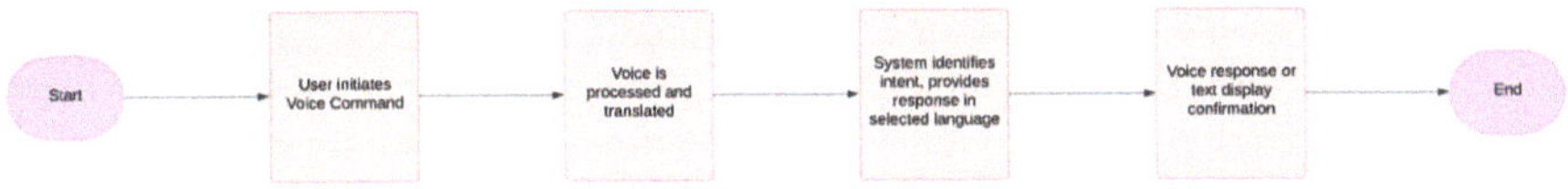

Fig. 2 Multilingual voice interaction module

and user-friendly. This structured workflow enables efficient and accurate handling of voice-based interactions, making MediAssist user-friendly and inclusive.

Chatbot Module

The chatbot interaction process in MediAssist as shown in Fig. 3. The user's engagement with the AI chatbot is seamless in the whole interaction. First, the interaction begins by the user reaching out and starting a chat session with the system. Then, the chatbot asks for the description of symptoms from the user to get the necessary information for analysis. The trained machine learning models—SVC, Naive Bayes, and Random Forest—analyze the symptoms and match the possible diagnosis through a medical database. Finally, the chatbot will recommend a suitable doctor or advise on further steps or even facilitate appointment scheduling. The process culminates with the user receiving actionable and context-aware guidance for an efficient and personalized experience.

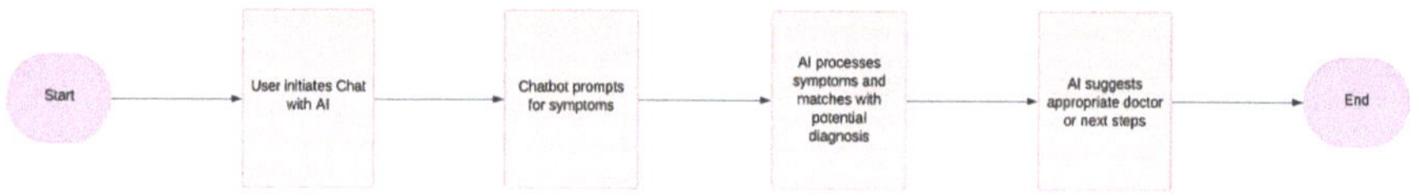

Fig. 3 Chatbot module for symptom diagnosis

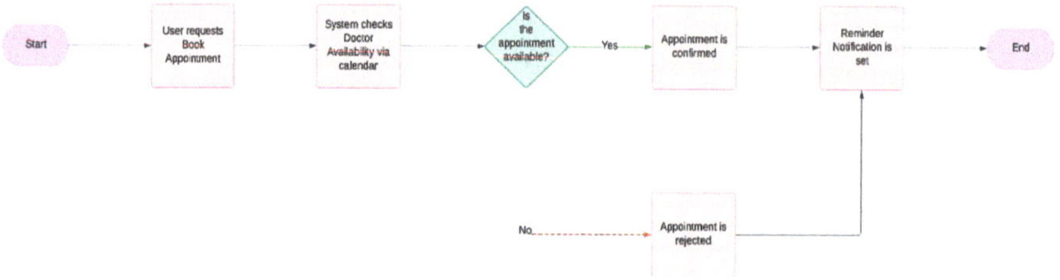

Fig. 4 Appointment booking module

Appointment Booking Module

Booking an appointment in MediAssist shown in Fig. 4 illustrates the workflow of smooth scheduling and appointment management. The user requests for a booking, and the system starts the scheduling module. It then checks if appointments are available, as it does not double-book or cause any conflicts through consulting its scheduling database or integrated calendar. If the slot is available, the appointment will be confirmed and booking details will be updated, or otherwise, it will notify the user that it cannot schedule the appointment. The system sends reminders through SMS, email, or in-app alert to remind the user on time. After sending the notification, the process will end, leaving the system prepared for further interaction or queries.

5.3 Use-Case Diagram

A Use-Case Diagram is a graphical representation of the functional requirement of a system, emphasizing interactions between users, or actors, and the system itself. It shows how different ways a user might interact with the system, and how the system might respond to that interaction. Understanding the use cases and their relationships by the development team ensures that the system meets its users' needs and functions properly. Key elements of a Use-Case Diagram are actors, use cases, and their relationship.

Doctor Use-Case Diagram

Use-Case Diagram in Fig. 5 illustrates the functionalities a doctor can do within a healthcare system are the following: make follow-up recommendations to patients, manage his/her availability by creating appointment times and schedules, viewing patient medical records to review histories and treatment plans, and communication with patients via secure means like email or messaging. The workflow begins when the doctor initiates action by selecting a use case and the system proceeds to process this request and furnish the relevant information or service. This diagram will represent the visual expression of how the doctor interacts with the system, especially in terms of the core services that it avails to ensure streamlined healthcare operations.

Patient Use-Case Diagram

Use-Case Diagram in Fig. 6 shows the relationship between a patient and a chatbot in a healthcare system. The actor patient represents the user, while the actor chatbot represents the system that provides different functionalities. Some of the main use cases are multi-language support, where the chatbot can assist the patient in several languages and functionalities like sending notifications such as appointment reminders or test results. Patients can view their medical records, book an appointment with a healthcare provider, and get support from the chatbot while booking appointments through guidance and information. The chatbot also does pre-diagnosis based on symptoms and provides preliminary advice. The diagram shows the associations between the patient and the system's functionalities; an "Include" relationship shows that "Assist with Appointment Booking" includes the "Book Appointment" use case, which ensures seamless interaction and service delivery.

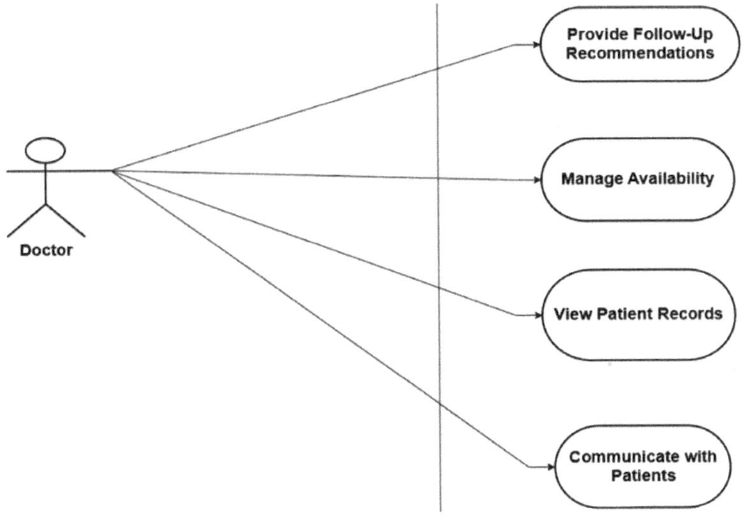

Fig. 5 Use-case diagram for doctor

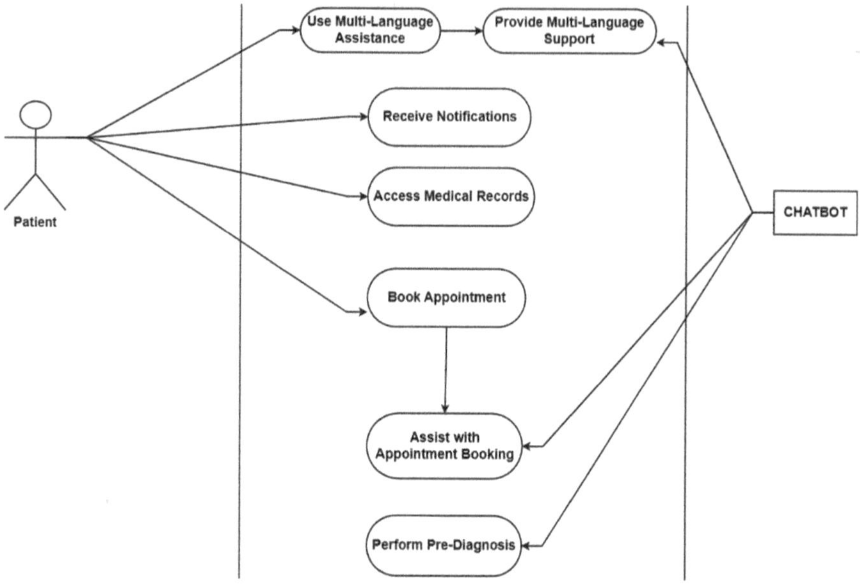

Fig. 6 Use-case diagram for patient

5.4 Sequential Diagram

Sequence diagrams describe the flow of interactions between system components. In Fig. 7, sequence diagram illustrates the step-by-step interactions between the patient, MediAssist system, AI chatbot, physician, and IPFS, describing a typical patient journey. It starts with patient registration or logs in to the MediAssist system that checks if the system is access to. The patient continues to arrange an appointment, the patient fills in the visit details in the system, and the doctor is notified if the availability is confirmed. If the patient also interacts with the AI chatbot to advise them about their symptoms, this is processed to issue an appropriate specialist referral. The consultation status is updated to "paid" and securely stored on the payment gateway. At the end of the consultation, the doctor uploads an e-prescription that is encrypted and securely stored in the IPFS for privacy and integrity. The system may notify the patient's information of the subsequent contract. This sequence is conducive to security transactions, proposals made by AI, and encrypted record management to ensure trouble-free and effective experience with medical treatment.

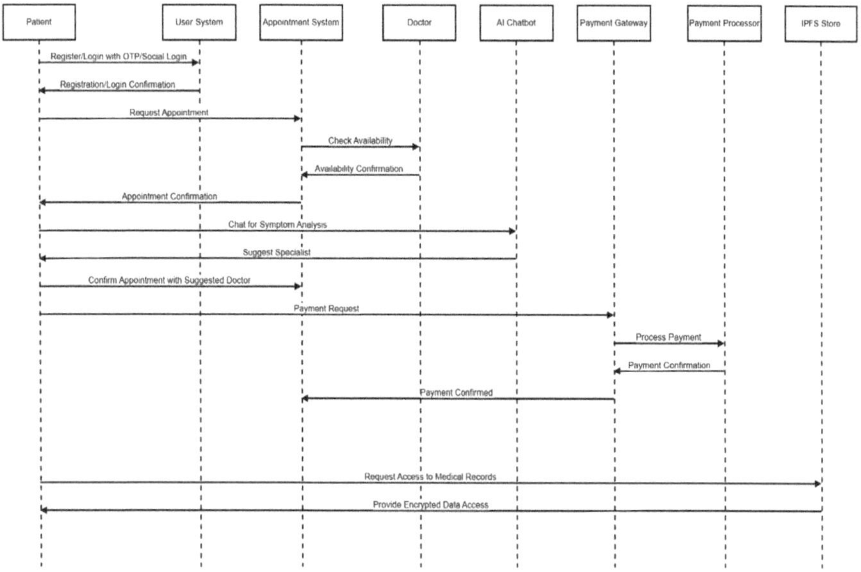

Fig. 7 Sequence diagram of end-to-end system

6 Implementation

6.1 Phase I (Project Initiation)

As moto of the project is recommend a specialist doctor based on the user symptoms, we need to first predict disease based on disease we can recommend doctors. So in order to achieve the goal, the first step of phase 1 involves collecting the desired dataset [16] and preprocessing by cleaning the data to make them ready for the analysis. From well-defined problem statement and a preprocessed dataset, appropriate algorithms are selected for sentimental analysis of symptoms as described by patients. It will predict trends like month-on-month increases or decreases in healthcare appointments or cases of diseases by analyzing data from various symptoms. At the stage of data preprocessing, the dataset is imported with the Pandas library, which removes columns that have no data present. The dataset is clean with no missing values and binary in format expressing every feature in terms of 0s and 1s. Another important classification check is to see if the target variable is well-balanced since the dataset for the solution could be imbalanced. This will be measured through a bar plot; the graph shows that each disease case spreads evenly across the data. Following this, it does not need extra techniques for balancing since it can balance out equally without this necessity. For this dataset, "prognosis," which is in object format, at its first stage, is a necessity to be changed and become acceptable in terms of fitting well to the needs for machine learning. That change has been made through label encoding, where each category of disease is mapped to a unique integer so that the model could understand and process the data correctly. All the above preprocessing steps ready the dataset for real-time disease classification.

Below is the Label Encoder snippet:

```
encoder = LabelEncoder()
data["prognosis"] = encoder.fit_transform(data["prognosis"])
```

6.2 Phase II (Project Planning)

When users describe their symptoms in any language, our system uses OpenAI's GPT-4.0 to analyze and extract relevant medical symptoms. It processes and refines the input for clarity before comparing it with a vast medical database to predict possible diseases. Advanced algorithms identify patterns and correlations, generating a list of likely conditions with probability scores. Based on the predictions, the system recommends a specialist doctor considering expertise, availability, and location. The process is seam-less, intuitive, and accessible in any language. Users receive real-time medical insights, enabling informed healthcare decisions. This AI-powered approach bridges the gap between symptom description and professional diagnosis.

We used Support Vector Classifier, Gaussian Naive Bayes Classifier, Random Forest Classifier for predicting disease for the symptoms analyzed by the NLP. These types of models are popular for disease predictions because their frameworks can be used to make sense of complex relationships inherent in medical data and classify patients. Since these types of models can work effectively with higher dimensional features, they can provide better prediction of health outcomes.

6.3 Phase III (Project Execution)

The dataset [16] was divided into training and test sets, using 80% for training and 20% for testing, which is a widely adopted guideline in the research community. No missing values were observed in the course of the data processing steps, and the "diseases" column was ready using label encoding. At first, this column contained text labels for different diseases; a Label Encoder was used to transform each disease label into a numeric value corresponding to it by giving it an index. These preprocessing techniques guaranteed that the data was properly formatted; in machine learning, the dataset must be complete and properly labeled. Regardless of whether it's for machine learning or deep learning, the training procedure does not penalize models based on variations in observation or prediction data.

The model will then be trained in $k - 1$ folds and checked or tested in the final fold. K-Fold Cross-Validation will complete k iterations so each fold will be tested once.

K-Fold Cross-Validation will be applied on the training dataset for the purpose of improving the generalization ability and robustness of the model. In the K-Fold

```
=============================================================
SVC
Scores: [0.9959432 1.         1.         1.         1.         1.         1
 1.         1.         1.        ]
Mean Score: 0.9995943204868155
=============================================================
Gaussian NB
Scores: [0.9959432 1.         1.         1.         1.         1.         1
 1.         1.         1.        ]
Mean Score: 0.9995943204868155
=============================================================
Random Forest
Scores: [0.9959432 1.         1.         1.         1.         1.         1
 1.         1.         1.        ]
Mean Score: 0.9995943204868155
```

Fig. 8 Result of K-fold cross-validation

Cross-Validation training dataset, each data point is used in training and validation multiple times on each iteration while training or modeling with $k - 1$ folds and testing or validating with the final fold k. The Cross-Validation results are depicted in Fig. 8.

Results and Performance

Each classifier was trained in parallel and was evaluated based on different metrics like accuracy, precision, and recall. In addition, the analysis of the confusion matrix has also been utilized in the determination of classification errors as well as the evaluation of overall performance. All three classifiers, namely Naive Bayes, Random Forest, and SVC, are tested on datasets that were divided as training and testing. All three resulted in an accuracy of 100% in both training and testing as shown in Fig. 9, and the accuracy, precision, and recall of all the models are shown in Fig. 10. Since the model is overfitted and should work for different unseen data, as we discussed above for robust application we take a mod of all three models shown in Eq. (1). From these results, one could clearly say that the preprocessing steps, the applied classifiers, and overall system were very effective in predicting disease prognosis. After predicting the disease, now specialist is been recommended to the user by looking in to the database which is the final motto.

$$\text{final\_preds} = \left[\text{stats.mode}\left([\text{i}, \text{j}, \text{k}]\right)[0] \text{ for i, j, k in zip(SVc\_preds, nb\_preds, rf\_preds)}\right] \tag{1}$$

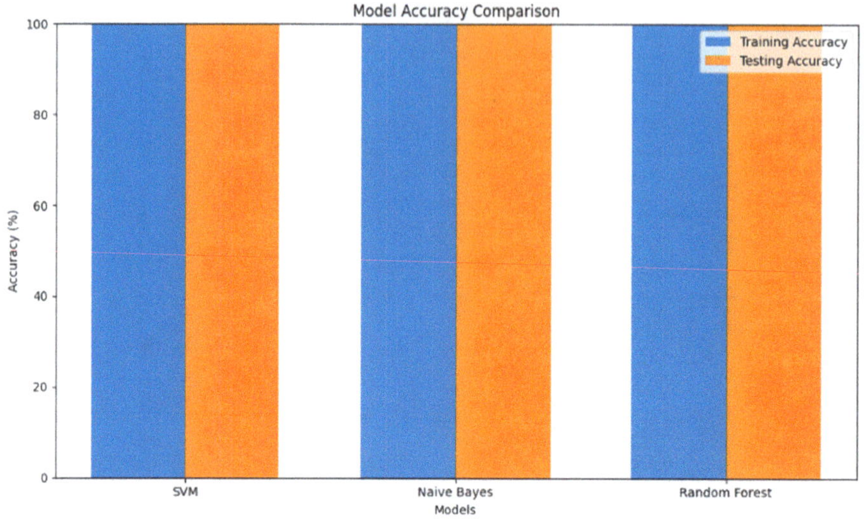

Fig. 9 Accuracy comparison of three models

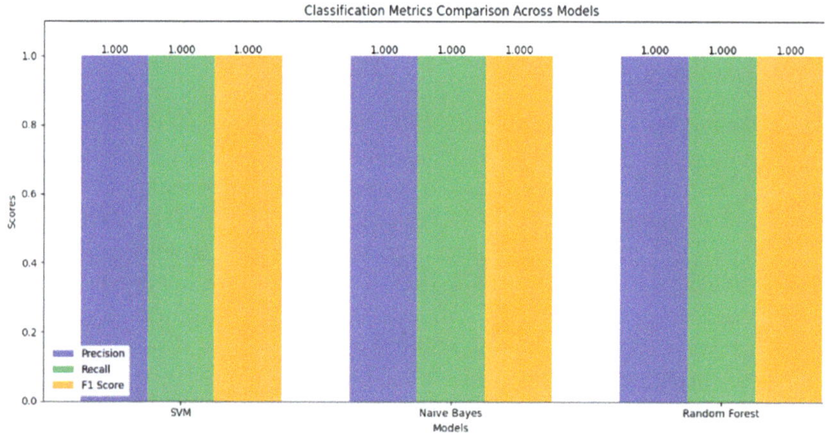

Fig. 10 Classification metrics comparison across model

6.4 Phase IV (Project Execution After Prediction)

Once the specialist doctor has been recommended to the patient based on the symptoms described. Then patient can check for additional feedbacks given by other patients and select a vacant slot comfortable for the patient enabling online consultancy, after making the required payment via gateway. The payment gateway model that will allow only one active user to select a slot at a given point of time, while blocking the concurrent user request in real time based on the above feature it is possible for MediAssist to handle concurrent request. After the successful completion of payment, the automatic remainder is triggered. The user can chat with the doctor via chats located in user dashboard. The patient can also update his/her profile by adding the allergies and previous medical reports if any. The medical reports are then stored in InterPlanetary File System (IPFS), and the corresponding IPFS hash is updated in user's database. Similarly after the diagnosis a medical report is generated by the doctor per patient and is uploaded as IPFS hash in users dashboard along with valuable feedback that is visible to user.

7 Testing

MediAssist is a health access system connecting patients with healthcare providers. The project has been built using React for frontend development, Node.js for backend support, and MongoDB and Firebase for database management. The testing report highlights the methodologies used and outcomes in testing the project to ensure the platform performs its intended functionality without any hiccups and follows the project's requirements while being user-friendly. Table 1 shows few examples of user-generated test cases pertaining to unit testing for containing both fail and pass cases.

The system ensures real-world usability by integrating patient feedback into the application and as a demonstration the screenshot has been included in Fig. 18. Collecting insights directly from both patients and doctors using this application helps to adopt agile process allowing the platform to be refined continuously for improved accessibility, accuracy, and efficiency.

8 Results

A classification report or a performance evaluation matrix gives an overview of the metrics by which the strength of the well-trained classification model can be quantified. For instance, its precision, recall, F1 score, and support explain the model's accuracy over its different classes.

In this report, we present the performance of the disease prediction model trained with the Random Forest (RF), SVC, and Naïve Bayes algorithm. The model is trained on the input features of patient medical records and symptom data. These input contains of textual descriptions of symptoms stored in structured data type. The classification results, as computed by Equation (2), reflect the performance of the model in predicting specific disease categories:

$$F1 = 2 * (\text{precision} * \text{recall})/(\text{precision} + \text{recall}), \tag{2}$$

Table 1 User-generated test cases

Test case ID	Test case description	Test steps	Expected result	Actual result	Pass/ fail
01	User registration	1. Enter valid user details in registration form 2. Submit registration form	User account created successfully	User account created successfully	Pass
02	Disease prediction	1. Enter symptoms in the prediction form 2. Submit the form for analysis	Predicted disease based on entered symptoms	Predicted disease based on entered symptoms	Pass
03	User login	1. Enter valid credentials	User logged into dashboard	User logged into dashboard	Pass
04	Voice to text (valid input)	1. Speak a clear sentence into the microphone 2. System converts voice input to text	Text accurately transcribed from voice input	Text accurately transcribed from voice input	Fail
05	Book appointment	1. Select a doctor. 2. Choose an available timeslot	Time slot selected. Appointment booked	Time slot selected. Appointment booked	Pass
06	Update doctor profile	1. Edit qualifications or experience	Changes saved successfully	Changes saved successfully	Pass

where:

- Precision = TP/(TP + FP).
- Recall = TP/(TP + FN).
- TP = True positives (correctly identified positive cases).

Performance comparison of four different models: SVC, Naive Bayes, Random Forest is fitted on complete dataset and evaluated on a binary classification task using both accuracy and F1-score as performance metrics is shown in Fig. 11, and by doing mod of three models in Eq. (2) as discussed implementation, it is given as combined model. The best model is the combined model with 100% accuracy and F1-score. SVC and Naive Bayes also have an accuracy of 100% but slightly lower F1-scores. Random Forest shows a minor drop in performance with 97% accuracy and F1-score.

Confusion Matrix

It is a table that shows how many were correct, not the predictions for each class. Figure 12 is the confusion matrix combined model. A confusion matrix is an evaluative visualization method to assess how well a classifier performs on a given classification task. Here, each row of the input matrix represents a correct or actual class name, and each column contains the predicted class. The diagonal line elements from the upper left corner indicate the number of cases of the correct classification

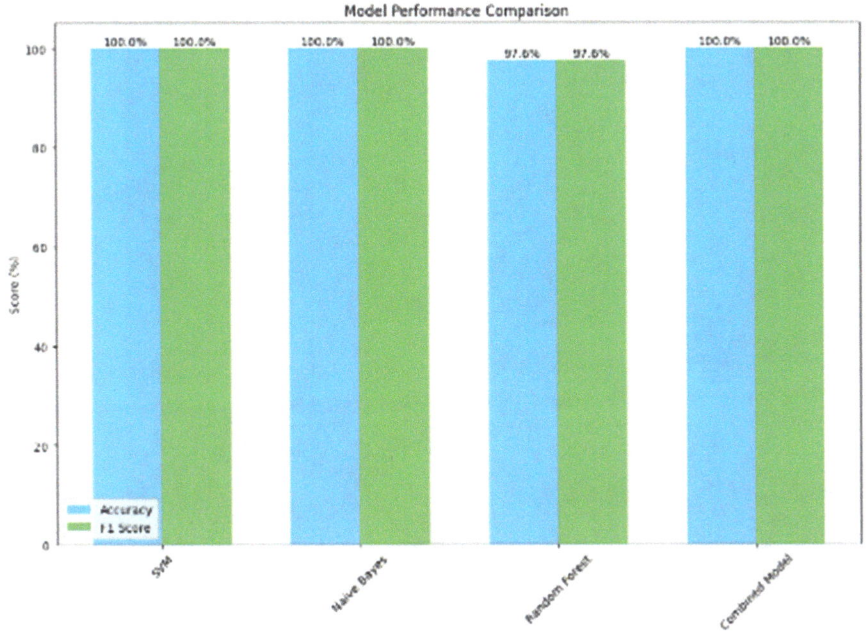

Fig. 11 F1-score comparison of all models

of each class and the ethomic elements denote the classification of errors. As we can see in Fig. 12 that since accuracy is 100%, only diagonal of confusion matrix is non zero, and rest is zero.

Model Agreement Analysis

Model agreement analysis evaluates the agreement of a set of multiple predictive models on the classification decisions. The agreement between models is analyzed as shown in Fig. 13, keeping in mind that this is for the specific prediction task. In view of this description, let's break down the model agreement analysis: Complete Agreement (97.6%): The largest section of the pie representing that in 96.1% of the cases the models in the analysis were making the same decision/classification. Partial Agreement (2.4%): A tiny section represents some of the cases where some models have agreed on the classification, while others did not. It might indicate that models are uncertain or disagreeing with each other. No agreement (0.0%) because we always that mod, so no disagreement. This depicts a model scenario whereby models are in sort of an agreement except in extremely rare cases of partial agreement or no agreement.

ROC Curve and AUC Score

ROC curve helps in visualizing the performance, while AUC provides an overall measure of model performance. With AUC = 1, it implies that all the three models

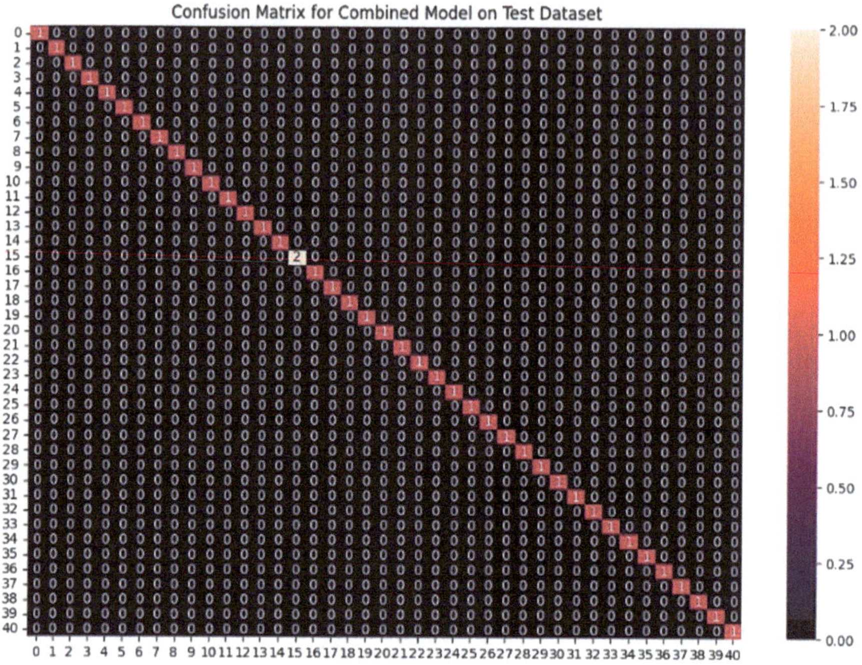

Fig. 12 Confusion matrix of combined model

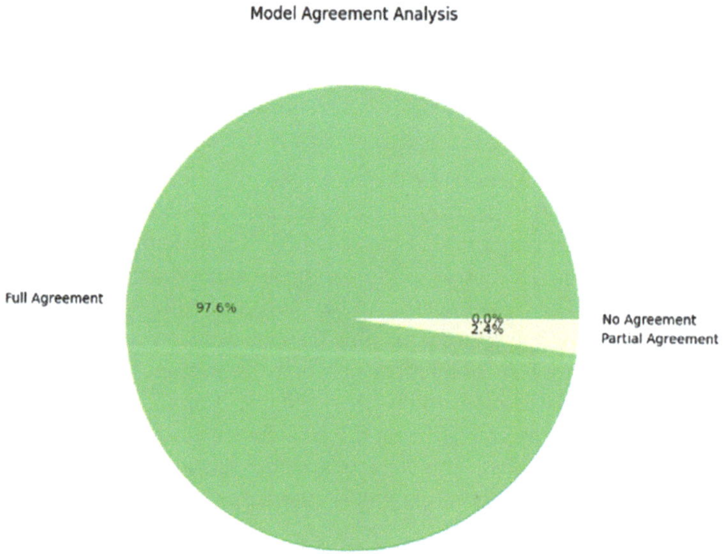

Fig. 13 Model agreement analysis

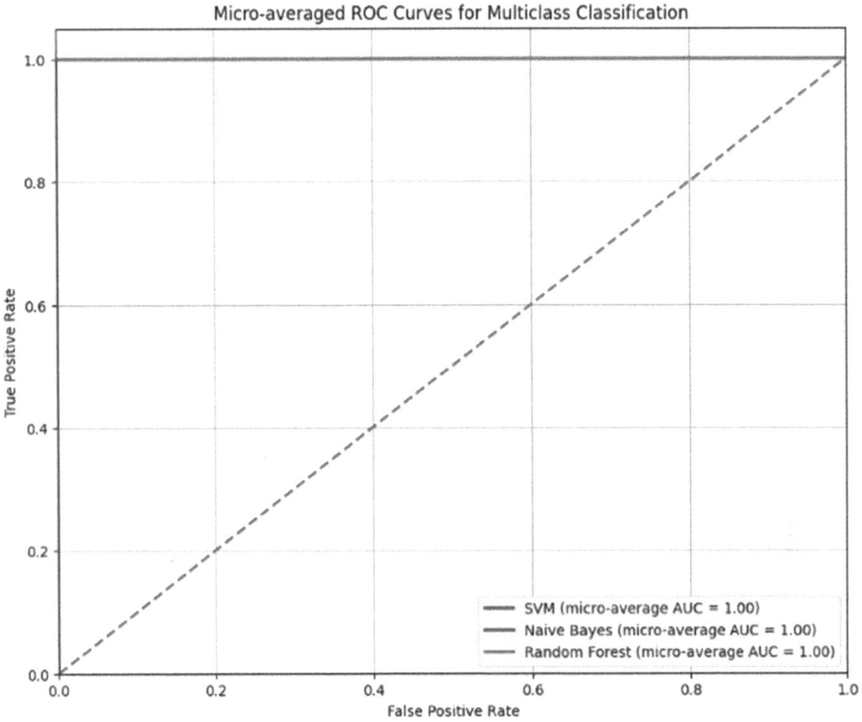

Fig. 14 ROC curve

used are perfect classifier and its ability to distinguish between positive and negative classes are shown in Fig. 14. ROC curve helps us to understand different decision thresholds on how well the model separates positive and negative classes.

Figures 15 and 17 show the result of disease predicted and based on the symptoms entered in user's desired language on the application developed. Then also the recommends the specialist doctor for the predicted disease. User can then select the doctor and take an appointment on his desired time slot provided by doctor as shown in Fig. 16.

9 Conclusion and Future Enhancement

Hence the proposed system not only preserves fuel consumption but at the same time saves the environment from being polluted as it minimizes traffic nearby to the hospitals enhancing sustainability and ensures paperless transactions resulting in green initiatives too while conserving both energy and environment at the same time. MediAssist will revolutionize traditional healthcare management systems by merging the latest available technologies, artificial intelligence, IPFS, and multi-language

Fig. 15 Prediction result in chatbot (Kan)

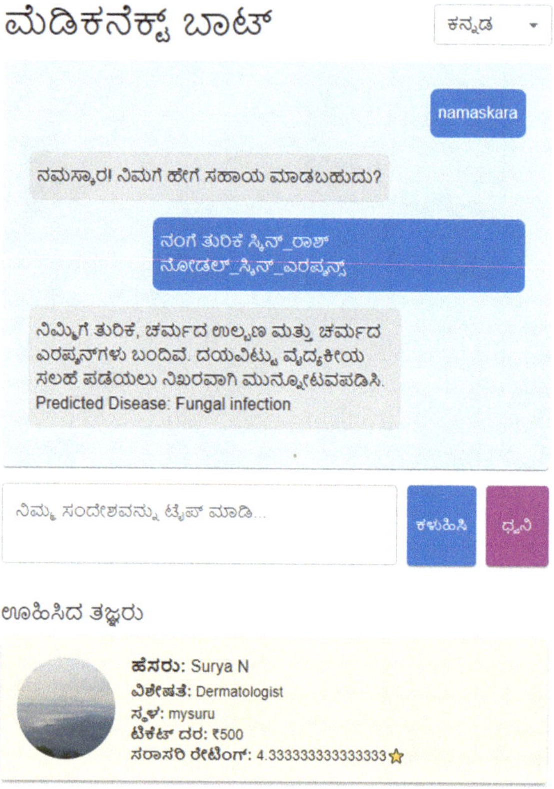

support, to achieve an inclusive and efficient platform while maintaining security. Key features including, AI-powered pre-diagnosis using a chatbot, multi-language voice assistance, secure e-payment systems, automated reminders, all designed to cater for a wide spread of users cut across the senior and non-technological adept users. The use of the IPFS technology ensures strong security for data and therefore gives patients control over their medical records while adhering to the set of regulations on privacy. It simplifies appointment bookings and reduces no-shows, easing the administrative burden that often healthcare providers encounter, thereby making it a whole package for healthcare modernization and bettering patient outcomes (Fig. 18).

Future developments will include use of advanced machine learning for the accurate prediction of diseases on large-scale real-world data from patients to avoid the problem of overfitting. Also implementing horizontal scaling with distributed architectures (such as microservices and containerization with Kubernetes) and leveraging cloud-based solutions can handle large amount of users enabling telemedicine support for remote consultations, real-time health monitoring through wearable devices, and extending several regional language support. Through constant improvement, MediAssist aims to build a more efficient, inclusive, and data-driven healthcare ecosystem for patients and doctors.

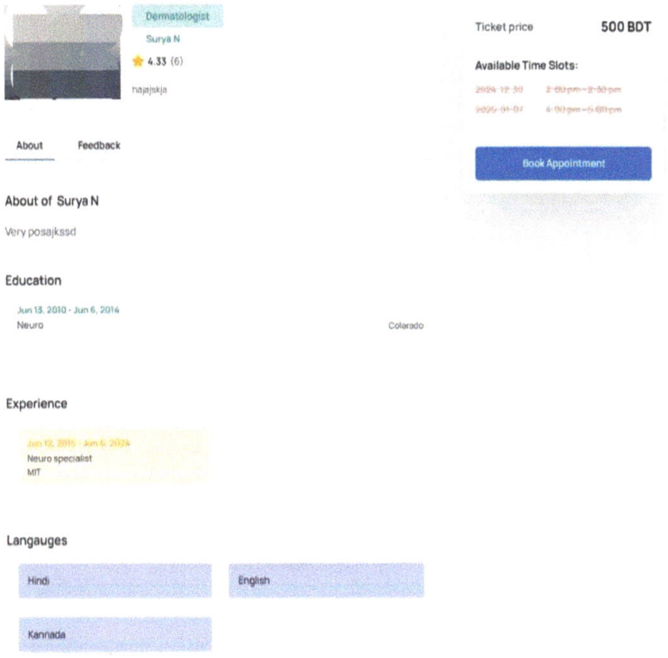

Fig. 16 Doctor details for appointment booking

Fig. 17 Prediction result in chatbot (Hin)

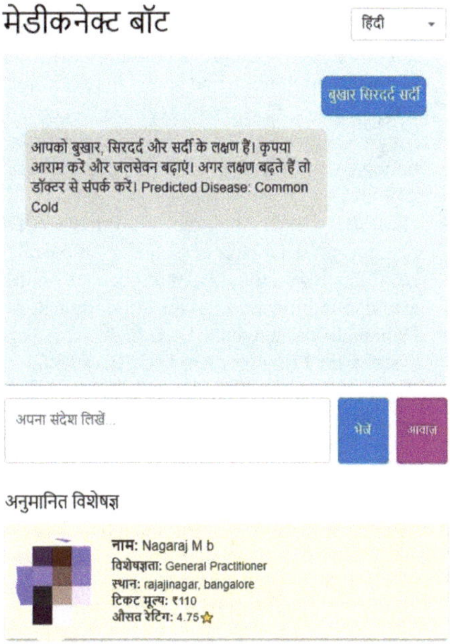

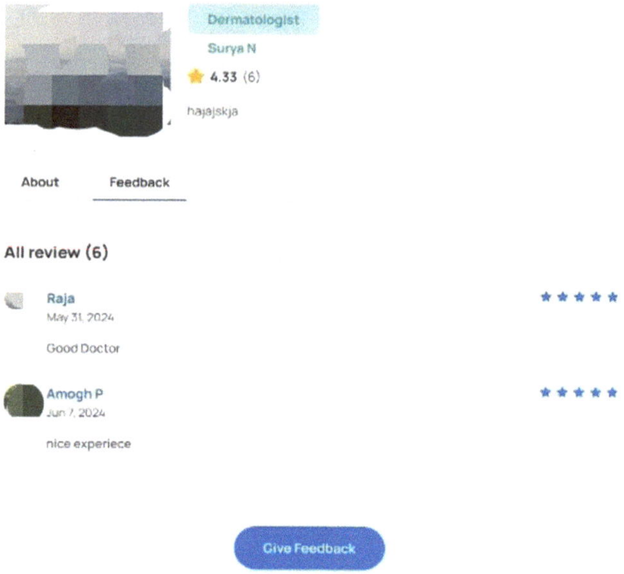

Fig. 18 Patient feedback

References

1. Mahanthi AN, Jacob TP (2023) Transforming traditional medical appointment booking with advanced technologies and user-centric design. In: 2023 3rd international conference on pervasive computing and social networking (ICPCSN). IEEE, pp 1656–1660
2. Subhi A, Jalal D, Samal M (2023) Doctor appointment system. Erbil, Kurdistan, Iraq
3. Singh A, Kaur IGP, Dabas C (2019) Get-a-doc: a doctor recommender system. In: 2018 7th international conference on reliability, Infocom technologies and optimization (ICRITO). IEEE, pp 219–223
4. Akilesh S, Sheik Abdullah A, Abinaya R, Dhanushkodi S, Sekar R (2024) A novel AI-based chatbot application for personalized medical diagnosis and review using large language models. In: 2023 International conference on research methodologies in knowledge management, artificial intelligence and telecommunication engineering (RMKMATE). IEEE, pp 1–5
5. Rallapalli V, Menghani D, Gallani H, Aasija G, Mane D (2022) Online doctor appointment system. Int J Eng Res Appl (IJERA) 12:48–52
6. Avhale SP, Ajabe WR, Chinchole PA, Changade MP, Bhil MN (2018) Doctor appointment online booking system. IJCRT 6(2). ISSN: 2320-2882
7. Jhansi Ida S, Neela K, Jayaram B, Indhumathi C, Kirubakaran D (2022) Doctor appointment booking system using content based filtering recommendation. Int J Mech Eng 7(1) (Kalahari J). ISSN: 0974-5823
8. Moulya S, Karthik NU, Priyanka M (2024) Mobile application for doctor appointment. JETIR 11(1)
9. Geeta SP (2022) Online doctor appointment booking system. Int Res J Modern Eng Technol Sci (IRJMETS) 04(07/July 2022). ISSN: 2582-5208
10. Nyayade S, Pawar K, Sonawane N, Patil A (2022) Online doctor's appointment system. Int J Adv Res Comput Commun Eng Impact Factor (IJARCCE) 11(3). ISSN(O) 2278-1021, ISSN (P) 2319-5940

11. Bharath H, Rahul N, Shylash S, Pious V (2020) Patient data management using IPFS. Int J Sci Res Public 10(7)
12. Banerjee K, Vinooth P (2024) Multiple disease prediction using ML and doctor recommendation by sentiment analysis. In: 2023 6th international conference on contemporary computing and informatics (IC3I). IEEE, pp 1469–1472
13. Himel GM, Hasan MS, Salsabil US, Islam MM (2024) MedLingua: a conceptual framework for a multilingual medical conversational agent. MethodsX. https://doi.org/10.1016/j.mex.2024.102614
14. Meza A, López G, Quesada L, Guerrero LA (2019) Architecture to design booking appointment applications for the smart personal assistant Alexa. In: 13th international conference on ubiquitous computing and ambient intelligence UCAmI 2019, Toledo, Spain
15. Dr. Bhutada S, Mahankali H, Chandupatla V, Narasimsetti SK (2023) Smart doctor's assistant-an advanced appointment booking system for hospitals. Int J Adv Res (IJAR)
16. https://www.kaggle.com/datasets/kaushil268/disease-prediction-using-machine-learning

A Portable UV–Vis Device for Polyethylene Microplastic Detection: Empowering Sustainable Water Monitoring

B. K. Vinay, M. Suchitra, T. R. Suranjan, S. Parameshwara, S. V. Panchami, and S. Anveesh

Abstract Aquatic ecosystems and human health are at risk due to microplastics (MP) in water, which pose a serious environmental concern. Particularly persistent contaminants that build up in water bodies and affect biodiversity and food chains include polyethylene (PE) MP. In response to this urgent need, we created a portable and reasonably priced tool that uses UV–Vis spectroscopy to identify PE MPs in water. Large-scale environmental monitoring is hampered by current detection techniques, which depend on costly and cumbersome laboratory settings. The proposed offers real-time detection of MPs at a distinctive absorbance peak of 230 nm by combining a UV light source, lenses, a 10 mm path length cuvette, a UV sensor and a microcontroller with a touch screen display and exhibited a Limit Of Detection (LOD) of 3.7966 ppm, a Limit of Quantification (LOQ) of 6.8381 ppm, coefficient of determination (R^2) of 0.9852 and Relative Standard Deviation (RSD) of 11.3719%. This advancement aids efforts to decrease MP pollution and maintain water quality for future generations by enabling thorough monitoring and improving affordability and portability.

Keywords Polyethylene microplastics · Sustainable detection · UV–vis spectroscopy · Portable device · Water quality · Environmental monitoring

---

B. K. Vinay (✉) · M. Suchitra · T. R. Suranjan · S. V. Panchami · S. Anveesh
Department of Electronics and communication, Vidyavardhaka College of Engineering, Mysuru, Karnataka, India
e-mail: vinaybk@vvce.ac.in

S. Parameshwara
Department of Electronics and Communication, The National Institute of Engineering, Mysuru, Karnataka, India

I. M. Jamadar et al. (eds.), *Integrating Sustainability in Modern Technologies*, Lecture Notes in Mechanical Engineering, https://doi.org/10.1007/978-981-95-9466-5_47

1 Introduction

Plastic's durability and versatility make them widely used in a variety of industries, including building, packaging and healthcare. It is expected that by 2050, the demand for plastics will increase production to 33 billion tonnes, exacerbating the pollution caused by plastic waste globally, as millions of metric tonnes flow into rivers and oceans [1]. Textiles, personal care items and bigger plastic deterioration are some of he sources of MP which are plastic particles smaller than 5 mm. They are bioavailable, ecologically disruptive and efficient carriers of dangerous compounds including pesticides and heavy metals, increasing environmental concerns due to their small size and extensive dissemination [2, 3].

MPs can enter the food chain through ingestion, breathing, or skin exposure, leading to potential health risks. Their detection in baby stool and breast milk raises significant concerns, particularly for at-risk populations. In addition to neurological and fertility effects, MPs may elicit an inflammatory response, oxidative stress and DNA damage elicited in chronic diseases such as cancer, heart diseases and autoimmune diseases. [4–6]. Aquatic environmental monitoring is important but conventional methods, filtration and sorting, are slow and not sensitive. Rapid and sensitive analysis is made possible by optical techniques such as UV–Vis spectroscopy, Raman spectroscopy and FTIR. UV–Vis spectroscopy is particularly notable for its ease of measurement [7–10].

To overcome the conventional methods, this research project aims to design a UV–Vis-based detection system on identifying PE MPs, which combines the UV light source, applicable sensors and a Raspberry Pi to achieve real-time, low-cost and scalable monitoring of MPs. This method provides a simple, on-site approach to plastic pollution for the simultaneous preservation of the aquatic ecosystem.

2 Literature Review

Optical approaches, such as Raman spectroscopy, FTIR imaging and UV imaging, offer high sensitivity and automation but face trade-offs in cost, scalability and precision [11]. A number of spectrometric and analytical methods have been developed to identify MP in water. The study by Mikac et al. [12], the remarkable sensitivity of surface-enhanced Raman spectroscopy to detect 1–4 μm polyethylene particles, as well as polystyrene particles as small as 350 nm, was demonstrated. However, obstacles including weak PE signals and restrictions on substrate design prevented wider use. Iri et al. [13] presented an affordable Raman spectrometer capable of identifying microplastics at concentrations as 0.015% w/v, with a linear response observed between 0.015% and 0.035% w/v. However, its use in highly contaminated areas was limited by signal saturation at higher doses. FTIR imaging was assessed by Chen et al. [14], who highlighted its capability to automatically detect microplastics without requiring any pre-sorting. Although efficiency was improved, the method's limitations while identifying particles smaller than 10 μm remained. The detection of MP by Jaruda and Gabayno [15] using UV imaging resulted in particle counts of up to 1,749 in surface water; nevertheless, they encountered difficulties differentiating between synthetic and organic particles and overlapping particles. These techniques demonstrate advancements while highlighting the compromises between scalability, cost and precision.

For the identification of MPs, UV–Vis spectroscopy provides a more straightforward, affordable option with encouraging outcomes. Using UV imaging and machine learning, Thammasanya et al. [16] achieved mAP scores of up to 35.7% and precision rates of 85.5–87.8%, but performance was impacted by limited datasets. Fan et al. [17] quantified PET MPs using UV/VIS spectrophotometry with an HITACHI U-3000 spectrometer, with excellent linearity ($R^2 = 0.9998$) and great precision (recovery ratios 94.33–100.02%) at a wavelength of 266 nm. Nevertheless, interference at 193 nm was observed in mixed samples, emphasizing the necessity of using different wavelengths for various polymers. The application of UV–Vis spectrophotometry for polystyrene nanoplastics via gold nanoparticle aggregation was proven by Hong et al. [18], who achieved a linear correlation for diameters between 350 and 880 nm, subject to certain ionic circumstances. In order to detect polyethylene MPs, Lee [19] used UV–Vis spectroscopy with Tween 80 surfactant having LOD of 12.9712 ppm and LOQ of 39.3067 ppm. Accurate quantification was achieved by calibration curves, while portability and cost issues persisted. Portable devices offer a promising balance between efficiency, affordability and ease of use, making them suitable for on-site detection [20].

These studies highlight the persistent challenge of developing portable, cost-effective and real-time tools for detecting PE microplastics in water. Gaps in environmental monitoring can be bridged using methods like UV–Vis, which when combined with portable devices, potentiates the prospect of environmental monitoring to be effective and economical. Our approach fills this void since we propose a portable device that uses UV visible spectroscopy for detection of PE MP in water.

3 Materials and Methods

3.1 Materials

PE MPs were obtained with a known density of 1.00 g/cc and a size of 38–45 μm^{-5} g (Cospheric) for this investigation. To improve MP suspension in water, sodium laurate, a surfactant was used (Sigma-Aldrich). The solvent employed throughout this study was deionized (DI) water. For calibration and validation, a tabletop UV–Vis spectrophotometer (SR-2XR250-25) was used to identify and quantify PE MPs. To assess the viability of field-scale applications, a portable UV sensor was used to integrate Raspberry pi.

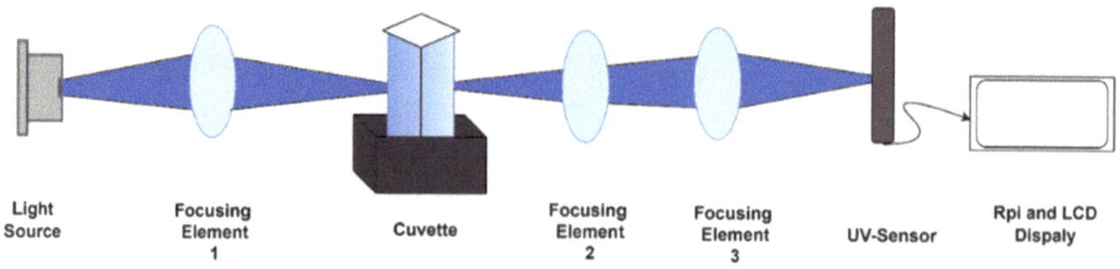

Fig. 1 Schematic diagram of portable UV–Vis spectrophotometer setup for PE MP detection

3.2 Device Design

The detecting device is made up of essential parts that maximize mobility and performance. The sample was illuminated using a 230 nm UV light source, and precise light transmission was ensured by collimating lenses and a 10 mm pathlength cuvette holder. The transmitted light's absorbance was measured by an UV sensor. A Raspberry Pi linked to a 7-inch LCD screen served as the system's control and display unit, facilitating effective data collection and real-time result visualization. Field applications and accurate measurements are made possible by the modular architecture as illustrated in (see Fig. 1).

3.3 Sample Preparation and Detection Protocol

Several critical components are organized in the setup to ensure accurate light measurement. Light emanating from a source that produces the necessary wavelength to travel through an optical pathway consisting of three focusing elements: two plano-convex lenses (Focusing Elements (FE) 1 and 3) and one plano-concave lens (FE 2). FE 1 focuses and directs the light towards the sample holder. The central FE 2 aids in evenly dispersing the light, while FE 3 channels the light to the UV sensor. This optical configuration provides a well-focused light pathway and reduces stray light, enhancing measurement precision. The sample containing the PE is placed in a cuvette and aligned with the path of the focused light.

PE MPs have a density of 1.00 g/cc, a predetermined quantity of MPs was mixed in 50 mL of deionized (DI) water using 100 ppm Sodium laurate as a surfactant to create a homogeneous suspension for analysis. To guarantee full Suspension, the mixture was agitated for half an hour. By diluting the stock solution with DI water, a concentration range of MPs, spanning from 5 to 500 ppm, was created. Quartz cuvettes were utilized for UV–Vis spectroscopy in order to prevent interference; they were completely cleaned with DI water. With an emphasis on the distinctive absorbance peak of PE MPs at 230 nm, the SR-2XR250-25 was calibrated within the wavelength range of 200–800 nm and displayed a linearity of 0.9997. The UV–Vis sensor built into the portable detection device was used to record absorbance data for each prepared sample and the reference solution.

A calibration curve was created by plotting absorbance values versus their corresponding concentrations, as illustrated in (Fig. 4a, b). The linearity of the response was evaluated to confirm a dependable relationship between absorbance and concentration. This confirmed the precision and efficacy of the portable device, which employs UV light for the accurate identification and measurement of PE microplastics.

4 Results and Discussion

4.1 Zelta Potential

To explore the behaviour of surface charges in PE MP, Zeta potential was assessed between pH values of 2 and 14. The findings revealed that PE microplastics generally carry a negative surface charge under typical environmental conditions, particularly in the pH range of 6 to 8. This negative charge promotes electrostatic repulsion, helping to maintain the stability of microplastics in suspension within aquatic systems. A shift to a positive surface charge was observed only under highly acidic conditions (pH below 3), which are uncommon in most natural water bodies (Fig. 2) illustrates the zeta potential measurements for PE MP.

This behaviour is significant as it has implications for the transport of PE microplastics within water environments, and how they interact with other contaminants. Finally, the negative zeta potential values under neutral to moderately alkaline pH confirming microplastics might be able to evade consolide and stay dispersed is

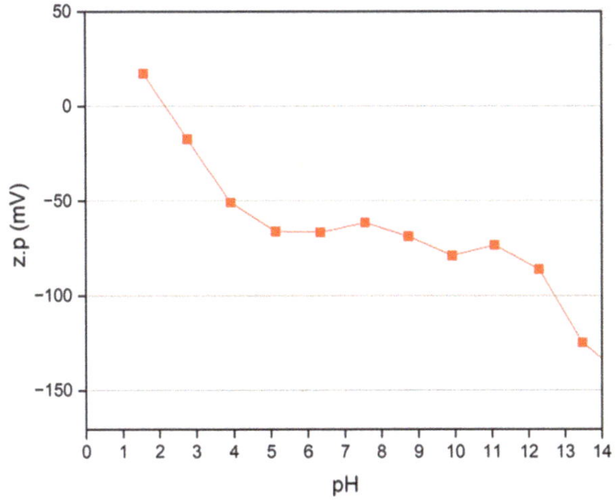

Fig. 2 Zeta potential of PE MP at different pH values, illustrating the change in surface charge as the pH varies

an important factor to understand their environmental behaviour and fate. Moreover, this characteristic influences how microplastics engage with charged substances, including natural organic matter and pollutants, which in turn affects their function in the transport and availability of contaminants within ecosystems.

4.2 Surfactants for PE Detection

PE MP must be effectively suspended in order to achieve accurate UV–Vis spectrometric detection. Among the surfactants tested, sodium laurate was identified as the most suitable, as it created a stable suspension of PE MPs at a concentration of 100 ppm. Not only does sodium laurate facilitate uniform dispersion, but it also contributes to generating a consistent absorbance peak at 230 nm, which is essential for UV–Vis analysis. The molecules of sodium laurate adhere to the surface of polyethylene microspheres due to the hydrophobic interactions between the polyethylene chains and the surfactant's hydrophobic tails. This interaction may lead to the formation of either a monolayer or multilayer coating on the polyethylene crystals.

The anionic surfactant Tween 80 showed a moderate level of efficacy in suspending PE MP. However its suspension performance was less reliable than Sodium laurate. Another surfactant, dodecyl sodium sulfonate (DSS), had some capacity to stabilize PE suspensions. While DSS provides adequate dispersion in controlled environments, its effectiveness in aqueous systems is limited due to weak adsorption on the PE surface. Similarly, the surfactant cetyltrimethylammonium bromide (CTAB) demonstrated poor performance in suspending PE microplastics, failing to achieve a stable suspension.

Scanning Electron Microscopy (SEM) images of PE MPs in the presence and absence of sodium laurate demonstrate the effect of a surfactant to the surface of the MPs. The SEM image was obtained from the Sodium laurate solution and PE MP which are dried and powdered (Fig. 3a) contains the PE MPs that had no treatment, showing no visible changes in the surface. PE MPs treated with sodium laurate, as depicted in (Fig. 3b), showed a coating of sodium laurate encasing the particles. SEM offers in-depth surface structure analysis, rendering it a crucial instrument for examining these changes. Energy Dispersive Spectroscopy (EDS) enhances SEM by validating the elemental makeup, affirming the presence of sodium laurate on the treated MPs and further supporting the efficacy of the surfactant treatment. For untreated PE MPs, EDS reveals a predominant carbon peak (81.48% by weight), affirming their hydrocarbon composition and oxygen presence (18.52% by weight) as illustrated in (Fig. 3c). The EDS spectrum of PE MPs treated with sodium laurate shown in (Fig. 3d) indicates the existence of carbon (68.58% by weight), oxygen (23.17% by weight), sodium (3.79% by weight), sulphur (1.26% by weight), chlorine (2.21% by weight) and titanium (0.99% by weight), which confirms the binding of sodium laurate to the PE surface. The development of this stabilizing layer improves the dispersibility of PE MPs in water, which is essential for detection using UV–Vis spectrometry.

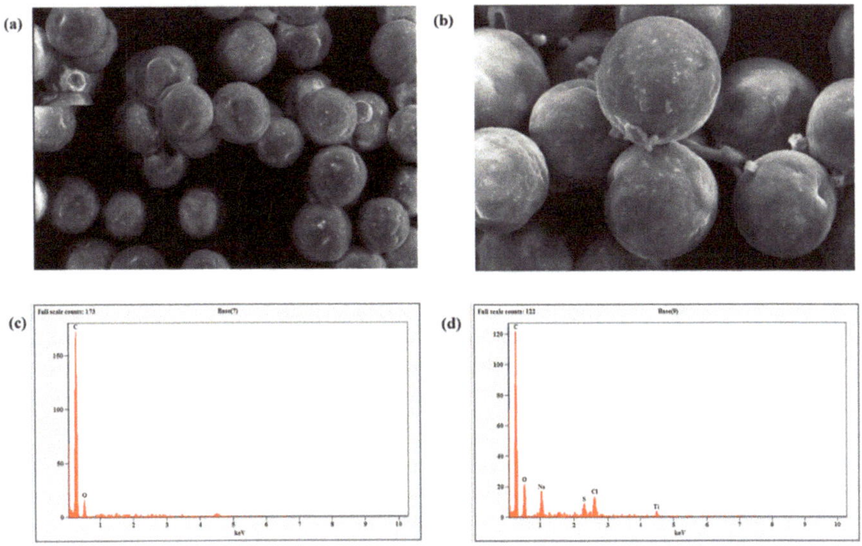

Fig. 3 **a** SEM image of untreated polyethylene (PE) microplastics (MPs), showing the typical surface morphology without any distinct modification. **b** SEM image of PE microplastics treated with sodium laurate, displaying a coating of sodium laurate around the particles. **c** EDS spectrum of untreated PE microplastics, confirming the dominant presence of carbon (81.48% by weight) and a hydrocarbon nature. **d** EDS spectrum of PE microplastics treated with sodium laurate

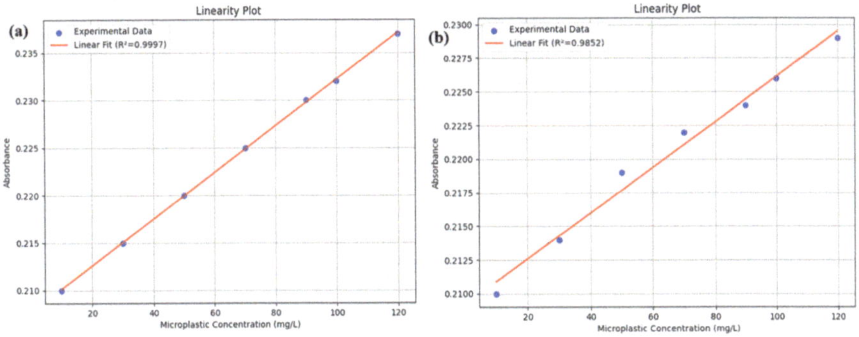

Fig. 4 Calibration curves for PE MP detection. **a** Calibration curve obtained using SR-2XR250-25. **b** Calibration curve for the proposed portable device, showcasing its performance and correlation with SR-2XR250-25 results

4.3 Calibration Method and Validation

MPs were quantified utilizing the calibration curve technique. Various concentrations of standard solutions were made. After measuring their absorbances, a calibration curve was created using the data. This aligns with Beer's rule, which states that absorbance is exactly proportional to analyte concentration in a solution and is determined by the equation

$$A = -\log T = \varepsilon BC, \tag{1}$$

$$T = \left(\frac{I_t}{I_o}\right), \tag{2}$$

where A is the absorbance, T is the transmittance, It is the intensity of transmitting light, I_o is the intensity of incident light, ε is molar absorption coefficient (M^{-1} cm^{-1}), C is the molar concentration (M) and B is the optical path length (cm).

According to the study's earlier part, sodium laurate peaks at 230 nm in wavelength. The absorbance at 230 nm is therefore employed. For the calibration curve drawing, six MP-contaminated water samples with varying amounts (10, 30, 50, 70, 100 and 120 ppm) are prepared.

The validation of UV–VIS spectroscopy was conducted with the LOD and LOQ

$$\text{LOD} = 3.3 * \left\{\frac{\text{SD}}{S}\right\}, \tag{3}$$

$$\text{LOQ} = 10 * \left\{\frac{\text{SD}}{S}\right\}, \tag{4}$$

where SD is the Standard Deviation of y-intercept and S is the Slope.

Linearity is the ability to obtain linear measurement readings in a specific range proportionate to analyte concentration. It can be analysed using the R^2 of the linear regression. The coefficient of determination gets closer to 1 the more linear it is.

4.4 Repeatability

Repeatability is a critical parameter that validates the reliability of the proposed device. It speaks about how consistently measurement results hold up when the same test is conducted again under the same circumstances. For this study, absorbance measurements of 50 ppm PE microplastic solutions were performed using the proposed device and SR-2XR250-25 as presented in (Fig. 5b). The SR-2XR250-25 ensure high repeatability our efforts were addressed by obtaining result as close as commercial device. RSD is a statistical measure of precision, indicating the

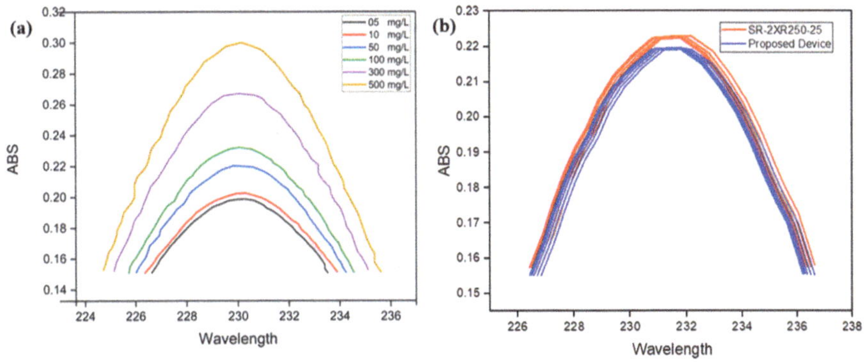

Fig. 5 **a** Absorbance peaks observed with Sodium laurate and varying concentrations of PE MPs. **b** Device repeatability curve between SR-2XR250-25 and proposed device with Sodium laurate and 50 ppm of PE MPs

percentage variation relative to the mean. The RSD values of 11.3730% for the spectrophotometer and 11.3719% for the proposed device demonstrate that the proposed device achieves precision comparable to the standard spectrophotometer. This highlights the proposed device's reliability and suitability for consistent absorbance measurements.

5 Conclusion and Future Aspects

This work presents a portable and environmentally friendly method for employing UV–Vis spectroscopy to find PE MPs in water. With a distinctive absorbance peak at 230 nm which is displayed in (Fig. 5a), the proposed device efficiently detects PE MPs and provides real-time, cost-effective and field-ready analysis. Table 1 shows the comparison of results between SR-2XR250-25 and proposed Device. By lowering resource requirements and costs, its small size overcomes the drawbacks of conventional large laboratory settings and guarantees accessibility for academics, policymakers and environmental agencies. The device is in line with sustainable and environmentally responsible practices thanks to the incorporation of the surfactant Sodium Laurate, which improves suspension stability while reducing secondary environmental concerns.

Table 1 Comparison of results

	LOD in ppm	LOQ in ppm	R^2	Slope	Intercept	RSD (50 ppm) (%)
SR-2XR250-25	2.0364	5.1711	0.9997	0.0002	0.2077	11.3730
Proposed device	3.7966	6.8381	0.9852	0.0002	0.2092	11.3719

This breakthrough aids international efforts to monitor and control MP contamination by fusing affordability, portability and precision. Its sensitivity and specificity are on par with those of traditional UV–Vis spectrometers, which encourages broad use for environmental preservation. To further strengthen its function in protecting aquatic ecosystems and water quality, future developments might concentrate on extending the device's detection capabilities to include other MP.

References

1. Williams AT, Rangel-Buitrago N (2022) The past, present, and future of plastic pollution. Marine Pollut Bull 176(113429). https://doi.org/10.1016/j.marpolbul.2022.113429
2. Rani L, Kaur G, Sood P, Kaushal J, Srivastav AL (2024) Remediation strategies for the removal of microplastics from the water. In: Srivastav AL, Grewal AS, Markandeya M, Pham TD (eds) Advances in pollution research: role of green chemistry in ecosystem restoration to achieve environmental sustainability. Elsevier, pp 191–200. https://doi.org/10.1016/B978-0-443-15291-7.00004-3
3. Campanale C, Massarelli C, Savino I, Locaputo V, Uricchio VF (2020) A detailed review study on potential effects of microplastics and additives of concern on human health. Int J Environ Res Public Health 17(4):1212. https://doi.org/10.3390/ijerph17041212
4. Winiarska E, Jutel M, Zemelka-Wiacek M (2024) The potential impact of nano- and microplastics on human health: Understanding human health risks. Environ Res 251(Part 2):118535. https://doi.org/10.1016/j.envres.2024.118535
5. Usman US, Abdull Razis AF, Shaari K, Amal MNA, Saad MZ, Mat Isa N, Nazarudin MF, Zulkifli SZ, Sutra J, Ibrahim MA (2020) Microplastics pollution as an invisible potential threat to food safety and security, policy challenges and the way forward. Int J Environ Res Public Health 17(24):9591. https://doi.org/10.3390/ijerph17249591
6. Wang C, Zhao J, Xing B (2021) Environmental source, fate, and toxicity of microplastics. J Hazard Mater 407:124357. https://doi.org/10.1016/j.jhazmat.2020.124357
7. Vinay BK, Ganesan S, Suranjan TR, Vivek BM (2025) Review of modern spectrometric techniques for monitoring cadmium in water and enhancing public health. Crit Rev Anal Chem 1–22. https://doi.org/10.1080/10408347.2025.2457391
8. Mariano MS, Tacconi S, Fidaleo M, Rossi M, Dini L (2021)Micro and nanoplastics identification: classic methods and innovative detection techniques. Front Toxicol 3:636640. https://doi.org/10.3389/ftox.2021.636640
9. Adhikari S, Kelkar V, Kumar R, Halden RU (2022) Methods and challenges in the detection of microplastics and nanoplastics: a mini-review. Polym Int 71(5):543–552. https://doi.org/10.1002/pi.6348
10. Randhawa JS (2023) Advanced analytical techniques for microplastics in the environment: a review. Bull Natl Res Cent 47:174. https://doi.org/10.1186/s42269-023-01148-0
11. BK V, Bagchi S, TR S (2025) Chromium detection in water using optical methods: a study of reagent and reagentless approaches. Crit Rev Anal Chem 1–38. https://doi.org/10.1080/10408347.2024.2419896
12. Mikac L, Rigó I, Himics L, Tolić A, Ivanda M, Veres M (2023) Surface-enhanced Raman spectroscopy for the detection of microplastics. Appl Surf Sci 608:155239. https://doi.org/10.1016/j.apsusc.2022.155239
13. Iri AH, Shahrah, M.H.A., Ali, A.M. et al. Optical detection of microplastics in water. Environ Sci Pollut Res 28:63860–63866. https://doi.org/10.1007/s11356-021-12358-2
14. Chen Y, Wen D, Pei J, Fei Y, Ouyang D, Zhang H, Luo Y (2020) Identification and quantification of microplastics using Fourier-transform infrared spectroscopy: current status and

future prospects. Curr Opin Environ Sci Health 18:14–19. https://doi.org/10.1016/j.coesh.2020.05.004

15. Jaruda RAJ, Gabayno JLF (2024) Detection of microplastics in surface water using a DSLR lens-based UV imaging system. J Phys: Conf Ser 2793:012018 (14th International Conference on Applied Physics and Mathematics, Taiwan). https://doi.org/10.1088/1742-6596/2793/1/012018
16. Thammasanya T, Patiam S, Rodcharoen E et al (2024) A new approach to classifying polymer type of microplastics based on Faster-RCNN-FPN and spectroscopic imagery under ultraviolet light. Sci Rep 14:3529. https://doi.org/10.1038/s41598-024-53251-5
17. Fan C, Huang Y-Z, Lin J-N, others (2022) Environ Technol Innov 28:102683
18. Hong J, Lee B, Park C et al (2022) Novel measurement method of determining PS nanoplastic concentration via AuNPs aggregation with NaCl. Korean J Chem Eng 39:2842–2848. https://doi.org/10.1007/s11814-022-1153-9
19. Lee H (2021) Quantification of microplastics using UV-VIS spectroscopy. M.S. thesis, Civil and Environmental Engineering, Seoul National University, Seoul, South Korea
20. BK V, TR S, Bagchi S (2025) Chromium contamination in water: a comparative analysis of detection methods with a focus on portability. Int J Environ Anal Chem 1–30. https://doi.org/10.1080/03067319.2025.2471388

Sustainable Automated Plastic Waste Segregation System Using IoT and Image Processing for Material Recovery Centers

Remya Jayachandran, Nidhishree, Anush R. Halappanavar, Laxman K. Budni, C. M. Abhishek, Charan S. Hegde, and S. Rupesh

Abstract The increasing volume of plastic waste has become a significant environmental concern, prompting the need for sustainable and automated waste management systems. Material recovery centers face challenges in segregating plastic waste based on type. This paper describes the design and development of an automated plastic waste segregation system based on grade classification, using NanoArduino and sensors for precise sorting. The system classifies plastic waste by its grade through image processing, enabling efficient recycling, with the NanoArduino as the central processing unit to control sorting mechanisms and overall operation. This system offers a cost-effective, automated solution for material recovery centers, promoting sustainable and efficient recycling with minimal manual intervention and errors.

Keywords Plastic grade · Recycle code · Sustainabilty · Object detection · Waste segregation · Automation

1 Introduction

Plastic waste accumulation is considered as one of the major environmental crises in today's world. A large portion of the millions of tons of plastic produced annually ends up polluting landfills, oceans, and various ecosystems. Efficient handling and recycling are essential for reducing pollution and protecting resources. However, the diverse nature of plastic waste makes sorting and recycling complex, as the quality of recycled products is highly dependent on the initial sorting of plastic

R. Jayachandran (✉) · Nidhishree · A. R. Halappanavar · L. K. Budni · C. M. Abhishek · C. S. Hegde
Department of Electronics and Communication Engineering, The National Institute of Engineering, Mysuru, Karnataka, India
e-mail: remyajayachandran@nie.ac.in

S. Rupesh
Bioenergy Research Information and Demonstration Center (BRIDC), P.E.S College of Engineering, Mandya, Karnataka, India

I. M. Jamadar et al. (eds.), *Integrating Sustainability in Modern Technologies*, Lecture Notes in Mechanical Engineering, https://doi.org/10.1007/978-981-95-9466-5_48

types. Traditional manual sorting is inefficient due to its labor-intensive nature and susceptibility to human error. The various plastic waste segregation systems utilizing the plastic material properties are reported in the literature [1–10]. Table 1 presents the various methods for sorting the plastic waste based on the type of polymer.

The comparison of various plastic waste segregation methods shows that each method has its own pros and cons. Due to the less manpower, automated segregation system is highly necessary for the timely sorting of the plastic wastes. Automated waste segregation systems offer a promising solution by improving the speed and precision of sorting, leading to higher quality recycled materials. The existing sorting methods have its own drawback that affects the overall performance of the segregation system. This paper presents a simple method to sort plastics on the basis of their

Table 1 Comparison of various plastic waste segregation methodologies

References	Method	Methodology	Advantages	Disadvantages
[9]	Manual segregation	Hand sorting by workers	Low cost, simple	Labor-intensive, inefficient
[1–3]	Mechanical segregation	Uses crushers, shredders, and sieves	Fast, scalable	High equipment cost, ineffective for some plastics
[6–8, 11]	Optical sorting	High-speed cameras and sensors for color/material detection	High accuracy	Expensive, may struggle with opaque plastics
[12]	Density separation	Uses water/air-based separation by density	Effective for PET versus PVC	Costly, requires specialized setup
[13]	Electrostatic separation	Charges plastics to separate based on attraction/repulsion	Useful for polystyrene versus polyethylene	High energy, limited plastic types
[14]	Friction-based sorting	Uses friction, air classifiers, or vibration	Cost-effective for low-value plastics	Inefficient for mixed plastic waste
[15]	Infrared (IR) sorting	Uses IR light to detect plastic types	Highly accurate and fast	Expensive, needs calibration, struggles with contamination.
[16]	Biological sorting	Uses microorganisms/enzymes for degradation	Eco-friendly	Slow, not industry-scalable
[17]	Chemical recycling	Converts plastics to monomers/chemicals via pyrolysis or solvolysis	Handles mixed plastics	Energy-intensive, costly
[18]	Magnetic separation	Uses magnets to remove metals	Good for eliminating metal contaminants	Not applicable to pure plastics

grade. The proposed method involves image processing to detect the plastic grade and microcontroller-based automation to segregate plastic waste based on plastic type. The camera module captures images of the plastic waste which is compared with the database, and a microcontroller analyses the data to activate sorting mechanisms that separate plastics into categories like PET, HDPE, and PVC. Image processing algorithm for feature extraction and comparison—SqueezeNet has been used for e-waste segregation which is reported in [11], achieving over 90% accuracy through extensive training and feature comparison. This algorithm compares the features extracted from the input image with the database and detects the e-waste. To classify plastics according to their recycle codes, advanced image processing algorithms are needed for object detection. The proposed automated waste segregation process is more efficient, economical, and precise compared to existing methods. This proposed design reduces the need for human labor and ensures higher quality recyclables, promoting sustainable waste management. Integrating sensors and microcontrollers enables real time, precise classification, ensuring accurate sorting, which improves the overall efficiency of recycling operations. This innovative approach promotes the use of recycled products, revolutionizes plastic waste handling, encourages environmental sustainability, and supports global efforts to tackle plastic pollution.

2 Methodology

Manual separation of the various grades of plastic in plastic waste is time-consuming in waste collection centers. Material Recovery Facilities (MRFs) are essential for sorting, processing, and recycling plastic waste but face challenges like labor shortages, inefficient sorting systems, contamination, such as food waste and non-recyclable materials mixed with recyclable plastics, etc. The diversity of waste streams, including plastics, metals, and paper, complicates the sorting efforts. The variety of plastic types (PET, HDPE, PVC, LDPE) with different properties require separate processing. The demand for recycled materials depends on the trends, regulations, and quality of the global market. Current sorting methods include near-infrared spectroscopy (NIR), multisensor hybrid sorting using NIR, optical and ultrasound sensors, robotic sorting with AI, and automated sorting with X-ray fluorescence (XRF). To address these issues, an automatic plastic grade detection system is proposed using image processing of the recycle code on plastic materials.

This paper presents an innovative method to classify the plastic based on the recycle code embedded or labeled over it. Figure 1 represents the different recycle codes which are labeled in plastic materials. Table 2 presents various plastic grades, uses and the recycling information. Based on this recycle code the plastic type can be determined manually. In MRCs, daily tones of plastics reach there for recycling. Due to the lack of man power and facilities, all these plastics are crushed together and made it in cubes. Figure 2 presents the proposed model of the plastic waste sorting system based on grade. The inlet to the conveyor belt is designed to align the waste one by one into the conveyor belt. The camera module is fixed on a vertical stand to capture the image of the input. The image captured by the camera module is send to the server for processing using deep learning algorithms. The algorithm is trained to detect the recycle symbol in the captured image. On detecting the recycle symbol, the system uses using optical character recognition (OCR) method to detect the number inside the recycle symbol which represents the type of plastic. This plastic grade information is transferred to the microcontroller—Arduino NanoUNO which is connected with the servo motors in the conveyor belt. The servomotor-controlled flaps are placed at a distance to push the specimen to the respective bin based on the plastic grade detection. The plastic grade which is not detected is collected at the terminal. The speed of the conveyor motor is set considering the response time of the image processing unit.

One of the major challenges is accurately detecting recycle codes in plastic waste sorting systems due to variations in size, position, and quality of the symbols across different companies. Fine-tuning image processing algorithms and using high-quality cameras with enhanced mobility can significantly improve detection accuracy. Moving cameras can scan different angles and capture symbols from various

Fig. 1 Recycle codes

Table 2 Plastic grades and recycling information

Grade	Plastic type	Applications	Recycling information
1	Polyethylene terephthalate (PET)	Soft drink and water bottles, food containers	Widely recycled into clothing, fleece, and packaging
2	High-density polyethylene (HDPE)	Milk jugs, detergent bottles, juice containers	Easily recycled into new containers, pipes, and plastic lumber
3	Polyvinyl chloride (PVC)	Pipes, flooring, clear food packaging	Less commonly recycled but used for tiles, flooring, and pipes
4	Low-density polyethylene (LDPE)	Grocery bags, bread bags, plastic wraps	Not widely recycled; can be repurposed into trash liners and tiles
5	Polypropylene (PP)	Yogurt containers, straws, food containers	Recyclable into brooms, brushes, and pallets
6	Polystyrene (PS)	Disposable cups, plastic cutlery, packaging peanuts	Hard to recycle but repurposed for insulation and low-grade products
7	Other (polycarbonate, acrylic)	Multi-layer food packaging, sunglasses, furniture	Difficult to recycle, but some types are reused in plastic lumber

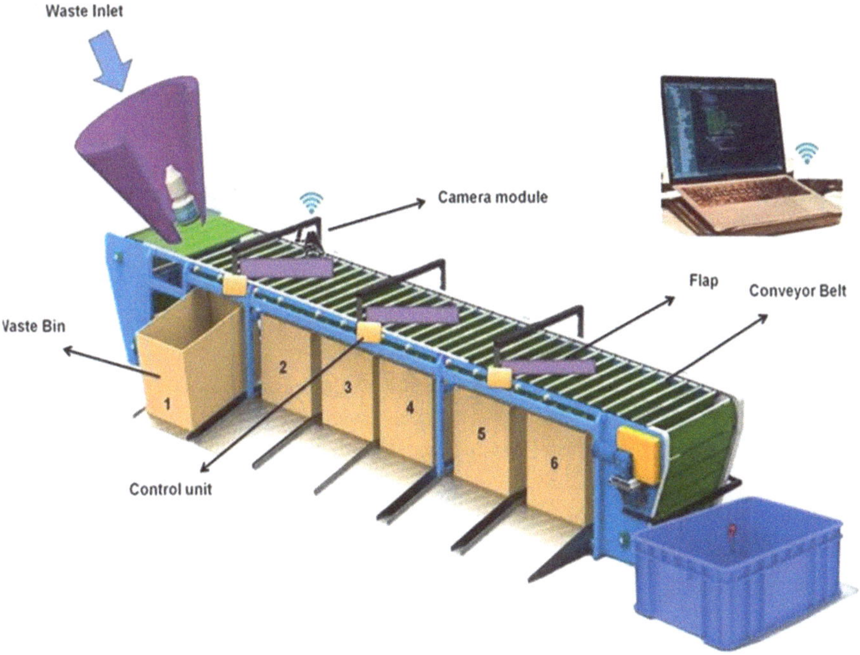

Fig. 2 Model of proposed plastic waste sorting system based on recycle code

perspectives, increasing the likelihood of correct detection. The proposed plastic waste sorting system for Material Recovery Centers (MRCs) could effectively use automated image processing to detect and categorize plastics by recycle codes, overcoming issues caused by manual sorting and variable plastic packaging. However, field surveys in Mysore have identified key challenges for MRCs, including limited manpower for sorting, lack of automated sorting systems, space constraints, and the need for timely sorting of plastics. The proposed system could help improve efficiency by automating the sorting process, while also allowing for scalable deployment depending on the facility's size and capacity. This would address the workforce shortage, enhance sorting accuracy, and optimize the use of space and time. Integrating artificial intelligence and machine learning models can further refine the system's ability to detect and sort a wide variety of plastic waste types based on recycle codes, making it a more efficient and sustainable approach for material recovery centers.

3 Methodology of Efficient Plastic Waste Segregation System with IoT and Image Processing

An efficient plastic waste segregation system that combines Internet of Things (IoT) and image processing can be highly effective in addressing the challenge of sorting plastic waste automatically. High-resolution cameras (such as RGB or infrared) are placed on the bins or along conveyor belts to capture images of the incoming waste. Cameras will scan the waste to classify it based on recycle code embedded on the plastic waste. AI-based image processing algorithms, typically using Convolution Neural Networks (CNNs), can process the captured images to identify and classify plastic items based on the recycle code as shown in Fig. 1. The image processing system can apply techniques like thresholding, edge detection, and object segmentation to segment the recycle code from the image and using optical character recognition (OCR), the algorithm will be able to detect the code from the recycle symbol. After image processing, actuators, air jets, or robotic arms can physically sort the waste into different bins based on the identified plastic type. The IoT platform will provide feedback to operators or users.

3.1 Plastic Grade Detection System Using Image Processing

Transfer Learning with Fine-tuning algorithm is utilized in this proposed system to detect the recycle code from the images of the plastic bottle. This methodology combines robust data preparation, advanced machine learning techniques, and efficient deployment strategies to achieve reliable and accurate plastic type classification. The plastic grade detection system has been designed with two main pipelines: the training pipeline and the prediction pipeline. In the training pipeline, the workflow begins with data collection, where a diverse set of images (plastic recycle code images) is sourced. This data is then organized into folders based on their respective grade types. Preprocessing plays a crucial role, which involves resizing all images to a uniform resolution of 224×224 pixels, rescaling pixel values to a [0, 1] range, and converting them into arrays compatible with TensorFlow. Advanced data augmentation techniques, such as random rotations, horizontal flips, and width/height shifts, are applied to expand the dataset and reduce overfitting during training artificially. After this, the dataset is divided into training and validation subsets, with an 80:20 ratio to ensure fair evaluation. The model training step uses MobileNetV2, a lightweight convolutional neural network pre-trained on ImageNet, as the backbone for transfer learning. The convolutional base of this model is frozen during initial training to leverage the pre-trained feature maps. Custom dense layers are added on top for classification. These layers include a global average-pooling layer, a dense layer with 128 units, a dropout layer with a rate of 0.5 to avoid overfitting and a final softmax layer to output probabilities for each class. The softmax activation function ensures that the outputs form a normalized probability distribution. The model is compiled using the Adam optimizer with a learning rate of 0.001, and categorical cross-entropy as the loss function. Early stopping is implemented to stop training if the validation accuracy does not improve after 10 consecutive epochs. Once the model is trained, it is evaluated on the validation set, and metrics like accuracy and loss are analyzed. After validation, the trained model and its corresponding class mapping (generated from the dataset's class indices) are saved for deployment.

In the Prediction Pipeline, the saved model (HDF5 format) and the class mapping (stored as a JSON file) are loaded into memory. Input images are preprocessed using the same steps as during training, ensuring consistency in dimensions and scaling. The preprocessed image is then fed into the trained model, which generates a probability distribution over all classes using the soft-max function. The class with the highest probability is selected as the prediction, and its confidence score is extracted. The system also supports batch predictions, allowing multiple images to be processed simultaneously. This ensures that the model's predictions are accurate, using the optimized weights and class mappings generated during training. Fig. 3 depicts the flowchart of the proposed plastic sorting system based on the type using recycle code embedded on it. The image processing part plays the major role in this proposed sorting system. Once the recycle code is detected by the algorithm, the information is passed to the hardware unit to control the motors attached to the conveyor belt. The hardware part is explained in the next section.

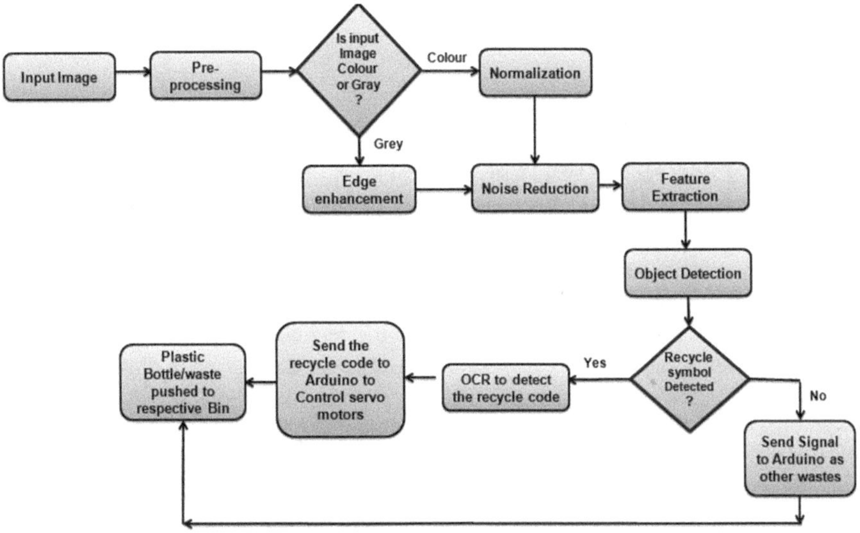

Fig. 3 Flowchart of proposed plastic waste sorting system based on recycle code

3.2 Hardware: Plastic Waste Segregation Based on Recycle Code

The plastic waste sorting system is designed with the NanoArduino microcontroller and a camera module strategically placed along the conveyor belt. Plastic. waste is introduced into the system via the funnel shaped inlet, and this allows only one bottle at a time to the conveyor belt. A camera module captures images of the incoming waste and transmits these images to the server. The server uses an image detection algorithm to process the images using Python. It extracts features from the image and compares them with a pre-existing database. If the algorithm identifies the recycling code and extracts the number, it sends a detection message to the microcontroller. The controller triggers the flap motor based on the detection message. IR sensors are placed in the bins to detect the level of wastes segregated. When the bin is filled above 75%, the alert signal is generated to inform the user. A horizontally aligned flap on the conveyor directs the plastic bottle/waste into a designated bin.

As the waste continues along the conveyor, the activated flap directs the detected waste-to the bin with appropriate recycle code. If the recycle code is not detected then the waste is conveyed to a general waste bin located at the end of the conveyor system. This proposed system efficiently categorizes waste into respective bins using a combination of visual and sensor-based detection technologies, enhancing both sorting accuracy and speed.

4 Results and Discussion

The proof of concept of the plastic waste segregation system and the sample of dataset used to train the model is shown in Figs. 4 and 5, respectively. The prototype is the scaled down version of the actual implementation. The proposed system is tested for bottles of height below 10 cm and width below 5 cm. The image of the bottle is captured using the android mobile camera and the image is saved in the cloud server. This image is preprocessed and the recycle code is detected from the image using the image processing algorithm. The dataset for each plastic grade is created, each grade with 100 images which is captured at various angles of different plastic bottles in which the recycle code is labeled. Before feeding the images into the model, they are resized to a fixed dimension of 224 × 224 pixels to match the model's input requirements and are normalized to ensure consistent pixel value ranges. Augmentation is performed in real time during training to provide the model with varied inputs, including rotated images, shifted images, and flipped images. This preprocessing pipeline ensures the dataset is well-prepared for training and validation, improving the overall performance of the model.

The accuracy chart representing the performance of plastic waste sorting system using object detection algorithm—Transfer Learning with Fine-tuning is shown in

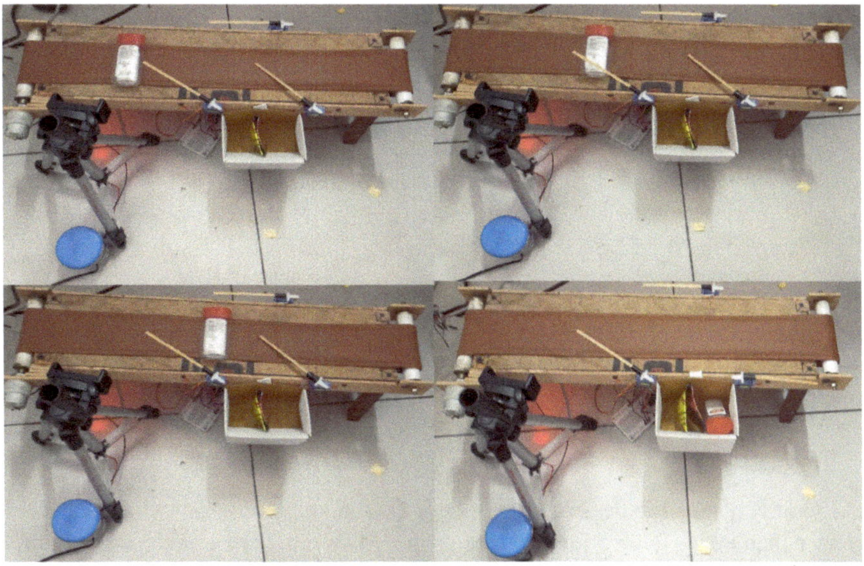

Fig. 4 Plastic waste sorting system based on recycle code using IoT and image processing

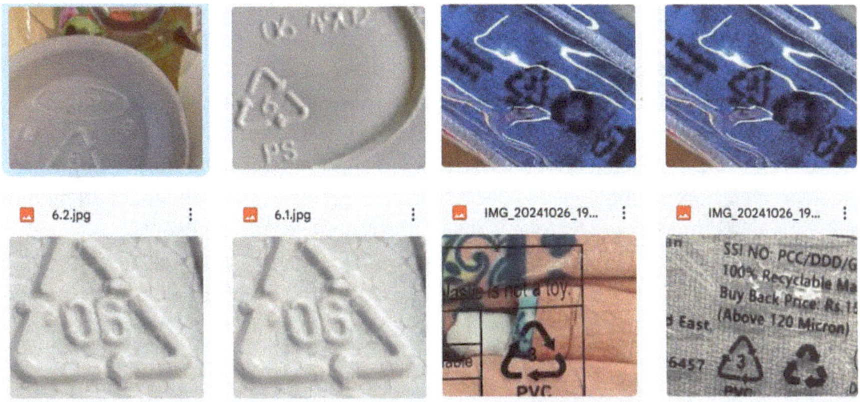

Fig. 5 Dataset of plastic with recycle code

Table 3 which is obtained using 100 trials. The chart depicts precision, recall, and F1-score for different plastic grades 1–7. $F1$-score combines both precision and recall to provide the performance of the algorithm used (Table 4).

The algorithm can provide precision, recall, and $F1$-scores of 95%, 92%, and 96%, respectively for plastics with recycle codes 1, 2, 3, 5, and 7. As the dataset is less for recycle code 6, the accuracy is less for plastics with recycle number 6. The dataset of a total of 1000 images is used in this algorithm of the developed system which is divided into training, validating, and testing sets. The algorithm parameters

Table 3 Accuracy chart of plastic waste segregation based on recycle code using image processing and IoT

Recycle code	Precision (%)	Recall (%)	$F1$-score (%)
1, 2, 3, 4, 5	95	97	96
6	92	89	90
Other wastes/7	94	95	94

Table 4 Detection accuracy and response time for plastic waste segregation

Plastic waste type.	Sensor type	Accuracy (%)	Response time (s)
Recycle code 1, 2	Image/camera (android)	95	3
Recycle code 3, 4	Image/camera (android)	95	4
Recycle code 5, 6	Image/camera (android)	98	4
Recycle code 7/other wastes	Image/camera (android)	90	6

are iteratively tuned to detect plastic grades 1–6 and other wastes with an accuracy greater than 90%. Table 5 presents the proposed system performance in real time. The method proposed in this paper efficiently detects and sorts plastics using image processing and IoT technologies. However, limited access to sufficient datasets for training the model has resulted in reduced accuracy for recycle code 6. The prototype is designed to handle bottles with dimensions below 10 cm in height and width. A key challenge is ensuring that bottles are placed one at a time and correctly oriented to capture the embedded recycle code. Future work can focus on enhancing system performance by incorporating a more advanced and suitable orientation mechanism. The real-time detection of plastic recycle code from the bottle is shown in Fig. 6; the recycle code 1 is detected from the image captured. Ensuring proper alignment of waste materials on the conveyor belt without overlap is a significant challenge in real-time detection, as it directly affects unobstructed sensor exposure.

Another difficulty lies in the lack of standardization in recycling symbol size and placement, which can vary widely and are often found at the bottom of plastic bottles. The accuracy and efficiency of the proposed system heavily depend on precise

Table 5 Real-time test results of the proposed plastic waste sorting system

Recycle code.	Bottles tested	Accurate detections	Accurate sortings	Detection accuracy (%)	Sorting success (%)
1	10	10	10	100	100
2	10	9	9	90	100
3	10	9	9	90	100
4	10	9	9	90	100
5	10	9	9	90	100
6	10	7	7	70	100
7/no code	10	10	10	100	100

Fig. 6 Real-time detection of recycle code from the plastic bottle

camera positioning. To ensure that bottles enter the conveyor belt one at a time, a funnel-shaped metal box with a slope between 45° and 60° (as shown in Fig. 2) is employed. The funnel outlet is designed to be smaller than 11 cm, effectively aligning bottles for individual processing. Additionally, system longevity can be improved through consistent sensor calibration, routine maintenance, and continuous real-time monitoring.

These challenges can be addressed by placing multiple cameras around the conveyor belt or an orientation system at fixed intervals to cover all possible angles (e.g., front, rear, sides, top, and bottom). Optionally include one or more cameras at adjustable angles for improved adaptability. Additionally, the proposed system can be enhanced by incorporating an initial stage specifically designed for a plastic bottle orientation system. In this setup, either a rotating camera or multiple cameras are arranged to capture images of the bottle from various angles. These images serve as input for an object detection algorithm, which identifies the recycle symbol. If the recycle symbol is detected within a specified time frame, the bottle is automatically directed onto the conveyor belt. Depending on the detected recycle code, the system activates a motor to sort the bottle into the appropriate bin.

5 Conclusion

An advanced automated plastic waste segregation system that utilizes IoT and image processing for detecting and classifying plastic waste, particularly for Material Recovery Facility (MRF) centers is presented in this work. The proposed system is developed based on the inputs received from the various MRF centers in Mysore and BIRDC Mandya. The proposed system employs an object detection image processing algorithm to identify recycle symbols and uses optical character recognition (OCR) to extract the number within the symbol. The proposed system achieves a detection accuracy of over 90% with a 4-seconds response time for plastic grades 1–5 and 7, maintaining a 10 cm separation between items on the conveyor belt. The proposed system aims to sort the plastics based on grade that enhances the quality of the recycled products. Future enhancements involve adopting advanced image processing algorithms like YOLO, setting up a multi-camera setup, and expanding the dataset to improve model training and accuracy. Integrating non-conventional energy sources can further optimize the system's sustainability.

Acknowledgments This project was funded by the Foundation for Advancement of Education and Research (FAER) Scholar Program 2023–24 and BRIDC Mandya. The authors express their sincere gratitude to the FAER Project Mentor, Dr. H. N. Chanakya, for his invaluable guidance and insights throughout the execution of this project. The authors also extend their heartfelt thanks to Vishakh J. K., Anagha M. A., Amaresh N., Trishna R., and Ashish R. for their significant contributions in data collection and prototype testing.

References

1. Idumah CI, Nwuzor IC (2019) Novel trends in plastic waste management. SN Appl Sci 1:1–14
2. Bernat K (2023) Post-consumer plastic waste management: from collection and sortation to mechanical recycling. Energies 16(8):3504
3. Pan D, Su F, Liu C, Guo Z (2020) Research progress for plastic waste management and manufacture of value-added products. Adv Compos Hybrid Mater 3:443–461
4. Santomasi G, Todaro F, Petrella A, Notarnicola M, Thoden van Velzen EU (2024) Mechanical recycling of PET multi-layer post-consumer packaging: effects of impurity content. Recycling 9(5):93
5. Lubongo C, Daej MAB, Alexandridis P (2024) Automated sorting technology for plastic waste. In: Reuse of plastic waste in eco-efficient concrete. Elsevier, pp 13–35
6. Tamin O, Moung EG, Dargham JA, Yahya F, Omatu S (2023) A review of hyperspectral imaging-based plastic waste detection state-of-the-arts. Int J Electr Comput Eng (IJECE) 13:3407–3419
7. Sio GA, Guantero D, Villaverde J (2022) Plastic waste detection on rivers using YOLOv5 algorithm. In: 2022 13th international conference on computing communication and networking technologies (ICCCNT). IEEE, pp 1–6
8. Hegde R, Patel S, Naik RG, Nayak SN, Shivaprakasha K, Bhandarkar R (2021) Underwater marine life and plastic waste detection using deep learning and Raspberry Pi. In: Advances in VLSI, signal processing, power electronics, IoT, communication and embedded systems: select proceedings of VSPICE, 2020. Springer, pp 263–272
9. Chawla K, Singh R, Singh J (2020) Segregation and recycling of plastic solid waste: a review. In: Advances in materials science and engineering: select proceedings of ICFMMP 2019, pp 205–221
10. Wen S, Yuan Y, Chen J (2023) A vision detection scheme based on deep learning in a waste plastics sorting system. Appl Sci 13(7):4634
11. Remya J, Rupesh S, Shaen K (2024) Intelligent solid waste sorting system using IoT and image processing. In: IEEE 2024 fourth Asian conference on innovation in technology (ASIANCON). IEEE, Pune, Maharashtra, India
12. Shukla P, Sharma PK, Pandey S, Kunwer R (2024) Optimizing municipal solid waste recovery through density-based waste plastic segregation. J Mater Cycles Waste Manag:1–21
13. Rybarczyk D, Jedryczka C, Regulski R, Sedziak D, Netter K, Czarnecka-Komorowska D, Barczewski M, Barański M (2020) Assessment of the electrostatic separation effectiveness of plastic waste using a vision system. Sensors 20(24):7201

14. Rodrigues B, Saron C (2022) Electrostatic separation of polymer waste by tribocharging system based on friction with PVC. Int J Environ Sci Technol 19(3):1293–1300
15. Adarsh U, Kartha V, Santhosh C, Unnikrishnan V (2022) Spectroscopy: a promising tool for plastic waste management. TrAC Trends Anal Chem 149:116534
16. Anitha R, Maruthi R, Sudha S (2022) Automated segregation and microbial degradation of plastic wastes: a greener solution to waste management problems. Glob Trans Proc 3(1):100–103
17. Schade A, Melzer M, Zimmermann S, Schwarz T, Stoewe K, Kuhn H (2024) Plastic waste recycling: a chemical recycling perspective. ACS Sustain Chem Eng 12(33):12270–12288
18. Zhang Y, Wang Q, Yalikun N, Wang H, Wang C, Jiang H (2023) A comprehensive review of separation technologies for waste plastics in urban mine. Resourc Conserv Recycl 197:107087

EmberEye: An Intelligent IoT and Cloud-Integrated Machine Learning Approach for a Sustainable Forest Fire Safety

Sneha Patro, Soumik Das, Srishti Kumari, Yuvraj Sonal, S. Gowrishankar, and A. Veena

Abstract Forest fires pose a significant threat to ecological balance and human safety, demanding advanced solutions for early identification and mitigation. This research introduces "EmberEye," an IoT-based automated system for detecting forest fires designed to monitor and classify fire risks in real time. The system employs a grid-based architecture with sensors measuring environmental parameters such as flame status, gas concentration, humidity, soil moisture, and temperature. Data from all sensor grids are transmitted via Long-Range (LoRa) communication to a central node, where it is processed using a trained Decision Tree Classifier. The system categorizes zones into Safe, Probable Fire, or Fire Detected Zone based on threshold conditions, with results displayed on a dynamic web interface featuring color-coded grids. Synthetic datasets representing realistic environmental conditions were used to achieve high prediction accuracy. EmberEye demonstrates scalability, low-power communication, and cost-effectiveness, providing a dependable technique for preventing forest fires and timely response, with potential applications in resource-constrained environments. As EmberEye ensures wildfire preventive measures, it enhances the adaptation to climate change, limits ecological harms,

S. Patro · S. Das · S. Kumari · Y. Sonal · S. Gowrishankar · A. Veena (✉)
Department of Computer Science and Engineering, Dr. Ambedkar Institute of Technology, Bengaluru, India
e-mail: veenaa1@acm.org

S. Patro
e-mail: patrosneha928@gmail.com

S. Das
e-mail: soumikjanka@gmail.com

S. Kumari
e-mail: imsrishti144@gmail.com

Y. Sonal
e-mail: imyuvrajsonal@gmail.com

S. Gowrishankar
e-mail: gowrishankarnath@acm.org

I. M. Jamadar et al. (eds.), *Integrating Sustainability in Modern Technologies*, Lecture Notes in Mechanical Engineering, https://doi.org/10.1007/978-981-95-9466-5_49

and promotes ecosystem recovery in a sustainable manner, therefore contributing to forest conservation. The energy efficiency of the sensor grids makes it possible to deploy the system in remote and off-grid sites. This paper thus highlights the potential of improved smart grid transformations and data connectivity to enhance the disaster response.

Keywords Internet of Things (IoT) · Wildfire prevention · Real-time surveillance · LoRa communication · Predictive analytics · Energy-conscious IoT systems · Data-driven risk prediction

1 Introduction

Forest fires are a significant global issue, causing severe environmental, economic, and social consequences. Over 56,000 wildfires occur annually in the United States alone, burning millions of acres of forested land [1]. Globally, these fires destroy ecosystems, reduce biodiversity, degrade air quality, and contribute to climate change. They endanger human lives, destroy properties, and impact local economies, especially in regions dependent on forest tourism, agriculture, and timber industries [2].

Early identification and observation of forest fires are challenging, particularly in vast forest areas and remote, inaccessible terrains. Traditional methods, such as aerial surveillance, satellite monitoring, and ground-based observation, are often slow, labor-intensive, and costly [3, 4]. These methods rely on manual inspection and are limited by weather conditions, making it challenging to identify fires early and respond effectively. Delays in detection and response times allow fires to spread uncontrollably, worsening the environmental and economic impact [5].

The Internet of Things (IoT) has emerged as a transformative technology capable of resolving these challenges. IoT enables deploying interconnected sensors to monitor parameters like temperature, humidity, and gas levels in real time[6]. This data is continuously transmitted and analyzed, providing a scalable, cost-effective, and a system for automating early warning of fires [7]. Integrated with machine learning, IoT can predict fire risks, ensuring early recognition and improving firefighting efficiency.

IoT-based identifications of forest fires [8] utilize sensors to monitor environmental parameters such as temperature, smoke, and humidity. These sensors transmit real-time data to a central system, enabling early detection and rapid alerts. This technology ensures continuous monitoring, improving response times and minimizing the impact of forest fires.

In this research, an IoT-based solution has been proposed for forest fire detection, consisting of a network of sensor grids, ESP32 microcontrollers, and LoRa communication [9] modules. The sensors continuously monitor environmental conditions, transmitting data to a central node for evaluation and processing. After that, this information is utilized to detect early signs of fire and trigger timely alerts. A cloud-based platform will store this data and provide real-time monitoring, offering access to the status of fire risks across large forest areas. The main objective of this study is to develop a scalable, cost-effective system that can improve fire detection and enhance the overall response to forest fires. The proposed solution will employ machine learning methods in order to analyze sensor data and predict fire risks, while providing real-time feedback through a user interface.

This research aims to develop an efficient and dependable system for detecting forest fires that employs Internet of Things technologies to monitor environmental factors using sensor grids and methods for machine learning to forecast fire risks accurately. The cloud-based solution enables real-time alerts, continuous monitoring, scalability, and efficient fire detection, enhancing forest resource management and public safety.

The rest of the paper is organized as follows: Related work is reviewed in Sects. 2 and 3 details the proposed design methodology. In fourth section, the paper elaborates on the results and outcomes of the applied methodology. Finally, the paper is concluded.

2 Related Work

Recent advancements in wildfire detection technologies have led to innovative methods to enhance detection accuracy and reduce false alarms. Pennypacker [10] introduced the Fire Urgency Estimator in Geosynchronous Orbit (FUEGO), an early-warning system using geosynchronous satellites with multi-spectral sensors and real-time image processing. FUEGO can efficiently detect small fires by employing rapid image subtraction and multi-spectral trend classification, addressing critical challenges in wildfire-prone areas and improving detection speed while minimizing false alarms.

Roy et al. [11] proposed a multi-sensor, multi-temporal change detection framework using Landsat-8 and Sentinel-2A satellites for mapping burned areas. By integrating random forest regression with synthetic data and temporal consistency checks, this method overcomes challenges like cloud cover, atmospheric variations, and sensor discrepancies, providing a robust solution for burned area estimation in complex environments. J. Wang presents an approach to identify forest fires using MODIS active fire products coupled with a number of environmental elements to enhance accuracy and support real-time monitoring and fire management [12].

A combined architecture for forest fire detection, developed by Dutta and Ghosh [13], merges separable convolutional neural networks (CNNs) using methods for digital image processing like segmentation and thresholding. Their computationally efficient model, trained on a large dataset of fire images, demonstrated high accuracy in diverse environments.

The use of unmanned aerial vehicles (UAVs) for forest fire monitoring was explored by Hossain [14], equipped with infrared cameras, GPS, and IMUs. UAVs offer continuous, precise surveillance, tracking fire spread, predicting direction, and providing real-time data to fire stations. Cooperative UAV systems offer more reliable data compared to single UAVs, aiding faster decision-making during emergencies.

Within one of the studies, Bosch et al. [15] concentrated on a ground-based identifying system that makes use of infrared (IR) signal processing. Their system optimized detection probability while minimizing false alarms, proving effective for fire detection over distances up to 10 km.

Jones [16] advanced Geostationary satellites for fire detection with the AHI-FSA and BAT algorithms, leveraging the Himawari-8 satellite for high-frequency detection every 10 minutes. These algorithms enhanced thermal anomaly detection at sub-pixel levels, outperforming traditional systems like MODIS and VIIRS in detection accuracy and response time. Alkhatib [17] reviewed several techniques for detecting forest fires, including satellite-based systems, optical sensors, and wireless sensor networks (WSNs). While satellite systems struggle with weak signals during early fire stages, WSNs provide low-cost, scalable solutions for monitoring environmental parameters such as temperature, humidity, and gas concentrations.

In their study, Pang et al. proposed FuF-Det [18], an innovative algorithm for target detection that combines UAV technology and machine vision to accurately detect early forest fires, even under foggy conditions, enhancing fire detection reliability in challenging environments. Chen [19], a GS-YOLOv5s for lightweight object detection algorithm, improves fire detection accuracy and reduces human-computer load in complex habitat backgrounds, which makes it suitable for real-time monitoring of fire.

In the context of these changes, the current project is an Internet of Things (IoT)-based identification of forest fires system which includes reliable data transfer, machine learning-based estimation, and monitoring of the environment. The proposed method uses on-ground sensors combined with LoRa for real-time data acquisition and transmission, particularly in vast forests, while satellite or UAV systems have drawbacks like coverage and cost. By analyzing temperature, humidity, soil moisture, gas conditions, and flame, the system uses machine learning to estimate the probabilities of a fire and how big it is.

The proposed project will be easy to implement in real time with instantaneous expansion and operation. LoRa radio mechanism ensures safe data transfer, use of wireless systems without any troubles, and enables information collection from remote areas. Like everything else, it also helps by giving information, thereby making it easier to plan and decide, as to how to use one's resources effectively.

This technique provides a new method of controlling and avoiding forest fires by tackling problems including hard-to-reach areas and slow reaction times. It's intended to slot in with what's already out there and to be able to be updated later. Its versatility allows use in any sector or site, offering a contemporary approach to a life-threatening issue.

3 Design Methodology

This study suggests a robust IoT-based technology for detecting forest fires that combines state-of-the-art sensing technologies, predictive modeling, and efficient communication protocols to overcome the challenges associated with traditional fire detection methods. The system consists of an integrated architecture with on-ground environmental sensors, LoRa modules for long-range communication, a centralized processing unit for data analysis, and a user interface for real-time monitoring and alerts. This methodology aims to address the limitations of existing solutions by providing a low-power, cost-effective, and scalable system capable of operating in remote regions with minimal infrastructure.

The primary objective of the suggested methodology is to enhance the early identification and forecasting of forest fires, thereby enabling timely responses and minimizing potential damage. The system is intended to monitor critical environmental factors, such as humidity, temperature, soil moisture, and smoke levels, and to utilize machine learning to analyze this data for fire risk assessment and prediction. By leveraging LoRa communication, the system ensures reliable data transmission in remote areas, overcoming connectivity challenges faced by satellite and conventional wireless networks.

The proposed architecture (Fig. 1) comprises multiple layers, each contributing to the system's functionality. The sensor layer is set up in a grid and has nodes with temperature, humidity, soil moisture, flame, and smoke sensors to monitor the environment completely. This information is transmitted by the communication layer to the central processing unit using low-power, long-range data transfer modules called LoRa.

The processing layer is formed of a cloud-based server, which pre-processes the incoming data, runs algorithms that predict fire risks, and sends alerts whenever present thresholds are exceeded. Last but not least, there is the application layer; it is the interface that brings to the user real-time data and visualization, including notifications about alarms, keeping stakeholders constantly monitoring the system.

The parts are engineered to ensure reliability and efficiency. Sensors for temperature and humidity such as Digital Humidity and Temperature Sensor (DHT11), soil moisture sensors that will indicate the moisture level present in soil, flame sensors for detection of emerging flames, and Metal Oxide Semiconductor Gas Sensor (MQ-2) for smoke detection capture important environmental data. These sensors are integrated with Espressif System Panel 32 (ESP32) microcontrollers for the acquisition and transmission of data while LoRa SX1278 modules offer robust communication over a long distance. Power jacks power the sensor nodes and enable them to operate with minimal or no disturbances in such areas.

LoRa technology underpins the communication framework, leveraging its long-range, low-power capabilities to ensure robust data transmission in remote forest environments. The network follows a star topology whereby each sensor node is directly connected to the central gateway. This setup ensures smooth communication while handling the challenges of thick forest areas effectively. It is, therefore,

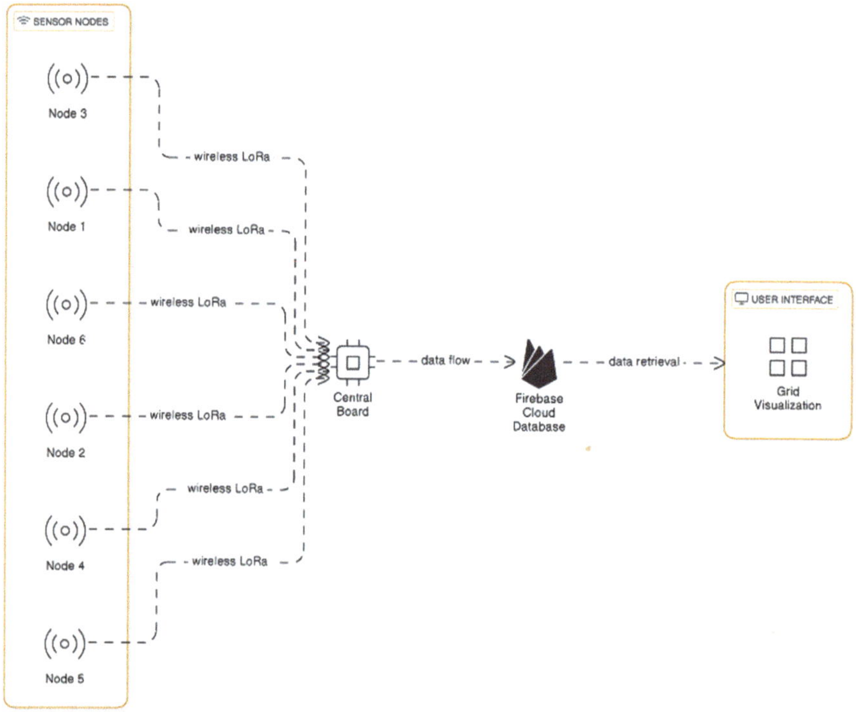

Fig. 1 A block diagram depicting overall system structure

imperative to consider software tools and frameworks in the system since they are vital in the performance of the system. Some of the technologies that have been used include Python for data preprocessing, machine learning, and back end; and React.js for the monitoring dashboard. The machine learning models which are implemented by using the scikit-learn library process the sensor data to assess the fire danger levels and detect possible fire growth.

The data is taken at some intervals. For this experimentation, the interval has been set to two seconds. The two-second interval ensures real-time data updates while keeping the system smooth and working well by avoiding too many requests. Sensors collect different environmental conditions. Then the taken data is sent to the cloud database for real-time analysis and storage for future references. The corresponding data handling steps in this system are as follows:

Threshold-Based Analysis: Comparing sensor readings with predetermined thresholds, for example, comparing high temperatures and smoke levels.

Advanced Analysis: The model acquires historical and real-time data from Firebase to predict potential fire outbreaks.

Results are visualized on a grid-based interface, utilizing color-coded indicators to signify varying risk levels (red for detected fire, yellow for potential spread, and green for safety). Real-time monitoring is achieved through a web-based dashboard that dynamically updates to display sensor data and highlight anomalies. The system generates alerts when specific thresholds are breached, notifying stakeholders via the different color-coded indicators on the grid-based interface.

The system incorporates error-correction protocols and data redundancy measures to maintain reliability and prevent data loss during transmission in adverse conditions. This methodology not only enhances the accuracy and timeliness of fire detection, but also offers a scalable and cost-effective solution for large-scale implementation. The integration of predictive modeling with IoT technology ensures that the system can proactively identify and respond to fire risks, ultimately reducing the impact of forest fires on ecosystems and human lives (Fig. 2).

Anticipated challenges include potential signal degradation of LoRa in densely forested regions, sensor inaccuracies under extreme weather conditions, and power limitations. Mitigation strategies involve deploying repeaters for enhanced signal coverage, periodic recalibration of sensors, and solar-powered solutions for uninterrupted functionality.

4 Results and Discussions

The proposed method for detecting forest fires was tested extensively in controlled environments to validate its effectiveness and reliability. The results demonstrate the system's ability to provide accurate, real-time monitoring, and early detection of potential fire risks. Data collected during these tests were analyzed both quantitatively and qualitatively, offering insights into the system's performance under different environmental conditions.

Figure 3 illustrates the complete system architecture, encompassing both the hardware components, including all the sensors and communication devices, as well as the virtual interface utilized for observation and control of the detection of forest fires system. The image (Fig. 4) illustrates all hardware components deployed on each grid, providing a comprehensive view of the system setup.

The machine learning model used in this study is a Decision Tree Classifier [20] that is designed to predict fire-prone conditions by analyzing environmental data such as temperature, humidity, soil moisture, flame status, and gas levels. The dataset (Fig. 5), consisting of over 5000 entries from real-world and simulated scenarios, underwent preprocessing to ensure data quality. Feature selection techniques were applied to identify key variables, and the model was trained using an 80–20 training-validation split.

Quantitative results highlight the system's accuracy and responsiveness. The machine learning model (Fig. 6) achieved an average precision of 0.53, recall of 0.62, and an $F1$-score of 0.57 in predicting fire risks based on sensor data with overall accuracy of 0.86 (Fig. 7). Detection latency was minimal, with alerts being generated within 5 s of threshold breaches.

Fig. 2 Data flow through the system

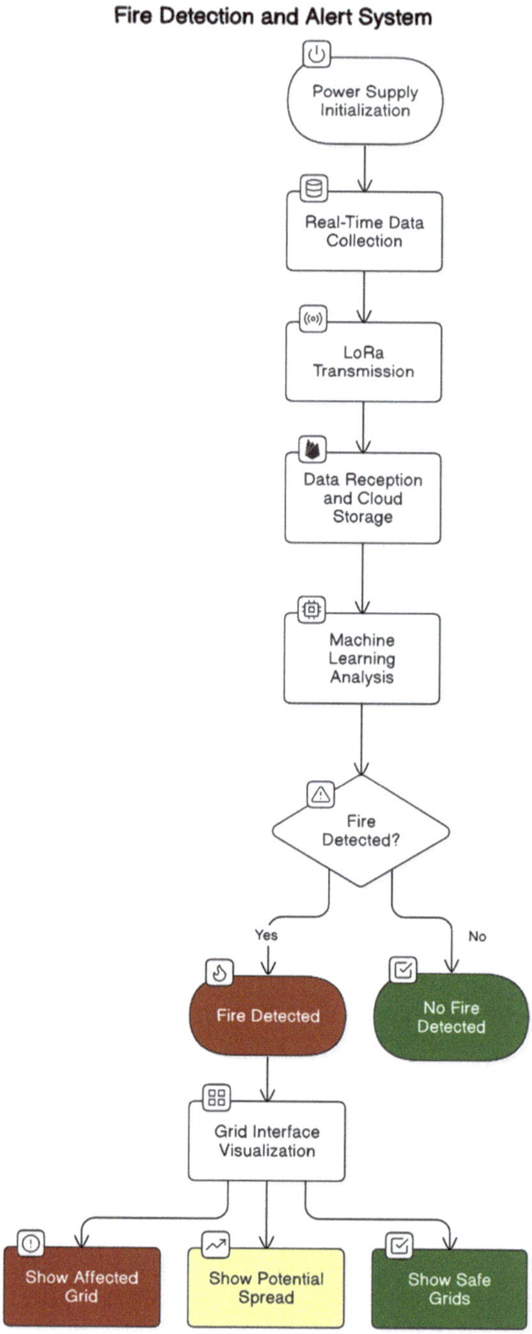

Fig. 3 System architecture showing hardware setup and virtual interface

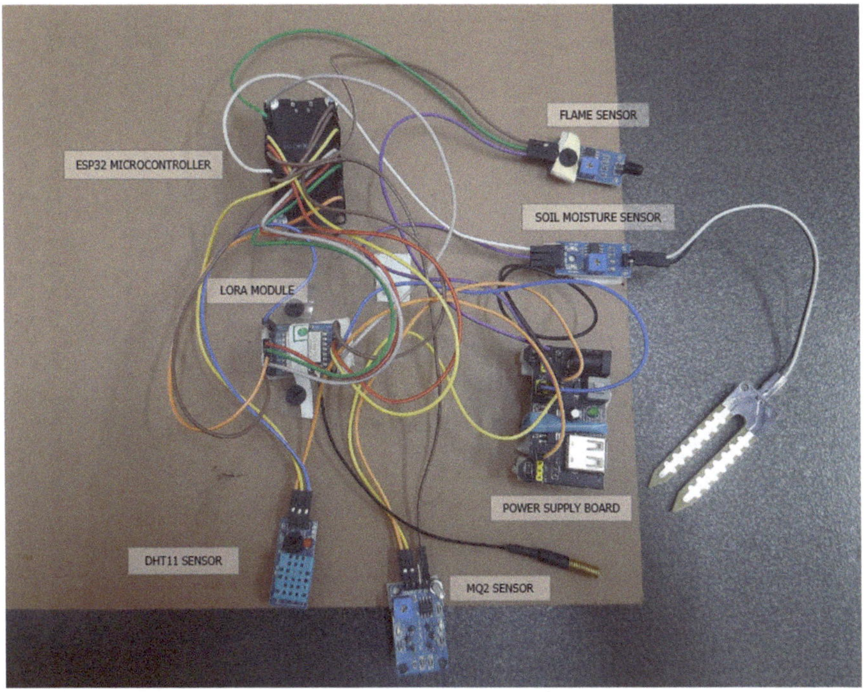

Fig. 4 Visuals of all the hardware components present on each grid

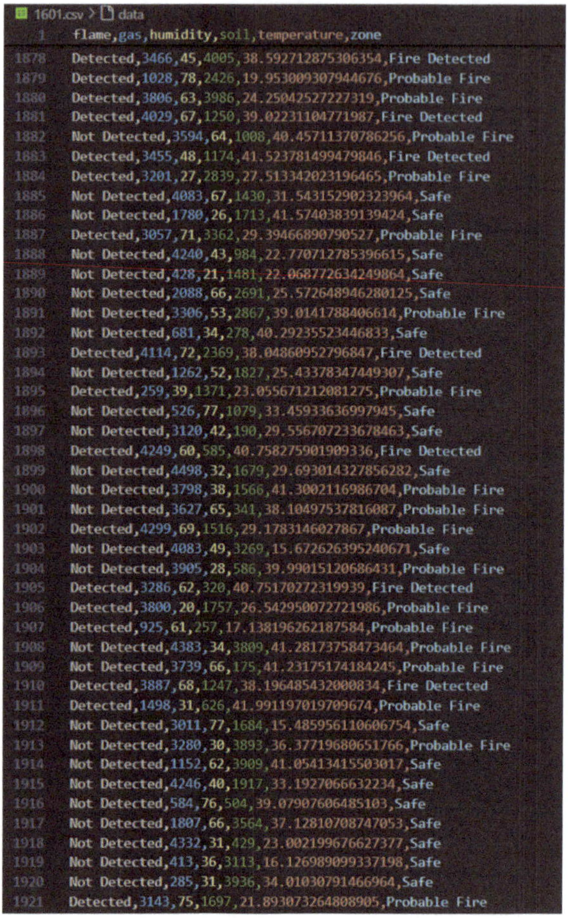

```
1601.csv > data
flame,gas,humidity,soil,temperature,zone
Detected,3466,45,4005,38.592712875306354,Fire Detected
Detected,1028,78,2426,19.953009307944676,Probable Fire
Detected,3806,63,3986,24.25042527227319,Probable Fire
Detected,4029,67,1250,39.02231104771987,Fire Detected
Not Detected,3594,64,1008,40.45711370786256,Probable Fire
Detected,3455,48,1174,41.523781499479846,Fire Detected
Detected,3201,27,2839,27.513342023196465,Probable Fire
Not Detected,4083,67,1430,31.543152902323964,Safe
Not Detected,1780,26,1713,41.57403839139424,Safe
Detected,3057,71,3362,29.39466890790527,Probable Fire
Not Detected,4240,43,984,22.770712785396615,Safe
Not Detected,428,21,1481,22.068772634249864,Safe
Not Detected,2088,66,2691,25.572648946280125,Safe
Not Detected,3306,53,2867,39.0141788406614,Probable Fire
Not Detected,681,34,278,40.2923552344683,Safe
Detected,4114,72,2369,38.04860952796847,Fire Detected
Not Detected,1262,52,1827,25.43378347449307,Safe
Detected,259,39,1371,23.055671212081275,Probable Fire
Not Detected,526,77,1079,33.45933636997945,Safe
Not Detected,3120,42,190,29.556707233678463,Safe
Detected,4249,60,585,40.7582759019093336,Fire Detected
Not Detected,4498,32,1679,29.693014327856282,Safe
Not Detected,3798,38,1566,41.3002116986704,Probable Fire
Not Detected,3627,65,341,38.10497537816087,Probable Fire
Detected,4299,69,1516,29.1783146027867,Probable Fire
Not Detected,4083,49,3269,15.672626395240671,Safe
Not Detected,3905,28,586,39.99015120686431,Probable Fire
Detected,3286,62,320,40.75170272319939,Fire Detected
Detected,3800,20,1757,26.542950072721986,Probable Fire
Detected,925,61,257,17.138196262187584,Probable Fire
Not Detected,4383,34,3809,41.28173758473464,Probable Fire
Not Detected,3739,66,175,41.23175174184245,Probable Fire
Detected,3887,68,1247,38.196485432000834,Fire Detected
Detected,1498,31,626,41.991197019709674,Probable Fire
Not Detected,3011,77,1684,15.485956110606754,Safe
Not Detected,3280,30,3893,36.37719680651766,Probable Fire
Not Detected,1152,62,3909,41.05413415503017,Safe
Not Detected,4246,40,1917,33.1927066632234,Safe
Not Detected,584,76,504,39.07907606485103,Safe
Not Detected,1807,66,3564,37.12810708747053,Safe
Not Detected,4332,31,429,23.002199676627377,Safe
Not Detected,413,36,3113,16.126989099337198,Safe
Not Detected,285,31,3936,34.01030791466964,Safe
Detected,3143,75,1697,21.893073264808905,Probable Fire
```

Fig. 5 Dataset key columns structure for forest fire detection

To store and manage real-time environmental data collected from sensors, Firebase Cloud Database [21] was utilized due to its scalability, real-time capabilities, and ease of integration. The database dynamically updates as sensors transmit data, providing a reliable and centralized storage system for analysis and monitoring.

The Firebase database (Fig. 8) contains key parameters such as flame, temperature, humidity, soil, and gas concentration, enabling a comprehensive overview of fire-prone conditions. The data is structured for easy retrieval, supporting both real-time monitoring and historical analysis for model training and evaluation.

The frontend interface (Fig. 9) provides a user-friendly visualization of real-time fire risk assessments for each grid. Using a color-coded system, the interface effectively communicates varying risk levels, with each grid dynamically updating based on the sensor data processed by the machine learning model.

```python
train_model.py > ...
import pandas as pd
from sklearn.model_selection import train_test_split
from sklearn.tree import DecisionTreeClassifier
from sklearn.metrics import accuracy_score, classification_report
import joblib

# Load the dataset
data = pd.read_csv('1601.csv')

# Drop columns with any missing values (if any)
data = data.dropna()

# Encode the target variable 'flame' - 'Detected' will be 1, 'Not Detected' will be 0
data['flame'] = data['flame'].apply(lambda x: 1 if x == 'Detected' else 0)
data['zone'] = data['zone'].apply(lambda y: 2 if y == 'Fire Detected' else (1 if y == 'Probable Fire' else 0))

# Feature columns (now including 'flame')
X = data[['flame', 'gas', 'humidity', 'soil', 'temperature']]  # Added 'flame' as a feature

# Target column ('zone')
y = data[['zone']]
print(data['zone'].value_counts())

# Split the data (80% for training, 20% for testing)
X_train, X_test, y_train, y_test = train_test_split(X, y, test_size=0.2, random_state=42)

# Initialize the classifier
clf = DecisionTreeClassifier(random_state=42)

# Train the classifier
clf.fit(X_train, y_train)

# Make predictions on the test set
y_pred = clf.predict(X_test)

# Evaluate the model
accuracy = accuracy_score(y_test, y_pred)
print(f'Accuracy: {accuracy * 100:.2f}%')
print('Classification Report:')
print(classification_report(y_test, y_pred))

# Save the trained model
joblib.dump(clf, 'fire_prediction_model_1601.pkl')
```

Fig. 6 Code snippet for training the fire detection model

```
PROBLEMS   OUTPUT   DEBUG CONSOLE   TERMINAL   PORTS

(venv) C:\Users\Yuvraj Sonal\Desktop\forest_fire_detection_2>python train_model.py
Accuracy: 86.00%
Classification Report:
                    precision    recall  f1-score   support

     Fire Detected       0.53      0.62      0.57        13
Probable Fire Zone       0.92      0.92      0.92        84
         Safe Zone       1.00      0.33      0.50         3

          accuracy                           0.86       100
         macro avg       0.82      0.62      0.66       100
      weighted avg       0.87      0.86      0.86       100

(venv) C:\Users\Yuvraj Sonal\Desktop\forest_fire_detection_2>
```

Fig. 7 Visualization of performance metrics

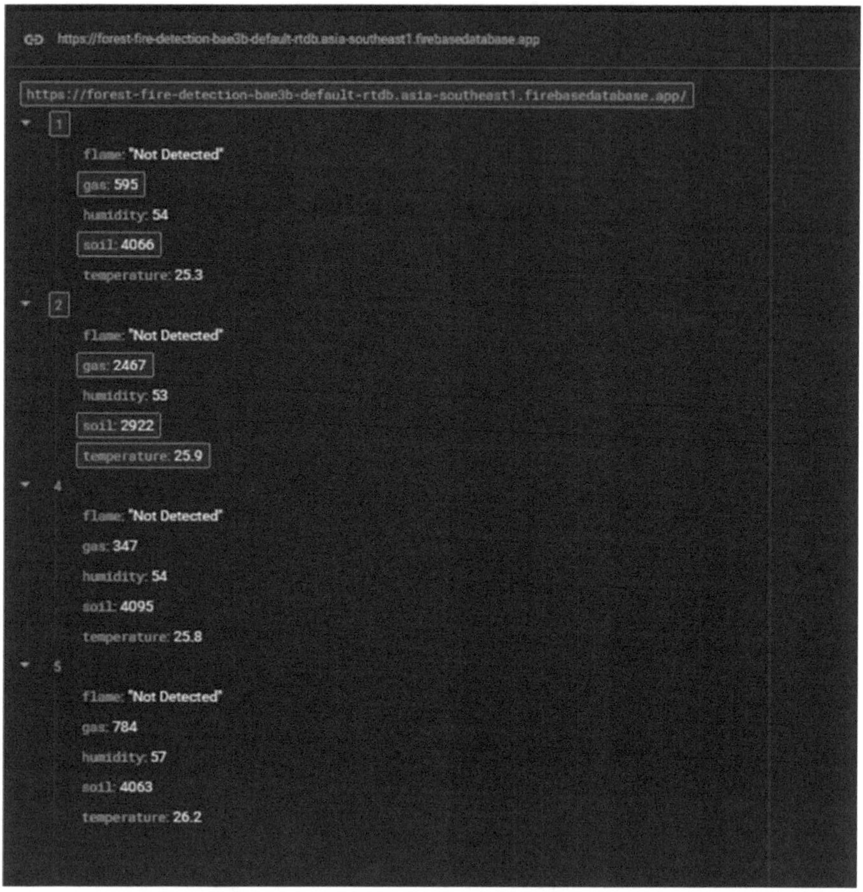

Fig. 8 Snapshot of firebase cloud database storing sensors data

Qualitative results were obtained by observing the system's response to simulated fire scenarios. The visual interface provided clear and actionable insights, with real-time updates on environmental parameters and predictive analytics displayed on the dashboard. Stakeholders reported high usability and effectiveness of the alert notifications, which facilitated timely decision-making during tests. The system effectively bridges the gap between early fire detection and timely response, outperforming traditional methods in accuracy, scalability, and cost-efficiency.

The real-world applicability of this system is evident from its ability to operate in remote regions with minimal infrastructure. The use of LoRa communication ensures scalability, making the system suitable for deployment across large forested areas. The modular design allows for easy expansion, with additional sensors and nodes seamlessly integrated into the network. These features, combined with the system's low operational costs, highlight its potential for widespread adoption in forest management and disaster prevention initiatives.

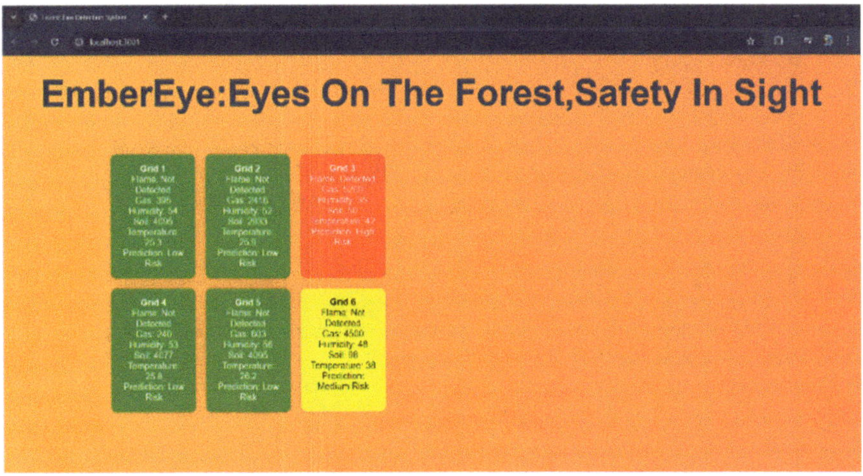

Fig. 9 Real-time detection interface with color-coded outputs showing fire risk levels

5 Conclusions

In this research, a technology has been proposed for detecting and tracking forest fires leveraging IoT devices, machine learning, and cloud computing to provide real-time, efficient, and accurate wildfire detection. The system uses various sensors to monitor environmental parameters, processes the data through machine learning methods for forecasting fire risks, and presents the findings on a color-coded interface. By integrating LoRa for communication, a reliable data transmission can be ensured even in remote areas, addressing the limitations of traditional systems. The proposed solution significantly enhances wildfire detection capabilities, reduces false alarms, and improves response times, contributing to more effective wildfire management.

While the proposed system offers substantial improvements, there are several areas for future development. The integration of additional sensor types, such as air quality and wind sensors, can provide more comprehensive monitoring. Future iterations could also explore the application of more advanced models for machine learning or deep learning techniques to improve prediction accuracy further. Additionally, expanding the system's scalability to monitor larger forested areas and integrating with satellite data for broader coverage could strengthen the system's overall effectiveness. Finally, incorporating automated response mechanisms, such as drone interventions or alert systems for authorities, could further enhance the system's capabilities in mitigating wildfire risks.

References

1. National Interagency Fire Center (2024) Fire statistics. National Interagency Fire Center. https://www.nifc.gov/fire-information/nfn
2. Food and Agriculture Organization (FAO) (2020) The impact of forest fire on forest biodiversity in the Indian Himalayas. FAO. https://www.fao.org/4/XII/0358-B1.htm
3. Abramov N, Emelyanova Y, Fralenko V, Khachumov V, Khachumov M, Shustova M, Talalaev A (2024) Intelligent methods for forest fire detection using unmanned aerial vehicles. Fire 7(3):89
4. Wang H, Zhang G, Yang Z, Xu H, Liu F, Xie S (2024) Satellite remote sensing false forest fire hotspot excavating based on time-series features. Rem Sens 16(13):2488
5. Huang MS, Wichmann B (2024) Machine learning estimates on the impacts of detection times on wildfire suppression costs. PLOS ONE 19(11)
6. Venugopal P, Babu J, Venugopal A, Suresh S (2023) IoT based fire detection system. Int J Adv Res Sci Commun Technol 3(5):816
7. Mouha R (2021) Internet of Things (IoT): definitions, challenges, and recent research directions. J Data Anal Inf Process 9:72–87
8. Kavitha M, Srinivasan R, Suganthy M, Mounika B (2023) IoT based forest fire detection and protection system. In: 2023 second international conference on augmented intelligence and sustainable systems (ICAISS), pp 123–127
9. Noreen U, Bounceur A, Clavier L (2017) A study of LoRa low power and wide area network technology. In: 2017 international conference on advanced technologies for signal and image processing (ATSIP). Fez, Morocco
10. Pennypacker CR, Jakubowski MK, Kelly M, Lampton M, Schmidt C (2013) FUEGO—fire urgency estimator in geosynchronous orbit—a proposed early-warning fire detection system
11. Roy DP, Huang H, Boschetti L, Giglio L, Yan L, Zhang HH, Li Z (2017) Landsat-8 and Sentinel-2 burned area mapping: a sensor harmonization approach. Rem Sens Environ 201:285–298
12. Wang J, Wang G, Qi J, Liu Y, Zhang W (2021) Research of forest fire points detection method based on MODIS active fire product. In: Proceedings of the 28th international conference on geoinformatics, pp 1–6
13. Dutta S, Ghosh S (2021) Forest fire detection using combined architecture of separable convolution and image processing. In: Proceedings of the 1st international conference on artificial intelligence and data analytics (CAIDA), pp 1–6
14. Anim Hossain FM, Zhang Y, Yuan C (2021) A survey on forest fire monitoring using unmanned aerial vehicles. Int J Rem Sens
15. Bosch SG, Vergara L (2011) A ground system for early forest fire detection based on infrared signal processing. Int J Rem Sens 32(17):4857–4870
16. Jones S (2021) Next generation fire detection from geostationary satellites. Rem Sens
17. Alkhatib AAA (2014) A review on forest fire detection techniques. Int J Electr Comput Sci
18. Pang Y, Wu Y, Yuan Y (2023) FuF-Det: an early forest fire detection method under fog. Rem Sens 15(23):5435
19. Chen Y, Wang T, Lin H (2023) Research on forest flame detection algorithm based on a lightweight neural network. Forests 14(12):2377
20. Mienye DI, Jere NR (2024) A survey of decision trees: concepts, algorithms, and applications. IEEE Access 99:1
21. Sharma D Fancy C (2022) Cloud storage security using firebase and Fernet encryption. Int J Eng Trends Technol 70(9):371–375

Comprehensive Analysis of Carbon Footprints, Emissions, and Sequestration in Different Sectors: A Way Forward for Sustainable Practices—An Investigation in the Pune Divisional Area

Rutuja Nitin Gawade, Nivedita Gogate, and Ramesh D. Dod

Abstract This research paper seeks to provide an understanding of carbon footprints in various sectors to assess the environmental effects and explore possible solutions for their reduction. The study concentrates on sectors like Agriculture, Land, Waste Management, Energy and Industry, Transport Infrastructure, and water resources sectors in the present study, carbon emissions of each sector are estimated and the existing measures to control the emission have also been assessed in the Pune Region (Maharashtra). Consequently, the life cycle assessment of this research shows areas that need to be focused on when it comes to carbon reduction. The results also indicate the need to adopt new technologies, improve the legal base, and encourage the use of environmentally friendly solutions in order to achieve the desired levels of CO_2 reduction. This study provides useful recommendations for policymakers, business entities, and scholars interested in the improvement of environmental standards and the fight against climate change. In line with the goals of this sustainability conference, the paper emphasizes the importance of collaborative efforts across various sectors and the adoption of innovative strategies to foster a low-carbon economy. It highlights the need for a unified approach that supports global initiatives focused on building climate resilience and promoting responsible environmental management.

Keywords Environmental sustainability · Carbon footprints · Agriculture · Waste management

R. N. Gawade (✉) · N. Gogate · R. D. Dod
Department of Civil Engineering, Dr. Vishwanath Karad MIT World Peace University, Pune, India
e-mail: rutujagawade29@gmail.com

N. Gogate
e-mail: nivedita.gogate@mitwpu.edu.in

R. D. Dod
e-mail: rameshdod@gmail.com

I. M. Jamadar et al. (eds.), *Integrating Sustainability in Modern Technologies*, Lecture Notes in Mechanical Engineering, https://doi.org/10.1007/978-981-95-9466-5_50

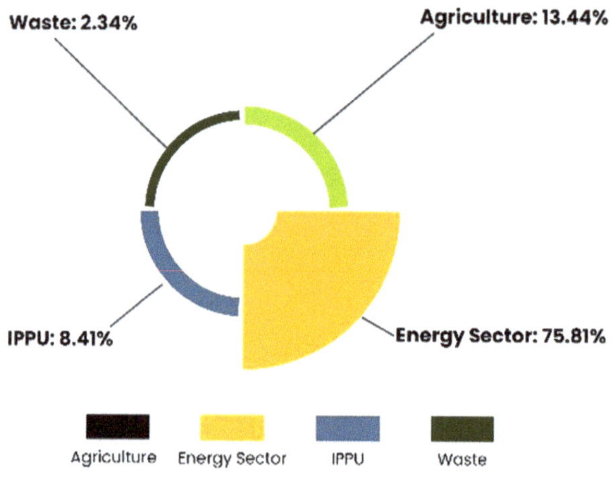

Fig. 1 The contribution of various sectors to the overall economy's emissions in 2019 (*Source* India Climate and Energy Dashboard 2019)

1 Introduction

1.1 Background and Motivation

The threats of climate change and environmental degradation have now become threats to sustainability hence calling for collective action in many fields. This paper will compare how various sectors including Agriculture and Land, Waste Management, Energy and Industries, Transport and Infrastructure, and water resource management sectors affect the environment in the Pune Region (Maharashtra) like any other urban area of the world. It is crucial to identify the carbon footprints of these sectors so that better strategies to minimize the impact can be devised and sustainable practices can be introduced (Fig. 1).

1.2 Significance of Work

The research paper seeks to dwell more on the identified gap of a lack of a comprehensive, systematic analysis and quantification of carbon emissions in various sectors in the Pune Region (Maharashtra). This can be done by evaluating pressures resulting from the major economic activities of the environment and evaluating the feasibility of measures taken in the process. The findings made in this research are useful for the practical application of decision-making practices that aim to improve the scope of environmental management and climate change mitigation in the region.

1.3 Research Objectives and Organization of Paper

This research objective is to study and seeks to conduct an analysis of different economic sectors involved in emitting carbon in the Pune Region (Maharashtra).

To achieve this, the study aims to.

- determine the amount of carbon emission covering Agriculture and Land uses, Waste disposal, Energy Generation and Industries, Transport and Infrastructure, and Water Resources,
- evaluate the effectiveness of current mitigation strategies in reducing carbon emissions,
- identify key areas for intensifying carbon reduction efforts, and
- propose pathways to sustainable practices for policymakers and stakeholders.

2 Literature Review

2.1 Work Reported

An analysis of the literature on the carbon footprint of different industries brings significant insight with regard to the impact of business on the environment, and how this may be improved.

Agriculture and Land

According to [1, 2], and other studies, the carbon emission from the agriculture sector is particularly high as a result of methane produced from livestock and nitrous oxide released from fertilizers. This is why sustainable land management practices are still crucial as they may assist in lowering these emissions. In their publications, Tilman et al. [3] and Vermeulen et al. [4] supported that with climate-smart agriculture and improved crop management, emissions can be minimized. The existing literature on emission reduction has given a better perception of how the reductions can be even more profound via diet change and optimal utilization of land as postulated by Garnett [5] and Foley et al. [6]. Robertson et al. [7] and Mosier et al. [8] have undertaken a discussion of the effects of nitrogen management and other facets of agriculture on the emission of greenhouse gases. Pimentel et al. [9] outline how various systems of agricultural production could be sustainable while Olesen et al. [10] discuss whether emissions could be reduced through organic farming.

Waste Management

Horne et al. [11] and Kumar et al. [12] also note carbon dioxide emissions of waste management and disposal with special reference to methane emissions from landfill sites. They point out that other decentralized technologies such as composting of waste and production of biogas could go a long way in reducing emissions. Levis et al. [13] and Rigamonti et al. [14] presenting a life cycle assessment of various processes of waste management have noted that recycling and energy recovery are favorable. Eriksson and Bisaillon [15] and Martinez-Sanchez et al. [16] describe the environmental effects of waste management and improvements. Cherubini et al. [17] provide a comparison of We technologies and Turner and Williams [18] focus on the management of landfill gas. Mazzant and Boboli [19] examine the economic incentives in waste management while Arena et al. [20] consider technologies for advanced waste treatment.

Energy and Industry

Following the same source, the IPCC [21] pointed out that it is now high time to shift to the use of renewable energy and reduce energy intensity in industries. Sathre and O'Connor [22] attempt to understand the ability of carbon capture and storage technologies in mitigation of emissions from industrial processes. Stern [23] and Pacala and Socolow [24] have given general views on the strategies required for deep cuts in the carbon intensity of the global energy sector. As IEA [25] and McCollum et al. [26] noted policy frameworks and technological advancement are key drivers of low-carbon energy systems. Rackstraw et al. [27] focus on the development of a low-carbon economy and Maranda and Wilkinson [28] analyze the relation between cleaner air and human health. The review of the literature indicates that Edenhofer et al. [29] have cantered their work on the economic consequences of climate policies while Van Vuuren et al. [30] have explored the scenarios for managing global warming.

Transport and Infrastructure

Davis et al. [31] and Kreutzer et al. [32] focus on the potential options for managing transport CO2 emissions: changing the transport mode share and land use transport strategies. These approaches are used in an attempt to reduce the environmental footprint of transport systems in urban centers. Litman [33] and Banister [34] highlight the use of public transport and non-motorized transport in the reduction of emissions. Sakamoto et al. [35] and Schwann et al. [36] have also supported the fact that policy measures and technological developments have made the transport sector possible. In their article, Ewing and Cervero [37] discuss the effects of urban form on travel behavior while Hickman and Banister [38] discuss sustainable mobility. Geels et al. [39] focus on transitions to low-carbon transport systems; Sperling and Gordon [40] examine the possibilities of electric vehicles.

Water Resource Management

Yang et al. [41] and Gonzalez-Eguino et al. [42] assess the carbon footprint of activities and tasks dealing with water resources and water treatment, which are energy intensive. Using less energy and implementing appropriate water management practices are two ways to reduce these emissions. Stokes and Horvath [43] and Sanders and Webber [44] also discuss the EW relationship, describing the dependencies between water usage and energy consumption. Cockerill et al. [44]; Plappally and Lienhard [45]; Siddiqi and Anadon [46] describe technologies and policies to reduce water system carbon footprint. As for the notion of water integration, Gallick

[47] offers meaningful information, and Kenway et al. [48] incorporate information regarding urban water infrastructure. In literature, Vörösmarty et al. [49] explored the impacts of water consumption on ecosystems of the global world, and Fisherman [50] addressed the issue of water scarcity and utilization.

Wiedmann and Minx [51] and Peters [52] provide good methodological papers that are helpful for the estimations and interpretations of the carbon footprint at the sector level. The global consumption and production-based analysis of carbon footprints are discussed in a wider sense by Harwich and Peters [53] and Tukker et al. [54] which supports the integrated approach to sustainability. In the work of Lenzen et al. [55], the authors analyze the implications of international trade on carbon emissions while Suh [56] describes the method of life cycle assessment. Le Quéré et al. [57] update the global carbon budget literature and Wiedmann et al. [58] present a systematic review of the carbon footprint studies. Munksgaard and Pedersen [59] limit themselves to discussing the carbon footprints of countries, whereas Minx and others (2009) address the issues related to carbon foot printing in mid-complexity systems.

2.2 Gap Area

- Even though there is a plethora of literature on carbon footprint analysis across various domains, there are a few areas that deserve further exploration. One giant hole in the literature is the absence of research examining areas such as the Pune Region (Maharashtra) and similar cities in India. This is an area that future research should consider by conducting research that focuses on these local areas. It would also be beneficial to have research that brings in concepts from various disciplines to develop full-fledged strategies for lowering carbon footprints and increasing sustainability (Fig. 2).

2.3 Gap in Approaches Used

Most of the research has focused on the analysis of sector specific techniques to determine the carbon footprints/emissions using approaches.

However, work based on real life data collection from various cities, i.e., ULB's and analysis is rarely reported further, the impact of various mitigation strategies, adopted sector-wise, on carbon emissions based on real-life data is not reported for Indian cities.

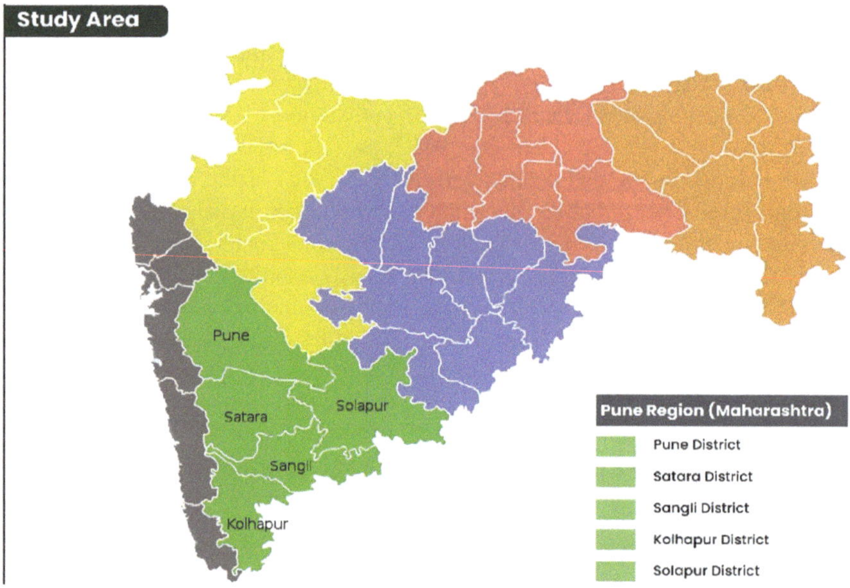

Fig. 2 Map showing Pune region (Maharashtra), i.e., study area

3 Research Methodology

1. **Initiation and Objective Setting**

To identify the research objectives and the research areas of focus, collaboration was made with the Divisional Commissioner Office, Pune. The following research objectives were developed: The first one was to explain what carbon footprints are and the second one was to demonstrate how they can be used for decision-making (Fig. 3).

2. **Stakeholder Engagement**

Consultative sessions were held with targeted representatives from the different sectors in the Pune Region (Maharashtra). Carbon emission information, sustainability goals, and policy considerations were also collected.

3. **Data Collection**

Continuous communication with the Divisional Commissioner Office was kept to secure data from government archives, environmental surveys, and industrial reports. The primary sources of information included energy use statistics, transport trends, waste production, industrial processes, and other factors that affect carbon emissions.

4. **Methodological Framework Development**

Altogether, the chosen approach and data sources were compatible with the research goals and context, and adequate input from stakeholders was incorporated into the

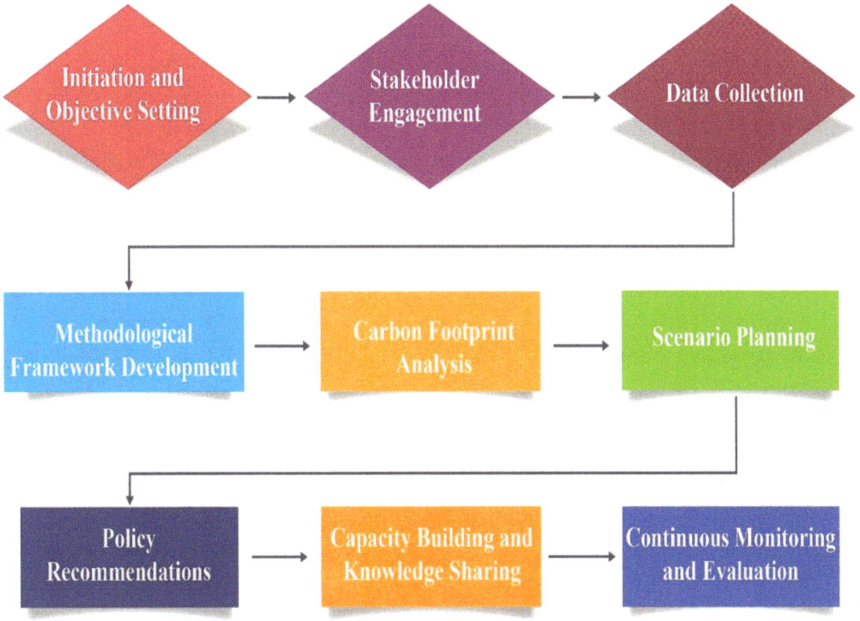

Fig. 3 Research methodology for the proposed study (Pictorial view)

methodological framework. Measures of data collection and analysis were developed to be accurate, and dependable and to maintain the applicability of the results.

5. **Carbon Footprint Analysis**

Standard techniques like life cycle assessment (LCA) were used to estimate the carbon footprint in various sectors of the Pune region (Maharashtra). Carbon emissions were computed by sector and key contributors to emissions such as energy, transport, and industries were identified.

6. **Scenario Planning**

Possible strategies for the reduction of carbon emissions were evaluated according to the available data and opinions of stakeholders as well as taking into account the existing policies. These were used to create carbon emissions scenarios that would illustrate the effects of various mitigation measures and policies on the emissions.

7. **Policy Recommendations**

Policy recommendations were made based on the findings of the carbon footprint assessment and the development of the subsequent scenarios. Recommendations were ranked based on their practicality, the effect they are expected to deliver, and levels of endorsement from stakeholders and submitted to the Divisional Commissioner Office.

8. **Capacity Building and Knowledge Sharing**

Seminars and trainings were conducted to educate the government personnel, industrialists, and other stakeholders in regard to the assessment and management of carbon footprints. The results of the research were presented in the form of reports, policy briefs, and dissemination events so as to ensure that the findings of the study are shared with the various stakeholders.

9. **Continuous Monitoring and Evaluation**

Procedures to track and assess the carbon emissions status and the efficacy of measures that were put in place were developed. The study design was regularly assessed and modified to reflect the availability of new data, new analytical techniques, and feedback from the stakeholders.

The following section was devoted to the presentation of the research methodology and its components: involvement of the Divisional Commissioner Office and the possible contribution to policy and practice in the Pune Region (Maharashtra). It was concluded that stakeholders' cooperation is crucial for maintaining carbon neutrality and achieving sustainable goals, which is why the author is grateful for the Divisional Commissioner's Office's assistance during the research process (Figs. 4 and 5).

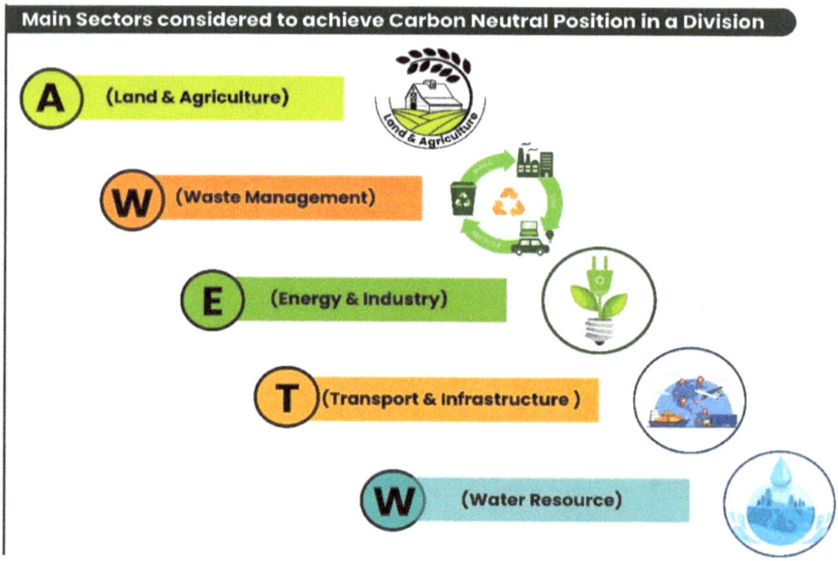

Fig. 4 Main sectors considered to achieve carbon neutral position in a division

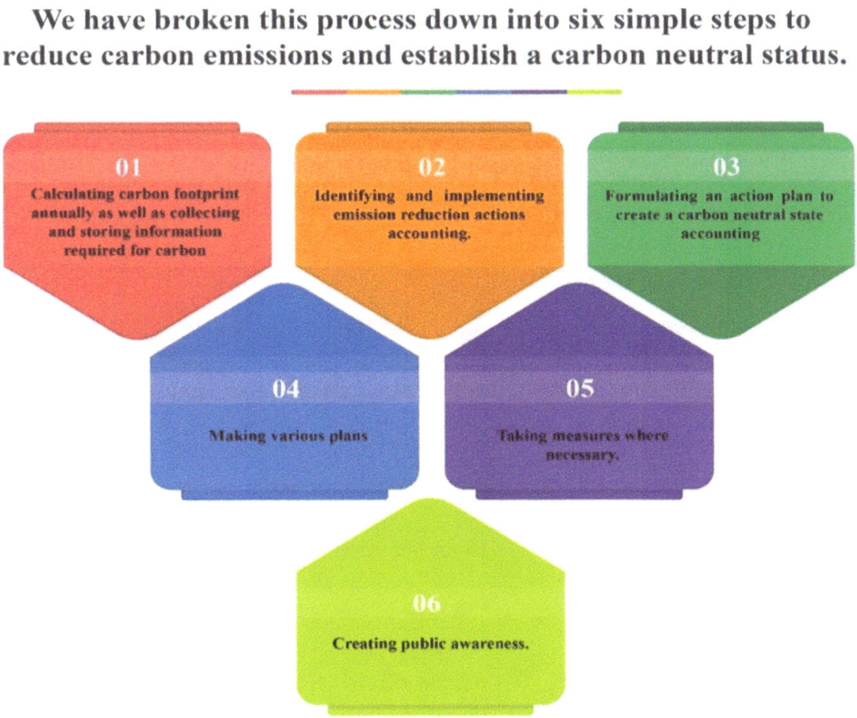

Fig. 5 Graphical representation of six simple steps to reduce carbon emissions and establish a carbon neutral position in a division

4 Data Analysis and Results

4.1 Analysis of Main Sectors Considered to Achieve Carbon Neutrality in the Pune Region (Maharashtra)

1. **Land and Agriculture**

The land sector has been identified as one of the main sources of carbon emissions; particularly deforestation. This process includes the cutting of trees and the change of land cover of forests to other uses such as agricultural or urban. This is because when trees are felled or burnt, the carbon content that is locked within the trees is liberated as carbon dioxide in the atmosphere. It is for this reason that afforestation is considered a vital tool for the sequestration of carbon and reducing emissions.

Farm operations, the use of transport in farming, farm equipment use, fertilizer, and pesticide application also lead to carbon emissions in agriculture. Further, the use of fossil fuels in the electricity-driven production of food and cash crops for export also contributes to these emissions. Some of the activities that contribute greatly to emissions include rice farming and animal farming which lead to the emission of methane, a very bad gas that contributes to the greenhouse effect.

Conversion of natural ecosystems including grasslands or wetlands to croplands or other forms of production can lead to emission of carbon sequestered in the soil and plants. This change often concerns drainage activities, which help to accelerate the process of disintegration of organic matter and consequently, enhance the level of carbon dioxide emission. Knowledge of these processes is crucial for comprehensively assessing the effects of LULC changes on carbon processes and developing efficient methods of their regulation.

- Based on the research analysis conducted for various sectors associated with carbon emissions in the Pune Region (Maharashtra), the following action plan is proposed and Implemented within division by divisional office for each sector to reduce carbon emissions (Table 1).

Sr. No.	Sectors	Description	Estimated total carbon reduction and sequestration
1	Land and agriculture	Tree plantation	16,406 ton sequestration
		Green area creation	3,985,108.02 tons of CO_2 sequestration
			797,021.6048 ton oxygen creation
		Organic compost	902,988 KG of CO_2 emission reduced and 193 ton CO_2 Sequestration

- Based on the research analysis conducted for various sectors associated with carbon emissions in the Pune Region (Maharashtra), the following graphs have been prepared for each sector. These graphs illustrate the estimated total carbon reduction, sequestration, and oxygen creation through the efforts undertaken by

Table 1 Actions taken in sector No. 1, i.e., land and agriculture

1. Land and agriculture	
Land	
1.1	Tree plantation and tree enumeration
1.2	Enhancing and developing green areas 1.2.1. Creation and restoration of grasslands 1.2.2. Biophilia—a green architecture 1.2.3. To create and maintain different parks • Creating biodiversity parks • Creating micro-ecosystems. (e.g., butterfly gardens)
Agriculture	
1.3	Rice farm management (planting recurrent crops and water management)
1.4	Proper management of crop residues (tillage)
1.5	Maximize the use of organic fertilizers
1.6	Cultivation of supplementary crops

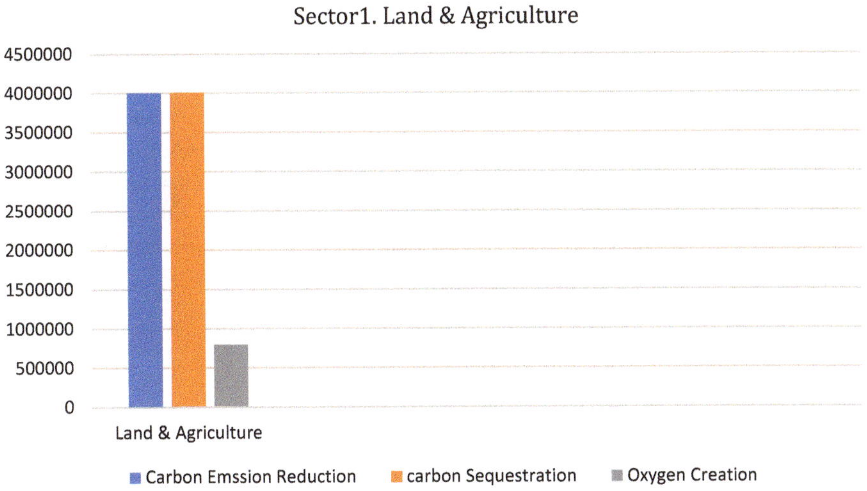

Graph 1 Estimated total carbon reduction/sequestration/oxygen creation through sector 1

the Division to achieve carbon neutrality within the defined parameters in the year 2023 (Graph 1).

2. **Waste Management**

Environmental action on solid waste management is also crucial in reducing Greenhouse effect. Approximately 16. Methane emissions originating from waste contribute to 6% of total emissions, making carbon–neutral practices of waste disposal and management alongside recycling important in reducing the carbon impact.

- **Wet Waste Management**: Reducing wet waste by utilizing fertilizer—an agricultural product, makes composting which is an effective carbon sequestration process. It is not only sequestering carbon in the soil but also ensuring that it does not get emitted to the atmosphere in the form of greenhouse gases. Organic waste items such as leaves, straw, hay, sawdust, and other brown materials have been found to have a high capacity for holding carbon, which makes composting even more effective. During composting, indigenous beneficial microorganisms use carbon sources for energy while nitrogen is used for the formation of proteins, which promote the decomposition process. Organic manure covering about 54% of organic carbon in its composition helps enhance the carbon strength in the soil as well as curtailing methane emission by undergoing aerobic breakdown. In the same way, the capability of each ton of organic manure to sequester carbon is about 0.5 tons of CO_2e, thus underlining its suitability for GHG reduction.
- **Dry Waste Management**: Recycled materials used in managing dry waste are more effective since the raw materials used in production have high-energy consumption and high carbon footprint. Choosing products made of recycled

paper, plastic, metal, and the like not only saves energy but also reduces waste products hence extending the concept of sustainable waste management as well as Greenhouse gas.

- **Wastewater Management**: In sewage management projects, measures such as covering sludge tanks have the possibility to minimize contribution to carbon emissions. This is because, in countries where the tanks are to be installed, methane generated by the process of sewage treatment can be trapped by covering the tanks and thus be stored as carbon. Furthermore, the covering of these tanks also eliminates the chances of gas leakage and hence minimizes the impact on the environment. However, when it comes to the management of wastewater, it is important to incorporate renewable energy sources for high energy-demanding operations. Municipal sewers refer to structures that convey wastewater from sources such as households, industries, and other establishments to WWTPs, and may consume substantial energy for pumping. This energy can be met using renewable energy sources to cut down the indirect carbon emission that comes with the transportation of wastewater. In addition, wastewater management projects should integrate new technologies and include wetlands in the efforts to reduce carbon emissions. New treatment methods may improve the productivity of the wastewater treatment process; wetlands also store carbon and thus can help reduce emissions.
- To achieve carbon- neutrality in wastewater management, it is essential to embrace sustainable sanitation services. Advanced technologies like sewage stabilization ponds and wetlands are environmentally friendly approaches to wastewater treatment, which are critical in achieving carbon neutrality for wastewater management. Therefore, these sustainable practices can be adopted to achieve carbon-free wastewater management for the promotion of environmental sustainability to reduce climate change impacts.
- **Sanitary Waste Management**: This makes the use of air pollution control devices a necessity in order to reduce the increase in carbon emissions that are associated with air pollution during sanitary waste disposal. These devices act as key tools in mitigating evil emissions, hence preserving the environment. Carbon emissions are high because the processes of treating sanitary wastes are energy-demanding; thus, renewable energy sources should be embraced. Carbon footprint can therefore be checked through the reduction of the use of fossil fuels by embracing renewable energy like solar or wind energy.

In addition, the application of gas cleaning installations, particularly scrubbers, is very crucial in minimizing the release of noxious gases. These systems are developed to strip acid gases and other undesirable gases from the flue gas produced during waste disposal processes, thus improving the quality of air. Therefore, in an attempt to reduce carbon emissions even more, using natural gas or biogas, which are considered clean fuels, is beneficial. These cleaner alternatives present feasible ways of minimizing the emission of carbon that is related to waste disposal processes. Furthermore, promoting the use of biodegradable products helps avoid the destruction of sanitary waste through the process of burning. Since biodegradable products

Table 2 Actions taken in the sector no. 2, i.e., waste management

Solid waste management	
2.1	1. Setting up of wet waste—organic fertilizer plant i. To produce biogas 2. Dry waste—setting up of dry waste recycling projects i. (Making durable goods from disposables) 3. Managing legacy waste 4. Solid waste management project using renewable energy 5. Creation of carbon sink areas at sites of solid waste management projects
Wastewater management (sewage treatment plant)	
2.2	1. Covering sludge tanks in sewage management projects 2. Utilization of renewable energy for various operational activities in wastewater management projects requires more energy 3. Use of advanced technology in sewage management projects as well as the creation of wetlands 4. Use of sustainable sanitation services in wastewater management
Sanitary waste (incineration)	
2.3	1. Using pollution free gas free equipment 2. Use of sanitary wares made from biodegradable materials

are used then the carbon emission is reduced, hence supporting the overall cause of emissions and environmental conservation.

- Based on the research analysis conducted for various sectors associated with carbon emissions in the Pune Region (Maharashtra), the following action plan is proposed and Implemented within division by divisional office for each sector to reduce carbon emissions (Table 2).

Estimated total carbon reduction/sequestration through work undertaken

2	Waste management	Dry waste	925.615 ton waste recycled
		Wet waste	902,988 KG of CO_2 carbon emissions reduced, 193 ton Co_2 Sequestration
		Sanitary waste	14.676 ton CO_2 carbon emissions reduced 0.77172 MLD CO_2

- Based on the research analysis conducted for various sectors associated with carbon emissions in the Pune Region (Maharashtra), the following graphs have been prepared for each sector. These graphs illustrate the estimated total carbon reduction, sequestration, and oxygen creation through the efforts undertaken by the Division to achieve carbon neutrality within the defined parameters in the year 2023 (Graph 2).

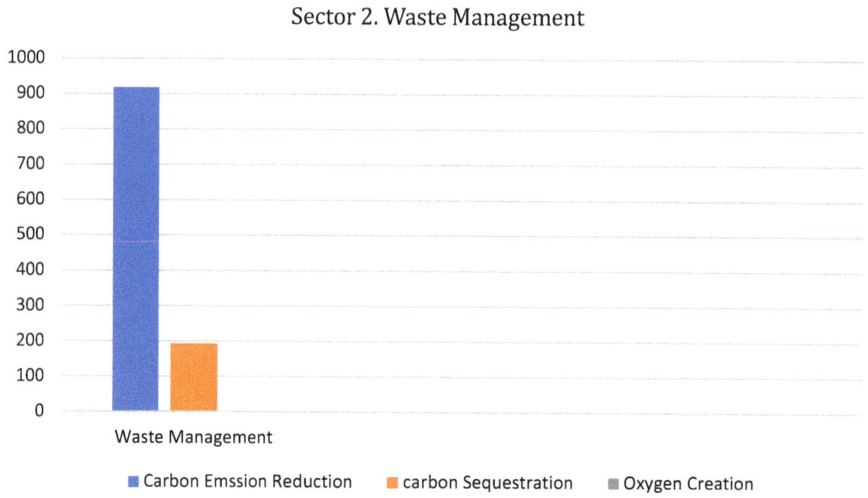

Graph 2 Estimated total carbon reduction/sequestration/oxygen creation through sector 2

3. Energy and Industry

3.1 Energy

Fossil energies primarily used in electricity production are coal, natural gas, and oil, which release carbon dioxide and other greenhouse gases into the environment. These increased carbon emissions make it imperative for the nation to switch to biofuels for power generation in a bid to reduce carbon emissions.

The use of carbon-free energy including solar and wind energy can be used in the creation of a carbon neutral status. Furthermore, hydropower forms another type of renewable energy source that would help mitigate carbon emissions in the long run. Certain types of FFPs have the possibility to add carbon capture and storage (CCS) technology with the purpose of reducing emissions. This advanced technology prevents carbon emissions from being emitted into the atmosphere and instead, deposits it in various forms underground in geological structures thus diminishing the amount of carbon emissions resulting from fossil fuel-generated electricity.

Biomass power plants are another source of electric power generation with minimal effects on the amount of carbon emissions. These plants use carbon–neutral inputs such as timber or crop residues and release carbon dioxide into the atmosphere as a part of the carbon cycle. Thus, the use of biomass power generation does not necessarily increase the amount of carbon dioxide in the atmosphere because new biomass can be produced to replace processed materials, making it carbon neutral.

3.2 Industry

In the industrial segment, the dependency on conventional sources of energy such as coal, natural gas, and oil is frequent for operating the equipment and for conducting numerous processes. Nonetheless, the burning of these fuels leads to the emission of carbon, which supports the prevalence of greenhouse gases in the atmosphere.

Table 3 Actions taken in sector no. 3, i.e., energy and industry

Energy and industry	
3.1	Biofuels should be used (in various processes as well as for transportation purposes)
3.2	Carbon–neutral raw materials should be used
3.3	Use of renewable energy
3.4	Installation of air filters in industrial area premises
3.5	Adoption of low carbon emitting electrical appliances like, 1. LED LIGHTS 2. USE OF SOLAR 3. Solar pumps 4. Biogas generation plant 5. Solar heater

Also, industries as a subsector of the energy and natural resources sector may directly emit greenhouse gases from processes. For instance, cement production causes emissions of carbon dioxide during the chemical process of converting limestone into clinker. Likewise, chemical production may emit gases like methane, and nitrous oxide, and thus the raw materials used in the process must be carbon neutral.

Industrial transport increases carbon emissions through diesel trucks and ships used in transporting products to and from industrial companies. To offset these emissions biofuels must be incorporated into the global mix and this can be the only way to effectively offset carbon emissions linked to industrial transport (Table 3).

Estimated total carbon reduction/sequestration through work undertaken

1.	Energy and industry	Solar and led lights	13,825.2 ton, CO_2 carbon emissions reduced

- Based on the research analysis conducted for various sectors associated with carbon emissions in the Pune Region (Maharashtra), the following graphs have been prepared for each sector. These graphs illustrate the estimated total carbon reduction, sequestration, and oxygen creation through the efforts undertaken by the Division to achieve carbon neutrality within the defined parameters in the year 2023 (Graph 3).

4 Transportation and Infrastructure

4.1 Transportation

Some biofuels such as ethanol and biodiesel can be characterized as carbon dioxide sinks since the plants, from which they are derived, remove this gas from the air during their growth. Carbon management and storage with the goal of using it for transportation can be achieved by carbon capture technology. They collect carbon emissions involved in transport activities and then effectively use them in different processes to reduce the impact on the environment.

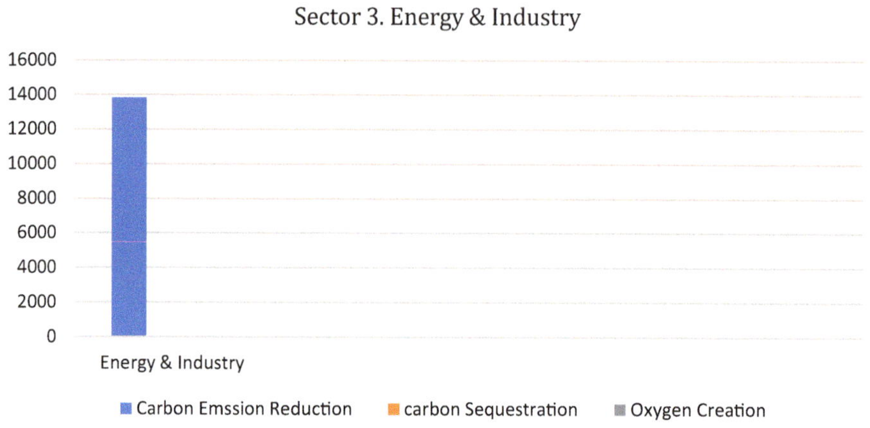

Graph 3 Estimated total carbon reduction/sequestration/oxygen creation through sector 3

Promoting the use of automobiles that do not emit carbon, like walking, cycling, using electric cars, and public transport is an excellent approach in the fight against the emission of carbon. These sustainable substitutes provide tangible measures to reduce the effects of environmental degradation in transportation.

Urban trees are very useful in carbon mitigation strategy and reduction since they reduce vehicular emissions and act as carbon reservoirs. Trees planted and preserved within urban areas regenerate carbon and therefore reduce the impact on the environment.

In addition, not leaving vehicles idling to save energy even for a short span of time like during red signal lights also reduces carbon emissions. This small but effective initiative saves fuel and decreases emissions to protect the environment and minimize people's carbon footprint.

4.2 **Infrastructure**

Building and transportation are significant sources of carbon emission through construction, functional, and maintenance processes. This involves the use of equipment and facilities such as the construction of roads, bridges, and airports; and the use of vehicles like cars, trucks, and airplanes that emit greenhouse gases including carbon dioxide and methane. To this end, there is a need for construction materials that are used for building constructions as well as transportation infrastructure materials to be made environmentally friendly by achieving carbon–neutral status in such processes.

Power generation facilities or power plants are one of the largest sources of carbon emissions if they are dependent on coal or natural gas. Substituting fossil fuels with biofuels is a possible way to counter these emissions. The concept of green spaces like parks, urban forests, and green roofs to be integrated into infrastructure can help reduce carbon emissions by sequestration through photosynthesis. Furthermore, the

Table 4 Actions taken in sector no.4, i.e., transportation and infrastructure

Transportation and infrastructure	
4.1	Air quality monitoring as well as installation of air filters in every signal area
4.2	Using carbon neutral fuel and Increasing footpaths and cycle tracks as well as using public transport
4.3	The number and frequency of public vehicles should be increased
4.4	Effective implementation of electric vehicles policy • Maximum use of electric vehicles • Charging stations for electric vehicles should be increased

integration of Carbon capture and storage technologies from industrial processes or power plants to infrastructure projects is effective in carbon storage.

Some of the activities that are necessary for carbon sequestration and environmental sustainability include rehabilitation, afforestation, reforestation activities, and planting of trees in construction projects.

The challenge of climate change is very real and to meet this challenge, it is crucial to push for sustainable infrastructure. This means that if sustainability is adopted as the key principle when constructing structures, climate change can be dealt with appropriately (Table 4).

Estimated total carbon reduction/sequestration through work undertaken

1.	Transport and Infrastructure	Use of electric vehicle	13.82 ton CO_2 carbon emissions reduced
		Cycle track and pedestrians	
		Air pollution control devices	Captured almost 10–20% of CO_2 daily

- Based on the research analysis conducted for various sectors associated with carbon emissions in the Pune Region (Maharashtra), the following graphs have been prepared for each sector. These graphs illustrate the estimated total carbon reduction, sequestration, and oxygen creation through the efforts undertaken by the Division to achieve carbon neutrality within the defined parameters in the year 2023 (Graph 4).

5. **Water Resource**

Coastal and inland water masses, including oceans, rivers, and lakes, release carbon dioxide and methane through natural processes like respiration and decomposition of organic matter and natural volcanic activity. However, factors associated with people like using fossil fuels and other products, deforestation, and agriculture increase carbon emissions from water sources.

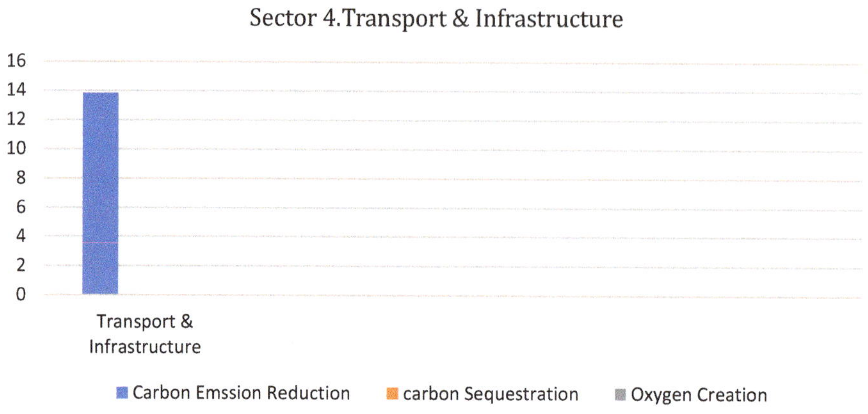

Graph 4 Estimated total carbon reduction/sequestration/oxygen creation through sector 4

With time, organic matter and carbon particles have the propensity to sink to the bottom of most water bodies thereby storing carbon in the sediments. Forests, particularly mangrove forests, tidal marshes, and seagrass beds, commonly known as "Blue Carbon" habitats, hold carbon both in their above-ground biomass and below-ground. Perspectives: The role of coastal ecosystems, particularly mangroves, grasses, and salt marshes in carbon sequestration—commonly known as "blue carbon" habitats—lies in both the plant tissues and the sediments. Furthermore, in the marine environment, the precipitation of calcium carbonate in the form of shells and other products of both biotic and abiotic processes contributes to the long-term sequestration of carbon.

Nevertheless, challenges persist. A number of human activities lead to increased nutrients in water bodies, which leads to a condition known as eutrophication, thereby fostering the growth of algae. When these algae are decomposed, carbon including methane is released to water and air and this underpins the need to address eutrophication. Secondly, carbon dissolved in the waters of the ocean absorbed from the atmosphere causes problems such as ocean acidification which is detrimental to sea life. Therefore, it is essential to mitigate the levels of atmospheric carbon to ensure that marine habitats are protected from the negative effects of climate change.

In conclusion, one has to agree with the fact that water bodies have their own natural way of regulating carbon but human beings unreasonably enhance the rate of carbon emissions, and therefore need to take proactive measures to reduce such effects on the aquatic environments (Table 5).

Estimated total carbon reduction/sequestration through work undertaken

1	Water resource	Creation of wetland and conservation of wetland	1776 ton of CO_2 sequestration

Table 5 Actions taken in the sector no. 5, i.e., water resources

Water resource	
5.1	Take preventive measures to prevent the occurrence of waterfowl and if it occurs, remove it and manage it properly
5.2	Creation and conservation of wetlands
5.3	Celebration of carbon–neutral festivals to prevent water pollution

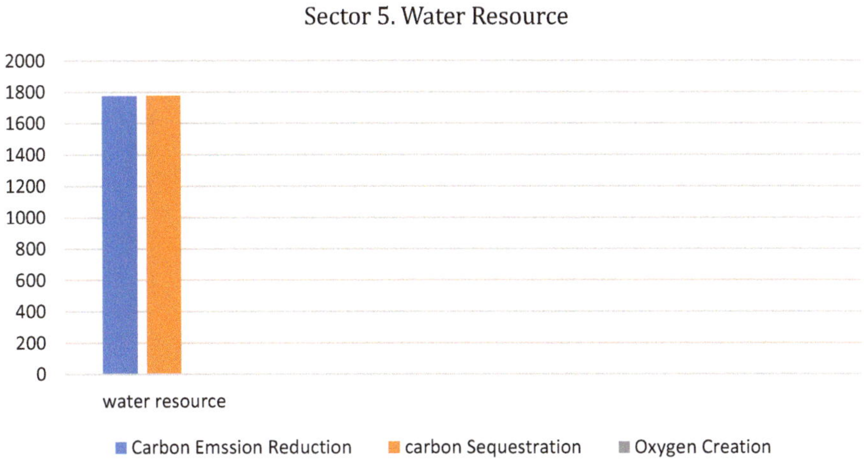

Graph 5 Estimated overall sequestration, reduction, and oxygen creation through sector 5

- Based on the research analysis conducted for various sectors associated with carbon emissions in the Pune Region (Maharashtra), the following graphs have been prepared for each sector. These graphs illustrate the estimated total carbon reduction, sequestration, and oxygen creation through the efforts undertaken by the Division to achieve carbon neutrality within the defined parameters in the year 2023 (Graph 5).

4.2 Summary of Estimated Carbon Reduction in All Sectors

See Table 6.

4.3 Results and Interpretation

This study reveals that the energy and transport sector is accountable for the largest share of carbon emissions in the Pune Region (Maharashtra) and this is followed by

Table 6 Estimated total carbon reduction/sequestration through work undertaken to make the Pune Division carbon neutral within the orbit

Sr. No.	Sectors	Description	Estimated total carbon reduction and sequestration
1	Land and agriculture	Tree plantation	16,406 ton sequestration
		Green area creation	3,985,108.02 tons of CO_2 sequestration
			797,021.6048 ton oxygen creation
		Organic compost	902,988 KG of CO_2 emission reduced and 193 ton CO_2 sequestration
2	Waste management	Dry waste	925.615 ton waste recycled
		Wet waste	902,988 KG of CO_2 carbon emissions reduced, 193 ton CO_2 sequestration
		Sanitary waste	14.676 ton CO_2 carbon emissions reduced 0.77172 MLD CO_2
3	Transport and infrastructure	Use of electric vehicle	13.82 ton CO_2 carbon emissions reduced
		Cycle track and pedestrians	
		Air pollution control devices	Captured almost 10–20% of CO_2 daily
4	Energy and industry	Solar and led lights	13,825.2 ton, CO_2 carbon emissions reduced
5	Water resource	Creation of wetland and conservation of wetland	1776 ton of CO_2 sequestration

Source Data collection from various Municipal councils under Pune Region (Maharashtra)

agriculture and waste management sectors. This therefore requires that more attention be shifted towards practicing within these sectors with the aim of reducing carbon emissions. Besides, it also highlights the significance of carbon storage processes such as afforestation and other conservation measures of land in controlling carbon emissions in the Pune Region (Maharashtra). This evidence supports the argument for adopting cross-sectoral systems in different sectors to achieve substantial reductions in carbon emissions and foster sustainability.

5 Contribution of the Paper

5.1 Contribution to Practice

- **Policy Informatics**: In sum, the research paper is beneficial to policymakers in that it highlights which sectors and what drivers most significantly contribute to carbon emissions. With reference to the areas that could potentially benefit from intervention, this study provides direction on how policy interventions could be implemented to reduce carbon footprints.

- **Strategic Planning**: The findings of the research paper can be useful in strategic management practices and activities as businesses and organizations can apply the findings in their practice. Carbon foot printing helps entities understand their emission levels and their impact, and in that way, it will be easier to spot the potential for emission reduction, resource efficiency, and sustainable advancement.
- **Community Engagement**: The research paper plays a significant part in contributing to the efforts toward community engagement through the provision of general information regarding carbon footprints and their impacts on the environment. Thus, the study not only makes research findings easily available to the public, in forms such as community workshops or educational programs but also enables individuals and communities to prevent carbon emissions.

In conclusion, the research paper has various strengths in theoretical contribution since it provides clarity in concepts, frameworks, and interdisciplinary knowledge. Further, its practical applications include policy making, strategic management, community involvement in sustainability initiatives, and thus the achievement of a sustainable environment toward the realization of a sustainable future.

6 Conclusion

In conclusion, these studies contribute to the understanding of carbon footprint across various sectors in the Pune Region (Maharashtra) region and thereby promote sustainability. Thus, through assembling and analyzing data systematically, the study contributes not only to the theoretical knowledge but also provides recommendations for policymakers, businesses, and communities within the region. These findings may be useful to policymakers who would like to develop policies that would help to address carbon footprints while it would also be useful to businesses who would like to know how they can minimize their emissions and make better use of their resources. Furthermore, it will help various community engagement strategies to enhance awareness and capacity within individuals to reach the goal of carbon neutrality. In sum, this study opens a way to further research and cooperation for the region to achieve the goal of environmental sustainability, therefore, more research should be conducted to support the progression towards a green future.

7 Limitations and Future Scope

7.1 Limitations

Data Availability: The first major limitation of this study is that the research is limited by the data that is available and the accuracy of the data. The estimates used for the amount of carbon that can be reduced and sequestered are based on measurements from various sources and hence if the data collected is not accurate or there is a lack of it, then the results are likely to be off.

Assumptions and Simplifications: It is also important to note that the analysis may have led to some assumptions and simplifications, which would affect the conclusions made. For example, when it comes to carbon sequestration rates, it is possible that they differ from one region to another depending on the factors in those regions, and these factors might not have been taken into account properly.

Implementation Challenges: Despite the availability of various actions to minimize carbon emissions as discussed in the proposed action plan, the actual implementation of such actions may be hindered by practicalities such as funding constraints, existing policy barriers, and community opposition. These challenges could easily reduce the impact of the proposed interventions.

Lack of Long-Term Assessment: The information derived from the research offers a projection of lowering carbon dioxide and carbon storage stemming from short-term management measures applied over a certain time frame. Yet, the durability and efficiency of these interventions are not clear and might require constant assessment and reinforcement.

7.2 Future Scope

Longitudinal Studies: Subsequent studies could conduct a cross-sectional examination to assess the effectiveness of the carbon reduction interventions put in place in the Pune Region (Maharashtra). Monitoring carbon emissions in the context of time would provide a useful understanding of how much impacts and mitigation measures are beneficial and how long they can be effective.

Integrated Modelling: Combining different types of simulation, like system dynamics or agent-based modeling could lead to a better understanding of the interdependencies between different sectors and the impact of the latter on carbon emissions. The synergy of these approaches would help policymakers design less flawed plans for the reduction of carbon emissions.

Stakeholder Engagement: Involving stakeholders across sectors such as government ministries, industries, academic institutions, and citizens could contribute to the development of policies with respect to carbon reduction and the effective implementation of those policies. In further studies, one may try to employ the participatory approach to develop sustainable interventions together with the stakeholders.

Technological Innovations: These new prospects for mitigation of carbon emissions in the Pune Region (Maharashtra) consist of further studies and advancements in renewable energy sources, carbon dioxide capture and sequestration, and efficient practices of agriculture. Possible future research directions include new technologies and their efficiency for large-scale implementation.

Policy Evaluation: Efficient assessment of current policies and regulations as to their capability to lower carbon emissions would be important for fine-tuning the policy framework in order to find out its weaknesses and strengths. Future work could, therefore, involve the evaluation of policy effects on carbon footprint and the documenting of policies that require more attention.

If these limitations are to be further addressed and other research directions are to be investigated, scholars and policymakers would be able to improve the understanding of carbon emissions in the Pune Region (Maharashtra) and design the corresponding strategies for attaining carbon neutrality and achieving environmental sustainability.

Acknowledgements We would like to express our heartfelt gratitude to the Municipal Administration Department, and Divisional Commissioner Office for their immense support and encouragement towards this research. Their commitment to attaining the objectives of environmental sustainability and carbon reduction in the Pune Region (Maharashtra) has been instrumental in the success of this research. We would like to thank the pertinent industry experts and stakeholder representatives for their insightful comments and valuable insights, which helped us to clarify and expand on the research findings. Their assistance has been instrumental in shaping our findings. Finally, we are grateful for the support of all the people and institutions involved in this research; the divisional office has been helpful, and its cooperation has been critical to our pursuit of carbon neutrality in the Pune Region (Maharashtra).

Declarations The authors declared no conflict of interest regarding the study.

References

1. Smith P et al Agriculture, forestry and other land use (AFOLU)
2. Farooq M, Sanaullah M, Nadeem F, Gogoi N, Arshad MS, Lal R (2018) Soil degradation and climate change in South Asia. In: Soil and climate. CRC Press
3. Global food demand and the sustainable intensification of agriculture | Request PDF. ResearchGate (2024). https://doi.org/10.1073/pnas.1116437108
4. (PDF) Climate change & food systems: assessing impacts and opportunities. ResearchGate. Accessed 11 Mar 2025. https://www.researchgate.net/publication/317170041_Climate_Change_Food_Systems_Assessing_Impacts_and_Opportunities
5. Where are the best opportunities for reducing greenhouse gas emissions in the food system (including the food chain)? | Request PDF. ResearchGate (2024). https://doi.org/10.1016/j.foodpol.2010.10.010

6. Foley JA et al (2011) Solutions for a cultivated planet. Nature 478(7369):337–342. https://doi.org/10.1038/nature10452
7. Robertson GP, Paul EA, Harwood RR (2000) Greenhouse gases in intensive agriculture: contributions of individual gases to the radiative forcing of the atmosphere. Science 289(5486):1922–1925. https://doi.org/10.1126/science.289.5486.1922
8. (PDF) Closing the global N_2O budget: nitrous oxide emissions through the agricultural nitrogen cycle. Accessed 11 Mar 2025. https://www.researchgate.net/publication/220048332_Closing_the_Global_N2O_Budget_Nitrous_Oxide_Emissions_through_the_Agricultural_Nitrogen_Cycle
9. (PDF) Environmental, energetic, and economic comparisons of organic and conventional farming systems. ResearchGate (2024). Accessed 11 Mar 2025. Available: https://www.researchgate.net/publication/271847517_Environmental_energetic_and_economic_comparisons_of_organic_and_conventional_farming_systems
10. (PDF) Impacts and adaptation of European crop production systems to climate change. ResearchGate (2025). https://doi.org/10.1016/j.eja.2010.11.003
11. Hoornweg D, Bhada-Tata P, Kennedy C (2013) Environment: waste production must peak this century. Nature 502(7473):615–617. https://doi.org/10.1038/502615a
12. (PDF) Challenges and opportunities associated with waste management in India. ResearchGate (2024). https://doi.org/10.1098/rsos.160764
13. Levis JW et al (2010) Greenhouse gas emissions from municipal solid waste management strategies. Environ Sci Technol 44(12):5298–5304
14. Rigamonti L et al (2009) Environmental impact evaluation of municipal solid waste management. Resour Conserv Recycl 54(2):64–78
15. Eriksson O, Bisaillon M (2011) Modelling multiple waste management systems. Waste Manage 31(12):2620–2630
16. Martinez-Sanchez V et al (2015) Life-cycle costing of food waste management in Denmark. Environ Sci Technol 49(8):4698–4707
17. Life cycle assessment (LCA) of waste management strategies: landfilling, sorting plant and incineration | Request PDF. ResearchGate (2025). https://doi.org/10.1016/j.energy.2008.08.023
18. Turner DA, Williams ID, Kemp S (2015) Greenhouse gas emission factors for recycling of source-segregated waste materials. Resour Conserv Recycl 105:186–197. https://doi.org/10.1016/j.resconrec.2015.10.026
19. Waste generation, waste disposal and policy effectiveness: evidence on decoupling from the European Union | Request PDF. ResearchGate (2024). https://doi.org/10.1016/j.resconrec.2008.07.003
20. A life cycle assessment of environmental performances of two combustion and gasification-based waste-to-energy technologies | Request PDF. Accessed 11 Mar 2025. Available: https://www.researchgate.net/publication/275342851_A_life_cycle_assessment_of_environmental_performances_of_two_combustion_and_gasification-based_waste-to-energy_technologies
21. IPCC Special Report, 'Global Warming of 1.5 °C'—WeForest. Accessed 11 Mar 2025. [Online]. https://www.weforest.org/blog/opinion/ippc-global-warming-15-c-special-report-6-oct-2018/?gad_source=1&gclid=CjwKCAjwvr--BhB5EiwAd5YbXkZd2KBCzPC1MsuPpTIho333u9N8G3ZvLL6O3G5m6aSd6SHuWWPZTxoCUwEQAvD_BwE
22. Seppälä J et al (2019) Effect of increased wood harvesting and utilization on required greenhouse gas displacement factors of wood-based products and fuels. J Environ Manage 247:580–587. https://doi.org/10.1016/j.jenvman.2019.06.031
23. 9780521700801_frontmatter.pdf. Accessed 11 Mar 2025. Available: https://assets.cambridge.org/97805217/00801/frontmatter/9780521700801_frontmatter.pdf
24. Pacala S, Socolow R (2004) Stabilization wedges: solving the climate problem for the next 50 years with current technologies. Science 305(5686):968–972. https://doi.org/10.1126/science.1100103
25. IEA—International energy agency. Accessed 11 Mar 2025. Available: https://www.iea.org/

26. Mccollum D et al (2013) Energy investments under climate policy: a comparison of global models. Clim Change Econ (CCE) 04(04):1–37
27. Rockström J, Gaffney O, Rogelj J, Meinshausen M, Nakicenovic N, Schellnhuber HJ (2017) A roadmap for rapid decarbonization. Science 355(6331):1269–1271. https://doi.org/10.1126/science.aah3443
28. Energy and health. 2. Electricity generation and health. Lancet | Request PDF ResearchGate (2024). https://doi.org/10.1016/S0140-6736(07)61253-7
29. (PDF) Transport (2025) Climate Change 2014: mitigation of climate change. In: Contribution of working group III to the fifth assessment report of the intergovernmental panel on climate change. Accessed 11 Mar 2025. Available: https://www.researchgate.net/publication/274897242_Transport_In_Climate_Change_2014_Mitigation_of_Climate_Change_Contribution_of_Working_Group_III_to_the_Fifth_Assessment_Report_of_the_Intergovernmental_Panel_on_Climate_Change
30. (PDF) The representative concentration pathways: an overview. Climatic Change. This issue. Accessed 11 Mar 2025. Available: https://www.researchgate.net/publication/227583841_The_representative_concentration_pathways_an_overview_Climatic_Change_This_issue
31. Consumption-based accounting of CO_2 emissions—PubMed. Accessed 11 Mar 2025. Available https://pubmed.ncbi.nlm.nih.gov/20212122/
32. Creutzig F, Fernandez B, Haberl H, Khosla R, Mulugetta Y, Seto KC (2016) Beyond technology: demand-side solutions for climate change mitigation. 01 Oct 2016. Social Science Research Network, Rochester, NY, 2859440. https://doi.org/10.1146/annurev-environ-110615-085428
33. (PDF) Sustainable transport: a state-of-the-art literature review. Accessed 11 Mar 2025. Available: https://www.researchgate.net/publication/365818929_Sustainable_Transport_A_State-of-the-Art_Literature_Review
34. The sustainable mobility paradigm | Request PDF. Accessed 11 Mar 2025. Available: https://www.researchgate.net/publication/223541451_The_sustainable_mobility_paradigm
35. Pathways to low-carbon transport—from technical and non-technical perspectives. Accessed 11 Mar 2025. Available: https://www.researchgate.net/publication/282579598_Pathways_to_Low-Carbon_Transport_-_From_Technical_and_Non-Technical_Perspectives
36. Scientific Research about climate change mitigation in transport: a critical review | Request PDF. Accessed 11 Mar 2025. Available: https://www.researchgate.net/publication/251629658_Scientific_Research_about_Climate_Change_Mitigation_in_Transport_A_Critical_Review
37. Travel and the built environment: a meta-analysis | request PDF. Accessed 11 Mar 2025. Available: https://www.researchgate.net/publication/275035324_Travel_and_the_Built_Environment_A_Meta-Analysis
38. Hickman R, Banister D (2014) Transport, climate change and the city. Routledge, London. https://doi.org/10.4324/9780203074435
39. (PDF) Strategies and instruments for low-carbon urban transport: an international review on trends and effects. ResearchGate (2024). https://doi.org/10.1016/j.tranpol.2012.07.003
40. Two Billion Cars: Driving Toward Sustainability | Request PDF. Accessed 11 Mar 2025. Available: https://www.researchgate.net/publication/227466960_Two_Billion_Cars_Driving_Toward_Sustainability
41. Analysis of Wastewater Reuse Potential in Beijing | Request PDF. Accessed 11 Mar 2025. Available: https://www.researchgate.net/publication/222301462_Analysis_of_Wastewater_Reuse_Potential_in_Beijing
42. (PDF) Climate Policy The nexus approach to water–energy–food security: an option for adaptation to climate change. ResearchGate (2024). Accessed 11 Mar 2025. Available: https://www.researchgate.net/publication/275152818_Climate_Policy_The_nexus_approach_to_water-energy-food_security_an_option_for_adaptation_to_climate_change
43. Stokes JR, Horvath A (2009) Energy and air emission effects of water supply. Environ Sci Technol 43(8):2680–2687. https://doi.org/10.1021/es801802h
44. (PDF) Water use in the US energy system: a national assessment and unit process inventory of water consumption and withdrawals. Accessed 11 Mar 2025. Available: https://www.researchgate.net/publication/325031477_Water_use_in_the_US_energy_system_A_national_assessment_and_unit_process_inventory_of_water_consumption_and_withdrawals

45. Energy requirements for water production, treatment, end use, reclamation, and disposal | Request PDF. ResearchGate (2024). https://doi.org/10.1016/j.rser.2012.05.022
46. (PDF) Water-Energy-Food Nexus in the Middle East and North African Countries (MENA). Accessed 11 Mar 2025. Available: https://www.researchgate.net/publication/347203245_Water-Energy-Food_Nexus_in_the_Middle_East_and_North_African_Countries_MENA
47. The Water Poverty Index: an international comparison. Accessed 11 Mar 2025. Available: https://ideas.repec.org/p/kee/kerpuk/2002-19.html
48. Non-conventional water resources and opportunities for water augmentation to achieve food security in water scarce countries | Request PDF. Accessed 11 Mar 2025. Available: https://www.researchgate.net/publication/222296599_Non-conventional_water_resources_and_opportunities_for_water_augmentation_to_achieve_food_security_in_water_scarce_countries
49. (PDF) Water use. Accessed 11 Mar 2025. Available: https://www.researchgate.net/publication/200043175_Water_use
50. (PDF) Crop evapotranspiration-Guidelines for computing crop water requirements-FAO Irrigation and drainage paper 56. Accessed 11 Mar 2025. Available: https://www.researchgate.net/publication/235704197_Crop_evapotranspiration-Guidelines_for_computing_crop_water_requirements-FAO_Irrigation_and_drainage_paper_56
51. Making the most of scarcity : accountability for better water management results in the Middle East and North Africa. Accessed 11 Mar 2025. Available: https://documents.worldbank.org/pt/publication/documents-reports/documentdetail/353971468280764676/making-the-most-of-scarcity-accountability-for-better-water-management-results-in-the-middle-east-and-north-africa
52. (PDF) Population momentum and the demand on land and water resources. Accessed 11 Mar 2025. Available: https://www.researchgate.net/publication/25456688_Population_momentum_and_the_demand_on_land_and_water_resources
53. Hoff H (2011) Understanding the Nexus. In: Background paper for the Bonn 2011 conference the water, energy and food security Nexus, Stockholm Environment Institute (SEI), Stockholm, Sweden. References. Scientific Research Publishing. Accessed 11 Mar 2025. Available: https://www.scirp.org/reference/ReferencesPapers?ReferenceID=2018811
54. OECD Environmental Outlook to 2050, OECD. Accessed 11 Mar 202. Available: https://www.oecd.org/en/publications/oecd-environmental-outlook-to-2050_9789264122246-en.html
55. Falkenmark M, Rockström J (2006) New paradigms in water resources planning and management. J Water Resour Plan Manag 132(3):129–132
56. Elimelech M, Phillip WA (2011) Future of seawater desalination: energy, technology, and environment. Science 333(6043):712–717
57. Wada Y et al (2011) Global depletion of groundwater resources. Geophys Res Lett 37(20):L20402
58. Postel SL, Richter BD (2003) Rivers for life: managing water for people and nature. Island Press
59. Molden D et al (2010) Optimism and caution in improving agricultural water productivity. Agric Water Manag 97(4):528–535

An Effective Approach to Identifying Plant Disease Through Image Processing and Machine Learning

Sandeep Sharma, Saruchi, Priyanka Garg, Pradeep Chauhan, Amandeep Kaur, and Harish Dutt Sharma

Abstract Sustainable practices can be achieved by integrating plant disease identification using image processing and machine learning. It is a smarter, more efficient, and eco-friendly agricultural system approach, which not only contributes to long-term environmental health but also plays a pivotal role in economic growth and social well-being. Another important requirement for automated and precise disease diagnosis in contemporary agriculture is addressed in this work, which proposes a deep learning-based system for disease detection in plants. Conventional visual inspection techniques can be subjective and consume a lot of time, which may cause delays in farmers' timely intervention and decision-making. The five main steps in the suggested technique are pre-processing and data collection, feature extraction, machine learning model construction, model evaluation and optimization, and deployment of the trained model into an intuitive real-time detection system. This

S. Sharma (✉)
Dept. of CSE, School of AI and Emerging Technologies, Lovely Professional University, Phagwara 144411, Punjab, India
e-mail: sanintel123@gmail.com

Saruchi
Computer Science and Engineering, Chandigarh University, Gharuan, Mohali 140413, Punjab, India
e-mail: ganpati.saruchi@gmail.com

P. Garg · P. Chauhan
Computer Science and Engineering, Shivalik College of Engineering, Dehradun 248197, Uttarakhand, India
e-mail: garg.priyanka1909@gmail.com

P. Chauhan
e-mail: pcpradeep0291@gmail.com

A. Kaur
Computer Science and Engineering, MMEC, MMDU Mullana, Ambala 133207, Haryana, India
e-mail: amandeepkaur@mmumullana.org

H. D. Sharma
School of Computer Engineering and Applications, Maya Devi University, Dehradun 248011, India
e-mail: sharma.harish106@gmail.com

I. M. Jamadar et al. (eds.), *Integrating Sustainability in Modern Technologies*, Lecture Notes in Mechanical Engineering, https://doi.org/10.1007/978-981-95-9466-5_51

methodology seeks to give farmers a dependable tool for early disease diagnosis by utilizing cutting-edge image processing and machine learning techniques. This will allow farmers to take proactive measures to safeguard crops and maximize agricultural productivity.

Keywords Image processing · Deep learning · Machine learning · AlexNet · AlexNetOWTBn · GoogLeNet · OverFeat · VGG

1 Introduction

Crop diseases affect agricultural production, food security, and economic stability, posing serious difficulties to farmers globally. The Food and Agriculture Organization (FAO) of the USA estimates that the plant-related diseases and pests in crop cause 20–25% of the world's crop production to be lost each year. Crop diseases can result in yield losses ranging from 10 to 40%. Conventional disease diagnosis techniques, which frequently depend on farmers visually inspecting samples, are labor- and time-intensive as well as subjective and prone to human error. Due to this, many farmers are facing troubles in identifying the crop diseases.

For farmers, creating a mobile application which is generally used to recognize the diseases that crops might encounter in the coming days using the latest technologies like image processing and machine learning algorithms is extensive. If the authors develop an application where a farmer can use it for the early recognition of the illness suffered by the crops and helps in the early diagnosis and in making informed decisions, it can help the farmers. The proposed program could handle and reduce the losses to maximum that might be caught because of the crop diseases, and also it can be helpful in utilizing the maximum resources and gradually helps in supporting the sustainable practices of the agriculture by providing the necessary knowledge and capability and helps in recognizing diseases and managing crops to the farmers. The dedicated approach for recognizing the plant diseases using deep learning and also using many trending technologies like image processing and machine learning models. Providing an effective, easy to use, and easy to adapt tool which can be helpful in recognizing the diseases early in the crop and also to treat the crop illness can lead to crop productivity to farmers which would be beneficial.

2 Literature Review

In the deep learning-based banana leaf disease classification, the author used Convolutional Neural Network (CNNs). This research is all about categorizing the leaves of banana using high-resolution image datasets. In this research it majorly says about the ability of deep learning principles in classifying and effectively recognizing the illness of the banana leaves [1].

The methodology for diagnosing the illness caused to the banana plants automatically. In this research, the team used the concepts like artificial neural networks and Z-transform characteristics for evaluating and calculating the diseases that might impact banana plants using the pictures. This research has got more accuracy in identifying the diseases in the banana plants and diagnosing the illness automatically [2].

The machine learning approaches are used in identifying the plant diseases and also help in diagnosing the illness of the plants. In this paper, the researchers also discussed both conventional and deep learning techniques in order to cure the plants illness. Also the researchers discussed the uses of the deep learning techniques, their uses, advantages, and also disadvantages. Also their study gave us insights of the existing models and further directions in the field of recognizing the plant diseases [3].

The image processing method is used to identify the disease on tomato leaves at its early stage. The researchers have also used many image processing techniques to analyze and visualize the tomato leaves based on their visual symptoms. This research says the ability to diagnose the diseases of the tomato leaves by using the image processing techniques at its early stages and also helps in producing more yield without any loss [4].

The detailed study of technologies that help in disease detection such as machine learning modules, conventional algorithms, and also deep learning techniques. The researchers have also provided the efficiency, benefits, and also drawbacks of the model. For students who are researching in this domain, this study gives valuable insights and more knowledge and also how machine learning can help to identify the disease at early stages and diagnose plant disease automatically [5].

The computer vision and machine learning techniques can be used to detect the plants diseases, and also researchers clearly explained about the benefits, drawbacks, and its potential applications. The researchers also mentioned that this paper was designed in a manner where it can be used as a resource for scholars, and also for the practitioners of agriculture. In addition to that, this paper also speaks about the difficulties that authors face in automated plant disease detection with the help of computer vision [6].

The Convolutional Neural Network (CNNs) for the automated rice disease diagnosis in their study "Diagnosis of Rice Diseases Using Convolutional Neural Networks." Their work speaks about the outcomes and the efficiency that the researchers have got by using the deep learning algorithms and also in precisely identifying, recognizing, and also classifying the diseases affecting rice crops based on image data [7].

The study majorly focuses on the plant leaf shapes and leaf color for examining the plants illness and categorizing it. Their paper also mentioned about the improvement that can be brought by combining these features with machine learning algorithms for obtaining more accuracy and precision and also for obtaining the promising outcomes in automating the disease diagnosis process [8].

This study majorly concentrated on the texture of analysis method. The authors preferred texture analysis methods in order to identify and categorize the diseases that are impacting the crops. Their study majorly focused on analyzing the textural information and also extracting the textural information from the images of maize plants. The researchers provided that using this strategy, the authors can automate the diagnosis process with the help of texture analysis tools [9].

This study demonstrates the use of machine learning to diagnose the disease automatically which is present in wheat crops. Their study promises that machine learning techniques can help in improving the automated disease diagnoses by recognizing and classifying the disease present in the wheat crops by analyzing the crop image data [10].

The thorough description and easy-to-use taxonomy of machine learning tools so that the plant community may accurately and quickly implement the best-practice guidelines and relevant ML techniques for different biotic and abiotic stress features [11].

The paper investigates the usage of isolated lesions and spots for the purpose. Without the need for more photos, the data's unpredictability is raised because every region has unique characteristics. This makes it possible to identify several illnesses that are affecting the same leaf. However, complete automation is still not possible due to the need for an appropriate symptom segmentation, which must be done by hand [12].

An accurate model has been built that can explain which visual symptoms are used to make predictions in addition to providing performance at the level of a qualified pathologist. It has been observed that the proposed approach is portable to other imaging circumstances and plants and may be used for a wide range of biotic and abiotic stressors [13].

Recent advancements in computer vision, enabled by deep learning and the growing global use of smartphones, have made it possible to diagnose diseases with the help of smartphones. The deep convolutional neural network was trained with a publically accessible dataset of 54,306 images of healthy and diseased plant leaves that were collected under controlled conditions in order to identify 14 crop types and 26 diseases [14].

The study focuses on a novel approach that combines deep neural networks based on leaf image categorization to create a model for plant disease recognition. The new training approach and the methodology used enable a quick and easy system implementation in practice. The created model can distinguish 13 different types of plant diseases from healthy leaves by differentiating between the leaves of the plant and their environment [15].

3 Research Methodology

Here the proposed methodology says about the phase-wise steps in order to identify the diseases of plant using image processing and machine learning technique. These steps include preprocessing the data, feature extraction, developing the machine learning models, model evaluation and optimization, and deployment. These are the phase-wise steps which should be followed in order to recognize the disease using image processing and machine learning models. Figure 1 illustrates the plant disease detection workflow using image processing and machine learning techniques.

Step 1: Data Pre-processing

- Image Acquisition: In this step, the authors need to take the picture of the crop using a mobile phone or camera which was used throughout the process for identifying and classifying the disease present in the crop.

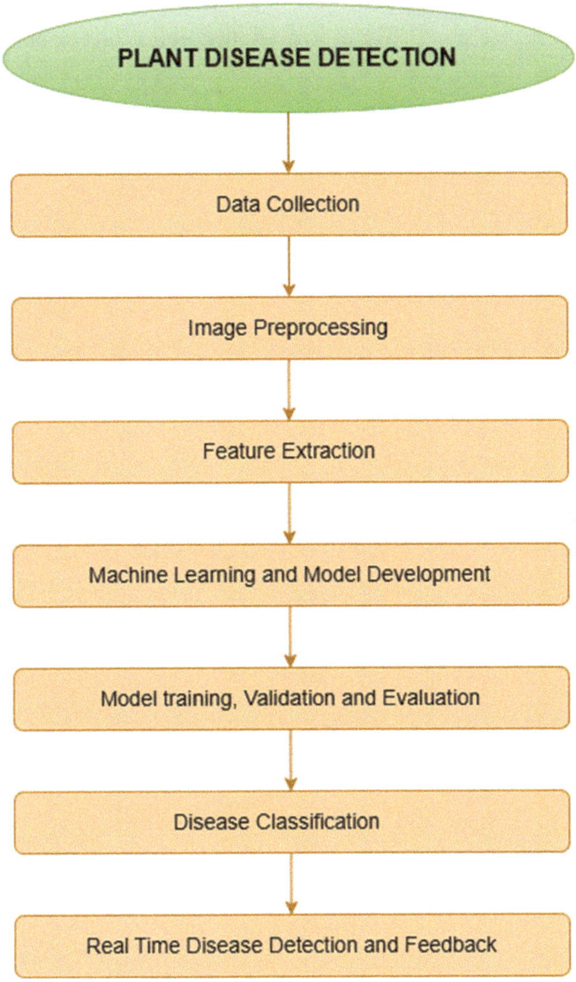

Fig. 1 Workflow of plant disease detection

- Data Augmentation: To get efficient results, the authors can make some transformations to the dataset by flipping, rotating, slicing, and also zooming so that it can fulfill the needs of the system for computer vision.
- Image Preparation: Preparation of the image is generally divided into two parts which are image enhancement which is used to improve the quality and contrast of image, and the second one is image segmentation which is used to differentiate between the normal and some unusual things.
- Image Enhancement: The contrast and quality of the image graph is an important parameter so that it can help in computer vision for better results. For improving the contrast and quality, authors generally use methods like histogram equalization.
- Image Segmentation: Image segmentation is the process where it helps in separating the segment of images into a healthy section and an unhealthy section.

Step 2: Extraction of the Features

- Feature extraction is the second step in this whole process where it helps in extracting the texture features and also the color features.
- Texture Analysis: Texture analysis is a process which the authors generally use to capture the spatial patterns in the pictures. For this, the authors have local binary patterns and Grey-Level Co-Occurrence matrix which are used in capturing the spatial patterns in the images.
- Color Analysis: In color analysis, the authors need to capture the differences between the healthy plant and diseased plant. For that, the authors need features like HSV, LAB histogram, or RGB to capture the color difference.

Step 3: Development of Machine Learning Models

The next step is to develop the machine learning models. In this step, the authors generally need to choose a model and then transfer learn the model and then train the model.

- Model Choice: As it is for picture-based disease identification, here the authors need to choose a model which is best for image classification. One such model is convolutional neural network where it has demonstrated the best performance in identifying the disease based on the picture.
- Model Training: For using model training, firstly the authors need to select the dataset which contains both healthy and unhealthy plants, and the authors need to choose a machine learning model and using that model the authors need to train the dataset.

Step 4: Model Evaluation and Optimization:

In model evaluation, the authors will evaluate the model based on the precision, accuracy, and F1 score and in optimization the authors will enhance the model's performance by using methods like random or grid search.

Step 5: Deployment:

- Deployment of the real-time detection system: Here the authors need to use the developed machine learning model which was trained to identify between the healthy and unhealthy leaves.
- Interface: The interface that the authors create should be user-friendly which should be easily understandable by the farmers and general users so that the researchers can easily upload the picture and get the required classification and information about the disease present in the leaves.

4 Results

The comparative result between both the original images and the pre-processed images using various models like AlexNet, AlexNetOWTBn, GoogLeNet, OverFeat, and VGG is shown below using the various graphical representations. In the comparison, it has been observed that the success rate of the disease prediction using various models in the case of original images is more than the one as compared to the pre-processed images as shown in Figs. 2 and 3.

In Figs. 4 and 5, it has been observed that the average error rate of the disease prediction using various models in the case of original images is less than the one as compared to that of pre-processed images.

In Figs. 6 and 7, it has been observed that the epoch value using various models in the case of original images is less than the one as compared to that of pre-processed images.

In Figs. 8 and 9, it has been observed that the time taken by epochs using various models in the case of original images is more than the one as compared to that of pre-processed images.

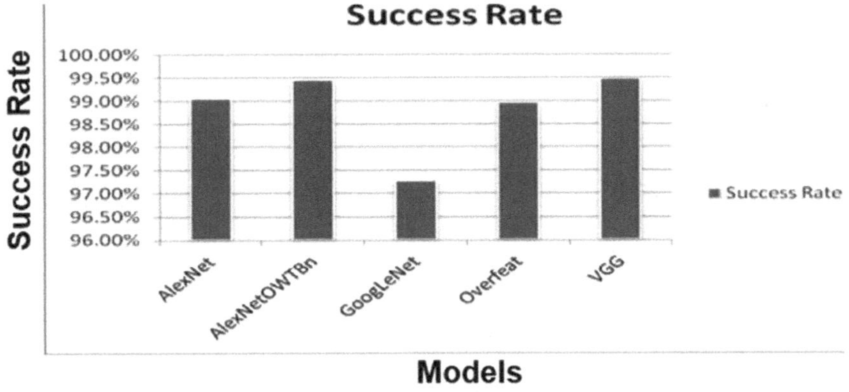

Fig. 2 Success rate of the original images by various models

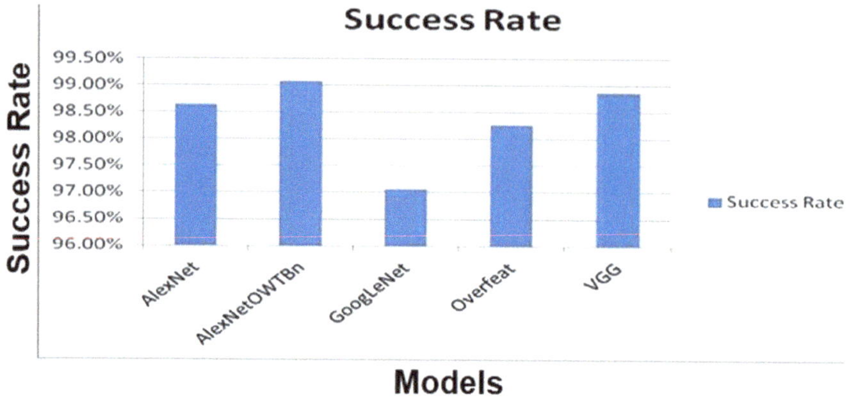

Fig. 3 Success rate of the pre-processed images by various models

Fig. 4 Average error of the original images by various models

5 Future Work

By using trending technologies like image processing, computer vision, and machine learning, the automatic disease recognition has a bright future. In future, the authors can improve the detecting algorithms, and the authors can combine the research with sensors and also Internet of Things so that it can reach wide audience who don't know to use computers. The authors can also improve the accuracy and efficiency of the model and can create a user-friendly chat box where farmers can chat with the machine and get to know more details about the disease that the leaves are suffering from. In order to create an effective and precise system for plant disease identification, this suggested methodology combines sophisticated image processing techniques to extract features with machine learning models based on deep learning for classifying the disease. This strategy can greatly increase early diagnosis and

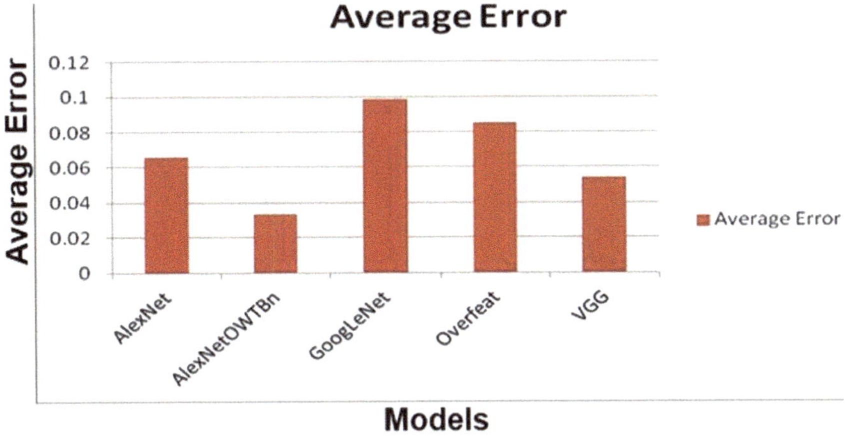

Fig. 5 Average error of the pre-processed images by various models

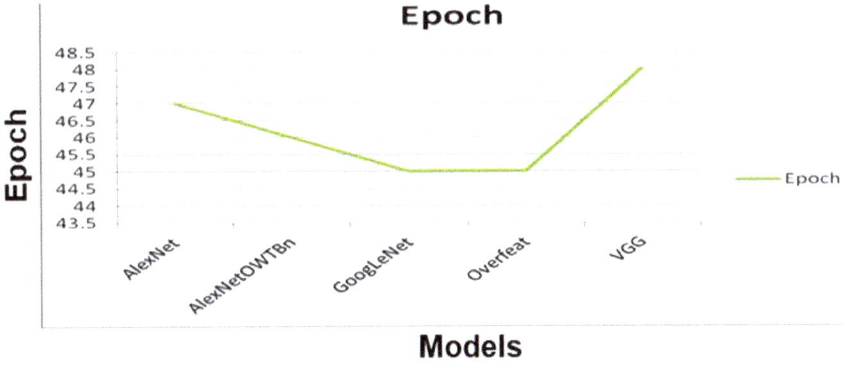

Fig. 6 Epoch value of the original images by various models

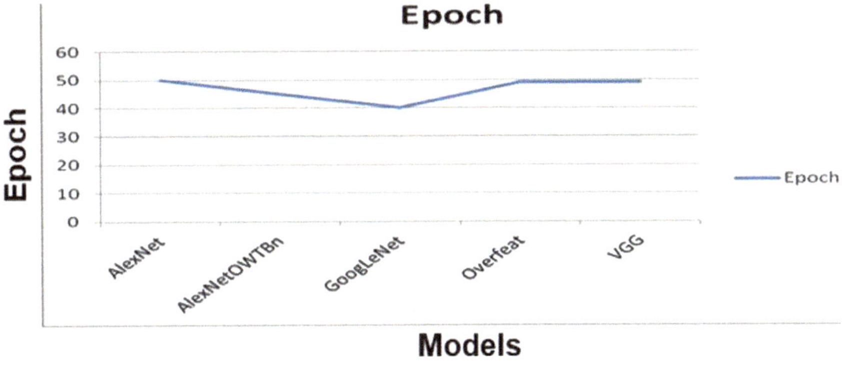

Fig. 7 Epoch value of the pre-processed images by various models

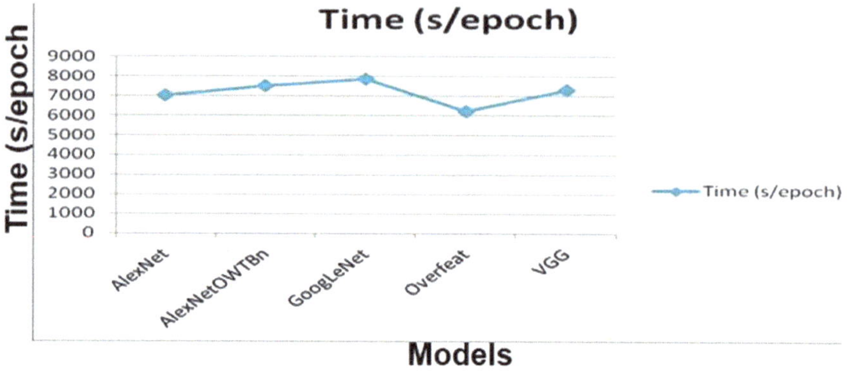

Fig. 8 Time taken to predict the original images by various models

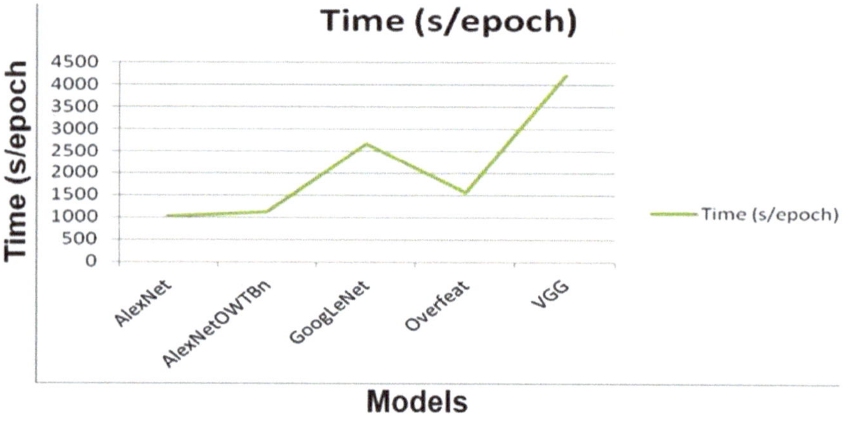

Fig. 9 Time taken to predict the pre-processed images by various models

intervention by utilizing the capabilities of contemporary technologies, which will increase agricultural output and decrease disease-related losses.

6 Conclusion

The conclusion is that integrating image processing and machine learning for plant disease detection is a revolutionary step toward sustainable agriculture. It automates the identification of diseases, enabling farmers to take proactive measures to prevent losses, enhance productivity, and maintain crop health efficiently. This research demonstrates the effectiveness of deep learning techniques in classifying plant diseases with high accuracy, significantly reducing human errors in traditional visual inspections.

The observations brought out in this research show that CNN models such as AlexNet and VGG are the best in identifying disease patterns with the help of image analysis. The real-time detecting and giving feedback ensure timely as well as informed decisions by farmers. The research method also forms a solid foundation for further developments in the area, including data preprocessing, feature extraction, training, and optimizing a model.

This will conclude that the adoption of automated disease detection systems does not only increase agricultural output but also aids in achieving the broader aim of food security. Further enhancements, such as IoT integration and real-time monitoring, will further drive the effectiveness of this approach to an unyielding position for modern agriculture.

References

1. Lu Z, Qin X, Wang Y et al (2017) Deep learning based banana leaf disease classification. Comput Electron Agrice
2. Kumar PV, Kumar SS, Raju MP et al (2019) Automated disease diagnosis for banana plants using z-transform features and artificial neural networks. Neural Comput Appl
3. Krizhevsky A, Sutskever I, Hinton GE (2018) A review of machine learning methods for leaf disease detection and classification. J Agric Inf
4. Waghmare SM, Patil PS, Deshmukh RB (2014) Identification of tomato leaf diseases using image processing techniques. Int J Comput Appl
5. Ghosal S, Bhat UA, Borah SVF et al (2020) Machine learning for plant disease detection: a review. Plants
6. Ferentinos A (2018) A survey on computer vision techniques for plant disease detection. Comput Electron Agric
7. Reddy TRM, Krishna A, Yadav RK (2016) Detection of rice diseases using convolutional neural networks. Int J Comput Appl
8. Mehmood H, Khan AM, Khan NY et al (2017) Machine learning approach for automated leaf disease detection using leaf shape and color features. Comput Electron Agric
9. Awada G, Awada MA (2015) Detection of maize diseases using texture analysis. Int J Agric Biol
10. Kumar R, Gupta A, Sharma A et al (2019) Application of machine learning algorithms for wheat disease detection. J Crop Improv
11. Singh A, Ganapathysubramanian B, Sarkar S et al (2016) Machine learning for high-throughput stress phenotyping in plants. Trends Plant Sci
12. Barbedo JGA (2019) Plant disease identification from individual lesions and spots using deep learning. Biosyst Eng
13. Ghosal S, Bhat UA, Jayaraj J et al (2018) An explainable deep machine vision framework for plant stress phenotyping. Proc Natl Acad Sci
14. Mohanty SP, Hughes DP, Salathé M (2016) Using deep learning for image-based plant disease detection. Front Plant Sci
15. Sladojevic S, Arsenovic M, Anderla A et al (2016) Deep neural networks based recognition of plant diseases by leaf image classification. Comput Ind

Experimental Study on Behaviour of Steel Fibre-Reinforced Concrete Beams Subjected to Combined Flexure, Shear and Torsion

T. H. Sadashiva Murthy, N. C. Balaji, Y. M. Manohar, and Mohammed Sakhib

Abstract This research deals with the study of behaviour of steel fibre-reinforced beams subjected to combined flexure, shear and torsion. The experimental study consists of 8 rectangular beams of the same dimensions 1700 × 150 ×120 mm. These beams were classified into 4 groups based on the volume fractions of steel fibres added, transverse and longitudinal reinforcement and concrete grade, and also the lever arm length was kept constant for all the beams. All the beams were designed to fail in regions of combined shear, flexure and torsion. According to the results of the tests done on hardened concrete, the best enhancing performance was provided by a beam reinforced with 1.5% steel fibres with increases in compressive and tensile strengths of 36.77% and 54.34% respectively. Diagonal tension cracks were observed in the side faces of the beam and flexural crack with new skew axis observed at top and bottom faces of the beam; the first cracks appeared after the control beam in all steel fibre-reinforced concrete (SFRC) beams. Furthermore, it was determined that hooked steel fibre improved beams' pre- and post-cracking behaviour. HSF 1.5% specimen was found to be the strongest beam in terms of the ultimate load with 43.75 per cent (average of two sets) enhancement was obtained compare to control beam. The specimen HSF 1% was in the second place, and HSF 0.5% was last. Therefore, the use 1.5% hooked end steel fibre enhanced torsional moment capacity by about 29.16%. The results of torsional moment versus angle of twist and load versus deflection for all specimens indicated a linear behaviour at the initial stages of loading (approximately 30–36% from the ultimate load), angle of twisting and deflection after that was raised gradually till the failure.

Keywords Combined bending · Shear and torsion · Steel fibres · SFRC

T. H. S. Murthy · N. C. Balaji (✉) · Y. M. Manohar · M. Sakhib
Department of Civil Engineering, The National Institute of Engineering, Mysuru, India
e-mail: balajinc@nie.ac.in

I. M. Jamadar et al. (eds.), *Integrating Sustainability in Modern Technologies*, Lecture Notes in Mechanical Engineering, https://doi.org/10.1007/978-981-95-9466-5_52

1 Introduction

When designing various types of reinforced concrete structures, torsion is a crucial factor to take into account. Although torsional stresses are common in reinforced concrete structures, they were typically disregarded by design engineers prior to 1960 [1]. The significant safety factor that was used in the flexural design of RC members was assumed to be sufficient to mitigate the minor torsional effects, but this assumption would cause many torsion-related failures.

Torsion is applied to a variety of structural elements, including space frames, curved girders, beams which are loaded eccentrically, exterior beams in buildings and circular staircases. When a member is not sufficiently reinforced for torsion, diagonal tension stresses are developed by the application of torsion, resulting in diagonal cracking, which leads to an unexpected brittle failure of the member. The outside surface of the beam will begin to crack spirally when the principal tensile stress of the beam exceeds the tensile stress of the beam.

In order to enhance the torsional capacity of reinforced concrete beam, adding steel fibres is one of the effective methods [2]. Reinforcement of steel fibres to concrete will improve the energy absorption capacity, flexural toughness, ductile behaviour prior to ultimate failure and crack arresting. It will mitigate the inherent weakness of concrete in resisting tension to some extent by including steel fibres.

Numerous researchers investigated torsion in beams under various conditions, including pure torsion in beams with fibres, combined torsion and flexure in beam strength with fibres, among others. Following are a few of these studies. Patil Premachand et al. [3] studied reinforced concrete beams with and without steel fibres that were subjected to pure torsion. The experimental program entails casting four 2000 mm × 150 mm × 150 mm reinforced concrete beams out of which one beam is without fibres and the remaining three beams were cast with increment of volume fraction of 0.5% like 0.5%, 1% and 1.5%. When compared to traditional RC beams, the cracking torsional strength and ultimate torsional strength of beam which is reinforced with 1.5% increased up to 54.85% and 72.10% respectively which is a very considerable rise in concrete strength.

Torsional behaviour and the mechanical efficiency of hybrid fibre-reinforced concrete beams were investigated experimentally by Saravanakumar et al. [4] who cast the specimens using Steel and Glass fibres with varied fibre volume fractions, including 0–1.5% with an incremental rate of 0.5%. By volume of concrete 0.5% each addition of steel and glass fibre increased the mechanical strength of conventional concrete by about 20% with the additions of fibres, the torsional behaviour of the SFRC beams continued to improve. The 1.5% SFRC beam showed a 33% maximum increase.

Karimipour et al. [5] compared how steel (SF) and polypropylene fibres (PPF) affected the torsional performance of square beams made of high-performance reinforced concrete (HPC). Nine HPC beams were cast and tested under torsion loading. SF and PPF were added at volumetric concentrations of 0–4%, with incremental rate of 1%. Experiments have shown that the torsion and torsion angles are improved with

PPF and SF, whereas the torsion performance of the HPC beam with PPF reinforcement is superior to that of the beam with SF reinforcement. In contrast, SF is more effective than PPF at increasing the torsional stiffness of HPC beams. As a result, the torsional strength was increased by roughly 122.3% and 85.2%, respectively, with the use of 4% PPF and SF.

2 Materials and Experimental Methods

2.1 Materials

For this research study OPC 53 grade cement in compliance with IS 1489-Part 1 [6] standards, as well as fine aggregates and coarse aggregates that complied with IS 2386-Part 3 [7] standards, and Conplast SP430, a superplasticizer based on sulfonated naphthalene polymers, was used to make concrete more workable. Hooked end steel fibre of 30 mm length 0.5 mm diameter having aspect ratio of 60 was used in this study. The manufacturer provided the fibres' properties, and the average ultimate tensile strength of steel fibres was 1100 MPa. Cubes measuring 150 mm in size for compressive strength and cylinders measuring 150 mm in diameter and 300 mm in length for split tensile strength were cast and tested after 28 days to determine the mechanical behaviour of SFRC. The RC beams of 1700 × 120 × 150 mm of dimensions were cast to study the torsional behaviour of SFRC beam.

2.2 Torsion Test on Beams

Tests were performed on beams made of grade M30 concrete having a dimension of 120 mm × 150 mm × 1700 mm (Fig. 1). Two 8-mm bars were provided as the beam's main reinforcement, along with two 8-mm hanger bars and 2-legged 8-mm stirrups at 100 mm c/c (Fig. 2). Figure 2 depicts the test setup. The moment arm's length was held constant at 0.25 m. The I-section member served as a two-point transfer for the two-point loading onto the beam. The I-section beam was placed diagonally over the lever arms. The steel I-section beam's middle received the load, which was then evenly distributed to both ends of the beam through the lever arm, and the torsional moment was applied to the ends of the beam. The applied force on each side and the distance between the moment arms were multiplied to determine the torque value (Tables 1 and 2; Figs. 3, 4, 5 and 6).

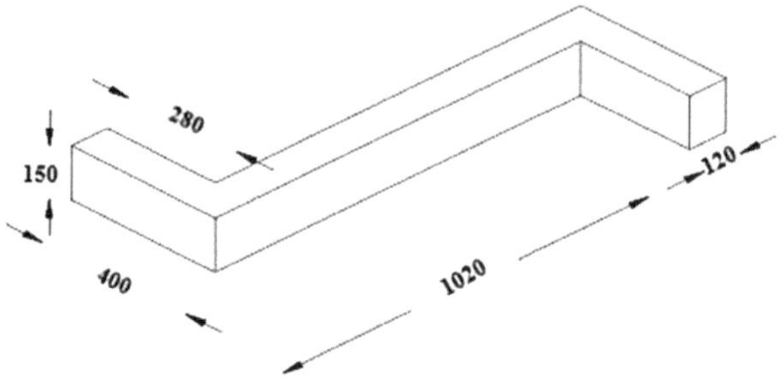

Fig. 1 Dimensions of the experimental specimen

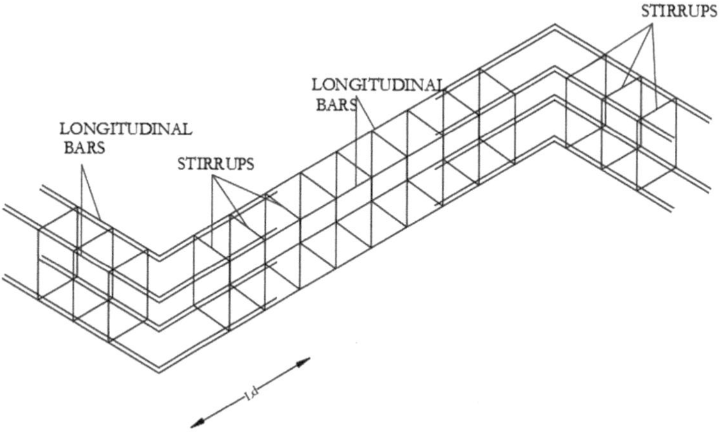

Fig. 2 Reinforcement details of the experimental specimen

Table 1 Compressive and split tensile strength of the specimens

Specimen designation	Compressive strength (N/mm^2)	Split tensile strength (N/mm^2)
CB 0%	38.75	3.11
0.5% HSF	44.88	3.81
1% HSF	49.06	4.24
1.5% HSF	53	4.8

Table 2 Nomenclature of the specimens

Specimens	Longitudinal reinforcement		Stirrups	Steel fibres volume fraction (%)
	Top	Bottom		
CB 0%	2-#8	2-#8	#8@100 mm c/c	0
HSF 0.5%	2-#8	2-#8	#8@100 mm c/c	0.5
HSF 1%	2-#8	2-#8	#8@100 mm c/c	1
HSF 1.5%	2-#8	2-#8	#8@100 mm c/c	1.5

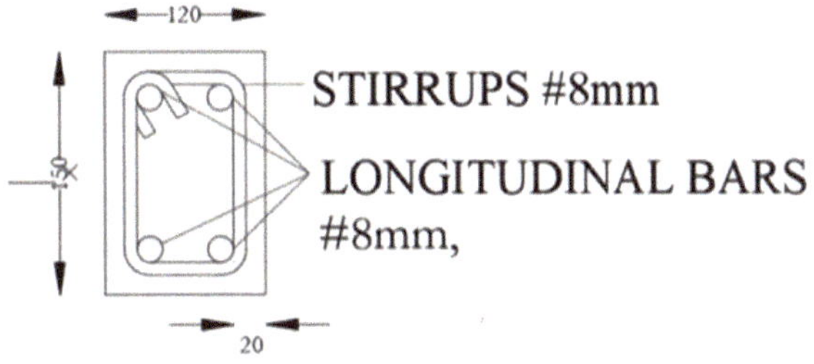

Fig. 3 Cross section of main beam

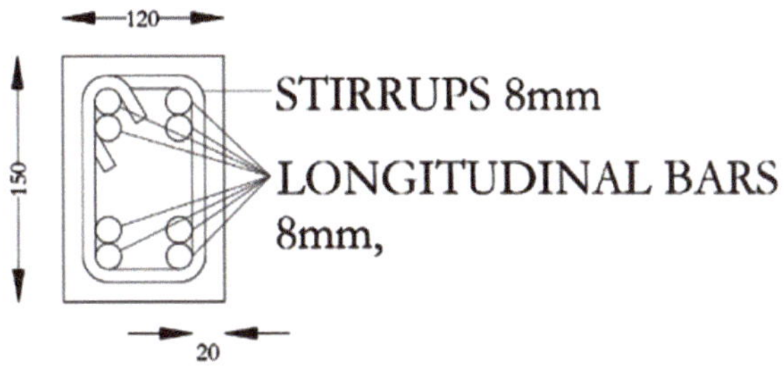

Fig. 4 Cross section of torsional arm

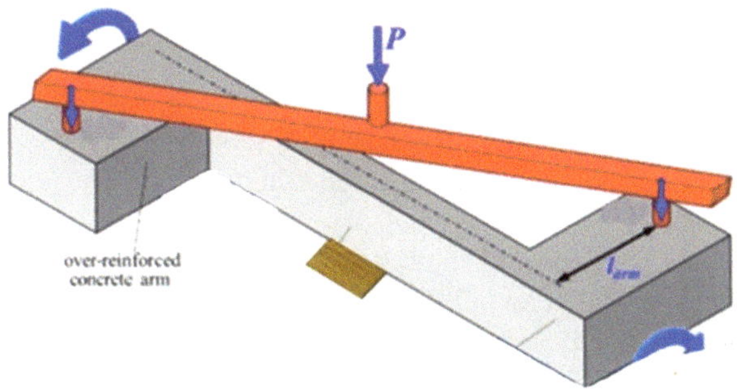

Fig. 5 Schematic picture of test setup

3 Results and Discussion

The present experimental work consists of four groups of beams which were reinforced with steel fibres with varying fibre volume fractions from 0 to 1.5% with an incremental rate of 0.5%; these beams were designated as CB0%, HSF 0.5%, HSF 1% and HSF 1.5%. The main objective was to study the effect of steel fibres on the beam which is subjected to combined effects of shear, flexure and torsion.

Fig. 6 Test setup of experimental specimen

3.1 *Torsional Moment—Angle of Twist*

Figure 7 shows the comparison of RC and SFRC set-1 beams; a linear behaviour is observed at the starting stages of loading; after that non-linear behaviour is observed until the failure.

The SFRC beams' torsional strength is increased as the percentage of fibres was added; therefore by adding fibres 0.5%, 1% and 1.5%, the first crack torsional strength increases by 29.41%, 38.23% and 50% respectively.

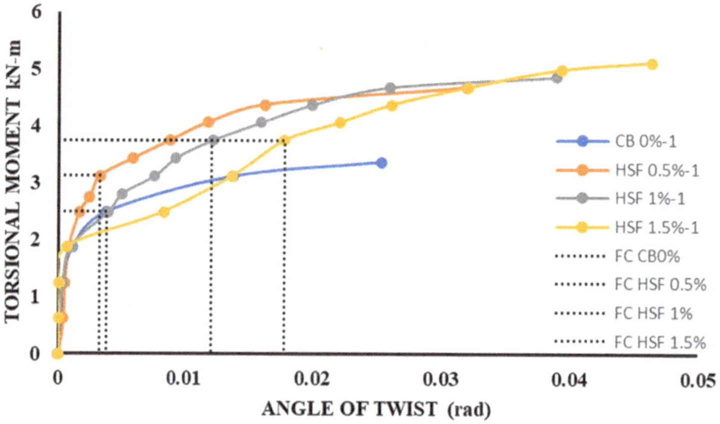

Fig. 7 Torsional moment versus angle of twist graph of set-1 beams

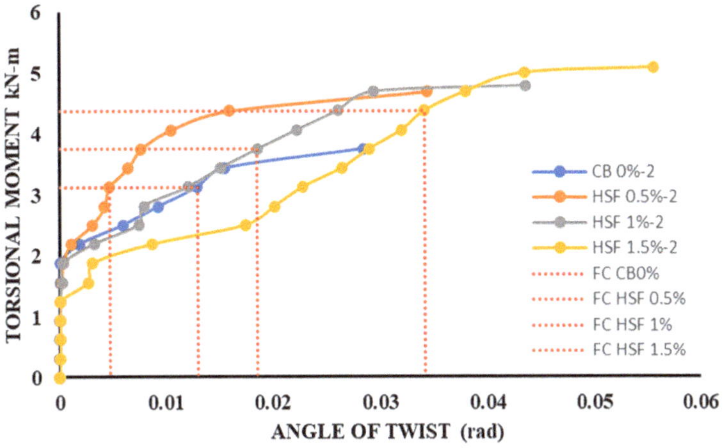

Fig. 8 Torsional moment versus angle of twist graph of set-2 beams

Figure 8 shows the comparison of RC and SFRC set-2 beams; a linear behaviour is observed at the starting stages of loading; after that non-linear behaviour is observed until the failure.

The SFRC beams' torsional strength is increased as the percentage of fibres was added; therefore by adding fibres 0.5%, 1% and 1.5% the first crack torsional strength increases by 23.68%, 26.31% and 34.21% respectively.

3.2 *Load-Deflection*

Figure 9 shows the comparison of RC and SFRC set-1 beams; a linear behaviour is observed at the initial stages of loading; after that non-linear behaviour is observed until the failure; as load increased, deformation starts to increase until the failure of the beam.

Specimen CB 0% shows lesser loading capacity than the SFRC beams. The load-carrying capacity of SFRC beams, the first crack is obtained at load of 20 kN and ultimate load of 27 kN after its first crack it manageable to take additional load of 7 kN a sudden failure is observed, by adding the fibres the toughness of the matrix getting in-creased as percentage of fibres content increased the failure time is post-poned, compared to the control beam CB 0% the SFRC beams strength increased by increased by approximately 27.2–43.33%. For all the cases taken into consideration, fibres did not significantly affect the elastic stiffness, though. At the non-linear stage, the effect of fibre addition became more noticeable; when fibres are used, beams can support a higher load at the same rate of deflection.

Figure 10 shows the comparison of RC and SFRC set-2 beams; a linear behaviour is observed at the initial stages of loading; after that non-linear behaviour is observed

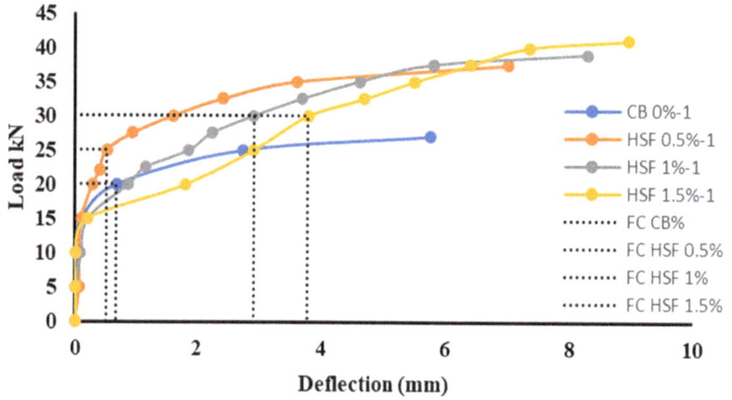

Fig. 9 Load versus deflection graph of set-1 beams

until the failure; as load increased, deformation starts to increase until the failure of the beam.

Specimen CB 0% shows lesser loading capacity than the SFRC beams. The load-carrying capacity of SFRC beams, the first crack is obtained at load of 25 kN and ultimate load of 30 kN after its first crack it manageable to take additional load of 5 kN a sudden failure is observed, by adding the fibres the toughness of the matrix getting in-creased as percentage of fibres content increased the failure time is postponed, compared to the control beam CB 0% the SFRC beams strength increased by increased by 25, 27.5 and 35.66%. For all the cases taken into consideration, fibres did not significantly affect the elastic stiffness, though. At the non-linear stage, the effect of fibre addition became more noticeable; when fibres are used, beams can support a higher load at the same rate of deflection.

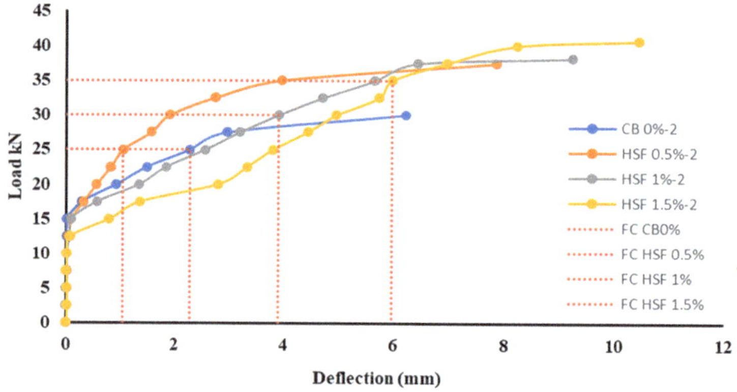

Fig. 10 Load versus deflection graph of set-2 beams

3.3 Crack Pattern

From Figs. 11 and 12, it is observed that the beams failed under flexure, the first crack occurred at the middle of the top face of the beam with a new skew axis, and diagonal tension crack was developed at Face B and Face D which rapidly propagates into the bottom and top faces beam like a spiral. The concrete at the bottom Face C became crushed as a result of cracks getting widened.

From Figs. 13, 14, 15, 16, 17 and 18, it is observed that all the SFRC beams follow a similar manner; they failed under predominant flexure, and the first crack is at the middle of the top Face A of the beam with a new skew axis and diagonal tension cracks appearing at the middle of the Face B and Face D and propagating towards top Face A and bottom Face B with respect to the longitudinal axis of the beam, at an angle of about 45 degrees, but the expansion of the diagonal tension crack is delayed as the percentage of fibre increased.

With further increase in the torsional moment, the diagonal tension cracks will appear at the bottom and top faces; these new cracks will join the existing diagonal

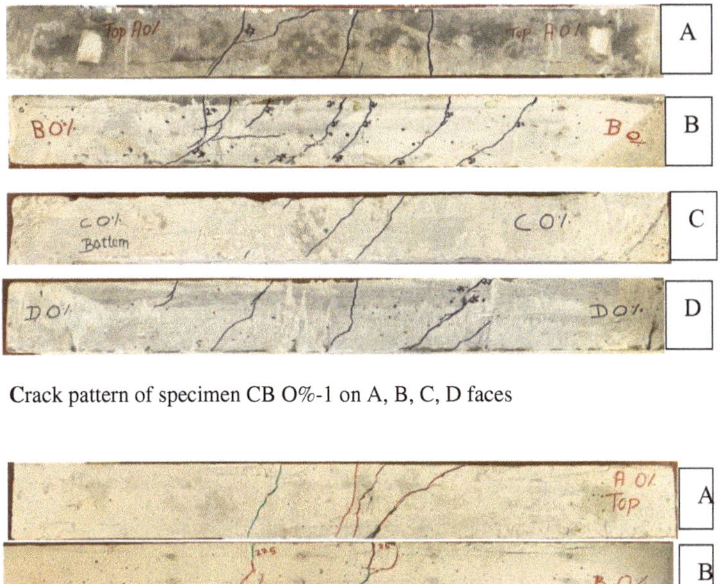

Fig. 11 Crack pattern of specimen CB O%-1 on A, B, C, D faces

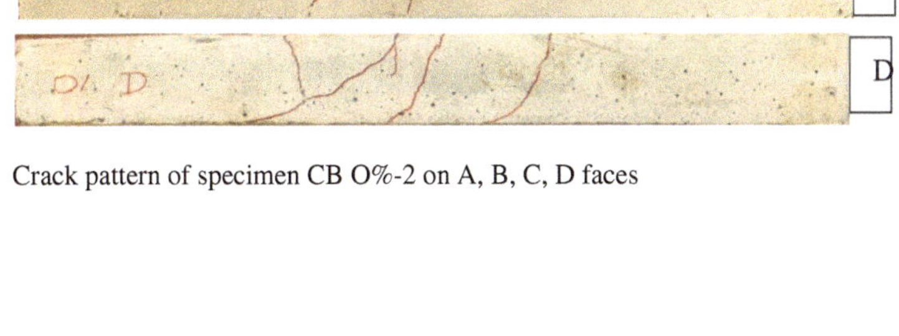

Fig. 12 Crack pattern of specimen CB O%-2 on A, B, C, D faces

Fig. 13 Crack pattern of specimen HSF 0.5%-1 on A, B, C, D faces

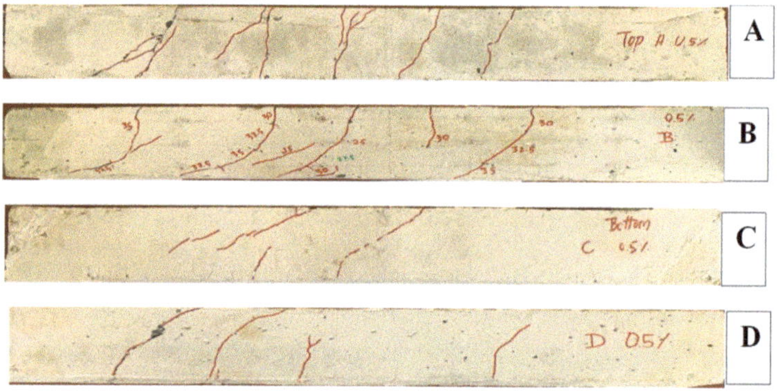

Fig. 14 Crack pattern of specimen HSF 0.5%-2 on A, B, C, D faces

Fig. 15 Crack pattern of specimen HSF 1%-1 on A, B, C, D faces

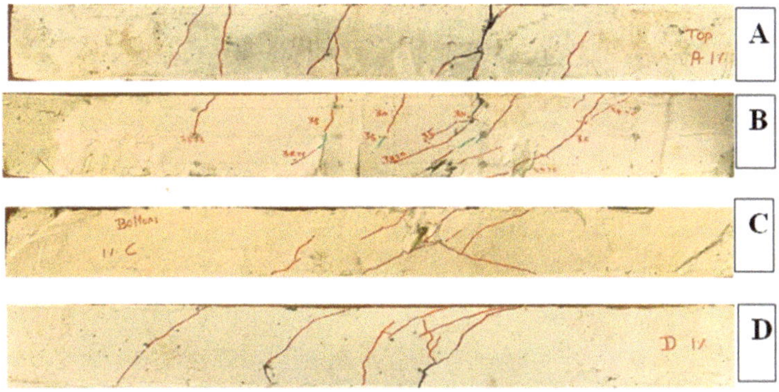

Fig. 16 Crack pattern of specimen HSF 1%-2 on A, B, C, D faces

Fig. 17 Crack pattern of specimen HSF 1.5%-1 on A, B, C, D faces

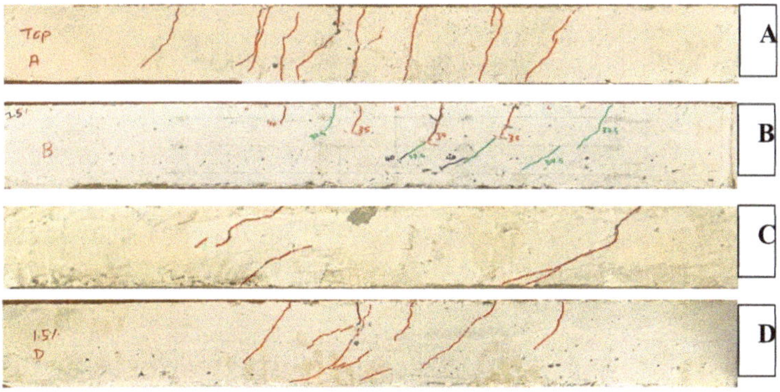

Fig. 18 Crack pattern of specimen HSF 1.5%-2 on A, B, C, D faces

tension cracks at Face B and Face D thus forming a spiral crack on the surfaces of beam with minimum spacings. A few of the diagonal cracks on faces A, B, D started growing excessively wide as the torque was increased further. The beams ultimately failed by crushing of concrete on the C face along a diagonal line connecting the ends of the primary crack.

An increase in fibre content doesn't change the mode of failure as can be seen from Figs. 13, 14, 15, 16, 17 and 18, when higher fibre content is used more closely and narrow cracks are observed.

3.4 Load-Crack Width

Due to SFRC beams' more ductile behaviour than RC beams, the first diagonal tension cracks in SFRC beams arrived later than in RC beams. The main diagonal tension crack appeared in the middle of the specimens, and a first crack was identified to measure its width in all specimens. For each loading increment, crack width was recorded. The specimens HSF 0.5%, HSF 1%, and HSF 1.5%, which showed up at 1.1 mm, 0.8 mm and 0.5 mm respectively, were followed by specimen CB 0%, whose crack width reached 1.5 mm to be the highest recorded crack width.

The experimental findings of this study clearly show that adding fibres to the concrete mix had a significant impact on delaying crack formation and its propagation and delaying the very first crack presence in each specimen by particular value (25, 27.5 and 35.66%). Figure 19 illustrates that the first crack in the HSF 1.5% beam that appeared at a total load of 35 kN, which is 35.66% higher than the load at which the control beam cracked. Furthermore, fibres enhance the post-crack behaviour of concrete due to their bridging effect. The beam is more load-resistant than the control beam (Table 3).

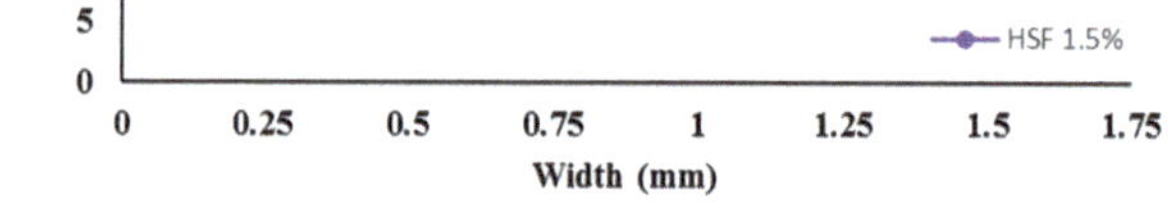

Fig. 19 Load versus crack width graph

Table 3 Summary of test results

Specimen name	Pu (kN)	Vu (kN)	Tu (kN-m)	Mu (kN-m)	δ (mm)	Θ (rad)
CB 0%-1	27	13.5	3.375	6.885	5.754	0.0253
CB 0%-2	30	15	3.75	7.65	6.218	0.0284
HSF 0.5%-1	37.5	18.75	4.687	9.562	7	0.0319
HSF 0.5%-2	37.5	18.75	4.687	9.562	5.366	0.0345
HSF 1%-1	39	19.5	4.875	9.945	8.286	0.0389
HSF 1%-2	38.25	19.125	4.781	9.753	9.238	0.0435
HSF 1.5%-1	41	20.5	5.125	10.455	8.954	0.0466
HSF 1.5%-2	40.7	20.35	5.087	10.378	10.439	0.055

Note Pu—ultimate load, Vu—shear force, Tu—torsional moment, Mu—bending moment, δ—deflection, θ—angle of twist

4 Conclusions

This experimental study consists of 8 rectangular beams of 1700 × 150 × 120 mm dimensions. These beams were classified into 4 groups based on the volume fractions of steel fibres added, transverse and longitudinal reinforcement, and concrete grade, and also the lever arm length was kept constant for all the beams. All the beams were designed to fail in regions of combined shear, flexure and torsion. The following conclusions can be drawn:

1. The addition of steel fibre positively affected the mechanical characteristics and significantly reduced brittleness.
2. The SFRC beams' crack resistance was enhanced by the steel fibre's tension stiffening mechanism, which also prevented crack growth and its propagations.
3. The ultimate load-carrying capacity of SFRC beam enhances approximately 38–51.85% as increase in the volume fraction of hooked end steel fibres.
4. Because fibres made concrete more ductile, all fibre-reinforced beams developed their major first cracks after the control beam (without fibres).
5. The addition of fibres increased the beam's stiffness slightly, but its effects became more noticeable at the non-linear stage when the beams could support a higher load at the same deflection rate.

References

1. Ju H, Lee DH, Hwang JH, Kang JW, Kim KS, Oh YH (2013) Torsional behavior model of steel-fiber-reinforced concrete members modifying fixed-angle softened-truss model. Compos Part B 45(1):215–231
2. Rao TG, Seshu DR (2003) Torsion of steel fiber reinforced concrete members. Cem Concr Res 33(11):1783–1788

3. Patil Premachand J, Kanase Jayant S (2016) Enhancement of torsional strength of reinforced concrete by application of steel fibres. Int J Innov Res Sci Technol 2(11)
4. Saravanakumar P, Sivakamidevi M, Meena K, Yamini SP (2021) An experimental study on hybrid fiber reinforced concrete beams subjected to torsion. Mater Today Proc
5. Karimipour A, De Brito J, Ghalehnovi M, Gencel O (2022) Torsional behaviour of rectangular high-performance fibre-reinforced concrete beams. In: Structures, vol 35. Elsevier, pp 511–519
6. IS 1489: Part 1-2015, Indian Standard Portland Pozzolana cement—specification part 1 fly ash based, fourth rev. BIS, New Delhi, India
7. IS 2386 (Part 3)—1963, Indian standard methods of test for aggregates for concrete, part 3: specific gravity, density, voids, absorption and bulking, BIS, New Delhi, India

Experimental Study on Behaviour of Steel Fibre-Reinforced Concrete Beam with Partial Replacement of Cement by Fly Ash Subjected to Combined Flexure, Shear and Torsion

T. H. Sadashiva Murthy, N. C. Balaji, Mohammed Sakhib, and Y. M. Manohar

Abstract The paper presented here deals with the experimental study of behaviour of steel fibre-reinforced concrete beams with partial replacement of fly ash subjected to combined flexure shear and torsion. The project has been taken up and an attempt has been made to study the combined loading effect of flexure, shear and torsion on steel fibre-reinforced concrete beams. The series consists of 8 beams of the same cross-sectional dimensions and same longitudinal shear reinforcement. Keeping optimized fly ash constant, the beams were divided into four groups based on the volume fraction of steel fibres added. Loading point, i.e., lever arm is kept constant for all the beams. The behaviour of the beam when first cracks occurs, cracking load, propagation of cracks at each load, cracking width and the ultimate failure of the specimen were observed carefully and recorded. Deflection and angle of twist were recorded using dial gauges at regular loading intervals.

Keywords Combined bending · Shear and torsion · Steel fibres · SFRC

1 Introduction

Concrete is a composite and brittle material which is most commonly used in construction. In concrete, cement is the most expensive ingredient which has been a significant source for CO_2 emissions by its production making approximately 2.4% of global CO_2 emissions from industrial and energy sources. One way to lower these emissions is to reduce the amount of cement and partially replace it [1]. Many research have shown that fly ash can decrease concrete's CO_2 emission and its cost while increasing its strength and workability. Depending on the type of fly ash, specification limits and geographic location, cement can be replaced by fly ash ranging

T. H. S. Murthy · N. C. Balaji (✉) · M. Sakhib · Y. M. Manohar
Department of Civil Engineering, The National Institute of Engineering, Mysuru, India
e-mail: balajinc@nie.ac.in

I. M. Jamadar et al. (eds.), *Integrating Sustainability in Modern Technologies*, Lecture Notes in Mechanical Engineering, https://doi.org/10.1007/978-981-95-9466-5_53

15–25% (most commonly) and when rapid setting time is not required it can be replaced by up to 40–60% which will help in reducing emissions. So fly ash is the greatest substitute since it can still offer high-strength concretes with the same strong mechanical properties as cement [1]. Concrete induces tensile stresses in which shrinkage is formed in both the plastic and hardened stages of concrete, and as a result crack occurs. In order to withstand these tensile stresses and prevent cracking, steel is reinforced in the concrete.

As a result, some beams are subjected to transverse loading through vertical shear and flexural moment [2]. Sometimes the entire structural behaviour may be controlled by the torsional response of the beams. Because of this, the shear and flexural behaviour of beams as well as their torsional behaviour should be researched and understood. By knowing the importance of the problem, many researchers studied torsional response of RC beams. Studies have been done on torsional behaviour of RC beam subjected to three different types of loading as follows:

1. Pure torsion.
2. Combined bending and torsion.
3. Combined bending, shear and torsion.

The earliest experimental studies found that adding steel fibres improves torsional behaviour of RC beams when subjected to pure torsion. Steel fibres of volume of fraction are varied to improve the ductility and torsional strength of the beams. Up to a certain volumetric level of steel fibre addition, the increase in strength reduces. With an increase in the percentage of steel fibres, the ability of fibre-reinforced concrete to absorb energy also increases linearly [3]. Some results found that 1.5% steel fibres gave the maximum cracking torsional strength and ultimate torsional strength [1]. Steel fibres made a positive impact even in the case of unreinforced concrete beams subjected to pure torsion which increased the torsional stiffness and torsional toughness of beams [4]. The mid span and then the area next to the supports, where the largest wide diagonal cracks have been created, are the most crucial areas that are strengthened in RC beams that are subjected to the combined effect of bending shear and torsion [4]. When compared to control beams, steel fibres increased the tensile strength up to 20–30% which also increases the torsional strength [5]. When RC beams subjected to combined bending shear and torsion increased the load-carrying capacity of SFRC by approximately 30–50% than the plain beam observed [6], it also concluded that the ultimate torsional strength is not affected by longitudinal reinforcement which mainly depends on properties of the concrete [7]. Some studied revealed that single type of reinforcement alone does not improve the torque-carrying capacity of the members beyond the first crack torque [3]. In order to improve torsional capacity of the beam, longitudinal and transverse reinforcement is provided in our study. Many investigations reveal that SFRC has attracted researchers the most due to improved mechanical properties. It is also necessary to evaluate the impact of torsional strength on SFRC beams partially replacing cement by fly ash. To effectively utilize this fly ash, it must be determined what percentage of fly ash should be used as a substitute for cement. There is very little well-documented information available on the behaviour of steel

Table 1 Mix design

Water	Cement	Fly ash	Fine aggregate	Coarse aggregate
0.48	339 kg	59.8 kg	729 kg	1123 kg

fibre-reinforced concrete beams with partially replacing cement by fly ash subjected to combined flexure, shear and torsion. Therefore, it is imperative to investigate the behaviour of steel fibre-reinforced concrete beams with partially replacing cement by fly ash subjected to combined loading effects.

2 Experimental Program

2.1 Materials

For the present study, beams were casted using concrete of the M30 grade. The properties of various concrete elements such as cement, M-sand, aggregate and granite powder have been discussed. In the present experimental study, cement of OPC-53 grade was used conforming to IS10262-2020 and having specific gravity (SG) of 3.12, fineness of 5.33% and normal consistency of 31%.

M-sand is used as fine-aggregate conforming to Zone-II and having SG of 2.61 and water absorption (WA) of 2%. The 20 mm downsize aggregate were used as coarse-aggregate and having SG of 2.66, WA of 0.5% conforming to IS:383–2015. Class "F" fly ash was commercially procured from a nearby fly ash processing industry in Mysuru. The superplasticizer used in our experimental study is CONPLAST SP430 having the specific gravity of 1.18. Fly ash has been sieved passing through 90 microns sieve. Then it is sieved through 90 μ to substitute cement in concrete production with specific gravity of 2.09%. Crimped steel fibres also called zigzag type fibres having aspect ratios of 60 corresponding to the length and diameter dimensions of 30/0.5 were used. It was procured from Stewols, Nagpur, Maharashtra. The tensile strength for the steel fibres given by the manufacturer was 1100 MPa. The mix design obtained is given in Table 1.

2.2 Optimization of Fly Ash

Different proportions of Class "F" fly ash replacing cement have been varied from 5, 10, 15, 20 and 25%. Compressive strength of concrete at 7 days and 30 days were tested. It was found that 15% of fly ash gave the maximum compressive strength of 20.8 and 32.5 kN/mm^2.

Fig. 1 Experimental setup

2.3 Test Specimens

The formwork is assembled according to size of the specimen (120 mm × 150 mm × 900 mm). Longitudinal reinforcement of 2-#8 mm diameter at top and 2-#8 mm diameter at bottom and stirrups were made of 8 mm diameter of 100 mm c/c spacing were used. A total of 8 beams were cast and tested as shown Fig. 1. For each specimen, three cubes of size 150 mm x150mm and cylinders of size 150 mm × 300 mm were cast and tested for compressive strength and split tensile strength.

2.4 Test Specimens

The experimental setup for testing of the beams is as shown in Fig. 2. The specimen testing were carried out in a 100 ton UTM. Load was transferred on to the beam by means of a steel channel section of 1.2 m length. The channel was supported on rollers on the torsional arm which produced the torque required for creating the condition of combined loading. The beam is supported at the centre of the span as shown in Fig. 1. The testing of the beams was carried out by applying loads in intervals of 5 KN up to the ultimate load. The loading of two-point nature was transferred on the beams through steel member. The DEMEC gauge was used to measure the strain. The readings were taken at different points by sticking studs on the front faces of the beams. Dial gauges were placed below the lever arm of the beam and along the length of the beam to measure angle of twist and deflection. At every interval, the DEMEC gauge readings and the dial gauge readings were observed and recorded. Cracks were marked during the loading procedure, and the corresponding loads were indicated on the specimen. The dial gauges were used to measure the deflection and angle of twist at different load intervals. Four dial gauges were used as shown in Figs. 1, 2, 3 and 4. The dial gauges 1 and 2 were used to measure the angle of twist, and dial gauges 3 and 4 was used to measure the deflection. Table 2 shows the designation of the specimens considered in the experimental work.

Fig. 2 Casted Z beams

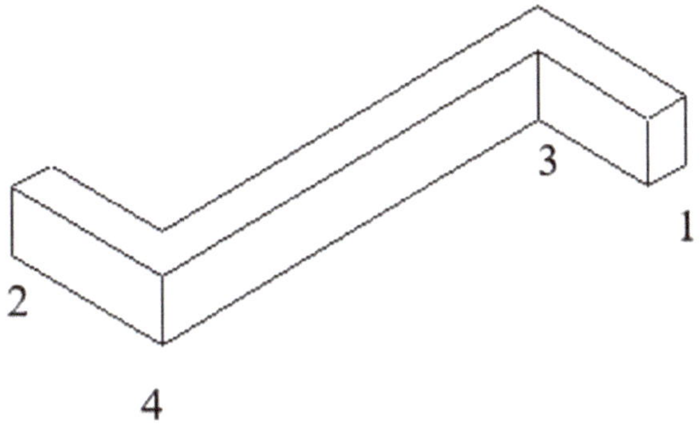

Fig. 3 Indication of dial gauges placement

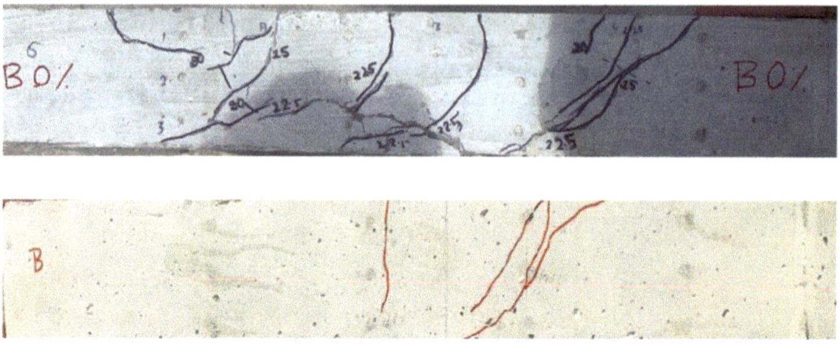

Fig. 4 Crack pattern of control beam specimen

Table 2 Designation of the specimens

Specimens	Percentage of steel fibres	Longitudinal reinforcement		Stirrups
		Top	Bottom	
CB	–	2-#8	2-#8	8 mm dia @100 mmc/c
CSF 0.5%	0.5%	2-#8	2-#8	8 mm dia @100 mmc/c
CSF 1%	1%	2-#8	2-#8	8 mm dia @100 mmc/c
CSF 1.5%	1.5%	2-#8	2-#8	8 mm dia @100 mmc/c

3 Results and Discussions

Tables 3 and 4 give the experimental values of shear force, bending moment, cracking load, ultimate load cracking torque, ultimate torque, deflection and angle of twist for the Specimen 1 and Specimen 2. Table 3 summarizes that when SFRC beams are compared with control beam, there is an increase in deflection, ultimate torque and angle of twist as volume fraction of steel fibres is increased. There is also a considerable increase in compressive strength and split tensile strength of concrete.

Table 3 Experimental data for test Specimen 1

Name of the specimen	% of steel fibres	Shear force (kN)	Bending moment (kN/m)	Cracking load (kN)	Cracking torque (kN-m)	Ultimate load (kN)	Ultimate torque (N-m)	Deflection (mm)	Angle of twist (rad)
Control beam	–	11.25	5.73	15	1.875	22.5	2.812	2.63	0.0189
CSF 0.5%	0.50%	15.75	8.03	25	3.125	31.5	3.937	5.09	0.0314
CSF 1%	1%	16.5	8.41	27	3.375	33	4.125	6.95	0.0446
CSF 1.5%	1.50%	18.75	9.43	30	0.75	37.5	4.687	7.54	0.0505

Table 4 Experimental data for test Specimen 2

Name of the specimen	% of steel fibres	Shear force (kN)	Bending moment (kN/m)	Cracking load (kN)	Cracking torque (kN-m)	Ultimate load (kN)	Ultimate torque (N-m)	Deflection (mm)	Angle of twist (rad)
CONTROL BEAM	–	10.12	5.16	15	1.875	20.25	2.531	2.48	0.0205
CSF 0.5%	0.50%	16.25	8.28	25	3.125	32.5	4.062	6.17	0.0386
CSF 1%	1%	18.25	9.3	30	3.75	36	4.562	7.69	0.043
CSF 1.5%	1.50%	19.02	9.702	32.5	4.026	38.05	4.756	8.39	0.0512

3.1 Crack Pattern

Figs. 4, 5, 6 and 7 show the crack pattern control beam, CSF0.5%, CSF1% and CSF1.5% specimens. The flexure cracks observed at the beginning in the middle of the top surface. The diagonal cracks developed side face of the beams propagates towards top face and bottom face at an inclination of approximately 45° to the longitudinal axis of the beam. As the percentage of steel fibre is increased, the growth of diagonal tension crack is delayed by absorbing energy and taking more torsional moment. From the Fig. 5, it can be seen that number of cracks is increased as the percentage of steel fibres is increased.

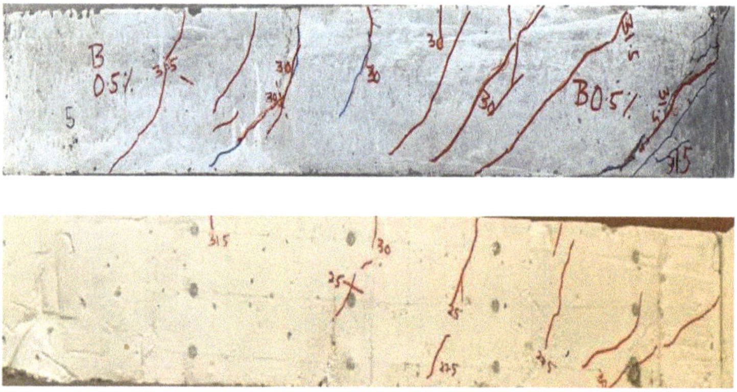

Fig. 5 Crack pattern of CSF0.5% specimen

Fig. 6 Crack pattern of CSF1% specimen

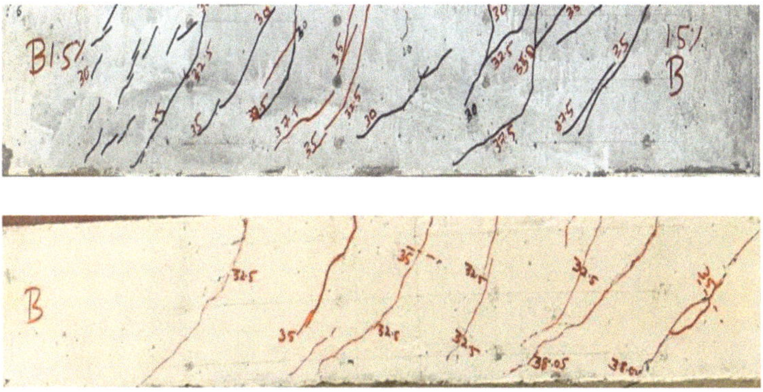

Fig. 7 Crack pattern of CSF1.5% specimen

3.2 *Load Versus Deflection*

Figures 8 and 9 show the comparison of FRC and SFRC beams for both specimens. It was observed that there is an increase in the deflection as the percentage of steel fibres is increased. A linear behaviour is observed at the initial stages of loading after that non-linear behaviour is observed until the failure of specimen, as load increased deformation start increases until the failure of the beam. Cracking load of different specimens is shown by a dashed line. Specimen CB 0% shows lesser loading capacity than the SFRC beams. The load-carrying capacity of SFRC beams, the first crack is obtained at load of 15 KN and ultimate load of 21.375 KN after its first crack it manageable to take additional load of 6.375 KN a sudden failure is observed, by adding the fibres the toughness of the matrix getting increased as percentage of fibres content increased the failure time is postponed, compared to the control beam the SFRC beams strength increased by approximately 50–80%.

3.3 *Torque Versus Angle of Twist*

From Figs. 10 and 11, it was observed that there is an increase in the angle of twist for Specimen 1 and Specimen 2 as the percentage of steel fibres is increased. Cracking load of different specimens is shown by a dashed line. From the result it was found that the angle of twist of volume fraction 0.5%, 1% and 1.5% steel fibres had increased by up to 60.6%, 122.35% and 158% respectively which was a noticeable increase in the angle of twist when compared to control beam. Percentage of steel fibres is increased, a linear behaviour is observed at the initial stages of loading after that non-linear behaviour.

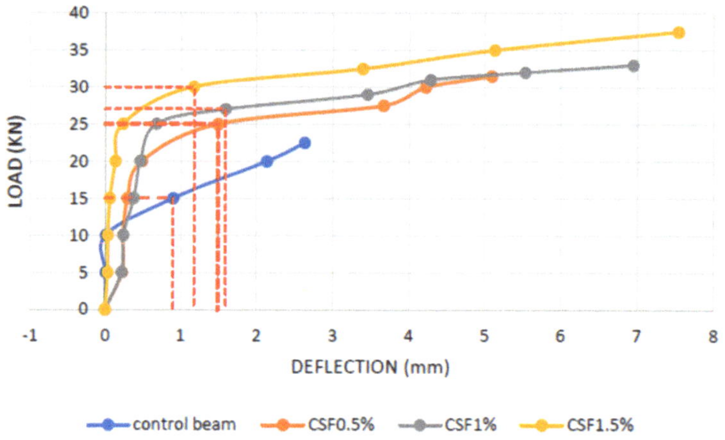

Fig. 8 Load versus deflection for Specimen 1

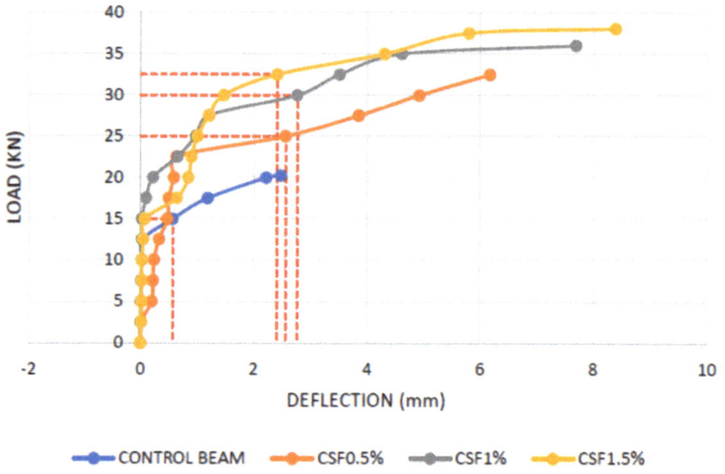

Fig. 9 Load versus deflection for Specimen 2

3.4 *Crack Width*

Figure 12 shows the load versus crack width of beams; it was noticed that SFRC beams exhibit more ductile behaviour than RC beam, thus the first diagonal tension cracks in SFRC beams appeared on the RC beam. A specific crack was defined to measure its width in all specimens, and the main diagonal tension crack was assigned to the middle section of the beams. Crack propagations during loading were recorded for each loading increment. As the load increment is increased, the control beam specimen recorded the widest crack at 0.6 mm, followed by the specimens CSF0.5%, CSF1% and CSF1.5% that arrived to 0.4 mm, 0.3 mm and 0.2 mm, respectively, while

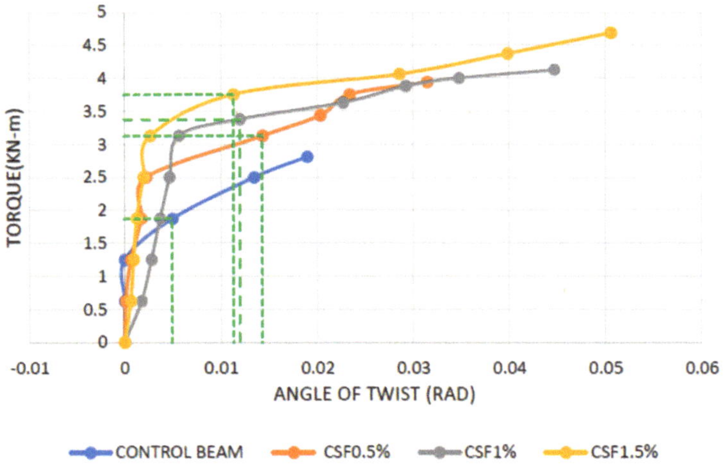

Fig. 10 Torque versus angle of twist for Specimen 1

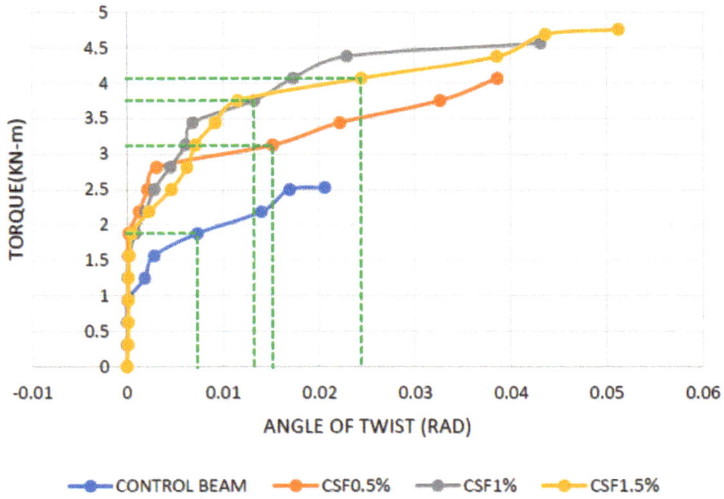

Fig. 11 Torque versus angle of twist for Specimen 2

the lowest crack width was noticed in HSF 1.5% specimen of 0.5 mm. As a result of their bridging effect, steel fibres improve the post-crack behaviour of concrete.

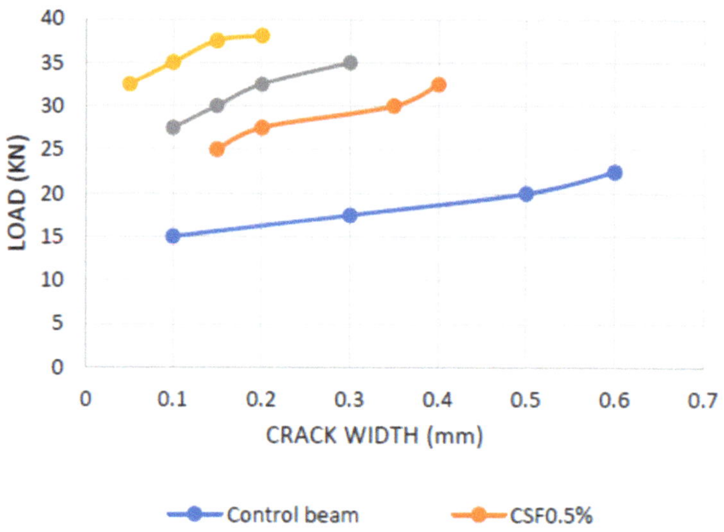

Fig. 12 Crack width

Table 5 Results

Name of the specimen	% of steel fibres	Compressive strength (N/mm²)		Split tensile strength (N/mm²)	
		7 days	28 days	7 days	28 days
Control beam	–	20.8	38.8	0.84	2.26
CSF 0.5%	0.5%	21.23	39.12	1.12	2.54
CSF 1%	1%	21.77	40.5	1.40	2.97
CSF 1.5%	1.5%	23.36	41.3	1.56	3.11

3.5 Compressive Strength and Split Tensile Strength Results

The split tensile strength and compressive strength at 7 days and 28 days for RC beams improved with additions of steel fibres at different volume fractions. From the given Table 5, 1.5% fraction of steel fibre addition gave the maximum compressive strength and maximum split tensile strength of 36.4 and 3.1 KN/mm^2.

3.6 Contribution of Steel Fibres to the Post-Elastic Response

The test results of Table 6 indicated that there is a significant improvement in the post-elastic behaviour of the specimens when they were subjected to maximum torsional moment. Fibre contribution is evaluated in terms of ductility μT, μT85MAX and

Table 6 Test results

Name of the specimen	% of steel fibres	Tcr (KNm)	VTcr (rad/m)	Elastic torsional stiffness (k) (kNm^2/rad)	Tmax	VTmax (rad/m)	v85Tmax (rad/m)	μT85 max	μTcr	μT
Specimen 1										
Control beam	–	1.875	0.0062	793.65	2.812	0.0189	0.035	1.85	5.64	3.048
CSF 0.5%	0.50%	3.125	0.0064	796.17	3.937	0.0314	0.058	1.84	9.06	4.906
CSF 1%	1%	3.375	0.008	605.811	4.125	0.0446	0.0825	1.84	10.31	5.575
CSF 1.5%	1.50%	3.75	0.01	594.05	4.687	0.0505	0.0934	1.85	9.34	5.05
Specimen 2										
Control beam	–	1.875	0.0053	731.707	2.531	0.0205	0.038	1.85	7.16	3.86
CSF 0.5%	0.50%	3.125	0.009	647.66	4.062	0.0386	0.0714	1.85	7.93	4.28
CSF 1%	1%	3.75	0.007	581.39	4.562	0.043	0.079	1.83	11.2	6.14
CSF 1.5%	1.50%	4.026	0.0074	634.76	4.756	0.0512	0.0947	1.84	12.79	6.91

μT85cr. From these values, it is shown that the advantage of using the steel fibres improving the post cracking torsional response is increased and also the torque rotation capabilities of control beam without losing its strength is increased.

4 Conclusions

The present study deals with the experimental behaviour of steel fibre-reinforced concrete members with partially replacing cement by fly ash subjected to combined loading effects of flexure, shear and torsion. Eight rectangular beams having the same cross section, reinforcing details and grade of concrete, with varying volume fractions of crimped steel fibres, designed to fail in combined loading are cast and tested till collapse. Dial gauges and a DEMEC gauge are used during testing to measure the deflections and strains. From the result obtained, the following conclusions can be drawn:

- The ultimate load-carrying capacity of steel fibre-reinforced concrete beam enhances with increase in the percentage of crimped steel fibre content.
- All fibre-reinforced beams developed their major first cracks after the control beam (without fibres) because fibres made concrete more ductile.

- The addition of fibres slightly increases the beam's stiffness, but its effects become more significant at the non-linear stage, where the beams can support a higher load at the same deflection rate.
- When compared to control beam, the ultimate torsional strength of the specimen CSF1.5% improved by 66.7%, which was very a significant improvement in concrete strength.
- Steel fibre addition improved crack growth resistance, resulting in a reduction in crack width.

References

1. Patil Premachand J, Kanase Jayant S (2016) Enhancement of torsional strength of reinforced concrete by application of steel Fibres. Int J Innov Res Sci Technol 2(11)
2. Abdullah MD (2018) Behavior of hollow and solid section of reinforced concrete beams under pure torsion strengthened by steel fiber. J Univ Babylon Eng Sci 26(9):344–355
3. Rao TG, Seshu DR (2006) Torsional response of fibrous reinforced concrete members: effect of single type of reinforcement. Constr Build Mater 20(3):187–192
4. Nath, P. and Sarker, P., 2011. Effect of fly ash on the durability properties of high strength concrete
5. Raut LL, Kulkarni DB (2014) Torsional strengthening of under reinforced concrete beams using crimped steel fiber. Int J Res Eng Technol 3(6):466–471
6. Patil SP, Sangle KK (2016) Tests of steel fibre reinforced concrete beams under predominant torsion. J Build Eng 6:157–162
7. Pant Avinash S, Parekar Suresh R (2009) Steel fiber reinforced concrete beams under bending, shear and torsion without web reinforcement. Int J Rec Trends Eng 1(6):86–88

Optimizing Energy Efficiency and Sustainability: A Case Study on Load Management and Solar Integration at an Engineering Institute

Najmuddin Jamadar, Ajinkya Sagare, and Suhani Jamadar

Abstract This study presents an energy audit conducted for an engineering institute, focusing on the impact of installing a 100 kW solar power plant to enhance sustainability by reducing energy consumption, costs, and environmental impact. The analysis reveals that integrating the solar plant can reduce grid electricity consumption by approximately 33%, yielding annual savings of ₹1,230,071. The payback period for the solar installation is estimated to be around 4.07 years, making it a financially viable and sustainable investment. In addition to cost savings, the solar plant will reduce the institute's carbon footprint by 61,179.50 kg of CO_2 annually and save 642,343.58 GJ of thermal energy. The study highlights the significance of renewable energy integration as a sustainable strategy for managing institutional energy demands while contributing to global environmental goals. This work emphasizes the importance of energy audits and renewable energy adoption to enhance energy efficiency, sustainability, and economic viability.

Keywords Solar integration · Energy consumption · Load management · Emission reduction

1 Introduction

Energy management in institutional settings, such as engineering institutes, is increasingly critical due to rising energy demands and environmental concerns. Energy audits serve as a fundamental tool in identifying inefficiencies, optimizing performance, and reducing operational costs [1, 2]. These audits provide a detailed analysis of current energy usage, allowing institutions to implement strategies for energy conservation and cost reduction [3, 4]. Sharma et al. highlight that energy audits play a crucial role in optimizing transformer capacities and ensuring long-term energy efficiency [5]. Similarly, Chen et al. discuss the methodologies for energy audits and their application to transformer capacity planning [6].

N. Jamadar (✉) · A. Sagare · S. Jamadar
Yashoda Technical Campus, Satara, Maharashtra, India

I. M. Jamadar et al. (eds.), *Integrating Sustainability in Modern Technologies*, Lecture Notes in Mechanical Engineering, https://doi.org/10.1007/978-981-95-9466-5_54

Effective load management is essential for maintaining transformer capacity and preventing system overloads. Accurate load forecasting and transformer capacity planning are crucial for ensuring the reliability of electrical systems [7, 8]. Previous studies emphasize the importance of load management strategies to enhance system performance and prevent potential failures [9, 10]. Ahmed et al. highlight the need for precise load forecasting and capacity planning to optimize transformer efficiency [11]. Additionally, Zhang et al. discuss how improper load management can lead to increased operational costs and system inefficiencies [12]. Further research by Zhang et al. provides insights into transformer efficiency and current best practices in load management [13].

The integration of renewable energy sources, such as solar power, offers a promising solution for reducing dependence on conventional energy and improving sustainability [14, 15]. Solar energy integration can significantly reduce energy costs and greenhouse gas emissions, as demonstrated by Khan and Kumar [16]. Furthermore, research by Liu et al. illustrates that solar energy integration can enhance system stability and contribute to long-term financial benefits [17]. The economic and environmental advantages of incorporating solar energy into institutional energy systems are well documented [18, 19]. Nguyen et al. explore various solar energy integration strategies for institutional energy systems, emphasizing their role in achieving sustainability [20]. Patel and Singh further discuss advanced methods for energy efficiency in large institutions [21]. Kim and Lee evaluate the benefits of load reduction and solar energy integration in institutional settings [22].

Accurate forecasting of future energy demands is essential for effective energy management and planning [23, 24]. Predictive models and simulations help estimate future load growth and assess the impact of energy efficiency measures [25]. Lee et al. suggest that advanced forecasting techniques can improve the accuracy of load predictions and support better decision-making for energy management [14]. The role of energy audits in optimizing transformer capacities and predicting future loads has been explored by Gupta et al. [23] and Singh et al. [24].

The significance of this study lies in analyzing the energy audit data of an engineering institute, focusing on load management, the impact of solar energy integration, and future load predictions. By examining current energy consumption patterns, evaluating load reduction strategies, and forecasting future energy needs, this study aims to provide insights into optimizing transformer capacity and enhancing overall energy efficiency [6, 25]. The results of this research will contribute to a deeper understanding of how energy audits can drive improvements in energy management practices and support sustainability goals.

The remaining part of the paper is organized as follows: Sect. 3 discuss materials and methods used for this case study. Section 3 represents results and discussion, and Sect. 4 concludes the case study outcomes.

2 Materials and Methods

2.1 *Study Area*

This study is conducted at an engineering institute that has undergone an energy audit to evaluate its energy consumption patterns and the effectiveness of potential energy-saving measures. The focus is on analyzing the historical energy usage data to determine the adequacy of the current transformer capacity and the potential impact of integrating a 100 kW solar energy system. The aim is to ensure that the transformer operates within safe limits while accommodating future load increases.

2.2 *Data Collection*

2.2.1 Energy Consumption Data

Energy consumption data shown in Table 1 represents a 12-month electricity consumption and billing record from June 2023 to May 2024. It shows monthly figures for units consumed (likely in kWh), bill demand (in KVA), and total bill amount (in INR). The consumption and costs vary throughout the year, with notable increases in the summer months, particularly in 2024. May 2024 had the highest consumption (30,815 units), demand (138 KVA), and bill amount (564,381 INR), while November 2023 had the lowest consumption (12,605 units) and bill amount (226,325 INR). The bill demand remained constant at 88 KVA for most months but increased in the last three months, coinciding with higher consumption and costs. This pattern suggests possible seasonal variations in electricity usage and a potential expansion of operations or increased activity in the recent months.

- **Units Consumed (kWh)**: The total kilowatt-hours of electricity consumed each month.
- **Reduced Units (kWh)**: The estimated reduction in energy consumption due to implemented or proposed energy efficiency measures.
- **New Units Consumed (kWh)**: The energy consumption after applying the reduction measures, calculated as the difference between the original units consumed and the reduced units.
- **Bill Demand (kVA)**: The maximum demand recorded on the utility bill, indicating the highest demand for electricity in kVA during the billing period.
- **Reduced kVA**: The estimated reduction in demand due to efficiency measures.
- **New Bill Demand (kVA)**: The new maximum demand after applying the reduction measures.
- **Bill Amount**: The total cost of energy consumption before applying reduction measures.

Table 1 Energy bill and consumption

Bill month	Units	Bill demand (KVA)	Bill amount (INR)
May-24	30,815	138	564,381
Apr-24	26,691	115	473,017
Mar-24	23,877	98	390,041
Feb-24	17,029	88	288,983
Jan-24	14,600	88	254,589
Dec-23	16,534	88	281,696
Nov-23	12,605	88	226,325
Oct-23	16,138	88	271,277
Sep-23	12,836	88	226,568
Aug-23	15,141	88	253,601
Jul-23	14,949	88	249,810
Jun-23	15,385	88	255,635

- **Reduced Bill Amount**: The estimated bill amount after applying efficiency measures.
- **New Bill Amount**: The adjusted bill amount after considering the reduction measures.

These data points provide a comprehensive view of energy usage, costs, and the impact of efficiency improvements as shown in Table 1.

2.2.2 Solar Energy Integration

To analyze the impact of integrating a 100 kW with existing 67 kW solar energy system, data on solar system capacity and performance were considered. This included the following:

- **Solar Capacity**: The total capacity of the solar energy system (167 kW).
- **Performance Factors**: Factors such as the capacity factor of the solar panels, which reflects their actual output compared to their maximum potential, and the operational hours of the solar system.

2.3 Analysis Methods

2.3.1 Load Analysis

Historical Load Data Analysis

Historical load data were analyzed to identify patterns and trends in energy consumption and demand as per Table 2. This data presents a detailed breakdown of electricity consumption for various items in an institution or facility over a 12-month period from June 2023 to May 2024. It lists 37 different electrical devices or systems, including lighting (CFL and LED tubes), HVAC equipment (fans and ACs), office equipment (PCs, printers), laboratory equipment, and various appliances. The consumption is measured monthly for each item, with the rightmost column showing the "Utilization Factor (%)" for each item. Air conditioning (AC) has the highest utilization at 22.41%, followed by PCs at 13.74% and fans at 11.17%. The data shows seasonal variations, with higher consumption in summer months (March–May) for most items. Some equipment, like lab equipment, show intermittent usage patterns. The detailed breakdown provides insights into the facility's energy consumption patterns, allowing for potential energy optimization strategies and identifying the most significant contributors to overall electricity usage.

Load Reduction Assessment

The impact of load reduction measures was assessed by comparing pre- and post-reduction data. This involved the following:

- **Reduction Calculation**: Determining the percentage reduction in energy consumption and demand achieved through efficiency measures.
- **Statistical Tests**: Applying paired t-tests to evaluate if the observed reductions were statistically significant. This helps ascertain whether the efficiency measures had a meaningful impact.

2.3.2 Transformer Capacity Analysis

The adequacy of the existing 200 kVA transformer was assessed by comparing its capacity with current and projected future loads. Key calculations included the following:

- **Current Load Analysis**: Reviewing current energy consumption and maximum demand to determine if the transformer operates within its safe limits.
- **Safe Operating Limits**: Calculating the safe operating range for the transformer, considering factors such as load factor and thermal limits.

To estimate the capacity required for future load, the following formula was used:

$$\text{Maximum Allowable Load (kW)} = \text{kVA} \times \text{Power Factor} \tag{1}$$

For a power factor of 0.9 and an updated transformer capacity of 108 kVA:

Table 2 Summary of equipment and their consumption in watts

Item	May-24	Apr-24	Mar-24	Feb-24	Jan-24	Dec-23	Nov-23	Oct-23	Sep-23	Aug-23	Jul-23	Jun-23	Utilization factor (%)
CFL tube	455	394	353	252	216	244	186	238	190	224	221	227	1.48
LED tube	651	564	505	360	309	349	266	341	271	320	316	325	2.11
FAN	3441	2983	2668	1903	1631	1846	1408	1803	1435	1693	1672	1721	11.17
AC	6900	5981	5351	3816	3270	3701	2823	3615	2877	3395	3352	3451	22.41
PC's	4232	3669	3282	2341	2006	2270	1731	2217	1765	2082	2056	2117	13.74
Printers	275	238	213	152	130	147	112	144	114	135	133	137	0.96
Network routers	448	388	347	248	212	240	183	234	186	220	217	224	1.03
Smart boards	136	118	106	75	65	73	56	71	57	66	66	80	0.43
Projector	424	336	320	231	172	218	182	121	210	0	0	0	0.1
Lab equipment 1 phase 5A	0	1686	1686	1686	1686	0	1686	1686	1686	1686	0	0	2.8
Lab equipment 1 phase 15A	0	3322	3322	3322	3322	0	3322	3322	3322	3322	0	0	5.94
Lab equipment 3 phase	0	2026	2026	2026	2026	0	2026	2026	2026	2026	0	0	3.75
Outdoor/outpost lightning	11	10	9	6	5	5	5	5	5	6	0	0	0.03
Refrigerator	51	44	39	28	24	27	21	26	21	25	24	25	0.18
Water pump	186	161	144	103	88	100	76	97	77	91	90	93	0.55
Inverters	60	52	47	32	28	32	24	31	25	29	29	30	0.22
Incinerator machine	18	16	14	10	9	10	9	7	9	9	9	9	0.05
Aqua filter plant	75	65	58	41	35	40	39	31	36	36	37	37	0.26
Water cooler	195	169	151	108	92	104	80	102	81	96	95	97	0.54

(continued)

Table 2 (continued)

Item	May-24	Apr-24	Mar-24	Feb-24	Jan-24	Dec-23	Nov-23	Oct-23	Sep-23	Aug-23	Jul-23	Jun-23	Utilization factor (%)
Flourmill	15	13	12	8	7	8	6	8	7	7	7	7	0.04
Sanitary napkin machine	1	1	1	1	1	1	1	1	1	1	1	1	0
Biometric	1	1	1	1	1	1	1	1	1	1	1	1	0
Ring bell	14	12	11	8	7	7	6	7	6	7	7	7	0.04
LED display board	20	17	16	11	9	8	10	10	10	10	10	10	0.06
Burwell motor	30	26	23	17	14	16	12	16	12	15	14	15	0.09
Xerox machine	112	97	87	62	53	60	46	59	47	55	54	56	0.37
Amplifier	16	14	12	9	8	7	8	8	8	8	8	8	0.05
Exhaust fan	21	18	16	12	10	11	9	10	10	10	10	10	0.06
UPS	2859	2478	2217	1582	1355	1534	1170	1498	1192	1407	1389	1430	8.29
Sound system	20	17	16	11	9	8	10	10	10	10	10	10	0.06
Cash count machine	8	7	6	4	4	4	3	4	3	4	4	4	0.02
Sandwich machine	15	13	12	8	7	8	6	8	7	7	7	7	0.04
Mixer	10	9	8	6	5	5	4	5	5	5	5	5	0.03
Halogen lamp	37	32	29	20	17	20	15	19	15	18	18	18	0.1
Geyser	190	165	147	105	90	102	78	99	79	92	90	92	0.53
Iron	15	13	12	8	7	8	6	8	7	7	7	7	0.04
Elevators	552	478	428	305	262	296	226	289	230	271	268	276	1.97

$$\text{Maximum Allowable Load (kW)} = 108\,\text{kVA} \times 0.9 = 97.2\,\text{kW}$$

3 Results and Discussion

3.1 Overview of Data and Initial Observations

The energy consumption data for the engineering institute from June 2023 to May 2024 were analyzed to assess the impact of energy efficiency measures and to evaluate transformer capacity and future load growth. The dataset includes monthly readings of units consumed, bill demand, and associated costs, before and after implementing load reduction measures.

Initial observations indicate a significant reduction in both energy consumption and bill demand after the implementation of energy efficiency measures. The average monthly reduction in units consumed was approximately 4000 kWh, and the reduction in bill demand was about 20 kVA.

3.2 Load Analysis Results

3.2.1 Historical Load Trends

Figure 1 presents the monthly energy consumption and demand over the study period. The data reveal a clear seasonal pattern, with higher energy consumption during the summer months and lower consumption in the winter:

- **Units Consumed**: The monthly energy consumption ranged from 12,605 kWh in November 2023 to 30,815 kWh in May 2024. The average monthly consumption was approximately 17,000 kWh.
- **Bill Demand**: The bill demand varied from 88 to 138 kVA, with the highest demand recorded in May 2024.

3.2.2 Impact of Load Reduction Measures

Table 3 summarizes the impact of load reduction measures on energy consumption and demand:

- **Reduction in Units Consumed**: The reduction in energy consumption ranged from 4160 kWh in November 2023 to 10,669 kWh in May 2024.
- **Reduction in Bill Demand**: The reduction in bill demand ranged from 29 to 46 kVA.

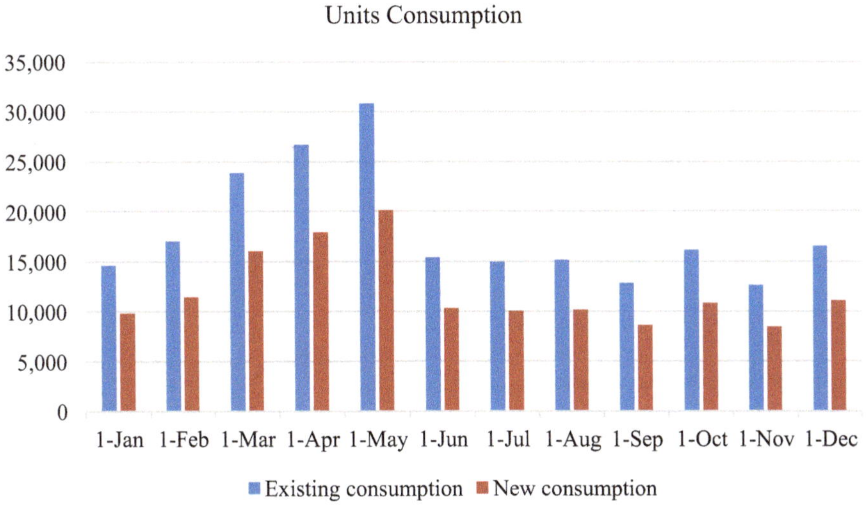

Fig. 1 Unit consumption over a year

These reductions led to a decrease in monthly energy bills, with reductions in bill amounts ranging from ₹90,665 in November 2023 to ₹226,946 in May 2024.

Table 3 Summary of reduction in demand and energy bill

Bill month	Units consumed	Reduced units	Reduction factor	New units consumed	Bill demand (kVA)	New bill demand (kVA)	Bill amount (INR)	New bill amount (INR)
May-24	30,815	10,669	0.33	20,146	138	92	564,381	378,138
Apr-24	26,691	8808	0.33	17,883	115	77	473,017	316,923
Mar-24	23,877	7879	0.33	15,998	98	66	390,041	261,928
Feb-24	17,029	5619	0.33	11,410	88	59	288,983	193,616
Jan-24	14,600	4818	0.33	9782	88	59	254,589	170,577
Dec-23	16,534	5456	0.33	11,078	88	59	281,696	188,536
Nov-23	12,605	4160	0.33	8445	88	59	226,325	151,638
Oct-23	16,138	5325	0.33	10,813	88	59	271,277	182,758
Sep-23	12,836	4236	0.33	8600	88	59	226,568	152,059
Aug-23	15,141	4997	0.33	10,144	88	59	253,601	170,656
Jul-23	14,949	4933	0.33	10,016	88	59	249,810	167,374
Jun-23	15,385	5077	0.33	10,308	88	59	255,635	171,649
			Annual bill				**3,735,923**	**2,505,852**
	Annual saving						**12,30,071**	

3.3 *Transformer Capacity Analysis*

3.3.1 Current Transformer Utilization

The existing transformer capacity of 200 kVA was evaluated against the current and reduced loads. The transformer was initially operating within its safe limits, with the maximum demand of 138 kVA in May 2024 being well within the 200 kVA capacity.

3.3.2 Future Capacity Projections

The future capacity needs were assessed considering an incremental load growth of 20 kW per year.

Maximum Allowable Load Calculation: For the updated transformer capacity of 108 kVA and a power factor of 0.9, the maximum allowable load was

$$\text{Maximum Allowable Load (kW)} = 108\,\text{kVA} \times 0.9 = 97.2\,\text{kW}$$

3.3.3 Load Extension Feasibility

The current load of 61.2 kW, including an estimated 20 kW per year growth, will reach the maximum allowable load in approximately 1.8 years.

3.4 *Impact of Solar Energy Integration*

3.4.1 Estimated Energy Savings

The integration of a 100 kW solar energy system is projected to significantly reduce grid electricity consumption. Assuming an output factor of 0.8 and 5 operational hours per day, the daily energy savings are as follows.

To estimate the impact of installing an additional 100 kW solar plant, we need to determine the potential reduction in units consumed, billed demand (kVA), and bill amount. We'll assume that the additional 100 kW solar plant will reduce the units consumed proportionally as shown in Table 3.

Assumptions:

1. Current Solar Plant Capacity: 67 kW.
2. New Solar Plant Capacity: 100 kW.
3. Total Solar Capacity: 67 kW + 100 kW = 167 kW.

4. Proportional Reduction: The additional solar plant will reduce the total units consumed by a proportionate amount based on its capacity relative to the total solar capacity.

Steps to Calculate Reduction:

1. Current Solar Capacity: 67 kW.
2. Additional Solar Capacity: 100 kW.
3. Total Solar Capacity: 167 kW.
4. Increase in Solar Capacity: 100 kW/67 kW = 1.49.

This means the total solar contribution will be 1.49 times the current contribution. Reduction Calculation:

- Reduction Factor: 1–1/1.49 = 0.331.

This means that the grid consumption will be reduced by approximately 33%. Example Calculation for May-24:

1. Units Consumed: 30,815.
2. Reduction in Units: 30,815 × 0.33 = 10,669.
3. New Units Consumed: 30,815 − 10,669 = 20,146.

Table 3 presents a comparative analysis of electricity consumption and billing over a 12-month period, showing the potential impact of energy reduction measures. For each month from June 2023 to May 2024, it displays the original units consumed, bill demand (kVA), and bill amount (INR), alongside projected figures after applying a consistent reduction factor of 0.33 (33%). This reduction results in significantly lower energy consumption, reduced bill demand, and, consequently, lower bill amounts each month. The most notable savings are seen in peak consumption months like May 2024, where the bill amount drops from 564,381 INR to 378,138 INR. The analysis concludes with an annual summary, showing that the original annual bill of 3,735,923 INR could potentially be reduced to 2,505,852 INR, resulting in substantial annual savings of 1,230,071 INR. This data highlights the significant financial benefits that could be achieved through implementing energy efficiency measures, potentially reducing energy consumption and costs by approximately one-third.

3.4.2 Cost of 100 kW Solar Installation

The cost of solar installation varies depending on the location, type of system, and market conditions. On average, the installation cost for a solar plant in India is around **₹40,000 to ₹60,000 per kW**. We'll take an average value of **₹50,000 per kW** for this calculation.

Installation Cost Calculation:

Cost of 100 kW Solar Plant = 100 × 50,000 = ₹50,00,000.

So, the estimated cost of installing a 100 kW solar plant is ₹50,00,000.

3.4.3 Payback Period Calculation

The payback period is calculated as the time it will take for the solar plant to pay for itself through the annual savings in electricity bills.

Payback Period Formula:

Payback Period (years) = Cost of Solar Installation/Annual Savings Substituting the values:

Payback Period = 50,00,000/12,30,071 = 4.07 years.

3.5 Effect on Transformer Capacity

The solar system will reduce the demand on the transformer by offsetting grid electricity. This reduction will prolong the transformer's service life by lowering its operational load.

To calculate how much additional load (in kVA) can be added before the 200 kVA transformer reaches its full capacity, we can subtract the current load from the transformer's rated capacity.

Step-by-Step Calculation:

1. **Current Load After Solar Installation**: The peak demand after solar installation is **92 kVA** (from May 2024 data).
2. **Transformer Capacity**: The transformer is rated for **200 kVA**.
3. **Available Capacity**:

$$\text{Available Capacity} = 200\,\text{kVA} - 92\,\text{kVA} = 108\,\text{kVA}$$

Safe Operating Range:

- Transformers typically perform best when operating in the range **70–80%** of their maximum capacity for efficiency and longevity.
- Using the **80%** rule of thumb: Recommended Capacity = 200 kVA × 0.8 = 160 kVA.
- Recommended Capacity = 200, kVA times 0.8 = 160 kVA.
- Available capacity under safe limits: 160 kVA – 92 kVA = 68 kVA.

3.6 Carbon Footprint and Thermal Energy Reduction

The carbon footprint reduction shown in Table 4 can be estimated using the average emission factor for electricity generation from fossil fuels. The emission factor varies by region, but a commonly used global average is **0.85 kg of CO_2 per kWh** for electricity generated from fossil fuel sources assuming grid power:

$$CO_2 \text{ Emission Reduction (kg)} = \text{kWh} \times \text{EF (kg } CO_2/\text{kWH)} \quad (3)$$

The average heat rate for coal plants is **8500 BTU/kWh** (British Thermal Units per kWh).

To convert BTUs to gigajoules (GJ), we use the conversion factor:

$$1\,\text{BTU} = 1.055 \times 10^{-3}\,\text{GJ}$$

Formula for Thermal Energy Savings (TES):

$$\text{TES (GJ)} = \text{kWh} \times \text{H (BTU/kWh)} \times 1.055 \times 10^{-3} \quad (4)$$

This data in Table 4 illustrates the environmental impact of energy reduction measures over a 12-month period from June 2023 to May 2024. It shows the monthly reduced electricity consumption in kilowatt-hours (kWh), the corresponding reduction in carbon dioxide (CO_2) emissions measured in kilograms, and the thermal energy savings in gigajoules (GJ). The reduction in electricity consumption ranges from 4160 kWh in November 2023 to 10,669 kWh in May 2024, aligning with seasonal variations in energy use. These reductions translate to significant environmental benefits, with CO_2 emission reductions varying from 3536 kg in November to 9068.65 kg in May. The thermal energy savings, which represent the amount of heat energy conserved, are also substantial, ranging from 37,222.40 to 95,482.64 GJ. This data underscores the considerable environmental advantages of implementing

Table 4 Monthly reduction in carbon footprint and thermal energy

Month	Reduced units (kWh)	CO_2 emission reduction (kg CO_2)	Thermal energy savings (GJ)
24-May	10,669	9068.65	95,482.64
24-Apr	8808	7486.80	78,847.56
24-Mar	7879	6697.15	70,492.31
24-Feb	5619	4776.15	50,243.68
24-Jan	4818	4095.30	43,080.67
23-Dec	5456	4637.60	48,827.16
23-Nov	4160	3536.00	37,222.40
23-Oct	5325	4526.25	47,662.88
23-Sep	4236	3600.60	37,597.26
23-Aug	4997	4247.45	44,315.52
23-Jul	4933	4193.05	43,747.81
23-Jun	5077	4315.45	45,021.69

energy efficiency measures, demonstrating how reducing electricity consumption can lead to substantial decreases in carbon emissions and conservation of thermal energy resources.

4 Conclusion

The energy audit conducted for the engineering institute has revealed significant potential for energy savings through the installation of a 100 kW solar power plant. By incorporating solar energy, the institution can reduce its grid electricity consumption by approximately 33%, resulting in a substantial decrease in energy costs. The analysis shows that the installation of the solar plant would save the institute around ₹1,230,071 annually in electricity bills, with an estimated payback period of approximately 4.07 years.

Additionally, the solar plant installation will contribute to environmental sustainability, reducing the carbon footprint by an estimated 61,179.50 kg of CO_2 annually and saving approximately 642,343.58 GJ of thermal energy each year. These outcomes align with global efforts to transition to renewable energy sources and reduce dependency on fossil fuels. Implementing the proposed solar plant will not only result in long-term financial savings but also contribute positively to the environment, making it a viable and impactful solution for the institution's energy needs.

The findings of this study underscore the importance of conducting regular energy audits and exploring renewable energy options to achieve both economic and environmental benefits. Further research can be conducted to optimize load management and explore additional energy-saving opportunities within the institution.

Disclosure of Interest There are no competing interests to declare.

Declaration of Funding No funding was received.

References

1. Ahmed S, Rahman MM, Karim MR (2019) Optimal load management strategies for improving transformer efficiency. IEEE Trans Power Deliv 34(2):850–860
2. Tognini G, Bianchini A (2019) Load forecasting and transformer capacity planning: a review of recent advances. Electr Power Syst Res 174:105908
3. Khedher NB, Feki M, Dammak N (2020) Energy audit and management: an essential tool for energy efficiency. J Clean Prod 244:118716
4. Singh R, Sharma A, Arora S (2020) Predictive analytics in energy management: a comprehensive review. J Energy Manag 45(7):45–58
5. Nguyen T, Ha T, Tran L (2022) Solar energy integration strategies for institutional energy systems. Renew Energy Technol J 9:320–331

6. Chen W, Wang Y (2023) Energy audit methodologies and their application to transformer capacity planning. J Electr Eng Technol 18(1):299–312
7. Zhang L, Li Y (2020) Solar energy integration and its impact on energy system stability. Energy Rep 6:1154–1162
8. Adamu I, Mahamud A (2021) Sustainable energy management: strategies and best practices. Energy Rep 7:345–356
9. Gupta A, Sharma N, Jain R (2021) Forecasting energy demands with advanced predictive models. Energy Rep 7:1047–1062
10. Patel S, Singh R (2021) Advanced methods for load forecasting and energy efficiency in large institutions. J Energy Resour Technol 143(9):090802
11. Khan MJ, Kumar A (2021) Economic and environmental benefits of solar energy integration in institutional settings. Renew Energy 167:1192–1203
12. Patel V, Kumar S (2021) Impact of renewable energy integration on power system stability. Renew Sust Energ Rev 139:110580
13. Liu C, Yang M, Wang J (2022) Cost-benefit analysis of energy efficiency measures in institutional buildings. Energy Econ 101:105410
14. Sharma P, Patel A (2021) Comprehensive review of energy audit practices and their impact on load management. Energy Rep 7:2917–2931
15. Sharma R, Patel P, Gupta D (2021) Energy audits and their role in optimizing transformer capacities. Electr Eng J 98(3):487–499
16. Sharma N, Gupta A (2021) Innovative strategies for energy audits and their implementation in institutional settings. Energ Effic 14:2001–2014
17. Singh A, Kumar V (2022) Energy efficiency and management strategies for educational institutions. J Sustain Energy 14(2):250–267
18. Lee J, Park Y, Kim H (2022) Advanced load forecasting techniques for effective energy management. Energy 236:121306
19. Kim D, Lee K, Choi J (2022) Enhanced forecasting methods for load prediction and energy management. IEEE Trans Power Syst 37(1):95–104
20. Zhang Y, Wang H, Zhou X (2022) Transformer efficiency and load management: Current practices and future directions. Int J Electr Power Energy Syst 125:106665
21. Gupta R, Khan M (2022) Renewable energy integration and its effect on load forecasting accuracy. Energy Convers Manag 252:115316
22. Kim J, Lee M (2022) Evaluating the benefits of load reduction and solar energy integration. Energy Rep 8:1984–1993
23. Nguyen M, Tran Q (2022) Energy management and forecasting techniques for future energy demands. Energy Manag J 45(4):361–375
24. Chen J, Xu B, Zhao Y (2023) Efficiency improvement through energy audits and load management. Energ Effic 16:569–584
25. Liu J, Zhang T, Xu Q (2022) Economic and environmental impacts of energy efficiency measures in educational institutions. Renew Energy 168:370–381